Role of Sample Return in Addressing Major Questions in Planetary Sciences

Mahesh Anand • Sara Russell • Yangting Lin •
Meenakshi Wadhwa • Kuljeet Kaur Marhas •
Shogo Tachibana
Editors

Role of Sample Return in Addressing Major Questions in Planetary Sciences

Previously published in *Space Science Reviews* in the Topical Collection "Role of Sample Return in Addressing Major Questions in Planetary Sciences"

Springer

Editors
Mahesh Anand
School of Physical Sciences
The Open University
Milton Keynes, UK

Sara Russell
Natural History Museum
London, UK

Yangting Lin
Institute of Geology and Geophysics
Chinese Academy of Sciences
Beijing, China

Meenakshi Wadhwa
School of Earth and Space Exploration
Arizona State University
Tempe, AZ, USA

Kuljeet Kaur Marhas
Physical Research Laboratory
Planetary Science Division
Ahmedabad, Gujarat, India

Shogo Tachibana
UTokyo Organization for Planetary and Space Science
University of Tokyo
Tokyo, Japan

ISSN 1385-7525 Space Sciences Series of ISSI
ISBN 978-94-024-2074-6

Cover Image: A composition of Mercury, Venus, Earth, Mars (courtesy of ESA), the Earth's Moon (courtesy of ESA/NASA), the Martian moon Phobos (courtesy of NASA/JPL-Caltech/University of Arizona), and the Asteroid Ryugu with the Spacecraft Hayabusa2 (courtesy of JAXA).

Printed on acid-free paper

This Springer imprint is published by the registered company Springer Nature B.V.
The registered company address is: Van Godewijckstraat 30, 3311 GX Dordrecht, The Netherlands

Contents

Preface to the Topical Collection: Role of Sample Return in Addressing Major Questions in Planetary Sciences

Mahesh Anand[1,2] · Sara Russell[2] · Yangting Lin[3] · Meenakshi Wadhwa[4] · Kuljeet Kaur Marhas[5] · Shogo Tachibana[6,7]

Major advances in our understanding of the origin and evolution of the inner Solar System have been made through analyses of samples from the Moon returned by the Apollo and Luna missions in 1960s and 1970s. More recently, the Stardust and Genesis missions sampled cometary grains and solar wind particles, respectively, revealing their elemental and isotopic compositions. Similarly, samples returned by the Hayabusa mission from the near-Earth asteroid 25143 Itokawa enabled insights into regolith evolution on minor bodies in the Solar System and provided the first direct link between meteorites and asteroids. As we write this preface, Hayabusa2 is on its return journey to Earth after having sampled asteroid 162173 Ryugu while OSIRIS-REx is preparing to collect samples from asteroid 101955 Bennu and Chang'E-5 mission is being readied for a robotic sample return from the Moon after a hiatus lasting over 4 decades. Indeed, the first phase of long awaited Mars Sample Return program is literally off the ground with the successful launch of Mars 2020 mission carrying Perseverance lander, which will aim to collect and cache Martian surface samples for a later mission to retrieve and return them to the Earth in early 2030s.

✉ M. Anand
Mahesh.Anand@open.ac.uk

1 Planetary and Space Sciences, School of Physical Sciences, Walton Hall, The Open University, Milton Keynes, MK7 6AA, UK

2 Department of Earth Sciences, The Natural History Museum, Cromwell Road, London SW7 5BD, UK

3 Key Laboratory of Earth and Planetary Physics, Institute of Geology and Geophysics, Chinese Academy of Sciences, Beijing, China

4 School of Earth and Space Exploration, Arizona State University, Tempe, AZ 85287, USA

5 Planetary Sciences Division, Physical Research Laboratory, Ahmedabad, Gujarat 380009, India

6 UTokyo Organization for Planetary and Space Science, The University of Tokyo, Tokyo, Japan

7 Institute of Space and Astronautical Science, Japan Aerospace Exploration Agency, Sagamihara, Japan

Finally, NASA's recently announced ARTEMIS program with the aim of landing the first woman and the next man on the Moon by 2024 heralds the next major phase of human exploration of the Solar System which will undoubtedly lead to the return of new samples from the Moon and any other celestial destinations visited by the humans in this endeavour.

Although lacking a parent body context, meteorites represent another type of 'returned' samples that are either 'finds' (collected from various locations on Earth, either through dedicated search efforts or through serendipity) or are 'falls' (observed to arrive on Earth, almost always without any forewarning but often in a spectacular fashion!). The past couple of years have witnessed a number of new meteorite falls, especially of carbonaceous chondrites, at multiple worldwide locations ranging from India, Indonesia and Costa Rica. Meteorite research has been invaluable in addressing a number of scientific topics related to Solar System formation and evolution, ranging from nucleosynthetic processes that contributed to materials forming the solar nebula through studies of pre-solar grains to processes (such as fluid-rock interactions and impacts) occurring on rocky parent bodies over the 4.5 Ga history of the Solar System. Indeed, some of the most exciting findings and developments of new concepts have been brought about as a result of painstaking research on meteorites in Earth-based laboratories.

Perhaps one of the most prolific outcomes from recent technological advances has been in our ability to send spacecraft missions (orbiters and landers) for remote and/or *in-situ* investigations of almost all major bodies in the Solar System, including the first audacious landings on a comet (67P/Churyumov–Gerasimenko) and on Saturn's moon, Titan. Consequently, there now exist global remote sensing datasets at various scales and resolutions for a number of planetary bodies that weren't available previously, and these have already played (and will continue to play) a vital role in the selection of targets for sample return missions. In addition, remote observations of planetary atmospheres are providing fascinating new insights into the nature and composition of gases, even those that are present in trace amounts (e.g., ppm or ppb levels). The recent discovery of phosphine gas in Venusian atmosphere has raised the tantalizing possibility of it being a potential biosignature and thus providing a major scientific driver for considering atmospheric sample return of Venus by future missions.

Notwithstanding the existence of a wealth of remote sensing and cosmochemical data on various Solar System objects (e.g., Moon, Mars, asteroids), significant gaps remain in our understanding of the origin and evolution of the Solar System. These outstanding science questions can only be addressed through high-precision and highly accurate measurements of returned samples for their mineralogical, chemical, and isotopic compositions from the macroscopic to the atomic scales using the most advanced laboratory instruments on Earth. Furthermore, as amply demonstrated by the Apollo and Luna missions, sample return provides us the critical 'ground truth' about the visited body, verifying and validating conclusions that can be drawn by remote sensing (both Earth-based and by spacecraft) and via landed instruments on other bodies (perhaps the best example being the application of a calibrated lunar crater chronology to the surfaces of other inner Solar System bodies, most notably Mars). Returned samples further enhance the scientific value of extraterrestrial materials such as meteorites and cosmic dust by providing additional clues about their parent bodies. Arguably, the biggest benefit arising from sample return missions is the fact that the returned samples can be preserved for decades and used by future generations of explorers to answer questions that haven't been framed yet using laboratory instruments that haven't even been imagined – fitting the metaphor 'the gift that keeps giving'.

In this context, a workshop was organised in February (5-9) 2018 at the International Space Science Institute (ISSI), Bern, Switzerland to review our current knowledge and understanding of Solar System formation and evolution and identify major outstanding science questions that remain to be answered through future sample return missions. This joint ISSI-Europlanet workshop was attended by 39 scientists (Fig. 1), representing a broad cross section of the international planetary science community. The career stages of attendees ranged from senior researchers to early stage researchers, including PhD students. The Workshop was structured around three main themes: (i) Initial conditions and workings of the Solar System, (ii) Planetary habitats, and (iii) Sample curation. The presentations and associated discussions in these themes were focused on: (1) reviewing major outstanding science questions in the planetary sciences that can only be answered via sample return missions, (2) identifying specific target body(ies) for answering specific science question(s), and (3) evaluating and considering sample curation facilities relevant for such sample return missions. An outcome from the discussions and debate held at the workshop has been the publication of 14 peer-reviewed papers in *Space Science Reviews* and subsequently regrouping of these papers under the Topical Collection 'Role of Sample Return in Addressing Major Questions in Planetary Sciences' and co-published as Volume 74 of the Space Sciences Series of ISSI (SSSI) by Springer Verlag.

The 2020s may usher in a new golden era of sample return. Although many such missions are spearheaded by individual space agencies, almost all of them have contributions from one or more international partners. This is an extremely positive development as it signifies that in the 21st century space exploration is envisaged as a common goal for humanity which can only be achieved through global cooperation. As scientific research on the Apollo and Luna samples continues to demonstrate, collaboration throughout the international science community is indispensable in order to make progress in science and the society.

The main purpose here is to present a collection of scientific papers reviewing the state-of-the-art in our understanding of the origin and evolution of the Solar System and then highlighting major gaps that can only be addressed by returning and analysing samples in Earth-based laboratories. The depth and breadth of the papers are likely to provide a firm basis for formulating strategies for future sample return missions by various space agencies.

Mezger et al. review our present understanding of the components that accreted to form the Earth – arguably the most extensively studied planetary body in the Solar System, especially through laboratory analysis of samples. They conclude that it is currently not possible to fully define and quantify the different chemical and isotopic materials that formed the Earth, because a major component seems missing in the extant collections of extraterrestrial samples. They identify Mercury and Venus as planets of great importance and interest in this respect.

Ireland et al. provide a thorough review of the remarkable utility of oxygen isotopic composition of extraterrestrial samples in elucidating some of the most fundamental planetary processes in the Solar System. Central to their argument is the observation that within the chondrite parent bodies, exchange between ^{16}O-poor fluids and relatively ^{16}O-rich solids created isotopic mixing lines. This model makes specific predictions for isotopic compositions of silicates and water ice throughout the Solar System. Although analysis of returned samples is the ultimate goal, these authors discuss deployment of a promising instrument for three oxygen isotope analysis using a landed mission which could be a precursor to a more challenging and ambitious sample return mission from the visited body.

Vander Kaaden et al. make a compelling case for sample return from Mercury – one of the least studied objects in the inner Solar System. According to these authors, a returned sample from the surface of Mercury would allow for transformative Solar System science, including new insights into space weathering processes on airless bodies and the geochemical behaviour of elements under extreme conditions.

Venus is sometimes referred to as the Earth's 'sister planet.' Greenwood and Anand argue that the single measurement ($\Delta^{17}O$) of a solid sample of Venus would have first order implications for our understanding of how large terrestrial planets such as our own Earth are built. They review the surface geology of Venus with the aim of identifying potential terrains that could be targeted by a robotic sample return mission. They make a strong case for a Venus sample return using a 'Grab and Go' strategy, elements of which have already been studied as part of various mission concept studies by NASA and ESA – so it may not be as outrageous as it may sound, initially!

As the only planetary body other than the Earth where humans have conducted field work and collected samples for laboratory analyses, the Moon has become a reference point for understanding planetary processes, serving multiple purposes (e.g., as a cornerstone, testbed, witness plate) for advancing our understanding of Solar System science. Tartèse et al. undertake a heroic task of distilling the vast amount of scientific information that has been derived from investigations of samples returned from the Moon over the past five decades. However, many new and fundamental questions remain which are difficult to address with currently available samples. For example, the abundance, distribution, and source(s) of volatiles on the Moon can only be fully addressed by sampling areas of the Moon not visited previously.

As the speculations grow that astrobiological signatures of the early Earth might be preserved in the lunar regolith, a similar possibility exists for the martian system in which the regolith on its largest moon, Phobos, could have preserved signatures of any life that might have existed on early Mars. Usui et al. describe the scientific objectives of a Japan Aerospace Exploration Agency (JAXA)-led Phobos sample return mission (MMX: Martian Moons eXploration), which aims to retrieve and return >10 g of Phobos regolith back to Earth in 2029.

The geological evidence suggesting that Mars was once a habitable planet, and could still harbour liquid water in its sub-surface, raises the tantalising possibility of answering the age-old question that humanity has faced, i.e., 'is there life elsewhere in the universe?'. As a result, Mars is currently the most intensively explored object in the Solar System with multiple ongoing and planned orbiter and landed missions. The next logical step in martian exploration is a Mars sample return (MSR) mission. Grady provides a review of the science rationale on behalf of the international Mars exploration community that should see the first mission in the MSR campaign, Mars 2020, landing at Jezero Crater in early 2021; samples are anticipated to return to Earth sometime after 2032.

Cockell et al. discuss the microbial ecology of volcanic rocks and hydrothermal systems on the Earth, relevant to developing a sampling strategy for martian samples of astrobiological importance. They suggest that a microbial ecology-driven strategy for investigating the habitability and presence of life on Mars is likely to yield the most promising sample set of the greatest use to the largest number of astrobiologists and planetary scientists.

Kleine et al. review one of the most vigorously debated recent topics in cosmochemistry wherein the isotopic dichotomy between non-carbonaceous (NC) and carbonaceous (CC) meteorites is thought to represent parent bodies from two genetically distinct reservoirs which were isolated within a few million years of the formation of the Solar System. The growth and/or migration of Jupiter subsequently resulted in inward scattering of CC bodies, which accounts for the co-occurrence of NC and CC bodies in the present-day asteroid belt. This is also reflected in the composition of the Earth's primitive mantle which has a mixed

NC–CC composition, indicating that Earth accreted CC bodies during the final stages of its growth, perhaps through the Moon-forming giant impactor.

Burbine and Greenwood make persuasive arguments for returning samples from a main-belt asteroid. They argue that returning samples from the main belt has the potential to test different planet migration models and the related geochemical and isotopic concept of a bimodal Solar System represented by the NC-CC dichotomy. They advocate a dedicated sample return mission to the dwarf planet (1) Ceres (recently visited by Dawn mission) as the best means of further exploring inherent Solar System variations and conclude that no other potential main-belt target is capable of providing as much insight into the early Solar System as Ceres.

Lin et al. adopt a different approach and make a strong case for developing concepts for sample return missions from asteroids that could reveal the first 10 Ma history of the Solar System, including the initial conditions, nebular processes, and the accretion of planetesimals. In order to 'see through' secondary processes that took place in planetesimals, such as thermal metamorphism, aqueous alteration, melting and differentiation, they favour targeting primordial objects such as enstatite chondrite-like or L-type Near-Earth asteroids.

Martin et al. review the similarities and differences for organic matter among various Solar System bodies such as comets, asteroids, meteorites, and dust particles in the context of detecting the most pristine organic material, either by on-site analysis or sample return missions. They also discuss the targets of potential future sample return missions, as well as the contamination controls that should be in place in order to successfully study pristine organic matter.

While staying with the topic of organic matter in the Solar System, Chan et al. review the state-of-the-art analytical protocols that are in use for analysing organics in extraterrestrial samples. One of the most challenging aspects of any sample return mission is the requirement to avoid or minimise terrestrial organic contamination. Despite various efforts in the design and execution of contamination control, it appears impossible to completely eliminate sources of organic contamination. Nevertheless, there have been some successes and lessons learned with regard to the identification of indigenous organic matter in the returned samples versus terrestrial contamination.

Sample curation is an indispensable element of any sample return mission plan. A well-planned sample curation strategy plays a vital role in ensuring the lasting scientific legacy of returned samples for several generations, as exemplified by the NASA's Astromaterials Research and Exploration Science (ARES) program. McCubbin et al. provide a detailed account of astromaterials acquisition and curation practices, especially in the context of advanced curation of astrobiologically sensitive samples that might be returned by future missions. These efforts require research and development of new innovative technologies and techniques for sample collection, handling, characterization, analysis, and curation of astromaterials. The primary result of advanced curation research is to both reduce and quantify contamination of astromaterials and preserve the scientific integrity of all samples from mission inception to secure delivery of samples to Earth-based laboratories for in-depth scientific analysis.

As we close this preface, we hope that the papers presented in this volume serve a reference document for formulating our strategies for future sample return missions in order to address outstanding scientific questions in planetary sciences. Our intention was not to provide an exhaustive list of target bodies that could be sampled by future missions but instead advocate an approach of being driven by major science questions that can only be addressed by returning samples from a specific body in the Solar System.

ISSI-Europlanet Workshop Participants (Feb 2018)

Acknowledgements We thank ISSI for hosting us in Bern during the workshop and are extremely grateful to all ISSI staff who provided vital logistical support to all participants, making their week in Bern a memorable one! We couldn't have brought this project to a successful completion without the guidance and support from Andrea Fischer (and her patience with the principal editor!) and the ISSI Directors over the past two years. We also thank Europlanet for co-funding this workshop.

Space Sci Rev (2020) 216:27
https://doi.org/10.1007/s11214-020-00649-y

Accretion of the Earth—Missing Components?

K. Mezger[1] · M. Schönbächler[2] · A. Bouvier[3]

Received: 5 February 2019 / Accepted: 19 February 2020 / Published online: 4 March 2020

Abstract Primitive meteorites preserve the chemical and isotopic composition of the first aggregates that formed from dust and gas in the solar nebula during the earliest stages of solar system evolution. Gradual increase in the size of solid bodies from dust to aggregates and then to planetesimals finally led to the formation of planets within a few to tens of million years after the start of condensation. Thus the rocky planets of the inner solar system are likely the result of the accumulation of numerous smaller primitive as well as differentiated bodies. The chemically most primitive known meteorites are chondrites and they consist mostly of metal and silicates. Chondritic meteorites are derived from distinct primitive planetary bodies that experienced only limited element fractionation during formation and subsequent differentiation. Different chondrite classes show distinct chemical and isotopic characteristics, which may reflect heterogeneities in the solar nebula and the slightly different pathways of their formation. To a first approximation the chemical composition of the bulk Earth bears great similarities to primitive meteorites. However, for some elements there are striking and significant differences. The Earth shows a much stronger depletion of the moderate to highly volatile elements compared to chondrites. In addition, mixing trends of specific isotopes reveal that the Earth is most enriched in *s*-process isotopes compared to all other analysed bulk solar system materials. It is currently not possible to fully define and quantify the different chemical and isotopic materials that formed the Earth, because a major component seems missing in the extant collections of extraterrestrial samples. Variations in nucleosynthetic isotope compositions as well as the strong depletion of moderately and strongly volatile elements points towards a source in the inner solar system for this missing material. It is conceivable that Venus and Mercury contain a much larger fraction of this

Role of Sample Return in Addressing Major Questions in Planetary Sciences
Edited by Mahesh Anand, Sara Russell, Yangting Lin, Meenakshi Wadhwa, Kuljeet Kaur Marhas and Shogo Tachibana

✉ K. Mezger

1 Institut für Geologie, Center for Space and Habitability, Universität Bern, Baltzerstrasse 1+3, 3012, Bern, Switzerland

2 Institute of Geochemistry and Petrology, ETH Zurich, 8092 Zurich, Switzerland

3 Bayerisches Geoinstitut, Universität Bayreuth, Bayreuth, 95447, Germany

missing component. Thus, for a complete reconstruction of the conditions that led to the formation of the inner solar system planets (Mercury to Mars) samples from the inner planets Venus and Mercury are of great interest and importance. High precision chemical and isotopic analyses in the laboratory of rocky material from inner solar system bodies could complete the knowledge on the chemical, isotopic and mineralogical make-up of the solar nebula just prior to planet formation and enhance our understanding of the evolution of the solar nebula in general and the formation of the rocky planets in particular.

Keywords Terrestrial planets · Accretion of Earth · Meteorites · Inner solar system

1 Introduction

The bulk composition of the Earth and that of the distinct terrestrial reservoirs is essential to understand and quantify the major differentiation processes that led to the present-day highly differentiated planet (e.g., McDonough and Sun 1995; Palme and O'Neill 2014). The bulk composition of the Earth is also required to reconstruct the formation of the Earth as a planet and to deduce the different processes that led to its specific chemical composition. Ultimately the material that formed the Earth originated from the solar nebula, which originally had the composition of the Sun. Starting from the solar nebula to the final accretion, the building material of the Earth underwent a variety of physical and chemical processes that led to the fractionation of elements. Key processes include condensation and evaporation of solids in the solar nebula, dust coagulation, evaporation of elements during collision of small bodies or during the magma ocean stage, accretion of solids to smaller bodies that grew to planetesimals and eventually to rocky planets like the Earth, and the late addition of bodies after the main phase of accretion (e.g., Chambers 2001; Day et al. 2012; Day 2015; Halliday 2014). Since solids formed at different distances from the Sun, different condensation histories and pathways of the building blocks have to be considered (e.g., Morbidelli et al. 2000; Ciesla 2010).

The asteroidal bodies that are the sources of most meteorites formed within the first few million years of the solar system; some were only minimally chemically modified after their formation (i.e. chondrite parent bodies), others differentiated significantly and segregated an iron core (i.e. achondrite parent bodies). The major groups of materials available for direct study in the laboratory are carbonaceous, ordinary and enstatite chondrites, iron meteorites, and achondrites such as eucrites or ureilites. A main constituent of the different chondrites are chondrules (e.g., Hewins 1997), which are sub-millimeter igneous spherules that can contain relict mineral grains that were not melted during chondrule formation. In addition to abundant chondrules that are surrounded by fine-grained matrix, some chondrites contain sulfides, metals as well as calcium–aluminum-rich inclusions (CAIs) and other refractory element enriched inclusions that formed as primary condensates from the solar nebula gas (e.g., Scott and Krot 2014; Yokoyama et al. 2017). The CAIs are particularly abundant in some carbonaceous chondrites (e.g., the CV group). The primitive chemical character has led to the idea that they could represent the building material of the terrestrial planets. The disparities in the chemical and isotopic composition of the rocky planets could reflect the different contributions of different known meteorite classes to their formation. Based on this assumption, it is possible to quantify the relative abundances of the different material that contributed to the growing Earth or other rocky planets whose bulk composition can be estimated (e.g., Sanloup et al. 1999; Lodders 2000; Fitoussi et al. 2016; Liebske and Khan 2019). However, as will be further shown in this contribution, a close investigation of the currently available chemical and isotopic data on samples from the rocky planets Earth and Mars and the different meteorite classes shows that it is impossible to account for all chemical and isotopic characteristics, particularly of the Earth, by combining different proportions of known materials from the solar system. In some respects, the Earth seems to be an apparent end member for many geochemical parameters among the rocky materials in the solar system studied so far (e.g., Wänke and Dreibus 1988; Dauphas et al. 2004; Burkhardt et al. 2011; Akram et al. 2013, 2015; Akram and Schönbächler 2016; Fischer-Gödde et al. 2015; Liebske and Khan 2019).

This contribution explores the chemical and isotope characteristics of different meteorites classes and their temporal evolution in order to evaluate their potential contributions to the formation of the Earth. Isotopes, element abundances and age constraints are used to evaluate, which known components contributed to the Earth and which terrestrial building blocks might be missing in the extant collection of extra-terrestrial materials. Particular questions addressed are: What are the characteristics of this potentially missing component and where might it be found within the solar system? Is the Earth possibly an end member in its chemical composition or does the Earth sample a common component of the inner solar system that is dominantly present in the inner solar system planets like Venus and Mercury? What is the timing of formation for planetesimals and planets, and is there a relationship with their nucleosynthetic compositions?

2 Compositional Constraints on the Building Blocks of Earth

Estimates for the abundances of the elements in bulk Earth and its different geochemical reservoirs (core, mantle, crust, hydro- and atmosphere) have been obtained using a variety of approaches (e.g., Allègre et al. 1995; McDonough and Sun 1995; Palme and O'Neill 2014). These estimates show that the Earth is a chemically highly differentiated planet that was accreted from heterogeneous and on average volatile element depleted (i.e. compared to the solar composition) materials with distinct isotopic and chemical compositions. These materials were added to the growing planet during different stages, while the planet was internally differentiating (e.g., Rubie et al. 2011). Chemical and isotopic diversity is also obvious in meteorites and their components that derive from primitive bodies of the solar system.

Due to the abundance of chondrites in the extant meteorite collection, their relatively primitive composition and first order similarities with estimates for the bulk Earth composition, chondrites have been traditionally used as a proxy for the material from which the Earth and the other rocky planets of the inner solar system formed (e.g., Wänke and Dreibus 1988; Schönbächler et al. 2010; Rubie et al. 2011; Fitoussi et al. 2016; Liebske and Khan 2019). However, chondrites may not be the only materials that contributed to the formation of the terrestrial planets. With the exception of CI chondrites, chondrites contain once molten materials, now preserved as spherical objects, i.e. chondrules, embedded in a fine-grained matrix. The chemically primitive chondrites, which represent a collection of materials from before or during before planet formation, may have played a key role in the early stages of planet formation. Most importantly, they are accessible for laboratory studies. In the global collection of meteorites, chondrites constitute the majority of samples with around 80% of all collected specimens (e.g., Bischoff and Geiger 1995; Gattacceca et al. 2019).

The chronology of primitive and differentiated materials from the early solar system has changed this traditional view (e.g., Kleine et al. 2005). The composition of the chondrites must reflect the composition of the reservoirs from which they accreted and because they accreted relatively late compared to some achondrite parent bodies, it is possible that these reservoirs were already modified by planetary formation and/or chondrule formation. It is therefore unlikely that the Earth formed by direct accretion of dust with a homogeneous composition from the solar nebula, but rather by the addition of smaller chemically distinct differentiated bodies that had element abundances different from the bulk solar system composition. These planetesimals were already differentiated into core, mantle and crust (e.g., Kleine et al. 2005; Kruijer et al. 2014). Some of these planetary crusts were also rapidly re-processed by impact melting producing further elemental fractionation (Frossard et al. 2019). An alternative planet formation scenario provides the relatively recently proposed mechanism of pebble accretion (see Johansen and Lambrechts 2017 for a review). In a protoplanetary gas disk, a planetary embryo can grow very quickly through accretion of small pebbles that formed further away from the Sun than the protoplanet and were captured by the protoplanet on their way towards the Sun.

Meteorites that are available for direct mineralogical, chemical and isotopic studies reveal a breadth of different materials that could potentially represent the building blocks for the Earth (e.g., Wasson 1985; Day 2015). Some meteorite bodies have compositions that are similar to the elemental composition of the Sun, like the rare CI meteorites, others are highly fractionated like iron meteorites that represent the cores of differentiated planetesimals. All these materials have in common that they are depleted to some degree in moderately to highly volatile elements. To a first approximation the degrees of depletion of the chemical elements in different primitive meteorite classes follow the respective condensation temperatures of the elements when condensing from a cooling solar gas (Lodders 2003; Wood et al. 2019) (Fig. 1). All materials from the solar system so far studied show various degrees of depletion for elements with 50% condensation temperatures below ca. 1300 K (e.g., Palme and O'Neill 2014). The different depletion patterns are characteristic for each meteorite class and by inferences their respective parent bodies (Fig. 1). The least depleted material is represented by the rare CI chondrites. They have lost atmophile elements, but are otherwise similar in composition to the Sun (e.g., Anders and Grevesse 1989; Asplund et al. 2009; Palme et al. 2014) and thus the primary solar nebula. For this reason, they are commonly used as a reference to which other materials are compared. Together with CI chondrites, the least differentiated materials are the carbonaceous, enstatite and ordinary chondrites. Their relative abundances of refractory elements are similar to the solar abundances for elements with a 50% condensation temperature above ca. 1300 K (Fig. 1), but they show different extents of depletion for the volatile elements (e.g., Palme and O'Neill 2014).

In order to reconstruct the quantity and composition of the possible components that formed the Earth from known materials, as represented by meteorites, it is essential to estimate the bulk Earth composition (BE) or at least the bulk silicate Earth composition (BSE). For this reason, several efforts have been made to obtain robust estimates for the abundances of the chemical elements in the Earth as a whole or in the silicate mantle, the volumetrically dominant chemical reservoir of the Earth (e.g., McDonough and Sun 1995; Allègre et al. 1999; Palme and Jones 2003; Palme and O'Neill 2014). These estimates can be compared with bulk analyses made to obtain the absolute element abundance for each meteorite class (e.g., Wasson and Kallemeyn 1988; Miyamoto et al. 2016; Clay et al. 2017; Braukmüller et al. 2018, 2019) (Fig. 1).

A comparison of the element abundances in different chondrite classes with the bulk composition of the Earth reveals general agreement among the refractory element ratios. All rocky materials known from the solar system have very similar relative abundances for refractory elements. However, there are striking differences in the moderately to highly volatile element abundances for different meteorite classes that show various degrees of depletion. These depletions start with the highly abundant elements Mg, Fe and Si and follow

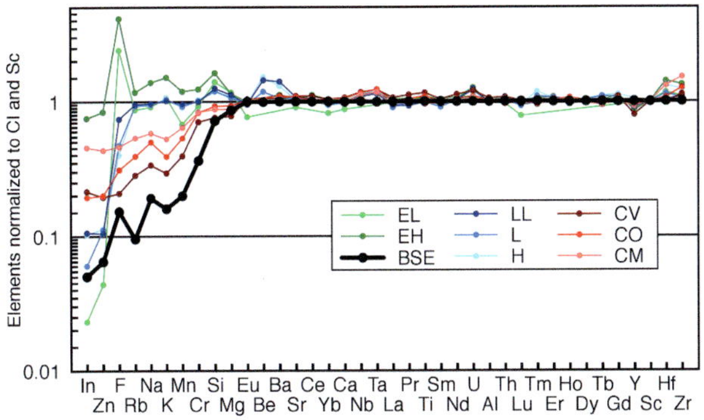

Fig. 1 Element abundances in different chondrites (carbonaceous chondrites: CM, CO, CV; ordinary chondrites: H, L, LL: enstatite chondrites: EH, EL) and the bulk silicate Earth (BSE). Compared to the different chondrites the Earth shows the strongest depletion in moderately to highly volatile elements. Elements (McDonough and Sun 1995; Wasson and Kallemeyn 1988) normalized to CI abundances (Anders and Grevesse 1989) and sorted according to their condensation temperature (Tc_{50}) (10^{-4} bar total pressure) (Lodders 2003)

the 50% condensation temperatures and thus reflect different thermal regimes in the feeding zones of the solids that subsequently accreted in different parent bodies (e.g., Palme and O'Neill 2014). This continuous and regular trend attests to a thermal cause for the depletion and may represent different places in the nascent solar system where these materials formed from a cooling and condensing nebula. Striking is the extreme depletion of volatile elements in the Earth; and compared to primitive meteorites, the Earth forms a compositional end member (Fig. 1). The chemical differences among the chondrites and the strong depletion of Earth in the more volatile elements are evidence that the Earth must have a major component that is more depleted than any known chondrite class (Wänke and Dreibus 1988; O'Neill and Palme 1998; Drake and Righter 2002; Schönbächler et al. 2010; Palme and O'Neill 2014; Palme and Zipfel 2016). Thus, for the formation of Earth, primitive but strongly volatile depleted material is needed to account for the observed missing volatile elements. The presence of such a depleted component in the Earth is supported by evidence from radiogenic isotope systems (e.g., Rb–Sr, U–Pb, Ag–Pd) that are sensitive to volatility related processes (e.g., Albarède 2009; Schönbächler et al. 2010; Peslier et al. 2017). It has been suggested that this strongly depleted material may be represented by achondrites, particularly angrites (e.g., Fitoussi et al. 2016). Although their abundances of moderately volatile lithophile elements make them a suitable component, the abundance pattern for the other moderately volatile lithophile elements does not show this depletion (e.g., Warren and Kallemeyn 1995). The volatile budget of the Earth must be a primary signature and cannot be the result of volatile loss by evaporation during accretion, because this would result in a step-function in the element depletion trend with a strong depletion below a certain threshold temperature and essentially no depletion above, which is not observed. The Earth shows a depletion in moderately volatile elements by a factor around 10 compared to the refractory elements (McDonough and Sun 1995; Fig. 1, 2). If this depletion occurred during condensation of the solids that formed the Earth or by evaporation during accretion into planetesimals that later formed the Earth, this should result in detectable stable isotope variations of these depleted elements. Such effects between the terrestrial planets are analytically barely resolvable (e.g., Nebel et al. 2011; Paniello et al. 2012). This depleted component was also most likely strongly reduced to account for the depletion of moderately siderophile elements in

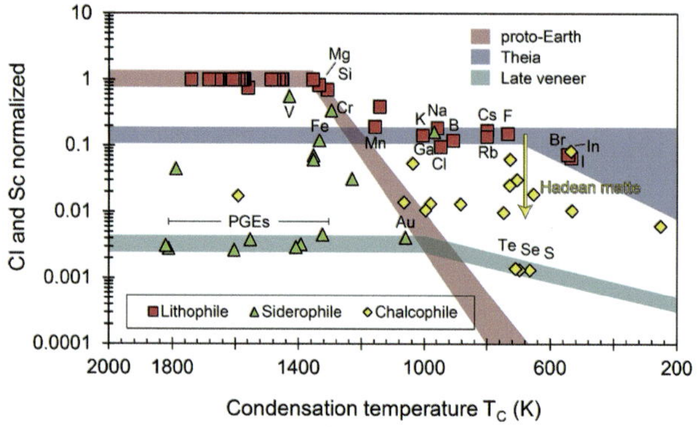

Fig. 2 Element abundances in bulk silicate Earth (BSE). Lithophile element abundances in the Earth can be modelled by a mixture of a ca. 85% of a component that was volatile element depleted (proto-Earth) and 15% of a component that was less volatile element depleted and oxidized (Theia). Siderophile and chalcophile elements are depleted in BSE due to metal core formation and segregation of a sulphide melt (Hadean matte; O'Neill 1991). Abundance of the platinum group elements (PGEs) is due to the addition of a late veneer after completion of core formation. Elements sorted according to their condensation temperature (Tc_{50}) (10^{-4} bar total pressure) (Lodders 2003). Modified after Maltese and Mezger (2020)

Earth's mantle, particularly Cr, which would otherwise be lithophile and not enter the metal melt during core formation (Wänke and Dreibus 1988). Such material has not been found in the meteorite collection yet. It may bear some similarities to enstatite or ordinary chondrites, however the latter have isotope compositions distinct from the bulk Earth (e.g., Warren 2011; Kruijer et al. 2019).

The similarity in the chemical and in isotopic composition of Earth and Moon has led to the suggestion that a giant impact is responsible for the formation of the Moon and at the same time completed the major accretion of Earth (e.g., Hartmann and Davis 1975; Benz et al. 1986; Wiechert et al. 2001; Canup and Asphaug 2001). This model implies that the present-day Earth consists of two major components that could have distinct compositions. Figure 2 shows the chondrite-normalized element abundances in bulk silicate Earth. Striking features of this element distribution in the silicate Earth are the relative chondritic abundances of the refractory lithophile elements and the less abundant moderately volatile lithophile elements, which are also present in relative chondritic abundances. The least abundant elements are the platinum group elements that also occur in relative chondritic abundances. Based on this observation, Wänke and Dreibus (1988) suggested that the Earth consists of two major chemically distinct components. The dominant component was strongly volatile depleted and highly reduced and the second was less volatile depleted and oxidized with nearly chondritic abundances of the volatile elements (e.g., Braukmüller et al. 2018, 2019). This model was further refined and modified (O'Neill 1991; Albarède 2009; Schönbächler et al. 2010; Albarède et al. 2013; Ballhaus et al. 2013) and is consistent with the following scenario: in a first stage highly reduced and volatile element depleted accreted to form proto-Earth, followed by a major chemical differentiation of the body into a metallic core and a silicate mantle. Subsequently an oxidized component was added to this proto-Earth (e.g., Ballhaus et al. 2013) followed by the segregation of a sulfide melt ("Hadean matte", O'Neill 1991). In a final stage a small amount of primitive material was added creating the late veneer and resulting in the relative chondritic abundances of the platinum group elements (e.g., Chou 1978). The two major components can be ascribed to a depleted proto-Earth that makes up ca. 85% of the present-day BSE and about 15% of a volatile richer component (e.g., O'Neill 1991). In terms of bulk composition of the Earth, the ratio of the two bodies is ca. 90:10, making the smaller body equivalent in size to Theia, the proposed body that impacted proto-Earth and led to the formation of the Moon (e.g., Canup and Asphaug 2001). A further implication of this accretion model is that proto-Earth was more volatile element depleted than known primitive meteorites and also more highly reduced. This may indicate that the material had a strong similarity to Mercury (Nittler et al. 2019), making this planet a prime target for a sample return missing to improve our understanding of the chemical and isotopic composition of the inner solar system during its early evolution (e.g., van der Kaaden et al. 2019).

3 Nucleosynthetic Isotope Constraints on Terrestrial Building Blocks

The missing volatile-depleted and reduced component indicated based on elemental concentrations can be further constrained by nucleosynthetic anomalies. Over the past decade, distinct nucleosynthetic anomalies in a variety of elements (Ca, Ti, Cr, Ni, Zr, Mo, Ru, Pd, Ba, Nd etc.) have been well characterised for the different meteorite parent bodies, Mars, and Earth. The only two major bodies with virtually identical mass-independent isotope compositions are Earth and Moon. These nucleosynthetic variations are powerful tools for provenance studies of planetary building blocks. The isotope variations stem from the heterogeneous distribution of presolar grains in the protoplanetary disc from which the planets formed. Primitive meteorites still today contain nm- to μm-sized presolar grains. These grains can possess extreme isotope compositions that were established in distinct stellar environments (e.g., AGB stars, novae) before the solar system formed (e.g., Zinner 2014; Hoppe et al. 2017). These presolar grains were gradually destroyed during parent body heating and are thus most abundant in very primitive meteorites of low metamorphic grade (e.g., Huss et al. 2003; Davidson et al. 2014). Some of the abundance variations in primitive meteorites and those between planetary bodies, however, result from incomplete homogenisation of this presolar dust in the solar nebula (Burkhardt and Schönbächler 2015). This heterogeneous distribution of presolar grains can account for the unique nucleosynthetic isotope compositions for many elements in the different meteorite parent bodies, Mars and the Earth (e.g., Cr, Ti, Mo, Zr; Dauphas et al. 2004; Trinquier et al. 2007, 2009; Leya et al. 2008; Qin et al. 2010; Burkhardt et al. 2011; Akram et al. 2015) (Figs. 3, 4). A preferred explanation for the origin of nucleosynthetic variations is selective thermal destruction of less robust presolar carriers (e.g., silicates), which occurred when dust was transported to hotter regions in the solar protoplanetary disc (Trinquier et al. 2009; Akram et al. 2015). Alternatively, it has been suggested that this isotope variability is a remnant of a heterogeneous distribution of this dust in the molecular cloud from which the solar system formed (Nanne et al. 2019). A recent study attributes these variations to mixing of dust that formed in the local interstellar medium with a second less abundant dust fraction (stardust) mainly consisting of presolar grains that formed around AGB-stars (Ek et al. 2019). This scenario also provides an explanation as to why s-process variations are generally smaller in heavy elements (e.g., Nd, W) compared to those in the mass region of Zr and Mo. Presolar grains only survive a limited time in the interstellar medium (Heck et al. 2020) and, therefore, were generally formed around higher metallicity stars than the remaining interstellar matter that on average has been synthesised over the whole galactic chemical evolution. Moreover, higher metallicity stars produce higher amounts of elements in the mass region of Zr and Mo compared to the heavy elements (e.g., Nd, W, Pt). This signal is imprinted in the presolar stardust component and thus leads to smaller anomalies in the heavy elements in the solar system bodies (Ek et al. 2019).

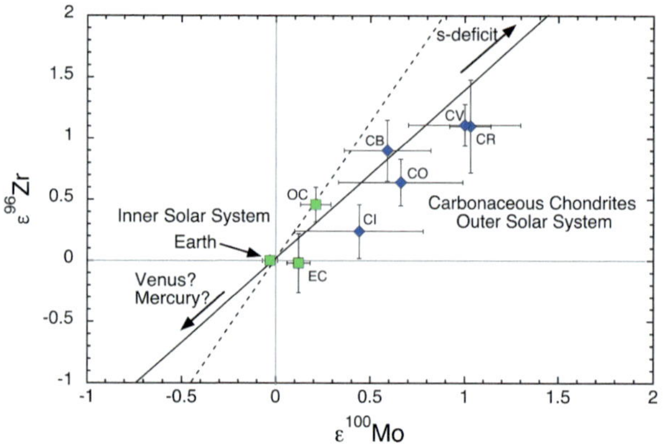

Fig. 3 Nucleosynthetic Zr and Mo isotope data for bulk meteorites and Earth are correlated despite their different geochemical properties (Zr: lithophile and Mo: moderately siderophile). The Earth shows the most extreme isotope composition closest to the *s*-process end member. Venus and Mercury may display even more extreme isotope compositions. Carbonaceous chondrites originating from further out in the solar system constitute the other end of the mixing array. Non-carbonaceous bodies are shown in green. Also shown are the mixing lines calculated for the addition of *s*-process Zr and Mo predicted from Arlandini et al. (1999) (dashed line, classical model) and Bisterzo et al. (2011) (black solid line) to a terrestrial composition. CI, CB, CO,CV, CR: different groups of carbonaceous chondrites, OC: ordinary chondrites, EC: enstatite chondrites. Figure updated from Akram et al. (2015). Zr data are from Akram et al. (2015), and Mo data are from Burkhardt et al. (2011) and Render et al. (2017)

However, independent of their exact origin, nucleosynthetic isotope variations are powerful tools to determine (i) genetic relationships of building blocks and planetary objects and (ii) the physio-chemical processes that influenced the planetary materials on their path from dust to planets. Several studies (e.g., Dauphas et al. 2004; Akram et al. 2015; Akram and Schönbächler 2016; Fitoussi et al. 2016; Render et al. 2017; Schiller et al. 2018; Liebske and Khan 2019) used nucleosynthetic isotope variations to trace the buildings blocks of the Earth. This is generally done by comparing the terrestrial isotope composition to those of meteorites, preferentially chondrites. It is important to note that depending on the geochemical behaviour of an element, its abundance in the silicate part of Earth (i.e. BSE) can be dominated by material delivered during different phases of accretion. Highly siderophile elements (HSEs), such as the platinum group elements, record the latest accretion history of Earth after core formation ceased. This is because previously delivered HSEs were stripped from the mantle into the Earth's core with high efficiency. With decreasing siderophile behaviour, the BSE composition reflects the integration of material accreted over larger parts of Earth's late accretion history (e.g., Mo isotopes; Dauphas 2017), while lithophile elements remained in the silicate part and reflect the average composition of the material delivered over the whole accretion history. Despite these distinct characteristics, elements with different geochemical behaviour in the mass region of Zr, Mo, Ru and Pd (Akram et al. 2013, 2015; Akram and Schönbächler 2016; Burkhardt et al. 2011; Dauphas et al. 2004; Fischer-Gödde et al. 2015; Ek et al. 2019) show correlated isotope effects indicating a two-component mixture with *s*-process material as one end member (Fig. 3). This implies that the average isotopic composition of accreted materials did not drastically change over time, although minor deviations are likely (e.g., for Mo isotopes; Budde et al. 2019). Strikingly, the Earth possesses the most *s*-process enriched composition for these elements and no mixture of known meteoritic materials can build the Earth. The known materials most similar to Earth are enstatite chondrites with isotopic compositions identical or very close to the Earth, when considering nucleosynthetic isotope variations and O isotopes (e.g., Akram et al. 2015; Akram and Schönbächler 2016; Bouvier and Boyet 2016; Burkhardt et al. 2016; Fukai and Yokoyama 2017; Dauphas 2017; Render et al. 2017; Boyet et al. 2018). Since enstatite chondrites are generally considered to originate from the inner solar system, this suggests that the Earth is mainly built from material of the inner solar system. However, while this meteorite group fits well in terms of isotopic composition, the group significantly differs from Earth in its bulk elemental abundances (see Sect. 2). Dynamical considerations and numerical models (e.g., O'Brien et al. 2006; Hansen 2009; Brasser 2013) also indicate that terrestrial planets mainly accreted material from their own local annulus. In dynamical simulations, materials from further away from the Sun (= volatile-rich material) are generally acquired in the later accretion stages (e.g., Morbidelli et al. 2000; O'Brien et al. 2014).

The observation that the Earth has the most s-process enriched composition so far identified for bulk planetary material, while members of the volatile-rich carbonaceous chondrite group belong to the most s-process depleted, severely limits the amount of carbonaceous chondrites that can be added to the Earth. Average estimates are around 10% (Fitoussi et al. 2016; Dauphas 2017; Budde et al. 2019), in good agreement with evidence from radiogenic isotopes for volatile element addition (up to 40%, Schönbächler et al. 2010). This volatile-rich carbonaceous chondrite material is added late during the main accretion history of the Earth to a more reduced composition likely typical for the inner solar system (Fig. 2). It hints to the so far elusive, reduced reservoir in the inner solar system (see Sect. 2).

However, it is possible that the observed trend of increasing s-process enriched compositions for bodies closer to the Sun (Fig. 3) continues further with Venus and Mercury. If so, material from inside the Earth's orbit can counterbalance volatile-rich, oxidised material from the outer solar system and therefore allows for much larger contributions of these materials. The most likely hosts with a significant portion of this volatile depleted and reduced component could be the inner solar system planets Mercury and Venus. They might contain larger amounts of this elusive material, if the solar nebula was indeed chemically zoned with stronger volatile depleted and s-process enriched material that formed closer to the Sun. These inner planets could preserve the missing link that is needed to form the Earth from materials that are derived from different regions of the solar system. Characterising this material would allow a better quantification of the accretion models for Earth that can describe the amount of different materials accreted in combination with concomitant differentiation of the growing planet. This illustrates that samples from Venus or Mercury play a fundamental role in answering the question of the building blocks of the Earth.

3.1 Evidence from the Dichotomy

The conclusions drawn from volatile elements concentrations and s-process anomalies are also supported by mass-independent variations in the O (mostly caused by irradiation processes. i.e. photo dissociation), Cr and Ti isotope compositions (nucleosynthetic anomalies) of different meteorites and the Earth. For O isotope analyses, samples from Venus and Mercury would help to address the O isotope homogeneity the inner solar system reservoir and compare this to Mars and the outer solar system, potentially tracing a signal of water in the disk (Krot et al. 2015).

In contrast to the expressed s-process anomalies (at first order produced by AGB stars) in the mass region of Zr, the nucleosynthetic Cr and Ti isotope compositions are generated by the heterogeneous distribution of dust formed in a supernovae environment (again at first order, less dominant contribution of other sources such as the s-process are likely;

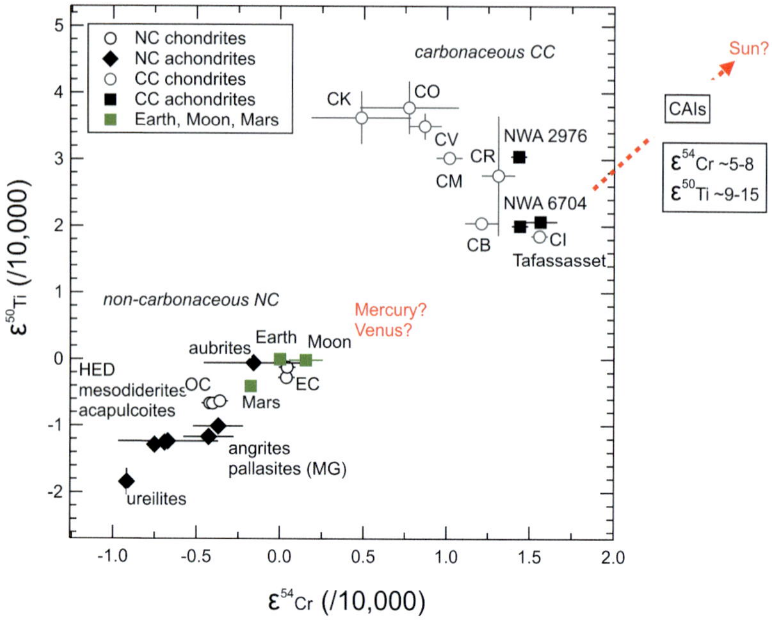

Fig. 4 ^{54}Cr vs. ^{50}Ti mass-independent isotope data in epsilon (parts per 10,000 deviation of the ratio from the respective terrestrial standards) of whole-rock chondrites and achondrites from the so-called carbonaceous and non-carbonaceous reservoirs (e.g., Warren 2011; Kruijer et al. 2019 for further details on their significance, originally discovered by Trinquier et al. 2007 and Leya et al. 2008), and Earth's mantle, lunar Apollo samples and Martian meteorites for Earth, Moon and Mars bulk compositions, respectively. Possible locations of Venus, Mercury, and the Sun are indicated but have not been measured yet. CAIs = calcium, aluminium-rich inclusions have reported ϵ^{54}Cr and ϵ^{50}Ti anomalies that range from ~5–8 and ~9–15, respectively (Larsen et al. 2018; Leya et al. 2009; Trinquier et al. 2009), and are thus represented, with the potential location of the Sun, off the chart. CC = carbonaceous chondrites and related groups, EC = enstatite chondrites (EH and EL groups), OC = ordinary chondrites (H, L and LL). Data sources from Trinquier et al. (2007), Leya et al. (2008, 2009), Göpel et al. (2015), Sanborn et al. (2019)

Schönbächler et al. 2011). Strikingly for all these elements (Ca, Ti, Cr and Ni), the Earth is not an endmember in the solar system, but falls at the end of the trend defined by Earth, Mars and non-carbonaceous (NC) meteorites (e.g., Steele et al. 2012; Schiller et al. 2018) (Fig. 4). This is often called the NC reservoir as compared to the carbonaceous chondrite (CC) reservoir, defined by CC. The two reservoirs are also referred to as solar system dichotomy and may have been separated by the early formation of Jupiter (Olsen et al. 2016; Kruijer et al. 2017; 2019; Alibert et al. 2018) or a ringed structure in the protoplanetary disc (Brasser and Mojzsis 2020). The CI chondrites fall on the extension of the trend of the NC material together with the Earth as the closest body to the NC reservoir (Fig. 4). Based on this observation, it was proposed that the Earth might have accreted significant amounts of CI materials through pebble accretion (Schiller et al. 2018). On the other hand, the data are also consistent with adding carbonaceous chondrite material to the Earth in the later history of Earth's main accretion phase through planetesimal accretion. The location of Mercury and Venus in Fig. 3 and 4 can potentially help to distinguish between the two scenarios (planetesimal accretion versus pebble accretion). For example, if the Cr and Ti isotope compositions of Mercury and Venus fall to the right of the Earth in Fig. 4, the addition of more CI-like material to Mercury and Venus can account for the Cr–Ti trend. If so, then the two planets must also fall between Earth and CI chondrites in the Zr–Mo graph (right to the Earth) (Fig. 3). If they fall to the left of the Earth, as indicated in Fig. 3, this does not allow for the accretion of more CI material and thus favours planetesimal accretion, where the material is not limited to CI chondrites. Instead, it would again indicate that Mercury and Venus have accreted material from an inner solar system reservoir that is enriched in the enigmatic strongly volatile depleted material from which also the Earth mainly accreted and from which no identified material is available in our meteorite collection.

3.2 Moon Formation

The similarity of Earth and the Moon in nucleosynthetic isotope compositions can be exploited to constrain the origin of the hypothetical impactor Theia that collided with the Earth to form the Moon. Recently, a new variation of the giant impactor scenario has been proposed, in which the Moon formed from an originally completely vaporized accretion disk that fully equilibrated with the silicate Earth (Synestia; Lock et al. 2018). In such a scenario, tracing of the impactor material using nucleosynthetic isotope anomalies is not possible. However, it is debated whether full equilibration indeed occurred. If it was not achived, comparison of elemental and isotopic measurements of the Earth and Moon today allows predicting the isotopic composition of the lunar impactor Theia using mass balance calculations. It requires the combination of isotopic data with the results of dynamic simulations that constrain the proportions of the impactor material that merged with the proto-Earth or mixed with terrestrial material to form the Moon. The prediction for the isotopic composition of the impactor can then be compared to data from solar system bodies to identify its origin (Akram and Schönbächler 2016). Such considerations yield that Theia most likely had a composition very similar to the Earth and, hence, must have formed in a similar orbit as the Earth (Dauphas et al. 2014; Meier et al. 2014; Melosh 2014; Akram and Schönbächler 2016). To fully exploit these models, however, knowledge of the compositions of Venus or Mercury to characterize the material that formed inside the Earth's orbit in the protoplanetary disk would be essential.

4 Constraints on the Reservoirs and Timescales of Formation for the Terrestrial Planets from Radiogenic Isotopes

4.1 Timescales of Planetary Accretion and Differentiation

The timescales of planet formation can be resolved with a precision of tens of thousands of years for processes that occurred 4.5 billion years ago using the daughter products of short-lived and now extinct radionuclides. Among the short-lived radionuclide isotope systems, 16 parent-daughter pairs have been used so far in cosmochemistry as high-resolution chronometers (Davis and McKeegan 2014). The parent isotopes were freshly synthesized within different stellar environments and injected shortly before or after the collapse of the solar nebula (e.g., Lugaro et al. 2014). They can be identified by finding a correlation of mass-independent isotopic variations of the radiogenic daughter products with the parent/daughter elemental ratio (e.g., Lee et al. 1976). Short-lived radioactive isotopes such as ^{26}Al ($T_{1/2} \approx 0.71$ Ma), ^{53}Mn ($T_{1/2} \approx 3.7$ Ma), ^{107}Pd ($T_{1/2} \approx 6.5$ Ma), ^{182}Hf ($T_{1/2} \approx 8.9$ Ma), ^{92}Nb ($T_{1/2} \approx 37$ Ma), and ^{146}Sm ($T_{1/2} \approx 103$ Ma) are principally used as high-resolution chronometers (down to a few 1,000 yr) and tracers for early planetary processes (e.g., Bizzarro et al. 2005; Kleine et al. 2009; Wadhwa et al. 2009; Boyet et al. 2010; Schönbächler et al. 2010; Bouvier et al. 2011; Iizuka et al. 2016; Pape et al. 2019). These isotope systems provide relative age information that can be placed on an absolute timescale using high-precision ages obtained from the long-lived U–Pb isotope system. For the latter a precision of 300,000 yr is now achievable. The decay products of short-lived radionuclides can be measured with high-precision and for some systems also with high spatial resolution by mass spectrometry (e.g., secondary ion mass spectrometry, SIMS; thermo-ionization MS, TIMS, multi-collector inductively coupled plasma MS, MC-ICPMS). They can be used as isotopic tracers and chronometers for the dynamic evolution of the nebula, if the parent isotopes were homogeneously distributed (e.g., Nyquist et al. 2009). Several short-lived radionuclides are also of interest to interpret isotopic anomalies identified in meteorites within a chronological context. This is the case of the decay of ^{247}Cm to ^{235}U ($T_{1/2} \approx 15.6$ Ma), which affects accurate ^{207}Pb–^{206}Pb age determinations of early solar system objects and processes (Brennecka and Wadhwa 2012; Tissot et al. 2016) relevant to planetary formation.

Chondrites come from asteroids that did not fully melt because these bodies formed relatively late (few Myr after CAIs) in the history of the solar system or derive from relatively small parent bodies. Most chondrules identified in chondrites (beside metal-rich CBs and large single OC, CR and CV chondrules dated by Pb–Pb; Bollard et al. 2017) formed 2–4 Myr after CAIs (e.g., Kita et al. 2013; Kita and Ushikubo 2012; Bollard et al. 2017; Pape et al. 2019) and therefore relatively late during the accretion period (Fig. 5) compared to the parent bodies of many achondrites. There is growing evidence of a variety of planetesimals that formed and differentiated within a short time interval after the condensation started, i.e. as little as 0.1 to 0.3 million years after the formation of CAIs based on iron meteorites (Kruijer et al. 2014). This time period is also within less than 2 Ma for silicate achondrite parent bodies (Barrat et al. 2015) and thus before the formation of most chondrules and the accretion of chondrite parent bodies (e.g., Hunt et al. 2018). Widespread mafic magmas formed on various parent bodies within the first 3 to 5 Myr (Fig. 5) as shown by old U–Pb and Al–Mg ages of basaltic meteorites from angrites, eucrites, diogenites, ureilites, primitive achondrites, or ungrouped meteorites (e.g., Goodrich et al. 2010; Bouvier et al. 2011; Srinivasan et al. 2018).

4.2 Timing of Formation of the Terrestrial Planets

Constraints from the Hf–W system indicate that 95% of the accretion of Mars took place within 5 Ma after the beginning of the solar system, contemporaneously with the formation of planetesimals and chondrules (e.g., Kleine et al. 2009; Dauphas and Pourmand 2011; Pape et al. 2019). The formation of the Earth was more protracted and can be divided into three main intervals (Fig. 5) with (i) the main accretion sequence from dust to oligarchic growth from planetesimals to planetary embryos merging into planets bringing the Earth to about 85 to 90% of its current volume over 30 to 100 Ma (e.g., Kleine et al. 2009), (ii) the Moon-formation event likely bringing about 10% of the Earth's mass between 60 to 100 Ma after the formation of the Solar System (Barboni et al. 2017; Kruijer et al. 2017; Maltese and Mezger 2020), and (iii) late accretion by impacts over 200 to 500 Ma after Earth's core formation bringing about 0.5% of its final mass (e.g., Day et al. 2016). The highly siderophile elements in the silicate Earth were added during this period forming the "late veneer" (Walker 2009; Bottke et al. 2010; Dauphas 2017).

The origin of materials, which may have been involved during this late stage, can be assessed through the measurements of highly siderophile elements in terrestrial and meteorite samples. The group of IAB irons uniquely share their Mo, Ru and Pd isotopic compositions with the silicate Earth (Burkhardt et al. 2011; Fischer-Gödde et al. 2015; Worsham et al. 2017; Ek et al. 2019). The only specimen of EL3 enstatite chondrite that has been measured

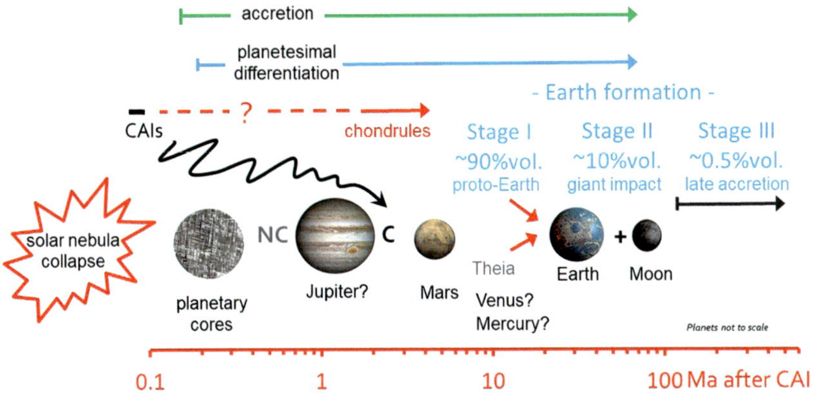

Fig. 5 Summary of timescales for the accretion and differentiation of planetary materials and their connection to Earth's formation (split here into 3 main stages, see also Fig. 2). After the collapse of the solar nebula, the oldest materials dated so far are CAIs, which mark the start of condensation, before their transportation and accretion within the carbonaceous chondrite parent body (C) within the outer regions of the protoplanetary disk. Accretion must start within 0.1 to 0.3 Ma and subsequent differentiation within 1 Ma after CAIs to explain the Hf–W systematics of iron meteorites (Kruijer et al. 2014; Hunt et al. 2018). The isolation of non–carbonaceous (NC) and carbonaceous (C) chondrite reservoirs with distinct isotopic signatures over 2–4 Ma has been linked to the formation of Jupiter within 1–2 Ma (Kruijer et al. 2017, 2019; Alibert et al. 2018). Although such a complete isolation based on Jupiter can be challenged (Brasser and Mojzsis 2020), Jupiter potentially provides a good barrier for at least CAI-sized particles (Desch et al. 2018)

so far is also identical in Ru isotopes to Earth's modern mantle (Fischer-Gödde et al. 2015). These results contrast with most carbonaceous chondrites, which have a range of compositions that limit their contribution to the late silicate Earth's HSE budget (Fischer-Gödde et al. 2015). They imply that carbonaceous chondrites, as strong candidates for the source of volatiles, would have to be involved in the earlier phase of the Earth's making process (i.e. Stage I to II in Fig. 5).

When combined with elemental and isotopic systematics, the chronology of solar system materials provides critical information about the evolution of isotopic reservoirs in space and time. It is thus important to obtain detailed chronological records with high time resolution for planetary processes, which took place within a few 10^4 to 10^7 years after the formation of CAIs, the oldest materials dated in the Solar System (e.g., Bouvier and Wadhwa 2010).

4.3 Constraints on Planetary Building Blocks from Radiogenic Isotope Systems

Due to their close similarities in major element abundances (e.g., Mg/Si, Al/Si) with the Earth's mantle, the carbonaceous chondrites have long been thought to represent the main components that make up the Earth (e.g., Jagoutz et al. 1979; Allègre et al. 1995). The refractory elements in carbonaceous, ordinary and enstatite chondrites have similar normalized abundances (Fig. 1). For this reason, the term Chondrite Uniform Reservoir was coined (CHUR) to represent the Bulk Earth's composition (DePaolo and Wasserburg 1976). Amongst several radio-isotopic systems used to trace and date planetary processes, the ^{176}Lu–^{176}Hf and ^{147,146}Sm–^{143,142}Nd systems have become extremely powerful because they are both composed of two refractory and lithophile elements with similar condensation temperatures and geochemical behaviour during igneous processes. The initial isotopic compositions of the terrestrial planets (e.g., Earth, Mars) have to be constrained to compare and interpret the isotope measurements obtained on Earth and planetary samples. The ratios of the refractory Sm–Nd and Lu–Hf systems are relatively homogeneous, with only a ~4% variation within unequilibrated chondrites (e.g., Bouvier et al. 2008). Variations in certain groups or metamorphic types of chondrites can be explained by secondary processes such as thermal metamorphism, or are due to the presence of exogenic components such as CAIs (Patchett et al. 2004; Bouvier et al. 2008; Martin et al. 2013; Debaille et al. 2017). There is no discernible difference in compositions for the long-lived systems ^{176}Lu–^{176}Hf and ^{147}Sm–^{143}Nd between the CHUR parameters and Earth's mantle array (Bouvier et al. 2008). However, when looking at the short-lived ^{146}Sm–^{142}Nd system, variations in initial ^{142}Nd abundances have been found between chondrite groups and the Earth's modern mantle composition. A ~20 part per million (ppm) difference in ^{142}Nd abundance was originally detected between chondrites and the Earth's mantle compositions (Boyet and Carlson 2005). It was interpreted as an early episode of global differentiation within the Earth (Boyet and Carlson 2005). However, additional measurements of stable and radiogenic Nd isotopes in a variety of chondrite groups and metamorphic types showed that the difference in the initial ^{142}Nd abundance of chondrites and the Earth is related to variable amounts of nucleosynthetic s-process isotopes, originally carried by dust grains within the disc, and not, as initially thought, due to Sm/Nd fractionation and radioactive decay (Bouvier and Boyet 2016; Burkhardt et al. 2016). Amongst the chondrite groups, the average composition of enstatite chondrites is closest to that of Earth with a small ^{142}Nd deficit of -10 ± 12 ppm (2SD), and even closer with -3 ± 5 ppm (2SD) when only considering the EL3 sub-type (Bouvier and Boyet 2016; Burkhardt et al. 2016; Boyet et al. 2018). A ~10 ppm ^{142}Nd gap between EC and BSE, if caused solely by ^{146}Sm decay, would correspond to a maximum of 3% fractionation of the parent/daughter isotope ratios during condensation and accretion of planetary bodies. Enstatite chondrites comprise several sub-groups, possibly originating from different parent bodies (Weyrauch et al. 2018). Amongst these, the rare EL3 chondrites best match the Earth's Nd composition. They may sample a different parent body compared to the other types of enstatite chondrites, which have been more commonly analysed for other isotopic systems. Nevertheless, it is likely that the different types of enstatite chondrites are genetically linked to each other (Boyet et al. 2018). The ordinary and carbonaceous chondrites are clearly distinct from Earth with anomalies of -16 ± 9 ppm (2SD) and -33 ± 13 ppm (2SD) in ε^{142}Nd, respectively (Boyet et al. 2018). Such variations have two potential implications: (1) the Earth formed from material(s) isotopically similar to enstatite chondrites, whereas ordinary and carbonaceous chondrites were only minor contributors, or (2) Earth accreted a larger variety of chondritic materials, which would require the contribution of a missing enriched composition with excesses in initial ^{142}Nd, a component that is currently missing from our collections (Burkhardt et al. 2016). This second implication is identical to that drawn from other isotopes that show larger s-process isotope anomalies (Zr, Mo, Ru and Pd, see above) than Nd.

In addition to the Earth, materials from other known planetary bodies have been also analysed for their ^{146}Sm–^{142}Nd systematics, including samples from the Moon (returned from Apollo and Luna missions and meteorites), meteorites from Mars, Vesta, and other unknown planetary bodies (e.g., angrites and mesosiderites) (e.g., Stewart et al. 1994; Foley et al. 2005; Boyet et al. 2010; Sanborn et al. 2015). The Moon has mass-dependent and mass-independent isotope compositions for many elements (e.g., O, Cr, Ti, Ca, Zr) that are identical within analytical precision or nearly indistinguishable from the Earth within a few tens of parts per million (e.g., Wiechert et al. 2001; Spicuzza et al. 2007; Hallis et al. 2010; Akram and Schönbächler 2016; Young et al. 2016; Greenwood et al. 2018) (Fig. 3). The Moon also appears to have a ^{146}Sm–^{142}Nd evolution identical to the Earth's mantle

(Sprung et al. 2013). Such identical isotopic compositions suggest that the proto-Earth and Theia, the impactor in the giant impact scenario that led to the formation of the Moon, may have formed from chemically similar building blocks or the material from both bodies was homogenized during the impact (e.g., Young et al. 2016; Akram and Schönbächler 2016). This interpretation is also supported by the similarity of nucleosynthetic elements in the two bodies. When considering giant impact scenarios and analytical precision for different isotopic system, some variations in composition for the impactor from the Earth are nevertheless allowed (e.g., Akram and Schönbächler 2016). Even further complexity arises from signatures from late accretion (stage III in Fig. 5) for O and W isotopes (e.g., Kruijer et al. 2014; Greenwood et al. 2018).

Mars formed before the Earth and the Moon (Fig. 5), with about 50% of its mass accreted within only 2 Myr after CAIs based on its ^{182}Hf–^{182}W and ^{60}Fe–^{60}Ni systematics (Dauphas and Pourmand 2011; Tang and Dauphas 2014; Barrat et al. 2015). It has generally closer isotopic similarities with ordinary chondrites (Trinquier et al. 2009; Warren 2011; Brasser et al. 2018) than Earth, and a positive Δ^{17}O of 0.318 ± 0.016‰ (Ali et al. 2016). From stable isotopic relationships, its initial ^{142}Nd abundance was deduced to be about -15 ppm compared to Earth, which would place Mars between enstatite and ordinary chondrite compositions, with a minor contribution from carbonaceous chondrites (Burkhardt et al. 2017a, 2017b).

The Howardite–Eucrites–Diogenite (HED) meteorite suite, sampling Vesta-type asteroids, is the most common group of achondrites with over 1200 eucrites alone available in the meteorite collections to this date. The HEDs are enriched in ^{16}O with a Δ^{17}O of -0.240 ± 0.014‰ (Greenwood et al. 2017) and depleted in ^{54}Cr and ^{50}Ti compared to Earth and Mars (Trinquier et al. 2007, 2009), suggesting accretion from materials that formed within the inner solar system, with lower contributions from carbonaceous chondrite materials. The initial ^{142}Nd abundance for Vesta was estimated from internal mineral isochrons of basaltic and cumulative eucrites (Boyet et al. 2010 and references therein). These meteorites have a similar isotopic evolution as angrites and in particular mesosiderites, which is also reflected in the similarity of U–Pb ages of zircons separated from these meteorites (Haba et al. 2019). Their ^{146}Sm–^{142}Nd evolution, when compared with the crystallization ages, is most consistent with a source that had a composition similar to ordinary chondrites (Gannoun et al. 2011). Thus, the comparison of the ^{146}Sm–^{142}Nd systematics in different materials from the solar system indicates that the Earth has an end member composition and cannot be the end product of mixing of known components. This result is in agreement with the constraints provided by the bulk chemical composition and the variations of nucleosynthetic isotope variation for Earth and extant meteorites.

In the search for planetary building blocks of the Earth, the chronology of planetary accretion and differentiation puts the focus on the investigation of achondrite meteorites to track the composition of precursors and early planetary formation reservoirs. In this search, many new and previously unknown types of primitive and differentiated achondrites have been recovered, primarily from the hot deserts. Some of these have the potential to represent the Earth's original main building material. Based on their mineralogical-geochemical characteristics and/or O-isotope compositions, many of these meteorites have been categorized as "ungrouped", because they do not appear to belong to any previously recognized achondrite types within the meteorite classification nomenclature. These include Northwest Africa (NWA) 7325 and paired meteorites, which were proposed to potentially originate from Mercury based on their elemental compositions. However, this is unlikely because their ancient age of 4563 Ma (within only 4–5 Myr after CAIs) suggests that they originated by re-melting of an early planetary crust that formed within 1.7 Myr of CAIs (Koefoed et al.

2016). Moreover, this crust had to be anorthositic in composition to explain the unique elemental composition of pyroxenes in NWA 7325 and paired meteorites (e.g., NWA 8486) (Frossard et al. 2019). Meteorite sample NWA 11119, an ungrouped achondrite, may be related to NWA 7325 based on its O isotopic composition (Srinivasan et al. 2018). It contains $\sim$20 vol.% of silica polymorphs mostly as tridymite. Such a high abundance of silica is generally found only in differentiated, evolved, crustal rocks on terrestrial planets. Its ^{26}Al–^{26}Mg age is the oldest of any silicate-rich achondrite: 4564.8 $\pm$ 0.3 Ma (Srinivasan et al. 2018). Silica-rich magmatism was thus taking place within $\sim$3 Ma after CAI formation. Both the NWA 7325 and NWA 11119 have O–Cr–Ti anomalies which would place them closer to the acapulcoite and ureilite formation reservoirs (Goodrich et al. 2017; Srinivasan et al. 2018), and therefore away from the innermost part of the protoplanetary disc where Mercury would be expected to be located (Fig. 4). Other ungrouped enstatite-rich achondrites also yielded identical O isotope compositions to Earth (e.g., Itqiy, NWA 5400), but differ in nucleosynthetic anomalies (Burkhardt et al. 2017a, 2017b). These meteorites may represent the crust and mantle of planetesimals and planets that formed in a similar accretion region as Earth before planetesimals were scattered by for example planet migration (e.g., the Grand tack, Raymond and Morbidelli 2014). As such, they may be remnants of Theia or the proto-Earth before the Moon-formation event took place. At present no connection between these meteorites and potential parent asteroids has been identified.

4.3.1 Conclusions

Bulk Earth has an extreme composition, when considering its element abundances as well as radiogenic initial and nucleosynthetic isotope compositions and comparing them with known materials from the solar system available for direct analysis in the form of meteorites. The carbonaceous and ordinary chondrite groups are clearly distinct from Earth's isotopic composition, while enstatite chondrites are the closest to Earth's mass-independent isotopic inventory (Figs. 2 and 3). However, due to their high Si abundance (Jagoutz et al. 1979), the composition of enstatite chondrites are quite distinct in their major element composition from that of Earth and thus cannot represent main precursors of the Earth's total mass. Possible condensation or vapour loss fractionation effects during accretion have been suggested to explain such discrepancies in the Si, Mg and Fe mass-dependent isotope compositions between Earth and chondrite compositions (Dauphas et al. 2015; Hin et al. 2017).

Currently, no mixing model can account for the observed composition of the Earth using known materials. The distinct nucleosynthetic variations observed in different meteorite classes or their components indicate that different regions in the early solar system had unique characteristic compositions. These differences in the isotopic compositions can be used as tracers for the region of origin within the disc for these meteorites and their parent bodies, before being transported and trapped within the asteroid belt (Akram et al. 2015; Akram and Schönbächler 2016; Bouvier and Boyet 2016; Burkhardt et al. 2016; Boyet et al. 2018). However, it is not well constrained where exactly each specific region was located in the protoplanetary disk. These regions can have varied in time and space. The chemical and isotopic diversity of known solar system materials and their difference to the Earth strongly points to a major missing component that possibly formed in the innermost regions of the solar system.

Most meteorites available for study are derived from the asteroid belt. Thus, only the regions beyond the Earth's orbit may be well represented in the meteorite collection. The CAIs are the earliest extant condensates of the solar nebula in our collection. They are considered to have formed close to the young Sun before being transported to the carbonaceous chondrite forming region (e.g., Nanne et al. 2019). Some CAIs share similar ^{142}Nd abundances to the Earth (Bouvier and Boyet 2016) suggesting a closer relationship to the inner solar system in terms of formation reservoirs compared to the accretion region of the carbonaceous chondrites, which may have been located beyond Jupiter (Kruijer et al. 2019). However, other isotope systematics are clearly distinct between the Earth and CAIs (e.g., Δ^{17}O, ^{50}Ti, ^{54}Cr ^{96}Zr; e.g., Clayton 2003; Leya et al. 2009; Trinquier et al. 2009; Akram et al. 2013) suggesting different carriers or formation mechanisms for these isotope effects.

Based on O-isotope abundances, meteorites sample only ca. 110 parent bodies (Greenwood et al. 2017) and many new objects from other reservoirs need to be studied to achieve further breakthroughs about the formation of Earth and the planets in the solar system. So far, no known group of chondrites satisfies the Earth's elemental and isotopic composition and it is likely the Earth formed from a variety of different materials. Based on their stable isotopic compositions achondrites (e.g., ureilites, HED meteorites) are remnants of distinct parent bodies, but they have not substantially contributed to the Earth. Materials that may constitute building blocks of the terrestrial planets likely formed within a short time interval of a few million years, but must also satisfy both elemental abundance and isotopic constraints.

In order to explain the particular isotopic composition of the Earth–Moon system with robust mixing models, samples from the innermost and outer solar system regions are needed to measure the range of chemical and isotopic variations within the solar system as a function of heliocentric distance. In search for the place and composition of the "missing component" that makes up the bulk of planet Earth, a sample return mission that provides material from inside the Earth's orbit would fill a current gap and help constrain models for the formation of rocky planets in general and the habitable Earth in particular. This illustrates that samples from Venus or Mercury can play a fundamental role in answering the question of the building blocks of the Earth. Models for the formation of the Earth would tremendously benefit from the knowledge of the compositions of Venus or Mercury to define the nature of the material that formed inside the Earth's orbit in the protoplanetary disk. This will lead to deeper understanding of the accretion of rocky planets and may help to understand how to build a habitable world.

Acknowledgements The paper benefited significantly from the comments and suggestions provided by three anonymous reviewers. KM and MS acknowledge support through the NCCR PlanetS grant 51NF40-141881. J. Pape and S. Maltese are thanked for providing figures.

References

W. Akram, M. Schönbächler, Zirconium isotope constraints on the composition of Theia and current Moon-forming theories. Earth Planet. Sci. Lett. **449**, 302–310 (2016)

W. Akram, M. Schönbächler, P. Sprung, N. Vogel, Zirconium–hafnium isotope evidence from meteorites for the decoupled synthesis of light and heavy neutron-rich nuclei. Astrophys. J. **777**, 169–181 (2013)

W. Akram, M. Schönbächler, S. Bisterzo, R. Gallino, Zirconium isotope evidence for the heterogeneous distribution of s-process materials in the solar system. Geochim. Cosmochim. Acta **165**, 484–500 (2015)

F. Albarède, Volatile accretion history of the terrestrial planets and dynamic implications. Nature **461**, 1227 (2009)

F. Albarède, C. Ballhaus, J. Blichert-Toft, C-T. Lee, B. Marty, F. Moynier, Q-Z. Yin, Asteroidal impacts and the origin of terrestrial and lunar volatiles. Icarus **222**, 44–52 (2013)

A. Ali, I. Jabeen, D. Gregory, R. Verish, N.R. Banerjee, New triple oxygen isotope data of bulk and separated fractions from SNC meteorites: evidence for mantle homogeneity of Mars. Meteorit. Planet. Sci. **51**, 981–995 (2016)

Y. Alibert, J. Venturini, R. Helled, S. Ataiee, R. Burn, L. Senecal, W. Benz, L. Mayer, C. Mordasini, S.P. Quanz, M. Schönbächler, The formation of Jupiter by hybrid pebble–planetesimal accretion. Nat. Astron. **2**(11), 873 (2018)

C.J. Allègre, J.P. Poirier, E. Humler, A.W. Hofmann, The chemical composition of the Earth. Earth Planet. Sci. Lett. **134**(3–4), 515–526 (1995)

C. Allègre, G. Manhès, E. Lewin, Chemical composition of the Earth and the volatility control on planetary genetics. Earth Planet. Sci. Lett. **185**, 49–69 (1999)

E. Anders, N. Grevesse, Abundances of the elements: meteoritic and solar. Geochim. Cosmochim. Acta **53**, 197–214 (1989)

C. Arlandini, F. Käppeler, K. Wisshak, R. Gallino, M. Lugaro, M. Busso, O. Straniero, Neutron capture in low-mass asymptotic giant branch stars: cross sections and abundance signatures. Astrophys. J. **525**, 886–900 (1999)

M. Asplund, N. Grevesse, A.J. Sauval, P. Scott, The chemical composition of the Sun. Annu. Rev. Astron. Astrophys. **47**, 481–522 (2009)

C. Ballhaus, V. Laurenz, C. Münker, R.O.C. Fonseca, F. Albarède, A. Rohrbach, M. Lagos, M.W. Schmidt, K-P. Jochum, B. Stoll, U. Weis, H.M. Helmy, The U/Pb ratio of the Earth's mantle—a signature of late volatile addition. Earth Planet. Sci. Lett. **362**, 237–245 (2013)

M. Barboni, P. Boehnke, B. Keller, I.E. Kohl, B. Schoene, E.D. Young, K.D. McKeegan, Early formation of the Moon 4.51 billion years ago. Sci. Adv. **3**, e1602365 (2017)

J.A. Barrat, R.C. Greenwood, A.B. Verchovsky, P. Gillet, C. Bollinger, J.A. Langlade, C. Liorzou, I.A. Franchi, Crustal differentiation in the early solar system: clues from the unique achondrite Northwest Africa 7325 (NWA 7325). Geochim. Cosmochim. Acta **168**, 280–292 (2015)

W. Benz, W.L. Slattery, A.G.W. Cameron, Origin of the Moon and the single-impact hypothesis I. Icarus **66**, 515–535 (1986)

A. Bischoff, T. Geiger, Meteorites for the Sahara: find locations, shock classification, degree of weathering and pairing. Meteorit. Planet. Sci. **30**, 113–122 (1995)

S. Bisterzo, R. Gallino, O. Straniero, S. Cristallo, F. Käppeler, The s-process in low-metallicity stars—II. Interpretation of high-resolution spectroscopic observations with asymptotic giant branch models. Mon. Not. R. Astron. Soc. **418**, 284–319 (2011)

M. Bizzarro, J.A. Baker, H. Haack, K.L. Lundgaard, Rapid timescales for accretion and melting of differentiated planetesimals inferred from 26Al-26Mg chronometry. Astrophys. J. **632**, L41–L44 (2005)

J. Bollard, J.N. Connelly, M.J. Whitehouse, E.A. Pringle, L. Bonal, J.K. Jørgensen, Å. Nordlund, F. Moynier, M. Bizzarro, Early formation of planetary building blocks inferred from Pb isotopic ages of chondrules. Sci. Adv. **3**(8), e1700407 (2017)

W.F. Bottke, R.J. Walker, J.M.D. Day, D. Nesvorny, L. Elkins-Tanton, Stochastic late accretion to Earth, the Moon, and Mars. Science **330**, 1527–1530 (2010)

A. Bouvier, M. Boyet, Primitive Solar System materials and Earth share a common initial ^{142}Nd abundance. Nature **537**, 399–402 (2016)

A. Bouvier, M. Wadhwa, The age of the Solar System redefined by the oldest Pb–Pb age of a meteoritic inclusion. Nat. Geosci. **3**, 637–641 (2010)

A. Bouvier, J.D. Vervoort, P.J. Patchett, The Lu–Hf and Sm–Nd isotopic composition of CHUR: constraints from unequilibrated chondrites and implications for the bulk composition of terrestrial planets. Earth Planet. Sci. Lett. **273**, 48–57 (2008)

A. Bouvier, L.J. Spivak-Birndorf, G.A. Brennecka, M. Wadhwa, New constraints on early Solar System chronology from Al–Mg and U–Pb isotope systematics in the unique basaltic achondrite Northwest Africa 2976. Geochim. Cosmochim. Acta **75**, 5310–5323 (2011)

M. Boyet, R.W. Carlson, ^{142}Nd evidence for early (>4.53 Ga) global differentiation of the silicate Earth. Science **309**, 576–581 (2005)

M. Boyet, R.W. Carlson, M. Horan, Old Sm–Nd ages for cumulate eucrites and redetermination of the solar system initial 146Sm/144Sm ratio. Earth Planet. Sci. Lett. **291**, 172–181 (2010)

M. Boyet, A. Bouvier, P. Frossard, T. Hammouda, M. Garçon, A. Gannoun, Enstatite chondrites EL3 as building blocks for the Earth: the debate over the ^{146}Sm–^{142}Nd systematics. Earth Planet. Sci. Lett. **488**, 68–78 (2018)
R. Brasser, The formation of Mars: building blocks and accretion time scale. Space Sci. Rev. **174**, 11–25 (2013)
R. Brasser, S.J. Mojzsis, The partitioning of the inner and outer Solar System by a structured protoplanetary disk. Nat. Astron. (2020). https://doi.org/10.1038/s41550-019-0978-6
R. Brasser, N. Dauphas, S.J. Mojzsis, Jupiter's influence on the building blocks of Mars and Earth. Geophys. Res. Lett. **45**, 5908–5917 (2018)
N. Braukmüller, F. Wombacher, D.C. Hezel, R. Escoube, C. Münker, The chemical composition of carbonaceous chondrites: implications for volatile element depletion, complementarity and alteration. Geochim. Cosmochim. Acta **239**, 17–48 (2018)
N. Braukmüller, F. Wombacher, C. Funk, C. Münker, Earth's volatile element depletion pattern inherited from a carbonaceous chondrite-like source. Nat. Geosci. **12**, 564–568 (2019)
G.A. Brennecka, M. Wadhwa, Uranium isotope compositions of the basaltic angrite meteorites and the chronological implications for the early Solar System. Proc. Natl. Acad. Sci. **109**, 9299–9303 (2012)
G. Budde, C. Burkhardt, T. Kleine, Molybdenum isotopic evidence for the late accretion of outer Solar System material to Earth. Nat. Astron. **3**, 736–741 (2019)
C. Burkhardt, M. Schönbächler, Intrinsic W nucleosynthetic isotope variations in carbonaceous chondrites: Implications for W nucleosynthesis and nebular vs. parent body processing of presolar materials. Geochim. Cosmochim. Acta **165**, 361–375 (2015)
C. Burkhardt, T. Kleine, F. Oberli, A. Pack, B. Bourdon, R. Wieler, Molybdenum isotope anomalies in meteorites: constraints on solar nebula evolution and origin of the Earth. Earth Planet. Sci. Lett. **312**, 390–400 (2011)
C. Burkhardt, L.E. Borg, G.A. Brennecka, Q.R. Shollenberger, N. Dauphas, T. Kleine, A nucleosynthetic origin for the Earth's anomalous ^{142}Nd composition. Nature **537**, 394–398 (2016)
C. Burkhardt, N. Dauphas, H. Tang, M. Fischer-Gödde, L. Qin, J.H. Chen, S.S. Rout, A. Pack, P.R. Heck, D.A. Papanastassiou, In search of the Earth-forming reservoir: mineralogical, chemical, and isotopic characterizations of the ungrouped achondrite NWA 5363/NWA 5400 and selected chondrites. Meteorit. Planet. Sci. **52**, 807–826 (2017a)
C. Burkhardt, T. Kruijer, L. Borg, T. Kleine, Nucleosynthetic heritage and origin of Mars, In: *Goldschmidt Abstracts* (2017b), p. 507
R.M. Canup, E. Asphaug, Origin of the Moon in a giant impact near the end of the Earth's formation. Nature **412**, 708–712 (2001)
J. Chambers, Making more terrestrial planets. Icarus **152**, 205–224 (2001)
C.L. Chou, Fractionation of siderophile elements in the Earth's upper mantle, in *Lunar and Planetary Science Conference Proceedings*, vol. 9 (1978), pp. 219–230
F.J. Ciesla, Radial transport in the solar nebula: implications for moderately volatile element depletions in chondritic meteorites. Meteorit. Planet. Sci. **43**, 639–655 (2010)
P.L. Clay, R. Burgess, H. Busemann, L. Ruzié-Hamilton, B. Joachim, J.M.D. Day, C.J. Ballentine, Halogens in chondritic meteorites and terrestrial accretion. Nature **551**, 614–618 (2017)
R.N. Clayton, Oxygen isotopes in the solar system. Space Sci. Rev. **106**, 19–32 (2003)
N. Dauphas, The isotopic nature of the Earth's accreting material through time. Nature **541**(7638), 521–524 (2017)
N. Dauphas, A. Pourmand, Hf-W-Th evidence for rapid growth of Mars and its status as a planetary embryo. Nature **473**, 489–492 (2011)
N. Dauphas, A.M. Davis, B. Marty, L. Reisberg, The cosmic molybdenum–ruthenium isotope correlation. Earth Planet. Sci. Lett. **226**(3–4), 465–475 (2004)
N. Dauphas, J.H. Chen, J. Zhang, D.A. Papanastassiou, A.M. Davis, C. Travaglio, Calcium-48 isotopic anomalies in bulk chondrites and achondrites: evidence for a uniform isotopic reservoir in the inner protoplanetary disk. Earth Planet. Sci. Lett. **407**, 96–108 (2014)
N. Dauphas, F. Poitrasson, C. Burkhardt, H. Kobayashi, K. Kurosawa, Planetary and meteoritic Mg/Si and δ^{30}Si variations inherited from solar nebula chemistry. Earth Planet. Sci. Lett. **427**, 236–248 (2015)
J. Davidson, H. Busemann, L.R. Nittler, C.M.D. Alexander, F.R. Orthous-Daunay, I.A. Franchi, P. Hoppe, Abundances of presolar silicon carbide grains in primitive meteorites determined by NanoSIMS. Geochim. Cosmochim. Acta **139**, 248–266 (2014)
A. Davis, K. McKeegan, Short-lived radionuclides and early solar system chronology, in *Meteorites and Cosmochemical Processes*. Treatise on Geochemistry, vol. 1 (2014), pp. 361–395
J.M. Day, Planet formation processes revealed by meteorites. Geol. Today **31**, 12–20 (2015)
J.M.D. Day, R.J. Walker, L. Qin, D. Rumble III., Late accretion as a natural consequence of planetary growth. Nat. Geosci. **5**, 614–617 (2012)

J.M.D. Day, A.D. Brandon, R.J. Walker, Highly siderophile elements in Earth, Mars, the Moon, and asteroids. Rev. Mineral. Geochem. **81**, 161–238 (2016)
V. Debaille, J. Van Orman, Q.-Z. Yin Y. Amelin, The role of phosphates for the Lu–Hf chronology of meteorites. Earth Planet. Sci. Lett. **473**, 52–61 (2017)
D. DePaolo, G. Wasserburg, Nd isotopic variations and petrogenetic models. Geophys. Res. Lett. **3**, 249–252 (1976)
S.J. Desch, A. Kalyaan, C.M.D. Alexander, The effect of Jupiter's formation on the distribution of refractory elements and inclusions in meteorites. Astrophys. J. Suppl. Ser. **238**(1), 11 (2018)
M. Drake, K. Righter, Determining the composition of the Earth. Nature **416**, 39–44 (2002)
M. Ek, A.C. Hunt, M. Lugaro, M. Schönbächler, The origin of *s*-process isotope heterogeneity in the solar protoplanetary disk. Nat. Astron. (2019). https://doi.org/10.1038/s41550-019-0948-z
M. Fischer-Gödde, C. Burkhardt, T.S. Kruijer, T. Kleine, Ru isotope heterogeneity in the solar protoplanetary disk. Geochim. Cosmochim. Acta **168**, 151–171 (2015)
C. Fitoussi, B. Bourdon, X. Wang, The building blocks of Earth and Mars: a close genetic link. Earth Planet. Sci. Lett. **434**, 151–160 (2016)
C.N. Foley, M. Wadhwa, L.E. Borg, P.E. Janney, R. Hines, T.L. Grove, The early differentiation history of Mars from ^{182}W-^{142}Nd isotope systematics in the SNC meteorites. Geochim. Cosmochim. Acta **69**, 4557–4571 (2005)
P. Frossard, M. Boyet, A. Bouvier, T. Hammouda, J. Monteux, Evidence for anorthositic crust formed on an inner solar system planetesimal. Geochem. Perspect. Lett. **11**, 28–32 (2019)
R. Fukai, T. Yokoyama, Neodymium isotope heterogeneity of ordinary and carbonaceous chondrites and the origin of non-chondritic 142Nd compositions in the Earth. Earth Planet. Sci. Lett. **474**, 206–214 (2017)
A. Gannoun, M. Boyet, H. Rizo, A. El Goresy, ^{146}Sm–^{142}Nd systematics measured in enstatite chondrites reveals a heterogeneous distribution of ^{142}Nd in the solar nebula. Proc. Natl. Acad. Sci. **108**, 7693–7697 (2011)
J. Gattacceca, A. Bouvier, J. Grossman, K. Metzler, M. Uehara, The meteoritical bulletin, No. 106. Meteorit. Planet. Sci. **54**, 469–471 (2019)
C.A. Goodrich, I.D. Hutcheon, N.T. Kita, G.R. Huss, B.A. Cohen, K. Keil, ^{53}Mn–^{53}Cr and ^{26}Al–^{26}Mg ages of a feldspathic lithology in polymict ureilites. Earth Planet. Sci. Lett. **295**, 531–540 (2010)
C.A. Goodrich, N.T. Kita, Q.-Z. Yin, M.E. Sanborn, C.D. Williams, D. Nakashima, S. Boyle, Petrogenesis and provenance of ungrouped achondrite Northwest Africa 7325 from petrology, trace elements, oxygen, chromium and titanium isotopes, and mid-IR spectroscopy. Geochim. Cosmochim. Acta **203**, 381–403 (2017)
C. Göpel, J.-L. Birck, A. Galy, J.-A. Barrat, B. Zanda, Mn–Cr systematics in primitive meteorites: insights from mineral separation and partial dissolution. Geochim. Cosmochim. Acta **156**, 1–24 (2015)
R.C. Greenwood, T.H. Burbine, M.F. Miller, I.A. Franchi, Melting and differentiation of early-formed asteroids: the perspective from high precision oxygen isotope studies. Chem. Erde **77**(1), 1–43 (2017)
R.C. Greenwood, J.-A. Barrat, M.F. Miller, M. Anand, N. Dauphas, I.A. Franchi, P. Sillard, N.A. Starkey, Oxygen isotopic evidence for accretion of Earth's water before a high-energy Moon-forming giant impact. Sci. Adv. **4**, eaao5928 (2018)
M.K. Haba, J.F. Wotzlaw, Y.J. Lai, A. Yamaguchi, M. Schönbächler, Mesosiderite formation on asteroid 4 Vesta by a hit-and-run collision. Nat. Geosci. **12**, 510–515 (2019)
A.N. Halliday, The origin and earliest history of the Earth, in *Treatise on Geochemistry*, vol. 2, ed. by A.M. Davis, second edn. (Elsevier, Amsterdam, 2014), pp. 509–557
L.J. Hallis, M. Anand, R.C. Greenwood, M.F. Miller, I.A. Franchi, S.S. Russell, The oxygen isotope composition, petrology and geochemistry of mare basalts: evidence for large-scale compositional variation in the lunar mantle. Geochim. Cosmochim. Acta **74**, 6885–6899 (2010)
B.M. Hansen, Formation of the terrestrial planets from a narrow annulus. Astrophys. J. **703**, 1131 (2009)
W.K. Hartmann, D.R. Davis, Satellite-sized planetesimals and lunar origin. Icarus **24**, 504–515 (1975)
P.R. Heck, J. Greer, L. Kööp, R. Trappitsch, F. Gyngard, H. Busemann, C. Maden, J.N. Ávila, A.M. Davis, R. Wieler, Lifetimes of interstellar dust from cosmic ray exposure ages of presolar silicon carbide. Proc. Natl. Acad. Sci. **117**(4), 1884–1889 (2020)
R.H. Hewins, Chondrules. Annu. Rev. Earth Planet. Sci. **25**, 61–83 (1997)
R.C. Hin, C.D. Coath, P.J. Carter, F. Nimmo, Y.-J. Lai, P.A.P. von Strandmann, M. Willbold, Z.M. Leinhardt, M.J. Walter, T. Elliott, Magnesium isotope evidence that accretional vapour loss shapes planetary compositions. Nature **549**, 511 (2017)
P. Hoppe, J. Leitner, J. Kodolányi, The stardust abundance in the local interstellar cloud at the birth of the solar system. Nat. Astron. **1**(9), 617 (2017)
A.C. Hunt, D.L. Cook, T. Lichtenberg, P.M. Reger, M. Ek, G.J. Golabek, M. Schönbächler, Late metal–silicate separation on the IAB parent asteroid: constraints from combined W and Pt isotopes and thermal modelling. Earth Planet. Sci. Lett. **482**, 490–500 (2018)

G.R. Huss, A.P. Meshik, J.B. Smith, C.M. Hohenberg, Presolar diamond, silicon carbide, and graphite in carbonaceous chondrites: implications for thermal processing in the solar nebula. Geochim. Cosmochim. Acta **67**, 4823–4848 (2003)
T. Iizuka, Y.-J. Lai, A. Akram, Y. Amelin, M. Schönbächler, The initial abundance and distribution of 92Nb in the Solar System. Earth Planet. Sci. Lett. **439**, 172–181 (2016)
E. Jagoutz, H. Palme, H. Baddenhausen, K. Blum, M. Cendales, D. Dreibus, B. Spettel, V. Lorenz, H. Wänke, The abundances of major, minor and trace elements in the Earth's mantle as derived from primitive ultramafic nodules, in *Proc. Lunar Planet. Sci. Conf.*, vol. 10 (1979), pp. 2031–2050
A. Johansen, M. Lambrechts, Forming planets via pebble accretion. Annu. Rev. Earth Planet. Sci. **45**, 359–387 (2017)
N.T. Kita, T. Ushikubo, Evolution of protoplanetary disk inferred from ^{26}Al chronology of individual chondrules. Meteorit. Planet. Sci. **47**, 1108–1119 (2012)
N.T. Kita, Q.-Z. Yin, G.J. MacPherson, T. Ushikubo, B. Jacobsen, K. Nagashima, E. Kurahashi, A.N. Krot, S.B. Jacobsen, 26Al-26Mg isotope systematics of the first solids in the early solar system. Meteorit. Planet. Sci. **48**, 1383–1400 (2013)
T. Kleine, K. Mezger, H. Palme, E.E. Scherer, C. Münker, Early core formation in asteroids and late accretion of chondrite parent bodies: evidence from ^{182}Hf-^{182}W in CAIs, metal-rich chondrites, and iron meteorites. Geochim. Cosmochim. Acta **69**, 5805–5818 (2005)
T. Kleine, M. Touboul, B. Bourdon, F. Nimmo, K. Mezger, H. Palme, S.B. Jacobsen, Q.-Z. Yin, A.N. Halliday, Hf–W chronology of the accretion and early evolution of asteroids and terrestrial planets. Geochim. Cosmochim. Acta **73**, 5150–5188 (2009)
P. Koefoed, Y. Amelin, Q.-Z. Yin, J. Wimpenny, M.E. Sanborn, T. Iizuka, A.J. Irving, U–Pb and Al–Mg systematics of the ungrouped achondrite Northwest Africa 7325. Geochim. Cosmochim. Acta **183**, 31–45 (2016)
A.N. Krot, K. Nagashima, C.M.O.D. Alexander, F.J. Ciesla, W. Fujiya, L. Bonal, Sources of water and aqueous activity on the chondrite parent asteroids, in *Asteroids IV* (2015), pp. 635–660
T.S. Kruijer, M. Touboul, M. Fischer-Gödde, K.R. Bermingham, R.J. Walker, T. Kleine, Protracted core formation and rapid accretion of protoplanets. Science **344**, 1150–1154 (2014)
T.S. Kruijer, C. Burkhardt, G. Budde, T. Kleine, Age of Jupiter inferred from the distinct genetics and formation times of meteorites. Proc. Natl. Acad. Sci. **114**, 6712–6716 (2017)
T.S. Kruijer, T. Kleine, L.E. Borg, The great isotopic dichotomy of the early Solar System. Nat. Astron. (2019). https://doi.org/10.1038/s41550-019-0959-9
K.K. Larsen, D. Wielandt, M. Bizzarro, Multi-element ion-exchange chromatography and high-precision MC-ICP-MS isotope analysis of Mg and Ti from sub-mm-sized meteorite inclusions. J. Anal. At. Spectrom. **33**, 613–628 (2018)
T. Lee, D.A. Papanastassiou, G.J. Wasserburg, Demonstration of ^{26}Mg excess in Allende and evidence for ^{26}Al. Geophys. Res. Lett. **3**, 109–112 (1976)
I. Leya, M. Schönbächler, U. Wiechert, U. Krähenbühl, A.N. Halliday, Titanium isotopes and the radial heterogeneity of the solar system. Earth Planet. Sci. Lett. **266**, 233–244 (2008)
I. Leya, M. Schönbächler, U. Krähenbühl, A.N. Halliday, New Titanium isotopes data for Allende and Efremovka CAIs. Astrophys. J. **702**, 1118–1126 (2009)
C. Liebske, A. Khan, On the principal building blocks of Earth and Mars. Icarus **322**, 121–134 (2019)
S.J. Lock, S.T. Stewart, M.I. Petaev, Z. Leinhardt, M.T. Mace, S.B. Jacobsen, M. Cuk, The origin of the Moon within a terrestrial synestia. J. Geophys. Res., Planets **123**(4), 910–951 (2018)
K. Lodders, An oxygen isotope mixing model for the accretion and composition of rocky planets, in *From Dust to Terrestrial Planets* (Springer, Dordrecht, 2000), pp. 341–354
K. Lodders, Solar system abundances and condensation temperatures of the elements. Astrophys. J. **591**, 1220–1247 (2003)
M. Lugaro, A. Heger, D. Osrin, S. Goriely, K. Zuber, A.I. Karakas, B.K. Gibson, C.L. Doherty, J.C. Lattanzio, U. Ott, Stellar origin of the ^{182}Hf cosmochronometer and the presolar history of solar system matter. Science **345**, 650–653 (2014)
A. Maltese, K. Mezger, The Pb isotope evolution of Bulk Silicate Earth: constraints from its accretion and early differentiation history. Geochim. Cosmochim. Acta **271**, 179–193 (2020)
C. Martin, V. Debaille, P. Lanari, S. Goderis, I. Vandendael, F. Vanhaecke, O. Vidal, P. Claeys, REE and Hf distribution among mineral phases in the CV–CK clan: A way to explain present-day Hf isotopic variations in chondrites. Geochim. Cosmochim. Acta **120**, 496–513 (2013)
W.F. McDonough, S.-S. Sun, The composition of the Earth. Chem. Geol. **120**, 223–253 (1995)
M.M. Meier, A. Reufer, R. Wieler, On the origin and composition of Theia: constraints from new models of the Giant Impact. Icarus **242**, 316–328 (2014)
H. Melosh, New approaches to the Moon's isotopic crisis. Philos. Trans. R. Soc., Math. Phys. Eng. Sci. **372**, 20130168 (2014)

H. Miyamoto, T. Niihara, T. Kuritani, P. Hong, J. Dohm, S. Sugita, Cluster analysis on the bulk elemental compositions of Antarctic stony meteorites. Meteorit. Planet. Sci. **51**, 906–919 (2016)

A. Morbidelli, J.E. Chambers, J.I. Lunine, J.M. Petit, F. Robert, G.B. Valsecchi, K.E. Cyr, Source regions and timescales for the delivery of water on Earth. Meteorit. Planet. Sci. **35**, 1309–1320 (2000)

J.A. Nanne, F. Nimmo, J.N. Cuzzi, T. Kleine, Origin of the non-carbonaceous–carbonaceous meteorite dichotomy. Earth Planet. Sci. Lett. **511**, 44–54 (2019)

O. Nebel, K. Mezger, W. van Westrenen, Rubidium isotopes in primitive chondrites: constraints on Earth's volatile element depletion and lead isotope evolution. Earth Planet. Sci. Lett. **lan305**, 309–316 (2011)

L. Nittler, N.L. Chabot, T.L. Grove, P.N. Peplowski, The chemical composition of Mercury, in *Mercury: The View After MESSENGER*, ed. by S.C. Solomon, L.R. Nittler, B.J. Anderson (Cambridge University Press, Cambridge, 2019), pp. 30–51

L.E. Nyquist, T. Kleine, C.Y. Shih, Y.D. Reese, The distribution of short-lived radioisotopes in the early solar system and the chronology of asteroid accretion, differentiation, and secondary mineralization. Geochim. Cosmochim. Acta **73**, 5115–5136 (2009)

D.P. O'Brien, A. Morbidelli, H.F. Levison, Terrestrial planet formation with strong dynamical friction. Icarus **184**(1), 39–58 (2006)

D.P. O'Brien, K.J. Walsh, A. Morbidelli, S.N. Raymond, A.M. Mandell, Water delivery and giant impacts in the 'Grand Tack' scenario. Icarus **239**, 74–84 (2014)

M.B. Olsen, D. Wielandt, M. Schiller, E.M. Van Kooten, M. Bizzarro, Magnesium and 54Cr isotope compositions of carbonaceous chondrite chondrules–insights into early disk processes. Geochim. Cosmochim. Acta **191**, 118–138 (2016)

H.S.C. O'Neill, The origin of the Moon and the early history of the Earth—a chemical model. Part 2: the Earth. Geochim. Cosmochim. Acta **55**, 1159–1172 (1991)

H.St.C. O'Neill, H. Palme, Composition of the silicate Earth: implications for accretion and core formation, in *The Earth's Mantle: Structure, Composition and Evolution—the Ringwood Volume*, ed. by I. Jackson (Cambridge University Press, Cambridge, 1998), pp. 3–126

H. Palme, A. Jones, Solar system abundances of the elements, in *Treatise on Geochemistry* (2003), pp. 41–61

H. Palme, H. O'Neill, Cosmochemical estimates of mantle composition, in *Treatise on Geochemistry*, vol. 3 (2014), pp. 1–38

H. Palme, J. Zipfel, The Earth contains a large fraction of material not represented by meteorites, in *47th Lunar and Planetary Science Conference* (2016), p. 2252

H. Palme, K. Lodders, A. Jones, Solar system abundances of the elements, in *Treatise on Geochemistry*, vol. 2, ed. by H.D. Holland, K.K. Turekian, second edn. (Elsevier, Oxford, 2014), pp. 15–36

R.C. Paniello, J.M. Day, F. Moynier, Zinc isotopic evidence for the origin of the Moon. Nature **490**, 376 (2012)

J. Pape, K. Mezger, A.-S. Bouvier, L. Baumgartner, Time and duration of chondrule formation: constraints from ^{26}Al–^{26}Mg ages of individual chondrules. Geochim. Cosmochim. Acta **244**, 416–436 (2019)

P.J. Patchett, J.D. Vervoort, U. Soderlund, V.J.M. Salters, Lu–Hf and Sm–Nd isotopic systematics in chondrites and their constraints on the Lu–Hf properties of the Earth. Earth Planet. Sci. Lett. **222**, 29–41 (2004)

A.H. Peslier, M. Schönbächler, H. Busemann, S.I. Karato, Water in the Earth's interior: distribution and origin. Space Sci. Rev. **212**, 743–810 (2017)

L. Qin, C.M.O.D. Alexander, R.W. Carlson, M.F. Horan, T. Yokoyama, Contributors to chromium isotope variation of meteorites. Geochim. Cosmochim. Acta **74**, 1122–1145 (2010)

R.A. Raymond, A. Morbidelli, The Grand Tack model: a critical review, in *Proceedings of the International Astronomical Union*, vol. 9 (2014), pp. 194–203

J. Render, M. Fischer-Gödde, C. Burkhardt, T. Kleine, The cosmic molybdenum–neodymium isotope correlation and the building material of the Earth. Geochem. Perspect. Lett. **3**, 170–178 (2017)

D.C. Rubie, D.J. Frost, U. Mann, Y. Asahara, T. Nimmo, T. Kyusei, P. Kegler, A. Holzheid, H. Palme, Heterogeneous accretion, composition and core-mantle differentiation of the Earth. Earth Planet. Sci. Lett. **301**, 31–42 (2011)

M.E. Sanborn, R.W. Carlson, M. Wadhwa, ^{147,146}Sm–^{143,142}Nd, ^{176}Lu–^{176}Hf, and ^{87}Rb–^{87}Sr systematics in the angrites: implications for chronology and processes on the angrite parent body. Geochim. Cosmochim. Acta **171**, 80–99 (2015)

M.E. Sanborn, J. Wimpenny, C.D. Williams, A. Yamakawa, Y. Amelin, A.J. Irving, Q.-Z. Yin, Carbonaceous achondrites Northwest Africa 6704/6693: milestones for early Solar System chronology and genealogy. Geochim. Cosmochim. Acta **245**, 577–596 (2019)

C. Sanloup, A. Jambon, P. Gillet, A simple chondritic model of Mars. Phys. Earth Planet. Inter. **112**, 43–54 (1999)

M. Schiller, M. Bizzarro, V.A. Fernandes, Isotopic evolution of the protoplanetary disk and the building blocks of Earth and the Moon. Nature **555**(7697), 507–510 (2018)

M. Schönbächler, R.W. Carlson, M.F. Horan, T.D. Mock, E.H. Hauri, Heterogeneous accretion and the moderately volatile element budget of Earth. Science **328**, 884–887 (2010)

M. Schönbächler, W.M. Akram, N.H. Williams, I. Leya, Nucleosynthetic heterogeneities of neutron-rich isotopes in calcium aluminum-rich inclusions and bulk solar system materials, in *Workshop on Formation of the First Solids in the Solar System*, vol. 1639 (2011), p. 9085

E.R.D. Scott, A. Krot, Chondrites and their components, in *Planets, Asteroids, Comets and the Solar System*, ed. by A.M. Davis, second edn. Treatise on Geochemistry, vol. 2 (Elsevier, Amsterdam, 2014), pp. 143–200

M.J. Spicuzza, J.M.D. Day, L.A. Taylor, J.W. Valley, Oxygen isotope constraints on the origin and differentiation of the Moon. Earth Planet. Sci. Lett. **253**, 254–265 (2007)

P. Sprung, T. Kleine, E.E. Scherer, Isotopic evidence for chondritic Lu/Hf and Sm/Nd of the Moon. Earth Planet. Sci. Lett. **380**, 77–87 (2013)

P. Srinivasan, D.R. Dunlap, C.B. Agee, M. Wadhwa, D. Coleff, K. Ziegler, R. Zeigler, F.M. McCubbin, Silica-rich volcanism in the early solar system dated at 4.565 Ga. Nat. Commun. **9**, 3036 (2018)

R.C. Steele, C.D. Coath, M. Regelous, S. Russell, T. Elliott, Neutron-poor nickel isotope anomalies in meteorites. Astrophys. J. **758**(1), 59–70 (2012)

B.W. Stewart, D.A. Papanastassiou, G.J. Wasserburg, Sm–Nd chronology and petrogenesis of mesosiderites. Geochim. Cosmochim. Acta **58**, 3487–3509 (1994)

H. Tang, N. Dauphas, ^{60}Fe–^{60}Ni chronology of core formation in Mars. Earth Planet. Sci. Lett. **390**, 264–274 (2014)

F.L.H. Tissot, N. Dauphas, L. Grossman, Origin of uranium isotope variations in early solar nebula condensates. Sci. Adv. **2**(3), e1501400 (2016)

A. Trinquier, J.L. Birck, C.J. Allèallègre, Widespread ^{54}Cr heterogeneity in the inner solar system. Astrophys. J. **655**, 1179–1185 (2007)

A. Trinquier, T. Elliott, D. Ulfbeck, C. Coath, A.N. Krot, M. Bizzarro, Origin of nucleosynthetic isotope heterogeneity in the solar protoplanetary disk. Science **324**, 374–376 (2009)

K.E. van der Kaaden, F.M. McCubbin, P.K. Byrne, N.L. Chabot, C.M. Ernst, C.L. Johnson, M.S. Thompson, Revolutionizing our understanding of the Solar System via sample return from Mercury. Space Sci. Rev. **215**, 49 (2019)

M. Wadhwa, Y. Amelin, O. Bogdanovski, A. Shukolyukov, G.W. Lugmair, P.E. Janney, Ancient relative and absolute ages for a basaltic meteorite: implications for timescales of planetesimal accretion and differentiation. Geochim. Cosmochim. Acta **73**, 5189–5201 (2009)

R.J. Walker, Highly siderophile elements in the Earth, Moon and Mars: update and implications for planetary accretion and differentiation. Chem. Erde **69**, 101–125 (2009)

H. Wänke, G. Dreibus, Chemical composition and accretion history of terrestrial planets. Philos. Trans. R. Soc., Math. Phys. Eng. Sci. **325**, 545–557 (1988)

P.H. Warren, Stable-isotopic anomalies and the accretionary assemblage of the Earth and Mars: a subordinate role for carbonaceous chondrites. Earth Planet. Sci. Lett. **311**, 93–100 (2011)

P.H. Warren, G.W. Kallemeyn, Angrites: a volatile-rich variety of asteroidal basalt (except for alkalis and gallium!). Meteoritics **30**, 593 (1995)

J.T. Wasson, *Meteorites: Their Record of Early Solar-System History* (WH Freeman and Co., New York, 1985). 274 pp.

J.T. Wasson, G.W. Kallemeyn, Compositions of chondrites. Philos. Trans. R. Soc., Math. Phys. Eng. Sci. **325**, 535–544 (1988)

M. Weyrauch, M. Horstmann, A. Bischoff, Chemical variations of sulfides and metal in enstatite chondrites—introduction of a new classification scheme. Meteorit. Planet. Sci. **53**, 394–415 (2018)

U. Wiechert, A.N. Halliday, D.C. Lee, G.A. Snyder, L.A. Taylor, D. Rumble, Oxygen isotopes and the Moon-forming giant impact. Science **294**, 345–348 (2001)

B.J. Wood, D.J. Smythe, T. Harrison, The condensation temperatures of the elements: a reappraisal. Am. Mineral. **104**, 844–856 (2019)

E.A. Worsham, K.R. Bermingham, R.J. Walker, Characterizing cosmochemical materials with genetic affinities to the Earth: genetic and chronological diversity within the IAB iron meteorite complex. Earth Planet. Sci. Lett. **467**, 157–166 (2017)

K. Yokoyama, Y. Kimura, C. Kaito, Experiments on condensation of calcium sulfide grains to demarcate environments for the formation of enstatite chondrites. ACS Earth Space Chem. **1**, 601–607 (2017)

E.D. Young, I.E. Kohl, P.H. Warren, D.C. Rubie, S.A. Jacobson, A. Morbidelli, Oxygen isotopic evidence for vigorous mixing during the Moon-forming giant impact. Science **351**, 493–496 (2016)
E. Zinner, 1.4—Presolar grains, in *Treatise on Geochemistry*, ed. by H.D. Holland, K.K. Turekian, second edn. (Elsevier, Oxford, 2014), pp. 181–213

Space Sci Rev (2020) 216:25
https://doi.org/10.1007/s11214-020-0645-3

Oxygen Isotopes and Sampling of the Solar System

Trevor R. Ireland[1] · Janaina Avila[1] · Richard C. Greenwood[2] · Leon J. Hicks[3] · John C. Bridges[3]

Received: 3 December 2019 / Accepted: 12 February 2020 / Published online: 3 March 2020

Abstract Oxygen is the dominant element in our planetary system. It is therefore remarkable that it shows substantial isotopic diversity both in mass-dependent fractionation, because it is a light element, and in mass-independent fractionation, primarily associated with variation in abundance of ^{16}O. On Earth, the primary variation in isotopic composition is related to temperature-dependent kinetic mass fractionation between hydrosphere and atmosphere. Meteorites provide samples of primitive bodies, that have not experienced melting, and planetesimals that have melted early in their history. Samples of Mars, Vesta, and the Moon are present in the meteorite collections. In meteorites, the cosmochemical fractionation related to the abundance of ^{16}O provides a useful classification scheme. Inclusions in chondrites show a large range in ^{16}O abundances from highly enriched (solar) through to compositions closer to terrestrial (planetary). The variability in ^{16}O appears originally to be related to predissociation and self-shielding of carbon monoxide likely in the primordial molecular cloud. Within the chondrite parent bodies, exchange between ^{16}O-poor fluids and relatively ^{16}O-rich solids created isotopic mixing lines. This model makes specific predictions for isotopic compositions of silicates and water ice throughout the solar system. One prediction, that the Earth should be isotopically heavier than the Sun, appears to be verified. But other tests based on oxygen isotopes within the solar system require either remote analysis or sample return missions. Remote analysis will require new instrumentation and analytical techniques to achieve the precision and accuracy required for three oxygen isotope analysis. Methodologies associated with cavity ring-down spectroscopy appear promising. Sample return appears viable only for the inner solar system including Mars and asteroids.

Role of Sample Return in Addressing Major Questions in Planetary Sciences
Edited by Mahesh Anand, Sara Russell, Yangting Lin, Meenakshi Wadhwa, Kuljeet Kaur Marhas and Shogo Tachibana

✉ T.R. Ireland

1 Research School of Earth Sciences, The Australian National University, Canberra, ACT 2601, Australia

2 Planetary and Space Sciences, School of Physical Sciences, The Open University, Milton Keynes MK7 6AA, United Kingdom

3 Space Research Centre, School of Physics and Astronomy, University of Leicester, Leicester LE1 7RH, United Kingdom

While sample return missions to either Venus or Mercury appear highly challenging, the scientific benefits are immense both in oxygen isotope characterisation, and in a variety of other geochemical analyses. Measurement of three oxygen isotopes throughout the solar system would further our concepts for formation of other solar systems, and give us insight into the general mechanisms of planetary system formation and the role of water in the formation and evolution of the chondrite parent bodies and planets.

Keywords Oxygen isotopes · Sample return · Remote analysis · Cosmochemistry

1 Introduction

Oxygen is a most remarkable element. It is the most abundant element on Earth. Combining with silicon it forms silicates, which are the most abundant minerals in the mantle and crust. The interrelationship with the oxidation state of iron potentially controls the core size of the terrestrial planets. Combining with hydrogen it forms water, a molecule which can single handedly define the habitability of a planet. Liquid water defines Earth. Water ice is common in the outer solar system. Condensation of water at the snow line of the early solar system potentially allowed the rapid formation of Jupiter. Understanding the processes that control oxygen in the materials of the solar system is therefore fundamental.

Another remarkable characteristic of oxygen, and the main focus of this paper, is the isotopic variability of oxygen in the solar system. Oxygen shows large mass dependent fractionation on Earth, especially between hydrosphere, atmosphere, and water precipitation. Calibration of these fractionations has allowed temperature proxies to be developed that allow insights into Earth's climate in the past.

In extraterrestial samples, including primitive and differentiated meteorites and their solar condensate components, variations associated with the abundance of ^{16}O are predominant. These can range from an 8% enrichment in ^{16}O to 20% deficits relative to terrestrial oxygen isotope ratio (e.g., Ireland 2012; Yurimoto 2018). The source of this isotopic diversity appears to be related to the chemistry of carbon monoxide rather than any particular nucleosynthetic carrier. Carbon monoxide can experience photodissociation under UV radiation, and self-shielding of the different isotopologues can create isotopic diversity (Bally and Langer 1982; Lada et al. 1994). One of the main products of this dissociation is water that is isotopically enriched in ^{17}O and ^{18}O. This mechanism yields testable predictions for the composition of the Sun, and for the compositions of silicates and water ice on outer solar system moons. A fundamental achievement of the NASA Genesis mission (Burnett et al. 2003) was to show that indeed the Earth is different from the Sun in a way that is consistent with the predissociation and self-shielding model (McKeegan et al. 2011).

The exploration of our solar system with respect to the nature and distribution of oxygen isotopes is therefore a priority in attempting to understand the formation processes of our solar system. Potentially this will guide our understanding of the formation of other exoplanetary systems.

The key to further our understanding of oxygen isotope compositions lies in the ability to either accomplish remote analysis of the three stable oxygen isotopes, or to bring samples back from solar system bodies. The analysis of three oxygen isotopes on Earth is enabled by the use of rather extreme chemical processing (use of highly reactive fluorine compounds) and analysis with rather heavy magnetic sector mass spectrometers. Neither of these aspects is particularly well suited to space flight. The alternative of a sample return mission is expensive, risky, and difficult. Still, if we are to realise an ambition to measure oxygen isotopes throughout the solar system, a combination of remote analysis and sample return missions will almost certainly be required.

This paper reviews the systematics of oxygen isotopes in our solar system including nomenclature, analysis protocols, isotopic processing on Earth as a guide to processes in the solar system, and the extent of isotopic diversity revealed in the solar system. The terrestrial oxygen isotope geochemistry section has been guided by the classic texts of Faure (1986) and Hoefs (2009). The extraterrestrial work is initiated from reviews of oxygen isotopes in the solar system by Ireland (2012), Greenwood et al. (2017), and Yurimoto et al. (2008). From these starting points we look at various possibilities to further our understanding of the solar system through remote analysis and sample return missions.

2 Oxygen Isotope Measurements

2.1 Nomenclature

Oxygen has three stable isotopes: ^{16}O (99.76%), ^{17}O (0.04%) and ^{18}O (0.20%), where the percentages in brackets refer to approximate overall terrestrial abundances. Oxygen isotope ratios show significant variation at the percent level in terrestrial materials as well as solar system materials. While a determination of absolute isotope abundances requires stringent measurement calibrations, in general usage an oxygen isotope composition is expressed as isotope ratios relative to the major isotope ^{16}O, i.e. $^{17}O/^{16}O$ and $^{18}O/^{16}O$.

Measured oxygen isotope ratios are then generally calibrated to a reference material and expressed as a deviation from that reference material in units of parts per thousand, i.e.,

$$\delta^{17}O = [(^{17}O/^{16}O)_{meas}/(^{17}O/^{16}O)_{std} - 1] * 1000\ (‰)$$

$$\delta^{18}O = [(^{18}O/^{16}O)_{meas}/(^{18}O/^{16}O)_{std} - 1] * 1000\ (‰)$$

where the subscript "meas" refers to the measured isotope ratio and "std" refers to the reference standard ratio.

2.2 Standards

The most common currently used reference material is V-SMOW, which is the Vienna Standard Mean Ocean Water standard. The original SMOW represented a mix of a number of ocean water samples from around the world that have been distilled to form pure water (Craig 1961). The limited supply of SMOW had been addressed by the International Atomic Energy Agency (based in Vienna) producing the replicate V-SMOW. The oxygen isotope ratios of this material are:

$$^{17}O/^{16}O = 0.0003799 \pm 8 \text{ (a ratio of approximately 1 in 2632 parts)}$$

$$^{18}O/^{16}O = 0.00200520 \pm 43 \text{ (a ratio of approximately 1 in 499 parts).}$$

The original determination for the absolute $^{18}O/^{16}O$ ratio of SMOW was carried out by Baertschi (1976) and this value is still taken as the reference for this ratio to fix the delta scale. The direct determination of $^{17}O/^{16}O$ of SMOW, or V-SMOW, is not a common practice. With the advent of dual-inlet multiple-collector mass spectrometers, relative differences between samples and standards are determined, rather than absolute values. Thus, the

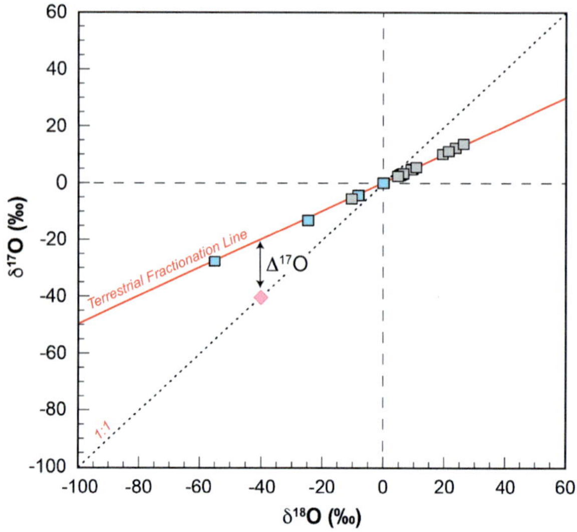

Fig. 1 The Oxygen three-isotope diagram. Oxygen isotope ratios are referenced to deviations from SMOW to generate the delta scale. Mass dependent fractionation related to kinetic isotope mass fractionation lies on the Terrestrial Fractionation Line of slope of approximate 1/2 (i.e. $\delta^{17}O = 0.5 * \delta^{18}O$). $\Delta^{17}O$ refers to the deviation of a composition from the mass dependent fractionation line, and is expressed as the distance to the TFL at the given $\delta^{18}O$ (i.e. $\Delta^{17}O = \delta^{17}O - 0.52 * \delta^{18}O$). For the red diamond at $\delta^{18}O = -40‰$ on the slope 1 reference line, the $\Delta^{17}O$ is $-19.2‰$. Oxygen isotope compositions of meteoritic water (blue squares) and terrestrial silicate and oxide minerals (grey squares) from Tanaka and Nakamura (2012) are consistent with mass dependent fractionation

$^{17}O/^{16}O$ ratio determined by Li et al. (1988) has an uncertainty of approximately 2‰, but this uncertainty is not propagated to measurements because the measurements are made in direct reference to the standard gas (e.g. Jabeen and Kusakabe 1997). A more complete discussion of the issues of calibration to SMOW are provided in Pack and Herwartz (2014) and Miller et al. (2015).

2.3 Mass Dependent Fractionation

Isotope compositions are known to change on Earth through mass dependent fractionation (MDF). For small degrees of MDF, a linear approximation may be used to describe the relationship between $\delta^{17}O$ and $\delta^{18}O$, with the slope being a function of the mass difference in the ratios, i.e. $(17 - 16)/(18 - 16) = 1/2$ (Fig. 1). If exact nuclidic masses are used, then the constant is 0.501. However, the interrelationship between $\delta^{17}O$ and $\delta^{18}O$ is not strictly constant and depends on the processes involved; Matsuhisa et al. (1978) noted a range in the fractionation factor between 0.50 and 0.53.

Typically in isotope geochemistry, the $^{17}O/^{16}O$ ratio is not measured. This is because of the functional relationship between $\delta^{17}O$ and $\delta^{18}O$, and also because of the increased difficulty of measuring $^{17}O/^{16}O$. The lower abundance of $^{17}O/^{16}O$ requires longer measurement times to achieve appropriate precision. Moreover, a common analytical mass spectrometric protocol involves the production and measurement of CO_2 where $^{18}O/^{16}O$ is represented by $^{12}C^{18}O^{16}O/\ ^{12}C^{16}O_2$. (i.e. mass 46/mass 44). However, in this case, mass 45 cannot be used for the $^{17}O/^{16}O$ ratio because of the isobaric $^{13}C^{16}O_2$ molecule interfering with $^{12}C^{17}O^{16}O$. Measurement of three oxygen isotopes requires the measurement of O (or O_2) directly, often released from silicate minerals with fluorine or fluorine compounds (i.e. F_2, BrF_5; Clayton and Mayeda 1963).

2.4 Mass Independent Fractionation

The purpose of measuring three oxygen isotopes is to ascertain if the mass dependence relationship holds, and if not, to identify the process responsible. Deviation from an assumed mass fractionation law is often referred to as Mass Independent Fractionation (MIF), although this particular expression has connotations of there being no relationship with mass dependent fractionation, which may, or may not be the case. A more generic expression is non-mass dependent fractionation and essentially expresses a departure from the accepted mass fractionation law.

This departure is quantified by calculating the residual $\Delta^{17}O$ (Fig. 1), where

$$\Delta^{17}O \approx \delta^{17}O - 0.52 * \delta^{18}O$$

In this case the value of 0.52 was adopted by the Chicago laboratory as an average value of the range found by Matsuhisa et al. (1978). Other laboratories have adopted different values, such as 0.5247 at the Open University based on a linearization method of a number of silicate standards (Miller 2002). However, Pack and Herwartz (2014) advocated a much steeper slope of 0.5305, which is the theoretical high temperature limit for equilibrium fractionation. It should be noted that at extremely high precision levels, such differences in fractionation factors will result in small shifts in $\Delta^{17}O$, and these will be systematically related to $\delta^{18}O$.

In the expression for $\Delta^{17}O$, the deviation of the composition from MDF is explicitly ascribed to the $^{17}O/^{16}O$ ratio. It should be remembered that departures from the MDF line are related to specific processes and reactions and that they may have a large MDF component as well as a MIF. Furthermore, while $\Delta^{17}O$ is a useful algebraic construct, it does not inform particularly well as to the cause of the shift away from the MDF line.

2.5 Analytical Techniques

2.5.1 Gas Source Mass Spectrometry

Gas source mass spectrometry (GSMS) is generally referred to as a bulk analytical technique because the sample is removed from its original context and chemically treated to release oxygen. The material can be solid, liquid, or gas, but the analytical methodology is based on the introduction of a gas (e.g. CO_2, O_2) to the mass spectrometer. However, for analysis of $\delta^{18}O$ and $\delta^{17}O$, oxygen must be used as the source gas because of the presence of unresolvable isobars with CO_2, and specifically with the major species $^{12}C^{16}O^{17}O$ and $^{13}C^{16}O_2$ at mass 45.

The controlled and complete release of oxygen from rocks is not straightforward (see review by Greenwood et al. 2017). Early chemical release of oxygen was based on "bombs", robust sealable containers made of a metal (e.g. Ni) that could be taken to elevated temperature (and thence internal pressure) to promote chemical reaction. Still, the temperature can only be taken to several hundred °C (e.g., ClF_3-HF at 430 °C or F_2-HF at 420 °C for periods of 6 to 20 hours, Baertschi and Silverman 1951) potentially resulting in incomplete reaction. Introduction of potential contaminants to the sample is an issue as hygroscopic NiF_2 is produced during heating of Ni bombs; NiF_2 can react with released sample oxygen compounds potentially compromising O isotopic ratios. The large surface area of the bombs can also contribute significantly to blank levels.

The development of laser fluorination techniques revolutionised oxygen isotope mass spectrometry (Sharp 1990). In this procedure, the sample is locally heated to extremely high temperatures (c. 1200 °C) by a focussed infrared laser, facilitating rapid reaction with the reaction agent e.g. F_2, BrF_5. The direct heating of the sample lowers blanks because the whole containment device is not heated.

The released oxygen is then cleaned up in a gas handling system before introduction to the mass spectrometer. A standard configuration is a dual inlet source that allows for rapid switching between a reference gas (the working standard) and the "unknown" gas, facilitating calibration to the V-SMOW reference scale. Mass spectrometry is generally carried out with a magnetic sector mass spectrometer typically at low resolution (c. 250 M/ΔM) in multiple collector configuration (^{16}O, ^{17}O, ^{18}O simultaneously collected in independent Faraday cup detectors). It is routinely possible to analyze 0.5 to 2 mg mineral and whole-rock samples with a precision in Δ^{17}O of better than ±0.03‰ (2σ) (Greenwood et al. 2017).

2.5.2 *Secondary Ion Mass Spectrometry*

Secondary ion mass spectrometry (SIMS) is based on a primary ion beam sputtering (ablating) a target mineral and analysis of the emitted secondary ions (Ireland 2014; Sangély et al. 2015). Sputtering with an energetic primary beam (of order 10–20 keV) results in ejection and ionisation of atoms and molecules from the target. The target is exclusively solid, and in particular, highly polished solid mineral species allow for the best reproducibility of isotope measurements. The sample must be coated with an electrically conductive material (C, Al, Au, etc.) so as to maintain the potential of the secondary ions being accelerated in the extraction field and passing into a double-focussing magnetic sector mass spectrometer.

For oxygen isotope analysis, a Cs^+ primary beam is typically used and bombardment with a positive ion beam, and extraction of negative ions (and electrons) causes sample charging. The charging effects can be neutralised with electrons delivered to the target either as a flood or as a targeted beam (e.g., Ávila et al. 2020; Kita et al. 2009).

For measurement of ^{18}O and ^{16}O, relatively low mass resolution is required to separate ^{17}OH from ^{18}O (M/ΔM $\approx$ 2000, where ΔM refers to the width of the peak at 10% peak height). This allows the slits used to define the ion beam, and in particular the source slit to be kept relatively wide allowing most of the O^- ion signal to be analysed. However, for three isotope measurement the measurement conditions are more complicated. In particular the presence of $^{16}OH^-$ adjacent to $^{17}O^-$ requires higher mass resolution (Fig. 2) because the $^{16}OH^-$ peak can be higher than the $^{17}O^-$, and particularly if hydrated species are being analysed. This requires higher mass resolution than predicted based on the nuclidic masses because of tailing of $^{16}OH^-$ under the $^{17}O^-$. To achieve this the source slit is narrowed restricting the transmission of O^- into the mass spectrometer. The first three oxygen isotope measurements were made on the small-magnet-radius Cameca ims-3f ion microscopes (McKeegan 1987), but now this type of analysis is largely carried out on large-radius ion microprobes where high precision is required (Ickert et al. 2008; Kita et al. 2010; Loiselle et al. 2019).

The detection system for oxygen isotope analysis is typically based around a multiple collector. For two isotope measurements (such as ^{18}O/^{16}O), Faraday cups in current mode are used, where the current is measured through high ohmic resistors (10^{11}–10^{12} Ω). The small amount of material consumed in an analysis (typically $\sim$ 0.2 to 1 ng) limits the ion signal and so for ^{17}O an electron multiplier can be used to improve signal to noise. However, the electron multiplier is subject to gain drift (and associated dead time drift), which

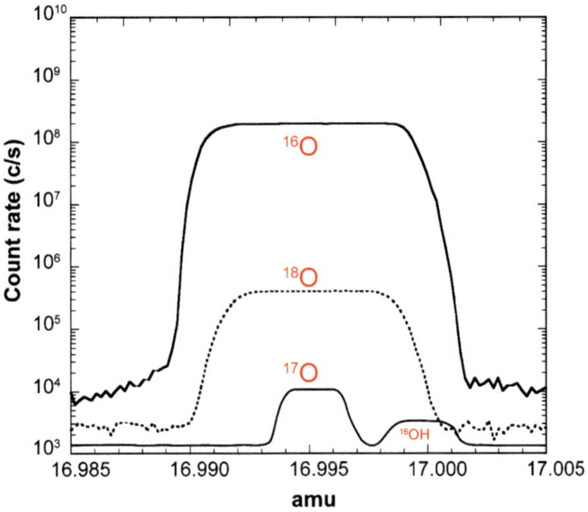

Fig. 2 Mass spectra of the oxygen isotope mass region. X-axis is determined by the $^{17}O^-$ mass position. For measurement of $^{18}O^-$ a mass resolution of ca. 2000 (M/ΔM) is required to resolve $^{18}O^-$ from the interferences of $^{17}OH^-$ and $^{16}OH_2^-$. However, to resolve $^{17}O^-$ from $^{16}OH^-$ a higher mass resolution of ca. 5000 (M/ΔM) is needed to completely resolve the hydrate contribution from $^{17}O^-$ with insignificant tailing contribution

can limit the accuracy of the measurement. Modification of a Faraday cup detection system where the feedback resistor is replaced by a capacitor has proven to be successfully to resolve subpermil variations in $\Delta^{17}O$ (Ireland et al. 2014; Loiselle et al. 2019).

2.5.3 Time-of-Flight Mass Spectrometry

An alternative to magnetic sector mass spectrometry is provided by time-of-flight mass spectrometry (TOF-MS). In this case the mass separation is enabled through the conversion of the extraction potential energy into kinetic energy. Lighter ions have higher velocity than heavier ions allowing time resolution of the individual species from a pulsed source (Wiley and McLaren 1955). The pulsed source can be an ion source (e.g. TOF-SIMS), or a laser, or any configuration that allows for rapid switching (ON/OFF) of the ion beam. The TOF spectrometer is based on having a drift length where the individual species can separate according to velocity. Detection of the signal is through an electron multiplier or a channel plate. Electron multipliers typically have very low background allowing for highly sensitive detection. Channel plates are composed of a resistive slab of material that has holes ("channels") that act as individual multipliers allowing multiple arrivals to be detected. Both detection systems are subject to dead time resulting in multiple coincident arrivals not being detected. The chief benefit of TOF-MS for space flight considerations is that the mass spectrometer is quite simple and a heavy magnetic sector mass spectrometer is not needed. On the other hand, the pulsing of the ion beam with a limited number of ions per pulse severely limits the number of ions collected and hence precision. Recently, new developments on the TOF-MS data acquisition method have been proposed to overcome the lack of high dynamic range. Kawai et al. (2018) presents a technique that combines simultaneously waveform-averaging and ion-counting data acquisition to measure target ions with a high concentration difference. The averaging mode and ion-counting mode are used simultaneously for high and low abundant ions, respectively.

For mass spectrometry on space missions, TOF-MS is a preferred method because it is relatively light and has a low energy consumption. The drift length can be straight and include electrostatic reflection, or can be configured in to a continuous flight path such as in a cyclotron.

2.5.4 Cavity Ring-Down Spectrometry

A novel technique for oxygen isotope analysis is based on the absorption of infrared radiation by water, or other O-bearing, molecules. Laser light is reflected in a cavity with a characteristic decay. If a gas is introduced, a faster decay proportional to abundance of the species occurs at relevant absorption lines. Absorption coefficients for the relevant lines have been determined to allow three isotope measurements in water and minerals (O'Keefe and Deacon 1988; Barker et al. 2011). It offers the possibility of a simple apparatus that is used in the field on Earth for triple oxygen isotope analysis. Even for saline waters, $\delta^{18}O$ measurements to the level of 0.1‰ levels have been achieved (Skrzypek and Ford 2014).

2.5.5 Accuracy and Precision

Precision relates to the level of uncertainty obtained in a given analysis. The fundamental limitation in mass spectrometry is typically taken as the number of ions available. According to Poisson counting statistics, the precision is limited to $1/\sqrt{N}$, where N is the number of counts. For an isotope ratio, it is largely determined by the numerator counts because the denominator is the major isotope. Precision can also be affected by noise in the system, for instance Johnson noise in the electrometer systems associated with Faraday cups.

Precision of oxygen isotope measurements is primarily determined by the amount of material available for analysis. This is typically a characteristic of the analytical technique where the procedure is optimised for a specific amount of material, which has been traded off with the ability to detect heterogeneity within a number of samples. For GSMS, this could be a specified amount of material such as a few mg, while for SIMS, the sample size is the amount of material sputtered during the course of a single spot, which might be as small as 1 ng for oxygen isotope analysis.

For GSMS, sample sizes are selected that optimise the precision that can be obtained through Faraday cup measurements. The signal strength can be adjusted to minimise electrometer noise and the analysis controlled to yield optimum precision and accuracy. Analytical time is not an issue. Ultimate precision at a level of 0.03‰ is attainable.

For SIMS, the spot size (and spot depth) control the amount of material in a single analysis. Precision for a given spot is typically of the order of 0.1‰ for $^{18}O/^{16}O$. For a $\Delta^{17}O$ measurement, the precision is poorer simply because of the low abundance of ^{17}O. For single spot analysis of spots of ca. 15 μm and a $^{16}O^-$ ion beam at 10^8 c/s, the ^{17}O count rate is 40000 c/s and so to get the minimum counts for 0.1‰ in ^{17}O (i.e. 10^8 counts) would take 2500 seconds. However, an analysis of this time would probably cause other changes in the ion emission, which would affect accuracy. A typical maximum analytical time for this type of SIMS analysis might be of order 1000 s, yielding a precision of order 0.3‰. Even then, this requires a stable primary ion beam as well as good stability in all other aspects of the instrument (electrical control, magnetic fields, etc.).

For TOF, the precision depends to a certain extent on the type of ionisation system, usually laser or focussed ion beam. The source controls the amount of material available as well as the ionisation. However, the main limitation for TOF spectrometers is the amount of material that can be included in a single pulse, and the time between pulses. Effectively, a TOF-MS spends most of its time waiting for the ions to arrive at the detector so that there is no interference between pulses of different species.

The *accuracy* of an analysis pertains as to how close the measurement comes to the true composition. A determination of accuracy is typically carried out by analysing reference materials, samples of known composition, and assessing reproducibility of those analyses.

It is extremely difficult to calibrate a measurement to determine absolute isotope abundances and this is generally regarded as unnecessary.

Accuracy is a particular issue for in situ techniques, where matrix composition may affect the ionisation yields and cause changes in the instrumental mass fractionation. As such, SIMS works best on well-defined matrices which have been characterised for chemistry and mineralogy (crystal structure) and for which well-matched standards and reference materials are available (Ireland 2014). SIMS cannot analyse mixtures particularly well. In some cases, atomic or molecular ion species can be used to assess the mix of two minerals and a calibration can be obtained, but in general SIMS measurements of complex matrices yield complex results.

GSMS has the ability through complete acid digestion (etc.) and release of oxygen to avoid any matrix effects. That is, for triple oxygen isotope measurements, only oxygen is ever introduced into the mass spectrometer and this can be run in reproducible conditions.

3 Terrestrial Oxygen Isotopes (Planet Earth)

Oxygen is present as a major component at Earth's surface: in silicate minerals in rocks, as free molecular oxygen in the atmosphere, and as water in the hydrosphere. These reservoirs, and their interactions, are part of the Earth system and can be characterised with oxygen isotopes. From a planetary science perspective, the detail available from the oxygen isotopic analysis of terrestrial materials is pretty much overwhelming. Furthermore, the Earth system as a whole is unique in the solar system in having surface water liquid, solid, and gas as coexisting phases. Effectively this is what makes Earth a habitable planet, and the resulting interactions in biogeochemistry are also evident.

3.1 Hydrosphere

The key processes of the water cycle are evaporation and condensation with fractionation between liquid and vapour (see Cappa et al. 2003). Water vapour that has evaporated from the surface of the ocean is enriched in ^{16}O, with $\delta^{18}O = -9.2‰$ (at 25 °C). However, this fractionation is strongly temperature dependent and ranges from +14‰ at −20 °C to +5‰ at 60 °C (Dansgaard 1964; Cappa et al. 2003). Condensation of water vapour to form precipitation enriches $^{18}O/^{16}O$ such that the composition is similar to the original seawater. The air mass carrying the water vapour becomes further enriched in the light isotope as condensation and precipitation proceed. If this is viewed as an open system, then the isotope compositions of vapour and liquid follow a trend defined by Rayleigh distillation (Dansgaard 1964). However, mixing of air masses probably result in compositions that are not strictly relatable via Rayleigh processing although the general trend of freshwater having lower $\delta^{18}O$ than seawater typically holds.

The strong temperature dependence of oxygen isotope composition of evaporating and condensing water is best viewed in terms of the change in oxygen isotope composition of water precipitation with latitude (Dansgaard 1964). At high latitudes, where the mean annual temperature can fall to below −30 °C, $\delta^{18}O$ is below −30‰ (Fig. 3). The precipitation record in ice cores allows model temperatures of sea surface temperatures to be estimated. These records can be traced back for hundreds of thousands of years into the past (Epstein et al. 1965; Johnsen et al. 1972; Siddall et al. 2003; Mayewski et al. 2004). While the ice records are restricted in time, other fossil proxies can then be used such as precipitation of silica, carbonate, and phosphate from seawater as occurs in various life forms (diatoms,

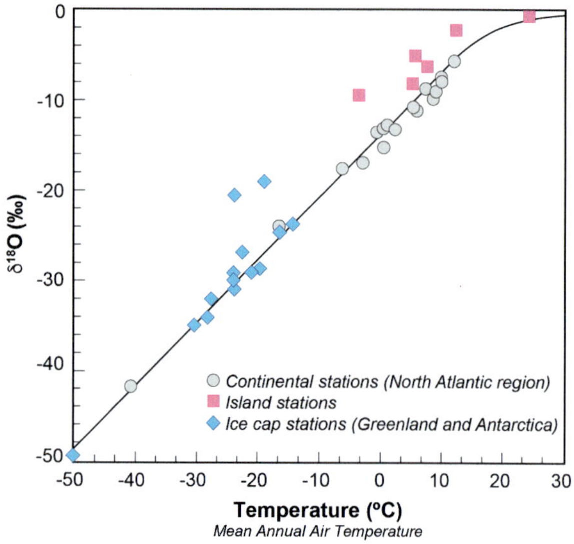

Fig. 3 Oxygen isotope compositions of meteoric waters are closely related to the annual mean temperature of the site of precipitation indicating a temperature related kinetic fractionation. This same effect can be used to model global sea water temperatures from ice cores. After Dansgaard (1964)

foraminifera, and conodonts respectively). Oxygen isotope analysis of fossil materials can then be used to estimate surface temperatures back through the Phanerozoic, provided the system remains closed to later re-equilibration with for example ground waters.

3.2 Atmosphere

Earth's oxygen-rich atmosphere is believed to have formed from progressive addition of oxygen through the Archean as a waste product of photosynthesising cyanobacteria (Holland 2006). Originally Earth appears to have had a reducing atmosphere, potentially similar to the N_2-CH_4 atmosphere found on Titan. Oxygen was initially taken up in oxidation of Fe^{2+} in the oceans with precipitation of magnetite forming Banded Iron Formations (Cloud 1973). In the Great Oxidation Event near the beginning of the Proterozoic, free oxygen became ultimately stable in the atmosphere, and led to stratospheric production of ozone (O_3) and concomitant shielding of Earth's surface from harmful UV radiation. This is reflected in the end of non-mass-dependent S isotope fractionations from SO_2 photodissociation that can only occur and be preserved in a low oxygen atmosphere (Farquhar et al. 2000; Pavlov and Kasting 2002). Detrital uraninite and pyrite, which are susceptible to weathering breakdown in oxidising conditions, disappeared from sedimentary sequences (see Johnson et al. 2014), and, deposition of large-scale Banded Iron Formations ceased (Cloud 1973).

Molecular oxygen (O_2) and stratospheric ozone (O_3) have distinct oxygen isotope compositions (Fig. 4) and the isotopic shift does not obey a mass dependent fractionation law (Heidenreich and Thiemens 1986; Thiemens 1999). Ozone oxygen isotope compositions can be enriched in $\delta^{17}O$ and $\delta^{18}O$ by over 10% (Schueler et al. 1990). It appears that this enrichment is related to stabilisation of asymmetric molecules, which favours those molecules with ^{17}O and ^{18}O. The fractionation lies on a line of slope 1 on the oxygen three isotope diagram.

3.3 Igneous Rocks

Oxygen is a major constituent in all of the principal mantle mineral phases: olivine $[(Mg,Fe)_2SiO_4]$, clinopyroxene $[Ca(Mg,Fe)Si_2O_6]$, orthopyroxene $[(Mg,Fe)SiO_3]$, and

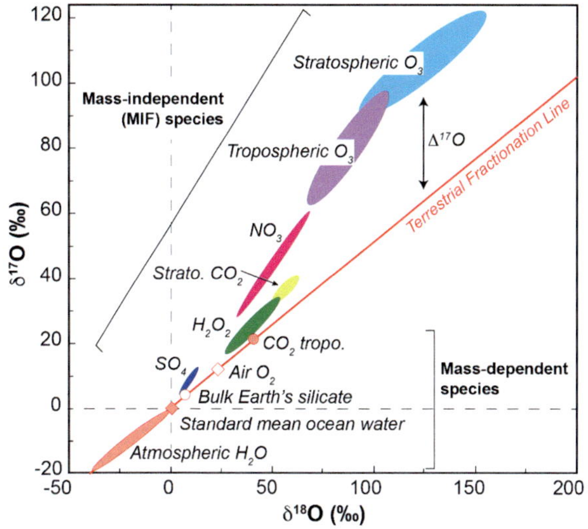

Fig. 4 Oxygen isotopic compositions in Earth's hydrosphere and atmosphere. Kinetic fractionation of oxygen isotopes follows mass dependent fractionation signified as the Terrestrial Fractionation Line. This includes Earth's oceans, surface level Air O_2 and bulk silicate compositions, and the large temperature related fractionation of atmospheric H_2O, which becomes isotopically light towards the poles. Mass independent fractionation occurs in the Earth's stratosphere and troposphere and is related to UV photolysis breaking oxygen molecules to form ozone. More energy states exist for asymmetric molecules leading to a build up of ^{17}O, ^{18}O-rich ozone. Equilibration of ozone with other species (CO_2, NO_2, SO_2) leads to MIF enrichments in these species as well. (Adapted from Thiemens 1999)

spinel [$(Mg, Fe)(Al, Cr)_2O_4$]. As such it constitutes around 50% of Earth's mantle. Earth's crust is a consequence of partial melting in the mantle and the preferential partitioning of incompatible elements into the melt. The Earth's crust has minerals with higher SiO_2 and alkali metals (Na, K) (quartz, feldspars) and lower Mg and Fe (both retained as compatible elements in the mantle). Oxygen isotopes also reflect this trend with quartz and feldspars having heavier oxygen (higher $\delta^{18}O$) leading to a trend of increasing $\delta^{18}O$ from ultramafic mantle rocks through basalts and gabbros and then granites and pegmatites (Fig. 5). In terms of oxygen isotopes, mantle rocks have $\delta^{18}O$ values in a narrow range from $+5$ to $+6.5‰$ relative to the SMOW reference, with granites and pegmatites ranging up to $+13‰$. The total range of $\delta^{18}O$ in igneous rocks at ca. 8‰ is not especially large but sufficient analytical resolution is available to distinguish the rocks according to igneous fractionation of specific minerals and the resultant bulk rocks. In igneous rocks not affected by atmospheric components, the isotopic mass fractionation appears to be entirely mass dependent and no $\Delta^{17}O$ effects are evident.

Oxygen isotopes are observed to partition into different phases in different proportions. The fractionations between coexisting minerals are consistent and can be shown to be temperature dependent. As such, a measurement of suitable phases can be used as a geothermometer (Taylor and Epstein 1962a,b).

3.4 Sedimentary Rocks

Clastic sedimentary rocks reflect the compositions of the particles that comprise the bulk rock. As such, sedimentary rocks can show heterogeneity reflecting differing proportions

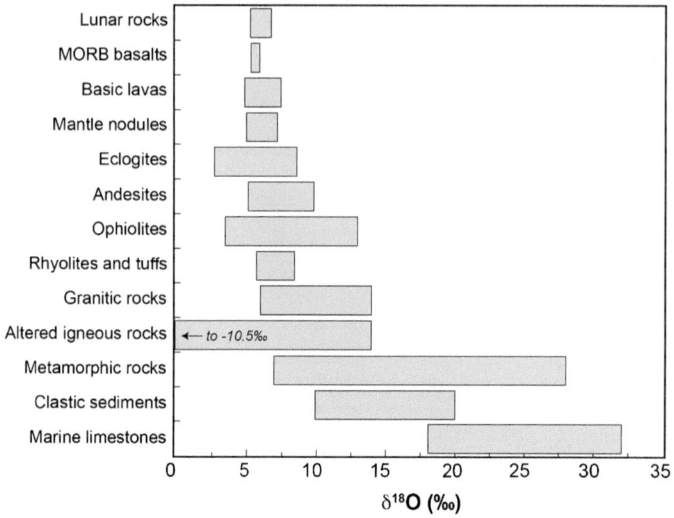

Fig. 5 Range of mass dependent fractionations (expressed as $\delta^{18}O$) in terrestrial reservoirs. Igneous rocks typically show a narrow range of compositions. Alteration of igneous rocks by meteoric water leads to isotopically light compositions. Biogeochemical reactions can lead to isotopically heavy fossils which are deposited as limestones. (Adapted from Hoefs 2009)

of differing source materials. Those source materials can be derived directly from igneous rocks, or can have more complex histories such as from metamorphic rocks. A characteristic of sedimentary rocks is their grain size, porosity, and permeability. The spacing between grains often allows the introduction of meteoric and/or marine water and so sedimentary rocks can be affected by low temperature reactions involving water and replacement of fine grained minerals.

A particular issue for sedimentary rocks is the formation of clay minerals. Notwithstanding the composition of the water reacting with silicate minerals to form clays, oxygen isotope fractionation factors at Earth surface temperatures for minerals such as montmorillonite and kaolinite are quite large at around +27‰ in terms of $\delta^{18}O$ (Savin and Epstein 1970).

Biological activity in the oceans can also fix oxygen in to a variety of phases such as silica, carbonates, and phosphates. Some of the species can reflect a near equilibrium with ocean water while others show variable compositions and systematic biases (i.e., kinetically induced MDF). As such, sediments composed of diatoms [SiO_2], coccoliths and foraminifera [$CaCO_3$], and conodont elements [$Ca_5(PO_4)_3(OH)$] can all reflect the changing oxygen isotopic composition of ocean water, which is typically responding to ice volume on Earth (Shackleton and Opdyke 1973).

3.5 Metamorphic Rocks

Metamorphic rocks can be mixtures of components from igneous, sedimentary, or metamorphic sources and inherit a variety of oxygen isotope signatures. This can include the extreme compositions of clay minerals and meteoric water. In a fully equilibrated metamorphic rock, the crystallisation of a new assemblage leads to a new fractionation of oxygen isotopes between the stable minerals. Equilibration temperatures in metamorphic rocks are lower and thence larger oxygen isotope fractionations are evident between minerals (Taylor and Epstein 1962b). These can also be used as geothermometers (Garlick and Epstein 1967).

However, the tendency of metamorphic systems to not be fully equilibrated, as well as late stage changes in fluid compositions, can cause perturbations.

3.6 The Earth System

Earth is a dynamic planet with 4.5 billion years of history. A lot of the events that led to the growth of Earth are probably the same as formed the other terrestrial planets. Earth's overall oxygen isotope composition was inherited from the planetesimals that contributed to it followed by efficient mixing and equilibration.

Earth is unique in the solar system in having solid, liquid, and gaseous water present. As such, water-rock interactions are important in geological processes. The formation of oceanic crust is followed by hydration reactions (serpentinisation) and subduction at convergent boundaries. Subduction is associated with dehydration, and then contribution to arc volcanoes, which release volatiles into the atmosphere, with some water likely recycled back into the mantle. The presence of water leads to a change in the physical nature of the mantle making it more plastic and amenable to convective overturn. This recycling has carried on over aeons and appears to be another unique aspect of Earth.

4 Extraterrestrial Oxygen

Oxygen is the most common element in the stony meteorites that come to Earth. Robert N. Clayton pioneered the characterisation of oxygen isotopes in extraterrestrial materials and his analytical prowess, thoroughness, and eye for detail offered insights into cosmochemical processes that still colour the field to this day (Clayton 1993, 2003). The discovery of three-oxygen isotope anomalies in meteorite inclusions was profound (Clayton et al. 1973). This completely changed our previous view of the hot homogeneous solar nebula (e.g. Grossman 1972). Isotopic anomalies were found in bulk meteorites, allowing oxygen isotopic compositions to become an essential tool in identifying clans of meteorites of common provenance, and effectively leading to the characterisation of meteorite parent bodies (Clayton et al. 1976).

While geochemical isotopic variations are dominated by mass dependent effects, cosmochemical oxygen isotope systematics in extraterrestrial materials are dominated by variability in ^{16}O. This was originally described as a mixing line in Allende inclusions between a ^{16}O-rich component ($\delta^{17}\mathrm{O} \approx \delta^{18}\mathrm{O} \approx -40‰$; $\Delta^{17}\mathrm{O} \approx -20$) and a fractionated component close to the terrestrial fractionation line, yielding a slope of approximately 0.94. This line is commonly referred to as CCAM (for Carbonaceous Chondrite Anhydrous Mixing; Clayton et al. 1977). Young and Russell (1998) subsequently defined a pure ^{16}O mixing line (slope 1) that includes the ^{16}O-enriched composition of the CAIs, as well as the compositions of the ordinary chondrites. Another representation of ^{16}O mixing in the early solar system is the PCM line (Primitive Chondrules Mineral line; Ushikubo et al. 2012), which is based on the dispersion observed in chondrules from the Acfer 094 carbonaceous chondrite.

4.1 Chondrite Meteorites and Their Components

4.1.1 Bulk Chondrites

Chondrites are regarded as the building blocks of the terrestrial planets. They are essentially cosmic breccias and are regarded as primitive in that they have never been melted and preserve materials of different provenance from the early solar system. A range of materials

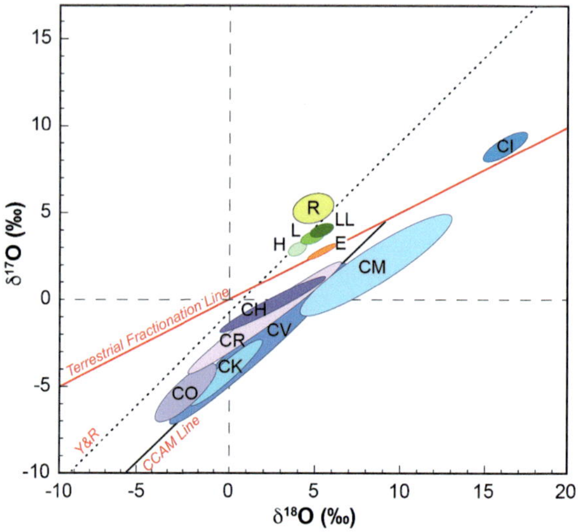

Fig. 6 Oxygen isotope compositions of chondrites are characteristic of different groups and are used for classification. Ordinary chondrites (H, L, LL) are enriched in δ^{17}O, δ^{18}O with a progressive increase in Δ^{17}O from H to L to LL. R chondrites have the highest Δ^{17}O of any of the chondrite groups, while Enstatite chondrites are not resolved from terrestrial. Carbonaceous chondrite compositions appear to be derived from a mixture of mass dependently fractionated oxygen (e.g. CI chondrites), and a ^{16}O-rich reservoir located on the CCAM mixing line. Reference lines shown are the terrestrial mass fractionation line, CCAM (Carbonaceous Chondrite Anhydrous Mineral) mixing line with slope of 0.95 (Clayton et al. 1977), and the ^{16}O fractionation line with slope = 1 (Young and Russell 1998)

exists in these breccias from high temperature inclusions through to the low temperature matrix, which has commonly experienced interaction with water. Chondrites are the most common type of meteorite falls, with Ordinary (O) chondrites by far the most common of these.

Ordinary chondrites have been split in to H, L, and LL groups based on the Fe metal abundance and compositions of the ferromagnesian silicates (see Van Schmus and Wood 1967). H chondrites have the highest overall Fe abundance, coupled with high Fe metal abundance and thence the lowest Fe substitution (i.e., as Fe^{2+}) in to ferromagnesian silicates. The L chondrites have lower total Fe abundances, lower Fe metal contents and higher Fe substitution into ferromagnesian silicates. The LL group are characterised by having the lowest total iron, low metal content, but the highest iron substitution in ferromagnesian silicates. Van Schmus and Wood (1967) used the fayalite content of olivine to distinguish the three groups with H (16–20% Fa), L (21–26%Fa) and LL (27–31% Fa).

The other chondrite types are the enstatite (E) chondrites, carbonaceous (C) chondrites, Rumuruti (R) chondrites, and Kakangari (K) chondrites. The enstatite chondrites are noteworthy in that they are extremely reduced (Van Schmus and Wood 1967); all Fe is present as metal and oxygen is in sufficient deficit that some metallic silicon is present in the Fe metal, and Ca is typically present in oldhamite (CaS). Carbonaceous chondrites have a large fraction of matrix material with a high volatile component (organics, water) and typically mostly oxidised Fe. Rumuruti (Bischoff et al. 2011) and Kakangari (Weisberg et al. 1996) chondrites are quite rare and have similarities to carbonaceous chondrites in terms of matrix abundance, but also have similarities to other chondrite types.

Oxygen isotope compositions of the chondrite families are presented in Fig. 6. The O chondrites show a progressive shift in terms of increasing δ^{17}O and δ^{18}O from H to L to LL, and hence increasing $\Delta^{17}O$ (Clayton et al. 1983, 1991). The changes are relatively small, but distinct. In terms of $\Delta^{17}O$, H chondrites are +0.7‰ with L chondrites at around +1.0‰, and LL chondrites +1.2‰. The R chondrites continue this trend with even higher $\Delta^{17}O$ of +2.0‰. E chondrites have compositions that appear to be coincident with the TFL (i.e. $\Delta^{17}O$ of 0‰; Clayton et al. 1984). The Kakangari chondrites have $\Delta^{17}O$ of −1.2 to −1.8‰, similar to the CR chondrites (Weisberg et al. 1996).

The carbonaceous chondrites show a wide range in oxygen isotopic compositions for the various subtypes. CI chondrites, have compositions of non-volatile elements close to solar abundances (Wang et al. 2019), but have strongly mass fractionated oxygen ($\delta^{18}O \approx +16$‰) and lie close to the TFL (Clayton and Mayeda 1999). CM chondrites also contain mass fractionated oxygen (up to +12‰), but lie significantly below the TFL indicating ^{16}O enrichment. CV chondrites lie close to the ^{16}O-mixing line with ranging down to $\delta^{17}O \approx \delta^{18}O$ of −4‰, with similar characteristics for the other carbonaceous chondrite groups. The bulk CI and CM carbonaceous chondrites appear to have been affected by alteration by water on their parent bodies.

4.1.2 *Chondrules and CAIs*

Chondrites contain high temperature inclusions. The most notable of these are chondrules, whose name provides the name of the rock type, chondrites. Chondrules are dominantly composed of ferromagnesian minerals (olivine, pyroxene), plus mesostasis (glass), some opaque Fe oxides (ilmenite, magnetite) and less commonly Fe metal and sulphide (Connolly and Jones 2016). The oxygen isotope compositions of chondrules in O chondrites lie on a trend of slope ~0.8 on three isotope plots (Bridges et al. 1999).

Chondrules in carbonaceous chondrites show a wider variety of ^{16}O enrichment compared to OC and EC. Populations of chondrules with ^{16}O-rich compositions are common. Clayton et al. (1983) found that chondrules in Allende ranged from $\delta^{18}O$ of +4 to −3‰. Ushikubo et al. (2012) reported two populations with $\delta^{17}O \approx \delta^{18}O \approx -4$‰ and −10‰ from the primitive Acfer 094 carbonaceous chondrite. However, these enrichments are small compared to highly forsteritic relic cores in chondrules, which can have $\delta^{17}O \approx \delta^{18}O \approx -50$‰ (Yurimoto and Wasson 2002). The most ^{16}O-enriched chondrule, found in the Acfer 214 carbonaceous chondrite, has $\delta^{17}O \approx \delta^{18}O \approx -80$‰ (Kobayashi et al. 2003).

Calcium, aluminium-rich inclusions (CAI) are the oldest materials formed in the solar system, and were the first objects to be found with large isotopic anomalies. Clayton et al. (1973) found Allende CV3 CAIs to have ^{16}O-rich compositions (i.e. $\delta^{17}O \approx \delta^{18}O \approx -40$‰; $\Delta^{17}O \approx -20$‰). The enrichment was observed to be mineral dependent with the most ^{16}O-enriched compositions found in pyroxene and spinel while melilite is closer to the TFL (Fig. 7). This has been related to thermal processing in the solar nebula (Ryerson and McKeegan 1994), or to parent body processes including aqueous alteration (e.g. Krot et al. 2019).

The ^{16}O-rich composition was initially interpreted as a nucleosynthetic component that came in to the solar system during the early stages, and potentially associated with other supernova debris products such as ^{26}Al (Clayton et al. 1977). However, the oxygen isotope anomalies are different from isotopic anomalies in other elements in that the variation is in the most abundant isotope and the range of variability is comparatively extremely large. Refractory minerals and inclusions from Murchison CM2 meteorite have been shown to have even larger anomalies, with up to 60‰ enhancements in ^{16}O (Fahey et al. 1987; Ireland et al. 1992; Liu et al. 2009; Kööp et al. 2016).

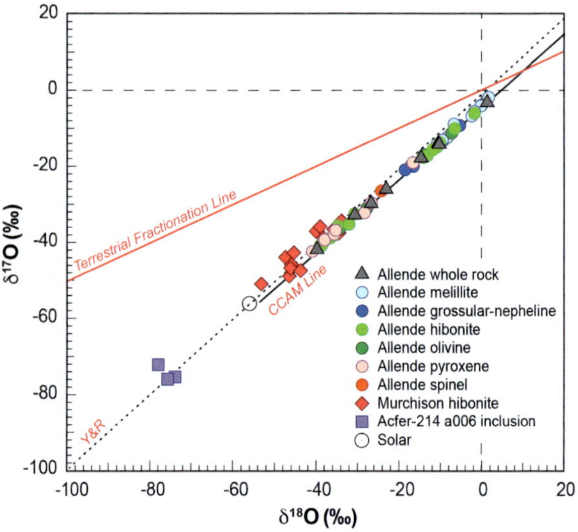

Fig. 7 Oxygen isotope compositions of CAIs are mineralogically controlled in Allende inclusions. Melilite and secondary minerals such as grossular and nepheline are ^{16}O-poor compared to olivine, pyroxene and spinel (Clayton et al. 1977). Whole rock (WR) measurements are also shown; other data are from density separates with dominant mineral indicated. The Allende mineral separates define CCAM, the Carbonaceous Chondrite Anhydrous Mixing line. Murchison hibonite compositions extend to an even more ^{16}O enriched compositions than Allende inclusions (Liu et al. 2009) and approach the solar oxygen isotope composition inferred from solar wind measurement (McKeegan et al. 2011). Also shown is the extreme ^{16}O enrichment measured thus far, that being in glass from the a006 inclusion in the Acfer 214 carbonaceous chondrite (Kobayashi et al. 2003). Terrestrial Fractionation Line shown, as well as a slope 1 reference line passing through the solar composition (broken line)

4.1.3 Mineralogical Controls on Oxygen Isotopic Compositions in Ordinary Chondrites

Bridges et al. (1995, 1999) used three isotope, laser fluorination analyses of mineral separates to show that there is a mineralogical control on the oxygen isotopic compositions of chondrules and clasts in ordinary chondrites. Silica polymorphs (notably cristobalite and tridymite), feldspar and feldspathic-composition glass all have relatively heavy δ^{18}O and δ^{17}O compared to olivines and pyroxenes (Fig. 8). The most ^{16}O-poor composition identified was $\delta^{18}\text{O} = +13‰$ and $\delta^{17}\text{O} = +9.5‰$, $\Delta^{17}\text{O} + 3.0‰$. Furthermore, both laser fluorination on mineral separates and subsequent SHRIMP analyses have shown that the largest isotopic differences between the co-existing ferromagnesian and feldspathic/silica phases have been preserved in the least equilibrated ordinary chondrites e.g. Semarkona, Krymka and Adrar 003 (Bridges et al. 1999; Bridges and Ireland 2015). Other ion probe studies confirm that olivines in ordinary chondrites have relatively ^{16}O-rich compositions, falling closer to the terrestrial fractionation line (e.g. Kita et al. 2010; Saxton et al. 2010). In order to explain this mineralogically controlled isotopic fractionation, Bridges et al. (1999) proposed that the diffusion of ^{16}O-poor water through the OC parent bodies would lead to the isotopic patterns observed. This is due to the propensity for framework silicates and glasses to allow relatively rapid exchange with a fluid. For instance, a high degree of isotopic equilibration can be attained rapidly for feldspar—H_2O exchange e.g. less than 100 years for a 10 mm grain of albite at 400 °C (Giletti et al. 1978). Silica polymorphs and feldspathic glass have

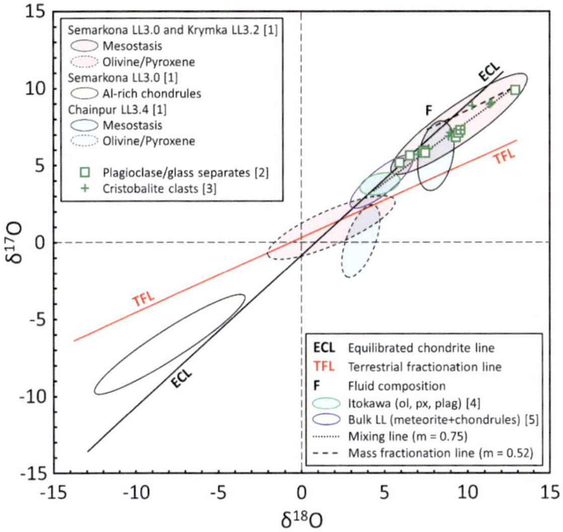

Fig. 8 Oxygen 3-isotope diagram of fluid-solid exchange in ordinary chondrites. Data from [1] Bridges and Ireland (2015); [2] Bridges et al. (1999); [3] Bridges et al. (1995); [4] Nakashima et al. (2013); [5] Clayton et al. (1991). The most highly thermally unequilibrated ordinary chondrites preserve a mineral specific oxygen isotopic fractionation and mass balance. Framework silicates and glass are ^{16}O-poor, whereas olivine, pyroxene are more ^{16}O-rich relative to bulk chondrule and clast isotopic compositions. This is the result of ^{16}O-poor fluid-^{16}O-rich solid exchange processes at low temperatures. We envisage the hydrous fluid reservoir originating on a slope 1.0 line, with mass dependent exchange between that and ^{16}O-rich solids

similar exchange properties. In contrast, gas-solid equilibration times for forsterite and low Ca pyroxene are many orders of magnitude longer.

Magnetite grains in Semarkona LL3.0 analysed by SIMS showed $\Delta^{17}O$ of up to $+5‰$, consistent with the presence of an ^{16}O-poor parent body fluid from which the magnetite formed by oxidation of metal (Choi et al. 1998). These mineralogical-isotopic studies highlight the importance of fluid process on the chondrite parent bodies in setting the oxygen isotopic compositions of chondrules and clasts. There is a large body of independent mineralogical evidence for the significance of fluid-mineral reactions in the OC parent bodes, including corroded chondrule mesostasis (Grossman et al. 2000), smectite clays and magnetite in chondrite matrix e.g. (Alexander et al. 1989; Brearley and Jones 1998; Hutchison et al. 1987). The oxygen isotopic data that show the greatest mineral specific fractionations, preserved in the least equilibrated ordinary chondrites, suggest that the OC isotopic mixing lines and associated aqueous activity predate thermal metamorphism and equilibration on the parent bodies.

The importance of low temperature fluid isotopic exchange processes on oxygen isotopes has been recognized for the carbonaceous chondrites as well, where the large extent of fluid activity—compared to the ordinary chondrites—has led to some minerals being partially equilibrated, leading to a slope 0.94 line on $\delta^{17}O$ $v.$ $\delta^{18}O$ plots (Krot et al. 2019). This is analogous to the slope 0.8 line in OC chondrules and clasts (Bridges et al. 1995, 1999). However, the added complexity of carbonaceous chondrites, with abundant CAIs and highly ^{16}O-rich components, described in the previous section, means that the relatively simple mixing lines preserved in the OCs are not always as easily defined.

We discuss the origin and isotopic composition of the associated water further in Sect. 7.3.2.

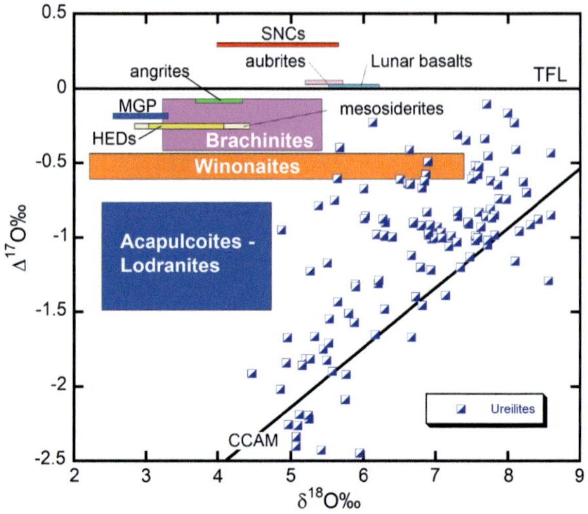

Fig. 9 Oxygen isotopic composition of planetary and asteroidal achondrites (modified after Greenwood et al. 2017). Coloured boxes: $\pm 2\sigma$ variation of group mean values (see Greenwood et al. 2012 for further details). Ureilite data: Clayton and Mayeda (1996), Bischoff et al. (2010), Rumble et al. (2010), Horstmann et al. (2012), Bischoff et al. (2014), Greenwood et al. (2017); SNC data: Franchi et al. (1999); Lunar basalt data: Greenwood et al. (2018); Aubrite data: Barrat et al. (2016); Acapulcoite-Lodranite clan: Greenwood et al. (2012); Brachinite data: Greenwood et al. (2012); Winonaite data: Greenwood et al. (2012); Abbreviations: MGP: main group pallasites, HEDs: Howardite-Eucrite-Diogenite suite, TFL: Terrestrial Fractionation Line, CCAM: Carbonaceous Chondrite Anhydrous Minerals line (Clayton et al. 1977; Clayton and Mayeda 1999)

4.2 Achondrites

Achondrites are meteorites that experienced variable degrees of high temperature processing, such that they lack primary "chondritic" components, including chondrules, CAIs, amoeboid olivine aggregates (AOAs) and matrix (Krot et al. 2014; Scott et al. 2015). Achondrites are derived both from asteroids and larger planetary bodies (specifically Mars and the Moon) and accordingly we refer to the former as asteroidal achondrites and the latter as planetary achondrites. The oxygen isotope composition of planetary and asteroidal achondrites is shown in Fig. 9.

4.2.1 Asteroidal Achondrites

Asteroidal achondrites tend to be divided into two broad subtypes: (i) primitive and (ii) differentiated achondrites (Weisberg et al. 2006; Krot et al. 2014; Scott et al. 2015). Primitive achondrites have near-chondritic bulk compositions, but non-chondritic textures and are thought to represent the products of low degrees of partial melting of a chondritic precursor material (Weisberg et al. 2006; Krot et al. 2014). Differentiated achondrites are compositionally more evolved and have well-developed igneous textures (Mittlefehldt et al. 1998; Krot et al. 2014). Differentiated achondrites are most likely derived from asteroids that experienced moderate to high degrees of partial melting (Krot et al. 2014; Scott et al. 2015).

Since primitive and differentiated achondrites show a complete gradation in terms of the extent to which they underwent melting and mobilization, there is disagreement about which individual groups should be assigned to each of these broad categories (Greenwood et al. 2017). Based on their heterogeneous oxygen isotopic compositions (Fig. 9), Greenwood et al. (2017) suggested that the acapulcoite-lodranite clan, winonaites and silicate-bearing IAB and IIICD irons, brachinites and ureilites are all best considered to be primitive achondrites. In contrast, due to their relatively homogeneous oxygen isotopic compositions (Fig. 9), Greenwood et al. (2017) included the angrites, aubrites, howardite-eucrite-diogenite suite (HEDs), mesosiderites and pallasites as differentiated achondrites.

In addition to achondrites that fit within the well-defined groups, an increasing number of both primitive and differentiated "ungrouped" achondrites are being identified. In a similar way to the ungrouped irons (Chabot and Haack 2006), ungrouped achondrites are of potential significance in that they increase the number of asteroidal bodies from which we have samples to investigate in the laboratory. Thus, Greenwood et al. (2020) estimated that in addition to the 15 to 17 parent bodies from which the primitive and differentiated "grouped" achondrites are derived, ungrouped achondrites may sample a further 28 to 34 parent bodies, significantly increasing the number of "igneous" asteroids from which we have samples.

Meteorites are linked to possible source asteroids through spectral similarities in the visible and near-infrared (Burbine 2016). However, in most cases it is not possible to make fully diagnostic meteorite-asteroid linkages, as most asteroids do not have sufficiently distinct reflectance spectra that would make any potential fit unambiguous (Burbine 2016). A further complication is that in most cases asteroid surfaces are likely affected by space weathering processes (Hapke 2001; Burbine 2016), which could cause a mismatch between the laboratory-derived and asteroidal spectra. Consequently, asteroids are generally linked to meteorites on the basis of taxonomic classes (DeMeo et al. 2009), rather than at the level of unique asteroidal bodies.

The HED-4 Vesta link appears to be a clear exception to the problem of identifying the asteroidal source of a particular group of meteorites. McCord et al. (1970) were the first to measure the reflectance spectra of 4 Vesta and match it to the eucrites. Since then there has been a general acceptance that the HED suite is derived from the Vesta family, which appears to be collisionally linked to asteroid 4 Vesta (McSween et al. 2011). The mesosiderites have also been linked to 4 Vesta (Greenwood et al. 2006, 2015), most recently by Haba et al. (2019), although no conclusive evidence to support this relationship was obtained by the NASA Dawn spacecraft (Peplowski et al. 2013).

4.2.2 Martian Meteorites and Mars Igneous Rocks

As of November 2019 there are 139 known martian meteorite samples (Meteoritical Bulletin database) of which about half are paired, i.e., there are about 65 distinct martian meteorites. The oxygen isotope composition for 98 individual martian meteorites is shown in Fig. 10a.

McSween (1984) first proposed that the shergottite-nakhlite-chassignite (SNC) clan of meteorites were derived from Mars. Prior to this, the energy of any potential impact on Mars had been deemed to be too small to allow ejection of martian material from the surface of Mars into an orbit that could bring them to Earth. The SNC meteorites are believed to come from Mars for three main reasons: geologically young ages, retention of trapped gases with composition similar to martian atmosphere, and high D/H ratios. Of these, the most decisive evidence was trapped noble gases from the martian atmosphere, as determined by the Viking spacecraft in 1976 (Pepin 1985).

The majority of the martian meteorites can also be further subdivided into groups, which include the shergottites (the most abundant, and in turn having basaltic, olivine-phyric and lherzolitic petrographic types), nakhlites (olivine clinopyroxenites), and the chassignites (dunites) (Fig. 10a). In addition, there is the ALH 84001 orthopyroxenite. A recently discovered meteorite NWA 7034 and its pairs including NWA 8114 form a new type of martian

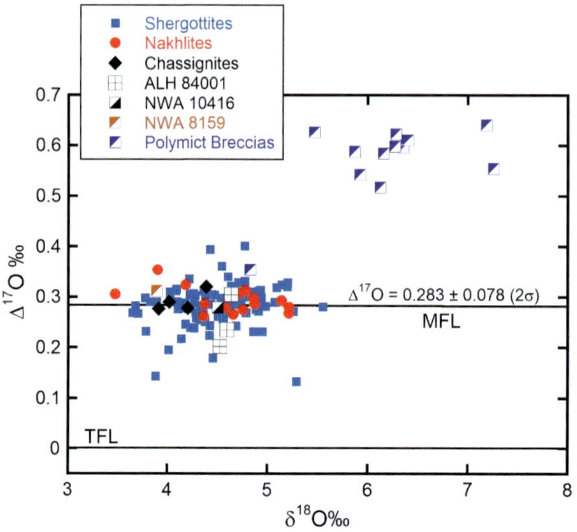

(a)

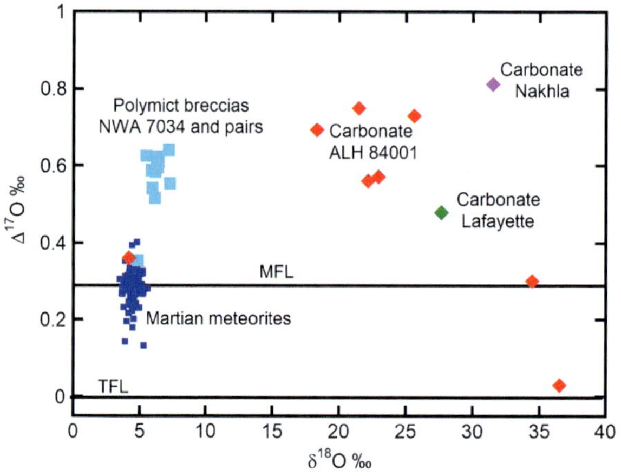

(b)

Fig. 10 (**a**) Oxygen isotopic composition of martian meteorites. MFL = Mars Fractionation Line; TFL = Terrestrial Fractionation Line. The plot is based on a compilation of data for 98 martian meteorites. Since $\Delta^{17}O$ for the original analyses was calculated using a variety of slope factors all the data in the plot has been recalculated using $\Delta^{17}O = \delta^{17}O - 0.525\delta^{18}O$. The MFL value of $\Delta^{17}O = 0.283 \pm 0.078$‰ (2σ) is based on the average of all the analyses plotted in (**a**) with the exception of the polymict breccias (NWA 7034 and pairs). This value is within error of that obtained by Franchi et al. (1999). Date Sources: Clayton and Mayeda (1996), Franchi et al. (1999), Rumble and Irving (2009), Basu Sarbadhikari et al. (2009), Irving et al. (2010), Filiberto et al. (2012), Agee et al. (2013), Llorca et al. (2013), Schwenzer et al. (2013), Wittmann et al. (2015), Ali et al. (2016), Meteoritical Bulletin Database. (**b**) The oxygen isotopic composition of martian meteorites shown in relation to the composition of carbonate extracted from ALH 84001, Nakhla and Lafayette (Farquhar et al. 1998; Farquhar and Thiemens 2000; Shaheen et al. 2015). Data sources for martian samples as per references in caption to (**a**). MFL = Mars Fractionation Line; TFL = Terrestrial Fractionation Line

meteorite, generally referred to as either a basaltic or polymict breccia (Fig. 10a) (Agee et al. 2013; Humayun et al. 2013; MacArthur et al. 2019).

In addition to petrographic-based classifications, the shergottites are subdivided geochemically in terms of their bulk REE abundances into Slightly, Moderately and Heavily Depleted groups, revealing a range of mantle source compositions for the associated primary basaltic melts (Bridges and Warren 2006). The major element compositions of the shergottites have frequently been used to represent a major compositional type representative of the martian crust e.g. McLennan (2003). The martian meteorites provide an essential comparison for igneous rocks identified by Mars landing missions. The Adirondack class basalts from Gusev Crater analysed by MER Spirit have a similar high Fe and low alkali composition (McSween et al. 2008). In contrast Gale Crater igneous float rocks are more alkali-rich than the shergottites and include feldspathic cumulates and trachybasalts (Cousin et al. 2017; Edwards et al. 2017; Sautter et al. 2015).

Martian meteorites cover a range of ages from nearly 4.5 Ga down to 130 Ma, although most ages fall within the Amazonian era between 150 and 650 Ma (Nyquist et al. 2001). However, recently, late-Amazonian <0.5 Ga (Darling et al. 2016; Moser et al. 2013) and early Amazonian >2.0 Ga (Herd et al. 2017; Lapen et al. 2017) ages for shergottite meteorites and the martian breccia NWA 7034 and pairs, have been identified. Zircons identified in martian breccia samples have been dated at 4.43–4.35 Gyr (Humayun et al. 2013; McCubbin et al. 2016) and a mineral isochron yielded a 4.39 Ga Sm-Nd age (Nyquist et al. 2016).

The $\delta^{18}O$ of martian meteorites falls in the range 3.4 to 5.7‰ (Fig. 10a) and is similar to that found in terrestrial basalts and gabbros. However, the triple oxygen isotope compositions of martian meteorites, excluding polymict breccias (see below), is highly distinctive with an average $\Delta^{17}O$ value of 0.283 ± 0.078‰ (2σ) (Fig. 10a). This value is within error of that obtained by Franchi et al. (1999). The relatively homogeneous oxygen isotope composition displayed by martian meteorites is consist with their derivation from a parent body that underwent large-scale melting and mixing. This evidence is consistent with the formation of a magma ocean on Mars early in its history (Zeff and Williams 2019).

A group of meteorites appear to be regolith breccias from Mars (Agee et al. 2013; Humayun et al. 2013). MacArthur et al. (2019) reported that the NWA 8114 martian breccia (paired with NWA 7034) is composed of mineral clasts, including pyroxenes, plagioclase and alkali feldspar, with minor volumes of iron oxides, and chlorapatite, as well as feldspar veins and aureoles. Pyroxene exsolution textures and feldspar cryptoperthite textures reveal slow cooling and a record of plutonic events, prior to the impact-related melting processes. An approximate ^{40}Ar-^{39}Ar maximum age of 1.13–1.25 Ga was found on an individual, separated, augite clast. Together with HRTEM (high-resolution transmission electron microscopy) evidence for oxidative breakdown of the pyroxene, this suggests the regolith within which shock melting occurred was an oxidizing, possibly water-rich environment.

These meteorites are also distinctive in having ferromagnesian minerals with $\Delta^{17}O$ of $+0.30$‰ but with bulk $\Delta^{17}O$ values of up $+0.65$‰ (Fig. 10a) (Ziegler et al. 2013). The oxygen isotope ratio of bulk martian breccia water shows it is not terrestrial contamination, with $\Delta^{17}O$ above the terrestrial fractionation line (Agee et al. 2013).

One of the main topics of research into martian meteorites is the presence of secondary mineral assemblages in the nakhlites and ALH 84001. As mentioned above, martian breccias also show isotopic and mineralogical signs of consolidation within a water-rich regolith environment. The 10 nakhlites contain variable amounts of a unique Fe-rich carbonate, ferric saponite, ferric serpentine and salts assemblage in veins within olivine and the mesostasis e.g. (Bridges et al. 2019; Hicks et al. 2014). Their truncation in places by fusion crust has established their martian origin (Gooding et al. 1991). A K-Ar age suggest formation at ages of ≤ 670 Ma (Swindle et al. 2010). Karlsson et al. (1992) and Farquhar and Thiemens (2000) demonstrated that the nakhlite secondary assemblage minerals have $\Delta^{17}O$ of +0.32‰ but they are not in mass-dependent equilibrium with the silicate minerals. Thus, they record a valuable record of atmosphere-crust interaction. The $\delta^{18}O$ of the carbonate in Nakhla and Lafayette is notably high being 31.5 and 27.6‰ respectively (Fig. 10b) (Farquhar and Thiemens 2000). The $\Delta^{17}O$ of carbonates in martian meteorites is significantly elevated with respect to bulk rock values of the igneous martian meteorites, but overlaps the values measured in the basaltic breccias (Fig. 10b). This evidence indicates that the hydrosphere, lithosphere and atmosphere on Mars are not in isotopic equilibrium (Farquhar and Thiemens 2000).

The other secondary mineral assemblage of bona fide martian origin is the Ca-Mg-Fe carbonate grains which comprise ~1% of ALH84001. The martian origin of these carbonate 'rosettes' was established on the basis of preterrestrial shear zones that cut them (Mittlefehldt 1994). The carbonate may have formed at lower temperatures (<50 °C) than the nakhlite carbonate (Bridges and Schwenzer 2012), but both in metastable, rapidly cooled conditions (Bridges et al. 2001, 2019). ALH 84001 experienced subsequent shock-induced recrystallization. These carbonate grains are hosts to putative evidence of ancient martian life (McKay et al. 1996), a stimulating hypothesis but which has not been generally accepted e.g. (Bradley et al. 1998; Treiman 2005; Treiman and Essene 2011).

The stable isotopic compositions of the ALH 84001 carbonates vary widely, for the most part in concert with their elemental compositions. The $\delta^{18}O$ of carbonates correlate with Mg/Fe (Leshin et al. 1998; Eiler et al. 2002; Holland et al. 2005; Saxton et al. 1998), as do their $^{13}C/^{12}C$ ratios (Niles et al. 2005). Both of these isotopic ratios suggest incorporation of C and O from the martian atmosphere. The carbonate $\Delta^{17}O$ is also characteristic of the martian atmosphere rather than its interior, at +0.03 to +0.75‰ (Fig. 10b; Farquhar et al. 1998; Shaheen et al. 2015). For a detailed review of carbonates and their isotopic compositions in martian meteorites see Bridges et al. (2019).

4.2.3 Lunar Samples and Meteorites

Between July 1969 and December 1972 six manned Apollo missions to the lunar surface successfully returned to Earth over two thousand samples of Moon rock, with a total mass of 382 kg. In addition to this material, over 300 g of surface samples were collected on the Moon and returned by three Russian robotic Luna probes during the period 1970 to 1976. Alan Hills 81005, collected in 1982 in Antarctica, was recognised by Brian Mason of the Smithsonian Institution as having characteristics similar to Apollo samples and was the first identified lunar meteorite (Marvin 1983). There are currently 394 individual lunar meteorite samples listed on the Meteoritical Bulletin Database, although many of these are likely to be paired.

The intense period of scientific activity that followed the Apollo landings led to the development of two key concepts concerning the formation and early evolution of the Moon. Based on the results of the initial examination of lunar rocks, it was suggested by two groups that a relatively deep magma ocean had been present on the Moon immediately after its formation (Smith et al. 1970; Wood et al. 1970). Later it was suggested independently by Hartmann and Davis (1975) and Cameron and Ward (1976) that the Moon formed as a consequence of a planetary-scale collision. These two key concepts remain central to current lunar research.

A collisional formation origin for the Moon was refined by Canup and Asphaug (2001). Their classic Moon-forming giant impact model was constrained by the present angular momentum of the Earth-Moon system and the small relative mass of the Moon's core ($\leq$ 3%

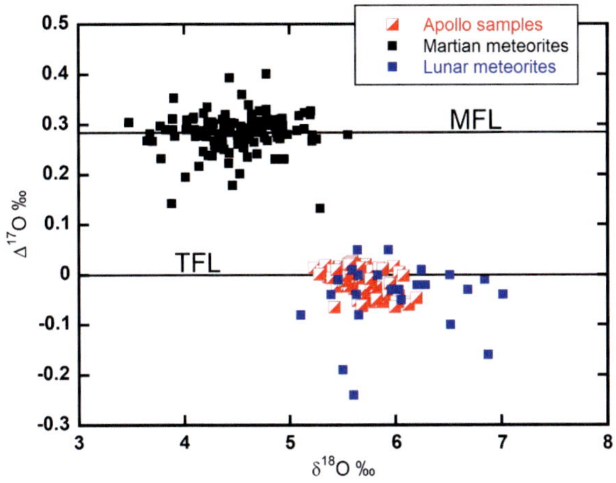

Fig. 11 Oxygen isotope compositions of lunar meteorites and Apollo samples shown in relation to martian meteorites (Fig. 9a). MFL = Mars Fractionation Line; TFL = Terrestrial Fractionation Line. Since $\Delta^{17}O$ for the original analyses was calculated using a variety of slope factors all the data in the plot have been recalculated using $\Delta^{17}O = \delta^{17}O - 0.525\delta^{18}O$. Data sources: Apollo samples—Wiechert et al. (2001), Spicuzza et al. (2007), Hallis et al. (2010), Herwartz et al. (2014), Young et al. (2016), Greenwood et al. (2018); Meteorite samples - Clayton and Mayeda (1996), Meteoritical Bulletin Database

of its total mass). Their model proposed a relatively low energy, oblique impact between the proto-Earth and a Mars-sized impactor, often referred to as Theia. Numerical simulations based on this model suggested that the Moon would have accreted from a debris disc containing approximately 70% impactor-derived material. As a result of this prediction, high-precision oxygen isotope analysis of lunar samples became an important test of the giant impact scenario. Given that Solar System bodies show significant oxygen isotopic variation with respect to $\Delta^{17}O$ (e.g. Fig. 6), the impactor would most likely have had a different oxygen isotopic composition to the proto-Earth. This signature should therefore be present in lunar rocks. However, high precision oxygen isotope studies have failed to detect significant differences between terrestrial and lunar rocks with respect to $\Delta^{17}O$ (Fig. 11) (Wiechert et al. 2001; Spicuzza et al. 2007; Hallis et al. 2010; Young et al. 2016; Greenwood et al. 2018). The small, ppm-level $\Delta^{17}O$ differences between terrestrial and lunar materials that have been detected (Herwartz et al. 2014; Greenwood et al. 2018), most likely relate to post formational processes (Greenwood et al. 2018).

Silicon and titanium isotope studies have also revealed that there are no significant differences between the Earth and Moon (Armytage et al. 2012; Zhang et al. 2012); while tungsten isotope studies indicate that both bodies initially had a closely similar ^{182}W composition (Dauphas et al. 2014). The lack of oxygen isotopic differences between the two bodies has been explained by a model involving post-impact isotopic equilibration within a hot melt-vapour disc (Pahlevan and Stevenson 2007).

A reassessment of the angular momentum constraints of the Earth-Moon system, has led to the formulation of new high-energy Moon-forming impact models, with essentially complete mixing of target and impactor-derived materials (Ćuk and Stewart 2012; Canup 2012; Lock et al. 2018). These scenarios would have involved near total isotopic homogenization. In contrast to such high energy models, it has been pointed out that the inner Solar system was likely to have been well mixed by the time of the giant impact and so the level of isotopic variability may have been small (Dauphas et al. 2014; Dauphas 2017).

As a consequence, Theia and proto-Earth might have been very close isotopically, hence relaxing the objections to the classic giant impact model (Dauphas 2017). This possibility raises the scientific importance, despite the technical difficulties involved, of sample return from either Venus or Mercury (Stevenson and Halliday 2014; Greenwood and Anand 2020). At present we have samples from relatively small (e.g. Moon), or distal (Mars) inner Solar System bodies. However, to test the possibility that the inner Solar System was isotopically well-mixed requires material from one of the other large inner Solar system bodies. Venus would be the preferred candidate being approximately the same size as Earth (Stevenson and Halliday 2014; Greenwood and Anand 2020).

As can be seen in Fig. 11, lunar meteorites show considerably more compositional scatter than lunar samples. This probably reflects a combination of factors, particularly terrestrial weathering, as well as analytical and calibration artefacts. In comparison, analyses of returned lunar samples show less variability. However, a compilation of lunar data of the type shown in Fig. 11 will overestimate the intrinsic level of isotopic variability due to calibration differences between labs (Rumble et al. 2007; Pack and Herwartz 2014; Miller et al. 2015, 2020). To date, returned lunar samples have been collected from a relatively restricted region corresponding to the Procellarum KREEP Terrane. It remains possible that samples recovered from other regions of the Moon will have oxygen isotopic compositions that are distinct from that of our existing sample collection. However, from our current perspective, as informed by the twin concepts of the giant impact and lunar magma ocean, this seems unlikely. Further sample return is required to test and refine these concepts.

4.3 Presolar Grains and Stellar Sources

The discovery of large variations in ^{16}O in CAIs was originally interpreted as representing admixture of grains from a distinctive nucleosynthetic source, likely a supernova. However, the lack of isotopic anomalies in different elements at the same scale as oxygen suggests that CAIs are not themselves of presolar origin. The discovery of hydrogen, carbon, and nitrogen isotope anomalies in interstellar dust particles (McKeegan et al. 1985) validated the notion that presolar material had survived in at least some materials coming from the primordial molecular cloud through the solar nebula to the planetary system. Ultimately the search for presolar grains based on progressive concentration of noble gases led to the discovery of a variety of carbonaceous carriers—SiC, graphite, and diamond, which were found to have distinctive C and N isotopic compositions (Anders and Zinner 1993; Zinner 1998).

These grains are small, really small. While the SiC grains can rarely get up to 30 μm, typically they, and the graphite grains, are less than 5 μm. The nanodiamonds are as they are named, with particle sizes of only a few nanometers. Despite these small sizes, isotopic analyses of multiple elements have been carried out on single grains of SiC (e.g., Ávila et al. 2012, 2013a, 2013b) and graphite (e.g., Jadhav et al. 2013). Then, the development of the Cameca nanoSIMS allowed the analyses of small areas (down to 50 nm across) within individual grains (Stadermann et al. 2005). However, as the size of the grains goes down, so does the absolute number of atoms that can be analysed. These presolar grains can be analysed because their isotopic anomalies are so large that even with the limitations of counting statistics they can be shown to be anomalous. Graphite, diamond, and silicon carbide are carbon-rich and this is the reason they survived the acid digestion used to concentrate them.

However, our solar system is oxygen-rich, not carbon-rich. Some indications of strange oxygen isotope compositions are found in carbon-rich grains. Stadermann et al. (2005) analysed submicron TiC crystals in a 10 μm graphite grain that appears to be sourced from a

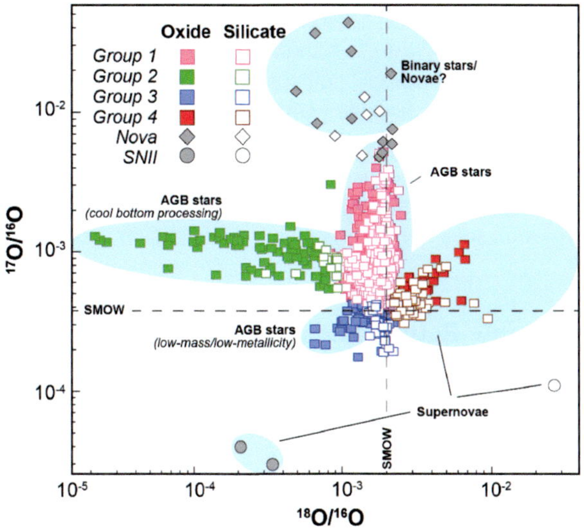

Fig. 12 Oxygen isotope compositions of presolar oxides and silicates. Most of the grains appear to be derived from AGB stars, but the grains show distinctive nucleosynthetic pathways labelled as Groups 1 to 4 of Nittler (1997). Compositions or rare supernovae grains, and compositions inferred to be from Binary stars or Novae are also shown. Data for presolar oxides and silicates from the Presolar Grain Database (https://presolar.physics.wustl.edu/presolar-grain-database/). Note logarithmic scale for the compositional range. Terrestrial SMOW composition shown for reference

supernova. These TiC inclusions had high oxygen concentrations and extreme $^{18}O/^{16}O$ enrichments (up to a factor of 20× solar), but curiously no $^{17}O/^{16}O$ anomalies. These oxygen isotope anomalies are completely different to the anomalies found in CAIs.

Notwithstanding the anomalies in the TiC, detailing oxygen isotope anomalies in silicates and oxides is of more relevance to the source of oxygen in our solar system. Some resilient oxide grains (hibonite, corundum, spinel) were found to have survived the acid treatment used to isolate the carbonaceous grains. Again, these grains have extremely large and variable O isotopic anomalies compared to any solar system materials (Nittler 1997; Nittler and Ciesla 2016). One of the signatures identified is a large excess of $^{17}O/^{16}O$, up to four times the terrestrial abundance, but with $^{18}O/^{16}O$ values similar to terrestrial (Group 1, Fig. 12). Other grains have large depletions in $^{18}O/^{16}O$ and enhanced $^{17}O/^{16}O$ ratios (Group 2). Grains from Group 3 show slight depletions in $^{17}O/^{16}O$ and $^{18}O/^{16}O$ values. These compositions have been interpreted largely as originating in red giant low-mass AGB stars. A small percentage of oxide and silicate grains have isotopic compositions pointing to an origin in Type II supernovae (Group 4). These grains are mostly ^{18}O-rich with variable $^{17}O/^{16}O$ ratios. Only two grains show large ^{16}O excess, the typical isotopic signature expected for supernova condensates (Nittler et al. 1998; Gyngard et al. 2010).

Discovering oxygen isotope anomalies in silicate minerals has been likened to finding a needle in a haystack, when your needle is also composed of straw. As such, the destructive techniques used to find carbonaceous presolar grains are completely inappropriate for identifying presolar silicate grains. The methodology adopted was through the painstaking examination of small regions of interplanetary dust particles with the Cameca NanoSIMS and chondrite matrix with the Cameca ims-1270 (Messenger et al. 2003; Nagashima et al. 2004). The search protocol was based on ion images of $^{16}O^-$, $^{17}O^-$ and $^{18}O^-$ and their reconstruction into isotope ratios. These images showed that extremely small regions were present in extraterrestrial materials and these had large oxygen isotopic anomalies. Many of the silicates have compositions in line with the oxide compositions, Nittler (1997) and are therefore likely related to dust ejected from AGB stars (Fig. 12).

The search for oxygen isotope anomalies has also been carried out based on isotopic anomalies in other elements. In seeking the source of ^{54}Cr isotopic anomalies that had been discovered during acid leaching of meteorite matrices, Qin et al. (2011) found sub-200 nm chromite grains with ^{54}Cr anomalies exceeding 1500‰. Despite the extreme ^{54}Cr anomalies, the oxygen isotopic compositions of these grains could not be resolved from terrestrial, as opposed to other oxide grains that were discovered during ion imaging.

As such, oxygen in the solar system appears to be mainly sourced from AGB stars, not supernovae. Other sources may exist, but the absence of grains might also suggest that these have either been fragmented coming in to the molecular cloud, or have otherwise been destroyed in solar nebula processing, leaving their chemical memory but not the grains.

5 Sample-Return Missions

Four extraterrestrial sample return missions have been successful, the targets being the Moon, the Sun, a comet, and an asteroid.

5.1 Apollo (and Luna)—Moon

Over 380 kg or rock and dust were delivered back to Earth by the Apollo missions of 1969 to 1972. In addition, around 0.3 kg of lunar soils were returned to Earth from the robotic Luna missions (USSR). The similarities of the oxygen isotope systematics of the Moon and Earth have been covered above in Sect. 4.2.3. The key issue here is that the recognition of lunar meteorites would not have been possible without the return of the Apollo samples. With progressive improvement in analytical techniques, the Apollo samples and lunar meteorites are revisited with a view to resolving the potential for ever decreasing differences. Herein lies the importance and massive advantage of sample return missions, samples can be retained and reanalysed as desired.

5.2 Genesis—Sun

The NASA Genesis mission was designed to obtain a sample of the solar wind and return it to Earth (Burnett et al. 2003). Genesis took off in 2003 to orbit the Sun-Earth Lagrange L1 point and exposed high-purity collectors to the solar wind. The solar wind is ionised atmosphere of the Sun accelerated to a velocity of ca. 400 km/s. With such energy, solar wind particles are implanted into the collectors and oxygen atoms reside some 50–100 nm below the collector surface and even less in non-concentrator samples as the concentrator boosted impact energy by typically 40 to 60 keV, depending on ionization level. A variety of collector compositions were flown so as to allow a variety of different elements to be analysed. An electrostatic solar wind concentrator was used to focus the solar wind and implant into four quadrants of an independent collector to increase the resulting concentration of selected light elements by 20 fold. Genesis returned to Earth in 2005. After atmospheric re-entry, it was intended that the capsule would be intercepted by a helicopter gathering the parachute so as to avoid possible terrestrial contamination. However, the capsule came in without the parachute release and crashed into the Utah desert, fragmenting the solar wind collectors (Burnett and Genesis Science Team 2011). Fortunately the concentrator targets largely remained intact. Such a crash and resulting contamination could have been the end of analysing the solar wind composition. The contamination introduced during the impact could, to first order, be removed by standard cleaning techniques, although persistent levels of microparticles still need to be avoided during analysis (McKeegan et al. 2011). Of a more

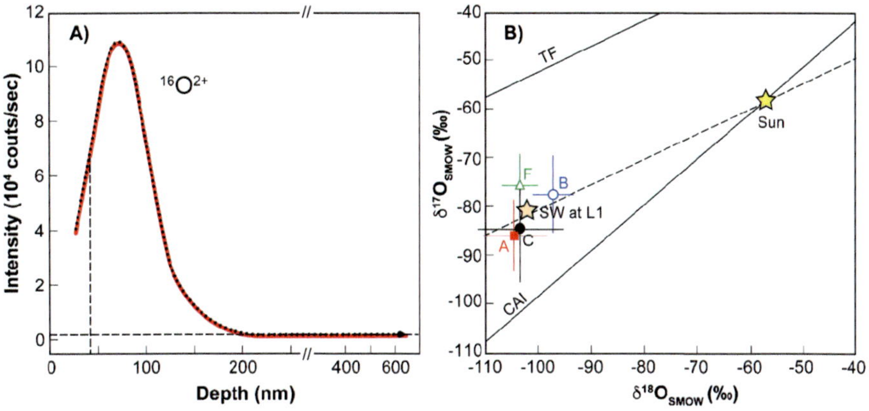

Fig. 13 MegaSIMS analysis of Genesis concentrator targets. Solar wind has a typical velocity of around 400 km/s resulting in the implantation of oxygen to a depth of around 50 nm. Additional acceleration from the electrostatic concentrator implants the solar wind at a higher energy. The depth profile obtained for $^{16}O^{2+}$ is consistent a mean implantation depth of around 80 nm (A). Oxygen isotope analyses across the concentrator, and corrected for variable electrostatic fractionation, indicate a composition for Solar Wind at L1 having a $\Delta^{17}O$ of approximately $-28‰$. Solar wind is known to be isotopically light (through mass dependent fractionation mechanism(s)) and so the inferred composition of the Sun lies at $\delta^{17}O \approx \delta^{18}O \approx -56‰$. After McKeegan et al. (2011)

insidious nature was a brown stain over the collectors. The brown stain was apparently related to glue outgassing from the spacecraft and being redeposited on to the clean collectors where it was affected by solar wind and UV radiation to form a very resistant film some 50 nm think. Analysis of the collectors was therefore compromised both by small adhering particles, likely related to terrestrial contamination, and the brown stain on the surface.

Oxygen isotope analysis of solar wind was the prime objective of the Genesis mission (Burnett et al. 2003). The energy of the solar wind results in implantation of oxygen below the surface of the collector. The means of analysing this oxygen are provided by an exceptional analytical instrument, MegaSIMS, which marries secondary ion mass spectrometry with accelerator mass spectrometry (McKeegan et al. 2004). This instrument was designed and constructed as part of the Genesis Mission. The issue in the analysis of solar wind oxygen is the high intrinsic concentration of hydrogen which results in a large OH^- fraction with SIMS analysis, which needs to be accommodated for the oxygen isotope analysis. The means of doing this is through accelerator mass spectrometry. The sputtered O^- and OH^- ions are recombined and passed into a gas stripper to produce an O^{2+} ion beam which is mass separated into the respective isotopes for analysis.

For the MegaSIMS oxygen isotope analysis, concentrator targets were used which had the highest concentration of oxygen. Progressive ablation of the sample demonstrated a rise in the O^{2+} counts consistent with the implantation of oxygen (Fig. 13a). Isotope ratios were determined from the depth profiles of the individual isotopes acquired simultaneously (with a multicollector). The oxygen is mass dependently fractionated as well as ^{16}O enriched, i.e. lying below the TFL (Fig. 13b). The $\delta^{18}O$ values of the solar wind are around $-100‰$. This mass dependent component appears to be related to a combination of gravitational settling in the solar photosphere as well as ionisation in the solar wind (see Burnett et al. 2003). The magnitude of the ^{16}O enrichment is approximately $56 \pm 3‰$ ($\delta^{17}O \approx \delta^{18}O \approx -56‰$; $\Delta^{17}O \approx -28‰$), which places it in a region similar to oxygen from refractory inclusions. This value exceeds the ^{16}O enrichment of Allende inclusions ($\Delta^{17}O \approx -20‰$; Clayton et al.

1977), but it does correspond to the extreme ^{16}O enrichments of Murchison hibonite grains (Δ^{17}O down to $\approx -30‰$; Ireland et al. 1992; Liu et al. 2009; Kööp et al. 2016). Thus, it is demonstrated that the Sun and Earth differ in their oxygen isotope compositions.

5.3 Stardust—Comet

The Stardust mission collected dust particles from a Jupiter Family Comet, 81P/Wild 2 (Brownlee et al. 2006). Comet Wild 2 was discovered in 1978, but it has been estimated that the comet had an encounter with Jupiter, approaching to within 0.0061 AU, in 1974 shifting the comet into a short-period orbit of 6.4 years within the inner solar system, with perihelion and aphelion distances of 1.49 AU and 5.24 AU respectively, and an inclination of 3.3° (Sekanina 2003). Thus it is probably a relatively fresh comet. The mission launched in 1999 and passed through the coma of comet Wild 2 (in 2006) with a differential velocity of around 6 km/s. Being a sample return mission, the spacecraft had to maintain a velocity that allowed it to return to Earth, and so matching the orbital speed of the comet was not an option (as opposed to the ESA Rosetta mission to comet P67 where a matched speed was required to allow landing of Philae). The high velocity of the spacecraft relative to the comet required a new type of collector that would allow for deceleration of particles without total vaporisation and destruction. The material chosen was aerogel, a lightweight glass composed of silica which approaches the density of air (Kistler 1931). Over 100 individual aerogel collectors were cast in an aluminium frame. Prior to the encounter, the detectors were faced backwards to avoid contamination.

Stardust passed through the Wild 2 coma on 02 Jan 2006. Remote sensing of the dust encounter was provided by the Dust Flux Monitor and the Cometary and Interstellar Dust Analyzer (CIDA). CIDA is a TOF-MS that was used to analyse particle compositions during the encounter (Kissel et al. 2004; Green et al. 2007). Most of the compositions appear to be dominated by organics. The CIDA instrument used on Stardust was identical to that used on Giotto and Vega 1 and 2 during the 1P/Halley comet encounter.

The Stardust sample capsule was returned to Earth on 14 Jan 2006 during Earth flyby. There was some worry about the re-entry because the same accelerometer timing system that had failed on Genesis was used on Stardust. However, the Stardust re-entry was nominal. The Stardust spacecraft was then redeployed as the NeXT mission to view the crater on comet Tempel created by the Deep Impact mission.

Examination of the Stardust collectors showed substantial (up to tens of μm in size) particles at the end of tracks. These included familiar inner solar system materials that looked like chondrule fragments, and even refractory inclusion-like materials (Brownlee et al. 2006) in elongated mm-length tracks which were classified as Type A tracks (Hörz et al. 2006; Burchell et al. 2008). In total, *Stardust* collected thousands of particles, 1–300 μm in size, totalling an estimated $\sim 3 \times 10^{-4}$ g in mass (Hörz et al. 2006; Burchell et al. 2008). Oxygen isotopic analysis of particles confirmed the inner solar system chondritic affinities with the chondrule fragments near the TFL, and the refractory particles enriched in ^{16}O relative to terrestrial (McKeegan et al. 2006; Bridges et al. 2012; Nakamura et al. 2008; Fig. 14). Zolensky et al. (2006), Joswiak et al. (2014), and Ishii et al. (2008) also reported a varied mineralogy, including olivines and low-Ca pyroxenes, Fe-Ni sulphides, Fe-Ni metal, and accessory phases. The presence of these particles has been interpreted as indicating substantial recycling of material from inner to outer solar system when the solar system was forming, and in particular, in the regions where comets accreted.

Type B, and C tracks are more bulbous and are thought to contain material analogous to fine grained chondrite matrix and organics (Burchell et al. 2008). In general, the large

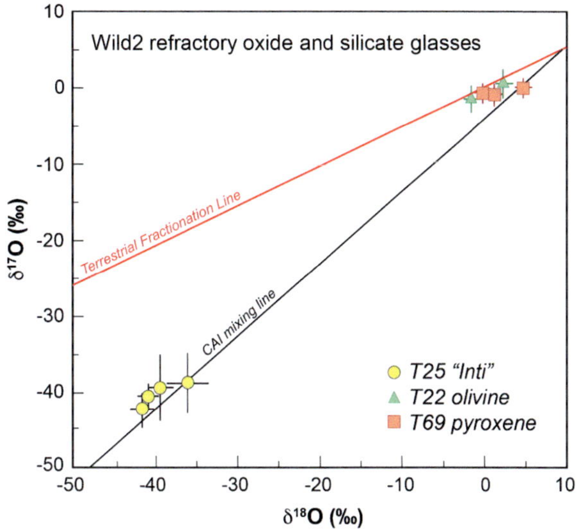

Fig. 14 Stardust grain fragments include materials that appear similar to chondrule and CAI derived minerals. Their oxygen isotope compositions is also similar to chondrule and CAI materials from chondrites suggesting that the comet forming region has been "polluted" with inner solar system materials likely transported out through jets or winds (Figure adapted from McKeegan et al. 2006)

particles at track ends have undergone less thermal processing during capture by the spacecraft from the coma, and so have been most useful for reconstructing the Wild2 mineralogy. The carbonaceous chondrite affinities are evident in the mineralogy. In addition to chondrule and CAI fragments, magnetite identified by Hicks et al. (2017) in Stardust tracks, reveals evidence for hydrous alteration on the cometary parent body.

In addition to the cometary particles, seven candidate interstellar grains were identified with the help of a 'Citizen Science' project by Westphal et al. (2014).

5.4 Asteroids

An asteroid sample return mission was, and is, premised on establishing the relationships between asteroids and meteorites. Infrared Spectroscopic characterisation of asteroids leads to two main types: S (17% abundance) and C (75%) (Tholen and Barucci 1989). The S-type asteroid spectrum has moderate albedo (c. 20%) and is further characterised by absorption features related to ferromagnesian silicate minerals (olivine, pyroxene). The C-type asteroid spectrum has a low albedo ($<10\%$) with no features apparent. It is therefore tempting to relate the S-class spectrum to meteorites rich in ferromagnesian minerals such as ordinary chondrites or pallasites, and the C-class asteroids to carbonaceous chondrites. However, while C-type asteroids are the most abundant, carbonaceous chondrites are relatively uncommon ($<5\%$). As such, there appears to be a flip in the apparent abundances of S- and C-class asteroids, and the abundance of stony meteorites and carbonaceous chondrites.

One successful sample return mission to an asteroid has been completed and two further missions are currently underway.

5.4.1 Hayabusa Mission to 25143 Itokawa (S Class Asteroid)

The Japanese Aerospace Exploration Agency (JAXA) Hayabusa mission began as the Muses-C mission, a technology demonstration to reach an asteroid and bring a sample back to Earth (Kawaguchi et al. 2008). This mission involved the use of autonomous navigation, ion engines, and a novel sampling device. Hayabusa was a 510 kg spacecraft about the size of a large refrigerator. It had a variety of imaging cameras (optical, infrared) as well as LIDAR (Light Detection and Ranging) altimeter. The sampling was based on the touchdown of a sample collection horn. When the horn touched the surface, a 5 g tantalum bullet was fired, and the resulting ejecta was to be transferred up the horn to the sample chamber. Immediately after touchdown, thrusters were activated to take the spacecraft away from the surface. As such, the touchdown and sampling is almost instantaneous and the operation is akin to a Peregrine falcon (Hayabusa in Japanese) hovering, catching prey, and returning to hovering.

The target asteroid was 1998SF36, which was subsequently formally renamed 25143 Itokawa. This asteroid was estimated to be around 500 m across and has an S-type reflectance spectrum. Hayabusa launched on 9 May 2003 with an August 2007 return planned. In the early stages of the voyage to Itokawa, two of the gyroscopic reaction wheels ceased operation. Spacecraft attitude was then only able to be controlled by the one reaction wheel and operation of the ion engine and thrusters. Despite this setback, Hayabusa arrived at Itokawa in September 2005.

Itokawa was observed to be a rubble pile asteroid ($535 \times 294 \times 209$ m) with many boulders apparent on the surface (Fujiwara et al. 2006; Saito et al. 2006). The asteroid is peanut shaped, apparently formed by a low speed merger of two objects. Following a program of imaging, the first touchdown was scheduled to occur in the smooth Muses Sea region of the asteroid, a region in the middle of the asteroid largely free of boulders. A target marker was dropped to guide the touchdown location. The spacecraft appeared to touch down on the surface of the asteroid before the interlocks on the sampling gun had been removed and so touchdown proceeded without the gun firing. In addition, the thrusters to power the spacecraft back to the home position did not operate and it appeared Hayabusa had not only touched down, it had actually landed on the surface of Itokawa. After establishing that the spacecraft was fully operational, the thrusters were used to take the spacecraft back to a hovering position. A second touchdown was scheduled and operation appeared nominal. However, after the touchdown an incident occurred that took the spacecraft out of communication with Earth. This appeared to be related to the chemical thrusters all firing, setting the spacecraft into chaotic tumbling. With the Hayabusa spacecraft out of control, it was decided that the August 2007 return was not possible, and so Hayabusa would stay near Itokawa for another orbit of the solar system. After 6 months, full communications with Hayabusa were recovered. Hayabusa then left to return to Earth. Issues with the ion engines and charge neutralisation required a reconfiguration, but overall the return journey was uneventful. However, it was felt that the extra 3 years in space may have affected the state of the battery on the return capsule that was required for the operation of the drogue parachute and the "beeper" used to locate the spacecraft. It was decided to separate the capsule and mother spacecraft as late as possible (to maximise battery power coming from the mother craft) leading to the spectacular reentry of both return capsule and space craft. The re-entry was nominal and the capsule was quickly located within 400 m of the nominal target at the Woomera Range in South Australia, and then recovered.

Around 150 grains were collected from the sample rooms of the return capsule in the first year, since then ca. 1000 grains have been recovered. This is consistent with the non-firing of the gun on Hayabusa, and the grains obtained simply reflect the slow speed impact of the spacecraft with the asteroid surface. The largest grains were hundreds of micrometers across but most are in the 10–30 μm size fraction. Petrographic examination determined that the grains were predominantly composed of olivine and pyroxene, and the chemistry is consistent with an equilibrated LL chondrite (Nakamura et al. 2011). The oxygen isotope composition was also consistent with L/LL chondrites (Yurimoto et al. 2011; Fig. 15). The

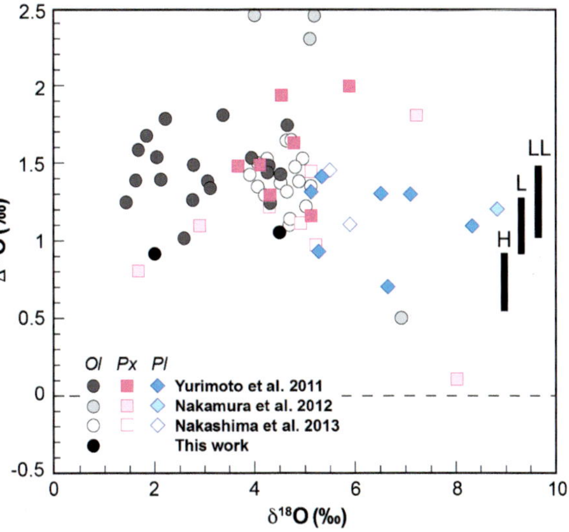

Fig. 15 Oxygen isotope composition of Itokawa grains including the Hayabusa preliminary examination team measurements by Yurimoto et al. (2011), Nakamura et al. (2012), Nakashima et al. (2013), and unpublished SHRIMP data. These analyses confirm the extraterrestrial compositions of these grains and the direct association of Itokawa with LL chondrites

exposure ages of the grains are young, less than 8 Ma based on ^{21}Ne accumulation and may reflect ongoing grinding of mineral surfaces on the asteroid, or assembly of the asteroid body (Nagao et al. 2011). It is established that Itokawa is an S-type asteroid and an LL ordinary chondrite parent body.

5.4.2 Hayabusa2 Mission to Ryugu (C Class Asteroid)

The success of the Hayabusa mission led to a swift turnaround and the building of Hayabusa 2. The short preparation time led to a similar retinue of instrumentation but with some added capability, and in particular to obtain close up images of the surface. The target asteroid of Hayabusa 2 is 1999JU3 now termed 162173 Ryugu. This is a C-type asteroid, which made it difficult to observe from Earth because of the low albedo.

Hayabusa 2 launched on 4 Dec 2014 and had an uneventful journey to Ryugu, arriving in June 2018. Ryugu is determined to have near equant dimensions at $1.04 \times 1.02 \times 0.88$ km (Watanabe et al. 2019). At home position, Ryugu appears as diamond shaped in profile and is close to the shape of a spinning top (i.e. looking down the rotation axis it appears to be round). Like Itokawa, Ryugu is a rubble pile asteroid with low density, and the surface is covered in a poorly sorted mix of pebbles to boulders (Watanabe et al. 2019; Sugita et al. 2019). Targets for the sample touchdown generally comprised locations that had the fewest boulders. After detailed imaging, a sampling touchdown was carried out in February 2019. All feedback from the touchdown indicates nominal operation and a successful sample recovery.

Hayabusa 2 also carried an experiment called the Small Carry-on Impactor (SCI). Here the objective was to make an artificial crater so that fresh sample free of space weathering could be obtained for analysis. The SCI was successfully deployed. A touchdown to the crater then followed in July 2019. At this stage, the sample capsule has been closed. Hayabusa left Ryugu in Dec 2019 for a Dec 2020 return to Earth at Woomera, South Australia.

5.4.3 OSIRIS-REx Mission to Bennu (C Class Asteroid)

Effectively in parallel with the JAXA Hayabusa 2 mission, the NASA OSIRIS-REx mission has targeted another C-class asteroid, 101955 Bennu. Bennu is the second spinning top rubble pile asteroid observed (Lauretta et al. 2019). Bennu is remarkably similar to Ryugu in terms of overall shape, although in detail there are differences in surface morphology. Touchdown evaluation is currently underway, with an expected sampling planned for December 2019. OSIRIS-REx will return to Earth in 2023.

6 Remote Analysis Missions (for Oxygen Isotopes)

Mass spectrometers have been flown frequently on space missions. However, the specific science goals of many of these mass spectrometers have been directed towards determining plasma elemental compositions for which low mass resolutions (tens to hundreds in terms of $M/\Delta M$) are suitable. The drawbacks of flying with mass spectrometers with higher mass resolving power and high sensitivity are related to instrument mass and power consumption. Missions related to determining plasma compositions are not covered here; see a review by Yokota (2018). The discussion below is not intended to be definitive, but rather to exemplify the range of techniques used in space missions to date. The main focus of this section is to examine the ability of mass spectrometers to determine oxygen isotopic compositions.

A remote mass spectrometer must either rely on the presence of plasma as the source of ions for analysis, or the ions must be generated from neutrals related to gas or dust. Ionisation can be active (e.g. electron bombardment), or can be "passive" (e.g. secondary ions caused by the impact of energetic particles). The environment around comets, in particular, is quite suitable for mass spectrometric analysis because of the abundant presence of gas and dust species with high relative velocity causing ionisation.

6.1 Halley's Comet

The return of comet 1P/Halley in 1986 generated an immense interest in the nature of comets. The Halley Armada consisted of five spacecraft: Giotto from the European Space Agency, two spacecraft from U.S.S.R. (Vega1 and Vega2), and two spacecraft from Japan (Sakigake and Suisei). The Giotto mission to Comet Halley included three mass spectrometers (for neutrals, ions and dust; Reinhard 1986). Giotto passed the nucleus of Halley on 14 March 1986 at a distance of less than 600 km and a relative velocity of 68 km/s. Balsiger et al. (1986) used the ion mass spectrometer to measure positive ion spectra during the encounter. The mass spectra are dominated by OH species (OH^+, H_2O^+ and H_3O^+). The dominant presence of these species makes it impossible to measure isotopologues to estimate O isotopic abundances. In addition, the $^{16}O^+$ signal at 1000 c/s is too low to provide accurate information on the $^{18}O^+$ abundance, or the $^{17}O^+$ abundance.

Kissel et al. (1986) reported dust impact data from the PUMA time-of-flight mass spectrometer on board the Vega 1 spacecraft. Dust particles impacted on to a silver conversion plate and positive ions generated are accelerated into the mass spectrometer. The mass spectra are quite detailed and complicated. The mass resolution appears to be around 50 (M/DM) allowing mass peaks at unity separation to be resolved below mass 50. The main species detected are H, C, O, with smaller peaks of Mg and Si of variable abundance, confirming the nature of comets as predominantly ice and silicate dust mixes. Mass resolution is not sufficient to resolve minor isotopes or molecular species from atomic.

The Cometary Impact Dust Analyzer (CIDA) is a similar instrument to the PUMA mass spectrometer that was flown on Stardust (Kissel et al. 2004). During the fly-through of the Wild 2 coma the particles were predominantly organic.

6.2 Rosetta at Comet 67P/Churyumov-Gerasimenko

The Rosetta mission to Comet 67P/Churyumov-Gerasimenko was the first mission to match the velocity of the comet in an attempt to place a lander (Philae) on the surface (Taylor et al. 2015). Comet 67P/C-G is a Jupiter-family comet with an orbital period of 6.45 years with a perihelion of 1.24 a.u. and so the velocity requirements were not as extreme as for longer period comets. Comet 67P/C-G is around 4 km across and Rosetta orbited for two years from 2014 to 2016. The closeup images of the comet are outstanding and include detailed images of pits associated with jets and other features associated with gas loss by sublimation (Sierks et al. 2015). On board Rosetta was a mass spectrometer system for analysing released gases. The Rosetta Orbiter Spectrometer for Ion and Neutral Analysis (ROSINA) consists of two mass spectrometers, being a double focussing mass spectrometer (DFMS), and a Reflectron time-of-flight (RTOF). The DFMS is particularly exciting because it has high mass resolution capabilities sufficient for separating H-O isotopologues with $M/\Delta M$ of 3000 (for ΔM at 1% peak height; Hässig et al. 2013). This enabled a determination of the D/H ratio (from the $HD^{16}O/H_2^{16}O$) of $5.3\ (\pm 0.7) \times 10^{-4}$ (Altwegg et al. 2015), approximately 3 times the terrestrial value, and precluding cometary water such as this as a major source of water for the Earth's oceans. The DFMS also had the capability to measure $^{18}O/^{16}O$ and $^{17}O/^{16}O$ through the measurement of H_2O species at masses 18, 19, 20 (viz. $H_2^{16}O$, $H_2^{17}O$, and $H_2^{18}O$). Low count rates lead to large uncertainties and the ratios could not be resolved from terrestrial with $^{17}O/^{16}O = 3.7\ (\pm 0.9) \times 10^{-4}$ and $^{18}O/^{16}O = 1.8\ (\pm 0.2) \times 10^{-3}$. The uncertainty in $^{17}O/^{16}O$ corresponds to approximately 250‰, approximately the entire range of oxygen found in stony materials in the solar system (Ireland 2012). Nevertheless, this represents the first three oxygen isotope analysis from an operating spacecraft.

The Philae lander also included two mass spectrometers. The COmetary SAmpling and Composition (COSAC) instrument was a combined gas chromatograph and time-of-flight mass spectrometer designed to determine volatile abundances and species, particularly of organic compounds (Goesmann et al. 2007). Most of the ingenuity (and mass) of this device lies in the gas handling and chromatograph system. The mass spectrometer is a comparatively simple TOF-MS designed for $M/\Delta M$ of 350 at mass 70.

The second mass spectrometer system, Ptolemy, was an instrument capable of measuring stable isotope ratios of key volatiles on the comet's nucleus (Wright et al. 2007). Ptolemy utilises an ion trap mass spectrometer that can be regulated to either long mass ranges (for compound characterisation), or short range for isotope compositions. In an ion trap, ions are first captured and then released into the cyclotron in a pulse for analysis. This is a similar concept to a Time-of-Flight MS except the ions are constrained in a circular orbit until ejection for analysis. The instrumentation has been miniaturised to a level where the entire instrument weighs only 5 kg. Samples were to be delivered to ovens where volatiles would be released and then purified before being analysed in the mass spectrometer. However, with the difficulties associated with Philae's landing, Ptolemy did not get the opportunity to operate in its intended fashion and only analysed gas emanating from the surface.

6.3 Cassini at Saturn

The Cassini spacecraft carried the Ion and Neutral Mass Spectrometer (INMS), a dual ion source quadrupole mass spectrometer. INMS was used to analyse the gas ejected from Enceladus (Waite et al. 2006). An electron impact source was used to ionise the gas. The mass spectrometer has effectively unit mass resolution and was capable of recognising the major gas species H_2O, CO_2, and CH_4, at masses 18, 44, and 16 respectively, although the interpretation of mass 28 is ambiguous in this type of analysis as being either CO or N_2.

6.4 Curiosity Rover on Mars

The Mars Curiosity Rover carries the SAM (Sample Analysis at Mars) system for determining chemical and isotopic compositions, particularly with reference to possible life habitats. Three instruments can be utilised as part of the characterisation including a Gas Chromatograph, a Quadrupole Mass Spectrometer, and a Tunable Laser Spectrometer. The total weight of this system is (a massive) 40 kg, but the capabilities are extensive. The quadrupole MS is similar to that carried on Cassini in that it uses electron bombardment and effectively has unit mass separation. Atmospheric samples can be directly input to the source (Mahaffy et al. 2013). Dust, soil, and rock samples can also be prepared for analysis through high temperature ovens (up to 1000 °C), as well as chemical solvents for wet chemistry experiments. This capability has been used for carrying out geochronological analysis of Mars surface materials (Cohen et al. 2019).

For the Mars atmosphere H, C, and O isotopes are dominated by atmospheric loss processes and water-rock reactions with its surface such as sulphate, clay and carbonate formation. SAM's measurements of the atmospheric isotope ratios of D/H ($\delta D = 4950 \pm 1080‰$), O ($\delta^{18}O = 48 \pm 5‰$) and C isotopes ($\delta^{13}C = 46 \pm 4‰$), have been interpreted on the basis that the martian reservoirs of CO_2 and H_2O were largely established by about 4 billion years ago, and that atmosphere was subsequently lost (Webster et al. 2013).

7 The Solar System

7.1 An Historical Perspective

In 1969, man landed on the moon and brought back the first samples collected from an extraterrestrial object. In that same year, two large carbonaceous chondrites landed on Earth making this a serendipitous year for obtaining samples relevant to our understanding of the solar system.

Fifty years ago, our view of the solar system was quite different to what it is now. It had been realized for some time that a key to understanding the provenance of our solar system was the search for isotopic anomalies produced during nucleosynthesis of the elements. B^2FH (Burbidge et al. 1957) showed that the abundances of the nuclides in the solar system could be described in eight nucleosynthetic processes and that the solar system elemental abundances had to be averages of multiple nucleosynthetic inputs. Hence, apparent isotopic anomalies are key to locating individual sources. However, prior to 1969, only a few elements had shown any isotopic variations and these were primarily for volatile species (H, and noble gases). Isotopic anomalies in the rock-forming elements were not known. One way of homogenising the elemental isotopic compositions is to completely evaporate everything in the solar nebula and then condense the rocky elements (Grossman 1972). Thus, the solar nebula was regarded by meteoriticists as extremely hot.

On 8 February 1969, two tons of the Allende (CV3) carbonaceous chondrite fell to Earth in Mexico. This was a propitious event in terms of having an abundance of extraterrestrial material available for testing analytical protocols for the upcoming lunar sample return. For this purpose a 2 kg aliquot of Allende was ground up and made available as a reference material by the Smithsonian Institution (Jarosewich et al. 1987). Allende also contained large (up to 25 mm across) and varied CAIs (Grossman 1975). Oxygen isotope analysis of these inclusions were the first indication of substantial isotopic heterogeneity in solar system materials (Clayton et al. 1973). With the ready availability of material, isotopic anomalies were sought in many other elements. Of particular interest were anomalies in ^{26}Mg correlated with ^{27}Al indicating the presence of live ^{26}Al in the early solar system with a $(^{26}\mathrm{Al}/^{27}\mathrm{Al})_0$ abundance of 5×10^{-5} (Lee et al. 1976). At this level, radioactive decay of ^{26}Al is a suitable heat source for planetesimals. Also of importance were anomalies in the Fe group of elements. Isotopic anomalies were first discovered in the heaviest isotope ^{50}Ti, suggesting the addition of neutron-rich material (Heydegger et al. 1979). The correlated excesses of ^{16}O, ^{26}Al, (and ^{50}Ti) supported the late stage SuperNova Injection model, where these nuclides were ejected from a supernova proximal to the early solar nebula, potentially causing the collapse of the molecular cloud leading to the solar nebula (Cameron and Truran 1977).

The Murchison (CM2) meteorite provided a different and perhaps complementary view to that of Allende (Ireland and Fegley 2000). On 28 September 1969, around 120 kg of Murchison were recovered in and around the township of Murchison, Victoria, Australia. The matrix of Murchison has seen aqueous alteration but not the metamorphism evident in the matrix of Allende. This cold matrix was found to contain a large variety of organic molecules, including amino acids. Murchison also contains its own suite of refractory inclusions. These are typically very small (less than 0.5 mm) and so analysis of these materials required a different approach to analysis of the larger Allende inclusions. They are also characterised by the common presence of a highly refractory mineral hibonite ($CaAl_{12}O_{19}$), which forms at higher temperatures than Allende CAI minerals and therefore might contain more of the initial heterogeneity of the solar system (Fahey et al. 1985). The isotopic analysis of these inclusions was possible because of the development of ion microprobes capable of in situ analysis while resolving various isobaric interferences including ^{48}Ca and ^{48}Ti ($M/\Delta M \approx 10000$; Zinner et al. 1986). These analyses showed a similar variability in oxygen isotope compositions from close to the terrestrial fractionation line to around a 6% enrichment in ^{16}O, and a similar maximum initial $(^{26}\mathrm{Al}/^{27}\mathrm{Al})_0$ of 5×10^{-5}. However, anomalies in Ti and Ca were greatly increased with excesses in ^{50}Ti and ^{48}Ca of up to +27% and +10% respectively (Ireland 1990). The lack of correspondence between ^{48}Ca-^{50}Ti anomalies and ^{16}O anomalies was used to argue against the late supernova injection model (Fahey et al. 1987).

7.2 Origins of Isotopic Anomalies

Variations in isotopic abundances (isotopic anomalies) were initially thought to be solely produced through nuclear processes, such as stellar nucleosynthesis, in addition to radioactive decay. However, a variety of chemical processes are now thought to produce isotope variability beyond that solely expected from mass dependent fractionation. This is particularly important for oxygen.

7.2.1 Nuclear Inheritance

The direct incorporation of dust grains ejected from other stars manifests itself as isotopically anomalous material in our solar system. These grains predate our solar system and so are often referred to as presolar grains. The source of these grains can be related to the big dust producers in the galaxy, namely AGB stars and supernovae (Zinner 2003).

In terms of oxygen isotope diversity in solar system materials, the most striking anomaly is the ^{16}O abundance variation seen in CAIs. However, a general mineral carrier for this anomaly has not been found. Most of the presolar oxide and silicate grains are distinctive and appear to be related to AGB stars, just like the carbonaceous grains (Nittler 1997; Floss and Haenecour 2016).

Moreover, the solar system oxygen isotope signature is striking in that it is planet Earth which is anomalous relative to the Sun (Clayton 2002; McKeegan et al. 2011). In this case, what would a nucleosynthetic carrier actually represent at the scale of these two bodies? The big issue with oxygen is that it is a major element in the solar system and that the variability is in the major isotope requiring exchange of a large number of atoms to manifest change.

7.2.2 Chemical Fractionations

The lack of a viable means of describing the solar system oxygen isotope composition with carriers of nucleosynthetic anomalies leaves us with the possibility that the anomalies are induced by chemical speciation and non-mass-dependent isotope fractionation.

The production of ozone in the atmosphere induces an isotope shift because the available energy levels for asymmetric molecules are more plentiful than for molecules of a single isotope. Such a mechanism can be readily envisaged for the solar system through various reactions, and was postulated very early in the interpretation of oxygen isotope anomalies (Thiemens and Heidenreich 1983; Thiemens 1999). However, making such a scheme operable in the early solar system brings up a number of challenges. For the oxygen-ozone exchange reaction, both species are in the gas phase. We see that the compositions are different because we can physically separate the oxygen from the ozone and measure them separately. So for CAI it is tempting to envisage the same mechanism because the isotopic composition ranges up to $+50‰$ for $\delta^{17}O$ and $\delta^{18}O$ in Earth's stratospheric oxygen, at the same scale as anomalies in CAI. However, in the case of CAI, we only have a solid phase. Hence any fractionation must be transferable to the solid phase and the gas then would be expected to be isotopically heavy.

7.2.3 Predissociation and Self-Shielding

While chemical fractionations along the lines of molecular asymmetry need more explicit development, another type of chemical fractionation is well known in the cosmos. Predissociation of CO by UV photons leads ultimately to dissociation of CO in to C and O (Bally and Langer 1982; Lada et al. 1994). The predissociation is subject to self-shielding because the photon stream is attenuated through absorption by the CO molecules. Furthermore, for a cold molecular cloud, the predissociation energy is quantised with a narrow energy window. The different isotopologues of CO, each have a different associated energy for predissociation. Thus photons relevant for the dissociation of $^{12}C^{16}O$ are absorbed at a short optical depth whereas photons for dissociating $^{12}C^{17}O$ and $^{12}C^{18}O$ can penetrate deeper into the core of the molecular cloud. Hence the core of a molecular cloud can have CO, which is highly enriched in ^{16}O, because $C^{17}O$ and $C^{18}O$ are dissociated and the ^{17}O and ^{18}O react with abundant H to eventually form ^{17}O-, ^{18}O-enriched water ice. This is consistent with what is observed in molecular clouds where the $C^{16}O/C^{18}O$ can be measured optically (Bally and Langer 1982).

The possibility of predissociation and self-shielding (PSS) in the early solar system was initially considered for molecular oxygen by Navon and Wasserburg (1985). These authors discounted predissociation and self-shielding because of the likely low abundance of free

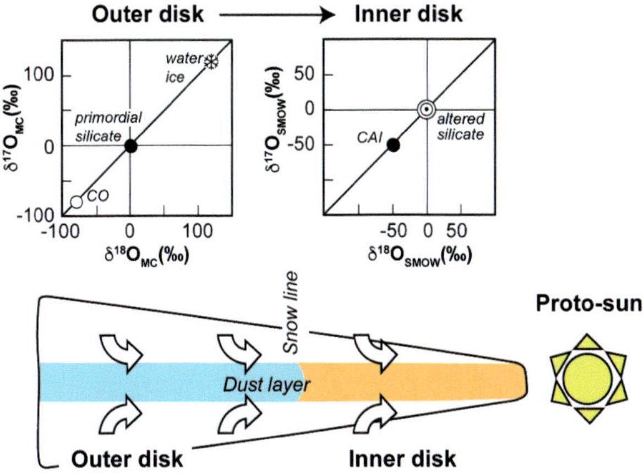

Fig. 16 The predissociation and self-shielding model of Yurimoto and Kuramoto (2004). In the primordial molecular cloud, UV photons cause predissociation of CO (i.e. CO -> C + O). The predissociation is subject to self-shielding because of the high relative abundance of $C^{16}O$, leading to rapid attenuation of photons that can predissociate $C^{16}O$. Predissociation of $C^{17}O$ and $C^{18}O$ can take place deeper in the molecular cloud and following reaction with H, can lead to water highly enriched in $H_2^{17}O$ and $H_2^{18}O$, while the associated CO is enriched in $C^{16}O$ (upper left). The primordial silicate composition is interpreted to represent the bulk molecular cloud composition and is also the starting CO composition ($\delta^{18}O$ of the molecular cloud, $\delta^{18}O_{MC} = 0$). When this assemblage arrives in the inner solar system, the water ice reacts with the original silicates to form an altered silicate reservoir, which represents the solids in the inner solar system including Earth ($\delta^{18}O$ of the inner solar system, $\delta^{18}O_{SMOW} = 0$). In normalising to the altered silicate composition of Earth, CAIs therefore appear anomalous, but they represent a remnant of the primordial dust composition. This model therefore suggests that the silicates in the inner solar system will differ from those of the outer solar system, and outer solar system silicates should reflect a composition close to CAIs. Figure based on Yurimoto and Kuramoto (2004)

gaseous oxygen. They did suggest CO as a more promising molecule. However, the concept languished until it was taken up by Robert Clayton (2002), who was the original advocate for a nucleosynthetic component carrying oxygen isotope anomalies (Clayton et al. 1973). Clayton (2002) suggested that PSS of CO occurred in the inner solar system and silicates reacted with the ^{17}O and ^{18}O preferentially released in the region of the terrestrial planets. Clayton (2002) importantly predicted that the solar composition would be relatively enriched in ^{16}O relative to Earth. However, Lyons and Young (2005) noted that an inner solar system locale is somewhat problematic because of the thermal line broadening that would occur in the inner solar system, which would affect the isotope selectivity and quickly erase the effects. Lyons and Young (2005) preferred a cold outer solar system location. Yurimoto and Kuramoto (2004) suggested that a molecular cloud setting was the most appropriate because this is where PSS is actually observed to occur. In their model, the ^{17}O and ^{18}O reacted with H to form OH and ultimately H_2O ice which would stick on dust grains (Fig. 16). On heating in the evolving solar system, water is released and reacts with dust to form an altered silicate reservoir, which is the process responsible for the terrestrial oxygen isotope composition.

One manifestation of this can be the large mineral specific oxygen isotopic fractionations and related mixing lines identified in ordinary chondrites (Sect. 4.1.1). The outer solar system would have silicates and oxides with the solar composition (i.e. similar to CAIs) while the water ice would be isotopically heavy (i.e. ^{17}O, ^{18}O-rich). Notwithstanding the specific location of the fractionation, these models all advocate a solar oxygen isotope composition which is enriched in ^{16}O, as has indeed been found by analysis of the Genesis targets (McKeegan et al. 2011).

7.3 Astrophysical Context

7.3.1 Formation of Solar Systems

Around 4568 million years ago, a molecular cloud was in the process of consolidation and forming new stellar systems. One of these stellar systems became our solar system. With the development of Hubble, Spitzer, Kepler, and other telescopes, we now have quite a detailed view of how stellar and planetary systems develop. Still, we cannot resolve down to inner solar system scales, nor get direct insight into the development of high temperature objects such as chondrules and CAIs around other stars. Understanding the formation of these objects is key to the formation of our solar system because the chemistry of the solar system is largely established at this time.

With the collapse of the molecular cloud, the solar nebula began to accrete and flatten into a disk due to conservation of angular momentum. Dust would be retained in Keplerian orbits but gas would be able to move quickly inwards to the growing protoSun. Dust would be accreted to form protoplanets and this must have occurred very early on if the gas giants were to be able to accrete H and He atmospheres before the gas disappeared into the Sun.

The present solar system shows a distinct difference between the inner solar system (terrestrial planets) and the outer solar system (gas and ice giants). This has been attributed to a snow line inside the orbit of Jupiter where water existed in the gas phase inside the line, and as water ice outside. In the simplest terms, a thermal gradient is indicated. This can be readily seen in calculations based on a minimum mass solar nebula. The temperatures in the disk are relatively low and are consistent with the presence of matrices in primitive meteorites that have not seen high temperatures. However, within that cold matrix are the high temperature chondrules and CAIs. Nowhere in a minimum mass nebula are temperatures high enough to melt silicates, which is required to form chondrules and CAIs (Salmeron and Ireland 2012b).

High temperature processing in the early solar system must be extensive but remains enigmatic. Chondrules are so prevalent in the solar system that they have to be formed in a general process of planetary systems. A full review is not warranted here but we will address some aspects that are pertinent to the thermal structure of the solar system that might be recorded. The key issue in this context is whether a thermal gradient can be seen in the processing of high temperature objects that would have implications for the temperature structure of the solar nebula.

CAIs are enriched in refractory elements and show characteristics of chemical fractionations at high temperatures such as overall high abundance of refractory elements, fractionation of ultra-refractory trace elements, and a lack of Fe metal. The lack, or low abundance, of Fe metal suggests a specific temperature range that CAIs formed within and that is above the condensation temperature of Fe; this is supported by the isotopic fractionation of Mg in CAIs (Richter et al. 2002). The lack of Fe^{2+}, except from secondary addition, suggests that CAIs formed in a reducing environment. The formation of CAIs must have occurred at extremely high temperature, and under quite restrictive temperature intervals such that the lanthanide elements could fractionate from each other. Furthermore, the isotopic variability in CAIs shows that they formed from precursors that weren't entirely homogenised. The most likely scenario is that CAIs formed close to the Sun and were then transported out into the accreting disk through jets, or winds, and/or turbulence.

The formation of chondrules appears to be quite distinct from the formation of CAIs. Chondrules appear to have formed in an oxidising region because of the presence of Fe^{2+}. They still require high temperatures to melt, but there is a distinct possibility that chondrules did not form in the same process/region as CAIs because CAIs appear to have formed under reducing conditions. A large number of the models for forming chondrules have been proposed.

Probably the mechanisms most widely accepted is that chondrules were melted in shock waves generated by gravitational instabilities in the protoplanetary disk (Desch and Connolly 2010). However, aggregates of appropriate size must be produced prior to entering the shock front. In addition, the nature of this process would suggest that partially shocked materials intermediate between the aggregates and chondrules should exist, but these do not appear to be common. Collisions between planetesimals is another likely scenario in the early solar system (Wakita et al. 2017), but again the record lacks the range of intermediate materials that might be expected if chondrules were formed by this process. Formation and transport in magnetohydrodynamic jets and winds have been modelled as suitable mechanisms to melt chondrules, either by direct solar radiation in close proximity to the Sun to for CAIs (Shu et al. 1996), or forming chondrules at 1 a.u. through friction generated between the wind and particles in a disc wind (Salmeron and Ireland 2012a,b). A drawback to these models is that chondrules would form and then have to be incorporated into matrix in rather specific proportions to yield chondritic abundances.

Chronology of the high temperature objects provides no definitive answers as to whether CAIs and chondrules formed in distinctly different processes. CAIs are the oldest material known from the early solar system and formed in a very narrow time window at 4567.3 ± 0.3 Myr; the oldest chondrules are close to the same age, but then record a much longer formation interval with ages up to ca. 4 Myr younger (Connelly et al. 2017). This time period is very similar to the observed duration of disks around young stars. It is also likely that Jupiter was largely formed before the chondrule forming event finished.

7.3.2 Oxygen Isotopes and the Structure of the Solar System

Oxygen isotopic compositions of chondrules and CAIs are similar in some aspects but different in others. Both indicate the presence of a ^{16}O-rich reservoir referred to as *solar*, and both indicate a reservoir closer to the TFL, referred to as *planetary* by Ireland (2012). These compositions are effectively those of the Sun, and the silicate dust composition, respectively.

CAIs and chondrules differ in the proportions of these components. CAI minerals hibonite, spinel, and pyroxene typically show oxygen isotope compositions with $\Delta^{17}O$ of -30 to $-20‰$ whereas melilite and anorthite can be variable but typically closer to the TFL. As such, these systematics are consistent with CAI formation close to the Sun with a solar gas composition mixing with solids during evaporation and condensation. It is debated as to whether the presence of melilite close to the TFL is a nebula product or a parent body product but either can be accommodated. The nebula could become more enriched as silicate evaporates, or, the parent body water could be equilibrated with its silicates.

As noted in Sect. 4.1, chondrules typically have compositions closer to the TFL than CAIs. Bulk compositions are within a few permil of TFL for $\Delta^{17}O$ for typical chondrule olivine in ordinary chondrites. Forsterite-rich relics are present, which frequently appear to have a greater ^{16}O-rich component in them (Kita et al. 2010). Chondrules in carbonaceous chondrites typically show larger ^{16}O enrichments. Noteworthy is chondrule a006 from the Acfer 214 carbonaceous chondrite, which has the largest ^{16}O excess ever recorded at $\Delta^{17}O \approx -40‰$ (Kobayashi et al. 2003). This composition is even more enriched in ^{16}O

than solar, which is not predicted for the primordial dust in the solar system. Potentially this is an important inclusion because its composition suggests reaction with a gas enriched in ^{16}O that might be associated with the otherwise lost CO reservoir.

In the PSS model, oxygen isotope compositions of H_2O-bearing species can be ^{17}O-^{18}O-rich (nebula water). Such compositions are uncommon but have been found. Ireland et al. (2006) found $\Delta^{17}O \approx +25‰‰$ in the surfaces of lunar metal grains, and while originally interpreted as related to solar wind, could be related to late stage water addition from cometary material. Sakamoto et al. (2007) documented a cosmic symplectite, an intergrowth of Fe, Ni-metal and sulphides apparently oxidised by water, in the Acfer 094 carbonaceous chondrite. This assemblage was found to have $\Delta^{17}O \approx +90‰$ and was interpreted as representing nebula water from the early solar system. Even heavier oxygen isotope ratios have come from organic-rich residues with $\Delta^{17}O$ up to $\approx 250‰$ (Hashizume et al. 2011).

The PSS model provides a context for understanding these oxygen isotope compositions in the solar system. The prediction that the Earth should be ^{17}O, ^{18}O-rich compared to the Sun appears to be borne out (Clayton 2002; McKeegan et al. 2011). PSS indicates that the nebula water composition becomes enriched in ^{17}O and ^{18}O, and the compositions above indicate that this isotopically heavy nebula water indeed appears to be present in the solar system. In the model of Yurimoto and Kuramoto (2004), this water reacts with dust to create an altered dust reservoir in the inner solar system. The corollary to this interpretation is that outer solar system dust and rocky material should be isotopically solar in the oxygen isotopes, like CAIs.

We have only a few indications of oxygen isotopes from the outer solar system. The analyses of particles from comets (McKeegan et al. 2006) determined that the same range in oxygen isotopes exists in the outer solar system as is found in the inner solar system. Interplanetary dust particles also show oxygen isotope compositions that have olivine close to the inner solar system planetary reservoir, and refractory minerals consistent with the CAI composition (McKeegan 1987). Even from this admittedly limited database, we are not seeing the variability in dust compositions that might be expected if outer solar system rocks were ^{16}O-rich relative to the inner solar system.

An alternative scenario could be postulated based on the uniformity of the planetary oxygen isotope composition, such as that indicated for Earth, Mars, and Vesta. These bodies have formed over large radial distances and yet their oxygen isotopes show only small subpermil differences. If these compositions were formed by interaction between the heavy nebula water reservoir and dust of solar composition, then it could be expected that larger variability in altered dust compositions would be apparent. The mixing of the solar and nebula water reservoirs, separated by around 250‰, would need to be thoroughly equilibrated with almost no variability in the mixing proportions to get such a specific oxygen isotope composition in planetary material (e.g. by the chondrule formation heating in the inner solar nebula; Yurimoto and Kuramoto 2004). An alternative interpretation might be that the planetary reservoir does represent the average dust from the molecular cloud and is only affected to a small degree by mixing with nebula water. Any interpretation of course awaits detailed analysis of materials including ice and rock from the outer solar system.

7.3.3 The Warren Gap

In assessing isotopic variability in planetary materials, Warren (2011) made an extraordinary observation that in plotting $\Delta^{17}O$ relative to $\delta^{54}Cr$, there are essentially two populations of material, with a distinct gap between (Fig. 17). This gap is perhaps the moreremarkable because there is no distinct gap in oxygen isotope compositions, and the Cr isotope compositions are near continuous in the plot. It is the correlation between oxygen and chromium

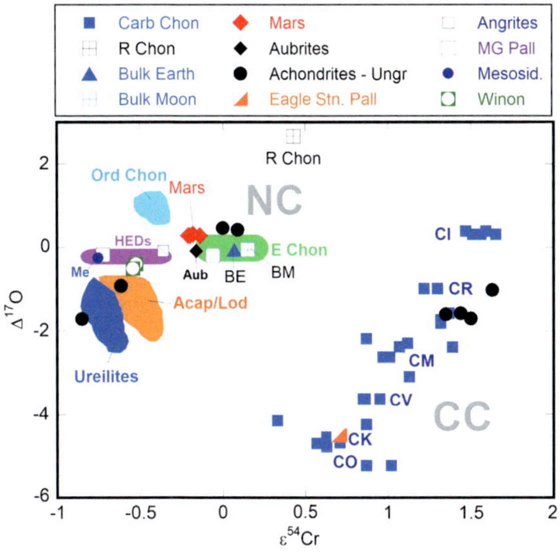

Fig. 17 A distinct separation exists between carbonaceous chondrites and non-carbonaceous chondrite-related meteorites. The separation of these compositions has recently been interpreted as due to the presence of Jupiter causing distinct inner and outer solar system reservoirs (Kruijer et al. 2017). However, this dichotomy is primarily represented by a shift in the chromium isotopic composition rather than oxygen. This is perhaps surprising given the large scale variability in oxygen isotope compositions in the inner solar system and particularly in inclusions in these meteorites. After Greenwood et al. (2017) and Warren (2011)

isotopes that produces the gap. The two groups are essentially related to ordinary chondrites and achondrites on one hand, versus carbonaceous chondrites on the other. The carbonaceous chondrites also show a positive correlation between δ^{54}Cr and Δ^{17}O indicating a fundamental relationship in this group between these two apparently unrelated elements.

An interpretation for this dichotomy has come forward based on the ALMA images of HL-Tau (ALMA Partnership 2015), that is that giant planets act as a road block for material accreting to the nebula (Dipierro et al. 2015). The HL-Tau image is effectively showing the dust distribution and it is clear the an annulus exists and that the most likely explanation is that a giant planet has accreted the dust coming in through the accretion disk and it acts as a block to material coming in to accrete on to any inner planets.

For our solar system, the explanation for the gap can then be posited in terms of the formation of Jupiter. Kruijer et al. (2017) used W and Mo isotopes to show that the gap exists for heavy elements and this can be interpreted as a difference in the accretion history of ordinary and carbonaceous chondrites, representing inner and outer solar system provenance. Once Jupiter had formed it would stop material flowing through the disk thereby allowing distinct reservoirs to exist. Scott et al. (2018) noted that refractory inclusions would come from the inner solar system via jets or winds and were aggregated into cosmic breccias in the outer solar system. Yet, the oxygen isotopic compositions of these outer solar system rocks remains largely "planetary" despite the potential presence of δ^{17}O, δ^{18}O rich water ice in the outer solar system. On the other hand, Yurimoto and Kuramoto (2004) predict isotopically light (solar) solids in the outer solar system. Such paradoxical relationships can only be resolved by sampling the outer solar system.

8 Outstanding Issues

Over a fifty year time frame, we have obtained close up images of all the planets, the dwarf planets—Ceres and Pluto, as well as an ever increasing number of asteroids and comets, and even a Kuiper Belt Object (Arrokoth). From a geologist's perspective these images show morphological features that can be described in terms of processes that have affected the surface. These processes are clearly many and varied, and have occurred in response to different temperatures, atmospheric pressures, and compositions. The radial change to first order in the compositions of planetary objects is quite striking in passing from the rocky and heavily cratered world of Mercury to the ice dominated Pluto. However, simply seeing the different morphologies is not enough. Every time an image is displayed, a planetary scientist wants to know the isotope compositions and formation and exposure ages. The only way of doing this is by remotely analysing it or by bringing a sample back to Earth.

8.1 Inner Solar System

8.1.1 Mercury

The visit by the orbital Messenger spacecraft allowed detailed mapping of the surface, as well as some chemical characterisation, which was limited by the low X-ray flare activity of the Sun. The upcoming Bepi-Colombo mission will revisit a number of the aspects discovered by Messenger but this remains a remote sensing mission. There will be no touchdown and no future touchdowns are planned.

8.1.2 Venus

The isotopic composition of Venus atmosphere was measured in two experiments during the Pioneer mission. During mission periapsis, upper atmosphere samples were analysed with the on-board neutral quadrupole mass spectrometer. It was found that $^{18}O/^{16}O$ and $^{17}O/^{16}O$ are within $\pm5\%$ uncertainty of terrestrial (Niemann et al. 1979). Similar results were obtained with the single focussing magnetic sector mass spectrometer. Hoffman et al. (1979) reported that the $^{18}O/^{16}O$ of CO_2 was found to be within 1% of terrestrial but with a large uncertainty of $\pm5\%$. The uncertainties of these measurements effectively cover the full range of terrestrial compositions and so refinement of these abundances is required.

Ten probes have landed on Venus. Of note were the 1978 Venera 9 and 10 spacecraft, which managed to transmit images of the Venusian surface back to Earth. Of most geochemical interest was the Venera 13 and 14 surface landers, which carried an XRF for soil analysis and a mass spectrometer for determining the atmospheric composition. Venus is covered by surface basalts of predicted age less than c. 500 Ma based on radar observations and crater counting. No current plans are in place to land and sample Venus, or to return a sample.

8.1.3 Mars

Mars is our most attractive planetary neighbour being readily accessible every two years with a relatively small Δv. Many landers have sampled the atmosphere and lithosphere with increasingly sophisticated methodologies. Despite all the successful missions, we are still left with a nagging question—Are SNC meteorites really from Mars? While the D/H and planetary gas compositions suggest they are, it won't be satisfactorily resolved in many geochemical minds until a rock demonstrably from Mars is shown to have $\Delta^{17}O$ of +0.30‰.

It is certain from lander, orbiter missions and secondary mineral assemblages in martian meteorites that Mars has likely had an active hydrological cycle in the past and so detailed comparisons with the terrestrial water cycle are eagerly anticipated. ExoMars and the NASA Mars2020 missions are the latest to head to Mars in the search for life. Mars2020 is expected to obtain and sequester a number of samples for a prospective future return to Earth by 2032.

8.1.4 Asteroids

One asteroid, the S-type 25103 Itokawa, has been visited and a sample collected and returned to Earth for analysis. The result was that Itokawa has been clearly identified as a parent body of equilibrated LL chondrite meteorites. The oxygen isotope compositions of the returned samples is uniform at around $\Delta^{17}O$ of +1.2‰.

Hayabusa 2 surveyed the asteroid Ryugu for a year and a half and took samples. It left the asteroid in November 2019. OSIRIS-REx arrived at Bennu in December 2018 and is currently mapping the asteroid in detail. Samples are expected to return in late 2020 and 2023 respectively. Returned samples will resolve some of the issues relating C-type asteroids to carbonaceous chondrites. Oxygen isotopes will provide a strong context for relating the mineralogies described to samples within the existing meteorite catalog.

In terms of a sample return mission to a well-defined and scientifically important achondritic parent body, asteroid 4 Vesta is undoubtedly a prime target. Remote sensing studies by the NASA Dawn spacecraft have already furnished a wealth of information about this body (Russell et al. 2015). With well over 2000 officially recognised specimens, the HED meteorites are by far and away the most important achondrite group (~75% of all achondritic meteorites). HEDs also have a diagnostic oxygen isotope composition (Greenwood et al. 2017). With the exception of planetary achondrites (see below), sample return from Vesta would validate, or otherwise, what appears to be the strongest meteorite-parent body relationship so far delineated. A returned sample would also help define the extent of oxygen isotope homogenization on Vesta. Basaltic achondrites with anomalous oxygen isotope compositions are presumed to be from differentiated asteroids other than Vesta (Greenwood et al. 2017). However, this remains an untested assumption and it is possible that Vesta is significantly more isotopically heterogeneous than is currently thought to be the case. A sample return mission to Vesta would directly address this issue.

8.1.5 Comets (Jupiter Class)

The Stardust mission performed a flyby with samples returned to Earth. These samples showed that comets were not as pristine as we thought, and oxygen isotopes showed that inner solar system material was transported out to the outer solar system when the comet parent bodies were forming.

The Rosetta-Philae mission to 67P/Churyumov-Gerasimenko provided a wealth of new data concerning the makeup of comets. However, the Philae touchdown did not proceed as planned. The anchor screws did not connect sending Philae bouncing over the surface in an uncontrolled manner. As such, high precision mass spectrometry of the comet surface was not obtained and analysing cometary ice remains a priority.

8.2 Outer Solar System

8.2.1 Gas and Ice Giants, and Pluto

The determination of isotopic compositions on Jupiter, Saturn, Uranus and Neptune is simply an issue of atmospheric sampling. No solid surfaces exist on these planets. Missions have mainly included flybys, but the atmospheric sampler on Galileo and the Huygens probe to the surface of Titan demonstrate that in situ analysis is possible.

8.2.2 Moons

The moons of the outer solar system planets offer a far more compelling case for geochemical analysis. From the basalt volcanoes on Io to the cryovolcanoes on Enceladus and Triton, much could be learned from detailed sampling concerning rock and water ice oxygen isotope compositions. From the Jupiter system alone, silicates from Io and ice compositions on Europa, Ganymede, Callisto would be of fundamental importance. The imaging of the surface of Titan shows a highly varied landscape and any mass spectrometric measurements on the surface of Titan would inform regarding the hydrocarbon cycle. The fly through of cryovolcanic ejecta on Enceladus indicated the presence of water ice.

8.2.3 Comets and KBOs

Obtaining samples from outer solar system comets and KBOs would provide a new constraint on the radial distribution of oxygen isotopes in the solar system. Ice alone would provide new constraints on the role of PSS in determining oxygen isotope compositions in the early solar system. Even a bulk sample of the silicates would prove invaluable in constraining the nature of dust and dust processing in the solar nebula.

9 Samples and Analysis

9.1 Priorities

The main priority to be addressed here is determining the oxygen isotope structure of our solar system. The PSS model provides a strong interpretative model for understanding oxygen isotope fractionations and obtaining analyses of water and silicate dust is essential. The clear element that is missing in this context is water and dust oxygen isotope compositions in the outer solar system. But, we also need better constraints on inner solar system planetary bodies.

The NASA mantra of "follow the water" is entirely attractive in this context. Water is the most common molecular compound in the solar system and its abundance reflects the gross thermal structure of the solar system. In terms of solar system structure, the ability to make oxygen isotope analyses is key. Water is routinely analysed by several forms of mass spectrometry, however, the key to unlocking the structure of our solar system is the ability to make three oxygen isotope measurements at high precision and accuracy. In addition, the coupled measurements of analysis of silicate dust is important in providing an understanding of the history of the solar system. These coupled measurements place stringent requirements on both remote analysis and sample return.

9.2 Remote Analysis

The issues concerning analysis of three oxygen isotopes relate to an integration of sample preparation and mass spectrometry. The key for remote analysis is to have a relatively light, low power consumption device capable of high resolution and high precision of triple oxygen isotopes.

Many of the systems that have been flown rely on the analysis of gas (and related ion) species that are formed in the outgassing of objects such as comets and outer solar system moons. Dust analysis has relied largely on the evaporation upon high energy collision with a conversion plate. However, in order to analyse silicate dust and to raise the signals to achieve high precision analysis, a means of direct sampling is required.

In terrestrial laboratories, the sample preparation has been based on obtaining a sample of pure oxygen to eliminate the isobaric interferences inherent in CO_2 mass spectrometry commonly used for $^{18}O/^{16}O$ analysis. This is achieved by reacting silicates with fluorine reagents (F_2, BrF_5), which are extremely difficult to handle. For mass spectrometry, the fundamental requirement is to analyse the $^{18}O/^{16}O$ ratio (at and abundance ratio of 1/500) and $^{17}O/^{16}O$ (at 1/2500). This may not seem a big difference in the ratio, but getting sufficient sample to generate a high signal to noise makes this quite a difficult measurement. In magnetic sector mass spectrometry, the ion beam can be brought into optimized collectors to allow simultaneous measurement of the three isotopes to high precision. This is used in gas source mass spectrometry and secondary ion mass spectrometry. However, a heavy magnetic sector mass spectrometer is not well suited to space flight.

Despite these obstacles, many elements of a successful triple oxygen isotope analysis have already flown on space flights. For water analysis, the double-focussing mass spectrometer in ROSINA was capable of getting to high mass resolution to separate water isotopologues (ca. 3000 $M/\Delta M$), but in doing so gaussian peak shapes were produced, which are not amenable to high precision/accuracy measurements. On Philae, the sample preparation system SD2 included an elaborate chemical laboratory with multiple ovens and reaction vessels (Finzi et al. 2007); unfortunately this could not be operated because of the landing difficulties experienced by Philae.

There are a number of specific modifications that could be made to improve versatility and reliability. Flying fluorine reagents is potentially problematic in terms of valve operation. An alternative methodology could be based on reaction of evaporated/pyrolyzed oxygen with graphite, which could be purified ^{12}C as was used in the Genesis wafers, to avoid the ^{13}C isobars in mass spectrometric analysis. This would potentially work well for both dust and ice analysis.

With chemical separation available, a low mass resolution magnetic sector mass spectrometer is capable of separating the oxygen-bearing ions, without the high resolution required for water ion analysis. A lighter magnetic sector mass spectrometer would be the best system for this type of analysis with the demonstrated capability of high precision and high accuracy.

TOF has been the preferred method of mass spectrometry in space because of low overall weight and a relatively simple electronic configuration. An example of an upcoming TOF-MS is the MASPEX instrument to fly on Europa Clipper (Brockwell et al. 2016). It has a flight weight of 8 kg. This TOF-MS is capable of high sensitivity and 30000 $M/\Delta M$. The main objective is to be able to identify different organic compounds based on their small mass differences. This then has the mass resolution capabilities for carrying out oxygen three isotopes, but obtaining high precision isotope measurements on TOF-MS systems remains a tricky prospect. Other forms of mass spectrometers based around time of flight include cyclotrons and the MULTUM II (a multi-turn TOF MS; Tonotani et al. 2016). But these are still based on bunches of a relatively small number of ions whereas high precision measurement requires a large number of ions to satisfy counting statistics.

Cavity Ring Down Spectrometry is a different type of measurement technology, based on optical spectroscopy and can be used in the field on Earth. Still, operation in space is a different story to operating on Earth and it remains to be seen if the instrumentation is sufficiently stable and robust to be capable of remote analysis. Ice measurement through water analysis is a definite possibility. If silicate oxygen could be released to form water, or CO_2, this would extend this capability.

9.3 Sample Return

The benefits of sample return in terms of analysis and sample documentation are immense. Hayabusa brought back 1500 grains of dust from Itokawa; this was sufficient to fully document a wide range of petrology, morphology, as well as isotopic systems. A timely approach to analysis can be carried out maximising the benefit of a single sample or indeed a single grain. Sample return allows optimised analytical conditions to get high precision and accuracy. Contamination can be assessed. Another important advantage of sample return is repeatability of measurements. Analysing a similar aliquot remotely is likely not possible with current technology.

The limitations to sample return are related to the logistics of bringing a sample capsule back to Earth. The asteroid return samples show that this is practical, and even the Stardust mission to a comet brought back the first solid mineral fragments thereof. However, these are all ostensibly inner solar system missions and with Δv back to Earth at a reasonable level.

The Hayabusa mission to Itokawa involved a Δv of only 0.5 km/s which was achievable through the ion engines. In coming back to Earth, the additional velocity was allowed for in the re-entry of the sample capsule. All sample return missions achieved so far are the same method. Even the return of Stardust followed this approach. The Earth re-entry already starts at 11 km/s and a small increment on top of this is not a major issue.

So a Mars return is feasible from the Earth reentry point of view. The difficulty for Mars sample return lies in the entry and exit from Mars itself. Getting samples from the Mars surface is currently being planned but remains a difficult and expensive logistical task.

For the asteroid belt, a potential Δv of 5 km/s becomes more daunting for direct re-entry. As such some form deceleration will likely be required either through engines, or through progressive atmospheric braking in Earth's or another planet's gravity well. The issue then becomes the duration of the mission. The Dawn mission lifted off in 2007 and took four years to get to Vesta; Dawn arrived at Ceres in 2015 and the mission was terminated in 2018. If a sample return mission was launched to Ceres a sample return would likely require at least 10 years depending on orbital alignment and profligacy of fuel expenditure.

It took Juno 5 years to reach Jupiter orbit. A mission to sample the Galilean Moons and return would be difficult not only for the entry and exit of Jupiter's gravitational well, but also the radiation coming from Jupiter. Add to this the issues of taking off from a sizeable planetary body. Potentially a mission to Callisto would be energetically more favourable but this would likely only result in the return of ice, and potentially contaminated with Trojan dust. A mission to Io, home of the furthest silicate volcanism from the Sun, would be highly desirable, but may be a step too far at this stage.

Inward from Earth are Venus and Mercury. A sample return from Venus involves many difficulties related to Venus mass, heavy atmosphere, surface heat, and then an uphill run home. Mercury might be easier on all fronts to land and take a sample, but it also represents a major difficulty in coming out of Mercury and coming back to Earth.

10 Concluding Remarks

Sampling extraterrestrial bodies is an essential ingredient in understanding the provenance of our solar system. While much material is available to us for free through meteorites, obtaining the details of location and history of the particular source body is essential to reconstruct distributions of materials in the early solar system. Still, meteorites have shown us a lot, and oxygen isotope analysis has proven to be a particularly powerful classification tool because individual parent bodies can have a unique identifier in their triple oxygen isotope compositions. The source of these anomalies are likely related to chemical speciation of oxygen isotope carriers and specifically silicate dust, nebular water and carbon monoxide through predissociation and self-shielding likely in the primordial molecular cloud. This is supported by the Genesis mission where the Earth has been shown to be enriched in ^{17}O and ^{18}O relative to the Sun. Another prediction of this model is that outer solar system silicates should be closer to solar oxygen in isotope composition and should be accompanied by isotopically heavy oxygen (^{17}O, ^{18}O-rich) water ice. There is also evidence in chondrites for ^{16}O-poor hydrous activity on the asteroidal parent bodies from the inner Solar System. However, mineralogical and oxygen isotopic analyses of material returned by the Stardust mission from the coma of 81P/Wild2 hint at the complexities recorded in material from the earliest Solar System. Wild2 grains have chondrite-like mineral assemblages and isotopic characteristics.

Obtaining remote triple oxygen isotope analyses of water in the outer solar system appears achievable in the medium term. This would allow some further context to be placed on the oxygen isotope systematics of the entire solar system. However, a silicate analysis would require a completely new chemical separation procedure, as well as new methodologies to measure the triple oxygen isotope abundance. Sample return from the outer solar system would allow a direct test of some of the predictions of the PSS model, but is energy and time expensive. Sample return from the inner solar system planets will allow material to be analysed with the best analytical capability and precision, as well as allowing characterisation of the samples with an arsenal of analytical techniques. Studies of martian meteorites have established that such samples will not only reveal the details of crustal oxygen and other isotopic reservoirs but also a record of atmospheric interactions.

Acknowledgements TRI and RCG acknowledge ISSI support to attend the Europlanet & International Space Science Institute Workshop on the *Role of Sample Return in Addressing Major Outstanding Questions in Planetary Sciences* held in Bern, Switzerland, 5-9 February 2018. We are grateful to the two reviewers whose suggestions significantly improved the manuscript and to the patience of Mahesh Anand. RCG is funded through a Consolidated Grant from the Science and Technology Facilities Council (STFC), UK Grant Number: ST/L000776/1. LJH and JCB are funded by STFC grant ST/S000429/1. TRI was supported by Australian Research Council Discovery Project Grant DP190102760. Thanks to all our friends and colleagues at ESA, JAXA, and NASA.

References

C.B. Agee, N.V. Wilson, F.M. McCubbin, K. Ziegler, V.J. Polyak, Z.D. Sharp, Y. Asmerom, M.H. Nunn, R. Shaheen, M.H. Thiemens, A. Steele, M.L. Fogel, R. Bowden, M. Glamoclija, Z. Zhang, S.M. Elardo,

Unique meteorite from early Amazonian Mars: Water-rich basaltic breccia Northwest Africa 7034. Science **339**, 780–785 (2013). https://doi.org/10.1126/science.1228858

C.M.O'D. Alexander, D.J. Barber, R. Hutchison, The microstructure of Semarkona and Bishunpur. Geochim. Cosmochim. Acta **53**, 3045–3057 (1989). https://doi.org/10.1016/0016-7037(89)90180-4

A. Ali, I. Jabeen, D. Gregory, R. Verish, N.R. Banerjee, New triple oxygen isotope data of bulk and separated fractions from SNC meteorites: Evidence for mantle homogeneity of Mars. Meteorit. Planet. Sci. **51**, 981–995 (2016). https://doi.org/10.1111/maps.12640

ALMA Partnership et al., The 2014 ALMA long baseline campaign: First results from high angular resolution observations toward the HL Tau region. Astrophys. J. Lett. **808**, L3 (2015). https://doi.org/10.1088/2041-8205/808/1/L3

K. Altwegg et al., 67P/Churyumov-Gerasimenko, a Jupiter family comet with a high D/H ratio. Science **347**, 1261952 (2015). https://doi.org/10.1126/science.1261952

E. Anders, E. Zinner, Interstellar grains in primitive meteorites: Diamond, silicon carbide, and graphite. Meteoritics **28**, 490–514 (1993). https://doi.org/10.1111/j.1945-5100.1993.tb00274.x

R.M.G. Armytage, R.B. Georg, H.M. Williams, A.N. Halliday, Silicon isotopes in lunar rocks: Implications for the Moon's formation and the early history of the Earth. Geochim. Cosmochim. Acta **77**, 504–524 (2012). https://doi.org/10.1016/j.gca.2011.10.032

J.N. Ávila, M. Lugaro, T.R. Ireland, F. Gyngard, E. Zinner, S. Cristallo, P. Holden, J. Buntain, S. Amari, A. Karakas, Tungsten isotopic compositions in Stardust SiC grains from the Murchison meteorite: Constraints on the s-process in the Hf-Ta-W-Re-Os region. Astrophys. J. **744**, 49–62 (2012). https://doi.org/10.1088/0004-637X/744/1/49

J.N. Ávila, T.R. Ireland, F. Gyngard, E. Zinner, G. Malmann, P. Holden, S. Amari, Barium isotopic compositions in large Stardust SiC grains from the Murchison meteorite: A cautionary tale of unresolved mass interferences. Geochim. Cosmochim. Acta **120**, 628–647 (2013a). https://doi.org/10.1016/j.gca.2013.03.039

J.N. Ávila, T.R. Ireland, M. Lugaro, F. Gyngard, E. Zinner, S. Cristallo, P. Holden, T. Rauscher, Europium s-process signature at close-to-solar metallicity in Stardust SiC grains from asymptotic giant branch stars. Astrophys. J. Lett. **768**, L18 (2013b). https://doi.org/10.1088/2041-8205/768/1/L18

J.N. Ávila, T.R. Ireland, P. Holden, P. Lanc, A. Latimore, N. Schram, J. Foster, I.S. Williams, L. Loiselle, B. Fu, High-precision, high-accuracy oxygen isotope measurements of zircon reference materials with the SHRIMP-SI. J. Geostand. Geoanal. Res. (2020, in press). https://doi.org/10.1111/ggr.12298

P. Baertschi, Absolute ^{18}O content of standard mean ocean water. Earth Planet. Sci. Lett. **31**, 341–344 (1976). https://doi.org/10.1016/0012-821X(76)90115-1

P. Baertschi, S.R. Silverman, The determination of relative abundances of the oxygen isotopes in silicate rocks. Geochim. Cosmochim. Acta **1**, 317–328 (1951). https://doi.org/10.1016/0016-7037(51)90006-3

J. Bally, W.D. Langer, Isotope-selective photodissociation of carbon monoxide. Astrophys. J. **255**, 143–148 (1982). https://doi.org/10.1086/159812

H. Balsiger, K. Altwegg, F. Bühler, J. Geiss, A.G. Ghielmetti, B.E. Goldstein, R. Goldstein, W.T. Huntress, W.-H. Ip, A.J. Lazarus, A. Meier, M. Neugebauer, U. Rettenmund, H. Rosenbauer, R. Schwenn, R.D. Sharp, E.G. Shelley, E. Ungstrup, D.T. Young, Ion composition and dynamics at comet Halley. Nature **321**, 330–334 (1986). https://doi.org/10.1038/321330a0

S.L.L. Barker, G.M. Dipple, F. Dong, D.S. Baer, Use of laser spectroscopy to measure the $^{13}C/^{12}C$ and $^{18}O/^{16}O$ compositions of carbonate minerals. Anal. Chem. **83**, 2220–2226 (2011). https://doi.org/10.1021/ac103111y

J-A. Barrat, R.C. Greenwood, K. Keil, M.L. Rouget, J.S. Boesenberg, B. Zanda, I.A. Franchi, The origin of aubrites: Evidence from lithophile trace element abundances and oxygen isotopic compositions. Geochim. Cosmochim. Acta **192**, 29–48 (2016). https://doi.org/10.1016/j.gca.2016.07.025

A. Basu Sarbadhikari, J.M.D. Day, Y. Liu, D. Rumble, L.A. Taylor, Petrogenesis of olivine-phyric shergottite Larkman Nunatak 06319: Implications for enriched components in martian basalts. Geochim. Cosmochim. Acta **73**, 2190–2214 (2009). https://doi.org/10.1016/j.gca.2009.01.012

A. Bischoff, M. Horstmann, A. Pack, M. Laubenstein, S. Haberer, Asteroid 2008 TC3—Almahata Sitta: A spectacular breccia containing many different ureilitic and chondritic lithologies. Meteorit. Planet. Sci. **45**, 1638–1656 (2010). https://doi.org/10.1111/j.1945-5100.2010.01108.x

A. Bischoff, N. Vogel, J. Roszjara, The Rumuruti chondrite group. Geochemistry **71**, 101–133 (2011). https://doi.org/10.1016/j.chemer.2011.02.005

A. Bischoff, M. Horstmann, J-A. Barrat, M. Chaussidon, A. Pack, D. Herwartz, D. Ward, C. Vollmer, S. Decker, Trachyandesitic volcanism in the early Solar System. Proc. Natl. Acad. Sci. USA **111**, 12689–12692 (2014). https://doi.org/10.1073/pnas.1404799111

J. Bradley, H. McSween, R. Harvey, Epitaxial growth of nanophase magnetite in martian meteorite Allan Hills 84001: Implications for biogenic mineralization. Meteorit. Planet. Sci. **33**, 765–773 (1998). https://doi.org/10.1111/j.1945-5100.1998.tb01682.x

A.J. Brearley, R.H. Jones, Chondritic meteorites, in *Planetary Materials*, ed. by J.J. Papike. Reviews in Mineralogy, vol. 36 (1998), pp. 3-1–3-398

J.C. Bridges, T.R. Ireland, Oxygen isotope analyses by SHRIMP of chondrules in highly unequilibrated L13 chondrites, in *Lunar and Planetary Science Conference XLVI* (2015), abstract #1674

J.C. Bridges, S.P. Schwenzer, The nakhlite hydrothermal brine on Mars. Earth Planet. Sci. Lett. **359–360**, 117–123 (2012). https://doi.org/10.1016/j.epsl.2012.09.044

J.C. Bridges, P.H. Warren, The SNC meteorites: Basaltic igneous processes on Mars. J. Geol. Soc., Lond. **163**, 229–251 (2006). https://doi.org/10.1144/0016-764904-501

J.C. Bridges, I.A. Franchi, R. Hutchison, A.D. Morse, J.V.P. Long, C.T. Pillinger, Cristobalite- and tridymite-bearing clasts in Parnallee (LL3) and Farmington (L5). Meteoritics **30**, 715–727 (1995). https://doi.org/10.1111/j.1945-5100.1995.tb01169.x

J.C. Bridges, I.A. Franchi, A.S. Sexton, C.T. Pillinger, Mineralogical controls on the oxygen isotopic compositions of UOCs. Geochim. Cosmochim. Acta **63**, 945–951 (1999). https://doi.org/10.1016/S0016-7037(98)00317-2

J.C. Bridges, D.C. Catling, J.M. Saxton, T.D. Swindle, I.C. Lyon, M.M. Grady, Alteration assemblages in martian meteorites: Implications for near-surface processes, in *Evolution of Mars*, ed. by R. Kallenbach, J. Geiss, W.K. Hartmann (Kluwer, Dordrecht, 2001), pp. 365–392. https://doi.org/10.1023/A:101196582 [also published in Space Sci. Rev. **96**, 365–392 (2001)]

J.C. Bridges, H.G. Changela, S. Nayakshin, N.A. Starkey, I.A. Franchi, Chondrule fragments from Comet Wild2: Evidence for high temperature processing in the outer solar system. Earth Planet. Sci. Lett. **341–344**, 186–194 (2012). https://doi.org/10.1016/j.epsl.2012.06.011

J.C. Bridges, L.J. Hicks, A. Treiman, Carbonates on Mars, in *Volatiles in the Martian Crust*, ed. by J. Filiberto, S.P. Schwenzer (Elsevier, Amsterdam, 2019), pp. 89–118. https://doi.org/10.1016/B978-0-12-804191-8.00005-2

T.G. Brockwell, K.J. Meech, K. Pickens, J.H. Waite, G. Miller, J. Roberts, J.I. Lunine et al., The mass spectrometer for planetary exploration (MASPEX), in *IEEE Aerospace Conference* (2016). https://doi.org/10.1109/AERO.2016.7500777

D. Brownlee et al., Comet 81P/Wild 2 under a microscope. Science **314**, 1711–1716 (2006). https://doi.org/10.1126/science.1135840

E.M. Burbidge, G.R. Burbidge, W.A. Fowler, F. Hoyle, Synthesis of the elements in stars. Rev. Mod. Phys. **29**, 547–650 (1957). https://doi.org/10.1103/RevModPhys.29.547

T. Burbine, *Asteroids: Astronomical and Geological Bodies*, Cambridge Planetary Science (Cambridge University Press, Cambridge, 2016). https://doi.org/10.1017/9781316156582

M.J. Burchell, S.A.J. Fairey, P. Wozniakiewicz, D.E. Brownlee, F. Hörz, A.T. Kearsley, T.H.S.P. Tsou, A. Westphal, S.F. Green, J.M. Trigo-Rodríguez, G. Domingúez, Characteristics of cometary dust tracks in Stardust aerogel and laboratory calibrations. Meteorit. Planet. Sci. **43**, 23–40 (2008). https://doi.org/10.1111/j.1945-5100.2008.tb00608.x

D.S. Burnett (Genesis Science Team), Solar composition from the Genesis Discovery Mission. Proc. Natl. Acad. Sci. **108**, 19147–19151 (2011). https://doi.org/10.1073/pnas.1014877108

D.S. Burnett, B.L. Barraclough, R. Bennett, M. Neugebauer, L.P. Oldham, C.N. Sasaki, D. Sevilla, N. Smith, E. Stansbery, D. Sweetnam, R.C. Wiens, Genesis discovery mission: Return of solar matter to Earth. Space Sci. Rev. **105**, 509–534 (2003). https://doi.org/10.1023/A:1024425810605

A.G.W. Cameron, J.W. Truran, The supernova trigger for formation of the solar system. Icarus **30**, 447–461 (1977). https://doi.org/10.1016/0019-1035(77)90101-4

A.G.W. Cameron, W.R. Ward, The origin of the Moon, in *7th Lunar and Planetary Science Conference*, vol. 7 (1976), pp. 120–122 (abstract)

R.M. Canup, Forming a Moon with an Earth-like composition via a giant impact. Science **338**, 1052–1055 (2012). https://doi.org/10.1126/science.1226073

R.M. Canup, E. Asphaug, Origin of the Moon in a giant impact near the end of the Earth's formation. Nature **412**, 708–712 (2001). https://doi.org/10.1038/35089010

C.D. Cappa, M.B. Hendricks, D.J. DePaolo, R.C. Cohen, Isotopic fractionation of water during evaporation. J. Geophys. Res., Atmos. **108**, 4525 (2003). https://doi.org/10.1029/2003JD003597

N.L. Chabot, H. Haack, Evolution of asteroidal cores, in *Meteorites and the Early Solar System II*, ed. by D.S. Lauretta, H.Y. McSween Jr. (University of Arizona Press, Tucson, 2006), pp. 747–771, 943 pp.

B.-G. Choi, K.D. McKeegan, A.N. Krot, J.T. Wasson, Extreme oxygen-isotope compositions in magnetite from unequilibrated ordinary chondrites. Nature **392**, 577–579 (1998). https://doi.org/10.1038/33356

R.N. Clayton, Oxygen isotopes in meteorites. Annu. Rev. Earth Planet. Sci. **21**, 15–149 (1993). https://doi.org/10.1146/annurev.ea.21.050193.000555

R.N. Clayton, Self-shielding in the solar nebula. Nature **415**, 860–861 (2002). https://doi.org/10.1038/415860b

R.N. Clayton, Oxygen isotopes in the solar system. Space Sci. Rev. **106**, 19–32 (2003). https://doi.org/10.1023/A:1024669l1

R.N. Clayton, T.K. Mayeda, The use of bromine pentafluoride in the extraction of oxygen from oxides and silicates for isotopic analysis. Geochim. Cosmochim. Acta **27**, 43–52 (1963). https://doi.org/10.1016/0016-7037(63)90071-1

R.N. Clayton, T.K. Mayeda, Oxygen isotope studies of achondrites. Geochim. Cosmochim. Acta **60**, 1999–2017 (1996). https://doi.org/10.1016/0016-7037(96)00074-9

R.N. Clayton, T.K. Mayeda, Oxygen isotope studies of carbonaceous chondrites. Geochim. Cosmochim. Acta **63**, 2089–2104 (1999). https://doi.org/10.1016/S0016-7037(99)00090-3

R.N. Clayton, L. Grossman, T.K. Mayeda, A component of primitive nuclear composition in carbonaceous meteorites. Science **182**, 485–488 (1973). https://doi.org/10.1126/science.182.4111.485

R.N. Clayton, N. Onuma, T.K. Mayeda, A classification of meteorites based on oxygen isotopes. Earth Planet. Sci. Lett. **30**, 10–18 (1976). https://doi.org/10.1016/0012-821X(76)90003-0

R.N. Clayton, N. Onuma, L. Grossman, T.K. Mayeda, Distribution of the pre-solar component in Allende and other carbonaceous chondrites. Earth Planet. Sci. Lett. **34**, 209–224 (1977). https://doi.org/10.1016/0012-821X(77)90005-X

R.N. Clayton, N. Onuma, Y. Ikeda, T.K. Mayeda, I.D. Hutcheon, E.J. Olsen, C. Molini-Velsko, Oxygen isotopic compositions of chondrules in Allende and ordinary chondrites, in *Chondrules and Their Origins*, ed. by E.A. King (1983), pp. 37–43

R.N. Clayton, T.K. Mayeda, A.E. Rubin, Oxygen isotopic compositions of enstatite chondrites and aubrites. J. Geophys. Res. **89**, 245–249 (1984). https://doi.org/10.1029/JB089iS01p0C245

R.N. Clayton, T.K. Mayeda, J.N. Goswami, E.J. Olsen, Oxygen isotope studies of ordinary chondrites. Geochim. Cosmochim. Acta **55**, 2317–2337 (1991). https://doi.org/10.1016/0016-7037(91)90107-G

P. Cloud, Paleoecological significance of the banded iron formation. Econ. Geol. **68**, 1135–1143 (1973). https://doi.org/10.2113/gsecongeo.68.7.1135

B.A. Cohen, C.A. Malespin, K.A. Farley, P.E. Martin, Y. Cho, P.R. Mahaffy, In situ geochronology on Mars and the development of future instrumentation. Astrobiology **19**(11), 1303–1314 (2019). https://doi.org/10.1089/ast.2018.1871

J.N. Connelly, J. Bollard, M. Bizzarro, Pb–Pb chronometry and the early Solar System. Geochim. Cosmochim. Acta **201**, 345–363 (2017). https://doi.org/10.1016/j.gca.2016.10.044

H.C. Connolly Jr., R.H. Jones, Chondrules: The canonical and noncanonical views. J. Geophys. Res., Planets **121**, 1885–1899 (2016). https://doi.org/10.1002/2016JE005113

A. Cousin, V. Sautter, V. Payré, O. Forni, N. Mangold, O. Gasnault, L. Le Deit, J. Johnson, S. Maurice, M. Salvatore, R.C. Wiend, P. Gasda, W. Rapin, Classification of igneous rocks analyzed by ChemCam at Gale crater, Mars. Icarus **288**, 265–283 (2017). https://doi.org/10.1016/j.icarus.2017.01.014

H. Craig, Standard for reporting concentrations of deuterium and oxygen-18 in natural waters. Science **133**, 1833–1834 (1961). https://doi.org/10.1126/science.133.3467.1833

M. Ćuk, S.T. Stewart, Making the Moon from a fast-spinning Earth: A giant impact followed by resonant despinning. Science **338**, 1047–1052 (2012). https://doi.org/10.1126/science.1225542

W. Dansgaard, Stable isotopes in precipitation. Tellus **16**, 436–468 (1964). https://doi.org/10.1111/j.2153-3490.1964.tb00181.x

J.R. Darling, D.E. Moser, I.R. Barker, K.T. Tait, K.R. Chamberlain, A.K. Schmitt, B.C. Hyde, Variable microstructural response of baddeleyite to shock metamorphism in young basaltic shergottite NWA 5298 and improved U–Pb dating of Solar System events. Earth Planet. Sci. Lett. **444**, 1–12 (2016). https://doi.org/10.1016/j.epsl.2016.03.032

N. Dauphas, The isotopic nature of the Earth's accreting material throughout time. Nature **541**, 521–524 (2017). https://doi.org/10.1038/nature20830

N. Dauphas, C. Burkhardt, P.H. Warren, F-Z. Teng, Geochemical arguments for an Earth-like Moon-forming impactor. Philos. Trans. R. Soc. Lond. A **372**, 20130244 (2014). https://doi.org/10.1098/rsta.2013.0244

F.E. DeMeo, R.P. Binzel, S.M. Slivan, S.J. Bus, An extension of the Bus asteroid taxonomy into the near-infrared. Icarus **202**, 160–180 (2009). https://doi.org/10.1016/j.icarus.2009.02.005

S.J. Desch, H.C. Connolly Jr., A model of the thermal processing of particles in solar nebula shocks: Application to the cooling rates of chondrules. Meteorit. Planet. Sci. **37**, 183–207 (2010). https://doi.org/10.1111/j.1945-5100.2002.tb01104.x

G. Dipierro, D. Price, G. Laibe, K. Hirsh, A. Cerioli, G. Lodato, On planet formation in HL Tau. Mon. Not. R. Astron. Soc. **453**, L73–L77 (2015). https://doi.org/10.1093/mnrasl/slv105

P.H. Edwards, J.C. Bridges, R.C. Wiens, R. Anderson, D. Dyar, M. Fisk, L. Thompson, P. Gasda, J. Filiberto, S.P. Schwenzer, D. Blaney, I. Hutchinson, Basalt-trachybasalt samples in gale crater, Mars. Meteorit. Planet. Sci. **52**, 2391–2410 (2017). https://doi.org/10.1111/maps.12953

J.M. Eiler, N. Kitchen, L. Leshin, M. Strausberg, Hosts of hydrogen in Allan Hills 84001: Evidence for hydrous martian salts in the oldest martian meteorite? Meteorit. Planet. Sci. **37**, 395–405 (2002). https://doi.org/10.1111/j.1945-5100.2002.tb00823.x

S. Epstein, R.P. Sharp, A.J. Gow, Six-year record of oxygen and hydrogen isotope variations in South Pole Firn. J. Geophys. Res. **70**, 1809–1814 (1965). https://doi.org/10.1029/JZ070i008p01809

A.J. Fahey, J.N. Goswami, K.D. McKeegan, E. Zinner, Evidence for extreme Ti-50 enrichments in primitive meteorites. Astrophys. J. Lett. **296**, L17–L20 (1985). https://doi.org/10.1086/184539

A.J. Fahey, J.N. Goswami, K.D. McKeegan, E.K. Zinner, O-16 excesses in Murchison and Murray hibonites—A case against a late supernova injection origin of isotopic anomalies in O, Mg, Ca, and Ti. Astrophys. J. Lett. **323**, L91–L95 (1987). https://doi.org/10.1086/185064

J. Farquhar, M.H. Thiemens, Oxygen cycle of the martian atmosphere-regolith system: $\Delta^{17}O$ of secondary phases in Nakhla and Lafayette. J. Geophys. Res., Planets **105**, 11991–11997 (2000). https://doi.org/10.1029/1999JE001194

J. Farquhar, M.H. Thiemens, T. Jackson, Atmosphere-surface interactions on Mars: $\Delta^{17}O$ measurements of carbonate from ALH 84001. Science **280**, 1580–1582 (1998). https://doi.org/10.1126/science.280.5369.1580

J. Farquhar, H. Bao, M. Thiemens, Atmospheric influence of Earth's earliest sulfur cycle. Science **289**, 756–758 (2000). https://doi.org/10.1126/science.289.5480.756

G. Faure, *Principles of Isotope Geology*, 2nd edn. (Wiley, New York, 1986), 589 pp.

J. Filiberto, E. Chin, J.M.D. Day, I.A. Franchi, R.C. Greenwood, J. Gross, S.C. Penniston-Dorland, S.P. Schwenzer, A.H. Treiman, Geochemistry of intermediate olivine-phyric shergottite Northwest Africa 6234, with similarities to basaltic shergottite Northwest Africa 480 and olivine-phyric shergottite Northwest Africa 2990. Meteorit. Planet. Sci. **47**, 1256–1273 (2012). https://doi.org/10.1111/j.1945-5100.2012.01382.x

A.E. Finzi, F.B. Zazzera, C. Dainese, F. Malnati, P.G. Magnani, E. Re, P. Bologna, S. Espinasse, A. Olivieri, SD2—How to sample a comet. Space Sci. Rev. **128**, 281–299 (2007). https://doi.org/10.1007/s11214-006-9134-6

C. Floss, P. Haenecour, Presolar silicate grains: Abundances, isotopic and elemental compositions, and the effects of secondary processing. Geochem. J. **50**, 3–25 (2016). https://doi.org/10.2343/geochemj.2.0377

I.A. Franchi, I.P. Wright, A.S. Sexton, C.T. Pillinger, The oxygen-isotopic composition of Earth and Mars. Meteorit. Planet. Sci. **34**, 657–661 (1999). https://doi.org/10.1111/j.1945-5100.1999.tb01371.x

A. Fujiwara, J. Kawaguchi, D.K. Yeomans, M. Abe, T. Mukai, T. Okada, J. Saito, H. Yano, M. Yoshikawa, D.J. Scheeres, O. Barnouin-Jha, A.F. Cheng, H. Demura, R.W. Gaskell, N. Hirata, H. Ikeda, T. Kominato, H. Miyamoto, A.M. Nakamura, R. Nakamura, S. Sasaki, K. Uesugi, The rubble-pile asteroid Itokawa as observed by Hayabusa. Science **312**, 1330–1334 (2006). https://doi.org/10.1126/science.1125841pmid:16741107

G.D. Garlick, S. Epstein, Oxygen isotope ratios in coexisting minerals of regionally metamorphosed rocks. Geochim. Cosmochim. Acta **31**, 181–214 (1967). https://doi.org/10.1016/S0016-7037(67)80044-9

B.J. Giletti, M.P. Semet, R.A. Yund, Studies in diffusion- III. Oxygen in feldspars: An ion microprobe determination. Geochim. Cosmochim. Acta **42**, 45–58 (1978). https://doi.org/10.1016/0016-7037(78)90215-6

F. Goesmann, H. Rosenbauer, R. Roll, C. Szopa, F. Raulin, R. Sternberg, G. Israel, U. Meierhenrich, W. Thiemann, G. Munoz-Caro, COSAC, the COmetary SAmpling and Composition experiment on Philae. Space Sci. Rev. **128**, 257–280 (2007). https://doi.org/10.1007/s11214-006-9000-6

J.L. Gooding, S.J. Wentworth, M.E. Zolensky, Aqueous alteration of the Nakhla meteorite. Meteoritics **26**, 135–143 (1991). https://doi.org/10.1111/j.1945-5100.1991.tb01029.x

S.F. Green, N. McBride, M.T.S.H. Colwell, J.A.M. McDonnell, A.J. Tuzzolino, T.E. Economou, B.C. Clark, Z. Sekanina, P. Tsou, D. Brownlee, Stardust Wild 2 dust measurements, in *Proc. "Dust in Planetary Systems" ESA SP-64*, ed. by H. Krueger, A. Graps (2007), pp. 35–44

R.C. Greenwood, I.A. Franchi, A. Jambon, J.A. Barrat, T.H. Burbine, Oxygen isotope variation in stony-iron meteorites. Science **313**, 1763–1765 (2006). https://science.sciencemag.org/content/313/5794/176

R.C. Greenwood, I.A. Franchi, J.M. Gibson, G.K. Benedix, Oxygen isotope variation in primitive achondrites: The influence of primordial, asteroidal and terrestrial processes. Geochim. Cosmochim. Acta **94**, 146–163 (2012). https://doi.org/10.1016/j.gca.2012.06.025

R.C. Greenwood, J-A. Barrat, E.R.D. Scott, H. Haack, P.C. Buchanan, I.A. Franchi, A. Yamaguchi, D. Johnson, A.W.R. Bevan, T.H. Burbine, Geochemistry and oxygen isotope composition of main-group pallasites and olivine-rich clasts in mesosiderites: Implications for the "Great Dunite Shortage" and HED-mesosiderite connection. Geochim. Cosmochim. Acta **169**, 115–136 (2015). https://doi.org/10.1016/j.gca.2015.07.023

R.C. Greenwood, T.H. Burbine, M.F. Miller, I.A. Franchi, Melting and differentiation of early-formed asteroids: The perspective from high precision oxygen isotope studies. Chem. Erde **77**, 1–43 (2017). https://doi.org/10.1016/j.chemer.2016.09.005

R.C. Greenwood, J-A. Barrat, M.F. Miller, M. Anand, N. Dauphas, I.A. Franchi, P. Sillard, N.A. Starkey, Oxygen isotopic evidence for accretion of Earth's water before a high-energy Moon-forming giant impact. Sci. Adv. **4**, eaao5928 (2018). https://doi.org/10.1126/sciadv.aao5928

R.C. Greenwood, M. Anand, What is the oxygen isotope composition of Venus? The scientific case for sample return from Earth's "sister" planet. Space Sci. Rev. (2020, accepted)

R.C. Greenwood, T.H. Burbine, I.A. Franchi, Linking asteroids and meteorites to the primordial planetesimal population. Geochim. Cosmochim. Acta (2020, accepted). https://doi.org/10.1016/j.gca.2020.02.004

L. Grossman, Condensation in the primitive solar nebula. Geochim. Cosmochim. Acta **36**, 597–619 (1972). https://doi.org/10.1016/0016-7037(72)90078-6

L. Grossman, Petrography and mineral chemistry of Ca-rich inclusions in the Allende Meteorite. Geochim. Cosmochim. Acta **39**, 433–454 (1975). https://doi.org/10.1016/0016-7037(75)90099-X

J.N. Grossman, C.M.O'D. Alexander, J. Wang, A.J. Brearley, Bleached chondrules: Evidence for widespread aqueous processes on the parent asteroids of ordinary chondrites. Meteorit. Planet. Sci. **35**, 467–486 (2000). https://doi.org/10.1111/j.1945-5100.2000.tb01429.x

F. Gyngard, E. Zinner, L.R. Nittler, A. Morgand, F.J. Stadermann, K.M. Hynes, Automated NanoSIMS measurements of spinel Stardust from the Murray meteorite. Astrophys. J. **717**, 107–120 (2010). https://doi.org/10.1088/0004-637X/717/1/107

M.K. Haba, J-F. Wotzlaw, Y-J. Lai, A. Yamaguchi, M. Schönbächler, Mesosiderite formation on asteroid 4 Vesta by a hit-and-run collision. Nat. Geosci. **12**, 510–515 (2019). https://doi.org/10.1038/s41561-019-0377-8

L.J. Hallis, M. Anand, R.C. Greenwood, M.F. Miller, I.A. Franchi, S.S. Russell, The oxygen isotope composition, petrology and geochemistry of mare basalts: Evidence for large-scale compositional variation in the lunar mantle. Geochim. Cosmochim. Acta **74**, 6885–6899 (2010). https://doi.org/10.1016/j.gca.2010.09.023

B. Hapke, Space weathering from Mercury to the asteroid belt. J. Geophys. **106**, 10039–10074 (2001). https://doi.org/10.1029/2000JE001338

W.K. Hartmann, D.R. Davis, Satellite-sized planetesimals and lunar origin. Icarus **24**, 504–514 (1975). https://doi.org/10.1016/0019-1035(75)90070-6

K. Hashizume, N. Takahata, H. Naraoka, Y. Sano, Oxygen isotope anomaly with a solar origin detected in meteoritic organics. Nat. Geosci. **4**, 165–168 (2011). https://doi.org/10.1038/NGEO1070

M. Hässig, K. Altwegg, H. Balsiger, J.J. Berthelier, U. Calmonte, M. Combi, J. De Keyser, B. Fiethe, S.A. Fuselier, M. Rubin, ROSINA/DFMS capabilities to measure isotopic ratios in water at comet 67P/Churyumov–Gerasimenko. Planet. Space Sci. **84**, 148–152 (2013). https://doi.org/10.1016/j.pss.2013.05.014

J.E. Heidenreich III, M.H. Thiemens, A non-mass-dependent oxygen isotope effect in the production of ozone from molecular oxygen: The role of molecular symmetry in isotope chemistry. J. Chem. Phys. **84**, 2129 (1986). https://doi.org/10.1063/1.450373

C.D.K. Herd et al., The Northwest Africa 8159 martian meteorite: Expanding the martian sample suite to the early Amazonian. Geochim. Cosmochim. Acta **218**, 1–26 (2017). https://doi.org/10.1016/j.gca.2017.08.037

D. Herwartz, A. Pack, B. Friedrichs, A. Bischoff, Identification of the giant impactor Theia in lunar rocks. Science **344**, 1146–1150 (2014). https://doi.org/10.1126/science.1251117

H.R. Heydegger, J. Foster, W. Compston, Evidence of a new isotopic anomaly from titanium isotopic ratios in meteoric materials. Nature **278**, 704–707 (1979). https://doi.org/10.1038/278704a0

L.J. Hicks, J.C. Bridges, S.J. Gurman, Ferric saponite and serpentine in the nakhlite martian meteorites. Geochim. Cosmochim. Acta **136**, 194–210 (2014). https://doi.org/10.1016/j.gca.2014.04.010

L.J. Hicks, J. MacArthur, J.C. Bridges, M. Price, J. Wickham-Eade, M. Burchell, G. Hansford, A. Butterworth, S.J. Gurman, S. Baker, Magnetite in Comet Wild 2: Evidence for parent body aqueous alteration. Meteorit. Planet. Sci. **52**, 2075–2096 (2017). https://doi.org/10.1111/maps.12909

J. Hoefs, *Stable Isotope Geochemistry*, 6th edn., vol. 285 (Springer, Berlin, 2009)

J.H. Hoffman, R.R. Hodges Jr., M.B. McElroy, T.M. Donahue, M. Kolpin, Composition and structure of the Venus atmosphere: Results from pioneer Venus. Science **205**, 49–52 (1979). https://doi.org/10.1126/science.205.4401.49

H. Holland, The oxygenation of the atmosphere and oceans. Philos. Trans. R. Soc. B **361**, 903–915 (2006). https://doi.org/10.1098/rstb.2006.1838

G. Holland, J.M. Saxton, I.C. Lyon, G. Turner, Negative $\delta^{18}O$ values in Allan Hills 84001 carbonate: Possible evidence for water precipitation on Mars. Geochim. Cosmochim. Acta **69**, 1359–1369 (2005). https://doi.org/10.1016/j.gca.2004.08.023

M. Horstmann, A. Bischoff, A. Pack, N. Albrecht, M. Weyrauch, H. Hain, L. Roggon, K. Schneider, Mineralogy and oxygen isotope composition of new samples from the Almahata Sitta strewn field, in *75th Annual Meeting of the Meteoritical Society, held August 12-17, 2012 in Cairns, Australia*. Published in Meteoritics and Planetary Science Supplement, id. 5052. http://adsabs.harvard.edu/abs/2012M&PSA..75.5052H

F. Hörz et al., Impact features on Stardust: Implications for Comet 81P/Wild 2 dust. Science **314**, 1716–1719 (2006). https://doi.org/10.1126/science.1135705

M. Humayun, A. Nemchin, B. Zanda, R.H. Hewins, M. Grange, A. Kennedy, J.-P. Lorand, C. Göpel, C. Fieni, S. Pont, D. Deldicque, Origin and age of the earliest martian crust from meteorite NWA 7533. Nature **503**, 513–516 (2013). https://doi.org/10.1038/nature12764

R. Hutchison, C.M.O'D. Alexander, D.J. Barber, The Semarkona meteorite: First recorded occurrence of smectite in an ordinary chondrite, and its implications. Geochim. Cosmochim. Acta **51**, 1875–1882 (1987). https://doi.org/10.1016/0016-7037(87)90178-5

R.B. Ickert, J. Hiess, I.S. Williams, P. Holden, T.R. Ireland, P. Lanc, N. Schram, J.J. Foster, S.W. Clement, Determining high precision, in situ, oxygen isotope ratios with a SHRIMP II: Analyses of MPI-DING silicate-glass reference materials and zircon from contrasting granites. Chem. Geol. **257**, 114–128 (2008). https://doi.org/10.1016/j.chemgeo.2008.08.024

T.R. Ireland, Presolar isotopic and chemical signatures in hibonite-bearing refractory inclusions from the Murchison carbonaceous chondrite. Geochim. Cosmochim. Acta **54**, 3219–3237 (1990). https://doi.org/10.1016/0016-7037(90)90136-9

T.R. Ireland, Oxygen isotope tracing of the solar system. Aust. J. Earth Sci. **59**, 225–236 (2012). https://doi.org/10.1080/08120099.2012.620626

T.R. Ireland, Ion microscopes and microprobes, in *Treatise on Geochemistry*, vol. 15, ed. by H.D. Holland, K.K. Turekian, 2nd edn. (Elsevier, Oxford, 2014), pp. 385–409

T.R. Ireland, B. Fegley Jr., The solar system's earliest chemistry: Systematics of refractory inclusions. Int. Geol. Rev. **42**, 865–894 (2000). https://doi.org/10.1080/00206810009465116

T.R. Ireland, E.K. Zinner, A.J. Fahey, T.M. Esat, Evidence for distillation in the formation of HAL and related hibonite inclusions. Geochim. Cosmochim. Acta **56**, 2503–2520 (1992). https://doi.org/10.1016/0016-7037(92)90205-W

T.R. Ireland, P. Holden, M.D. Norman, J. Clarke, Isotopic enhancements of ^{17}O and ^{18}O from solar wind particles in the lunar regolith. Nature **440**, 776–778 (2006). https://doi.org/10.1038/nature04611

T.R. Ireland, N. Schram, P. Holden, P. Lanc, J. Ávila, R. Armstrong, Y. Amelin, A. Latimore, D. Corrigan, S. Clement, J.J. Foster, W. Compston, Charge-mode electrometer measurements of S-isotopic compositions on SHRIMP-SI. Int. J. Mass Spectrom. **359**, 26–37 (2014). https://doi.org/10.1016/j.ijms.2013.12.020

A.J. Irving, S.M. Kuehner, C.D.K. Herd, M. Gellissen, R.L. Korotev, I. Puchtel, R.J. Walker, T. Lapen, D. Rumble, Petrologic, elemental and multi-isotopic characterization of permafic olivine-phyric shergottite northwest Africa 5789: A primitive magma derived from depleted martian mantle, in *41st Lunar and Planetary Science Conference* (2010), abstract #1547

H.A. Ishii, J.P. Bradley, Z.R. Dai, M. Chi, A.T. Kearsley, M.J. Burchell, N.D. Browning, F. Molster, Comparison of Comet 81P/WIld 2 dust with interplanetary dust from comets. Science **319**, 447–450 (2008). https://doi.org/10.1126/science.1150683

I. Jabeen, M. Kusakabe, Determination of $\delta^{17}O$ values of reference water samples VSMOW and SLAP. Chem. Geol. **143**, 115–119 (1997). https://doi.org/10.1016/S0009-2541(97)00109-5

M. Jadhav, E.K. Zinner, S. Amari, T. Maruoka, K.K. Marhas, R. Gallino, Multi-element isotopic analyses of presolar graphite grains from Orgueil. Geochim. Cosmochim. Acta **113**, 193–224 (2013). https://doi.org/10.1016/j.gca.2013.01.018

E. Jarosewich, R.S. Clarke Jr., J.N. Barrows, in *Allende Meteorite Reference Sample*. Smithsonian Contributions to the Earth Sciences, vol. 27 (1987), pp. 1–49. https://doi.org/10.5479/si.00810274.27.1

S. Johnsen, W. Dansgaard, H. Clausen et al., Oxygen isotope profiles through the Antarctic and Greenland ice sheets. Nature **235**, 429–434 (1972). https://doi.org/10.1038/235429a0

J.E. Johnson, A. Gerpheide, M.P. Lamb, W.W. Fischer, O_2 constraints from Paleoproterozoic detrital pyrite and uraninite. Geol. Soc. Am. Bull. **126**, 813–830 (2014). https://doi.org/10.1130/B30949

D.J. Joswiak, D. Nakashima, D.E. Brownlee, G. Matrajt, T. Ushikubo, N.T. Kita, S. Messenger, M. Ito, Terminal particle from Stardust track 130: Possible Al-rich chondrule fragment from comet Wild 2. Geochim. Cosmochim. Acta **144**, 277–298 (2014). https://doi.org/10.1016/j.gca.2014.08.017

H.R. Karlsson, R.N. Clayton, E.K. Gibson, T.K. Mayeda, Water in SNC meteorites—Evidence for a martian hydrosphere. Science **255**, 1409–1411 (1992). https://doi.org/10.1126/science.11537889

J. Kawaguchi, A. Fujiwara, T. Uesugi, Hayabusa—Its technology and science accomplishment summary and Hayabusa-2. Acta Astronaut. **62**, 639–647 (2008). https://doi.org/10.1016/j.actaastro.2008.01.028

Y. Kawai, T. Hondo, K.R. Jensen, M. Toyoda, K. Terada, Improved quantitative dynamic range of time-of-flight mass spectrometry by simultaneously waveform-averaging and ion-counting data acquisition. J. Am. Soc. Mass Spectrom. **29**, 1403–1407 (2018). https://doi.org/10.1007/s13361-018-1967-1

J. Kissel et al., Composition of comet Halley dust particles from Vega observations. Nature **321**, 280–282 (1986). https://doi.org/10.1038/321280a0

J. Kissel, F.R. Krueger, J. Silén, B.C. Clark, The Cometary Impact Dust Analyzer (CIDA). Science **304**, 1774–1776 (2004). https://doi.org/10.1126/science.1098836

S.S. Kistler, Coherent expanded aerogels and jellies. Nature **127**, 741 (1931). https://doi.org/10.1038/127741a0

N. Kita, T. Ushikubo, B. Fu, J. Valley, High precision SIMS oxygen isotope analysis and the effect of sample topography. Chem. Geol. **264**, 43–57 (2009). https://doi.org/10.1016/j.chemgeo.2009.02.012

N.T. Kita, H. Nagahara, S. Tachibana, S. Tomomura, M.J. Spicuzza, J.H. Fournelle, J.W. Valley, High precision SIMS oxygen three isotope study of chondrules in LL3 chondrites: Role of ambient gas during chondrule formation. Geochim. Cosmochim. Acta **74**, 6610–6635 (2010). https://doi.org/10.1016/j.gca.2010.08.011

S. Kobayashi, H. Imai, H. Yurimoto, New extreme ^{16}O-rich reservoir in the early solar system. Geochem. J. **37**, 663–669 (2003). https://doi.org/10.2343/geochemj.37.663

L. Kööp, A.M. Davis, D. Nakashima, C. Park, A.N. Krot, K. Nagashima, T.J. Tenner, P.R. Heck, N.T. Kita, A link between oxygen, calcium and titanium isotopes in ^{26}Al-poor hibonite-rich CAIs from Murchison and implications for the heterogeneity of dust reservoirs in the solar nebula. Geochim. Cosmochim. Acta **189**, 70–95 (2016). https://doi.org/10.1016/j.gca.2016.05.014

A.N. Krot, K. Keil, E.R.D. Scott, C.A. Goodrich, M.K. Weisberg, Classification of meteorites and their genetic relationships, in *Meteorites and Cosmochemical Processes, vol. 1, Treatise on Geochemistry*, ed. by A.M. Davis, 2nd edn. (Elsevier, Oxford, 2014), pp. 1–63

A.N. Krot, K. Nagashima, K. Fintor, E. Pal-Molnar, Evidence for oxygen-isotope exchange in refractory inclusions from Kaba (CV3.1) carbonaceous chondrite during fluid-rock interaction on the CV parent asteroid. Geochim. Cosmochim. Acta **246**, 419–435 (2019). https://doi.org/10.1016/j.gca.2018.11.002

T.S. Kruijer, C. Burkhardt, G. Budde, T. Kleine, Dating the formation of Jupiter. Proc. Natl. Acad. Sci. **114**, 6712–6716 (2017). https://doi.org/10.1073/pnas.1704461114

C.J. Lada, E.A. Lada, D.P. Clemens, J. Bally, Dust extinction and molecular gas in the dark cloud IC 5146. Astrophys. J. **429**, 694–709 (1994). https://doi.org/10.1086/174354

T.J. Lapen et al., Two billion years of magmatism recorded from a single Mars meteorite ejection site. Sci. Adv. **3**(2), e1600922 (2017). https://doi.org/10.1126/sciadv.1600922

D.S. Lauretta et al., The unexpected surface of asteroid (101955) Bennu. Nature **568**, 55–60 (2019). https://doi.org/10.1038/s41586-019-1033-6

T. Lee, D.A. Papanastassiou, G.J. Wasserburg, Demonstration of ^{26}Mg excess in Allende and evidence for ^{26}Al. Geophys. Res. Lett. **3**, 41–44 (1976). https://doi.org/10.1029/GL003i001p00041

L.A. Leshin, K.D. McKeegan, P.K. Carpenter, R.P. Harvey, Oxygen isotopic constraints on the genesis of carbonates from martian meteorite ALH84001. Geochim. Cosmochim. Acta **62**, 3–13 (1998). https://doi.org/10.1016/S0016-7037(97)00331-1

W. Li, N. Baoling, J. Dequi, Z. Qinglian, Measurement of the absolute abundance of oxygen-17 in VSMOW. Chin. Sci. Bull. **33**, 1610–1613 (1988). https://doi.org/10.1360/sb1988-33-19-1610

M.C. Liu, K.D. McKeegan, J.N. Goswami, K.K. Marhas, S. Sahijpal, T.R. Ireland, A.M. Davis, Isotopic records in CM hibonites: Implications for timescales of mixing of isotope reservoirs in the solar nebula. Geochim. Cosmochim. Acta **73**, 5051–5079 (2009). https://doi.org/10.1016/j.gca.2009.02.039

J. Llorca, J. Roszjar, J. Roszjar, J.A. Cartwright, A. Bischoff, U. Ott, A. Pack, S. Merchel, G. Rugel, L. Fimiani, P. Ludwig, J.V. Casado, D. Allepuz, The Ksar Ghilane 002 shergottite—The 100th registered martian meteorite fragment. Meteorit. Planet. Sci. **48**, 493–513 (2013). https://doi.org/10.1111/maps.12074

S.J. Lock, S.T. Stewart, M.I. Petaev, Z. Leinhardt, M.T. Mace, S.B. Jacobsen, M. Ćuk, The origin of the Moon within a terrestrial synestia. J. Geophys. Res., Planets **123**, 910–951 (2018). https://doi.org/10.1002/2017JE005333

L. Loiselle, P. Holden, J.N. Ávila, P. Lanc, J.C. Bridges, J.L. MacArthur, T.R. Ireland, The O isotope composition of martian meteorites using SHRIMP SI: Evidence of multiple reservoirs in silicate minerals of the regolith breccia Northwest Africa 8114, in *50th Lunar and Planetary Science Conference* (2019), abstract #2648

J.R. Lyons, E.D. Young, CO self-shielding as the origin of oxygen isotope anomalies in the early solar nebula. Nature **435**, 317–320 (2005). https://doi.org/10.1038/nature03557

J. MacArthur, J.C. Bridges, L.J. Hicks, R. Burgess, K. Joy, M.J. Branney, G.M. Hansford, S.H. Baker, S.P. Schwenzer, S.J. Gurmand, N.R. Stephen, E.D. Steer, J.D. Piercy, T.R. Ireland, Mineralogical constraints on the thermal history of martian regolith breccia Northwest Africa 8114. Geochim. Cosmochim. Acta **246**, 267–298 (2019). https://doi.org/10.1016/j.gca.2018.11.026

P.R. Mahaffy, C.R. Webster, S.K. Atreya, H. Franz, M. Wong, P.G. Conrad, D. Harpold, J.J. Jones, L.A. Leshin, H. Manning, T. Owen, R.O. Pepin, S. Squyres, M. Trainer (MSL Science Team), Abundance and isotopic composition of gases in the martian atmosphere from the curiosity rover. Science **341**, 263–266 (2013). https://doi.org/10.1126/science.1237966

U.B. Marvin, The discovery and initial characterization of Allan Hills 81005: The first lunar meteorite. Geophys. Res. Lett. **10**(9), 775–778 (1983). https://doi.org/10.1029/GL010i009p00775

Y. Matsuhisa, J.R. Goldsmith, R.N. Clayton, Mechanism of hydrothermal crystallization of quartz at 250 °C and 15 kbar. Geochim. Cosmochim. Acta **42**, 173–182 (1978). https://doi.org/10.1016/0016-7037(78)90130-8

P. Mayewski et al., Holocene climate variability. Quat. Res. **62**(3), 243–255 (2004). https://doi.org/10.1016/j.yqres.2004.07.001

T.B. McCord, J.B. Adams, T.V. Johnson, Asteroid Vesta: Spectral reflectivity and compositional implications. Science **168**, 1445–1447 (1970). https://doi.org/10.1126/science.168.3938.1445

F.M. McCubbin et al., Geologic history of martian regolith breccia Northwest Africa 7034: Evidence for hydrothermal activity and lithologic diversity in the martian crust. J. Geophys. Res., Planets **121**, 2120–2149 (2016). https://doi.org/10.1002/2016JE005143

D.S. McKay, E.K. Gibson, K.L. Thomas-Keprta, H. Vali, C.S. Romanek, S.J. Clemett, X.D. Chillier, C.R. Maechling, R.N. Zare, Search for past life on Mars: Possible relic biogenic activity in martian meteorite ALH84001. Science **273**, 924–930 (1996). https://doi.org/10.1126/science.273.5277.924

K.D. McKeegan, Oxygen isotopes in refractory stratospheric dust particles: Proof of extraterrestrial origin. Science **237**, 1468–1471 (1987). https://doi.org/10.1126/science.237.4821.1468

K.D. McKeegan, R.M. Walker, E. Zinner, Ion microprobe isotopic measurements of individual interplanetary dust particles. Geochim. Cosmochim. Acta **49**, 1971–1987 (1985). https://doi.org/10.1016/0016-7037(85)90091-2

K.D. McKeegan, C.D. Coath, P.H. Mao, G. Jarzebinski, D. Burnett, A high energy secondary ion mass spectrometer for the analysis of captured solar wind, in *36th Lunar and Planetary Science Conference* (2004), abstract #2000

K.D. McKeegan et al., Isotopic compositions of cometary matter returned by Stardust. Science **314**, 1724–1728 (2006). https://doi.org/10.1126/science.1135992

K.D. McKeegan, A. Kallio, V. Heber, G. Jarzebinski, P. Mao, C. Coath, T. Kunihiro, R. Wiens, J. Nordholt, R. Moses, D. Reisenfeld, A. Jurewicz, D. Burnett, The oxygen isotopic composition of the Sun inferred from captured solar wind. Science **332**, 1528–1532 (2011). https://doi.org/10.1126/science.1204636

S.M. McLennan, Large-ion lithophile element fractionation during the early differentiation of Mars and the composition of the martian primitive mantle. Meteorit. Planet. Sci. **38**, 895–904 (2003). https://doi.org/10.1111/j.1945-5100.2003.tb00286.x

H.Y. McSween, SNC meteorites: Are they martian rocks? Geology **12**, 3–6 (1984). https://doi.org/10.1130/0091-7613(1984)12%3C3:SMATMR%3E2.0.CO;2

H.Y. McSween, S.W. Ruff, R.V. Morris, R. Gellert, G. Klingelhöfer, P.R. Christensen, T.J. McCoy, A. Ghosh, J.M. Moersch, B.A. Cohen, A.D. Rogers, C. Schröder, S.W. Squyres, J. Crisp, A. Yen, Mineralogy of volcanic rocks in Gusev Crater, Mars: Reconciling Mössbauer, Alpha Particle X-Ray Spectrometer, and Miniature Thermal Emission Spectrometer spectra. J. Geophys. Res. **113**, E06S04 (2008). https://doi.org/10.1029/2007JE002970

H.Y. McSween, D.W. Mittlefehldt, A.W. Beck, R.G. Mayne, T.J. McCoy, HED meteorites and their relationship to the geology of Vesta and the dawn mission. Space Sci. Rev. **163**, 141–174 (2011). https://doi.org/10.1007/s11214-010-9637-z

S. Messenger, L.P. Keller, F.J. Stadermann, R.M. Walker, E. Zinner, Samples of stars beyond the solar system: Silicate grains in interplanetary dust. Science **300**, 105–108 (2003). https://doi.org/10.1126/science.1080576

M.F. Miller, Isotopic fractionation and the quantification of Δ^{17}O anomalies in the oxygen three-isotope system—An appraisal and geochemical significance. Geochim. Cosmochim. Acta **66**, 1881–1889 (2002). https://doi.org/10.1016/S0016-7037(02)00832-3

M.F. Miller, R.C. Greenwood, I.A. Franchi, Comment on "The triple oxygen isotope composition of the Earth mantle and understanding Δ^{17}O variations in terrestrial rocks and minerals" by A. Pack and D. Herwartz [Earth Planet. Sci. Lett. **390**, 138–145 (2014)]. Earth Planet. Sci. Lett. **418**, 181–183 (2015). https://doi.org/10.1016/j.epsl.2014.12.026

M.F. Miller, A. Pack, I.N. Bindeman, R.C. Greenwood, Standardizing the reporting of Δ^{17}O data from high precision oxygen triple-isotope ratio measurements of silicate rocks and minerals. Chem. Geol. (2020, accepted). https://doi.org/10.1016/j.chemgeo.2019.119332

D.W. Mittlefehldt, ALH84001, a cumulate orthopyroxenite member of the martian meteorite clan. Meteorit. Planet. Sci. **29**, 214–221 (1994). https://doi.org/10.1111/j.1945-5100.1994.tb00673.x

D.W. Mittlefehldt, T.J. McCoy, C.A. Goodrich, A. Kracher, Non-chondritic meteorites from asteroidal bodies, in *Planetary Materials*, ed. by J.J. Papike. Reviews in Mineralogy, vol. 36 (Mineralogical Society of America, Chantilly, 1998), pp. 4.1–4.195

D.E. Moser et al., Solving the martian meteorite age conundrum using micro-baddeleyite and launch-generated zircon. Nature **499**, 454–457 (2013). https://doi.org/10.1038/nature12341

K. Nagao et al., Irradiation history of Itokawa regolith material deduced from noble gases in the Hayabusa samples. Science **333**, 1128–1131 (2011). https://doi.org/10.1126/science.1207785

K. Nagashima, A.N. Krot, H. Yurimoto, Stardust silicates from primitive meteorites. Nature **428**, 921–924 (2004). https://doi.org/10.1038/nature02510

T. Nakamura, T. Noguchi, A. Tsuchiyama, T. Ushikubo, N.T. Kita, J.W. Valley, M.E. Zolensky, Y. Kakazu, K. Sakamoto, E. Mashio, K. Uesugi, T. Nakano, Chondrule-like objects in short-period Comet 81P/Wild 2. Science **321**, 1664–1667 (2008). https://doi.org/10.1126/science.1160995

T. Nakamura et al., Itokawa dust particles: A direct link between S-type asteroids and ordinary chondrites. Science **333**, 1113–1116 (2011). https://doi.org/10.1126/science.1207758

E. Nakamura, A. Makishima, T. Moriguti, K. Kobayashi, R. Tanaka, T. Kunihiro, T. Tsujimori, C. Sakaguchi, H. Kitagawa, T. Ota, Y. Yachi, T. Yada, M. Abe, A. Fujimura, M. Ueno, T. Mukai, M. Yoshikawa, J. Kawaguchi, Space environment of an asteroid preserved on micrograins returned by the Hayabusa spacecraft. Proc. Natl. Acad. Sci. USA **109**, E624–E629 (2012). https://doi.org/10.1073/pnas.1116236109

D. Nakashima, N.T. Kita, T. Ushikubo, T. Noguchi, T. Nakamura, J.W. Valley, Oxygen three-isotope ratios of silicate particles returned from asteroid Itokawa by the Hayabusa spacecraft: A strong link with equilibrated LL chondrites. Earth Planet. Sci. Lett. **379**, 127–136 (2013). https://doi.org/10.1016/j.epsl.2013.08.009

O. Navon, G.J. Wasserburg, Self-shielding in O_2—A possible explanation for oxygen isotopic anomalies in meteorites? Earth Planet. Sci. Lett. **73**, 1–16 (1985)

H.B. Niemann, R.E. Hartle, W.T. Kasprzak, N.W. Spencer, D.M. Hunten, G.R. Carignan, Venus upper atmosphere neutral composition: Preliminary results from the pioneer Venus orbiter. Science **203**, 770–772 (1979). https://doi.org/10.1126/science.203.4382.770

P.B. Niles, L. Leshin, Y. Guan, Microscale carbon isotope variability in ALH84001 carbonates and a discussion of possible formation environments. Geochim. Cosmochim. Acta **69**, 2931–2944 (2005). https://doi.org/10.1016/j.gca.2004.12.012

L.R. Nittler, Presolar oxide grains in meteorites. AIP Conf. Proc. **402**, 59–82 (1997). https://doi.org/10.1063/1.53320

L.R. Nittler, F. Ciesla, Astrophysics with extraterrestrial materials. Annu. Rev. Astron. Astrophys. **54**, 53–93 (2016). https://doi.org/10.1146/annurev-astro-082214-122505

L.R. Nittler, C.M.O'.D. Alexander, J. Wang, X. Gao, Meteoritic oxide grain from supernova found. Nature **393**, 222 (1998). https://doi.org/10.1038/30377

L.E. Nyquist, C.-Y. Shih, D.D. Bogard, Ages and geologic histories of martian meteorites, in *Chronology and Evolution of Mars*, ed. by R. Kallenbach, J. Geiss, W.K. Hartmann (Kluwer, Dordrecht, 2001), pp. 105–164

L.E. Nyquist, C-Y. Shih, F.M. McCubbin, A.R. Santos, C.K. Shearer, Z.X. Peng, P.V. Burger, C.B. Agee, Rb-Sr and Sm-Nd isotopic and REE studies of igneous components in the bulk matrix domain of martian breccia Northwest Africa 7034. Meteorit. Planet. Sci. **51**, 483–498 (2016)

A. O'Keefe, D.A.G. Deacon, Cavity ring-down Optical Spectrometer for absorption measurements using pulsed laser sources. Rev. Sci. Instrum. **59**, 2544 (1988). https://doi.org/10.1063/1.1139895

A. Pack, D. Herwartz, The triple oxygen isotope composition of the Earth mantle and understanding $\Delta^{17}O$ variations in terrestrial rocks and minerals. Earth Planet. Sci. Lett. **390**, 138–145 (2014). https://doi.org/10.1016/j.epsl.2014.01.017

K. Pahlevan, D.J. Stevenson, Equilibration in the aftermath of the lunar-forming giant impact. Earth Planet. Sci. Lett. **262**, 438–449 (2007). https://doi.org/10.1016/j.epsl.2007.07.055

A.A. Pavlov, J.F. Kasting, Mass-independent fractionation of sulfur isotopes in archean sediments: Strong evidence for an anoxic archean atmosphere. Astrobiology **2**, 27–41 (2002). https://doi.org/10.1089/153110702753621321

R. Pepin, Meteorites: Evidence of martian origins. Nature **317**, 473–475 (1985). https://doi.org/10.1038/317473a0

P.N. Peplowski, D.J. Lawrence, T.H. Prettyman, N. Yamashita, D. Bazell, W.C. Feldman, L. Le Corre, T.J. McCoy, V. Reddy, R.C. Reedy, C.T. Russell, M.J. Toplis, Compositional variability on the surface of 4 Vesta revealed through GRaND measurements of high-energy gamma rays. Meteorit. Planet. Sci. **48**, 2252–2270 (2013). https://doi.org/10.1111/maps.12176

L. Qin, L.R. Nittler, C.M.O'D. Alexander, J. Wang, F.J. Stadermann, R.W. Carlson, Extreme ^{54}Cr-rich nano-oxides in the CI chondrite orgueil—Implications for a late supernova injection into the solar system. Geochim. Cosmochim. Acta **75**, 629–644 (2011). https://doi.org/10.1016/j.gca.2010.10.017

R. Reinhard, The Giotto encounter with comet Halley. Nature **321**, 313–318 (1986). https://doi.org/10.1038/321313a0

F. Richter, A.M. Davis, D.S. Ebel, A. Hashimoto, Elemental and isotopic fractionation of Type B calcium-, aluminum-rich inclusions: Experiments, theoretical considerations, and constraints on their thermal evolution. Geochim. Cosmochim. Acta **66**, 521–540 (2002). https://doi.org/10.1016/S0016-7037(01)00782-7

D. Rumble, A.J. Irving, Dispersion of oxygen isotopic compositions among 42 martian meteorites determined by laser fluorination: Evidence for assimilation of (ancient) altered crust, in *40th Lunar and Planetary Science Conference* (2009), abstract #2293

D. Rumble, M.F. Miller, I.A. Franchi, R.C. Greenwood, Oxygen three-isotope fractionation line in terrestrial silicate minerals: An inter laboratory comparison of hydrothermal quartz and eclogitic garnet. Geochim. Cosmochim. Acta **71**, 3592–3600 (2007). https://doi.org/10.1016/j.gca.2007.05.011

D. Rumble, M.E. Zolensky, J.M. Friedrich, P. Jenniskens, M.H. Shaddad, The oxygen isotope composition of Almahata Sitta. Meteorit. Planet. Sci. **45**, 1765–1770 (2010). https://doi.org/10.1111/j.1945-5100.2010.01099.x

C.T. Russell, H.Y. McSween, R. Jaumann, C.A. Raymond, The dawn mission to Vesta and ceres, in *Asteroids IV*, ed. by P. Michel, F.E. DeMeo, W.F. Bottke (University of Arizona Press, Tucson, 2015), pp. 419–432. https://muse.jhu.edu/chapter/1705177

F.J. Ryerson, K.D. McKeegan, Determination of oxygen self-diffusion in åkermanite, anorthite, diopside, and spinel: Implications for oxygen isotopic anomalies and the thermal histories of Ca-Al-rich inclusions. Geochim. Cosmochim. Acta **58**, 3713–3734 (1994). https://doi.org/10.1016/0016-7037(94)90161-9

J. Saito et al., Detailed images of asteroid 25143 Itokawa from Hayabusa. Science **312**, 1341–1344 (2006). https://doi.org/10.1126/science.1125722

N. Sakamoto, Y. Seto, S. Itoh, K. Kuramoto, K. Fujino, K. Nagashima, A.N. Krot, H. Yurimoto, Remnants of the early solar system water enriched in heavy oxygen isotopes. Science **317**, 231–233 (2007). https://doi.org/10.1126/science.1142021

R. Salmeron, T.R. Ireland, Formation of chondrules in magnetic winds blowing through the proto-asteroid belt. Earth Planet. Sci. Lett. **327–328**, 61–67 (2012a). https://doi.org/10.1016/j.epsl.2012.01.033

R. Salmeron, T.R. Ireland, The role of protostellar jets in star formation and the evolution of the early solar system: Astrophysical and meteoritical perspectives. Meteorit. Planet. Sci. **47**, 1922–1940 (2012b). https://doi.org/10.1111/maps.12029

L. Sangély, B. Boyer, E. de Chambost, N. Valle, J.-N. Audinot, T. Ireland, M. Wiedenbeck, J. Aleon, H. Jungnickel, J.-P. Barnes, P. Bienvenu, U. Breuer, Secondary ion mass spectrometry, in *Sector Field Mass Spectrometry for Elemental and Isotopic Analysis* (The Royal Society of Chemistry, London, 2015), pp. 439–499. Chap. 15

V. Sautter, M.J. Toplis, R.C. Wiens, A. Cousin, C. Fabre, O. Gasnault, S. Maurice, O. Forni, J. Lasue, A. Ollila, J.C. Bridges, N. Mangold, S. Le Mouélic, M. Fisk, P.-Y. Meslin, P. Beck, P. Pinet, L. Le Deit, W. Rapin, E.M. Stolper, H. Newsom, D. Dyar, N. Lanza, D. Vaniman, S. Clegg, In situ evidence for continental crust on early Mars. Nat. Geosci. **8**, 605–609 (2015). https://doi.org/10.1038/ngeo2474

S.M. Savin, S. Epstein, The oxygen and hydrogen isotope geochemistry of clay minerals. Geochim. Cosmochim. Acta **34**, 25–42 (1970). https://doi.org/10.1016/0016-7037(70)90149-3

J. Saxton, I. Lyon, G. Turner, Correlated chemical and isotopic zoning in carbonates in the martian meteorite ALH84001. Earth Planet. Sci. Lett. **160**, 811–822 (1998). https://doi.org/10.1016/S0012-821X(98)00129-0

J.M. Saxton, I.C. Lyon, G. Turner, Oxygen isotopes in forsterite grains from Julesburg and Allende: Oxygen-16-rich material in an ordinary chondrite. Meteorit. Planet. Sci. **33**, 1017–1027 (2010). https://doi.org/10.1111/j.1945-5100.1998.tb01708.x

B. Schueler, J. Morton, K. Mauersberger, Measurement of isotopic abundances in collected stratospheric ozone samples. Geophys. Res. Lett. **17**, 1295–1298 (1990). https://doi.org/10.1029/GL017i009p01295

S.P. Schwenzer, R.C. Greenwood, S.P. Kelley, U. Ott, A.G. Tindle, R. Haubold, S. Herrmann, J.M. Gibson, M. Anand, S. Hammond, I.A. Franchi, Quantifying noble gas contamination during terrestrial alteration in martian meteorites from Antarctica. Meteorit. Planet. Sci. **48**, 929–954 (2013). https://doi.org/10.1111/maps.12110

E.R.D. Scott, K. Keil, J.L. Goldstein, E. Asphaug, W.F. Bottke, N.A. Moskovitz, Early impact history and dynamical origin of differentiated meteorites and asteroids, in *Asteroids IV*, ed. by P. Michel, F. DeMeo, W.F. Bottke (University of Arizona Press, Tucson, 2015), pp. 573–595

E.R.D. Scott, A.N. Krot, I.S. Sanders, Isotopic dichotomy among meteorites and its bearing on the protoplanetary disk. Astrophys. J. **854**, 164 (2018). https://doi.org/10.3847/1538-4357/aaa5a5

Z. Sekanina, A model for comet 81P/Wild 2. J. Geophys. Res., Planets **108**, 8112 (2003). https://doi.org/10.1029/2003JE002093

N. Shackleton, N. Opdyke, Oxygen isotope and palaeomagnetic stratigraphy of equatorial Pacific core V28-238: Oxygen isotope temperatures and ice volumes on a 10^5 year and 10^6 year scale. Quat. Res. **3**, 39–55 (1973). https://doi.org/10.1016/0033-5894(73)90052-5

R. Shaheen, P.B. Niles, K. Chong, C.M. Corrigan, M.H. Thiemens, Carbonate formation events in ALH84001 trace the evolution of the martian atmosphere. Proc. Natl. Acad. Sci. **112**, 336–341 (2015). https://doi.org/10.1073/pnas.1315615112

Z.D. Sharp, A laser-based microanalytical method for the in situ determination of oxygen isotope ratios in silicates and oxides. Geochim. Cosmochim. Acta **54**, 1353–1357 (1990). https://doi.org/10.1016/0016-7037(90)90160-M

F.H. Shu, H. Shang, T. Lee, Toward an astrophysical theory of chondrites. Science **271**, 1545–1552 (1996). https://doi.org/10.1126/science.271.5255.1545

M. Siddall, E. Rohling, A. Almogi-Labin et al., Sea-level fluctuations during the last glacial cycle. Nature **423**, 853–858 (2003). https://doi.org/10.1038/nature01690

H. Sierks et al., On the nucleus structure and activity of comet 67P/Churyumov-Gerasimenko. Science **347**, aaa1044 (2015). https://doi.org/10.1126/science.aaa1044

G. Skrzypek, D. Ford, The stable isotope analysis of saline water samples on a cavity ring–down spectroscopy instrument. Environ. Sci. Technol. **48**, 2827–2834 (2014). https://doi.org/10.1021/es4049412

J.V. Smith, A.T. Anderson, R.C. Newton, E.J. Olsen, P.J. Wyllie, A petrologic model for the Moon based on petrogenesis, experimental petrology, and physical properties. J. Geol. **78**, 381–405 (1970). https://doi.org/10.1086/627537

M.L. Spicuzza, J.M.D. Day, L.A. Taylor, J.W. Valley, Oxygen isotope constraints on the origin and differentiation of the Moon. Earth Planet. Sci. Lett. **253**, 254–265 (2007). https://doi.org/10.1016/j.epsl.2006.10.030

F. Stadermann, T.K. Croat, T.J. Bernatowicz, S. Amari, S. Messenger, R.M. Walker, E. Zinner, Supernova graphite in the NanoSIMS: Carbon, oxygen and titanium isotopic compositions of a spherule and its TiC sub-components. Geochim. Cosmochim. Acta **69**, 177–188 (2005). https://doi.org/10.1016/j.gca.2004.06.017

D.J. Stevenson, A.N. Halliday, The origin of the Moon. Philos. Trans. R. Soc., Math. Phys. Eng. Sci. **372**, 20140289 (2014). https://doi.org/10.1098/rsta.2014.0289

S. Sugita et al., The geomorphology, color, and thermal properties of Ryugu: Implications for parent-body processes. Science **364**, eaaw0422 (2019). https://doi.org/10.1126/science.aaw0422

T.D. Swindle, A.H. Treiman, D.J. Lindstrom, M.K. Burkland, B.A. Cohen, J.A. Grier, B. Li, E.K. Olson, Noble gases in iddingsite from the Lafayette meteorite: Evidence for liquid water on Mars in the last few hundred million years. Meteorit. Planet. Sci. **35**, 107–115 (2010). https://doi.org/10.1111/j.1945-5100.2000.tb01978.x

R. Tanaka, E. Nakamura, Determination of ^{17}O-excess of terrestrial silicate/oxide minerals with respect to Vienna Standard Mean Ocean Water (VSMOW). Rapid Commun. Mass Spectrom. **27**, 285–297 (2012). https://doi.org/10.1002/rcm.6453

H.P. Taylor, S. Epstein, Relationship between ^{18}O/^{16}O ratios in coexisting minerals of igneous and metamorphic rocks: Part 1: Principles and experimental results. Geol. Soc. Am. Bull. **73**, 461–480 (1962a). https://doi.org/10.1130/0016-7606(1962)73[461:RBORIC]2.0.CO;2

H.P. Taylor, S. Epstein, Relationship between ^{18}O/^{16}O ratios in coexisting minerals of igneous and metamorphic rocks: Part 2. Application to petrologic problems. Geol. Soc. Am. Bull. **73**, 675–693 (1962b). https://doi.org/10.1130/0016-7606(1962)73[675:RBORIC]2.0.CO;2

M.G.G.T. Taylor et al., Rosetta begins its Comet Tale. Science **347**, 387 (2015). https://doi.org/10.1126/science.aaa4542

M. Thiemens, Mass-independent isotope effects in planetary atmospheres and the early solar system. Science **283**, 341–345 (1999). https://doi.org/10.1126/science.283.5400.341

M. Thiemens, J.E. Heidenreich III, The mass-independent fractionation of oxygen: A novel isotope effect and its possible cosmochemical implications. Science **219**, 1073–1075 (1983). https://doi.org/10.1126/science.219.4588.1073

D.J. Tholen, M.A. Barucci, Asteroid taxonomy, in *Asteroids II; Proceedings of the Conference*, Tucson, AZ, Mar. 8–11, 1988 (A90-27001 10-91) (University of Arizona Press, Tucson, 1989), pp. 298–315. https://doi.org/10.1016/0012-821X(85)90030-5

A. Tonotani, K. Bajo, S. Itose, M. Ishihara, K. Uchino, H. Yurimoto, Evaluation of multi-turn time-of-flight mass spectrum of laser ionization mass nanoscope. Surf. Interface Anal. **48**, 1122–1126 (2016). https://doi.org/10.1002/sia.6112

A.H. Treiman, Olivine and carbonate globules in ALH84001: A terrestrial analog, and implications for water on Mars, in *36th Lunar and Planetary Science Conference XXXVI* (2005), abstract #1107

A.H. Treiman, E.J. Essene, Chemical composition of magnetite in martian meteorite ALH 84001: Revised appraisal from thermochemistry of phases in Fe–Mg–C–O. Geochim. Cosmochim. Acta **75**, 5324–5335 (2011). https://doi.org/10.1016/j.gca.2011.06.038
T. Ushikubo, M. Kimura, N.T. Kita, J.W. Valley, Primordial oxygen isotope reservoirs of the solar nebula recorded in chondrules in Acfer 094 carbonaceous chondrite. Geochim. Cosmochim. Acta **90**, 242–264 (2012). https://doi.org/10.1016/j.gca.2012.05.010
W.R. Van Schmus, J.A. Wood, A chemical-petrologic classification for the chondritic meteorites. Geochim. Cosmochim. Acta **31**, 747–754 (1967). https://doi.org/10.1016/S0016-7037(67)80030-9
J.H. Waite Jr., M.R. Combi, W.-H. Ip, T.E. Cravens, R.L. McNutt Jr., W. Kasprzak, R. Yelle, J. Luhmann, H. Niemann, D. Gell, B. Magee, G. Fletcher, J. Lunine, W.-L. Tseng, Cassini ion and neutral mass spectrometer: Enceladus plume composition and structure. Science **311**, 1419–1422 (2006). https://doi.org/10.1126/science.1121290
S. Wakita, Y. Matsumoto, S. Oshino, Y. Hasegawa, Planetesimal collisions as a chondrule forming event. Astrophys. J. **834**, 125 (2017). https://doi.org/10.3847/1538-4357/834/2/125
H. Wang, C. Lineweaver, T.R. Ireland, The volatility trend of protosolar and terrestrial elemental abundances. Icarus **328**, 287–305 (2019). https://doi.org/10.1016/j.icarus.2019.03.018
P. Warren, Stable-isotopic anomalies and the accretionary assemblage of the Earth and Mars: A subordinate role for carbonaceous chondrites. Earth Planet. Sci. Lett. **311**, 93–100 (2011). https://doi.org/10.1016/j.epsl.2011.08.047
S. Watanabe et al., Hayabusa2 arrives at the carbonaceous asteroid 162173 Ryugu—A spinning top—Shaped rubble pile. Science **364**, 268–272 (2019). https://doi.org/10.1126/science.aav8032
C.R. Webster, P.R. Mahaffy, G.J. Flesch, P.B. Niles, J.H. Jones, L.A. Leshin, S.K. Atreya, J.C. Stern, L.E. Christensen, T. Owen, H. Franz, R.O. Pepin, A. Steele (the MSL Science Team), Isotope ratios of H, C, and O in CO_2 and H_2O of the martian atmosphere. Science **341**, 260–263 (2013). https://doi.org/10.1126/science.1237961
M.K. Weisberg, M. Prinz, R.N. Clayton, T.K. Mayeda, M.M. Grady, I. Franchi, C.T. Pillinger, G.W. Kallemeyn, The K (Kakangari) chondrite grouplet. Geochim. Cosmochim. Acta **60**, 4253–4263 (1996). https://doi.org/10.1016/S0016-7037(96)00233-5
M.K. Weisberg, T.J. McCoy, A.N. Krot, Systematics and evaluation of meteorite classification, in *Meteorites and the Early Solar System II*, ed. by D.S. Lauetta, H.Y. McSween Jr. (University of Arizona Press, Tucson, 2006), pp. 19–52
A.J. Westphal et al., Evidence for interstellar origin of seven dust particles collected by the Stardust spacecraft. Science **345**, 786–791 (2014). https://doi.org/10.1126/science.1252496
U. Wiechert, A.N. Halliday, D-C. Lee, G.A. Snyder, L.A. Taylor, D. Rumble, Oxygen isotopes and the Moon-forming Giant Impact. Science **294**, 345–348 (2001). https://doi.org/10.1126/science.1063037
W.C. Wiley, I.H. McLaren, Time-of-flight mass spectrometer with improved resolution. Rev. Sci. Instrum. **26**, 1150 (1955). https://doi.org/10.1063/1.1715212
A. Wittmann, R.L. Korotev, B. Jolliff, A.J. Irving, D.F. Moser, I. Barker, D. Rumble, Petrography and composition of martian regolith breccia meteorite Northwest Africa 7475. Meteorit. Planet. Sci. **50**, 326–352 (2015). https://doi.org/10.1111/maps.12425
J.A. Wood, J.S. Dickey, U.B. Marvin, B.N. Powell, Lunar anorthosites. Science **167**, 602–604 (1970). https://doi.org/10.1126/science.167.3918.602
I.P. Wright, J. Barber, G.H. Morgan, A.D. Morse, S. Sheridan, D.J. Andrews, J. Maynard, D. Yau, T. Evans, M.R. Leese, J.C. Zarnecki, B.J. Kent, N.R. Waltham, M.S. Whalley, S. Heys, D.L. Drummond, R.L. Edeson, E.C. Sawyer, R.F. Turner, C.T. Pillinger, Ptolemy—An instrument to measure stable isotopic ratios of key volatiles on a cometary nucleus. Space Sci. Rev. **128**, 363–381 (2007). https://doi.org/10.1007/s11214-006-9001-5
S. Yokota, Isotope mass spectrometry in the solar system exploration. Mass Spectrom. **7**, S0076 (2018). https://doi.org/10.5702/massspectrometry.S0076
E.D. Young, S.S. Russell, Oxygen reservoirs in the early solar nebula inferred from an Allende CAI. Science **282**, 452–455 (1998). https://doi.org/10.1126/science.282.5388.452
E.D. Young, I.E. Kohl, P.H. Warren, D.C. Rubie, S.A. Jacobson, A. Morbidelli, Oxygen isotopic evidence for vigorous mixing during the Moon-forming giant impact. Science **351**, 493–496 (2016). https://doi.org/10.1126/science.aad0525
H. Yurimoto, Oxygen isotopes, in *Encyclopedia of Geochemistry: A Comprehensive Reference Source on the Chemistry of the Earth*, ed. by W.M. White (Springer, Cham, 2018), pp. 1129–1135
H. Yurimoto, K. Kuramoto, Molecular cloud origin for the oxygen isotope heterogeneity in the solar system. Science **305**, 1763–1766 (2004). https://doi.org/10.1126/science.1100989
H. Yurimoto, J.T. Wasson, Extremely rapid cooling of a carbonaceous-chondrite chondrule containing very ^{16}O-rich olivine and a ^{26}Mg-excess. Geochim. Cosmochim. Acta **66**, 4355–4363 (2002). https://doi.org/10.1016/S0016-7037(02)01218-8

H. Yurimoto, A.N. Krot, B. Choi, J. Aléon, T. Kunihiro, A.J. Brearley, Oxygen isotopes of chondritic components. Rev. Mineral. Geochem. **68**, 141–186 (2008)
H. Yurimoto et al., Oxygen isotopic compositions of asteroidal materials returned from Itokawa by the Hayabusa mission. Science **333**, 1116–1119 (2011). https://doi.org/10.1126/science.1207776
G. Zeff, Q. Williams, Fractional crystallization of martian magma oceans and formation of a thermochemical boundary layer at the base of the mantle. Geophys. Res. Lett. **46**, 10997–11007 (2019)
J. Zhang, N. Dauphas, A.M. Davis, I. Leya, A. Fedkin, The proto-Earth as a significant source of lunar material. Nat. Geosci. **5**, 251–255 (2012). https://doi.org/10.1038/ngeo1429
K. Ziegler, Z.D. Sharp, C.B. Agee, The unique NWA 7034 martian meteorite: Evidence for multiple oxygen isotope reservoirs, in *44th Lunar and Planetary Science Conference* (2013), abstract #2639
E. Zinner, Stellar nucleosynthesis and the isotopic composition of presolar grains from primitive meteorites. Annu. Rev. Earth Planet. Sci. **26**, 147–188 (1998). https://doi.org/10.1146/annurev.earth.26.1.147
E.K. Zinner, Presolar grains, in *Treatise on Geochemistry*, vol. 1, ed. by A.M. Davis, H.D. Holland, K.K. Turekian (Elsevier, Amsterdam, 2003), pp. 17–39. https://doi.org/10.1016/B0-08-043751-6/01144-0
E.K. Zinner, A.J. Fahey, J.N. Goswami, T.R. Ireland, K.D. McKeegan, Large ^{48}Ca anomalies are associated with ^{50}Ti anomalies in Murchison and Murray hibonites. Astrophys. J. Lett. **311**, L103–L106 (1986). https://doi.org/10.1086/184807
M. Zolensky et al., Mineralogy and petrology of Comet 81P/Wild 2 nucleus samples. Science **314**, 1735–1739 (2006). https://doi.org/10.1126/science.1135842

Space Sci Rev (2019) 215:49
https://doi.org/10.1007/s11214-019-0614-x

Revolutionizing Our Understanding of the Solar System via Sample Return from Mercury

Kathleen E. Vander Kaaden[1] · Francis M. McCubbin[2] · Paul K. Byrne[3] · Nancy L. Chabot[4] · Carolyn M. Ernst[4] · Catherine L. Johnson[5,6] · Michelle S. Thompson[7]

Received: 7 May 2019 / Accepted: 4 October 2019 / Published online: 7 November 2019

Abstract Data from Mariner 10, MESSENGER, and ground-based telescopic observations have facilitated great advancements towards understanding the geochemistry, geology, internal structure, exosphere, and magnetosphere of Mercury. However, there are critical science questions that can be only addressed via examination of a sample in Earth-based laboratories, where numerous highly sensitive analytical measurements are possible. Collecting a sample from the surface of Mercury and bringing it to Earth for in-depth analysis would allow for transformative Solar System science to be conducted, examining aspects of our Solar System such as the evolution of the protoplanetary disk, space weathering on airless bodies, the geochemical behavior of elements at extreme conditions, and the origin and distribution of volatiles across the terrestrial planets. Furthermore, our knowledge of Mercury's differentiation and geochemical processes, chronology and geologic evolution, tectonism and geomechanical properties, and past and ongoing magnetism would be greatly advanced via analysis of a sample from Mercury. Although there are ample challenges and knowledge gaps associated with sample return from Mercury in terms of both spacecraft requirements and material requirements for curatorial facilities, a sample from the planet would be an

Role of Sample Return in Addressing Major Questions in Planetary Sciences
Edited by Mahesh Anand, Sara Russell, Yangting Lin, Meenakshi Wadhwa, Kuljeet Kaur Marhas and Shogo Tachibana

✉ K.E. Vander Kaaden
kathleen.e.vanderkaaden@nasa.gov

1 Jacobs, NASA Johnson Space Center, Mail Code XI3, Houston, TX, USA

2 NASA Johnson Space Center, Mail Code XI, 2101 NASA Parkway, Houston, TX, USA

3 Planetary Research Group, Department of Marine, Earth, and Atmospheric Sciences, North Carolina State University, Raleigh, NC, USA

4 The Johns Hopkins University Applied Physics Laboratory, Laurel, MD, USA

5 Department of Earth, Ocean and Atmospheric Sciences, University of British Columbia, Vancouver, British Columbia, Canada

6 Planetary Science Institute, Tucson, AZ, USA

7 Department of Earth, Atmospheric, and Planetary Sciences, Purdue University, West Lafayette, IN, USA

invaluable scientific resource for generations to come, enabling the most sophisticated measurements to be brought to bear for decades and helping to truly unlock the mysteries of our Solar System.

Keywords Mercury · Sample return · Geochemistry · Oxygen fugacity · Space weathering · Exploration

1 Introduction

Our knowledge of the formation and evolution of the Solar System is constantly changing with each new set of spacecraft data, meteorite analysis, and sample collected from another planetary body. Given the vast amount of data we have from Mars from orbiters (e.g., Rogers and Hamilton 2015), landers (e.g., Arvidson et al. 1989), rovers (e.g., Ming et al. 2008), and meteorite samples (e.g., McSween 1994), increasing our understanding of the formation, makeup, and evolution of Mercury through similar efforts will help to constrain models of the lateral heterogeneities of the inner planetary disk and allow for questions relating to transformative Solar System science to be answered (Table 1). Although we have greatly increased our knowledge of Mercury with the return of data from the NASA MErcury Surface, Space ENvironment, GEochemistry, and Ranging (MESSENGER) mission (Solomon et al. 2001; Solomon et al. 2019), there are still outstanding, fundamental questions that can only be answered via high-precision analyses conducted in Earth-based laboratories.

The first flybys of Mercury were conducted in 1974 and 1975 by the Mariner 10 spacecraft. This initial set of analyses from Mercury yielded images from $\sim 45\%$ of the planet, detection of a magnetic field, and measurements of elements such as H, He, and O in the exosphere, along with determination other physical characteristics of the planet (Broadfoot et al. 1974; Broadfoot et al. 1976; Chase et al. 1974; Murray 1975; Ness et al. 1975; Ness et al. 1974; Ogilvie et al. 1974; Simpson et al. 1974). In addition to these flyby data, ground-based analyses of telescopic data showed evidence for volatile elements such as Na and K in the atmosphere of Mercury as well as radar-reflective deposits at its poles (Harmon and Slade 1992; Potter and Morgan 1985, 1986; Slade et al. 1992). With this limited dataset, however, much information about Mercury still had to be inferred. Fortunately, NASA's MESSENGER mission provided the first in-depth study of the innermost planet.

Orbiting Mercury from 2011 to 2015, instruments onboard the MESSENGER spacecraft were able to image the entirety of the planet and so provide the first global view of Mercury. Coupling this imagery with geochemical instruments onboard MESSENGER, we have developed a better understanding of the geochemical terranes associated with the surface materials on the planet and the unique nature of Mercury's geochemical state compared with the other terrestrial planets (e.g., Peplowski et al. 2015b; Vander Kaaden et al. 2017; Weider et al. 2015). MESSENGER instruments also provided data that facilitated great advancements in our understanding of the geology, internal structure, exosphere, and magnetosphere of Mercury (e.g., Denevi et al. 2009; Hauck et al. 2013; Johnson et al. 2016; Killen 2016; Margot et al. 2019). These past mission datasets show that Mercury is a volatile-rich planet, has extensive polar deposits, exhibits a north-south asymmetry in its global magnetic field, experienced prolonged global contraction, and has a dynamic magnetosphere as well as a seasonal exosphere. While the immense amount of data collected by MESSENGER revealed Mercury as a geochemical end-member among the terrestrial planets, the new knowledge gained and ultimate ending of the mission left many unanswered questions that necessitate additional exploration of the planet (Tables 1 and 2).

Table 1 Transformative Solar System Science Questions. Questions that would be addressed via sample return from Mercury and high-precision measurements in Earth-based laboratories. Questions are not listed in order of priority but are representative of the scope of knowledge that would be gained

Science question	High-precision laboratory measurements
What can we learn about the composition of the Sun from Mercury's regolith?	Geochemical analyses
What are the primary building blocks for terrestrial planets in the inner Solar System?	Geochemical and mineralogical analyses
What can the composition of Mercury tell us about exoplanet formation?	Geochemical, mineralogical, and isotopic analyses
What does the present metal:silicate ratio of Mercury tell us about the chemical composition across the inner Solar System?	Geochemical and isotopic analyses
What do the volatile abundances of Mercury tell us about volatile distribution of the inner Solar System?	Geochemical and isotopic analyses
If Mercury is an example of a planetary embryo, were there lateral heterogeneities in the protoplanetary disk with distance from the Sun?	Geochemical and isotopic analyses
Is Mercury enriched or depleted in nucleosynthetic anomalies that are abundant in other Solar System objects?	Isotopic analyses
Are there any Mercurian meteorites in our current collections?	Geochemical, mineralogical, isotopic analyses, age dating, coordinated analyses at the nano-scale, comparison with meteorite collection
What are the radiogenic ages of some of Mercury's surface materials and what does this tell us about ongoing processes in the inner Solar System?	Age dating of crater material
What is the chronology of Mercury's geological evolution and how does it compare to other bodies such as the Moon?	Age dating of crater material
What are the microstructural, chemical, and optical signatures of space weathering closer to the Sun?	Coordinated analyses at the nano-scale

2 Future Exploration of Mercury

2.1 BepiColombo

Launched on October 20, 2018, the joint ESA/JAXA dual-orbiter spacecraft BepiColombo is due to arrive at Mercury in December of 2025 and is the most ambitious effort yet attempted to explore Mercury (Benkhoff et al. 2010; Rothery et al. 2010; Milillo et al. 2010). The Mercury Planetary Orbiter (MPO) includes 11 scientific instruments and is set to study the surface, the internal composition of the planet, the magnetosphere, and the exosphere in unprecedented detail (Fraser et al. 2010; Huovelin et al. 2010; Mitrofanov et al. 2010; Flamini et al. 2010; Hiesinger et al. 2010; Orsini et al. 2010; Chassefière et al. 2010; Glassmeier et al. 2010; Gunderson and Thomas 2010; Iafolla et al. 2010). The second spacecraft, Mio (formerly the Mercury Magnetospheric Orbiter), will characterize Mercury's dynamic magnetosphere and its interaction with the solar wind plasma and interplanetary magnetic field, as well as sodium emissions and dust monitoring (Nogami et al. 2010; Saito et al. 2010; Yoshikawa et al. 2010; Kasaba et al. 2010; Baumjohann et al. 2010). This interdisciplinary approach will provide insight into the origin and evolution of Mercury, by simultaneously conducting a global characterization of the planet and investigating its inte-

Table 2 Transformative Mercury Science Questions. Questions that would be addressed via sample return from the planet and high-precision measurements in Earth laboratories. Questions are not listed in order of priority but are representative of the scope of knowledge that would be gained

Science question	High-precision laboratory measurements
What is the mineralogy of the largest volcanic plains unit on the planet, Borealis Planitia?	Mineralogical analyses
What is the composition of the low-reflectance material on the planet, hypothesized to be the primary crust?	Geochemical and mineralogical analyses
What are the carriers of remanant magnetism on Mercury?	Geochemical, mineralogical and magnetic properties analyses
How did Mercury differentiate to form its large core and how did this process compare to the other terrestrial planets?	Geochemical, mineralogical, isotopic analyses
What are the composition constraints on Mercury's interior units, such as its solid core, liquid core, mantle, and potential other layers?	Geochemical, mineralogical, isotopic analyses
What is the composition of the hollows, and how did they form?	Geochemical and isotopic analyses
Did Mercury undergo a giant impact early in its history or did it accrete with the metal–silicate ratio it has?	Geochemical and isotopic analyses
Are the circum-Caloris plains volcanic or impact in origin?	Geochemical and isotopic analyses
What is the age of major units on Mercury like Caloris basin or the northern volcanic plains?	Age dating and isotopic analyses
When did widespread effusive volcanism cease on Mercury?	Age dating and isotopic analyses
What are the mechanical properties of Mercury's crust?	Geochemical, isotopic analyses, petrographic analysis, age dating, coordinated analyses at the nano-scale
How much of the surface has been space weathered and how does this change our understanding of the geochemistry of the surface materials?	Coordinated analyses at the nano-scale
What are the microstructural, chemical, and optical signatures of space weathering in samples from Mercury?	Coordinated analyses at the nano-scale
How can the data from MESSENGER and eventually BepiColombo be refined with new ground-truth data?	Geochemical, mineralogical, isotopic analyses, age dating, coordinated analyses at the nano-scale

rior, surface, exosphere, and magnetosphere (Benkhoff et al. 2010). The more circular, and lower altitude orbit of MPO will allow for higher resolution measurements of the southern hemisphere of Mercury that will likely reveal further complexities in the chemical and physical properties of the planet. BepiColombo will build upon the successes of MESSENGER to continue advancing our understanding of the planet's magnetic field, volcanic activity, shadowed craters, history of global contraction, unique surface features like hollows, enigmatic geochemical composition of different terranes, particularly in the southern hemisphere, dynamic magnetosphere, and the evolution of its exosphere (Benkhoff et al. 2010).

2.2 Landed Science

Although there are not any currently planned missions to Mercury beyond BepiColombo, the continued exploration of the planet has strong support within the scientific community (e.g., Byrne et al. 2018a). Furthermore, following the general strategy of exploration of other planets, this continued exploration of Mercury should be conceived as a multi-mission, multi-generational effort (e.g., a sequence comprising flyby, orbiter, lander/rover, and sample return). Guided by the heritage of mission capabilities sent to Mars, the next step in Mercury exploration should be to send a lander to the planet (Byrne et al. 2018a). There are several major aspects of Mercury's character and evolution for which significant knowledge gaps exist, but where our current understanding could be dramatically improved with data acquired from the planet's surface via *in situ* landed science. Specifically, a Mercury lander would greatly advance our understanding of the planet's geochemical makeup, its interior structure, geological evolution, the present-day processes at work there, and even the planet's polar volatile inventory (Byrne et al. 2018a). *In situ* compositional and petrological measurements could be obtained to understand the nature, origin, and abundance of Mercury's low-reflectance material (LRM; see Klima et al. 2018), the mineralogy of the planet's varied surface materials, and the composition of diffuse deposits interpreted to be pyroclastic in nature. By including a seismometer, heat probe, and magnetometer on the lander, our understanding of the interior structure of the planet, present-day level of seismicity, heat flux observations related to the core dynamo, and characterization of the electrical and conductivity structure of the crust and mantle would be greatly enhanced. *In situ* geochronological measurements of the surface materials and investigations of the remanent magnetization would provide additional information regarding the age of materials on Mercury's surface, surface geological processes and evolution, and the history of interior melt production and dynamo generation. Furthermore, *in situ* imaging of the surface would return useful information regarding the physical properties of the regolith, including grain size, shape, and mechanical strength. In addition, a lander visiting one of Mercury's poles would have the unique opportunity to provide *in situ* measurements to understand the origin and composition of the volatile compounds there, the purity of the ice in these deposits, as well as the physical and mechanical properties (e.g., volume, grain size, strength) of the volatiles themselves. Although challenging, the feasibility of conducting landed science at Mercury has greatly increased over the last few decades with new engineering capabilities (e.g., the Mars Science Laboratory mission) and with a better understanding of the Mercurian surface enabled by MESSENGER data. Given the long cruise phase necessitated by Mercury's proximity to the Sun, it is crucial that planning for missions to land on the surface of Mercury and conduct *in situ* measurements of it surface materials are underway as soon as possible to avoid any lapse in our continued exploration of the Sun's closest planetary neighbor (Byrne et al. 2018a). Furthermore, as discussed in Sect. 6, a lander to the surface of Mercury is crucial to prepare for an eventual sample return mission.

2.3 Sample Return

Although the data from remote-sensing missions provide a wealth of knowledge regarding the physical and chemical characteristics of a planetary body, and *in situ* analyses of the surface materials will transform our current understanding of Mercury, there are critical science questions that can best be addressed via examination of a sample in Earth-based laboratories, where numerous highly sensitive analytical measurements are possible (Tables 1 and 2). The remainder of this paper details the vast amount of information and transformative Solar System science that can be achieved via sample return from Mercury. It is crucial that a landed mission be carried out prior to sample return in order to assess engineering constraints, determine appropriate sample site location(s), and facilitate the requirements for curation and handling of a Mercurian sample. As demonstrated by ongoing analysis of Apollo lunar samples (e.g., Niihara et al. 2019), a sample from Mercury would be an invaluable scientific resource for generations to come, enabling the most sophisticated measurements to be brought to bear for decades, helping to truly unlock the mysteries of our Solar System.

3 A New Window into the Solar System

The history of our Solar System is recorded within the astromaterials that it comprises. At present, we have samples of Earth, the Moon, Mars, and numerous asteroids and comets. We do not have recognized samples from Mercury or Venus, nor of any of the outer Solar System planets, although some of the asteroids and most of the cometary materials are thought to originate in the outer Solar System, beyond the orbit of Jupiter (Budde et al. 2016; Kruijer et al. 2017; Nanne et al. 2019). The limited subset of materials we have to construct the history of our Solar System results in many unanswered questions (Table 1). Sample return from Mercury is essential to answer broad questions about the lateral heterogeneities with distance from the Sun of the inner Solar System, space weathering on airless bodies, the behavior of elements at extreme conditions, and the origin and distribution of volatiles in the inner Solar System (Table 1). Furthermore, a sample from Mercury would enable comparative planetology studies, and would yield transformative science relevant to the origin, timing, and composition of planetary building blocks in the Solar System.

3.1 Insights into the Protoplanetary Disk

Analyses of astromaterials, the current structure of our Solar System, and observations of young protostellar disks have helped to develop dynamical models of the earliest stages of disk evolution and planet formation. Based on these models, the Solar System likely started as a protoplanetary disk consisting of dust and gas with the Sun at its center. This dust coalesced to form pebble-sized objects that accreted and collapsed by dust pile-up or through streaming instabilities to form planetesimal-sized bodies (Bitsch et al. 2015; Gonzalez et al. 2017; Johansen and Lambrechts 2017; Johansen and Youdin 2007; Youdin and Goodman 2005). These planetesimals combined further into planetary embryos in a somewhat orderly fashion through a period of oligarchic growth (Chambers 2004; Chambers and Wetherill 1998; Raymond et al. 2009; Walsh et al. 2011). These planetary embryos may have had narrow feeding zones within the disk that spanned about 0.01 AU (Chambers 2004; Chambers and Wetherill 1998). Once most of the dust and gas dissipated, the planetary embryos started to interact with each other and collide, forming planets (Chambers 2004; Chambers and Wetherill 1998; Raymond et al. 2009; Walsh et al. 2011).

At present, we have four terrestrial planets and their associated moons in the inner Solar System along with numerous asteroids, thought to be relict planetesimals that survived the epoch of planet formation. The relict planetesimals provide the earliest history of our Solar System (Bouvier et al. 2011; Brennecka and Wadhwa 2012; Kleine et al. 2009; Kruijer et al. 2014, 2017; Schrader et al. 2017; Srinivasan et al. 2018; Wadhwa et al. 2009), forming within the first 0.5–8 Ma after the formation of the first solids in the protoplanetary disk (i.e., CAI in CV meteorites dated at 4567.30 ± 0.16 Ma before the present; Connelly et al. 2012). Of the four terrestrial planets, Mercury and Mars are of a relatively small size such that they likely comprise only 1–2 planetary embryos (Chambers 2004). In contrast, Earth and Venus are likely comprised of tens of planetary embryos (Chambers 2004). If Mars and Mercury are relict planetary embryos or are even composed of a small number of embryos, they can provide a crucial record of the period of

Solar System evolution between the formation of planetesimals and the formation of planets. Moreover, given the limited feeding zones of planetary embryos, samples of Mars and Mercury can provide unique insights into lateral heterogeneities within the inner portion of the protoplanetary disk. If their current positions are indicative of their location of formation, samples from these planets would yield information about the conditions in the innermost and outermost regions of the terrestrial-planet forming region of the protoplanetary disk.

Martian meteorites have already shown that Mars is unique among the terrestrial planets with a very narrow primary accretion window spanning from 2–10 Ma (Dauphas and Pourmand 2011; Tang and Dauphas 2014) and at most 2–20 Ma (Bouvier et al. 2018) after CAIs, consistent with its status as a planetary embryo. Importantly, however, the bulk composition of Mars indicates that it formed from a feeding zone within the protoplanetary disk that is chemically distinct from the Earth and Moon (Dauphas 2017; Schiller et al. 2018; Warren 2011). In particular, the composition of its nucleosynthetic isotopic anomalies (i.e., ε^{54}Cr, ε^{50}Ti, ε^{62}Ni, and μ^{48}Ca) indicate its bulk composition was dominated by building blocks from both enstatite chondrite-type and ordinary chondrite-type materials (Fig. 1 and Brasser et al. 2017; Dauphas 2017). In contrast, the Earth and Moon are dominated by enstatite chondrite-type materials (Dauphas 2017). Moreover, the information available from remote observations of Venus and Mercury indicate that their building blocks are similar to Earth and dominated by enstatite chondrite-type materials (Dauphas 2017; Filiberto 2014; Nittler et al. 2019). Samples from Mercury would provide a direct test of this hypothesis through analyses of Δ^{17}O, ε^{54}Cr, ε^{50}Ti, ε^{62}Ni, and μ^{48}Ca in Mercurian material (Fig. 1). Although only minimal amounts of sample are required to make such measurements (e.g., $<$ mg), the importance of returning as large of a sample as is feasible cannot be overstated because more mass allows for additional analyses, repeat measurements, and curation for future analyses. Regardless of the results from those analyses, samples from Mercury would provide unique constraints on the geochemistry and isotopic compositions of primary planetary building blocks available in the inner portion of the protoplanetary disk. The residual planetesimals that have been sampled (i.e., meteorites) may not represent all of the primary planet-building materials that were available at the time of planet formation, and samples from Mercury could lend insights into previously unsampled Solar System building blocks. For example, the compositions of materials from ureilites, asteroid 4 Vesta, and the angrite parent body cannot be constructed from known undifferentiated materials (i.e., chondrites), which is not surprising given that these materials accreted prior to the formation of chondritic parent bodies (as summarized by McCubbin and Barnes 2019). The compositions of these materials illustrate that temporal variations in the composition of the protoplanetary disk also need to be considered in understanding the compositions of planetary bodies, including Mercury, but samples of Mercury are needed to address these questions.

The comparison of samples from Mercury and Mars would provide crucial constraints on lateral chemical heterogeneities in the protoplanetary disk at the time of planet formation. Aside from Mars, Mercury is the only remaining relict planetary embryo from which information about the planet-formation epoch of our Solar System history can be attained. Samples from Mercury would therefore help to refine models of protoplanetary disk chemistry and the available building blocks of the terrestrial planets. Sample return from Mercury could provide important insights into the origin of nucleosynthetic isotope anomalies and early Solar System chronology.

Currently, we do not have any recognizable mercurian meteorites in our collections; however, there is potential that we do have a mercurian meteorite that has been misclassified as a different highly reduced meteorite group (Love and Keil 1995). Not only do dynamical calculations suggest that mercurian meteorites should exist at a factor of

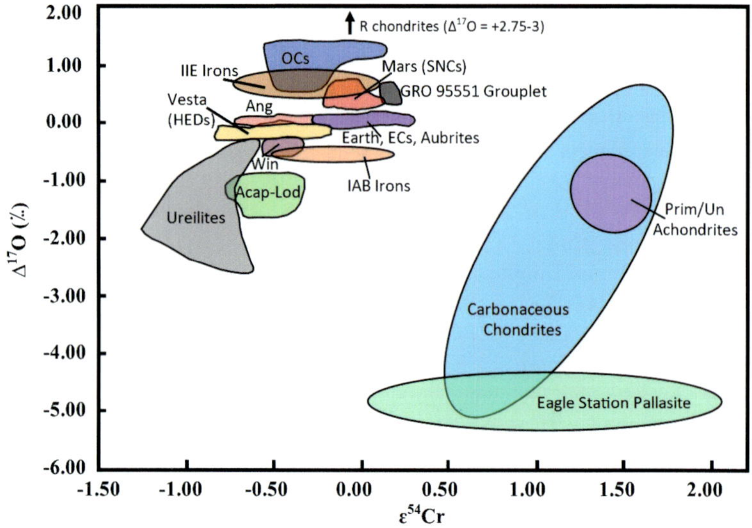

Fig. 1 Analyzing a sample from Mercury and determining where it lies on this plot would allow for a better understanding of the chemical heterogeneities in the proto-planetary disk. Abbreviations include: OC—ordinary chondrites, Acap-Lod—acapulcoite-lodranite, Win—winonaites, Ang—angrites, EC—enstatite chondrite, GRO—Grosvenor Mountains Antarctica, Prim/Un—primitive and ungrouped. This figure is modified after Sanborn et al. (2019), and includes data from Dey et al. (2019), Huyskens et al. (2019), and references therein

2–3 fewer on Earth compared with martian meteorites, but the lack of geochemical information for Mercury likely contributes to our inability to distinguish mercurian meteorites from meteorites of other highly reduced planetary bodies (e.g., enstatite achondrites/aubrites or slowly cooled enstatite chondrite impact melts; Gladman and Coffey 2009; Love and Keil 1995; McCubbin and McCoy 2016; Udry et al. 2019). Given the data collected during the MESSENGER mission, a meteorite from Mercury would likely contain albitic plagioclase, Mg-rich orthopyroxene, and forsteritic olivine as major phases, as well as graphite and exotic sulfides as accessory phases (Vander Kaaden et al. 2017; McCubbin and McCoy 2016). Identification of a mercurian meteorite in our current collections, ahead of a sample return mission, would lead to improved expectations of surface characteristics and instrument selection.

3.2 Space Weathering on Airless Bodies

Space weathering modifies the surfaces of airless bodies such as Mercury via micrometeorite impacts and solar wind irradiation. Together, these processes alter the microstructure, chemical composition, and reflectance properties of surface material (e.g., Domingue et al. 2014; Hapke 2001; Pieters and Noble 2016). The effects of space weathering complicate the interpretation of remotely sensed data and the subsequent characterization of planetary surfaces. Our understanding of space weathering across the inner Solar System is based predominantly on the analysis of returned samples from the lunar surface and from asteroid Itokawa (e.g., Keller and McKay 1993, 1997; Noguchi et al. 2014, 2011; Thompson et al. 2014). From these studies, we have learned to recognize the microstructural and chemical signatures of space weathering: amorphous grain rims arising from solar wind irradiation, melt features and vapor deposits resulting from impacts and sputtering, vesiculated textures, and

Fe-bearing nanoparticles. These nanoparticles exhibit compositional variability in returned samples, with phases including Fe-metal, -oxides, and -sulfides (Burgess and Stroud 2018; Keller and McKay 1997; Noguchi et al. 2014, 2011; Thompson et al. 2016). The effect of these nanoparticles on the spectral properties of surface material is pronounced, typically causing reddening (an increase of reflectance with increasing wavelength), darkening (a reduction of overall reflectance), and attenuation of characteristic absorption bands (Pieters and Noble 2016).

However, the space weathering environment at Mercury is substantially different than at the bodies from which we already have returned samples in two important ways: 1) the composition of Mercury's surface is not fully characterized; and 2) the intensity of solar wind irradiation and the flux and velocity of micrometeorites impacting the surface is greater. Compositionally, the low Fe content (less than 2 wt.%), high S content (up to 4 wt.%), and localized carbon abundance (up to 4 wt.% above the local mean in the LRM) likely have considerable effects on the development of space-weathering features in Mercury's surface material (Evans et al. 2012; Klima et al. 2018; McCubbin et al. 2017; Nittler et al. 2011; Weider et al. 2014). Analyses of returned samples from the Moon and asteroid Itokawa have demonstrated the importance that Fe and S play in controlling the development of microchemical space weathering features. Metallic Fe nanoparticles dominate the surfaces of lunar soil grains and are responsible for the reddened, darkened, often featureless spectral properties of mature lunar soils. In contrast, Fe-sulfide nanoparticles are common in particles from Itokawa, where sulfur-bearing phases (e.g., troilite) are abundant, and the influence that these nanoparticle compositions have on changes in surface sample spectral effects is not yet well understood (Noguchi et al. 2014, 2011). With lower Fe but higher S contents expected on the surface of Mercury than on other airless bodies, returned sample analysis would enable the exploration of a new parameter space for the development of microstructural and chemical signatures of space weathering and, in turn, their influence on optical characteristics. Similarly, we have a very limited understanding of the role of C in the space weathering process; early results from experimental work performed for the weathering of carbonaceous chondrites has suggested that changes in organic microstructure and functional group chemistry can contribute to substantial alterations in spectral reflectance features (Matsuoka et al. 2015; Moroz et al. 2004). Examining the effect of C-bearing phases on the overall space weathering of airless surfaces, including Mercury, would be relevant for many carbonaceous surfaces across the Solar System.

In addition to having a unique surface composition, Mercury experiences a different space weathering environment than the Moon or near-Earth asteroids. Although the composition and velocity of the solar wind at Mercury is likely the same as in near-Earth space, the flux of ions, which varies with distance from the Sun, is higher. Without a dense atmosphere, the surface of Mercury is susceptible to experiencing this high flux of solar wind irradiation. However, the planet's magnetic field (discussed in Sect. 4.4) may limit the amount of radiation reaching the surface, creating an environment unlike the surfaces of other planetary bodies from which samples have been returned (Domingue et al. 2014). In addition to differences in solar wind flux reaching the surface, the population of micrometeorites at Mercury also differs in comparison to previously visited planetary bodies. With higher impact fluxes and velocities, the regolith on Mercury is expected to overturn more quickly and contain more impact glasses and agglutinate grains than Itokawa or even the surface of the Moon (Borin et al. 2009; Cintala 1992). Examining the likely influence of impacts and solar wind irradiation together suggests that micrometeorite bombardment may be the primary process driving the development of space weathering features on Mercury's surface. Returning samples from Mercury would provide an end-member dataset by enabling the analysis of materials with unique starting compositions that have experienced probable extreme weathering rates, dominated by constituent processes unlike any other airless surface. To enable the proper interpretation of remote sensing data for airless bodies across the Solar System, a fundamental understanding of how space weathering operates across all surfaces is crucial.

3.3 Geochemical Behavior of Elements Under Highly Reduced Conditions

The oxygen fugacity (fO_2), or partial pressure of O_2 gas that is in equilibrium with a magmatic system, plays an important role in the physicochemical properties of that system, because fO_2 has a substantial impact on the partitioning behavior and solubility of elements in various phases. The reducing nature of Mercury has raised many questions regarding the geochemical behavior of lithophile elements in magmas at low oxygen fugacity. In fact, interpretations of magmatic processes on Mercury are difficult at present because the number of experimental studies relevant to the extremely low fO_2 and high sulfur content of Mercury are somewhat limited (e.g., Chabot et al. 2014; Malavergne et al. 2007, 2014, 2004; McCoy et al. 1999; Namur et al. 2016a, 2016b; Ricolleau et al. 2011; Rose-Weston et al. 2009; Vander Kaaden and McCubbin 2016; Vogel et al. 2018). At such reducing conditions, elements are likely to deviate from their typical geochemical behavior displayed at higher oxygen fugacity (e.g., Foley et al. 2005; McCoy et al. 1999; McCubbin et al. 2012; Vander Kaaden and McCubbin 2016).

For example, lithophile elements such as Mg and Ca display chalcophile behavior in some highly reduced chondrites and achondrites, forming the minerals oldhamite [(Ca,Mg,Fe,)S] and niningerite [(Mg,Fe,Mn)S] (McCoy et al. 1999), and Si and P display siderophile behavior with elevated abundances in Fe-rich metals in those same meteorites (Keil 2010; Ringwood 1961). Other than the highly reduced enstatite chondrites and achondrites, we do not have many natural examples of what to expect with respect to element behavior in such oxygen-starved systems (e.g., Udry et al. 2019; Fogel 2005). Samples of Mercury would provide unique insights into the distributions of major, minor, and trace elements among metal, sulfide, silicate, and oxide phases. Mercury samples would also fill a critical gap in our understanding of the behavior of elements under reducing conditions in complex geological systems. Furthermore, when combined with the geochemical behavior of elements under more oxidizing conditions, these data would help to constrain models of element distribution and behavior over a wide range of conditions relevant to exoplanetary systems. Finally, constraints on the elemental behavior exhibited by samples from Mercury is a prerequisite to using those samples to constrain the bulk composition of the planet, including that of its core, mantle, and crust (see Sect. 4.1).

3.4 The Storage of H_2O in Planetary Interiors

Mercury's status as a planetary embryo could have important implications for constraining H_2O delivery to the inner Solar System given that the period of planetary embryo formation was complete by about 10 Ma after the formation of CAIs (Chambers 2004; Dauphas and Chaussidon 2011). In fact, the interior of Mars (i.e., the only other planetary embryo among the inner Solar System bodies) may preserve the H-isotopic composition of its primary building blocks (McCubbin and Barnes 2019), and the interior of Mercury may preserve a similar record of its respective precursor materials. Mercury has surface features indicating past explosive volcanism, consistent with the presence of magmatic volatiles in its interior (Kerber et al. 2009, 2011; Thomas et al. 2014;

Weider et al. 2016). The surface of Mercury is also enriched in moderately volatile elements such as K, Na, S, and Cl (Evans et al. 2015; Nittler et al. 2011; Peplowski et al. 2011, 2014; Vander Kaaden et al. 2017; Weider et al. 2015; Zolotov et al. 2013). However, enrichments in moderately volatile elements are not typically a good indicator of the abundances of more volatile components, such as H_2O (McCubbin and Barnes 2019; Sarafian et al. 2017). In fact, the highly reducing conditions of Mercury's interior may limit the abundance of H_2O, given the effect of oxygen fugacity on the fugacity ratio of H_2O/H_2 via the reaction $H_2 + 1/2O_2 \longleftrightarrow H_2O$ (e.g., Hirschmann et al. 2012; Sharp et al. 2013). If the computed fO_2 range of Mercury's mantle is an accurate predictor of the molecular H_2O/H_2 ratio of Mercury's mantle, it implies an H_2O/H_2 ratio in the range of 0.034–0.002 over a temperature range of 1300–1800 K (McCubbin and Barnes 2019). Samples of Mercury would enable a test of this hypothesis, and would provide crucial insights into storage mechanisms of both H_2 and H_2O in the interiors of reduced planetary bodies. When coupled with knowledge of H-storage in more oxidized planetary bodies, storage mechanisms of H in reduced planetary bodies would have important implications for the storage of H in exoplanets that may exhibit a range of fO_2 similar to those for the terrestrial planets.

4 A New Window into Mercury's Formation and Evolution

Sample return from Mercury is essential to address broad questions about the lateral heterogeneities, as a function of distance to the Sun, of the inner Solar System, the impact of space weathering on airless bodies, the behavior of elements at extreme conditions, and the origin and distribution of volatiles in the inner Solar System (Table 1). However, a sample from Mercury is also crucial to increasing our understanding of the geochemistry, mineralogy, geology, tectonism, and magnetism of the planet itself, and our efforts to begin to answer questions about its formation and evolution as a whole (Table 2).

4.1 Differentiation and Geochemical Evolution

Several instruments onboard MESSENGER collected geochemical data from the surface of Mercury, including a gamma ray spectrometer (GRS), neutron spectrometer (NS), and X-ray spectrometer (XRS), but these data are limited in terms of spatial extent, resolution, and the elements analyzed. Furthermore, MESSENGER did not include instruments that allowed for the examination of the oxidation state of the measured elements, nor their host mineral(s), although BepiColombo will shed light on these properties. Such information is crucial to refining our understanding of the thermal and magmatic evolution of Mercury including bulk composition, crust formation, volcanism, smelting events, and the mineralogy and petrology of the planet (e.g., Byrne et al. 2016; McCubbin et al. 2017; Vander Kaaden et al. 2017; Vander Kaaden and McCubbin 2015, 2016). MESSENGER data indicate Mercury is enriched in moderately volatile elements relative to Earth (Evans et al. 2015; Nittler et al. 2011; Peplowski et al. 2011), was formed under highly reducing conditions (McCubbin et al. 2012; Zolotov et al. 2013), and has a carbon-rich surface (Murchie et al. 2015; Peplowski et al. 2016; Vander Kaaden and McCubbin 2015). The graphite flotation crust hypothesis (Fig. 2) has been supported by measurements on the surface of Mercury, mainly within the LRM excavated from depth, which suggests that the measured carbon is likely endogenic rather than exogenic (Klima et al. 2018; Murchie et al. 2015; Peplowski et al. 2015a, 2016). As the majority of Mercury's surface is covered by volcanic materials (e.g., Denevi et al. 2009),

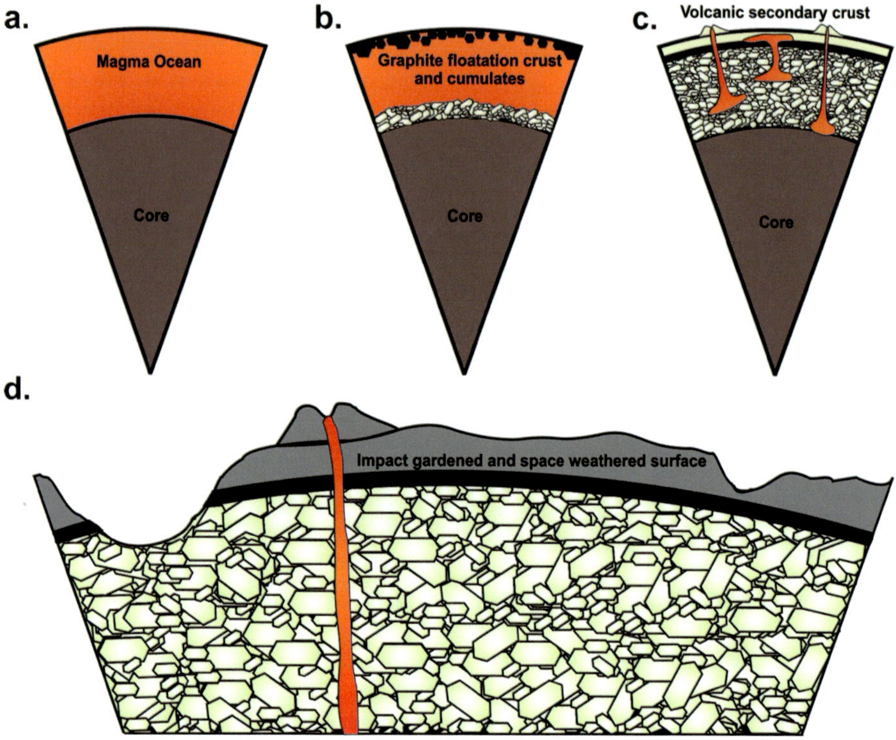

Fig. 2 Mercury's differentiation and early crust formation has been proposed to be very different than that of Earth, the Moon, or Mars. A sample collected at Mercury and sent to the Earth would test these hypotheses, just as lunar and martian samples have been used to unravel the geochemical evolution of these bodies. This figure is modified after Vander Kaaden and McCubbin (2015)

eruption through a C-rich layer has implications for the oxidation state of elements measured at the surface and the resulting mineralogy (e.g., McCubbin et al. 2017). Measurements of the mineralogy and siderophile element abundances in a sample from Mercury would enable modeling of the core formation process on this highly reduced planet, as has been done for Earth, Mars, and other planetary bodies. Such measurements would directly test the proposed hypothesis for Mercury's geochemical evolution and provide key new insight into the conditions and composition involved during Mercury's differentiation.

In addition to its unique geochemical nature, Mercury comprises proportionally more core by volume than any of the other terrestrial planets, with a core mass fraction of 0.70–0.77 (computed based on the range of core density from 6700–7260 kg/m^3, range of mantle density from 2800–3600 kg/m^3, the planet radius of 2439.36 km, and the core radius estimate of 2024 km; Margot et al. 2019). The origin of the planet's large core and volatile-rich composition are key to understanding the formation history of Mercury and, by extension, its bulk composition. There are two endmember scenarios for the origin of Mercury's metal-rich composition: 1) enrichment of metal through loss of silicate material by an early giant impact (Benz et al. 2007), or 2) the metal–silicate ratio is a primordial feature of Mercury that reflects the nature of locally sourced material in the protoplanetary disk (Burbine et al. 2002; Taylor and Scott 2004). However, each of the two proposed formation scenarios above imposes complicating factors for constraining the bulk composition of Mercury, which has been inferred from MESSENGER data to be similar to enstatite chondrites (Nittler et al.

2019). Moreover, neither formation model can explain all our observations. For example, the impact-stripping model may be difficult to reconcile with Mercury's volatile-rich composition (Asphaug and Reufer 2014). In contrast, $\geq$ 18.5 wt% Si in Mercury's core is needed if the metal–silicate ratio is a primordial feature and reflects a bulk chondritic composition, which would imply an oxygen fugacity that is far below current estimates (i.e., 3.2–4.3 log units below the iron-wüstite buffer; McCubbin et al. 2017) and inconsistent with other MESSENGER geochemical observations (Vander Kaaden et al. 2017).

Samples from Mercury would help to constrain the bulk composition of Mercury beyond that possible from orbital data, and returned samples would enable both the hypotheses for Mercury's origin, and its purported enstatite chondrite-like bulk composition, to be tested. For example, the exact mineralogy of the sample could be fully characterized, greatly expanding the compositional measurements to include trace elements, isotopes, and mineralogy at microscopic levels. The presence or absence of key mineral phases (e.g., Si-metal and graphite) would directly test existing hypotheses for the history of Mercury and lend insight into the early evolution of the planet. Understanding the oxidation state of elements and minerals present on the surface would help to refine oxygen fugacity measurements and the compositional relationship between materials pre- and post-eruption through a C-rich layer. Moreover, better constraints on Mercury's bulk composition, including siderophile trace element data, would further enable comparative planetology studies, lending important insights into the planetary building blocks from which Mercury was assembled and its subsequent differentiation and expression of these materials on the surface.

4.2 Chronology and Geologic Evolution

The geologic time scale of Mercury is segmented into five time-stratigraphic eras. From oldest to youngest, these are the pre-Tolstojan, Tolstojan, Calorian, Mansurian, and Kuiperian (Spudis 1985; Spudis and Guest 1988). These systems (Fig. 3) were originally defined as analogs to the lunar time-stratigraphic system, assuming lunar impact flux models could be applied to Mercury (Shoemaker and Hackman 1962; Wilhelms 1987). More recent analysis using MESSENGER data has led to revised age estimates for these transitions. The ages of the Tolstoj and Caloris basins (stratigraphic markers for the beginnings of the Tolstojan and Calorian, respectively) were updated based on improved crater counts obtained from images better suited for identifying crater morphology (Denevi et al. 2019; Ernst et al. 2017). The ages of the Mansurian and Kuiperian eras were revised by examining the spatial statistics of craters classified as having formed in these two eras (Banks et al. 2016). Given these better-refined estimates from higher-resolution datasets, the shift in the positions of the time-stratigraphic boundaries has implications for the geologic history of Mercury. In particular, the Calorian system, defined as lasting from 3.9–3.3 Ga, may have lasted much longer than previously thought (possibly extending to < 2 Ga). Furthermore, the emplacement ages of some large expanses of volcanic material (e.g., Borealis Planitia, Caloris interior plains) could shift by up to a billion years. These updated age constraints have major implications for models of the cooling of the planet and subsequent volcanic activity. However, these dates are based on models derived from crater statistics, rather than from measured radiometric age constraints, and are not supported by independent data from Mercury.

The values used for deriving Solar System absolute model ages were obtained from radiogenic analysis of the Apollo samples (e.g., Schaeffer and Husain 1974). However, there are different factors that must be taken into account between impacts at the Moon and at Mercury, such as surface properties and bombardment flux. Determining a radiometric age

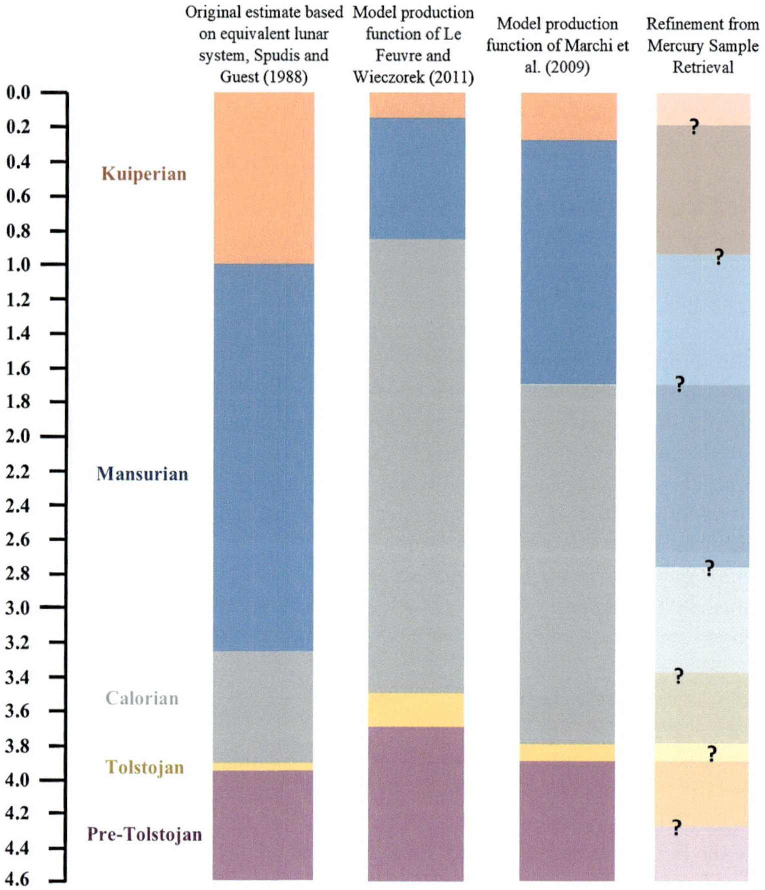

Fig. 3 Age constraints from a Mercurian sample would help to refine the geologic timescale for the planet as well as the impactor flux of the inner Solar System. This figure is modified after Denevi et al. (2019) using data from Spudis and Guest (1988), Le Feuvre and Wieczorek (2011), and Marchi et al. (2009)

date for even a single terrain on Mercury via sample return would provide an absolute constraint on the timeline for that planet (Fig. 3), while greatly improving our understanding of Mercury's geologic history in particular, and of the impact flux of the Solar System in general. An absolute age date obtained via radiogenic analysis of a Mercurian sample would tell us whether the current models, extrapolated from lunar data and used to interpret the geologic history of Mercury, are correct or, more likely, if they need refining. Furthermore, such data would place localized timing constraints on important events in Mercury's history. On a much broader scale, obtaining absolute age dates for materials from Mercury would provide an age anchor in the inner Solar System that we currently lack. Given the complexity of impact records and the interpretation of ages from different materials and by different systems, sample site selection will be key to allow for the greatest knowledge gain from returned materials. By combining absolute age data for lunar materials with an absolute date for a mercurian sample, it would be possible to assess the overall bombardment history of the inner Solar System and the variation of flux and crater density of these planetary bodies from Mercury to Mars.

Similar to the timescale used for Earth's history, each stratigraphic era of Mercury's timescale has representative geologic events that define that portion of the sequence. For example, the pre-Tolstojan and Tolstojan eras were dominated by impact bombardment and impact and volcanic resurfacing to form what has been defined as the "intercrater plains" (Whitten et al. 2014). There is very little evidence on the surface of Mercury of representative, pre-Tolstojan crust. Furthermore, there are no areas of the surface that are as heavily cratered as the Moon by craters > 20 km in diameter. Model age estimates of Mercury's oldest terrains based on craters > 25 km in diameter yield ages of ∼ 4.1 Ga, suggesting global resurfacing of the planet up to that time (Marchi et al. 2013). However, the areal density of craters > 65 km in diameter shows areas on Mercury with equivalent densities to the oldest surfaces on the Moon (Denevi et al. 2019). This finding suggests a partial or patchy resurfacing history, whereby the smaller craters were erased, but some larger basins were at least partially retained. As such, it remains possible that some portions of the surface of Mercury could be older than ∼ 4.1 Ga. The Calorian era was dominated by the formation of the planet's smooth plains and continued impact bombardment (albeit at a lower flux). With the end of the Calorian, the majority of effusive volcanism had ceased (e.g., Byrne et al. 2016). However, limited pyroclastic volcanism continued into the Mansurian and Kuiperian (e.g., Jozwiak et al. 2018; Thomas et al. 2015). The Mansurian and Kuiperian eras appear to correspond to less geologic activity, with crater formation the dominant process throughout these eras. The sublimation of surface materials (e.g., perhaps the mechanism responsible for "hollows"; Blewett et al. 2013) occurred at least in the Kuiperian (and may be still occurring today). Acquiring absolute age data for these regions in particular would be instrumental in refining the geologic timescale for Mercury.

4.3 Tectonism and Geomechanical Properties

The history of tectonism on Mercury is dominated by global contraction—a process in which the planetary volume has decreased as the interior cooled through time. The results of this process manifest on Mercury as a planet-wide population of crustal shortening structures (Strom et al. 1975), most of which boast relief of a few hundred meters but the largest of which stand several kilometers high and are hundreds of kilometers long (Byrne et al. 2018b). These tectonic structures cross-cut virtually all major surface units (Banks et al. 2015); on the basis of crater statistics, most of these units have collective ages of 3.9–3.5 Gyr (e.g., Denevi et al. 2013; Head et al. 2011; Whitten et al. 2014) and so it follows that most tectonic deformation associated with global contraction took place after around 3.5 Ga. Notably, major plains volcanism on Mercury ended at around this time (Byrne et al. 2016), with only minor effusive and explosive activity continuing thereafter (e.g., Jozwiak et al. 2018; Thomas et al. 2014; Kerber et al. 2009). Such a relationship between tectonic and volcanic activity is consistent with predictions for a planet undergoing global contraction (cf. Solomon 1977).

The acquisition of a sample from the surface of Mercury would provide valuable insight into the tectonic evolution and behavior of the planet. For example, the mineralogical properties of a sample would in turn yield information regarding its mechanical characteristics, which play important roles in how tectonic deformation is manifested. A difference in mechanical properties has been proposed to account for the apparent contrast in tectonic styles between the volcanic smooth plains units on Mercury and the older, more heavily cratered "intercrater plains" units: shortening structures in the smooth plains tend to be smaller and more distributed than the larger, more isolated tectonic features that formed in the older plains (e.g., Byrne et al. 2018b). This difference may reflect the influence of a lower fracture density for the volcanic strata that comprise the smooth plains on the structures that formed there, in contrast to the likely more complex, fractured, and mechanically weaker interior makeup of the older plains, which results from a longer history of impact bombardment and volcanic resurfacing (cf. Mueller and Golombek 2004).

Thus, the physical condition of a sample would yield useful information regarding the mechanical characteristics of Mercury's crust. Many planetary tectonic analyses assume representative values for the properties of the crust (e.g., strength, cohesion, rigidity, etc.) that are taken from laboratory samples, but such samples tend not to be representative of the true state of a planet's outer rigid shell (e.g., Klimczak et al. 2015; Schultz 1993). Sustained impact bombardment (e.g., Wieczorek et al. 2013) and thermal stresses (Beuthe 2010; Eppes et al. 2015) can serve to fracture the crust, resulting in rocks that are substantially weaker than those assessed in the laboratory (Klimczak et al. 2015). Appraising the physical state of the sample in terms of fracture density, porosity, permeability, etc. (cf. Bieniawski 1989) would return invaluable information on the bulk mechanical properties of the crust in general, with commensurate insight into the surfaces of other planetary bodies for which samples are not available. Importantly, this information could be acquired for any sample taken at any location on Mercury, with the broader geological context of that sample serving as the basis for extrapolating those findings to other parts of the surface. As a result, efforts to understand the mechanical properties of Mercury with an acquired sample do not face any particular constraints or requirements in terms of from where that sample should be sourced.

4.4 Past and Ongoing Magnetism

The MESSENGER mission led to a wealth of discoveries regarding Mercury's magnetic field. In addition to characterizing the global dynamo field (Johnson et al. 2019 and references therein), and the surrounding magnetospheric environment (Korth et al. 2019; Slavin et al. 2019), magnetic field data obtained at altitudes below ~ 50 km toward the end of the orbital mission led to the discovery of weak, spatially localized signals from magnetized rocks (e.g., Hood 2016; Johnson et al. 2015, 2016, 2019). The geographical distribution of magnetization (Fig. 4) shows that much of the northern hemisphere crust is weakly magnetized. Stronger magnetizations are associated with the region around Mercury's Caloris basin, and with some, but not all, other impact craters. Similar to rocks on Earth, some of this magnetization is likely induced in iron-bearing minerals in the crust by Mercury's present field, though iron is in low abundances on Mercury's surface, but the strongest magnetizations point to at least some of the magnetization being remanent, a relic of an ancient dynamo field (Johnson et al. 2015, 2019).

Although the discovery of crustal magnetization and the inferred geographical distribution over the northern hemisphere are robust, fundamental issues remain. First, the depth extent of magnetization and the iron mineralogy(ies) responsible for the magnetization are unknown. Second, remanent magnetization places tight constraints on the thermal evolution of the planet because it requires a dynamo field present at the time of acquisition of remanence (~ 3.7 Ga; Johnson et al. 2019). However, because the remanent contribution to the magnetization is difficult to quantify from remotely sensed data alone, and because there is no overall correlation of magnetization with geological units of specific ages (Hood 2016; Johnson et al. 2015, 2019), both the timing and strength of an ancient field are poorly constrained. The strongly reduced nature of Mercury's crustal rocks (e.g., McCubbin et al. 2012; Nittler et al. 2019; Zolotov et al. 2013) suggests iron metal or iron alloys are likely magnetic carriers, although more exotic mineralogies such as iron carbides and iron silicides are possible. Sample return would enable the identification of the magnetic phases and laboratory

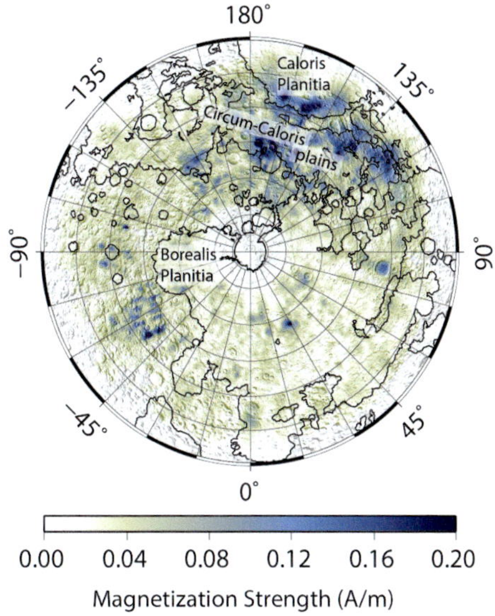

Fig. 4 A map of magnetization strength overlain on a shaded-relief map of Mercury's northern hemisphere, with the black contour lines outlining smooth plains units. The magnetization strength shown here assumes a 10-km-thick magnetized layer. In order to understand the mineralogical hosts of magnetism, a sample should ideally be collected from an area shaded in darker blues. The figure is modified after Hauck and Johnson (2019)

experiments to establish their magnetic properties. Knowledge of these phases would enable direct comparisons with enstatite meteorites thought to contain magnetic mineralogies analogous to Mercury. Although magnetic measurements on such meteorites have been made, the library of such data is very limited (Rochette et al. 2008). Key properties with which to understanding the origin (i.e., induced versus remanent versus shock from impacts) of the magnetization are magnetic susceptibility (low-field and thermal) and the Curie temperature. Measurements of these properties would enable accurate characterization of the magnetization induced in the present-day magnetic field. Furthermore, direct measurement of the remanent magnetization in a sample would constrain the strength of the ancient field, as has been possible from Apollo samples for the Moon (Tikoo et al. 2017; Weiss and Tikoo 2014; Wieczorek et al. 2012).

5 Potential Sampling Locations for Mercury Sample Return

Although there are ample locations for future exploration of Mercury via sample return, we broadly describe a few potential sampling site options and the scientific rationale behind their selection. Figure 5 shows an enhanced color map of Mercury, illustrating the diversity of potential sampling locations and identifying potential site categories of interest. In reality, however, a sample from anywhere on the planet has the potential to transform our current understanding of Mercury and the terrestrial planets in our Solar System.

One potential sampling site location is the low reflectance material on Mercury, hypothesized to contain components of the planet's primary crust (Fig. 2 and Vander Kaaden and McCubbin 2015). Analyses to understand the elemental compositions, mineralogy, geochemical behavior of elements, isotopic abundances, and radiometric ages, among other measurements, would help to answer numerous scientific questions about this material and the primary crust on Mercury including: is this material the planet's primary crust? What is the composition of the LRM? How does this composition compare with the other terrestrial planets? How old is the primary crust of Mercury? What were the geochemical conditions

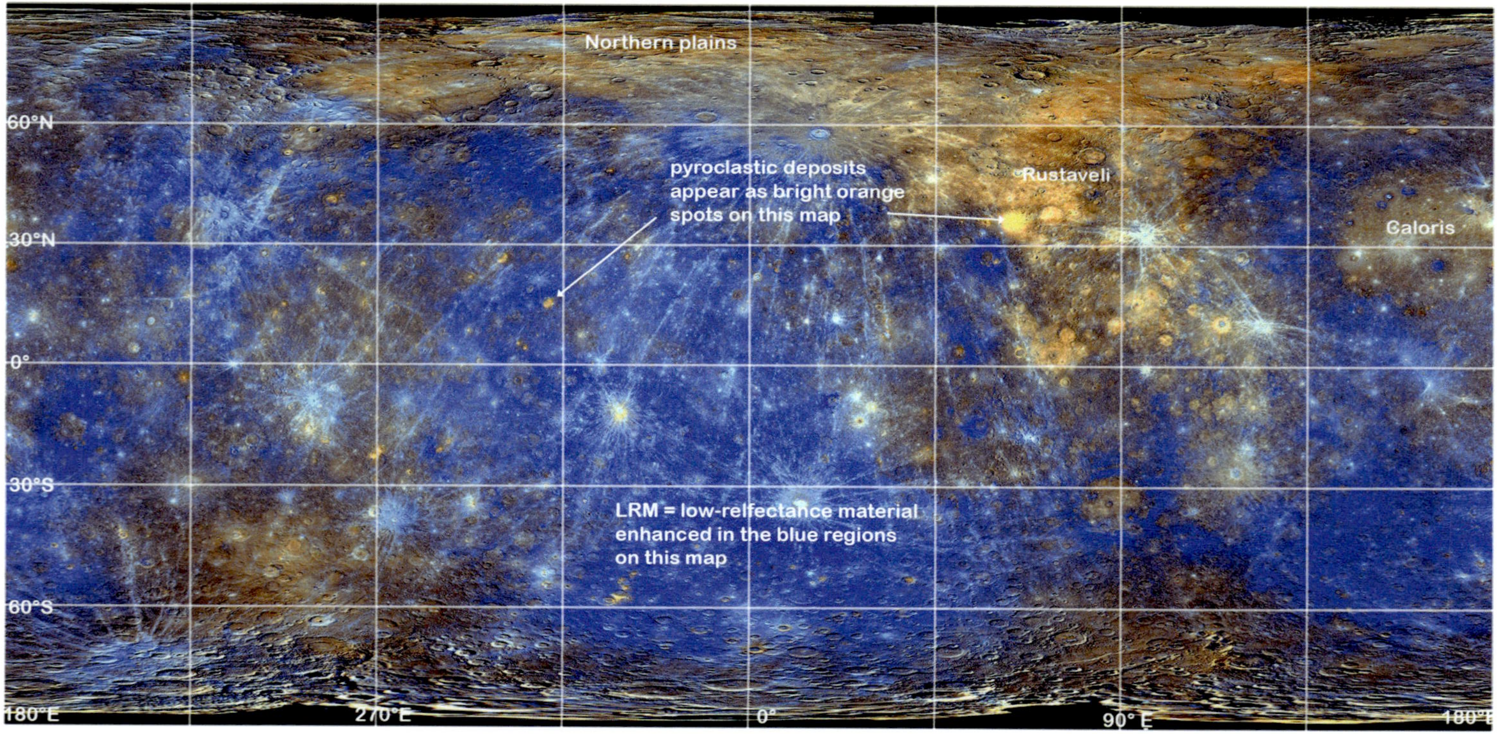

Fig. 5 Enhanced color map of Mercury, showing the variety and diversity of potential sampling site locations. The first principal component is in the green channel, the second principal component is in the red channel, and the ratio of the 430 nm/1000 nm filters in the blue channel

during Mercury's core formation? What volatiles are present in this material and what do they tell us about the composition of the material from which Mercury formed? What role does C play in controlling the development of space weathering features on airless surfaces?

A second potential sample site to further understand the volcanic and volatile nature of Mercury is one of the planet's pyroclastic deposits (Goudge et al. 2014; Kerber et al. 2011). Measurements similar to those that would be made of LRM would address questions including: What are the dominant volatile species associated with pyroclastic eruptions? What are the major, trace, and rare earth element compositions of the deposits? What is the mineralogy and petrology associated with pyroclastic deposits? What are the radiometric ages of these deposits?

Understanding the stratigraphy of, and attributing absolute chronological measurements to, major geologic events on Mercury is crucial for characterizing not only the timing of events on Mercury but the inner Solar System impact flux itself. The major geologic terrain types on Mercury are volcanic plains and impact craters/basins (Peplowski et al. 2015b). Volcanic plains consist of both the older, more cratered, intercrater plains as well as the younger, less cratered, smooth plains (e.g., Denevi et al. 2013, 2019; Ostrach et al. 2015). These terrain types were identified from Mariner 10 images (which covered only 45% of the surface), and additional geologic terrains were identified from the full-coverage MESSENGER images. However, with the global coverage and higher resolution of the MESSENGER images, the distribution of these various units has been updated and clues to their origins revealed. For example, the expansive nature of Borealis Planitia were seen in MESSENGER images showing that these plains cover $\sim 12\%$ of the northern hemisphere and are volcanic nature (Head et al. 2011; Ostrach et al. 2015). Sampling of a smooth plains unit (e.g., Denevi et al. 2013; Head et al. 2011), such as the vast Borealis Planitia, would provide us with invaluable insight regarding the timing of large volcanic events throughout Mercury's geochemical evolution, as well as the mechanical properties of relatively younger geologic materials on the planet's surface.

In addition, there is evidence of ancient, extremely large (e.g., Caloris-sized) basins that have been completely filled by volcanic materials, adding to the evidence of major and widespread resurfacing of older portions of the planet's surface. Many of Mercury's larger "young" basins have also been infilled by volcanism; in some places, these infilling deposits have been estimated to be 2–4 km thick (Ernst et al. 2015). Sampling of the interior portion of the Caloris basin (e.g., Besse et al. 2015; Ernst et al. 2015), for instance, would place vital age constraints on this large impact event and key stratigraphic marker. Most of what we know about the timing of Mercury's evolution hinges on extrapolations from lunar chronological constraints, and so any absolute marker for Mercury has the potential to revolutionize our understanding of the timing of important events on the innermost planet. Sampling of sites such as Caloris and volcanic plains units can help answer questions such as: what is the timing and evolution of major geologic events on Mercury? What do the ages of these events tell us about the influx of materials to the inner Solar System? When did massive eruptions like those that produced Borealis Planitia take place, and what do these ages tell us about the thermal and geologic evolution of Mercury?

To answer questions about Mercury's history of magnetism (Sect. 4.3), an ideal sampling location would currently be more strongly magnetized than its surrounding materials (e.g., Fig. 4). Potential sites to understand Mercury's remanant magnetism include regions around or within the Caloris basin, select impact craters showing strong magnetization in their interior (e.g., Rustaveli), or those impact craters showing strong magnetizations in their ejecta blankets (e.g., Rustaveli, Stieglitz; Hood et al. 2018). A sample from these locations would allow us to answer questions including: what are the dominant mineral hosts of magnetization in these regions? What is the strength of the ancient magnetic field?

Given the global distribution of tectonic features, and evidence of space weathering across the surface of Mercury, a sample collected from anywhere on the planet would be able to shed light on these processes. Although some potential sample site locations were described above, any sample collected from Mercury would be instrumental in enhancing our understanding of the planet's formation and evolution. Similarly, regardless of where the returned sample is from ideally this material would contain a diversity of materials with which we could address a wide range of scientific questions. Moreover, the returned sample ought to be of sufficient size that materials could be preserved for future analysis when technology has far exceeded our abilities at the time of sample acquisition.

6 Curation and Flight Hardware Requirements for Mercury Sample Return

The successful return of samples from Mercury would provide many new and unique challenges, and those challenges do not cease once the samples arrive safely on Earth. Contamination concerns for the samples would be part of every decision made for the materials used to fabricate and machine the flight hardware, and special consideration would need to be given to any sample-intimate hardware that comes into direct contact with the samples, including the curation facilities that would receive and process the samples once they arrive on Earth. Each sample return mission has a contamination knowledge plan that is designed to trace back and verify any potential contaminants from the flight hardware or curatorial facilities that might have made their way into the samples (Dworkin et al. 2018; McCubbin et al. 2019). This practice holds true for previous sample return missions, and the continued success of this practice has reinforced its importance to the process going forward (McCubbin et al. 2019). At present, the paucity of verifiable mineralogical information for the surface of Mercury presents a high level of risk for choosing appropriate sample containment requirements for both the flight hardware and the curatorial facilities.

If the surface of Mercury is dominated by typical silicate, oxide, sulfide, and Fe-rich metals that are present in many meteorites (Namur and Charlier 2017; Vander Kaaden et al. 2017), special consideration of sample containment on the flight hardware and in the curatorial facilities may not be necessary. However, if the surface of Mercury is also rich in highly reduced metals produced through low-pressure smelting reactions (e.g., McCubbin et al. 2017), the surface materials would be incompatible with certain spacecraft components and common curatorial laboratory infrastructure. Moreover, the nature of the surface materials would also affect sample containment requirements. For example, the presence of highly reduced metal particles in the sample would present a risk of exothermic reactions with Earth's atmosphere if the samples were not collected and sealed under vacuum or purged with an inert gas prior to Earth arrival. Such a composition would pose a risk both to maintaining the pristine nature of the samples but also to the safety of the sample processors and scientists working on the samples. The choice of vacuum-sealing or purging with inert gas also has important implications for the sample handling requirements in the curation facility. For example, the sample processing cabinet for JAXA's Hayabusa samples operates under vacuum (Yada et al. 2014), and astromaterials at NASA's Johnson Space Center are typically stored and processed in gloveboxes under dry (< 10 ppm H_2O) gaseous nitrogen (N_2) purge (McCubbin et al. 2019). In a similar vein, any sample collected from the Mercurian surface would lose its induced magnetization, although this property could be calculated *a posteriori* through measurements of susceptibility in Earth-based laboratories. For those measurements to reflect the magnetization on Mercury's surface, sample collection would need to be completed with non-magnetic tools and the sample would have to be stored in a magnetically shielded container for return to Earth, as well as in the sample return facility on Earth. To mitigate these risks and fill the knowledge gap on the nature of Mercury surface materials, a Mercury lander to collect mineralogical information on the surface of Mercury is needed before sample containment requirements can be developed for a Mercury sample return mission (Byrne et al. 2018a).

7 Challenges Associated with Mercury Sample Return

Our continued exploration of Mercury through both landed science and by bringing to Earth a sample of the planet for laboratory analyses will be difficult. Perhaps the greatest challenge of sample return from Mercury is the large amount of launch energy and associated change in velocity required for not only reaching Mercury and landing on its surface, but for lofting a sample back into space, accelerating it to escape velocity, and having it successfully intercepted and brought to Earth (Hauck et al. 2010). Additionally, the large fluctuation in temperature on Mercury ($> 600^\circ C$ over the course of a Mercury day) challenges the longevity of instruments on its surface. These temperatures place tight time constraints on any analyses as well as collection and return efforts. Considerations about the topography of the region being explored, as well as the impact of crustal properties for landing on the surface, need to be highly scrutinized. Furthermore, the sampling mechanism required (e.g., scooping) has likely yet to be developed, though can potentially build off of what will be used for future Mars sample return missions. Although there are numerous risks and challenges associated with landing on the surface of the Sun's closest planetary neighbor, the amount of knowledge gain from a single piece of Mercury far outweighs the challenges and risks. With the speed of technological advancements seen not just with MESSENGER and BepiColombo, but also with sample return missions such Hayabusa 1 and 2, OSIRIS-REx, and future Mars sample return missions such as Mars2020, we are confident that it is possible to develop the technology in the coming decades with which to undertake the grand challenge of collecting a sample from Mercury's surface and returning it safely to Earth.

8 Conclusions

To ensure the continued exploration of the innermost planet, mission concepts to explore Mercury from both landed science and sample return need to be underway now. Because of the requirement to shed velocity on approach to Mercury from Earth, a mission dispatched to the planet takes several years to reach its destination, comparable to the cruise duration for an outer planets mission (Byrne et al. 2018a). Time is therefore of the essence. Lander science and sample return from Mercury would facilitate transformative Solar System science that would place new and vital constraints on the building blocks and thermochemical evolution of Mercury and the terrestrial planets. Given the substantial investment of resources that would be needed to collect a sample from Mercury and bring it to Earth, it is critical that all challenges and risks are assessed and overcome prior to any sample return attempts, and that the material returned from the planet be diverse and curated appropriately. Any delay in future exploration efforts could have detrimental impacts on the continued exploration of Mercury and, with it, our understanding of our Solar System at large.

Acknowledgements The authors thank Dave Blewett, Steve Hauck, and Erwan Mazarico for fruitful discussions about the future exploration of Mercury and all members of the MESSENGER team for providing us with the necessary data to start thinking about Mercury sample return in detail. KVK and FMM acknowledge support from NASA's Planetary Science Research Program. PKB acknowledges support from North Carolina State University. NLC acknowledges support from DDAP grant #NNX15AK89G. CLJ acknowledges support from the Natural Sciences and Engineering Research Council of Canada. MST acknowledges support from Purdue University. We thank three anonymous reviewers for their helpful and thorough reviews of this manuscript as well as the editors for the handling of this manuscript.

Publisher's Note Springer Nature remains neutral with regard to jurisdictional claims in published maps and institutional affiliations.

References

R.E. Arvidson, J.L. Gooding, H.J. Moore, The martian surface as imaged, sampled, and analyzed by the Viking landers. Rev. Geophys. **27**, 39–60 (1989)

E. Asphaug, A. Reufer, Mercury and other iron-rich planetary bodies as relics of inefficient accretion. Nat. Geosci. **7**, 564–568 (2014)

M.E. Banks, Z. Xiao, T.R. Watters, R.G. Strom, S.E. Braden, C.R. Chapman, S.C. Solomon, C. Klimczak, P.K. Byrne, Duration of activity on lobate-scarp thrust fauls on Mercury. J. Geophys. Res., Planets **120**, 1751–1762 (2015)

M.E. Banks, Z. Xiao, S.E. Braden, S. Marchi, C.R. Chapman, N.G. Barlow, C.I. Fassett, Revised age constraints for Mercury's Kuiperian and Mansurian systems, in *Lunar and Planetary Science Conference*, The Woodlands, TX (2016), p. #2943

W. Baumjohann, A. Matsuoka, W. Magnes, K-H. Glassmeier, R. Nakamura, H. Biernat, M. Delva, K. Schwingenschuh, T. Zhang, H-U. Auster, K-H. Fornacon, I. Richter, A. Balogh, P. Cargill, C. Carr, M. Dougherty, T.S. Horbury, E.A. Lucek, F. Tohyama, T. Takahashi, M. Tanaka, T. Nagai, H. Tsunakawa, M. Matsushima, H. Kawano, A. Yoshikawa, H. Shibuya, T. Nakagawa, M. Hoshino, T. Tanaka, R. Kataoka, B.J. Anderson, C.T. Russell, U. Motschmann, M. Shinohara, Magnetic field investigation of Mercury's magnetosphere and the inner heliosphere by MMO/MGF. Planet. Space Sci. **58**, 279–286 (2010)

J. Benkhoff, J. van Casteren, H. Hayakawa, M. Fujimoto, H. Laakso, M. Novara, P. Ferri, H.R. Middleton, R. Ziethe, BepiColombo—comprehensive exploration of Mercury: mission overview and science goals. Planet. Space Sci. **58**, 2–20 (2010)

W. Benz, A. Anic, J. Horner, J.A. Whitby, The origin of Mercury. Space Sci. Rev. **132**, 189–202 (2007)

S. Besse, A. Doressoundiram, J. Benkhoff, Spectroscopic properties of explosive volcanism within the Caloris basin with MESSENGER observations. J. Geophys. Res., Planets **120**, 2102–2117 (2015)

M. Beuthe, East-west faults due to planetary contraction. Icarus **209**, 795–817 (2010)

Z.T. Bieniawski, *Engineering Rock Mass Classifications* (Wiley, New York, 1989)

B. Bitsch, M. Lambrechts, A. Johansen, The growth of planets by pebble accretion in evolving protoplanetary discs. Astron. Astrophys. **582**, A112 (2015)

D.T. Blewett, W.M. Vaughan, Z. Xiao, N.L. Chabot, B.W. Denevi, C.M. Ernst, J. Helbert, M. D'Amore, A. Maturilli, J.W. Head, S.C. Solomon, Mercury's hollows: Constraints on formation and composition from analysis of geological setting and spectral reflectance. J. Geophys. Res., Planets **118**, 1–20 (2013)

P. Borin, G. Cremonese, F. Marzari, M. Bruno, S. Marchi, Statistical analysis of micrometeoroids flux on Mercury. Astron. Astrophys. **503**, 259–264 (2009)

A. Bouvier, L.J. Spivak-Birndorf, G. Brennecka, M. Wadhwa, New constraints on early Solar System chronology from Al-Mg and U-Pb isotope systematics in the unique basaltic achondrite Northwest Africa 2976. Geochim. Cosmochim. Acta **75**, 5310–5323 (2011)

L.C. Bouvier, M.M. Costa, J.N. Connelly, N.K. Jensen, D. Wielandt, M. Storey, A.A. Nemchin, M.J. Whitehouse, J.F. Snape, J.J. Bellucci, F. Moynier, A. Agranier, B. Gueguen, M. Schonbachler, M. Bizzarro, Evidence for extremely rapid magma ocean crystallization and crust formation on Mars. Nature **558**, 586–589 (2018)

R. Brasser, S.J. Moizsis, S. Matsumura, S. Ida, The cool and distant formation of Mars. Earth Planet. Sci. Lett. **468**, 85–93 (2017)

G.A. Brennecka, M. Wadhwa, Uranium isotope compositions of the basaltic angrite meteorites and the chronological implications for the early Solar System. Proc. Natl. Acad. Sci. USA **109**, 9299–9303 (2012)

A.L. Broadfoot, S. Kumar, M.J.S. Blelton, M.B. McElroy, Mercury's atmosphere from Mariner 10: preliminary results. Science **185**, 166–169 (1974)
A.L. Broadfoot, D.E. Shemansky, S. Kumar, Mariner 10: Mercury atmosphere. Geophys. Res. Lett. **3**, 577–580 (1976)
G. Budde, C. Burkhardt, G.A. Brennecka, M. Fischer-Godde, T.S. Kruijer, T. Kleine, Molybdenum isotopic evidence for the origin of chondrules and a distinct genetic heritage of carbonaceous and non-carbonaceous meteorites. Earth Planet. Sci. Lett. **454**, 293–303 (2016)
T.H. Burbine, T.J. McCoy, L.R. Nittler, G.K. Benedix, E.A. Cloutis, T.L. Dickinson, Spectra of extremely reduced assemblages: Implications for Mercury. Meteorit. Planet. Sci. **37**, 1233–1244 (2002)
K.D. Burgess, R.M. Stroud, Coordinated nanoscale compositional and oxidation state measurements of lunar space-weathered material. J. Geophys. Res., Planets **123**, 2022–2037 (2018)
P.K. Byrne, L.R. Ostrach, C.I. Fassett, C.R. Chapman, B.W. Denevi, A.J. Evans, C. Klimczak, M.E. Banks, J.W. Head, S.C. Solomon, Widespread effusive volcanism on Mercury likely ended by about 3.5 Ga. Geophys. Res. Lett. **43**, 7408–7416 (2016)
P.K. Byrne, D.T. Blewett, N.L. Chabot, S.A. Hauck II., E. Mazarico, K.E. Vander Kaaden, White paper on the case for Landed Mercury Exploration and the Timely Need fora Mission Concept Study (2018a). https://bit.ly/2GqsmC2
P.K. Byrne, C. Klimczak, A.M.C. Şengör, The tectonic character of Mercury, in *Mercury: The View After MESSENGER*, ed. by S.C. Solomon, L.R. Nittler, B.J. Anderson (Cambridge University Press, Cambridge, 2018b)
N.L. Chabot, E.A. Wollack, R.L. Klima, M.E. Minitti, Experimental constraints on Mercury's core composition. Earth Planet. Sci. Lett. **390**, 199–208 (2014)
J.E. Chambers, Planetary accretion in the inner Solar System. Earth Planet. Sci. Lett. **223**, 241–252 (2004)
J.E. Chambers, G.W. Wetherill, Making the terrestrial planets: N-body integrations of planetary embryos in three dimensions. Icarus **136**, 304–327 (1998)
S.C. Chase, E.D. Miner, D. Morrison, G. Munch, G. Neugebauer, M. Schroeder, Preliminary infrared radiometry of the night side of Mercury from Mariner 10. Science **185**, 142–145 (1974)
E. Chassefière, J.-L. Maria, J.-P. Goutail, E. Quemerais, F. Leblanc, S. Okano, I. Yoshikawa, O. Korablev, V. Gnedykh, G. Naletto, P. Nicolosi, M.-G. Pelizzo, J.-J. Correia, S. Gallet, C. Hourtoule, P.-O. Mine, C. Montaron, N. Rouanet, J.-B. Rigal, G. Muramaki, K. Yoshioka, O. Kozlov, V. Kottsov, P. Moisseev, N. Semena, J.-L. Bertauz, M.-Th. Capria, J. Clarke, G. Cremonese, D. Delcourt, A. Doressoundiram, S. Erard, R. Gladstone, M. Grande, D. Hunten, W. Ip, V. Izmodenov, A. Jambon, R. Johnson, E. Kallio, R. Killen, R. Lallement, J. Luhmann, M. Mendillo, A. Milillo, H. Palme, A. Potter, S. Sasaki, D. Slater, A. Sprague, A. Stern, N. Yan, PHEBUS: A double ultraviolet spectrometer to observe Mercury's exosphere. Planet. Space Sci. **58**, 201–223 (2010)
M.J. Cintala, Impact-induced thermal effects in the lunar and mercurian regoliths. J. Geophys. Res., Planets **97**, 947–973 (1992)
J.N. Connelly, M. Bizzarro, A.N. Krot, A. Nordlund, D. Wielandt, M.A. Ivanova, The absolute chronology and thermal processing of solids in the solar protoplanetary disk. Science **338**, 652–655 (2012)
N. Dauphas, The isotopic nature of the Earth's accreting material through time. Nature **541**, 521–524 (2017)
N. Dauphas, M. Chaussidon, A perspective from extinct radionuclides on a young stellar object: The Sun and its accretion disk. Annu. Rev. Earth Planet. Sci. **39**, 351–386 (2011)
N. Dauphas, A. Pourmand, Hf–W–Th evidence for rapid growth of Mars and its status as a planetary embryo. Nature **473**, 489–492 (2011)
B.W. Denevi, M.S. Robinson, S.C. Solomon, S.L. Murchie, D.T. Blewett, D.L. Domingue, T.J. McCoy, C.M. Ernst, J.W. Head, T.R. Watters, N.L. Chabot, The evolution of Mercury's crust: A global perspective from MESSENGER. Science **324**, 613–618 (2009)
B.W. Denevi, C.M. Ernst, H.M. Meyer, M.S. Robinson, S.L. Murchie, J.L. Whitten, J.W. Head, T.R. Watters, S.C. Solomon, L.R. Ostrach, C.R. Chapman, P.K. Byrne, C. Klimczak, P.N. Peplowski, The distribution and origin of smooth plains on Mercury. J. Geophys. Res., Planets **118**, 891–907 (2013)
B.W. Denevi, C.M. Ernst, L.M. Prockter, M.S. Robinson, The geologic history of Mercury, in *Mercury: The View After MESSENGER*, ed. by S.C. Solomon, L.R. Nittler, B.J. Anderson (Cambridge University Press, Cambridge, 2019)
S. Dey, Q.Z. Yin, M.E. Sanborn, K. Ziegler, T.J. McCoy, Planetary genealogy of iron meteorites and pallasites using e54Cr-D17O isotope systematics, in *50th Lunar and Planetary Science Conference*, The Woodlands, TX (2019), p. #2977
D.L. Domingue, C.R. Chapman, R.M. Killen, T.H. Zurbuchen, J.A. Gilbert, M. Sarantos, M. Benna, J.A. Slavin, D. Schriver, P.M. Trávníček, T.M. Orlando, A.L. Sprague, D.T. Blewett, J.J. Gillis-Davis, W.C. Feldman, D.J. Lawrence, G.C. Ho, D.S. Ebel, L.R. Nittler, F. Vilas, C.M. Pieters, S.C. Solomon, C.L. Johnson, R.M. Winslow, J. Helbert, P.N. Peplowski, S.Z. Weider, N. Mouawad, N.R. Izenberg, W.E. McClintock, Mercury's weather-beaten surface: Understanding Mercury in the context of lunar and asteroidal space weathering studies. Space Sci. Rev. **181**, 121–214 (2014)

J.P. Dworkin, L.A. Adelman, T. Ajluni, A.V. Andronikov, J.C. Aponte, A.E. Bartels, E. Beshore, E.B. Bierhaus, J.R. Brucato, B.H. Bryan, A.S. Burton, M.P. Callahan, S.L. Castro-Wallace, B.C. Clark, S.J. Clemett, H.C. Connolly Jr., W.E. Cutlip, S.M. Daly, V.E. Elliott, J.E. Elsila, H.L. Enos, D.F. Everett, I.A. Franchi, D.P. Glavin, H.V. Graham, J.E. Hendershot, J.W. Harris, S.L. Hill, A.R. Hildebrand, G.O. Jayne, R.W. Jenkens Jr., K.S. Johnson, J.S. Kirsch, D.S. Lauretta, A.S. Lewis, J.J. Loiacono, C.C. Lorentson, J.R. Marshall, M.G. Martin, L.L. Matthias, H.L. McLain, S.R. Messenger, R.G. Mink, J.L. Moore, K. Nakamura-Messenger, J.A. Nuth III., C.V. Owens, C.L. Parish, B.D. Perkins, M.S. Pryzby, C.A. Reigle, K. Righter, B. Rizk, J.F. Russell, S.A. Sandford, J.P. Schepis, J. Songer, M.F. Sovinski, S.E. Stahl, K. Thomas-Keprta, J.M. Vellinga, M.S. Walker, OSIRIS-REx contamination control strategy and implementation. Space Sci. Rev. **214**, 19 (2018)

M.-C. Eppes, A. Willis, J. Molaro, S. Abernathy, B. Zhou, Cracks in Martian boulders exhibit preferred orientations that point to solar-induced thermal stress. Nat. Commun. **6**, 6712 (2015)

C.M. Ernst, B.W. Denevi, O.S. Barnouin, C. Klimczak, N.L. Chabot, J.W. Head, S.L. Murchie, G.A. Neumann, L.M. Prockter, M.S. Robinson, S.C. Solomon, T.R. Watters, Stratigraphy of the Caloris basin, Mercury: Implications for volcanic history and basin impact melt. Icarus **250**, 413–429 (2015)

C.M. Ernst, B.W. Denevi, L.R. Ostrach, Updated absolute age estimates for the Tolstoj and Caloris basins, in *Mercury, Lunar and Planetary Science Conference XLVIII*, The Woodlands, TX (2017), p. #2934

L.G. Evans, P.N. Peplowski, E.A. Rhodes, D.J. Lawrence, T.J. McCoy, L.R. Nittler, S.C. Solomon, A.L. Sprague, K.R. Stockstill-Cahill, R.D. Starr, S.Z. Weider, W.V. Boynton, D.K. Hamara, J.O. Goldsten, Major-element abundances on the surface of Mercury: Results from the MESSENGER Gamma-Ray Spectrometer. J. Geophys. Res., Planets **117**, E00L07 (2012)

L.G. Evans, P.N. Peplowski, F.M. McCubbin, T.J. McCoy, L.R. Nittler, M.Y. Zolotov, D.S. Ebel, D.J. Lawrence, R.D. Starr, S.Z. Weider, S.C. Solomon, Chlorine on the surface of Mercury: MESSENGER gamma-ray measurements and implications for the planet's formation and evolution. Icarus **257**, 417–427 (2015)

J. Filiberto, Magmatic diversity on Venus: Constraints from terrestrial analog crystallization experiments. Icarus **231**, 131–136 (2014)

E. Flamini, F. Capaccioni, I. Colangeli, G. Cremonese, A. Doressoundiram, J.L. Josset, Y. Langevin, S. Debei, M.T. Capria, M.C. De Sanctis, L. Marinangeli, M. Massironi, E. Mazotta Epifani, G. Naletto, P. Palumbo, P. Eng, J.F. Roig, A. Caporali, V. Da Deppo, S. Erard, C. Federico, O. Forni, M. Sgavetti, G. Filacchione, L. Giacomini, G. Marra, E. Martellato, M. Zusi, M. Cosi, C. Bettanini, L. Calamai, M. Zaccariotto, L. Tommasi, M. Dami, J. Ficai Veltroni, F. Poulet, Y. Hello (The SIMBIO-SYS Team), SIMBIO-SYS: The spectrometer and imagers integrated observatory system for the BepiColombo planetary orbiter. Planet. Space Sci. **58**, 125–143 (2010)

R.A. Fogel, Aubrite basalt vitrophyres: The missing basaltic component and high-sulfur silicate melts. Geochim. Cosmochim. Acta **69**, 1633–1648 (2005)

C.N. Foley, M. Wadhwa, L.E. Borg, P.E. Janney, R. Hines, T.L. Grove, The early differentiation history of Mars from W-182-Nd-142 isotope systematics in the SNC meteorites. Geochim. Cosmochim. Acta **69**, 4557–4571 (2005)

G.W. Fraser, J.D. Carpenter, D.A. Rothery, J.F. Pearson, A. Martindale, J. Huovelin, J. Treis, M. Anand, M. Anttila, M. Ashcroft, J. Benkoff, P. Bland, A. Bowyer, A. Bradley, J. Bridges, C. Brown, C. Bulloch, E.J. Bunce, U. Christensen, M. Evans, R. Fairbend, M. Feasey, F. Giannini, S. Hermann, M. Hesse, M. Hilchenbach, T. Jorden, K. Joy, M. Kaipiainen, I. Kitchingman, P. Lechner, G. Lutz, A. Malkki, K. Muinonen, J. Naranen, P. Portin, M. Prydderch, J. San Juan, E. Sclater, E. Schyns, T.J. Stevenson, L. Struder, M. Syrjasuo, D. Talboys, P. Thomas, C. Whiford, S. Whitehead, The Mercury Imaging X-ray Spectrometer (MIXS) on BepiColombo. Planet. Space Sci. **58**, 79–95 (2010)

B. Gladman, J. Coffey, Mercurian impact ejecta: Meteorites and mangle. Meteorit. Planet. Sci. **44**(2), 285–291 (2009)

K.-H. Glassmeier, H.-U. Auster, D. Heyner, K. Okrafka, C. Carr, G. Berghofer, B.J. Anderson, A. Balogh, W. Baumjohann, P. Cargill, U. Christensen, M. Delva, M. Doughtery, K.-H. Fornacon, T.S. Horbury, E.A. Lucek, W. Magnes, M. Mandea, A. Matsuoka, M. Matsushima, U. Motschmann, R. Nakamura, Y. Narita, H. O'Brien, I. Richter, K. Schwingenschuh, H. Shibuya, J.A. Slavin, C. Sotin, B. Stoll, H. Tsunakawa, S. Vennerstrom, J. Vogt, T. Zhang, The fluxgate magnetometer of the BepiColombo Mercury Planetary Orbiter. Planet. Space Sci. **58**, 287–299 (2010)

J.F. Gonzalez, G. Laibe, S.T. Maddison, Self-induced dust traps: overcoming planet formation barriers. Mon. Not. R. Astron. Soc. **467**, 1984–1996 (2017)

T.A. Goudge, J.W. Head, L. Kerber, D.T. Blewett, B.W. Denevi, D.L. Domingue, J.J. Gillis-Davis, K. Gwinner, J. Helbert, G.M. Holsclaw, N.R. Izenberg, R.L. Klima, W.E. McClintock, S.L. Murchie, G.A. Neumann, D.E. Smith, R.G. Strom, Z. Xiao, M.T. Zuber, S.C. Solomon, Global inventory and characterization of pyroclastic deposits on Mercury: New insights into pyroclastic activity from MESSENGER orbital data. J. Geophys. Res., Planets **119**, 635–658 (2014)

K. Gunderson, N. Thomas, BELA receiver performance modeling over the BepiColombo mission lifetime. Planet. Space Sci. **58**, 309–318 (2010)
B. Hapke, Space weathering from Mercury to the asteroid belt. J. Geophys. Res., Planets **106**, 10039–10073 (2001)
J.K. Harmon, M.A. Slade, Radar mapping of Mercury: full-disk images and polar anomalies. Science **258**, 640–642 (1992)
S.A. Hauck II., C.L. Johnson, Mercury: Inside the iron planet. Elements **15**, 21–26 (2019)
S.A. Hauck II., D.A. Eng, G.J. Tahu, Mercury Lander Mission Concept Study. NASA Science Solar System Exploration (2010). https://solarsystem.nasa.gov/studies/199/mercury-lander-mission-concept-study/
S.A. Hauck II., J.L. Margot, S.C. Solomon, R.J. Phillips, C.L. Johnson, F.G. Lemoine, E. Mazarico, T.J. McCoy, S. Padovan, S.J. Peale, M.E. Perry, D.E. Smith, M.T. Zuber, The curious case of Mercury's internal structure. J. Geophys. Res., Planets **118**, 1204–1220 (2013)
J.W. Head, C.R. Chapman, R.G. Strom, C.I. Fassett, B.W. Denevi, D.T. Blewett, C.M. Ernst, T.R. Watters, S.C. Solomon, S.L. Murchie, L.M. Prockter, N.L. Chabot, J.J. Gillis-Davis, J.L. Whitten, T.A. Goudge, D.M.H. Baker, D.M. Hurwitz, L.R. Ostrach, Z. Xiao, W.J. Merline, L. Kerber, J.L. Dickson, J. Oberst, P.K. Byrne, C. Klimczak, L.R. Nittler, Flood volcanism in the northern high latitudes of Mercury revealed by MESSENGER. Science **333**, 1853–1856 (2011)
H. Hiesinger, J. Helbert (MERTIS Co-I Team), The Mercury Radiometer and Thermal Infrared Spectrometer (MERTIS) for the BepiColombo mission. Planet. Space Sci. **58**, 144–165 (2010)
M.M. Hirschmann, A.C. Withers, P. Ardia, N.T. Foley, Solubility of molecular hydrogen in silicate melts and consequences for volatile evolution of terrestrial planets. Earth Planet. Sci. Lett. **345**, 38–48 (2012)
L.L. Hood, Magnetic anomalies concentrated near and within Mercury's impact basins: Early mapping and interpretation. J. Geophys. Res., Planets **121**, 1016–1025 (2016)
L.L. Hood, J.S. Oliveira, V. Galluzzi, D.A. Rothery, Investigating sources of Mercury's crustal magnetic field: Further mapping of MESSENGER magnetometer data. J. Geophys. Res., Planets **123**(10), 2647–2666 (2018)
J. Huovelin, R. Vainio, H. Andersson, E. Valtonen, L. Alha, A. Malkki, M. Grande, G.W. Fraser, M. Kato, H. Koskinen, K. Muinonen, J. Naranen, W. Schmidt, M. Syrjasuo, M. Anttila, T. Vihavainen, E. Kiuru, M. Roos, J. Peltonen, J. Lehti, M. Talvoija, P. Portin, M. Prydderch, Solar Intensity X-ray and particle Spectrometer (SIXS). Planet. Space Sci. **58**, 96–107 (2010)
M.H. Huyskens, M.E. Sanborn, Q.Z. Yin, Y. Amelin, P. Koefoed, Chronology of carbonaceous achondrites from the outer solar system, in *50th Lunar and Planetary Science Conference*, The Woodlands, TX (2019), p. #2736
V. Iafolla, E. Fiorenza, C. Lefevre, A. Morbidini, S. Nozzoli, R. Peron, M. Persichini, A. Reale, F. Santoli, Italian Spring Accelerometer (ISA): A fundamental support to BepiColombo Radio Science Experiments. Planet. Space Sci. **58**, 300–308 (2010)
A. Johansen, M. Lambrechts, Forming planets via pebble accretion. Annu. Rev. Earth Planet. Sci. **45**, 359–387 (2017)
A. Johansen, A. Youdin, Protoplanetary disk turbulence driven by the streaming instability: Nonlinear saturation and particle concentration. Astrophys. J. **662**, 627–641 (2007)
C.L. Johnson, R.J. Phillips, M.E. Purucker, B.J. Anderson, P.K. Byrne, B.W. Denevi, J.M. Feinber, S.A. Hauck II., J.W. Head, H. Korth, P.B. James, E. Mazarico, G.A. Neumann, L.C. Philpott, M.A. Siegler, N.A. Tsyganenko, S.C. Solomon, Low-altitude magnetic field measurements by MESSENGER reveal Mercury's ancient crustal field. Science **348**, 892–895 (2015)
C.L. Johnson, L.C. Philpott, B.J. Anderson, H. Korth, S.A. Hauck II, D. Heyner, R.J. Phillips, R.M. Winslow, S.C. Solomon, MESSENGER observations of induced magnetic fields in Mercury's core. Geophys. Res. Lett. **43**, 2436–2444 (2016)
C.L. Johnson, B.J. Anderson, H. Korth, R.J. Phillips, L.C. Philpott, Mercury's internal magnetic field, in *Mercury: The View After MESSENGER*, ed. by S.C. Solomon, L.R. Nittler, B.J. Anderson (Cambridge University Press, Cambridge, 2019)
L.M. Jozwiak, J.W. Head, L. Wilson, Explosive volcanism on Mercury: Analysis of vent and deposit morphology and modes of eruption. Icarus **302**, 191–212 (2018)
Y. Kasaba, J.-L. Bougeret, L.G. Blomberg, H. Kojima, S. Yagitani, M. Moncuquet, J.-G. Trotignon, G. Chanteur, A. Kumamoto, Y. Kasahara, J. Lichtenberger, Y. Omura, K. Ishisaka, H. Matsumoto, The Plasma Wave Investigation (PWI) onboard the BepiColombo/MMO: First measurement of electric fields, electromagnetic waves, and radio waves around Mercury. Planet. Space Sci. **58**, 238–278 (2010)
K. Keil, Enstatite achondrite meteorites (aubrites) and the histories of their asteroidal parent bodies. Chem. Erde **70**, 295–317 (2010)
L.P. Keller, D.S. McKay, Discovery of vapor deposits in the lunar regolith. Science **261**, 1305–1307 (1993)
L.P. Keller, D.S. McKay, The nature and origin of rims on lunar soil grains. Geochim. Cosmochim. Acta **61**, 2331–2341 (1997)

L. Kerber, J.W. Head, S.C. Solomon, S.L. Murchie, D.T. Blewett, L. Wilson, Explosive volcanic eruptions on Mercury: Eruption conditions, magma volatile content, and implications for interior volatile abundances. Earth Planet. Sci. Lett. **285**, 263–271 (2009)

L. Kerber, J.W. Head, D.T. Blewett, S.C. Solomon, L. Wilson, S.L. Murchie, M.S. Robinson, B.W. Denevi, D.L. Domingue, The global distribution of pyroclastic deposits on Mercury: The view from MESSENGER flybys 1–3. Planet. Space Sci. **59**, 1895–1909 (2011)

R.M. Killen, Pathways for energization of Ca and Mg in Mercury's exosphere. Icarus **268**, 32–36 (2016)

T. Kleine, M. Touboul, B. Bourdon, F. Nimmo, K. Mezger, H. Palme, S.B. Jacobsen, Q.-Z. Yin, A.N. Halliday, Hf-W chronology of the accretion and early evolution of asteroids and terrestrial planets. Geochim. Cosmochim. Acta **73**, 5150–5188 (2009)

R.L. Klima, B.W. Denevi, C.M. Ernst, S.L. Murchie, P.N. Peplowski, Global distribution and spectral properties of low-reflectance material on Mercury. Geophys. Res. Lett. **45**, 2945–2953 (2018)

C. Klimczak, P.K. Byrne, S.C. Solomon, A rock-mechanical assessment of Mercury's global tectonic fabric. Earth Planet. Sci. Lett. **416**, 82–90 (2015)

H. Korth, B.J. Anderson, C.L. Johnson, J.A. Slavin, J.M. Raines, T.H. Zurbuchen, Structure and configuration of Mercury's magnetosphere, in *Mercury: The View After MESSENGER*, ed. by S.C. Solomon, L.R. Nittler, B.J. Anderson (Cambridge University Press, Cambridge, 2019)

T.S. Kruijer, M. Touboul, M. Fischer-Godde, K.R. Bermingham, R.J. Walker, T. Kleine, Protracted core formation and rapid accretion of protoplanets. Science **344**, 1150–1154 (2014)

T.S. Kruijer, C. Burkhardt, G. Budde, T. Kleine, Age of Jupiter inferred from the distinct genetics and formation times of meteorites. Proc. Natl. Acad. Sci. USA **114**, 6712–6716 (2017)

M. Le Feuvre, M.A. Wieczorek, Nonuniform cratering of the Moon and a revised crater chronology of the inner Solar System. Icarus **214**, 1–20 (2011)

S.G. Love, K. Keil, Recognizing mercurian meteorites. Meteoritics **30**, 269–278 (1995)

V. Malavergne, J. Siebert, F. Guyot, L. Gautron, R. Combes, T. Hammouda, S. Borensztajn, D. Frost, I. Martinez, Si in the core? New high-pressure and high-temperature experimental data. Geochim. Cosmochim. Acta **68**, 4201–4211 (2004)

V. Malavergne, S. Berthet, K. Righter, Formation of CaS-MgS in enstatite chondrites and achondrites as a function of redox conditions and temperature: constraints on their evolution in a planetesimal and in a proto-planet, in *38th Lunar and Planetary Science Conference*, Lunar and Planetary Institute, Houston (2007), p. #1737

V. Malavergne, P. Cordier, K. Righter, F. Brunet, B. Zanda, A. Addad, T. Smith, H. Bureau, S. Surblé, C. Raepsaet, E. Charon, R.H. Hewins, How Mercury can be the most reduced terrestrial planet and still store iron in its mantle. Earth Planet. Sci. Lett. **394**, 186–197 (2014)

S. Marchi, S. Mottola, G. Cremonese, M. Massironi, E. Martellato, A new chronology for the Moon and Mercury. Astron. J. **137**, 4936–4948 (2009)

S. Marchi, C.R. Chapman, C.I. Fassett, J.W. Head, W.F. Bottke, R.G. Strom, Global resurfacing of Mercury 4.0–4.1 billion years ago by heavy bombardment and volcanism. Nature **499**, 59–61 (2013)

J.-L. Margot, S.A. Hauck II., E. Mazarico, S. Padovan, S.J. Peale, Mercury's internal structure, in *Mercury: The View After MESSENGER*, ed. by S.C. Solomon, L.R. Nittler, B.J. Anderson (Cambridge University Press, Cambridge, 2019), pp. 85–113

M. Matsuoka, T. Nakamura, Y. Kimura, T. Hiroi, R. Nakamura, S. Okumura, S. Sasaki, Pulse-laser irradiation experiments of Murchison CM2 chondrite for reproducing space weathering on C-type asteroids. Icarus **254**, 135–143 (2015)

T.J. McCoy, T.L. Dickinson, G.E. Lofgren, Partial melting of the Indarch (EH4) meteorite: A textural, chemical, and phase relations view of melting and melt migration. Meteorit. Planet. Sci. **34**, 735–746 (1999)

F.M. McCubbin, J.J. Barnes, Origin and abundances of H_2O in the terrestrial planets, Moon, and asteroids. Earth Planet. Sci. Lett. **526**, 115771 (2019)

F.M. McCubbin, T.J. McCoy, Expected geochemical and mineralogical properties of meteorites from Mercury: Inferences from MESSENGER data, in *79th Annual Meeting of the Meteoritical Society* (2016), p. #6242

F.M. McCubbin, M.A. Riner, K.E. Vander Kaaden, L.K. Burkemper, Is Mercury a volatile-rich planet? Geophys. Res. Lett. **39**, L09202 (2012)

F.M. McCubbin, K.E.V. Kaaden, P.N. Peplowski, A.S. Bell, L.R. Nittler, J.W. Boyce, L.G. Evans, L.P. Keller, S.M. Elardo, T.J. McCoy, A low O/Si ratio on the surface of Mercury: Evidence for silicon smelting? J. Geophys. Res., Planets **122**, 2053–2076 (2017)

F.M. McCubbin, C.D.K. Herd, T. Yada, A. Hutzler, M.J. Calaway, J.H. Allton, C.M. Corrigan, M.D. Fries, A.D. Harrington, T.J. McCoy, J.L. Mitchell, A.B. Regberg, K. Righter, C.J. Snead, K.T. Tait, M. Zolensky, R.A. Zeigler, Advanced curation of astromaterials for planetary science. Space Sci. Rev. (2019). https://doi.org/10.1007/s11214-019-0615-9

H.Y. McSween, What we have learned about Mars from SNC meteorites. Meteoritics **29**, 757–779 (1994)

A. Milillo, M. Fujimoto, E. Kallio, S. Kameda, F. Leblanc, Y. Narita, G. Cremonese, H. Laakso, M. Laurenza, S. Massetti, S. McKenna-Lawlor, A. Mura, R. Nakamura, Y. Omura, D.A. Rothery, K. Seki, M. Storini, P. Wurz, W. Baumjohann, E.J. Bunce, Y. Kasaba, J. Helbert, A. Sprague, et al. (Hermean Environment WG), The BepiColombo mission: An outstanding tool for investigating the Hermean environment. Planet. Space Sci. **58**, 40–60 (2010)

D.W. Ming, R. Gellert, R.V. Morris, R.E. Arvidson, J. Brueckner, B.C. Clark, G. Klingelhöfer, Geochemical properties of rocks and soils in Gusev crater, Mars: Results of the Alpha Particle X-Ray Spectrometer from Cumberland Ridge to Home Plate. J. Geophys. Res., Planets **113**, E12 (2008)

I.G. Mitrofanov, A.S. Kozyrev, A. Konovalov, M.L. Litvak, A.A. Malakhov, M.I. Mokrousov, A.B. Sanin, V.I. Tret'ykov, A.V. Vostrukhin, Y.I. Bobrovnitskjj, T.M. Tomilina, I. Gurvits, A. Owens, The Mercury Gamm and Neutron Spectrometer (MGNS) on board the Planetary Orbiter of the BepiColombo mission. Planet. Space Sci. **58**, 116–124 (2010)

L. Moroz, G. Baratta, G. Strazzulla, L. Starukhina, E. Dotto, M.A. Barucci, G. Arnold, E. Distefano, Optical alteration of complex organics induced by ion irradiation: 1. Laboratory experiments suggest unusual space weathering trend. Icarus **170**, 214–228 (2004)

K. Mueller, M. Golombek, Compressional structures on Mars. Annu. Rev. Earth Planet. Sci. **32**, 435–464 (2004)

S.L. Murchie, R.L. Klima, B.W. Denevi, C.M. Ernst, M.R. Keller, D.L. Domingue, D.T. Blewett, N.L. Chabot, C. Hash, E. Malaret, N.R. Izenberg, F. Vilas, L.R. Nittler, J.W. Head, Orbital multispectral mapping of Mercury using the MESSENGER Mercury Dual Imaging System: Evidence for the origins of plains units and low-reflectance material. Icarus **254**, 287–305 (2015)

B.C. Murray, The Mariner 10 pictures of Mercury: an overview. J. Geophys. Res. **80**, 2342–2344 (1975)

O. Namur, B. Charlier, Silicate mineralogy at the surface of Mercury. Nat. Geosci. **10**, 9–13 (2017)

O. Namur, B. Charlier, F. Holtz, C. Cartier, C. McCammon, Sulfur solubility in reduced mafic silicate melts: Implications for the speciation and distribution of sulfur on Mercury. Earth Planet. Sci. Lett. **448**, 102–114 (2016a)

O. Namur, M. Collinet, B. Charlier, T.L. Grove, F. Holtz, C. McCammon, Melting processes and mantle sources of lavas on Mercury. Earth Planet. Sci. Lett. **439**, 117–128 (2016b)

J.A.M. Nanne, F. Nimmo, J.N. Cuzzi, T. Kleine, Origin of the non-carbonaceous-carbonaceous meteorite dichotomy. Earth Planet. Sci. Lett. **511**, 44–54 (2019)

N.F. Ness, K.W. Behannon, R.P. Lepping, Y.C. Whang, K.H. Schatten, Magnetic field observations near Mercury: preliminary results from Mariner 10. Science **185**, 151–160 (1974)

N.F. Ness, K.W. Behannon, R.P. Lepping, Y.C. Whang, The magnetic field of Mercury. J. Geophys. Res. **80**, 2708–2716 (1975)

T. Niihara, S.P. Beard, T.D. Swindle, L.A. Schaffer, H. Miyamoto, D.A. Kring, Evidence for multiple 4.0–3.7 Ga impact events within the Apollo 16 collection. Meteorit. Planet. Sci. **54**, 675–698 (2019)

L.R. Nittler, R.D. Starr, S.Z. Weider, T.J. McCoy, W.V. Boynton, D.S. Ebel, C.M. Ernst, L.G. Evans, J.O. Goldsten, D.K. Hamara, D.J. Lawrence, R.L. McNutt, C.E. Schlemm, S.C. Solomon, A.L. Sprague, The major-element composition of Mercury's surface from MESSENGER X-ray spectrometry. Science **333**, 1847–1850 (2011)

L.R. Nittler, N.L. Chabot, T.L. Grove, P.N. Peplowski, The chemical composition of Mercury, in *Mercury: The View After MESSENGER*, ed. by S.C. Solomon, L.R. Nittler, B.J. Anderson (Cambridge University Press, Cambridge, 2019), pp. 30–51

K. Nogami, M. Fujii, h. Ohashi, T. Miyachi, S. Sasaki, S. Hasegawa, H. Yano, H. Shibata, T. Iwai, S. Minami, S. Takechi, E. Grun, R. Srama, Development of the Mercury dust monitor (MDM) onboard the BepiColombo mission. Planet. Space Sci. **58**, 108–115 (2010)

T. Noguchi, T. Nakamura, M. Kimura, M.E. Zolensky, M. Tanaka, T. Hashimoto, M. Konno, A. Nakato, T. Ogami, A. Fujimura, M. Abe, Incipient space weathering observed on the surface of Itokawa dust particles. Science **333**, 1121–1125 (2011)

T. Noguchi, M. Kimura, T. Hasimoto, M. Konno, T. Nakamura, M.E. Zolensky, R. Okazaki, M. Tanaka, A. Tsuchiyama, A. Nakato, T. Ogami, Space weathered rims found on the surfaces of the Itokawa dust particles. Meteorit. Planet. Sci. **492**, 188–214 (2014)

K.W. Ogilvie, J.D. Scudder, R.E. Hartle, G.L. Siscoe, H.S. Bridge, A.J. Lazarus, J.R. Asbridge, S.J. Bame, C.M. Yeates, Observations at Mercury encountered by the plasma science experiment on Mariner 10. Science **185**, 145–151 (1974)

S. Orsini, S. Livi, K. Torkar, S. Barabash, A. Milillo, P. Wurz, A.M. Di Lellis, E. Kallio (SERENA Team), SERENA: A suite of four instruments (ELENA, STROFIO, PICAM, and MIPA) on board BepiColombo-MPO for particle detection in the Hermean environment. Planet. Space Sci. **58**, 166–181 (2010)

L.R. Ostrach, M.S. Robinson, J.L. Whitten, C.I. Fassett, R.G. Strom, J.W. Head, S.C. Solomon, Extent, age, and resurfacing history of the northern smooth plains on Mercury from MESSENGER observations. Icarus **250**, 602–622 (2015)

P.N. Peplowski, L.G. Evans, S.A. Hauck, T.J. McCoy, W.V. Boynton, J.J. Gillis-Davis, D.S. Ebel, J.O. Goldsten, D.K. Hamara, D.J. Lawrence, R.L. McNutt, L.R. Nittler, S.C. Solomon, E.A. Rhodes, A.L. Sprague, R.D. Starr, K.R. Stockstill-Cahill, Radioactive elements on Mercury's surface from MESSENGER: Implications for the planet's formation and evolution. Science **333**, 1850–1852 (2011)

P.N. Peplowski, L.G. Evans, K.R. Stockstill-Cahill, D.J. Lawrence, J.O. Goldsten, T.J. McCoy, L.R. Nittler, S.C. Solomon, A.L. Sprague, R.D. Starr, S.Z. Weider, Enhanced sodium abundance in Mercury's north polar region revealed by the MESSENGER Gamma-Ray Spectrometer. Icarus **228**, 86–95 (2014)

P.N. Peplowski, D.J. Lawrence, L.G. Evans, R.L. Klima, D.T. Blewett, J.O. Goldsten, S.L. Murchie, T.J. McCoy, L.R. Nittler, S.C. Solomon, R.D. Starr, S.Z. Weider, Constraints on the abundance of carbon in near-surface materials on Mercury: Results from the MESSENGER Gamma-Ray Spectrometer. Planet. Space Sci. **108**, 98–107 (2015a)

P.N. Peplowski, D.J. Lawrence, W.C. Feldman, J.O. Goldsten, D. Bazell, L.G. Evans, J.W. Head, L.R. Nittler, S.C. Solomon, S.Z. Weider, Geochemical terranes of Mercury's northern hemisphere as revealed by MESSENGER neutron measurements. Icarus **253**, 346–363 (2015b)

P.N. Peplowski, R.L. Klima, D.J. Lawrence, C.M. Ernst, B.W. Denevi, E.A. Frank, J.O. Goldsten, S.L. Murchie, L.R. Nittler, S.C. Solomon, Remote sensing evidence for an ancient carbon-bearing crust on Mercury. Nat. Geosci. **9**, 273–276 (2016)

C.M. Pieters, S.K. Noble, Space weathering on airless bodies. J. Geophys. Res., Planets **121**, 1865–1884 (2016)

A.E. Potter, T.H. Morgan, Discovery of sodium in the atmosphere of Mercury. Science **229**, 651–653 (1985)

A.E. Potter, T.H. Morgan, Potassium in the atmosphere of Mercury. Icarus **67**, 336–340 (1986)

S.N. Raymond, D.P. O'Brien, A. Morbidelli, N.A. Kaib, Building the terrestrial planets: Constrained accretion in the inner Solar System. Icarus **203**, 644–662 (2009)

A. Ricolleau, Y. Fei, A. Corgne, J. Siebert, J. Badro, Oxygen and silicon contents of Earth's core from high pressure metal-silicate partitioning experiments. Earth Planet. Sci. Lett. **310**, 409–421 (2011)

A.E. Ringwood, Silicon in the metal phase of enstatite chondrites and some geochemical implications. Geochim. Cosmochim. Acta **25**, 1–13 (1961)

P. Rochette, J. Gattacceca, L. Bonal, M. Bourot-Denise, V. Chevrier, J.-P. Clerc, G. Consolmagno, L. Folco, M. Gournelle, T. Kohout, L. Pesonen, E. Quirico, L. Sagnotti, A. Skripnik, Magnetic classification of stony meteorites: 2. Non-ordinary chondrites. Meteorit. Planet. Sci. **43**, 959–980 (2008)

A.D. Rogers, V.E. Hamilton, Compositional provinces of Mars from statistical analyses of TES, GRS, OMEGA, and CRISM data. J. Geophys. Res., Planets **120**, 62–91 (2015)

L. Rose-Weston, J.M. Brenan, Y.W. Fei, R.A. Secco, D.J. Frost, Effect of pressure, temperature, and oxygen fugacity on the metal-silicate partitioning of Te, Se, and S: Implications for Earth differentiation. Geochim. Cosmochim. Acta **73**, 4598–4615 (2009)

D. Rothery, L. Marinangeli, m. Anand, J. Carpenter, U. Christensen, I.A. Crawford, M.C. De Sanctis, E.M. Epifani, S. Erard, A. Frigeri, G. Fraser, E. Hauber, J. Helbert, H. Hiesinger, K. Joy, Y. Langevin, M. Marrironi, A. Milillo, I. Mitrofanov, K. Muinonen, J. Naranen, C. Pauselli, p. Potts, J. Warell, P. Wurz, Mercury's surface and composition to be studied by BepiColombo. Planet. Space Sci. **58**, 21–39 (2010)

Y. Saito, J.A. Sauvaud, M. Hirahara, S. Barabash, D. Delcourt, T. Takashima, K. Asamura (BepiColombo MMO/MPPE Team), Scientific objectives and instrumentation of Mercury Plasma Particle Experiment (MPPE) onboard MMO. Planet. Space Sci. **58**, 182–200 (2010)

M.E. Sanborn, J. Wimpenny, C.D. Williams, A. Yamakawa, Y. Amelin, A.J. Irving, Q.Z. Yin, Carbonaceous achondrites Northwest Africa 6704/6693: Milestones for early Solar System chronology and genealogy. Geochim. Cosmochim. Acta **245**, 577–596 (2019)

A.R. Sarafian, E.H. Hauri, F.M. McCubbin, T.J. Lapen, E.L. Berger, S.G. Nielsen, H.R. Marschall, G.A. Gaetani, K. Righter, E. Sarafian, Early accretion of water and volatile elements to the inner Solar System: Evidence from angrites. Philos. Trans. R. Soc. A, Math. Phys. Eng. Sci. **375**(2094), 20160209 (2017)

O.A. Schaeffer, L. Husain, Chronology of lunar basin formation, in *Proceedings of the Fifth Lunar Conference* (1974), pp. 1541–1555

M. Schiller, M. Bizzarro, V.A. Fernandes, Isotopic evolution of the protoplanetary disk and the building blocks of Earth and the Moon. Nature **555**, 507 (2018)

D.L. Schrader, K. Nagashima, A.N. Krot, R.C. Ogliore, Q.Z. Yin, Y. Amelin, C.H. Stirling, A. Kaltenbach, Distribution of Al-26 in the CR chondrite chondrule-forming region of the protoplanetary disk. Geochim. Cosmochim. Acta **201**, 275–302 (2017)

R.A. Schultz, Brittle strength of basaltic rock masses with applications to Venus. J. Geophys. Res. **98**, 10883–10895 (1993)

Z.D. Sharp, F.M. McCubbin, C.K. Shearer, A hydrogen-based oxidation mechanism relevant to planetary formation. Earth Planet. Sci. Lett. **380**, 88–97 (2013)

E.M. Shoemaker, R.J. Hackman, Stratigraphic basis for a lunar time scale, in *The Moon* (Academic Press, New York, 1962), pp. 289–300

J.A. Simpson, J.H. Eraker, J.E. Lamport, P.H. Walpole, Electrons and protons accelerated in Mercury's magnetic field. Science **185**, 160–166 (1974)
M.A. Slade, B.J. Butler, D.O. Muhleman, Mercury radar imaging: evidence for polar ice. Science **258**, 635–640 (1992)
J.A. Slavin, D.N. Baker, D.J. Gershman, G.C. Ho, S.M. Imber, S.M. Krimigis, T. Sundberg, Mercury's dynamic magnetosphere, in *Mercury: The View After MESSENGER*, ed. by S.C. Solomon, L.R. Nittler, B.J. Anderson (Cambridge University Press, Cambridge, 2019)
S.C. Solomon, The relationship between crustal tectonics and internal evolution in the Moon and Mercury. Phys. Earth Planet. Inter. **15**, 135–145 (1977)
S.C. Solomon, R.L. McNutt Jr., R.E. Gold, M.H. Acuña, D.N. Baker, W.V. Boynton, C.R. Chapman, A.F. Cheng, G. Gloeckler, J.W. Head, S.M. Krimigis, W.E. McClintock, S.L. Murchie, S.J. Peale, R.J. Phillips, M.S. Robinson, J.A. Slavin, D.E. Smith, R.G. Strom, J.I. Trombka, M.T. Zuber, The MESSENGER mission to Mercury: scientific objectives and implementation. Planet. Space Sci. **49**, 1445–1465 (2001)
S.C. Solomon, L.R. Nittler, B.J. Anderson, *Mercury: The View After MESSENGER* (Cambridge University Press, Cambridge, 2019)
P.D. Spudis, A Mercurian chronostratigraphic classification, in *Reports of the Planetary Geology and Geophysics Program NASA Technical Memorandum* (1985), pp. 595–597
P.D. Spudis, J.E. Guest, Stratigraphy and geologic history of Mercury, in *Mercury* (University of Arizona Press, Tucson, 1988), pp. 118–164
P. Srinivasan, D.R. Dunlap, C.B. Agee, M. Wadhwa, D. Coleff, K. Ziegler, R. Zeigler, F.M. McCubbin, Silica-rich volcanism in the early solar system dated at 4.565 Ga. Nat. Commun. **9**, 3036 (2018)
R.G. Strom, N.J. Trask, J.E. Guest, Tectonism and volcanism on Mercury. J. Geophys. Res. **80**, 2478–2507 (1975)
H. Tang, N. Dauphas, Fe-60-Ni-60 chronology of core formation in Mars. Earth Planet. Sci. Lett. **390**, 264–274 (2014)
G.J. Taylor, E.R.D. Scott, Mercury, in *Treatise on Geochemistry: Meteorites, Comets and Planets*, ed. by A.M. Davis (Elsevier, Amsterdam, 2004), pp. 477–485
R.J. Thomas, D.A. Rothery, S.J. Conway, M. Anand, Long-lived explosive volcanism on Mercury. Geophys. Res. Lett. **41**, 6084–6092 (2014)
R.J. Thomas, D.A. Rothery, S.J. Conway, M. Anand, Explosive volcanism in complex impact craters on Mercury and the Moon: Influence of tectonic regime on depth of magmatic intrusion. Earth Planet. Sci. Lett. **431**, 164–172 (2015)
M.S. Thompson, R. Christoffersen, T.J. Zega, L.P. Keller, Microchemical and structural evidence for space weathering in soils from asteroid Itokawa. Earth Planets Space **66**, 89 (2014)
M.S. Thompson, T.J. Zega, P. Becerra, J.T. Keane, S. Byrne, The oxidation state of nanophase Fe particles in lunar soil: Implications for space weathering. Meteorit. Planet. Sci. **51**, 1082–1095 (2016)
S.M. Tikoo, B.P. Weiss, D.L. Shuster, C. Suavet, H. Wang, T.L. Grove, A two-billion-year history for the lunar dynamo. Sci. Adv. **3**, e1700207 (2017)
A. Udry, Z.E. Wilbur, R.R. Rahib, F.M. McCubbin, K.E. Vander Kaaden, T.J. McCoy, K. Ziegler, J. Gross, C. DeFelice, L. Combs, B.D. Turrin, Reclassification of four aubrites as enstatite chondrite impact melts: Potential geochemical analogues for Mercury. Meteorit. Planet. Sci. **54**, 785–810 (2019)
K.E. Vander Kaaden, F.M. McCubbin, Exotic crust formation on Mercury: Consequences of a shallow, FeO-poor mantle. J. Geophys. Res., Planets **120**, 195–209 (2015)
K.E. Vander Kaaden, F.M. McCubbin, The origin of boninites on Mercury: An experimental study of the northern volcanic plains lavas. Geochim. Cosmochim. Acta **173**, 246–263 (2016)
K.E. Vander Kaaden, F.M. McCubbin, L.R. Nittler, P.N. Peplowski, S.Z. Weider, E.A. Frank, T.J. McCoy, Geochemistry, mineralogy, and petrology of boninitic and komatiitic rocks on the mercurian surface: Insights into the mercurian mantle. Icarus **285**, 155–168 (2017)
A.K. Vogel, E.S. Jennings, V. Laurenz, D.C. Rubie, D.J. Frost, The dependence of metal-silicate partitioning of moderately volatile elements on oxygen fugacity and Si contents of Fe metal: Implications for their valence states in silicate liquids. Geochim. Cosmochim. Acta **237**, 275–293 (2018)
M. Wadhwa, Y. Amelin, O. Bogdanovski, A. Shukolyukov, G.W. Lugmair, P. Janney, Ancient relative and absolute ages for a basaltic meteorite: Implications for timescales of planetesimal accretion and differentiation. Geochim. Cosmochim. Acta **73**, 5189–5201 (2009)
K.J. Walsh, A. Morbidelli, S.N. Raymond, D.P. O'Brien, A.M. Mandell, A low mass for Mars from Jupiter's early gas-driven migration. Nature **475**, 206–209 (2011)
P.H. Warren, Stable-isotopic anomalies and the accretionary assemblage of the Earth and Mars: A subordinate role for carbonaceous chondrites. Earth Planet. Sci. Lett. **311**, 93–100 (2011)
S.Z. Weider, L.R. Nittler, R.D. Starr, T.J. McCoy, S.C. Solomon, Variations in the abundance of iron on Mercury's surface from MESSENGER X-Ray Spectrometer observations. Icarus **235**, 170–186 (2014)

S.Z. Weider, L.R. Nittler, R.D. Starr, E.J. Crapster-Pregont, P.N. Peplowski, B.W. Denevi, J.W. Head, P.K. Byrne, S.A. Hauck II., D.S. Ebel, S.C. Solomon, Evidence for geochemical terranes on Mercury: Global mapping of major elements with MESSENGER's X-Ray Spectrometer. Earth Planet. Sci. Lett. **416**, 109–120 (2015)

S.Z. Weider, L.R. Nittler, S.L. Murchie, P.N. Peplowski, T.J. McCoy, L. Kerber, C. Klimczak, C.M. Ernst, T.A. Goudge, R.D. Starr, N.R. Izenberg, R.L. Klima, S.C. Solomon, Evidence from MESSENGER for sulfur- and carbon-driven explosive volcanism on Mercury. Geophys. Res. Lett. **43**, 3653–3661 (2016)

B.P. Weiss, S.M. Tikoo, The lunar dynamo. Science **346**, 1198 (2014)

J.L. Whitten, J.W. Head, B.W. Denevi, S.C. Solomon, Intercrater plains on Mercury: Insights into unit definition, characterization, and origin from MESSENGER datasets. Icarus **241**, 97–113 (2014)

M.A. Wieczorek, B.P. Weiss, S.T. Stewart, An impactor origin for lunar magnetic anomalies. Science **335**, 1212–1215 (2012)

M.A. Wieczorek, G.A. Neumann, F. Nimmo, W.S. Kiefer, G.J. Taylor, H.J. Melosh, R.J. Phillips, S.C. Solomon, J.C. Andrews-Hanna, S.W. Asmar, A.S. Konopliv, F.G. Lemoine, D.E. Smith, M.M. Watkins, J.G. Williams, M.T. Zuber, The crust of the Moon as seen by GRAIL. Science **339**, 671–675 (2013)

D.E. Wilhelms, The geologic history of the Moon. USGS Professional Paper, #1348 (1987)

T. Yada, A. Fujimura, M. Abe, T. Nakamura, T. Noguchi, R. Okazaki, K. Nagao, Y. Ishibashi, K. Shirai, M.E. Zolensky, S. Sandford, T. Okada, M. Uesugi, Y. Karouji, M. Ogawa, S. Yakame, M. Ueno, T. Mukai, M. Yoshikawa, J. Kawaguchi, Hayabusa-returned sample curation in the Planetary Material Sample Curation Facility of JAXA. Meteorit. Planet. Sci. **49**, 135–153 (2014)

I. Yoshikawa, O. Korablev, S. Kameda, D. Rees, H. Nozawa, S. Okano, V. Gnedykh, V. Kottsov, k. Yoshioka, G. Murakami, F. Ezawa, G. Cremonese, The Mercury sodium atmospheric spectral imager for the MMO spacecraft of Bepi-Colombo. Planet. Space Sci. **58**, 224–237 (2010)

A.N. Youdin, J. Goodman, Streaming instabilities in protoplanetary disks. Astrophys. J. **620**, 459–469 (2005)

M.Y. Zolotov, A.L. Sprague, S.A. Hauck, L.R. Nittler, S.C. Solomon, S.Z. Weider, The redox state, FeO content, and origin of sulfur-rich magmas on Mercury. J. Geophys. Res., Planets **118**(1), 138–146 (2013)

Space Sci Rev (2020) 216:52
https://doi.org/10.1007/s11214-020-00669-8

What is the Oxygen Isotope Composition of Venus? The Scientific Case for Sample Return from Earth's "Sister" Planet

Richard C. Greenwood[1] **· Mahesh Anand**[1,2]

Received: 15 October 2019 / Accepted: 3 April 2020 / Published online: 11 May 2020

Abstract Venus is Earth's closest planetary neighbour and both bodies are of similar size and mass. As a consequence, Venus is often described as Earth's sister planet. But the two worlds have followed very different evolutionary paths, with Earth having benign surface conditions, whereas Venus has a surface temperature of 464 °C and a surface pressure of 92 bar. These inhospitable surface conditions may partially explain why there has been such a dearth of space missions to Venus in recent years.

The oxygen isotope composition of Venus is currently unknown. However, this single measurement ($\Delta^{17}O$) would have first order implications for our understanding of how large terrestrial planets are built. Recent isotopic studies indicate that the Solar System is bimodal in composition, divided into a carbonaceous chondrite (CC) group and a non-carbonaceous (NC) group. The CC group probably originated in the outer Solar System and the NC group in the inner Solar System. Venus comprises 41% by mass of the inner Solar System compared to 50% for Earth and only 5% for Mars. Models for building large terrestrial planets, such as Earth and Venus, would be significantly improved by a determination of the $\Delta^{17}O$ composition of a returned sample from Venus. This measurement would help constrain the extent of early inner Solar System isotopic homogenisation and help to identify whether the feeding zones of the terrestrial planets were narrow or wide.

Determining the $\Delta^{17}O$ composition of Venus would also have significant implications for our understanding of how the Moon formed. Recent lunar formation models invoke a high energy impact between the proto-Earth and an inner Solar System-derived impactor body, Theia. The close isotopic similarity between the Earth and Moon is explained by these models as being a consequence of high-temperature, post-impact mixing. However, if

Role of Sample Return in Addressing Major Questions in Planetary Sciences
Edited by Mahesh Anand, Sara Russell, Yangting Lin, Meenakshi Wadhwa, Kuljeet Kaur Marhas and Shogo Tachibana

✉ R.C. Greenwood
r.c.greenwood@open.ac.uk

1 Planetary and Space Sciences, School of Physical Sciences, The Open University, Walton Hall, Milton Keynes MK7 6AA, UK

2 Department of Earth Sciences, The Natural History Museum, Cromwell Road, London, SW7 5BD, UK

Earth and Venus proved to be isotopic clones with respect to $\Delta^{17}O$, this would favour the classic, lower energy, giant impact scenario.

We review the surface geology of Venus with the aim of identifying potential terrains that could be targeted by a robotic sample return mission. While the potentially ancient tessera terrains would be of great scientific interest, the need to minimise the influence of venusian weathering favours the sampling of young basaltic plains. In terms of a nominal sample mass, 10 g would be sufficient to undertake a full range of geochemical, isotopic and dating studies. However, it is important that additional material is collected as a legacy sample. As a consequence, a returned sample mass of at least 100 g should be recovered.

Two scenarios for robotic sample return missions from Venus are presented, based on previous mission proposals. The most cost effective approach involves a "Grab and Go" strategy, either using a lander and separate orbiter, or possibly just a stand-alone lander. Sample return could also be achieved as part of a more ambitious, extended mission to study the venusian atmosphere. In both scenarios it is critical to obtain a surface atmospheric sample to define the extent of atmosphere-lithosphere oxygen isotopic disequilibrium. Surface sampling would be carried out by multiple techniques (drill, scoop, "vacuum-cleaner" device) to ensure success. Surface operations would take no longer than one hour.

Analysis of returned samples would provide a firm basis for assessing similarities and differences between the evolution of Venus, Earth, Mars and smaller bodies such as Vesta. The Solar System provides an important case study in how two almost identical bodies, Earth and Venus, could have had such a divergent evolution. Finally, Venus, with its runaway greenhouse atmosphere, may provide data relevant to the understanding of similar less extreme processes on Earth. Venus is Earth's planetary twin and deserves to be better studied and understood. In a wider context, analysis of returned samples from Venus would provide data relevant to the study of exoplanetary systems.

Keywords Venus · Sample return · Oxygen isotopes · Planetary protection · Exoplanets · Mars · Moon · Giant impact · Grab and Go · Bimodal Solar System · Inner Solar System · Tessera terrains

1 Introduction

1.1 Earth's Twin Sister?

Venus (Fig. 1) is often viewed as Earth's sister planet, or perhaps more accurately its ugly sister. This familial comparison stems from their approximately equal mass and physical size (Taylor et al. 2018). But compared to the relatively benign and hospitable surface conditions that prevail on Earth, Venus is the stuff of nightmares. The planet has a surface temperature of 464 °C (Williams 2018a), sufficient to melt lead, tin and zinc! Its broiling atmosphere is composed principally of carbon dioxide (96.5%), with a minor amount of nitrogen (3.5%) and only a trace of water vapour (0.002%). The pressure at the surface is a spacecraft crushing 92 bars. To add to this hellish vision, its deep, thick clouds are composed principally of sulphuric acid, mixed with water (approx. 85% H_2SO_4; 15% H_2O) (Titov et al. 2018).

At least in terms of the prevailing surface conditions, Venus appears to be a dead, uninhabitable world. From a human perspective, it is widely perceived to be an apocalyptic warning about the ultimate consequences of a runaway greenhouse effect (Leconte et al. 2014). Apart from speculative science fiction-style scenarios, no serious space agency plans

Fig. 1 Global view of the surface of Venus constructed from Magellan synthetic aperture radar mosaics and Pioneer Venus Orbiter data. Simulated colour used to enhance small-scale structure. Image produced by the Solar System Visualization project and the Magellan science team JPL. (Image credit: NASA/JPL)

exist to land humans on the surface of Venus (Taylor 2014), in sharp contrast to NASA's long term strategic goal for Mars (National Space Policy 2010).

Venus is Earth's nearest planetary neighbour, being 38.2 million kilometres away at their closest approach, compared with 55.7 million kilometres for Earth and Mars (Williams 2018a, 2018b). This proximity makes space travel to Venus slightly shorter than to Mars, such that the European Space Agency's (ESA) Venus Express spacecraft took 153 days to arrive at its destination, compared to 201 days for Mars Express. Historically, Venus and Mars have received similar levels of attention from international space agencies, with both being targeted by 47 spacecraft (Williams 2020a, 2020b). Taylor et al. (2018) estimate that about 60% of the probes sent to Venus have largely succeeded in their missions; a rate that is similar to that for Mars (Planetary Society 2020). There are eight active missions to the red planet and only one, Japan Aerospace Exploration Agency's (JAXA) Akatsuki orbiter (Nakamura et al. 2016), currently studying Venus. In addition, there are four imminent missions to Mars (ESA's Exo Mars Rover, National Aeronautics and Space Administration's (NASA) Mars 2020 mission, Emirates Mars Mission, China Mars 2020). In contrast, there are currently no firm commitments from any international space agency to build a spacecraft to study Venus (Glaze and Garvin 2018; Glaze et al. 2018). However, this situation may be about to change. In May 2018 ESA selected the Venus orbiter mission EnVision (Ghail et al. 2017) as one of three finalists for its medium-class M5 Cosmic Vision programme. Final selection is due in 2021, with a possible launch in 2032. Two of the four missions selected in February 2019 for further evaluation as part of the NASA Discovery Programme are Venus remote sensing missions: DAVINCI+ (Garvin et al. 2020) and VERITAS (Smrekar et al. 2019).

A renewed interest in Venus exploration is long overdue. The perception of Venus as a "pariah" planet does not stand up to closer scientific scrutiny. Thus, while Venus may currently be an uninhabitable world, the situation may have been very different in the relatively recent past. It has been suggested that Venus was once an Earth-like planet with oceans (Hashimoto et al. 2008) and that evidence of erosion by flowing liquid water remains detectable on surface rocks (Khawja et al. 2020). This interpretation is consistent with the proposal that Venus maintained a habitable climate until about 700 Myr ago (Way et al. 2016). There is also the potential possibility that life may still be present on Venus within the cloud deck, at altitudes of between 51 and 62 km (Dartnell et al. 2015).

There is a growing recognition that developing realistic models for the long-term climatic and geological evolution of Venus will help to improve our understanding of why the Earth has remained habitable for most its existence (e.g. Driscoll and Bercovici 2013; Kodama et al. 2019), despite a progressive increase in solar energy input; an issue generally referred to as the "faint young Sun paradox" (Feulner 2012). The luminosity of the Sun will continue to increase in the future, with stark, albeit very long-term, implications for Earth's habitability (Wolf and Toon 2014). At some stage in its later evolution the Earth may have a climate that resembles present day Venus. The proximity of Earth and Venus and their apparent climatic dichotomy may be just that, "apparent". Venus could simply be at a more advanced stage on an evolutionary route that Earth will one day follow. While this is little more than speculation, it is certainly the case that the study of Venus has implications for our understanding of terrestrial planet evolution in its widest sense (Wilson et al. 2019). A related scientific development has been the recognition that understanding the different evolutionary paths of Earth and Venus has important implications for the search for habitable, Earth-like exoplanets (Angelo et al. 2017; Kane et al. 2018, 2019).

1.2 The Scientific Case for Sample Return from Venus

Space exploration is by its very nature a hugely expensive enterprise. In order to justify the very large sums of public money that are necessarily involved, it is important that the scientific objectives of any space mission are of primary significance. Here we argue that a sample return mission to the surface of Venus, despite the significant costs involved, represents an urgent and first-order scientific challenge. What is at stake is our understanding of how the Solar System formed and the mechanism by which the planets, particularly the inner terrestrial planets, including our own, accreted and evolved. At the heart of the scientific case is the question posed by the title of this article, namely: What is the oxygen isotope composition of Venus?

We are far from being the first to suggest that sample return from Venus would be a high priority scientific objective (Rodgers et al. 2000; Basilevsky et al. 2007; Stevenson and Halliday 2014). In 2014, The Royal Society convened an international discussion meeting entitled Origin of the Moon. In the published overview of the meeting, Stevenson and

Table 1 Mass of inner Solar System bodies

	Mass (kg)	%
Mercury	3.30×10^{23}	2.78
Venus	4.87×10^{24}	40.97
Earth	5.97×10^{24}	50.22
Moon	7.35×10^{22}	0.62
Mars	6.41×10^{23}	5.39
Asteroid belt	2.30×10^{21}	0.02
Total	1.19×10^{25}	100.00

Data Sources: Planetary Fact Sheet (Williams 2019a); Asteroid Fact Sheet (Williams 2019b)

Halliday (2014) stated: "One question that recurred during discussion at the Royal Society conference on 'The Origin of the Moon' was that of what single piece of information the speakers would want to know to move the science forward. A recurring answer was the composition of Venus. It would tell us if the Earth and Moon are unusual in being so similar isotopically."

Table 1 demonstrates why speakers at The Royal Society conference were so keen to obtain samples from Venus. Excluding material from our own planet, we currently have samples from Mars, the Moon and the asteroid belt (McSween 1994; Udry et al. 2019; Greenwood et al. 2020; Ireland et al. 2020). However, these bodies represent little more than 6% of the mass of the inner Solar System. If you include Earth that jumps to just over 56%. But with a returned sample from Venus we would have material from bodies that represent over 97% of the inner Solar System's mass. As discussed later in this article, a venusian sample would almost certainly transform our understanding of how the inner Solar System was built. This could be put another way. If sample return was based solely on how representative a body was of the inner Solar System, would we choose to go first to Mars (5.4% relative mass) or Venus (41% relative mass)? Of course we are going to Mars for reasons other than its rather diminutive size, which is just as well. However, as we attempt to demonstrate in this article, there are also very compelling reasons why we should not ignore Venus (Glaze and Garvin 2018).

It is perhaps surprising that more effort and resources are not currently being expended trying to understand the formation and early evolution of our "twin" planet. As we discuss in later sections, each new discovery about Venus has implications for our Earth. The absence of plate tectonics, or lack of a global magnetosphere are puzzling in a planet which is roughly the same size and composition as Earth (Wilson et al. 2019). Venus has an atmosphere, but it is nothing like ours. A runaway greenhouse effect on Venus is a troubling reminder that our own atmosphere may be heating up, only in our case human activity is at least partially to blame. Are we heading in the same direction as Venus? Research on Venus has a compare and contrast feel about it! Venus may be a nightmare world, but we ignore it at our peril (Goldblatt and Watson 2012).

Retrieving material from the surface of Venus and returning it to Earth is not just technically feasible, but in many cases important components for such a mission have already been tested successfully, either directly in the venusian environment, or under simulated conditions (Rodgers et al. 2000; Rehnmark et al. 2017; Shibata et al. 2017; Glaze et al. 2018; Wilson et al. 2019). We discuss here, not only the overwhelming scientific case for a sample return mission to Venus, but also outline various practical scenarios for achieving this goal on a realistic timescale and at a cost that is no greater than currently funded missions to Mars.

In the first part of this article we look at why essentially a single measurement of the bulk oxygen isotope composition of Venus is of critical importance in understanding the formation and early evolution of the Solar System. We then review some key features of Venus, including its surficial geology, with the aim of defining areas which might potentially be favourable for sample recovery. Next, we consider possible configurations for a sample return mission. Finally, we review the range of analytical techniques that could be applied to the characterisation and analysis of these precious extraterrestrial materials once they have been brought back to Earth.

2 Why is an Oxygen Isotope Analysis of Venus so Important?

2.1 What's so Great About Oxygen?

Spectroscopic measurements of the solar photosphere show that oxygen is the third most abundant element in the Solar System, after hydrogen and helium (Lodders 2003). Of equal importance, from the perspective of planetary formation processes, is that oxygen is a major mineral-forming element and makes up nearly 46 wt.% of the Bulk Silicate Earth (Javoy et al. 2010). All the current indications are that Venus has essentially the same bulk composition as the Earth (Treiman 2009). In addition to its Solar System and Bulk-Earth abundances, a further critical feature of oxygen is the fact that it readily combines with hydrogen to form water. Water played an essential role in the evolution of early life (Nisbet and Sleep 2001), is widespread throughout the Solar System (Raymond and Izidoro 2017), and experiments indicate that it may have promoted the fast accretion of early planetary bodies by helping dust grains stick together (Gundlach and Blum 2015). It has been shown that a significant proportion of the water in the Solar System was inherited from its parent molecular cloud (Cleeves et al. 2014).

2.2 Oxygen Isotopes – Notation

Oxygen has three stable isotopes: ^{16}O, ^{17}O, ^{18}O, with average terrestrial abundances of 99.757%, 0.038% and 0.205% respectively (Rosman and Taylor 1998). For analytical reasons, absolute abundances of the three isotopes are not generally measured and instead compositions are determined as isotope ratios ($^{17}O/^{16}O$ and $^{18}O/^{16}O$) (Ireland et al. 2020). Oxygen isotopic analyses are generally reported in standard δ notation, where $\delta^{18}O$ is calculated relative to the international standard Vienna Standard Mean Ocean Water (VSMOW) as $\delta^{18}O = [(^{18}O/^{16}O_{\text{sample}})/(^{18}O/^{16}O_{\text{VSMOW}}) - 1] \times 1000$ (‰) and similarly for $\delta^{17}O$ using the $^{17}O/^{16}O$ ratio.

Oxygen is readily fractionated by a variety of chemical and physical processes, with the magnitude of this variation being a function of the masses of the isotopes and as a consequence is termed mass-dependent fractionation. Thus, for any particular process the $^{18}O/^{16}O$ ratio will vary by approximately twice as much as the $^{17}O/^{16}O$ ratio. When oxygen isotope analyses for terrestrial samples are plotted on a three-isotope diagram, with $\delta^{18}O$ plotted as the abscissa (Fig. 2), the resultant line of slope ~0.52 (Matsuhisa et al. 1978) is commonly referred to as the terrestrial fractionation line (TFL). Deviations from this reference line are conventionally expressed as: $\Delta^{17}O = \delta^{17}O - 0.52\delta^{18}O$ (e.g. Clayton and Mayeda 1988). $\Delta^{17}O$ can be expressed more accurately using the linearized format of Miller (2002):

$$\Delta^{17}O = 1000 \ln\left(1 + \delta^{17}O/1000\right) - \lambda 1000 \ln\left(1 + \delta^{18}O/1000\right)$$

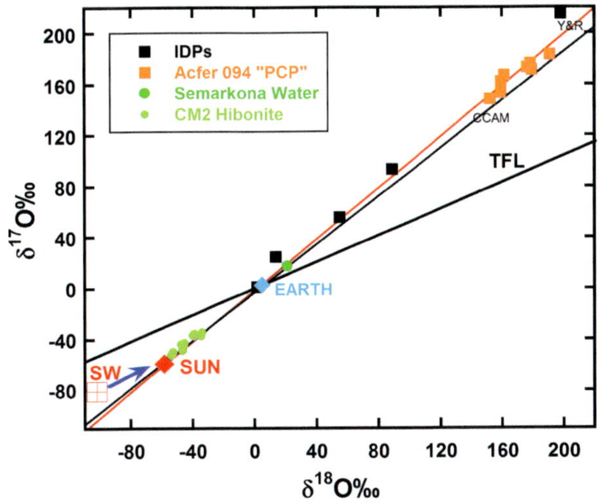

Fig. 2 Oxygen three-isotope plot showing the principal Solar System reservoirs. The diagram is essentially a mixing line between a ^{16}O-rich endmember (the Sun) and a more poorly constrained ^{16}O-poor endmember, as detected in the matrix of Acfer 094 (Sakamoto et al. 2007) and some IDPs (Starkey et al. 2014) and possibly reflecting the composition of primordial water (Sakamoto et al. 2007). The composition of the Sun is derived from Genesis solar wind (SW) measurements projected onto the CCAM line (McKeegan et al. 2011). Refractory solids (e.g. hibonite, corundum) have ^{16}O-rich compositions (Liu et al. 2009; Bodénan et al. 2014), consistent with formation from a gas of solar composition (Krot et al. 2009). Also shown is the inferred water composition from Semarkona (Choi et al. 1998). Mixing lines: CCAM (Clayton et al. 1977), Y&R (Young and Russell 1998)

Values for λ used in the literature for meteorites and related extraterrestrial samples vary between approximately 0.5247 and 0.5305 (Pack and Herwartz 2014; Greenwood et al. 2017; Miller et al. 2020).

2.3 High-Precision Oxygen Isotope Measurements – Sample Return or In Situ Analysis?

There are currently two principal laboratory techniques for oxygen isotope analysis of rocks and minerals: (i) infrared laser-assisted fluorination and (ii) secondary ion mass spectrometry (SIMS) (Greenwood et al. 2017; Ireland et al. 2020). Laser fluorination is a bulk technique and so unlike SIMS is not capable of spot analysis, however, it has the important advantage of achieving high levels of precision. Thus, based on repeat analysis of their internal obsidian standard ($n = 39$), Greenwood et al. (2018b) obtained the following precision: $\pm 0.05‰$ for δ^{17}O; $\pm 0.09‰$ for δ^{18}O, and $\pm 0.02‰$ for Δ^{17}O (2σ). At the present time and for the foreseeable future, this level of precision is unobtainable by instrumentation that can realistically be operated on a spacecraft. Amongst the highest levels of precision so far achieved for oxygen isotope analysis by in situ spacecraft measurements are those made by the Curiosity rover at Gale Crater using a tunable laser spectrometer (Webster et al. 2013). These measurements were made on atmospheric CO_2 and had the following precision: $\pm 5‰$ (2SEM) for δ^{17}O and $\pm 5‰$ (2SEM) for δ^{18}O. Other oxygen isotope measurements made by spacecraft-based instruments generally have lower precision than those achieved by Curiosity (Alday et al. 2019). While these measurements represent significant scientific achievements, their precision is much lower than can be achieved by laboratory-based laser fluorination. As will be discussed later in this paper, there is a realistic possibility that Venus will have a lithospheric $\Delta^{17}O$ composition that is extremely close to that of Earth's. An in situ analysis by a spacecraft-based instrument would not be capable of resolving such differences and instead sample return is the only viable option.

2.4 The Origin of Mass-Independent Oxygen Isotope Variation in Solar System Materials

The discovery by Clayton and co-workers (Clayton et al. 1973) that oxygen in carbonaceous chondrite meteorites, and its constituent high-temperature components, displays mass-independent variation was a major breakthrough in planetary sciences and provided cosmochemists with a powerful new tool to investigate early Solar System processes (Greenwood et al. 2017). Clayton et al. (1973) showed that rather than plotting on a mass fractionation line with a slope close to 0.5, carbonaceous chondrites and constituent inclusions, plot along a line with a slope close to 1 (Fig. 2). As such variation does not show any mass dependency, it is generally referred to as mass-independent fractionation. Clayton et al. (1973) initially suggested that slope 1 variation in meteorites resulted from an early injection into the Solar System of a hypothetical component composed almost exclusively of ^{16}O. Experimental work undertaken by Thiemens and Heidenreich (1983) demonstrated a possible alternative mechanism when it was found that ozone formation was associated with a slope 1 oxygen isotope anomaly.

Failure to identify the putative ^{16}O-rich component led Clayton (2002) to suggest that early Solar System mass-independent variation might have been caused by photodissociation of CO, followed by self-shielding. Clayton (2002) further proposed that the site of this process was the x-wind spot close to the young Sun (Shu et al. 2001). Other sites that have been proposed for the operation of this process are, the edge of the solar nebula (Lyons and Young 2005), or the precursor molecular cloud from which the Solar System formed (Yurimoto and Kuramoto 2004; Yurimoto et al. 2007).

In the years since the breakthrough discovery of Clayton et al. (1973), extensive analysis of meteorites and their components has shown that they generally have distinct compositions in oxygen three-isotope space (Clayton 2003, 2006; Franchi 2008; Greenwood et al. 2017) (Fig. 3). Planets are no exception and the two such bodies from which we have samples, Mars and Earth, have distinct oxygen isotope ($\Delta^{17}O$) compositions (+0.3‰ for Mars and 0‰ for Earth). (Fig. 3). The Moon has essentially the same $\Delta^{17}O$ composition as the Earth for reasons which most likely reflect its mode of formation in a giant impact event (Wiechert et al. 2001; Spicuzza et al. 2007; Hallis et al. 2010; Young et al. 2016; Greenwood et al. 2018b). However, an alternative view for the Earth-Moon isotopic similarity is that it reflects the near-identical planetary feeding zones from which the proto-Earth and the giant impactor accreted (Dauphas et al. 2014; Dauphas 2017). As will be discussed in further detail in the following sections, what is required to solve these controversies is a sample from Venus (Stevenson and Halliday 2014).

2.5 The Bimodal Solar System

2.5.1 How are Planets Built?

At the largest scale, the Solar System has a distinct bimodal structure, comprising an inner zone of rocky terrestrial planets and an outer region of giant planets (Morbidelli et al. 2015). The processes which led to this dichotomy remain poorly understood (Chambers 2016; Walsh and Levison 2019; Weidenschilling 2019). One possibility is that the giant planets

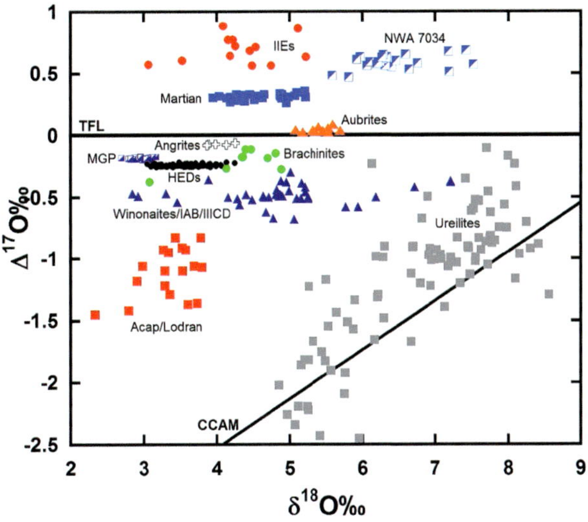

Fig. 3 Oxygen isotope diagram of planetary and achondritic meteorites (Modified after Greenwood et al. 2017)

grew large because of the presence of additional components, such as ices and organics that were essentially not available in the inner Solar System (Chambers 2016; Morbidelli 2018; Weidenschilling 2019). But such a simple explanation does not take into account the rapid timescales involved. Protoplanetary discs typically exist for relatively short time periods of only about 6 Myr or less (Haisch et al. 2001) and without other factors being involved, accretion in the outer Solar System would have been too slow to explain the existence of giant planets (Thommes et al. 2003).

Planetary growth models start with dust grains in a gaseous disc. Interactions with the gas give the grains varying velocities that promote frequent collisions. Initially at least, grains stick together and form larger and larger aggregates (Blum 2018; Weidenschilling 2019). This process has been termed collisional coagulation (Weidenschilling 2019). However, there appears to be a size limit to this process, generally referred to as the "metre-size barrier" (Morbidelli 2018). Essentially, this is a reflection of the fact that above a certain grain-size, mutual collisions are destructive, rather than constructive. The actual limit varies according to the radial distance from the proto-Sun, being approximately mm-sized in the inner disc and cm-sized or greater in the outer disc (Morbidelli 2018). Two processes have been proposed to overcome this difficulty: i) turbulent concentration and ii) streaming instabilities (Weidenschilling 2019). The first of these mechanisms involves the formation of dense clumps of chondrule-sized particles in regions of lower vorticity within a turbulent velocity field (Cuzzi et al. 2010; Weidenschilling 2019). Such dense clumps become susceptible to gravitational collapse, leading to the formation of bodies that significantly exceed the metre-size limit (Cuzzi et al. 2010). Where a disc is non-turbulent, clumping can still take place via streaming instabilities (Youdin and Goodman 2005; Morbidelli 2018; Weidenschilling 2019). This involves zones with higher amounts of solid interacting with gas, such that the inwards drift of these denser clumps is reduced compared to isolated particles. As a consequence, isolated particles which lie outwards from such a denser zone will be spiralling in at a faster rate and so tend to merge with the denser clump. This produces a positive feedback with the clump gradually increasing its mass. This will eventually result in its contraction due to self-gravity and result in the formation of a planetesimal (Morbidelli 2018).

Following the emergence of a planetesimal population, planet formation is classically considered to proceed via a series of discrete stages. This commences with "runaway growth" in which the largest planetesimal in a local region grows at a faster rate than its smaller neighbours. Once this body attains about half the local mass it begins to perturb the orbits of its smaller neighbours. As a result, relative velocities increase, leading to a process termed "oligarchic growth" (Walsh and Levison 2019). At the end of this phase a bimodal population is formed consisting of relatively few, large planetary embryos and a much larger number of remaining planetesimals. The final step involves planet-building via energetic impacts between embryos and is termed the "giant impact" or "chaotic growth" stage (Walsh and Levison 2019).

The classic stages of planetary growth outlined above may not occur rapidly enough to account for the formation of the outer giant planets (Morbidelli 2018). Instead, once planetesimals formed, further growth may have involved a process known as pebble accretion (Lambrechts and Johansen 2012; Johansen and Lambrechts 2017). This model is based on the premise that newly-formed planetesimals were embedded within a disc of gas containing small solid particles, or "pebbles". These particles would be strongly coupled to the gas and hence undergo drift with respect to the planetesimals. As pebbles collide and are incorporated into the growing planetesimal the accretion zone around the body is continuously replenished by new solids and hence accretion rates are significantly greater than predicted by classical growth models. The distinction between the inner and outer Solar System may have taken hold at this stage (Morbidelli et al. 2015). Pebbles inward of the snow line may have been less massive than those outside it. The result is that sub-Moon-sized planetesimals in the inner Solar System only attained Mars-sized dimensions, whereas in the outer Solar System they reached 20 Earth masses (Morbidelli et al. 2015). Initial planetesimal size may also have been an important factor in controlling the size of bodies that eventually formed in the inner disc, with inside-out accretion being favoured by larger-sized primordial planetesimals (Walsh and Levison 2019). The growth of Jupiter may have taken place by a complex sequence, initially involving a rapid phase of pebble-accretion, then a slower stage of planetesimal accretion, with a final stage of runaway gas accretion (Alibert et al. 2018).

2.5.2 Isotopic Evidence for a Bimodal Solar System

The Solar System is not only bimodal in terms of its overall structure but also shows a clear dichotomy at the level of its bulk and isotopic composition. Based on data from a number of earlier studies (e.g. Trinquier et al. 2007; Qin et al. 2010), Warren (2011) showed that Solar System materials have a bimodal composition with respect to a number of isotopic systems (Δ^{17}O, ε^{54}Cr, ε^{50}Ti, ε^{62}Ni). Thus, on a plot of ε^{54}Cr vs. Δ^{17}O two groups are present (Fig. 4). One comprises the carbonaceous chondrites and a minor subset of achondrites. The other group is defined by all other Solar System materials, including planetary-derived samples (Mars, Earth, and Moon), ordinary and enstatite chondrites and a wide range of achondrites (main-group pallasites, HEDs, ureilites, aubrites). Warren (2011) suggested that the carbonaceous chondrite group (CC) may comprise material formed in the outer Solar System and the non-carbonaceous chondrite group (NC) in the inner Solar System. The gap between the NC and CC groups has been referred to as "The Warren Gap" (e.g. Voosen 2018). It has now been demonstrated that the isotopes of a wide group of elements display the dichotomy seen in Fig. 4 (Warren 2011; Budde et al. 2016; Van Kooten et al. 2016; Kruijer et al. 2017, 2020; Burkhardt et al. 2019). Iron meteorites also display a similar bimodal distribution of nucleosynthetic anomalies (e.g. Budde et al. 2016).

With the notable exception of oxygen, isotopic mass-independent variation in Solar System materials, which is not caused by either spallation or radioactive decay, is the result

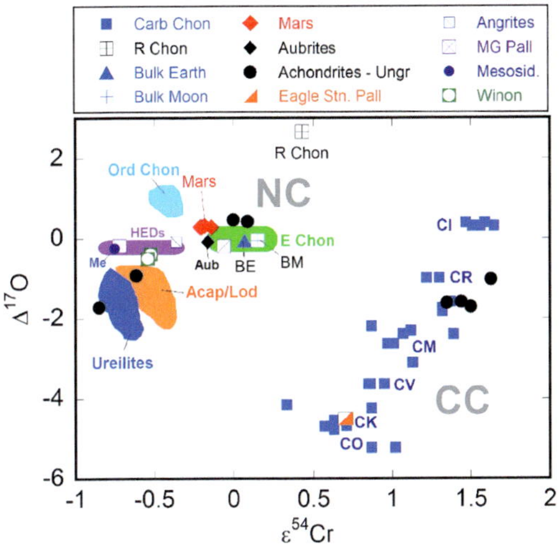

Fig. 4 Plot of ε^{54}Cr vs. Δ^{17}O for a range of planetary materials. Carbonaceous chondrite (CC) and non–carbonaceous (NC) groups show a clear separation. The NC group includes the ordinary chondrites (Ord Chon), enstatite chondrites (E Chon), R chondrites (R Chon), Bulk Earth (BE), Bulk Moon (BM), martian meteorites (Mars), ureilites, mesosiderites (ME), acapulcoite – lodranite suite (Acap/Lod), aubrites (Aub), howarite-eucrite-diogenite suite (HEDs). The CC group includes all the carbonaceous chondrites as well as the Eagle Station pallasites and various ungrouped achondrites (Modified after Warren 2011; Greenwood et al. 2017, 2020)

of nucleosynthetic processes that took place in the precursor stars from which the Solar System formed (Dauphas and Schauble 2016; Scott et al. 2018 Burkhardt et al. 2019; Nanne et al. 2019). The isotopic variation observed in Solar System materials likely reflects the heterogeneous distribution of the presolar carrier grains (Kruijer et al. 2020) and suggests that the mixing and homogenisation processes that took place in the precursor molecular cloud and later in the protoplanetary disc, were not sufficient to erase these anomalies (Burkhardt et al. 2019; Nanne et al. 2019; Kruijer et al. 2020).

In contrast, oxygen mass-independent isotopic variation likely had a non-nucleosynthetic origin (Clayton 2002). The most widely accepted formation model involves selective UV photo-dissociation of CO, either in the presolar giant molecular cloud, or in the solar nebula itself (Clayton 2002; Yurimoto and Kuramoto 2004; Lyons and Young 2005). The oxygen isotope anomalies produced by this process would have become incorporated into different phases, including water ice, gas and dust (Yurimoto and Kuramoto 2004). Preservation of such isotopic differences may again be a reflection of incomplete homogenization in the protosolar nebula (Ireland et al. 2020).

Evidence from the Stardust sample return mission (Brownlee 2014) suggests that large-scale mixing of gas and dust was taking place during the earliest stages of Solar System evolution. In view of this, an important and not yet fully resolved question is why the NC-CC dichotomy was not erased at an early stage due to vigorous mixing. In terms of how the anomaly was initially established, recent studies have suggested that it may reflect a change in the composition of the fresh infalling material that was feeding the proto-solar nebula (Burkhardt et al. 2019; Nanne et al. 2019; Kruijer et al. 2020). Calcium Aluminium-rich inclusions (CAIs) are generally considered to have formed at an early stage in nebular evolution and at a location that was close to the proto-Sun (Shu et al. 2001; Ireland et al. 2020). CAIs thus preserve a record of the isotopic composition of the earliest infall to the proto-solar disc. The evidence from CAIs is that this early material was enriched in nuclides produced in neutron-rich stellar environments. This infall material, including CAIs, would have been transported outwards through viscous spreading (Kruijer et al. 2020). Following this early stage of disc accretion, it is suggested that there was a change in the composition of the infalling solids, which became depleted in neutron-rich nuclides. These later solids would have tended to accumulate in the inner part of the disk and have diluted the isotopic signature originally present there (Kruijer et al. 2020). Thermal processing of solids in the hotter inner part of the disc (Scott et al. 2018), may not have been a major contributory process in the establishment of the NC-CC dichotomy (Nanne et al. 2019). One explanation that has been advanced to explain the preservation of the NC-CC dichotomy is that it can be related to the early, rapid accretion of Jupiter, which during the later stages of its growth would have formed a barrier separating the inner and outer Solar System regions (Kruijer et al. 2017, 2020).

The role of Jupiter in preserving the NC-CC dichotomy has been disputed by Brasser and Mojzsis (2020). They suggest that Jupiter's accretion took place at too slow a rate to have been a significant barrier to lateral mixing within the disc. Brasser and Mojzsis (2020) instead invoke a pressure maximum in the disc close to the present location of Jupiter and suggest that it was this feature, rather than the planet itself, which prevented mixing between the inner and outer disc. In a companion paper (Burbine and Greenwood 2020), a sample return mission to Ceres is proposed as a means of investigating in detail the origin and early evolution of the CC group.

Bimodality is also evident in the bulk composition of extraterrestrial samples. There is a well-established dichotomy between carbonaceous chondrites on the one hand and ordinary and enstatite chondrites on the other (Weisberg et al. 2006; Krot et al. 2014). Carbonaceous chondrites generally contain abundant CAIs, predominantly plot below the terrestrial fractionation line (TFL) on an oxygen three-isotope diagram and show evidence of having accreted water ice into their parent bodies (e.g. Grimm and McSween 1989; Trigo-Rodríguez et al. 2019). In contrast, ordinary and enstatite chondrites generally show anhydrous characteristics, plot on or above the TFL and have a relatively low CAI content (Weisberg et al. 2006; Krot et al. 2014).

2.6 Is the Inner Solar System Heterogeneous or Homogeneous? Sorting Out the Giant Impact Once and for All?

The question as to whether the inner Solar System is isotopically heterogeneous or homogeneous would appear at first glance to be fairly straightforward. Based on the evidence of Fig. 4 and similar plots of ε^{50}Ti vs. ε^{54}Cr (Warren 2011; Scott et al. 2018) it would seem that the inner Solar System (NC) group, while distinct from the carbonaceous chondrite (CC) group, still displays a significant level of heterogeneity. If such lithologies are representative of the material that formed in the inner Solar System then mixing and homogenisation was far from complete. However, there are problems with this analysis. Apart from the Earth, Moon and Mars, the other non-carbonaceous chondrite lithologies plotted in Fig. 4 are all asteroidal in origin and hence derived from very small bodies. In fact, all of the mass in the asteroid belt is estimated to represent just 0.04% that of the Earth (Pitjeva and Pitjev 2018) (see also Table 1) and so can hardly be considered to be representative. Although much more massive than the asteroid belt, Mars is anomalously small compared to either Earth or Venus (Table 1). One of the drivers for the formulation of the Grand Tack model was to explain the relatively small size of Mars (Walsh et al. 2011, 2012). As a result of the inward and then outward migration of Jupiter, the Grand Tack model predicts that the feeding zone for Mars was significantly impoverished, hence its small size. This suggests that Mars is also not fully representative of the inner Solar System, which leaves just Earth and the Moon in terms of relatively large bodies from the inner zone of the Solar System from which we already have samples.

While both the Earth and Moon are almost identical in terms of their $\Delta^{17}O$ compositions, as discussed above, this may be the result of very high energy mixing in the wake of a giant impact (Wiechert et al. 2001; Spicuzza et al. 2007; Hallis et al. 2010; Young et al. 2016; Greenwood et al. 2018b). However, an alternative hypothesis, which helps to maintain the basic tenants of the canonical giant impact model (low-energy, oblique impact) (Canup and Asphaug 2001), postulates an impactor, which was essentially isotopically identical to the proto-Earth (Dauphas et al. 2014; Dauphas 2017). Evidence from multiple isotopic systems suggests that during a significant portion of its accretion Earth was fed by compositionally homogeneous enstatite chondrite-like material (Dauphas 2017). This implies that the inner Solar System may have been well-mixed prior to the Giant Impact. To test such a hypothesis we need a sample from another large inner Solar System body, either Venus or Mercury. While it would be optimal to have samples from both planets and this should be a long term strategic, exploration goal, a choice would need to be made in the short term. Mercury appears to have an anomalously high iron content that may reflect a complex history of multiple giant impacts with loss of crustal and mantle material (Chau et al. 2018). It is also much smaller than Venus (Table 1). Hence, Venus must be regarded as a more representative inner Solar System sample return target.

Aubrites are the achondrites with the closest oxygen isotope composition to Earth with respect to $\Delta^{17}O$ (Greenwood et al. 2018b). Using aubrites as a proxy for the giant impactor, Greenwood et al. (2018b) were able to demonstrate that, at high precision, a clear oxygen isotopic difference between the Earth and Moon should be observed if the impact conditions were those defined by the canonical model (Canup and Asphaug 2001). The magnitude of this difference would far exceed the maximum 4 ppm isotopic difference actually measured in lunar and terrestrial samples by Greenwood et al. (2018b). It was concluded by Greenwood et al. (2018b) that the giant impact was significantly more energetic than the canonical model suggests and most likely resembled the situation envisaged by the "Synestia" simulations of Lock et al. (2016) and Lock and Stewart (2017). However, analysis of another large inner Solar System body is needed to test this further. As pointed out by Stevenson and Halliday (2014), we don't know whether "the Earth and Moon are unusual in being so similar isotopically". It is also possible that the inner Solar System experienced a very advanced level of homogenisation and that the giant impactor which formed our Moon was an example of such a body. If the impactor was more Earth-like than aubrites in terms of its isotopic composition then the canonical giant impact model may represent a more realistic scenario than is currently believed. A sample from Venus would undoubtedly help to resolve this controversy.

2.7 What was the $\Delta^{17}O$ Structure of the Inner Solar System?

With the notable exception of meteorites from Mars, the Moon and Vesta, it is generally very difficult to link asteroids and meteorites (Greenwood et al. 2020). A further complication is the prediction of the Grand Tack model that the material present in the asteroid belt is not indigenous to this region, but instead may have formed at varying locations throughout the early Solar System (Walsh et al. 2011, 2012). However, as pointed out by Goodrich and O'Brien (2014) in the case of the ureilite parent body, knowing the heliocentric distance at which an asteroid accreted is a critical piece of information in trying to understand early

Solar System evolution. A concept that is frequently discussed in meteorite studies is the proposition that the $\Delta^{17}O$ of parent bodies varies sequentially with heliocentric distance (e.g. Rubin and Wasson 1995; Kallemeyn et al. 1996). In its broadest sense this concept suggests that parent bodies with $\Delta^{17}O$ values close to or above the TFL e.g. Mars, Earth, ordinary and enstatite chondrites, R chondrites, may have formed closer to the Sun than those with negative $\Delta^{17}O$ values, such as the carbonaceous chondrites. While in an overall sense this concept may have value and appears to be a forerunner of the bimodal view of the Solar System discussed in Sect. 2.5, in detail the oxygen isotopic evolution of the solar nebula was complex (Yurimoto et al. 2007; Lyons et al. 2009) and a simple and systematic variation in $\Delta^{17}O$ with heliocentric distance is likely to be an over simplification. However, without a sample from another large inner Solar System body we have no way of testing whether there is any relationship between $\Delta^{17}O$ and heliocentric distance.

The later stages of terrestrial planet formation are thought to involve the accretion of several hundred Moon-sized bodies (Kaib and Cowan 2015). The final $\Delta^{17}O$ value of a planet is therefore likely to be the weighted average of the $\Delta^{17}O$ values of these contributing bodies (Young et al. 2016). In theory then, the values for the Earth and Venus could be significantly different, just as they are for Earth and Mars (Fig. 3). On the other hand, if the inner Solar System underwent an early stage of homogenisation and Mars is an outlier, as predicted by the Grand Tack model (Walsh et al. 2011, 2012), then Earth and Venus could have nearly identical $\Delta^{17}O$ values. It is impossible to discriminate between these two possibilities without a sample from Venus. In conclusion, the $\Delta^{17}O$ value for Venus would likely have profound implications both for the formation of our own Solar System and by extrapolation for other exoplanetary systems (e.g. Torres and Fressin 2019).

3 Venus: A Brief Geological Overview

In this section we provide a brief overview of venusian geology, focussing in particular on the structure and composition of its surface. The aim of this section is to suggest potential terrains which could be sampled by a robotic mission to Venus.

3.1 Planetary Parameters

Venus is Earth's closest planetary neighbour and the two bodies show a number of striking similarities in terms of their overall size and structure. Venus is slightly smaller than Earth having a relative mass, volume and mean radius compared to our planet of 81.5%, 85.7% and 95%, respectively (Williams 2018a). The uncompressed density of Venus is nearly identical to that of Earth and hence both planets are considered to have similar internal structures and compositions (Taylor et al. 2018). Venus is situated 0.72 AU from the Sun and in theory resides within the Solar System's habitable zone (Byrne et al. 2019). Based on the results of modelling studies, Venus is thought to have a large metallic core, broadly similar to that of Earth (Zhang and Zhang 1995). However, unlike Earth, Venus does not have a significant global magnetic field (Jacobson et al. 2017). The reason for this is poorly understood. Venus has a 243 day retrograde rotation, the slowest for any planet. However, in theory this would still be sufficient to generate an Earth-like magnetic field (Stevenson 2003; O'Rourke et al. 2018). One possible reason for the lack of a core dynamo is that "normal" planetary accretion processes result in a compositionally stratified core which is gravitationally stable and hence does not undergo convection, which may be the case for Venus (Jacobson et al. 2017). In contrast, Earth experienced a late-stage, Moon-forming, giant impact that may have homogenized its core, resulting in the onset of thermal convection (Jacobson et al. 2017). The possible lack of a late stage giant impact is only one of a range of scenarios that might explain the absence of a core dynamo on Venus (Smrekar et al. 2018).It is also possible that the core of Venus may already have totally crystallised or that the magnetic dynamo may be episodic in nature (Smrekar et al. 2018).

Although the present day venusian atmosphere is essentially dry, the primordial water content of the planet was probably similar to that of Earth (Taylor et al. 2018). The D/H ratio of water in the venusian atmosphere is between 95 and 120 times greater than that on Earth (Krasnopolsky et al. 2013), consistent with models that Venus started with a water inventory similar to Earth's and that this was then lost over time to space (Donahue et al. 1982; Greenwood et al. 2018a). Hydrogen and oxygen can rapidly escape from a planetary atmosphere following photolysis of water under conditions of strong X-ray and ultraviolet radiation (Guo 2019). Kasting and Pollack (1983) indicate a timescale of about a billion years for Venus to have lost one terrestrial ocean equivalent of water by hydrodynamic escape, whereas Guo (2019) suggests an early and much more rapid process of atmospheric desiccation.

Tectonic processes on Venus are considered to be very different to those operating on Earth (Turcotte 1993; Ghail 2015). A random distribution of craters across the surface of the planet is taken as evidence that the crust of Venus acts like a single rigid plate that is gradually thickening due to conductive cooling (Turcotte 1993). This tectonic regime is generally referred to as a "stagnant lid" (Ghail 2015) and may be a consequence of the high viscosity of the venusian mantle, due to low concentrations of water (Nimmo and McKenzie 1998). In contrast to Earth, Venus appears to experience global resurfacing events, the last being 300 to 600 Myr ago (Nimmo and McKenzie 1998). The heat flux on Venus is currently about half that expected from radiogenic heat production, suggesting that Venus is currently heating up internally (Nimmo and McKenzie 1998).

3.2 Surface Geology

The following account of the surface geology of Venus is derived mainly from Basilevsky and McGill (2007); Ivanov and Head (2011) and Taylor et al. (2018). The venusian surface is dominated by volcanic plains ($\sim$80% of total area), which show varying degrees of deformation. The plains have been divided into a number of distinct terrains, which include: (i) densely fractured plains, (ii) ridge belts, (iii) regional plains, (iv) shield plains, (v) lobate plains and flows, (vi) smooth plains (Basilevsky and McGill 2007; Ivanov and Head 2011).

Densely fractured plains occupy 3 to 5% of the planetary surface and form small isolated areas up to 200 km in diameter, elevated a few hundred metres above the regional plains. As their name suggests, densely fractured terrains are composed of intensely fractured material. Ridge belts are slightly elevated above the regional plains and comprise material that has been deformed into elongate ridges that tend to run subparallel to the belt margins (Fig. 6). Ridge belts comprise 3 to 5% of the venusian surface. Regional plains tend to have smooth surfaces with flow-like structures suggesting that they may represent basaltic lavas of low viscosity (Basilevsky and McGill 2007) (Fig. 7). A characteristic feature of the regional plains is the development of narrow, but hundreds of kilometre long, "wrinkle ridges" (Fig. 7). Regional plains occupy 50 to 60% of the planetary surface. Shield plains are areas that consist of coalescing larger volcanic shield volcanoes, each a few hundred kilometres across, which in turn are covered by smaller, 5 to 15 km diameter shield volcanoes. Shield plains comprise 10 to 15% of the surface of Venus. Lobate plains and flows generally

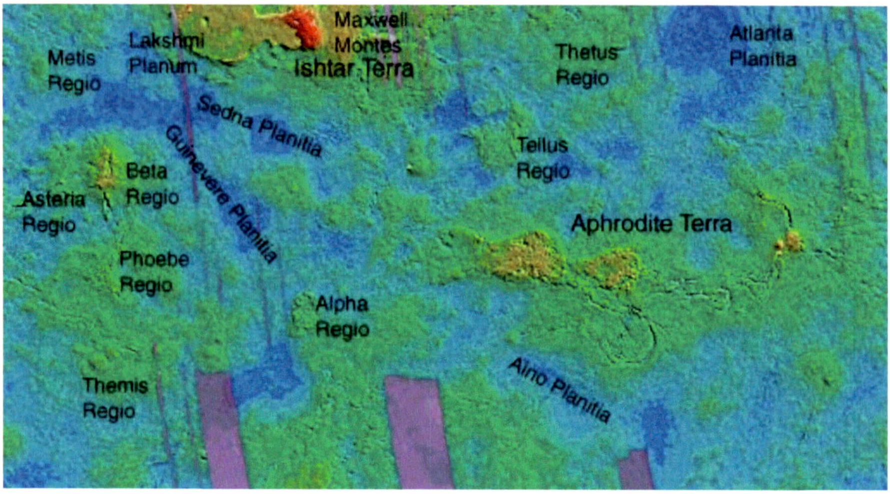

Fig. 5 Altimeter topographic map of Venus constructed using data from the Magellan orbiter. Map is displayed in Mercator projection. Unmapped areas shown in grey. Red areas highest and blue lowest. Synthetic shadowing added to enhance visibility of features. Credit NASS/JSC

have a distinct cuspate morphology and are sometimes associated with volcanic structures. Lobate plains and flows comprise about 10% of the planet's surface and show variable radar brightness, suggesting different surface textures. Smooth plains appear dark in radar images and occupy only a few percent of the venusian surface.

Distinct from the plains are elevated areas, known as tessera terrains which occupy about 8% of the surface of Venus (Basilevsky and McGill 2007; Ivanov and Head 2011). The two largest "highland" areas dominated by tessera terrains are Ishtar Terra in the north and the somewhat larger, more southerly, Aphrodite Terra (Fig. 5). Tessera terrains are the most highly deformed regions on Venus and are characterised by numerous intersecting ridges and grooves, which are typically a few kilometres wide and tens of kilometres long (Basilevsky and McGill 2007; Whitten and Campbell 2019). Ridge features are considered to have formed by compressional tectonics and the grooves by extension. Tessera terrains appear to represent the oldest units on Venus and tend to appear bright on Magellan radar images (Fig. 8). The composition of the material forming the tessera terrains is unknown, with alternative suggestions that they may be either basaltic or feldspathic (anorthosite/granite) in nature (Basilevsky and McGill 2007; Gilmore et al. 2017). Tessera terrains are often associated with mountain belts, as is clearly seen in the case of Ishtar Terra, which contains the highest mountain on Venus, Maxwell Montes (Fig. 5). The mountain belts are considered to have been formed by compressional tectonics and may be of similar age to the tessera terrains (Basilevsky and McGill 2007). The absolute age of either the mountain belts or tessera terrains is unknown.

Some of the landforms found on Venus are generally considered to be unique to the planet. These include flat-topped volcanic features known as "pancake-domes" or "lava domes" or "farra" (Fig. 9). These can be up to 70 km diameter and rise up to 1 km above the surrounding plains. There are also star-like features termed "novae" (Fig. 10) and structures which comprise both radial and concentric fractures, termed "arachnoids" based on their resemblance to spider's webs (Fig. 11). Circular rings sometimes surrounded by a depression are termed "coronae" (Fig. 12). All of these landforms are likely to be volcanic in origin.

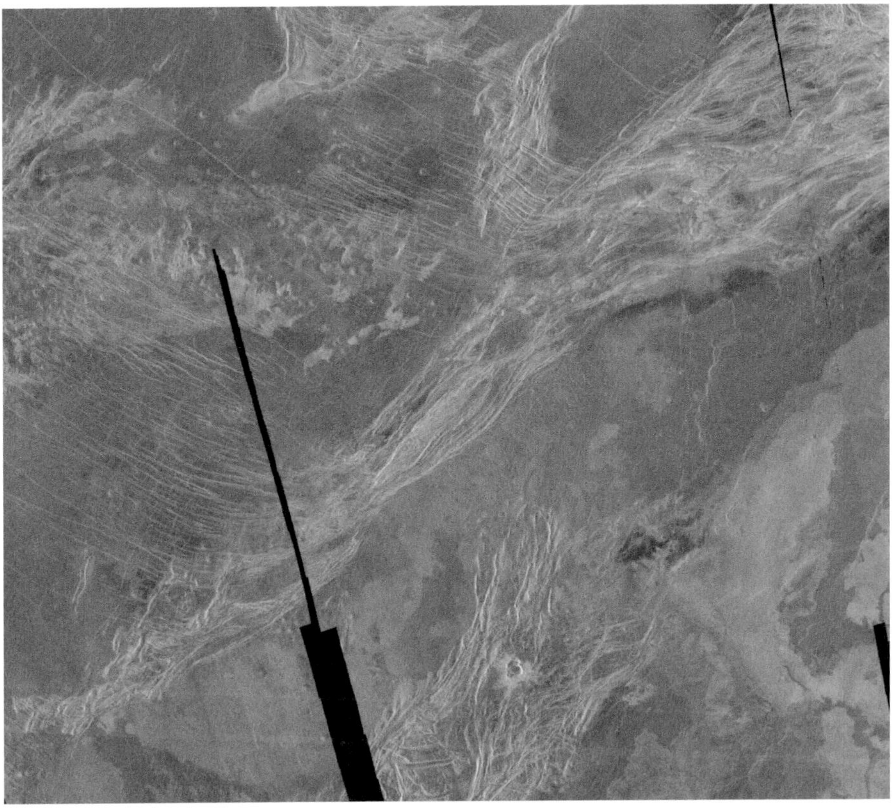

Fig. 6 Magellan full resolution radar mosaic of the Lavinia region of Venus. The mosaic is centred at 50 degrees south latitude, 345 degrees east longitude, and spans 540 km north to south and 900 km east to west. A diverse set of geologic features are shown on the image. The bright area running from the upper right to the lower left is interpreted as part of a belt of ridges, formed by compression and thickening of the upper layers of the planet. The areas between ridges suggest flooding by smoother lavas. The varied texture of the lavas can be seen in the mottled appearance of the plains, which are cut by the ridges. Brighter, rougher flows are also common. The particularly bright flows in the lower right corner are the northern extension of Mylitta Fluctus. The bright ridges adjacent to Mylitta Fluctus at the bottom centre of the image also appear to have been affected by the volcanic activity. Some of these bright features have been interpreted as down-dropped areas roughly 5 km wide. This would imply a region of extension where the crust has been pulled apart and so was more easily flooded by the later lava flows. The thinner fractures running from the upper left seem to end at the ridge belt in the centre of the mosaic. These thinner fractures are a continuation of a pattern seen throughout much of Lavinia and suggest a pattern of compression over a very large region. (Image credit: NASA/JSC. Caption modified from original JSC version)

The dense cloud cover on Venus means that surface temperatures show only slight variations and as a result surface winds are slow and not effective agents of erosion. This coupled with the lack of water, may explain why there is such a diversity of surface features on Venus (Basilevsky and McGill 2007).

Volcanism appears to be the dominant process shaping the surface of Venus (Taylor et al. 2018). There is evidence that Venus remains volcanically active. The Venus Express orbiter detected transient high emissivity anomalies associated with the Ganiki Chasma rift zone in the northern Atla Regio region, which are thought to represent recent eruptions of surface lava (Shalygin et al. 2015; Gilmore et al. 2017).

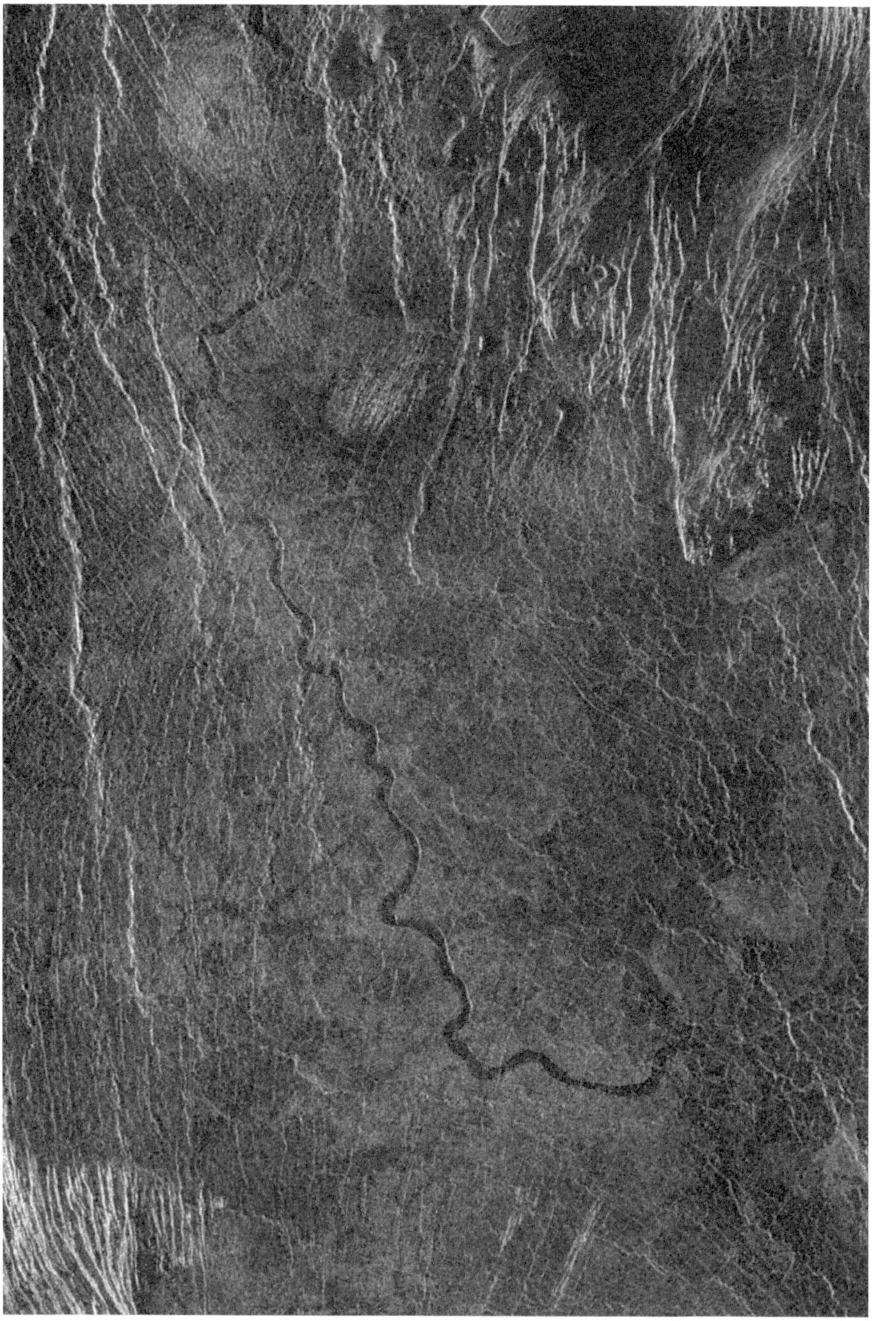

Fig. 7 Full resolution radar mosaic from Magellan at 49 degrees south latitude, 273 degrees east longitude. The area shown has dimensions of 130 by 190 km. The sinuous channel at the centre of the image is approximately 200 km long by 2 km wide. Similar channel-like features are common on the plains of Venus. In some places they appear to have been formed by lava which may have melted or thermally eroded a path over the plains' surface. Most channels are partly buried by younger lava plains, making their sources difficult to identify. (Image credit: NASA/JSC. Caption modified from original JSC version)

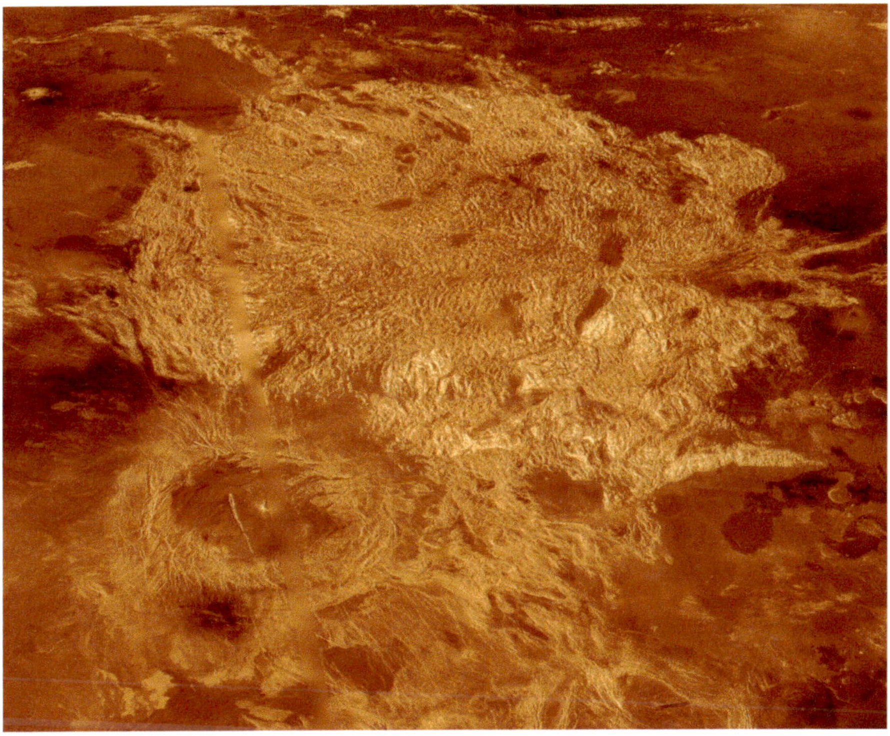

Fig. 8 A three-dimensional perspective view of the Alpha Regio region, which is a topographic upland approximately 1300 km across, centred on 25 degrees south latitude, 4 degrees east longitude. The radar-bright area of Alpha Regio is characterized by multiple sets of intersecting structural features, such as ridges, troughs, and flat-floored fault valleys that together form a polygonal outline. Directly south of the complex ridged terrain is a large ovoid-shaped feature named Eve. (Image credit: NASA/JPL Caption modified from original JSC version http://photojournal.jpl.nasa.gov/catalog/PIA00481)

Almost 1000 impact craters are evenly spread across the surface of Venus. Most are in almost pristine condition and this has led to the proposition that Venus underwent a global resurfacing event 300 to 600 Myr ago. This is thought to reflect the lack of plate tectonics on Venus (Turcotte 1993; Ghail 2015). In contrast to the continuous subduction of crustal material that occurs at destructive plate boundaries on Earth, Venus experiences a cyclic process whereby its mantle gradually heats up until the crust is weakened and subduction then takes place on an enormous scale. No crater on Venus is less than 3 km diameter. This probably reflects filtering by the thick atmosphere which slows down any incoming bolides that are less than about 50 m in diameter.

3.3 What Samples to Collect and Where?

It is clear from the previous sections that the surface of Venus displays great structural diversity and to an unknown extent may also be compositionally diverse. However, the surface geology does appear to be dominated by the products of basaltic volcanism, with the possible exception of the tessera terrains (Basilevsky and McGill 2007). An important first order issue for any sample return mission would be to decide what sort of terrain should be targeted to gain the maximum amount of information about the planet. The various plains units

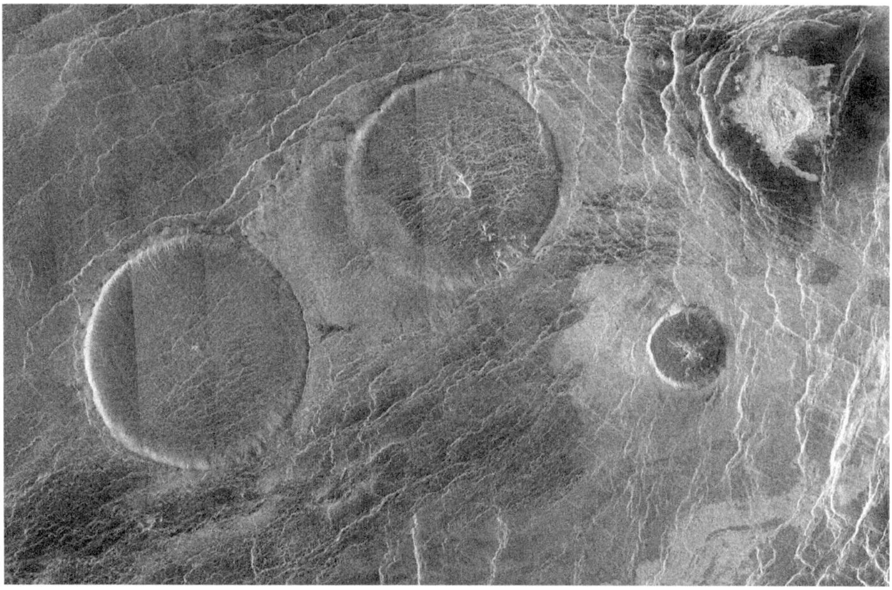

Fig. 9 Magellan full resolution mosaic, centred at 12.3 north latitude, 8.3 degrees east longitude, showing an area 160 km by 250 km in the Eistla region of Venus. The prominent circular features are volcanic domes, 65 km in diameter with broad, flat tops less than one km in height. Sometimes referred to as 'pancake' domes, they represent a unique category of volcanic edifice on Venus formed by the extrusion of viscous lava. The cracks and pits commonly found in these features may result from cooling and withdrawal of lava. (Image credit: NASA/JPL. Caption modified from original JSC version)

comprise 80% of the planet's surface, so that a sample collected from these areas would clearly be the most representative. On the other hand, there is some evidence to suggest that the tessera terrains are both older and compositionally distinct from the other units on Venus (Gilmore et al. 2019). A single sample from such a terrain might prove to be a Rosetta stone, capable of yielding an enormous amount of information about the evolution of Venus. But from one perspective it doesn't really matter where the returned venusian samples are collected. As discussed earlier, the most important single piece of information we seek to acquire from returned samples is their $\Delta^{17}O$ composition and this should be the same no matter what terrain they are collected from.

In their study of potential landing sites on Venus, Basilevsky et al. (2007) recommended that for lander missions undertaking "in situ" analysis, "true" tessera terrains would be of highest importance, with other terrains of secondary importance. Interestingly, they switched this priority for sample return missions and suggested that the highest priority target were the regional plains with wrinkle ridges, with the tessera terrains only of secondary importance. Basilevsky et al. (2007) consider that as the regional plains are the most widespread unit on Venus, full geochemical and isotopic analysis of this material back on Earth would provide the maximum amount of information about the processes operating in the mantle and crust of the planet.

The extent to which individual terrains have experienced surface weathering is a further important constraint on landing site selection. As pointed out by Taylor et al. (2018), the high pressures and temperatures that prevail at the surface of Venus and the reactive species present in its atmosphere, such as SO_2, HCl and HF, are likely to result in significant levels of surface-atmosphere interaction. This problem may be compounded

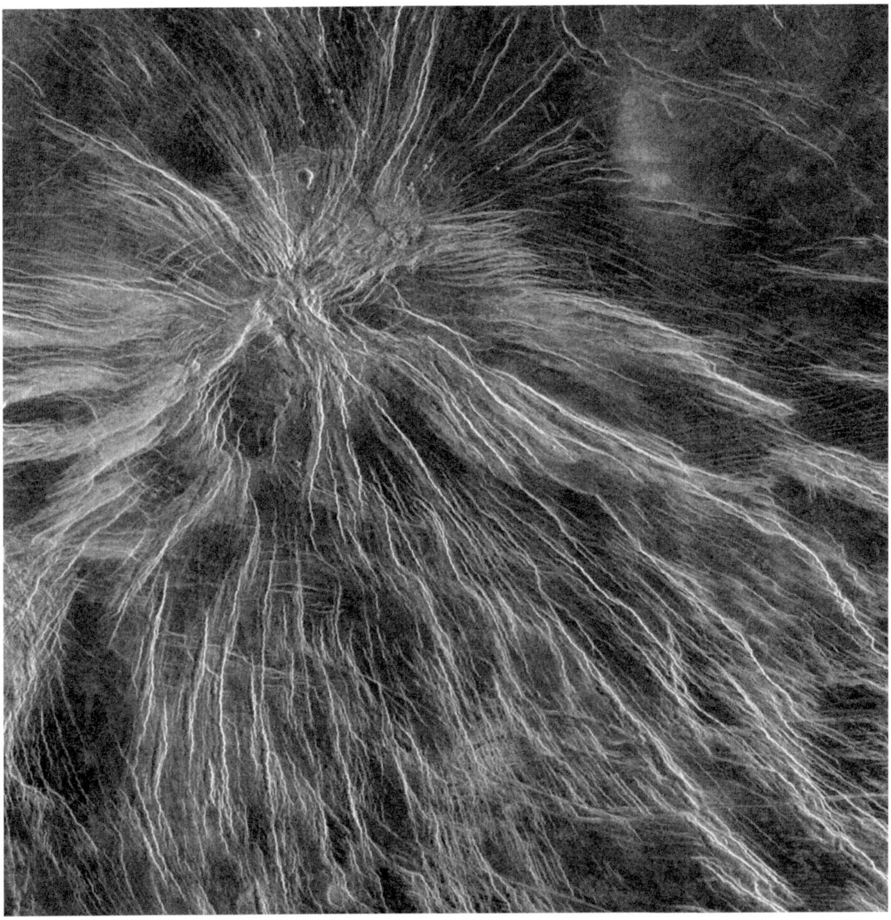

Fig. 10 Magellan image of a nova, a radial network of grabens, in Themis Regio, Venus. There have been about 50 novae identified on Venus, which consist of closely spaced graben radiating from a central area. This nova is about 250 km in diameter. (Image credit: NASA/JPL. Caption modified from original JSC version)

by the fact that the atmosphere and lithosphere on Venus are unlikely to be in isotopic equilibrium, as is also the case for Earth and Mars (e.g. Thiemens et al. 1995; Farquhar et al. 1998). It is therefore possible that surface samples may have been isotopically modified as a result of extensive exchange with the atmosphere. Three strategies could be adopted to mitigate this problem. Firstly, the landing site would need to be in an area covered in relatively young, and as a consequence, fresh basaltic lavas. The volcanically active Ganiki Chasma region might be one possibility. Secondly, a drill sample of reasonable depth needs to be collected (Figs. 13, 14). Finally, a surface atmospheric sample should be returned in order to define the oxygen isotopic composition of the lower atmosphere. However, while it is certainly important to pay due attention to the issue of surface-atmosphere interactions, it should also be noted that the images returned by the Venera spacecraft suggest that appropriate sample site selection would significantly alleviate this potential problem (Fig. 13).

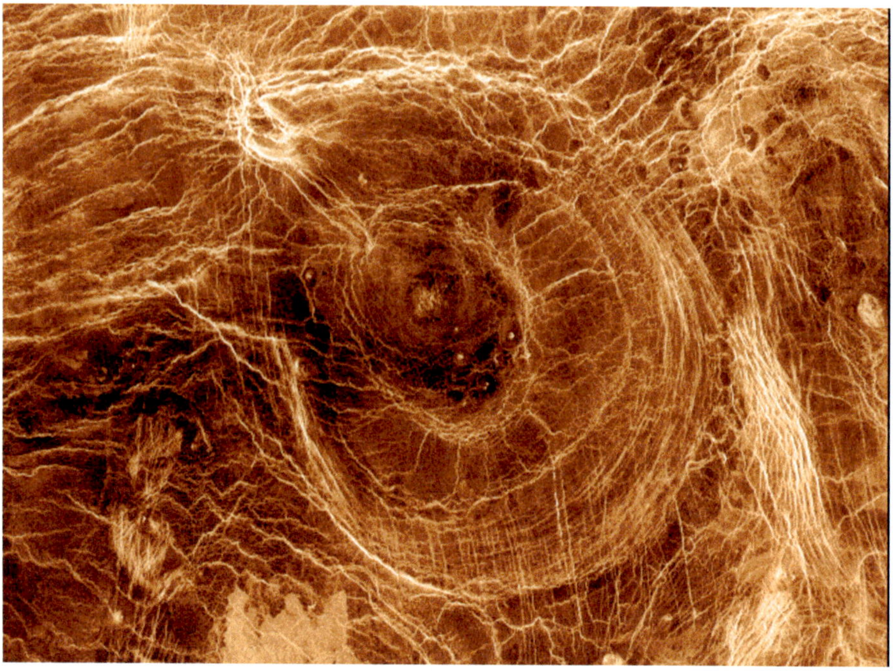

Fig. 11 Arachnoids are large structures of unknown origin that have been found only on the surface of Venus. Arachnoids get their name from their resemblance to spider-webs. They appear as concentric ovals surrounded by a complex network of fractures and can have diameters of up to 200 km. Over 30 arachnoids have been identified on Venus. (Image credit: NASA/JPL. Caption modified from original JSC version)

4 A Robotic Sample Return Mission to Venus

4.1 Studying the Surface and Atmosphere of Venus – Past Missions and Mission Proposals

Past missions to study the atmosphere and surface of Venus have been summarised by Basilevsky et al. (2007) and Taylor et al. (2018). The NASA flyby mission Mariner 2 was the first spacecraft to study Venus, arriving in December 1962. Between 1967 and 1983 a series of thirteen successful Venera missions (Venera 4 to 16) were launched by the Soviets to variously study the atmosphere and surface of Venus (Fig. 13) (Basilevsky et al. 2007). The NASA Pioneer Venus Project consisted of two spacecraft launched in 1978. One of these spacecraft was an orbiter and the other deployed four small descent probes into the venusian atmosphere, one of which reached the surface and continued to operate for over an hour. Vega 1 and Vega 2 were Soviet missions designed to study Venus and then Halley's Comet (Dolgopolov et al. 2012). Both spacecraft combined lander and balloon modules, which were inserted into the venusian atmosphere in June 1985. The Vega 1 lander experienced a large atmospheric shock during descent and so failed to operate successfully on the surface. In contrast, Vega 2 undertook surface observations for approximately an hour and successfully analysed sample material. The balloon systems deployed be Vega 1 and 2 both transmitted atmospheric data successfully for about 46 hours; a duration defined by their battery lifetimes (Sagdeev et al. 1986). ESA's Venus Express orbiter operated between 2006

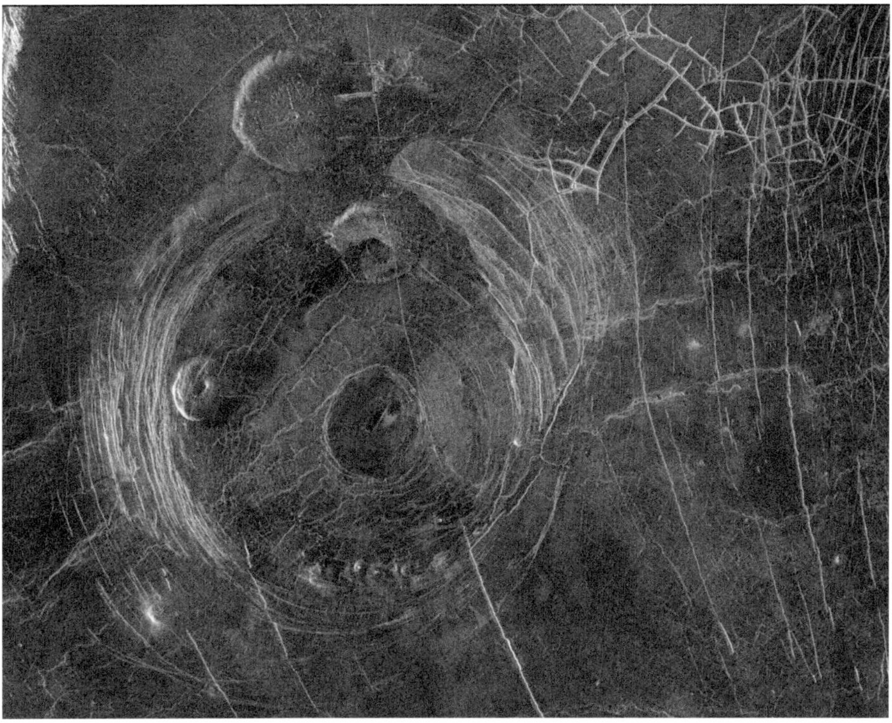

Fig. 12 Magellan radar image showing a region approximately 300 km across, centred on 59 degrees south latitude, 164 degrees east longitude and located in a vast plain to the south of Aphrodite Terra. The large circular structure near the centre of the image is a corona, approximately 200 km in diameter and provisionally named Aine Corona. Just north of Aine Corona is a 'pancake' dome about 35 km in diameter. Another pancake dome is located inside the western parts of the annulus of the corona fractures. Complex fracture patterns like the one in the upper right of the image are often observed in association with coronae and various volcanic features. (Image credit: NASA/JPL Caption modified from original JSC version)

Fig. 13 Surface image of Venus taken by the Soviet Venera 9 lander in October 1975. The angular boulders suggest only a limited degree of interaction with the venusian atmosphere. It seems likely that fresh material could be obtained by limited drilling of one of the larger, angular boulders. (Image source: Wikipedia/Ted Stryk/Planetary.org)

and 2014 and undertook detailed observations of the venusian atmosphere (Taylor et al. 2018). The JAXA Akatsuki orbiter is the only spacecraft currently studying Venus (Nakamura et al. 2016). Launched in 2010, it initially failed to enter orbit around Venus, but was later recovered and began observations of the atmosphere in 2015 (Nakamura 2019).

Fig. 14 Schematic diagram showing the possible stages of a "Grab and Go" sample recovery mission from the surface of Venus. This scenario is based on the detailed proposals of Rodgers et al. (2000). The original spacecraft would consist of an orbiter and a lander module. The orbiter (1) passes over the landing site every 93 mins. Once the lander has detached from the orbiter (2) it descends to the surface slowed by parachutes and possibly also rocket motors. On the surface sampling operations take place (3). The lander makes use of multiple sampling devices, including a scoop, drill and "vacuum cleaner". Sample material is transferred to the ascent stage (4) which consists of a three part rocket (individual stages not shown). The ascent rocket is lifted to a height of 66 km by a helium balloon (5). The balloon is jettisoned and the ascent rocket (6) takes the sample back to the orbiter (7). In an extended mission scenario, a cloud-level balloon platform (8) (Chassefière et al. 2009; Wilson et al. 2019) could also be deployed, which would carry a suite of instruments to undertake detailed analysis of the venusian atmosphere. Data from the cloud-level balloon platform would be transmitted to the orbiter

A wide range of mission proposals and feasibility studies have been undertaken that involve at least some form of direct compositional analysis of venusian samples, either gaseous or solid. Most commonly, such analyses are to be carried out remotely, with a more limited number of studies proposing a return of material to Earth. Here we briefly review various proposals that involve sample return.

A detailed assessment study of how sample return from the surface of Venus could be achieved was undertaken by ESA as an in-house activity within the Directorate of Scientific Programmes (Coradini et al. 1998; Taylor et al. 1999). The mission plan envisaged two spacecraft, the first to be launched being a combined orbiter and Earth return capsule. The second spacecraft comprised a combined lander and ascent module (Whitcomb et al. 2002). Planned surface operations included collection of 100 g of rock by drilling, 100 g of dust using a "vacuum cleaner" device and 100 g of surface atmosphere. Either during the descent or ascent phases, additional atmospheric sampling was planned, involving collection of 250 ml of gas at altitudes of 15 km, 40 km and 65 km (Coradini et al. 1998). A feasibility study for a "Grab and Go" style sample return mission from the surface of Venus has also been undertaken by Rodgers et al. (2000) and is discussed in detail in Sect. 4.4. Multiple configurations to undertake both atmospheric and surface sample return have been outlined by Shibata et al. (2017).

Venera-D is a proposed Russian mission to Venus, which would comprise an orbiter and lander (Gregg et al. 2019). While the main lander is expected to only survive for two to three hours on the surface it would incorporate a longer-duration module that would operate for between two to three months (Gregg et al. 2019). A comprehensive concept mission involving both atmospheric and surface sampling has recently been proposed by Valentian et al. (2019) in response to the ESA Voyage 2050 white paper call. The proposed mission would involve either two or three separate launches carrying a variety of probes, including two orbiters, one of which brings back the samples to Earth, a balloon sampling platform and multiple landers for surface sample collection.

4.2 How Much Sample Material Needs to be Collected?

Defining the minimum amount of material that would need to be returned to Earth to fulfil the mission's principal scientific goals is an issue of paramount importance. If the only objective was to determine the $\Delta^{17}O$ composition of the venusian crust, then about 100 mg of material would probably suffice. High-precision laser fluorination analysis is normally undertaken on aliquots of approximately 2 mg and a return sample size of about 100 mg would ensure that the material could be adequately characterised for major and trace elements, as well as providing for multiple oxygen isotope analyses to be performed at a number of laboratories.

However, as set out below, a wide range of analytical techniques should also be used on these precious materials in order to glean as much information as possible about the origin and evolution of Venus. In other words, and within reason, the more material you can bring back, the more techniques you can apply and consequently the more you are likely to learn about Venus. In Table 2 an estimate is given of the sample mass required for a series of techniques that might be undertaken on a returned sample. Based on these estimates a sample mass of 9.4 g would be enough to provide very detailed information about the age and composition of the venusian lithosphere. The list of techniques given in Table 2 is far from exhaustive, but serves to demonstrate that a considerable amount of scientific data could be collected on an approximelty10 g sample. In fact, the allowances assigned to each of the techniques listed in Table 2 are on the generous side and by carful curation management the required amount of material could be reduced considerably. Thus, a 10 g nominal sample mass can accommodate all the essential techniques required to characterise the returned material.

However, bringing back just enough material to undertake measurements as they exist at the time of the mission does not provide any legacy for future generations. One of the great successes of the Apollo program was that a very large amount of material was returned from the Moon, with a significant fraction kept in clean conditions for future scientists to work on. Just defining a sample size on the basis of current constraints is not a good long-term strategy. Consequently, the question about how much material needs to be returned to Earth can be reconfigured based on engineering and budgetary perspectives to: how much material can be collected from the surface of Venus within "reasonable" cost and logistic constraints? One way to address this issue is to look at other sample return missions.

Between 1970 and 1976 the Soviet Union launched three successful remote sampling missions to the lunar surface (Luna 16, Luna 20 and Luna 24) (Robinson et al. 2012). Luna 16 returned a 35 cm-long core sample with an approximate mass of 101 g, Luna 20 returned 52 g of material and Luna 24 returned a 160 cm-long core containing 170 g of material (Robinson et al. 2012). While the logistics of returning a sample from the Moon are much less complex than would be the case for Venus, the Luna missions do at least demonstrate that remote sampling of a large-sized planetary body is a realistic option. The Luna 24 mission made use of a rotary percussion corer, demonstrating that such technology is a viable way to collect a large-sized sample from a planetary surface. Tests with a similar type

Table 2 Sample mass estimates for various analytical techniques

Method	Sample mass	Example study[@]
Initial characterisation		
4 × 1 cm^2 polished sections	1.2 g	Stephant et al. (2019)
ICP-MS (major and traces)*	1.0 g	Barrat et al. (2016)
Dating studies		
Sm-Nd, Lu-Hf, Pb-Pb	2.5 g	Bouvier et al. (2008)
Mn-Cr (bulk)	0.1 g	Trinquier et al. (2008)
Mn-Cr (minerals)	0.5 g	Trinquier et al. (2008)
Rb-Sr	0.5 g	Shih et al. (2005)
In situ dating (SIMS)	Polished sections	Bellucci et al. (2015), Stephant et al. (2019)
Stable isotope analyses		
H isotopes bulk**	0.1 g	Alexander et al. (2012)
C isotopes bulk**	0.1 g	Alexander et al. (2012)
N isotopes bulk**	0.1 g	Alexander et al. (2012)
O isotopes bulk**	0.1 g	Greenwood et al. (2017, 2018b)
S isotopes bulk	0.5 g	Franz et al. (2014)
In situ stable isotopes	polished sections	Stephant et al. (2019)
Ti Cr isotopes*		
Cr	0.1 g	Bonnand et al. (2016)
Ti	0.1 g	Goodrich et al. (2017)
Mo Ru isotopes		
Ru	1.0 g	Hopp et al. (2020)
Mo	1.0 g	Hopp et al. (2020)
Highly siderophile elements (HSE)*		
HSE and Re-Os isotopes	0.5 g	Dhaliwal et al. (2017)
Total	9.4 g	

[@] Example studies are given to show how the quoted amount of material has been used as part of an analogous extraterrestrial study

*Mass will allow multiple complete reruns at different labs

**Sample masses shown are what is generally considered to be a representative fraction. The amount of material consumed for each individual replicate is considerably less than 100 mg and so these sample masses should allow for reruns at different labs

of drill designed to operate on Venus showed that it could core to a depth of 4.5 cm in 9.5 minutes and was able to recover 25 g of material (Rehnmark et al. 2017). Due to Venus' hostile surface conditions the collection phase of the sample return mission would have to be of short duration, perhaps no more than about an hour in total. Using a combination of drilling and surface grabbing (see Sect. 4.3.1), it would seem feasible to obtain at least 50 g of material. This would certainly be sufficient to allow all the analytical techniques outlined in Sect. 5 and Table 2 to be undertaken.

A number of other remote sample collection missions have been undertaken and have successfully returned extraterrestrial material to Earth. These include NASA's Stardust mission to Comet Wild 2 (Brownlee 2014), NASA's Genesis mission to collect solar wind material (Jurewicz et al. 2003) and JAXA's Hayabusa mission to asteroid Itakowa (Nakamura et al. 2013). While these missions returned extremely limited amounts of material, they clearly demonstrate that major scientific breakthroughs can be achieved, even with very small samples (e.g. Matzel et al. 2010; McKeegan et al. 2011; Jin and Bose 2019). Current asteroid sample return missions OSIRIS-REx (Lauretta et al. 2017) and Hayabusa2 (Watanabe et al. 2017) are planning to return > 60 g and > 1 g of surface materials from asteroids Bennu and Ryugu, respectively. Even with the more limited amounts of material likely to be returned by Hayabusa2 from asteroid Ryugu, a full suite of isotopic and geochemical studies are envisaged (Watanabe et al. 2017). This again suggests that return of a 10 g nominal sample from Venus would provide sufficient material to support an ambitious analytical campaign. However, from a technical standpoint, there seems no reason why a much larger amount of material, for example 100 g, could not be returned from Venus. Such a sample size would provide both for immediate analytical work and also serve as a longer-term resource. It is worth noting again (see Sect. 4.1) that the ESA Venus sample return study (Coradini et al. 1998) envisaged a return mass of 300 g, consisting of 100 g of rock, 100 g of unconsolidated soil and 100 g of surface atmosphere.

Due to the likelihood that surface samples will have interacted, at least to some extent, with the harsh venusian atmosphere (Sect. 3.3), it is important that a surface atmospheric sample is collected to define the degree of atmosphere-lithosphere oxygen isotope disequilibrium. In addition, aliquots of gas could also be collected at higher levels in the atmosphere during either the lander ascent or descent phases. As a consequence of the high atmospheric pressures that prevail close to the surface of Venus, sampling adequate amounts of gas for high-precision oxygen and carbon analysis back on Earth would be relatively straightforward and requires only sample tanks of moderate volume. Thus, a 1 dm^3 collection tank at the surface would passively collect about 1.5 moles of atmosphere ($\sim$65 g). This quantity of gas would be more than sufficient for a very large number of high-precision carbon, oxygen and nitrogen analyses using gas source mass spectrometry.

4.3 How Would We do a Sample Return Mission to Venus?

Retrieving samples from the surface of Venus and successfully returning them to Earth would be a technically challenging project and as a consequence relatively expensive to implement. However, while Venus represents an extremely hostile environment in which to operate a spacecraft, it has a mission failure rate that is not significantly different to that for Mars (sect. 1.1). In addition, many of the required technologies for sample return have already been tried and tested during past missions to Venus (Rodgers et al. 2000; Basilevsky et al. 2007; Dolgopolov et al. 2012; Rehnmark et al. 2017; Shibata et al. 2017; Glaze et al. 2018; Wilson et al. 2019). Here we outline two possible strategies to acquire and return surface samples from Venus. The first, and lower cost mission configuration, involves a lander with, or without, an orbiter, to transport the sample back to Earth. The second, more costly scenario, would be as part of an integrated mission to study in detail the atmosphere of Venus, but with a subsidiary lander to collect a small surface sample.

4.3.1 A "Grab and Go" Sample Return Scenario

Coradini et al. (1998) and Rodgers et al. (2000) provide scenarios outlining how a "Grab and Go" venusian sample return mission might be accomplished. The two studies share many features in common and so the summary scenario presented here is a composite derived from these earlier proposals. A schematic diagram summarising the various stages of the mission are given in Fig. 14. The spacecraft would be two part, comprising an orbiter-return module and a lander-ascent stage. The full payload would be delivered to Venus by a single conventional launch, as proposed by Rodgers et al. (2000). Aerocapture by Venus could be facilitated by the use of a ballute, a hybrid of a balloon and parachute. A subsequent aerobraking manoeuvre would then put the spacecraft into a final equatorial and circular orbit.

Subsequent to orbital insertion, the lander would detach from the orbiter, with its velocity reduced using a small rocket motor. It would descend to the surface in about 1.5 hours. The dense venusian lower atmosphere facilitates a slow descent onto the surface and this would be further aided by the use of parachutes (Fig. 14). As proposed by Coradini et al. (1998), a limited number of atmospheric samples could be taken either during the descent or subsequent ascent stages. Once the lander reaches the surface, sampling would be undertaken rapidly using multiple methodologies, including drilling, scooping and possibly also a vacuum cleaner-type system (Fig. 14) (Coradini et al. 1998). Due to the high prevailing temperatures, surface operations would need to be conducted rapidly, probably lasting in total not much more than 1 hour. As discussed above, collection of a surface atmospheric sample would be an essential part of the mission in order to define the extent of oxygen isotopic disequilibrium between the atmosphere and lithosphere.

The initial ascent of the lander would be achieved using a helium balloon system, lifting the payload to a height of 66 km. A three-stage solid rocket system would then take the sample payload into orbit, where it would be captured by the orbiter. Return of the orbiter plus payload to Earth is a challenging procedure due to the large mass of Venus. A solar electric propulsion (SEP) system is preferred by Rodgers et al. (2000) over conventional chemical propulsion because of the large mass advantage of the SEP system (approximately 30%). Return of the sample payload to Earth would take about 3 years.

Significant advances have recently been made in sounding rocket technology, with the three stage JAXA SS-520-5 becoming the smallest rocket to place a satellite into Earth orbit in February 2018 (Graham 2018). The SS-520-5 measured 9.54 m and weighed 2600 kg. Incorporating a larger-sized sounding rocket within the lander module would have the potential to eliminate the need for a separate orbiter. Based on the mass estimates given by Rodgers et al. (2000), a single spacecraft incorporating a larger-sized sounding rocket within the lander assembly should weigh no more than about 5000 kg, which is slightly less than the Cassini spacecraft. Incorporating a rocket that could return the samples directly to Earth would eliminate the need for a potentially risky orbital rendezvous sample transfer manoeuvre.

4.3.2 An Extended Sample Return – Atmospheric Analysis Mission Scenario

A "Grab and Go" mission similar to that outlined by Rodgers et al. (2000) could also form part of an extended mission to study the atmosphere of Venus (Fig. 14). Wilson et al. (2019) have proposed a "strawman" L-class mission comprising an orbiter, cloud-level balloon platform and (optional) Russian-style descent probe. To facilitate sample return, the descent lander would need to be configured along the lines suggested by Rodgers et al. (2000). This mission could involve the collection of a comprehensive suite of atmospheric samples. As discussed in Sect. 5.3, there is a possibility that life may be present on Venus within the cloud layer at altitudes of between 51 km (65 °C) and 62 km (−20 °C) (Dartnell et al. 2015). This could be tested by returning atmospheric samples collected at the appropriate altitude.

Atmospheric data could also be collected remotely using a balloon platform, with the data transferred to the orbiter. In this case only the surface sample would be returned to Earth. In view of current concerns about global warming there can be no doubt that a combined atmospheric and surface return mission would have a higher profile than just a surface sample collection mission with limited atmospheric sampling, as outlined in Sect. 4.3.1 above.

4.4 How Much Would it Cost?

Rodgers et al. (2000) estimate that a Venus sample return mission would require a budget in excess of a NASA Discovery mission, but would cost significantly less than Flagship missions. Most NASA Discovery missions fall in a cost range of between 300 and 800 million dollars, whereas recent Flagship missions to Mars, such as Mars 2020 or Mars Science Laboratory are in the 2 to 2.5 billion dollar cost bracket. On this basis we could put a tentative cost for a Venus "Grab and Go" sample return mission at about 1.5 billion dollars. A more extensive mission, involving a significant atmospheric sampling program would clearly be more expensive. Exactly how much more such a mission would cost would be dependent on the exact range of additional analytical tasks that were undertaken. As a guide, the NASA Venus Flagship Study Report (Venus Science & Technology Definition Team 2009) details a range of possible architectures, with estimated costs that vary from 1.6 to 3 billion dollars. The recommended option for further detailed evaluation included an orbiter, two mid-level orbital platforms and two landers and had an estimated cost of 2.7 billion dollars. On this basis 1.5 billion dollars for a "Grab and Go" mission seems generally realistic. In view of the importance of the science case for sample return from Venus, and despite the relatively high costs involved, such a mission ought to be given a very high priority.

An important additional consideration is the accumulating financial disparity between research on Mars compared to that on Venus. With a significant long-term program and extensive on-going activities, the budget for Mars research now dwarfs that for Venus. However, it is far from clear that this disparity can be justified from a scientific perspective. Mars is not a good analogue planet for Earth and its formation and early evolution are not representative of those that built the larger terrestrial planets, such as Earth and Venus. In contrast, Venus and Earth appear to have formed in closely similar ways. The more we learn about Venus, the more we learn about Earth (Wilson et al. 2019). From a scientific perspective the balance of funding needs to be redressed and research on Venus should be given significantly more emphasise than it is at present (Glaze and Garvin 2018; Wilson et al. 2019).

5 The Analytical Campaign Following Venus Sample Return

5.1 Overview

In view of the scientific value of returned samples from Venus, the analytical campaign following return to Earth would be a major undertaking and require detailed planning and preparation. This program would have the advantage of drawing on the experience gained from the lunar, Stardust and Hayabusa missions and hopefully also from Hayabusa2 and OSIRIS REx. In addition, the detailed planning that has been carried out for Mars sample return should provide a template for designing the necessary analytical protocols and strategies (Mars Exploration Program Analysis Group (MEPAG), 2012). In fact we would envisage the Venus sample analysis campaign following a very similar strategy to that outlined for Mars.

5.2 Clean Receiving Facility and Initial Handling Procedures

An important and critical component of the analytical program would be the development of a clean receiving laboratory. This might not necessarily be a brand new facility, but it would need to have been totally isolated and purged of any potential sources of terrestrial contamination prior to the return of actual Venus samples. Using analogue materials, appropriate protocols would need to be developed in order to establish initial characterisation and manipulation procedures. However, the experience gained from previous sample return campaigns should help to streamline this development process (Mars Exploration Program Analysis Group (MEPAG), 2012).

As envisaged for the return of martian material, samples from Venus would need to be initially screened using a non-invasive and non-destructive technique, such as X-ray computed tomography (XCT) (Welzenbach et al. 2019). This would potentially provide significant benefits in establishing the three-dimensional structure and textural variation of the returned materials, which would be of assistance in planning subsequent phases of the analytical campaign. However, there have been recent suggestions that XCT can cause permanent changes to sample materials (Sears et al. 2018) and so further evaluation of this technique is required (Welzenbach et al. 2019).

5.3 Planetary Protection Issues

From a planetary protection perspective, spacecraft operations in the venusian environment are currently classified as category II by the Committee on Space Research (COSPAR). This category covers: "All types of missions (gravity assist, orbiter, lander) to a target body where there is significant interest relative to the process of chemical evolution and the origin of life, but where there is only a remote chance that contamination carried by a spacecraft could compromise future investigations" (COSPAR 2019). Compared to operations on Mars (category III – IV), the harsh venusian environment clearly poses less of a threat in terms of both forward and backward contamination. Samples collected from the upper regolith layers of Venus are unlikely to contain a viable biological component and should an organism be transported by the spacecraft from Earth it would not be viable in such an environment.

Despite the relatively low level of threat posed by Venus, prudence dictates that planetary protection issues would need to be taken seriously as part of the planning and preparation for a sample return mission. It now appears likely that Venus possessed surface conditions favourable for life in the past and these conditions may have persisted until as little as 700 Myr ago (Way et al. 2016). If life did evolve on Venus then it may still persist in various niche environments. It is theoretically possible that extremophiles could be present close to the venusian surface, at a minimum depth of about 10 m, where temperatures could be as low as 200 °C (Joseph 2019). However, due to time constrains, drill coring operations are unlikely to penetrate more than a few tens of centimetres below the surface.

The need for planetary protection is more acute for any atmospheric samples, in view of the theoretical possibility that life could be present within the lower cloud layer on Venus (Limaye et al. 2018). Based on a combination of temperature and pH constraints, Dartnell et al. (2015) estimated that extremophiles could live at altitudes of between 51 km (65 °C) and 62 km (−20 °C). The threat from such organisms may not just arise during atmospheric sampling activities. If the spacecraft and any collected materials are not properly protected, contamination could take place during the ascent phase following surface operations. There is also the possibility of forward contamination during the descent phase and so spacecraft build operations would need to be undertaken using protocols that minimised this threat.

5.4 Initial Classification and Characterisation Studies

Either with or without the use of XCT, initial characterisation work would start with the controlled removal of the samples from their enclosing collection containers, followed by weighing and initial imaging. This work would need to be undertaken based on the best practice approaches developed for other extraterrestrial sample suites (Allen et al. 2011). Suitable sub-samples would then need to be selected and prepared for mineralogical and petrographic study. This would involve the preparation of polished sections and grain mounts. These would be used to determine the textural and lithological diversity present in the samples using microbeam and related analysis techniques (quantitative scanning electron microscopy, electron microprobe, transmission electron microscopy). Other techniques that might be employed in these initial stages include: Raman spectroscopy, atomic force spectroscopy, ultraviolet-visible-near infrared (UV-Vis-NIR) microscopy, Electron Backscatter Diffraction (EBSD) and cathodoluminescence.

5.5 Bulk Major-Element, Trace-Element and Isotopic Analysis

As is the case for terrestrial, lunar and martian samples, bulk major- and trace-element analysis, generally now undertaken by Inductively Coupled Plasma Mass Spectrometry (ICP-MS) and related techniques (Jenner et al. 1990), would provide important constraints on the origin and evolution of returned samples from Venus. However, prior to crushing and dissolution for ICP-MS, cut surfaces of samples can first be studied by micro X-ray Fluorescence (μXRF) scanning techniques. Secondary Ion Mass Spectrometry (SIMS) and Laser Ablation-Inductively Coupled Plasma-Mass Spectrometer (LA-ICP-MS) are in situ techniques that have been successfully employed in the study of martian meteorites (e.g. Lorand et al. 2018; Ireland et al. 2020) and would also have distinct advantages for the analysis of returned venusian samples, In addition, bulk isotopic data, particularly for Cr, Ti, Ni, W, Ca and Mo would permit a comparison between Venus and other Solar System bodies (Warren 2011; Scott et al. 2018). This data would be relevant to the proposition that Venus formed from essentially identical precursor material to that of the Earth (Sects. 2.3 and 2.4).

5.6 Stable Isotope Analysis

In this article we have highlighted the importance of undertaking oxygen isotope analysis of returned venusian samples. However, other stable isotope measurements (hydrogen, carbon, nitrogen, sulphur) would also yield critical information relevant to the evolution of Venus' early hydrosphere, lithosphere and atmosphere. Hydrogen isotope analysis would be particularly informative concerning the fate of water on Venus and surface sample measurements would be helpful in interpreting any data collected during atmospheric analysis (Grinspoon 2017). The venusian atmosphere contains very little water (about 30 ppm) and has a D/H ratio about 150 times higher than the oceans on Earth (Greenwood et al. 2018a). Based on the high D/H ratio, it has been suggested that Venus once contained similar amounts of water to Earth, but that this was subsequently lost to space (Donahue et al. 1982). D/H measurements of material collected from the tessera terrains would be particularly important in view of recent suggestions that these may have formed in the presence of surface water (Mueller et al. 2017). In terms of its carbon and nitrogen isotopic composition, Venus appears to have similar values to Earth (Greenwood et al. 2018a). High levels of relatively unstable sulphur species in the atmosphere of Venus may be due to ongoing volcanic replenishment mechanisms (Greenwood et al. 2018a). However, it has also been suggested that Venus is significantly less active volcanically than Earth (Mikhail and Heap 2017). Combining the information obtained by stable isotope analysis of both returned surface and atmospheric samples from Venus should allow, for the first time, the development of quantitative circulation models for the planet and so constrain the nature of the ongoing interactions between its mantle, crust and atmosphere (Taylor et al. 2018).

5.7 The Importance of Dating Studies

Dating studies undertaken on the returned samples would be particularly important in view of the fact that there is currently no data relevant to the absolute ages of the various terrains identified on Venus (Taylor et al. 2018). Significant information is available on the relative ages of these surface units (Taylor et al. 2018) and there is general agreement that the tessera terrains represent the oldest exposed lithologies on the planet (e.g. Whitten and Campbell 2019). As discussed in Sect. 3.3, sample return from a tessera terrain would be valuable, although there is a significant risk that such ancient regions have experienced protracted surface-atmosphere interactions. Dating studies of returned tessera material should provide ages for the oldest geological events recorded by extant venusian rocks. Depending on the results, this data would help to constrain the likely tectonic regime on Venus. For example, do the tessera terrains represent continental crust as old as that on Earth or are they something completely different? A significant amount of comparative geology has already been undertaken for Venus using remote sensing data (Basilevsky et al. 2007). The dating of just one sample from the surface of Venus would help to place this information into a comparative context and would as a result enormously increase the value of the datasets already available. It is generally suggested that plate tectonics does not operate on Venus, but the manner in which the planet loses its heat is poorly constrained (Taylor et al. 2018). Dating of a single surface sample would certainly help to clarify significantly this important issue.

5.8 Noble Gas Analysis

Glaze and Garvin (2018) and Wilson et al. (2019) have highlighted the need for a complete suite of noble gas measurements to be made during sampling of the venusian atmosphere. As is the case for Earth and Mars, this data would help constrain any episodes of major degassing and atmospheric blow-off (Wilson et al. 2019). It would also be important to undertake such noble gas measurements on returned surface samples, which would then provide a basis for comparison with similar data from Earth and Mars (Peron et al. 2018; Ott et al. 2019).

5.9 Other Analysis Techniques

In addition to those outlined above, a wide range of other techniques could be used to interrogate samples returned from the surface of Venus. In particular, synchrotron-based techniques such as X-ray Absorption Near Edge Spectroscopy (XANES) would be useful in identifying the different oxidation states of species present in the samples (e.g. Nakada et al. 2019), while high-resolution X-ray diffraction studies would provide detailed structural information (e.g. Hicks et al. 2014). In addition, there is no doubt that new analytical techniques are likely to be available by the time samples from Venus are available for terrestrially-based researchers!

6 Some Final Thoughts and Speculations

While it is not possible to predict the $\Delta^{17}O$ composition of the returned venusian rock samples, we can at least examine the possible implications of various end-member scenarios (Fig. 15). A well-mixed, essentially homogeneous inner Solar System should lead to a venusian lithosphere with a $\Delta^{17}O$ that is close to that of the Earth (point 1 in Fig. 15) with a $\Delta^{17}O$ composition of ~0 ‰. In such a homogenised inner Solar System, Mercury might also have a similar $\Delta^{17}O$ to Venus and Earth, leaving Mars as the odd man out. If this does turn out to be the correct scenario then it would have significant implications for our understanding of the Moon-forming giant impact. The impactor, Theia was likely to have been of inner Solar System origin (Dauphas et al. 2014; Dauphas 2017) and so also have a $\Delta^{17}O$ composition close to 0 ‰. As a consequence, a well-mixed inner Solar System would mitigate many of the isotopic objections to the classic giant impact model of Canup and Asphaug (2001).

As discussed in Sect. 2.6, it has often been suggested that $\Delta^{17}O$ may vary with heliocentric distance. Currently we have samples from only two of the four terrestrial planets. Using the $\Delta^{17}O$ values for Earth and Mars, a linear relationship with distance from the Sun (AU) can be used to predict the $\Delta^{17}O$ values for Venus (Fig. 15, point 2, $\Delta^{17}O = -0.16$‰) and Mercury ($\Delta^{17}O = -0.35$‰). If Venus had such a composition it would essentially put an end to the classic giant impact model for the origin of the Moon (Canup and Asphaug 2001) and instead provide strong evidence in favour of higher-energy impact scenarios (Greenwood et al. 2018b; Lock and Stewart 2017). Perhaps more significantly, it would suggest that terrestrial planet-building took place on a more localized scale than would be the case for a totally homogenised inner Solar System.

A third possibility is that the $\Delta^{17}O$ composition of Venus doesn't have a value close to zero, nor does it fall on any simple linear mixing line. This would imply that building terrestrial planets may be a relatively chaotic process, particularly during the final "giant impact" stage (Walsh and Levison 2019). Even if the $\Delta^{17}O$ composition of Venus appears to be unrelated in any straightforward way to its neighbouring planets, surprises are still possible. As depicted in Fig. 15, Venus might turn out to have a $\Delta^{17}O$ value of -0.24‰, essentially identical to that of Vesta (Greenwood et al. 2017; Ireland et al. 2020). It was suggested by Bottke et al. (2006) that Vesta may be a main belt "interloper", scattered into its present position from a location within the terrestrial planet region. In addition, Vesta is

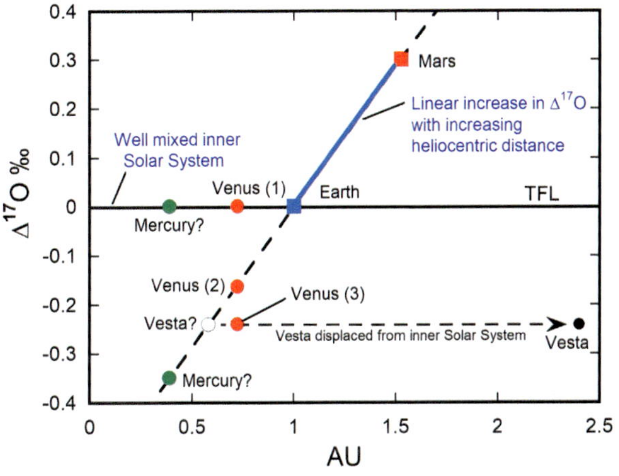

Fig. 15 Speculative Venus $\Delta^{17}O$ compositions based on different models for the formation of the inner Solar System. See Sect. 6 for further discussion

highly alkali-depleted and has a potassium isotopic composition ($\delta^{41}K = +0.36 \pm 0.16‰$) that is even heavier than the Moon (Tian et al. 2019). In the case of the Moon, its heavy potassium isotopic composition compared to terrestrial values was explained by Wang and Jacobsen (2016) as being the result of partial condensation from vapour in the aftermath of the giant impact. Vesta may represent a natural satellite that formed in a similar way to the Moon and was later displaced from its companion planet. It is possible that Vesta originally formed as a moon of Venus. Although this scenario is speculative, it remains unclear why Venus has no natural satellite, whereas the Earth does. It has been suggested that this reflects loss of an earlier-formed moon due to close encounters with other bodies (Brooks and Jacobson 2019).

7 Conclusions

In this paper we have highlighted the scientific imperative for determining the $\Delta^{17}O$ composition of Venus. This single measurement has the potential to radically change our understanding of how the early Solar System evolved. In particular, it would provide answers to the question: How well-mixed was the terrestrial planet region? If the Earth and Venus prove to be isotopic clones it could potentially alter our understanding of such events as the Moon-forming giant impact. There is already clear evidence that the early Solar System was bimodal in composition. A returned sample from Venus would be indispensable in studying and quantifying this phenomena.

In addition to this oxygen isotope data, a returned sample from the surface of Venus would also furnish a wealth of scientific information relevant to the evolution of the inner Solar System. Analysis of samples from Venus would deepen our understanding of comparative planetology, providing a firm basis for assessing similarities and differences with the evolution of early-Earth and Mars. In a wider context, analysis of returned samples from Venus will provide data relevant to the study of exoplanetary systems. Our Solar System provides us with an important case study in how two almost identical bodies, Earth and Venus, could have had such a divergent evolution. Finally, Venus, with its runaway greenhouse atmosphere, may provide data relevant to the understanding of similar less extreme processes on Earth. Venus is Earth's planetary twin and deserves to be better studied and understood. While humans will probably never set foot on its surface, this is not a good reason for neglecting the scientific study of such an important and nearby planet.

Acknowledgements Planetary science research studies at the Open University are supported by a Consolidated Grant from the UK Science and Technology Facilities Council (STFC), Grant number: ST/P000657/1. The authors would like to thank the Guest Editor Professor Sara Russell and the two anonymous reviewers for their helpful comments and constructive advice. We are grateful to Colin Wilson for discussions and advice about past missions to Venus and mission proposals.

References

J. Alday, C.F. Wilson, P.G.J. Irwin, K.S. Olsen, L. Baggio, F. Montmessin, A. Trokhimovskiy, O. Korablev, A.A. Fedorova, D.A. Belyaev, A. Grigoriev, A. Patrakeev, A. Shakun, Oxygen isotopic ratios in Martian water vapour observed by ACS MIR on board the ExoMars Trace Gas Orbiter. Astron. Astrophys. **630**, A91 (2019). https://doi.org/10.1051/0004-6361/201936234

C.M.OD. Alexander, R. Bowden, M.L. Fogel, K.T. Howard, C.D.K. Herd, L.R. Nittler, The provenances of asteroids, and their contributions to the volatile inventories of the terrestrial planets. Science **337**, 721–723 (2012). https://doi.org/10.1126/science.1223474

Y. Alibert, J. Venturini, R. Helled, S. Ataiee, R. Burn, L. Senecal, W. Benz, L. Mayer, C. Mordasini, S.P. Quanz, M. Schönbächler, The formation of Jupiter by hybrid pebble-planetesimal accretion. Nat. Astron. **2**, 873–877 (2018). https://doi.org/10.1038/s41550-018-0557-2

C. Allen, J. Allton, G. Lofgren, K. Righter, M. Zolensky, Curating NASA's extraterrestrial samples—past, present, and future. Chem. Erde **71**, 1–20 (2011). https://doi.org/10.1016/j.chemer.2010.12.003

I. Angelo, J.F. Rowe, S.B. Howell, E.V. Quintana, M. Still, A.W. Mann, B. Burningham, T. Barclay, D.R. Ciardi, D. Huber, S.R. Kane, Kepler-1649b: an exo-venus in the solar neighborhood. Astron. J. **153**(4), 162 (2017). https://doi.org/10.3847/1538-3881/aa615f

J.A. Barrat, R.C. Greenwood, K. Keil, M.L. Rouget, J.S. Boesenberg, B. Zanda, I.A. Franchi, The origin of aubrites: evidence from lithophile trace element abundances and oxygen isotope compositions. Geochim. Cosmochim. Acta **192**, 29–48 (2016). https://doi.org/10.1016/j.gca.2016.07.025

A.T. Basilevsky, G.E. McGill, Surface evolution of Venus exploring Venus as a terrestrial planet, in *Geophys. Monogr. Ser.*, vol. 176, ed. by L.W. Esposito, E.R. Stofan, T.E. Cravens (Am. Geophys. Union, Washington, 2007), pp. 23–43

A.T. Basilevsky, M.A. Ivanov, J.W. Head, M. Aittola, J. Raitala, Landing on Venus: past and future. Planet. Space Sci. **55**, 2097–2112 (2007). https://doi.org/10.1016/j.pss.2007.09.005

J.J. Bellucci, A.A. Nemchin, M.J. Whitehouse, J.F. Snape, P. Bland, G.K. Benedix, The Pb isotopic evolution of the Martian mantle constrained by initial Pb in Martian meteorites. J. Geophys. Res., Planets **120**, 2224–2240 (2015). https://doi.org/10.1002/2015JE004809

J. Blum, Dust evolution in protoplanetary discs and the formation of planetesimals. What have we learned from laboratory experiments? Space Sci. Rev. **214**, 52 (2018). https://doi.org/10.1007/s11214-018-0486-5

J-D. Bodénan, N.A. Starkey, S.S. Russell, I.P. Wright, I.A. Franchi, An oxygen isotope study of wark-lovering rims on type A CAIs in primitive carbonaceous chondrites. Earth Planet. Sci. Lett. **401**, 327–336 (2014). https://doi.org/10.1016/j.epsl.2014.05.035

P. Bonnand, H.M. Williams, I.J. Parkinson, B.J. Wood, A.N. Halliday, Stable chromium isotopic composition of meteorites and metal–silicate experiments: implications for fractionation during core formation. Earth Planet. Sci. Lett. **435**, 14–21 (2016). https://doi.org/10.1016/j.epsl.2015.11.026

W.F. Bottke, D. Nesvorný, R.E. Grimm, A. Morbidelli, D.P. O'Brien, Iron meteorites as remnants of planetesimals formed in the terrestrial planet region. Nature **439**, 821–824 (2006). https://doi.org/10.1038/nature04536

A. Bouvier, J. Blichert-Toft, J.D. Vervoort, P. Gillet, F. Albarède, The case for old basaltic shergottites. Earth Planet. Sci. Lett. **266**, 105–124 (2008). https://doi.org/10.1016/j.epsl.2007.11.006

R. Brasser, S.J. Mojzsis, The partitioning of the inner and outer Solar System by a structured protoplanetary disk. Nat. Astron. (2020). https://doi.org/10.1038/s41550-019-0978-6

J. Brooks, S.A. Jacobson, Losing moons: the gravitational influence of close encounters on satellite orbits. AAS Division on Dynamical Astronomy meeting #50, id. 302.05. Bull. Am. Astron. Soc. **51**, 5 (2019). https://113qx216in8z1kdeyi404hgf-wpengine.netdna-ssl.com/wp-content/uploads/2019/06/dda50_abstracts.pdf#abs302.05

D. Brownlee, The stardust mission: analyzing samples from the edge of the Solar System. Annu. Rev. Earth Planet. Sci. **42**, 179–205 (2014). https://doi.org/10.1146/annurev-earth-050212-124203

G. Budde, C. Burkhardt, G.A. Brennecka, M. Fischer-Gödde, T.S. Kruijer, T. Kleine, Molybdenum isotopic evidence for the origin of chondrules and a distinct genetic heritage of carbonaceous and non-carbonaceous meteorites. Earth Planet. Sci. Lett. **454**, 293–303 (2016). https://doi.org/10.1016/j.epsl.2016.09.020

T.H. Burbine, R.C. Greenwood, Exploring the Bimodal Solar System via Sample Return from the Main Asteroid Belt: The Case for Revisiting Ceres. Space Sci. Rev. (2020). https://doi.org/10.1007/s11214-020-00671-0

C. Burkhardt, N. Dauphas, U. Hans, B. Bourdon, T. Kleine, Elemental and isotopic variability in Solar System materials by mixing and processing of primordial disk reservoirs. Geochim. Cosmochim. Acta **261**, 145–170 (2019). https://doi.org/10.1016/j.gca.2019.07.003

P.K. Byrne, R.C. Ghail, M.S. Gilmore, S.E. Smrekar, A.H. Treiman, C.F. Wilson, S.C. Solomon, The exploration of Venus: current understanding and open questions, in *50th Lunar and Planetary Science Conference*. LPI Contribution, vol. 2132 (2019). Abstract 2853. https://www.hou.usra.edu/meetings/lpsc2019/pdf/2853.pdf

R.M. Canup, E. Asphaug, Origin of the Moon in a giant impact near the end of the Earth's formation. Nature **412**, 708–712 (2001). https://doi.org/10.1038/35089010

J.E. Chambers, Pebble accretion and the diversity of planetary systems. Astrophys. J. **825**, 63 (2016). https://doi.org/10.3847/0004-637X/825/1/63

E. Chassefière et al., European Venus Explorer (EVE): an in-situ mission to Venus. Exp. Astron. **23**, 741–760 (2009). https://doi.org/10.1007/s10686-008-9093-x

A. Chau, C. Reinhardt, R. Helled, J. Stadel, Forming Mercury by giant impacts. Astrophys. J. **865**, 35 (2018). https://doi.org/10.3847/1538-4357/aad8b0

B-G. Choi, K.D. McKeegan, A.N. Krot, J.T. Wasson, Extreme oxygen-isotope compositions in magnetite from unequilibrated ordinary chondrites. Nature **392**, 577–579 (1998). https://doi.org/10.1038/33356

R.N. Clayton, Self-shielding in the solar nebula. Nature **415**, 860–861 (2002). https://doi.org/10.1038/415860b

R.N. Clayton, Oxygen isotopes in meteorites, in *Treatise on Geochemistry*, vol. 1, ed. by A.M. Davis (Elsevier, Amsterdam, 2003), pp. 129–142. https://doi.org/10.1016/B0-08-043751-6/01063-X

R.N. Clayton, Meteorites and their parent asteroids. Science **313**, 1743–1744 (2006). https://doi.org/10.1126/science.1133137

R.N. Clayton, T.K. Mayeda, Formation of ureilites by nebular processes. Geochim. Cosmochim. Acta **52**, 1313–1318 (1988). https://doi.org/10.1016/0016-7037(88)90286-4

R.N. Clayton, L. Grossman, T.K. Mayeda, A component of primitive nuclear composition in carbonaceous chondrites. Science **182**, 485–488 (1973). https://doi.org/10.1126/science.182.4111.485

R.N. Clayton, N. Onuma, L. Grossman, T.K. Mayeda, Distribution of the pre-solar component in Allende and other carbonaceous chondrites. Earth Planet. Sci. Lett. **34**, 209–224 (1977). https://doi.org/10.1016/0012-821X(77)90005-X

I.I. Cleeves, E.A. Bergin, C.M.O'D. Alexander, F. Du, D. Graninger, K.I. Oberg, T.J. Harries, The ancient heritage of water ice in the Solar System. Science **345**, 1590–1593 (2014). https://doi.org/10.1126/science.1258055

Committee on Space Research, The International Planetary Protection Handbook (2019). http://pposs.org/planetary-protection-handbook/

M. Coradini, G. Scoon, J-P. Lebreton, Venus sample return: assessment study report. M. Coradini, G. Scoon, J.-P. Lebreton. [S.l.]: European Space Agency, 1998. ESA/(Observatory Publications) (1998)

J.N. Cuzzi, R.C. Hogan, W.F. Bottke, Towards initial mass functions for asteroids and Kuiper Belt Objects. Icarus **208**, 518–538 (2010). https://doi.org/10.1016/j.icarus.2010.03.005

L.R. Dartnell, T.A. Nordheim, M.R. Patel, J.P. Mason, A.J. Coates, G.H. Jones, Constraints on a potential aerial biosphere on Venus: I. Cosmic rays. Icarus **257**, 396–405 (2015). https://doi.org/10.1016/j.icarus.2015.05.006

N. Dauphas, The isotopic nature of the Earth's accreting material through time. Nature **541**, 521–524 (2017). https://doi.org/10.1038/nature20830

N. Dauphas, E.A. Schauble, Mass fractionation laws, mass-independent effects, and isotopic anomalies. Annu. Rev. Earth Planet. Sci. **44**, 709–783 (2016). https://doi.org/10.1146/annurev-earth-060115-012157

N. Dauphas, C. Burkhardt, P. Warren, F-Z. Teng, Geochemical arguments for an Earth-like Moon-forming impactor. Philos. Trans. R. Soc. Lond. A **372**, 20130244 (2014). https://doi.org/10.1098/rsta.2013.0244

J.K. Dhaliwal, J.M.D. Day, C.A. Corder, K.T. Tait, K. Marti, N. Assayag, P. Cartigny, D. Rumble III., L.A. Taylor, Early metal-silicate differentiation during planetesimal formation revealed by acapulcoite and lodranite meteorites. Geochim. Cosmochim. Acta **216**, 115–140 (2017). https://doi.org/10.1016/j.gca.2017.06.042

V.P. Dolgopolov, K.M. Pichkhadze, K.G. Sukhanov, The Vega Project: a space mission to Venus and Halley's comet. Space Sci. Rev. **46**, 568–577 (2012). https://doi.org/10.1134/S0038094612070106

T.M. Donahue, J.H. Hoffman, R.R. Hodges, A.J. Watson, Venus was wet—a measurement of the ratio of deuterium to hydrogen. Science **216**, 630–633 (1982). https://doi.org/10.1126/science.216.4546.630

P. Driscoll, D. Bercovici, Divergent evolution of Earth and Venus: influence of degassing, tectonics, and magnetic fields. Icarus **226**, 1447–1464 (2013). https://doi.org/10.1016/j.icarus.2013.07.025

J. Farquhar, M.H. Thiemens, T. Jackson, Atmosphere-surface interactions on Mars: $\Delta^{17}O$ measurements of carbonate from ALH 84001. Science **280**, 1580–1582 (1998). https://doi.org/10.1126/science.280.5369.1580

G. Feulner, The faint young Sun problem. Rev. Geophys. **50**, RG2006 (2012). https://doi.org/10.1029/2011RG000375

I.A. Franchi, Oxygen isotopes in asteroidal materials. Rev. Miner. **68**, 345–397 (2008). https://doi.org/10.2138/rmg.2008.68.13

H.B. Franz, S-T. Kim, J. Farquhar, J.M.D. Day, R.C. Economos, K.D. McKeegan, A.K. Schmitt, A.J. Irving, J. Hoek, J. Dottin III., Isotopic links between atmospheric chemistry and the deep sulphur cycle on Mars. Nature **508**, 364–368 (2014). https://doi.org/10.1038/nature13175

J. Garvin, G. Arney, S. Getty, N. Johnson, W. Kiefer, R. Lorenz, M. Ravine, C. Malespin, C. Webster, B. Campbell, N. Izenberg, V. Cottini, DAVINCI+: deep atmosphere of Venus investigation of noble gases, chemistry and imaging, plus, in *51st Lunar Planet. Sci. Conf.* (2020). Abstract number 2599. https://www.hou.usra.edu/meetings/lpsc2020/pdf/2599.pdf

R. Ghail, Rheological and petrological implications for a stagnant lid regime on Venus. Planet. Space Sci. **113–114**, 2–9 (2015). https://doi.org/10.1016/j.pss.2015.02.005

R. Ghail, C. Wilson, T. Widemann, L. Bruzzone, C. Dumoulin, J. Helbert, R. Herrick, E. Marcq, P. Mason, P. Rosenblatt, A. Carine Vandaele, L-J. Burtz, EnVision: understanding why our most Earth-like neighbour is so different (2017). Eprint arXiv:1703.09010

M. Gilmore, A. Treiman, J. Helbert, S. Smrekar, Venus surface composition constrained by observation and experiment. Space Sci. Rev. **212**, 1511–1540 (2017). https://doi.org/10.1007/s11214-017-0370-8

M.S. Gilmore, J.L. Whitten, R. Perkins, J.F. Brossier, The ancient environments of Venus as recorded by Tessera terrain, in *American Geophysical Union, Fall Meeting 2019* (2019). Abstract #P51G-3445. https://ui.adsabs.harvard.edu/abs/2019AGUFM.P51G3445G/abstract

L.S. Glaze, J.B. Garvin, Exploring Venus: never give up, never surrender, in *49th Lunar and Planetary Science Conference*. LPI Contribution, vol. 2083 (2018). Abstract 2024. https://www.hou.usra.edu/meetings/lpsc2018/pdf/2024.pdf

L.S. Glaze, C.F. Wilson, L.V. Zasova, M. Nakamura, S. Limaye, Future of Venus research and exploration. Space Sci. Rev. **214**, 89 (2018). https://doi.org/10.1007/s11214-018-0528-z

C. Goldblatt, A.J. Watson, The runaway greenhouse: implications for future climate change, geoengineering and planetary atmospheres. Philos. Trans. R. Soc., Math. Phys. Eng. Sci. **370**, 4197–4216 (2012). https://doi.org/10.1098/rsta.2012.0004

C. Goodrich, D.P. O'Brien, Where did the ureilite parent body accrete? Constraints from chemical and isotopic compositions, in *American Astronomical Society, DPS Meeting #46, id. 304.03* (2014). https://ui.adsabs.harvard.edu/abs/2014DPS....4630403G/abstract

C.A. Goodrich, N.T. Kita, Q-Z. Yin, M.E. Sanborn, C.D. Williams, D. Nakashima, M.D. Lane, S. Boyle, Petrogenesis and provenance of ungrouped achondrite northwest Africa 7325 from petrology, trace elements, oxygen, chromium and titanium isotopes, and mid-IR spectroscopy. Geochim. Cosmochim. Acta **203**, 381–403 (2017). https://doi.org/10.1016/j.gca.2016.12.021

W. Graham, Japanese sounding rocket claims record-breaking orbital launch. NASAspaceflight.com (2018). https://www.nasaspaceflight.com/2018/02/japanese-rocket-record-borbital-launch/

R.C. Greenwood, T.H. Burbine, M.F. Miller, I.A. Franchi, Melting and differentiation of early-formed asteroids: the perspective from high precision oxygen isotope studies. Chem. Erde **77**, 1–43 (2017). https://doi.org/10.1016/j.chemer.2016.09.005

J.P. Greenwood, S. Karato, K.E. Vander Kaaden, K. Pahlevan, T. Usui, Water and volatile inventories of Mercury, Venus, the Moon, and Mars. Space Sci. Rev. **214**, 92 (2018a). https://doi.org/10.1007/s11214-018-0526-1

R.C. Greenwood, J-A. Barrat, M.F. Miller, M. Anand, N. Dauphas, I.A. Franchi, P. Sillard, N.A. Starkey, Oxygen isotopic evidence for accretion of Earth's water before a high-energy Moon-forming giant impact. Sci. Adv. **4**, 5928 (2018b). https://doi.org/10.1126/sciadv.aao5928

R.C. Greenwood, T.H. Burbine, I.A. Franchi, Linking asteroids and meteorites to the primordial planetesimal population Geochim. Cosmochim. Acta (2020, in press). https://doi.org/10.1016/j.gca.2020.02.004

T.K.P. Gregg, L. Zasova, T. Economou, N. Eismont, M. Gerasimov, D. Gorinov, N. Ignatiev, M. Ivanov, I. Khatuntsev, O. Korablev, T. Kremic, K. Jessup, S. Limaye, I. Lomakin, A. Martynov, A. Ocampo, Venera-D: a potential long-lived mission to explore Venus' surface, atmosphere and plasma environment, in *50th Lunar Planet. Sci. Conf.*. LPI Contrib., vol. 2132 (2019). Abstract number 1738. https://www.hou.usra.edu/meetings/lpsc2019/pdf/1738.pdf

R.E. Grimm, H.Y. McSween Jr., Water and the thermal evolution of carbonaceous chondrite parent bodies. Icarus **82**, 244–280 (1989). https://doi.org/10.1016/0019-1035(89)90038-9

D.H. Grinspoon, The need for new in situ measurements to understand the climate, geology and evolution of Venus, in *American Geophysical Union, Fall Meeting* (2017). Abstract #P52A-08. https://ui.adsabs.harvard.edu/abs/2017AGUFM.P52A..08G/abstract

B. Gundlach, J. Blum, The stickiness of micrometer-sized water-ice particles. Astrophys. J. **798**, 34 (2015). https://doi.org/10.1088/0004-637X/798/1/34

J.H. Guo, The effect of photoionization on the loss of water of the planet. Astrophys. J. **872**, 99 (2019). https://doi.org/10.3847/1538-4357/aaffd4

K.E. Haisch Jr., E.A. Lada, C.J. Lada, Disk frequencies and lifetimes in young clusters. Astrophys. J. Lett. **553**, L153 (2001). https://doi.org/10.1086/320685
L.J. Hallis, M. Anand, R.C. Greenwood, M.F. Miller, I.A. Franchi, S.S. Russell, The oxygen isotope composition, petrology and geochemistry of mare basalts: evidence for large-scale compositional variation in the lunar mantle. Geochim. Cosmochim. Acta **74**, 6885–6899 (2010). https://doi.org/10.1016/j.gca.2010.09.023
G.L. Hashimoto, M. Roos-Serote, S. Sugita, M.S. Gilmore, L.W. Kamp, R.W. Carlson, K.H. Baines, Felsic highland crust on Venus suggested by Galileo near-infrared mapping spectrometer data. J. Geophys. Res. **113**(E9), E00B24 (2008). https://doi.org/10.1029/2008JE003134
L.J. Hicks, J.C. Bridges, S.J. Gurman, Ferric saponite and serpentine in the nakhlite martian meteorites. Geochim. Cosmochim. Acta **136**, 194–210 (2014). https://doi.org/10.1016/j.gca.2014.04.010
T. Hopp, G. Budde, T. Kleine, Heterogeneous accretion of Earth inferred from Mo-Ru isotope systematics. Earth Planet. Sci. Lett. **534**, 116065 (2020). https://doi.org/10.1016/j.epsl.2020.116065
T.R. Ireland, J. Avila, R.C. Greenwood, L.J. Hicks, J.C. Bridges, Oxygen isotopes and sampling of the Solar System. Space Sci. Rev. (2020). https://doi.org/10.1007/s11214-020-0645-3
M.A. Ivanov, J.W. Head, Global geological map of Venus. Planet. Space Sci. **59**, 1559–1600 (2011). https://doi.org/10.1016/j.pss.2011.07.008
S.A. Jacobson, D.C. Rubie, J. Hernlund, A. Morbidelli, M. Nakajima, Formation, stratification, and mixing of the cores of Earth and Venus. Earth Planet. Sci. Lett. **474**, 375–386 (2017). https://doi.org/10.1016/j.epsl.2017.06.023
M. Javoy, E. Kaminski, F. Guyot, D. Andrault, C. Sanloup, M. Moreira, S. Labrosse, A. Jambon, P. Agrinier, A. Davaille, C. Jaupart, The chemical composition of the Earth: enstatite chondrite models. Earth Planet. Sci. Lett. **293**, 259–268 (2010). https://doi.org/10.1016/j.epsl.2010.02.033
G.A. Jenner, H.P. Longerich, E. Jackson, B.J. Fryer, ICP-MS — a powerful tool for high-precision trace-element analysis in Earth sciences: evidence from analysis of selected U.S.G.S. reference samples. Chem. Geol. **83**, 133–148 (1990). https://doi.org/10.1016/0009-2541(90)90145-W
Z. Jin, M. Bose, New clues to ancient water on Itokawa. Sci. Adv. **5**, 8106 (2019). https://doi.org/10.1126/sciadv.aav8106
A. Johansen, M. Lambrechts, Forming planets via pebble accretion. Annu. Rev. Earth Planet. Sci. **45**, 359–387 (2017). https://doi.org/10.1146/annurev-earth-063016-020226
R.G. Joseph, Life on Venus and the interplanetary transfer of biota from Earth. Astrophys. Space Sci. **364**(11), 191 (2019). https://doi.org/10.1007/s10509-019-3678-x
A.J.G. Jurewicz et al., The genesis solar-wind collector materials. Space Sci. Rev. **105**, 535–560 (2003). https://doi.org/10.1023/A:1024469927444
N.A. Kaib, N.B. Cowan, The feeding zones of terrestrial planets and insights into Moon formation. Icarus **252**, 161–174 (2015). https://doi.org/10.1016/j.icarus.2015.01.013
G.W. Kallemeyn, A.E. Rubin, J.T. Wasson, The compositional classification of chondrites: VII. The R chondrite group. Geochim. Cosmochim. Acta **60**, 2243–2256 (1996). https://doi.org/10.1016/0016-7037(96)88430-4
S.R. Kane, A.J. Ceja, M.J. Way, E.V. Quintana, Climate modelling of a potential ExoVenus. Astrophys. J. **869**, 46 (2018). https://doi.org/10.3847/1538-4357/aaec68
S.R. Kane, G. Arney, D. Crisp, S. Domagal-Goldman, L.S. Glaze, C. Goldblatt, D. Grinspoon, J.W. Head, A. Lenardic, C. Unterborn, M.J. Way, K.J. Zahnle, Venus as a laboratory for exoplanetary science. J. Geophys. Res., Planets **124**, 2015–2028 (2019). https://doi.org/10.1029/2019JE005939
J.F. Kasting, J.B. Pollack, Loss of water from Venus. I - Hydrodynamic escape of hydrogen. Icarus **53**, 479–508 (1983). https://doi.org/10.1016/0019-1035(83)90212-9
S. Khawja, R.E. Ernst, C. Samson, P.K. Byrne, R.C. Ghail, Erosional features on tessera terrains Venus, in *51st Lunar Planet. Sci. Conf.* (2020). Abstract number 1161. https://www.hou.usra.edu/meetings/lpsc2020/pdf/1161.pdf
T. Kodama, H. Genda, R. O'ishi, A. Abe-Ouchi, Y. Abe, Inner edge of habitable zones for Earth-sized planets with various surface water distributions. J. Geophys. Res., Planets **124**, 2306–2324 (2019). https://doi.org/10.1029/2019JE006037
V.A. Krasnopolsky, D.A. Belyaev, I.E. Gordon, G. Li, L.S. Rothman, Observations of D/H ratios in H_2O, HCl, and HF on Venus and new DCl and DF line strengths. Icarus **224**, 57–65 (2013). https://doi.org/10.1016/j.icarus.2013.02.010
A.N. Krot, Y. Amelin, P. Bland, F.J. Ciesla, J. Connelly, A.M. Davis, G.R. Huss, I.D. Hutcheon, K. Makide, K. Nagashima, L.E. Nyquist, S.S. Russell, E.R.G. Scott, K. Thrane, H. Yurimoto, Q-Z. Yin, Origin and chronology of chondritic components: a review. Geochim. Cosmochim. Acta **73**, 4963–4997 (2009). https://doi.org/10.1016/j.gca.2008.09.039
A.N. Krot, K. Keil, E.R.D. Scott, C.A. Goodrich, M.K. Weisberg, Classification of meteorites and their genetic relationships, in *Meteorites and cosmochemical processes; Treatise on Geochemistry, vol. 1*, ed. by A.M. Davis 2nd edn. (Elsevier, Amsterdam, 2014), pp. 1–63

T.S. Kruijer, C. Burkhardt, G. Budde, T. Kleine, Thorsten Age of Jupiter inferred from the distinct genetics and formation times of meteorites. Proc. Natl. Acad. Sci. **114**, 6712–6716 (2017). https://doi.org/10.1073/pnas.1704461114

T.S. Kruijer, T. Kleine, L.E. Borg, The great isotopic dichotomy of the early Solar System. Nat. Astron. **4**, 32–40 (2020). https://doi.org/10.1038/s41550-019-0959-9

M. Lambrechts, A. Johansen, Rapid growth of gas-giant cores by pebble accretion. Astron. Astrophys. **544**, A32 (2012). https://doi.org/10.1051/0004-6361/201219127

D.S. Lauretta et al., Sample return from Asteroid (101955) Bennu. Space Sci. Rev. **212**, 925–984 (2017). https://doi.org/10.1007/s11214-017-0405-1

J. Leconte, F. Forget, B. Charnay, B. Wordsworth, A. Pottier, Increased insolation threshold for runaway greenhouse processes on Earth-like planets. Nature **504**, 268–271 (2014). https://doi.org/10.1038/nature12827

S.J. Limaye, R. Mogul, D.J. Smith, A.H. Ansari, G.P. Słowik, P. Vaishampayan, Venus' spectral signatures and the potential for life in the clouds. Astrobiology **18**, 1181–1198 (2018). https://doi.org/10.1089/ast.2017.1783

M-C. Liu, K.D. McKeegan, J.N. Goswami, K.K. Marhas, S. Sahijpal, T.R. Ireland, A.M. Davis, Isotopic records in CM hibonites: implications for timescales of mixing of isotope reservoirs in the solar nebula. Geochim. Cosmochim. Acta **73**, 5051–5079 (2009). https://doi.org/10.1016/j.gca.2009.02.039

S.J. Lock, S.T. Stewart, The structure of terrestrial bodies: impact heating, corotation limits, and synestias. J. Geophys. Res. (2017). https://doi.org/10.1002/2016JE005239

S.J. Lock, S.T. Stewart, M.I. Petaev, Z.M. Leinhardt, M. Mace, S.B. Jacobsen, M. Ćuk, A new model for lunar origin: equilibration with Earth beyond the hot spin stability limit, in *LPSC*, vol. 47 (2016). Abstract #2881. https://www.hou.usra.edu/meetings/lpsc2016/pdf/2881.pdf

K. Lodders, Solar system abundances and condensation temperatures of the elements. Astrophys. J. **591**, 1220–1247 (2003). https://doi.org/10.1086/375492

J-P. Lorand, R.H. Hewins, M. Humayun, L. Remusat, B. Zanda, C. La, S. Pont, Chalcophile-siderophile element systematics of hydrothermal pyrite from martian regolith breccia NWA 7533. Geochim. Cosmochim. Acta **241**, 134–149 (2018). https://doi.org/10.1016/j.gca.2018.08.041

J.R. Lyons, E.D. Young, CO self-shielding as the origin of oxygen isotope anomalies in the early solar nebula. Nature **435**, 317–320 (2005). https://doi.org/10.1038/nature03557

J.R. Lyons, E.A. Bergin, F.J. Ciesla, A.M. Davis, S.J. Desch, K. Hashizume, J-A. Lee, Timescales for the evolution of oxygen isotope compositions in the solar nebula. Geochim. Cosmochim. Acta **73**, 4998–5017 (2009). https://doi.org/10.1016/j.gca.2009.01.041

Mars Exploration Program Analysis Group, Planning for Mars returned sample science: final report of the MSR end-to-end international science analysis group (E2E-iSAG). Astrobiology **12**, 175–230 (2012). https://doi.org/10.1089/ast.2011.0805

Y. Matsuhisa, J.R. Goldsmith, R.N. Clayton, Mechanisms of hydrothermal crystallization of quartz at 250°C and 15 kbar. Geochim. Cosmochim. Acta **42**, 173–182 (1978). https://doi.org/10.1016/0016-7037(78)90130-8

J.E.P. Matzel, H.A. Ishii, D. Joswiak, I.D. Hutcheon, J.P. Bradley, D. Brownlee, P.K. Weber, N. Teslich, G. Matrajt, K.D. McKeegan, G.J. MacPherson, Constraints on the formation age of cometary material from the NASA Stardust Mission. Science **328**, 483–486 (2010). https://doi.org/10.1126/science.1184741

K.D. McKeegan, A.P.A. Kallio, V.S. Heber, G. Jarzebinski, P.H. Mao, C.D. Coath, T. Kunihiro, R.C. Wiens, J.E. Nordholt, R.W. Moses, D.B. Reisenfeld, A.J. Jurewicz, D.S. Burnett, The oxygen isotopic composition of the Sun inferred from captured solar wind. Science **332**, 1528–1532 (2011). https://doi.org/10.1126/science.1204636

H.Y. McSween Jr., What we have learned about Mars from SNC meteorites. Meteoritics **29**(6), 757–779 (1994). https://doi.org/10.1111/j.1945-5100.1994.tb01092.x

S. Mikhail, M.J. Heap, Hot climate inhibits volcanism on Venus: constraints from rock deformation experiments and argon isotope geochemistry. Phys. Earth Planet. Inter. **268**, 18–34 (2017). https://doi.org/10.1016/j.pepi.2017.05.007

M.F. Miller, Isotopic fractionation and the quantification of ^{17}O anomalies in the oxygen three-isotope system: an appraisal and geochemical significance. Geochim. Cosmochim. Acta **66**, 1881–1889 (2002). https://doi.org/10.1016/S0016-7037(02)00832-3

M.F. Miller, A. Pack, I.N. Bindeman, R.C. Greenwood, Standardizing the reporting of $\Delta^{17}O$ data from high precision oxygen triple-isotope ratio measurements of silicate rocks and minerals. Chem. Geol. **532**, 119332 (2020). https://doi.org/10.1016/j.chemgeo.2019.119332

A. Morbidelli, *Accretion Processes, Oxford Research Encyclopedia, Planetary Science* (Oxford University Press, USA, 2018). https://doi.org/10.1093/acrefore/9780190647926.013.32

A. Morbidelli, M. Lambrechts, S. Jacobson, B. Bitsch, The great dichotomy of the Solar System: small terrestrial embryos and massive giant planet cores. Icarus **258**, 418–429 (2015). https://doi.org/10.1016/j.icarus.2015.06.003

N.T. Mueller, C. Tsang, D.C. Nunes, J. Helbert, M.D. Dyar, S.E. Smrekar, Near infrared multispectral mapping of Venus supports the hypothesis that tessera plateau material was formed in the presence of surface water, in *American Geophysical Union, Fall Meeting 2017* (2017). Abstract #P53A-2645. https://ui.adsabs.harvard.edu/abs/2017AGUFM.P53A2645M/abstract

R. Nakada, T. Usui, M. Ushioda, Y. Takahashi, Micro-Xanes determination of oxidation states of V and Fe in olivine-hosted glass inclusion and groundmass of Yamato 980459, in *Ninth International Conference on Mars*. LPI Contrib., vol. 2089 (2019). Abstract number 6133. https://www.hou.usra.edu/meetings/ninthmars2019/pdf/6133.pdf

M. Nakamura, Venus Observation by Akatsuki: 2015–2019. EPSC Abstracts 13, EPSC-DPS2019-391-3 (2019). https://meetingorganizer.copernicus.org/EPSC-DPS2019/EPSC-DPS2019-391-3.pdf

T. Nakamura et al., Itokawa dust particles: a direct link between S-type asteroids and ordinary chondrites. Science **333**, 1113–1116 (2013). https://doi.org/10.1126/science.1207758

M. Nakamura et al., AKATSUKI returns to Venus. Earth Planets Space **68**, 75, 10 (2016). https://doi.org/10.1186/s40623-016-0457-6

J.A.M. Nanne, F. Nimmo, J.N. Cuzzi, T. Kleine, Origin of the non-carbonaceous–carbonaceous meteorite dichotomy. Earth Planet. Sci. Lett. **511**, 44–54 (2019). https://doi.org/10.1016/j.epsl.2019.01.027

National Space Policy of the United States of America (2010). https://history.nasa.gov/national_space_policy_6-28-10.pdf

F. Nimmo, D. McKenzie, Volcanism and tectonics on Venus. Annu. Rev. Earth Planet. Sci. **26**, 23–53 (1998). https://doi.org/10.1146/annurev.earth.26.1.23

E.G. Nisbet, N.H. Sleep, The habitat and nature of early life. Nature **409**, 1083–1091 (2001). https://doi.org/10.1038/35059210

J.G. O'Rourke, C. Gillmann, P. Tackley, Prospects for an ancient dynamo and modern crustal remanent magnetism on Venus. Earth Planet. Sci. Lett. **502**, 46–56 (2018). https://doi.org/10.1016/j.epsl.2018.08.055

U. Ott, T.D. Swindle, S.P. Schwenzer, Chap. 3 - Noble Gases in Martian Meteorites: Budget, Sources, Sinks, and Processes in Volatiles in the Martian Crust (2019), pp. 35–70. https://doi.org/10.1016/B978-0-12-804191-8.00003-9

A. Pack, D. Herwartz, The triple oxygen isotope composition of the Earth mantle and understanding $\Delta^{17}O$ variations in terrestrial rocks and minerals. Earth Planet. Sci. Lett. **390**, 138–145 (2014). https://doi.org/10.1016/j.epsl.2014.01.017

S. Peron, M. Moreira, A. Agranier, Origin of light noble gases (He, Ne, and Ar) on Earth: a review. Geochem. Geophys. Geosyst. **9**, 979–996 (2018). https://doi.org/10.1002/2017GC007388

E.V. Pitjeva, N.P. Pitjev, Masses of the Main Asteroid Belt and the Kuiper Belt from the motions of planets and spacecraft. Astron. Lett. **44**, 554–566 (2018). https://doi.org/10.1134/S1063773718090050

Planetary Society website, Missions to Mars, Planetary Society (2020). http://www.planetary.org/explore/space-topics/space-missions/missions-to-mars.html

L. Qin, C.M.O'D. Alexander, R.W. Carlson, M.F. Horan, T. Yokoyama, Contributions to chromium isotope variations of meteorites. Geochim. Cosmochim. Acta **74**, 1122–1145 (2010). https://doi.org/10.1016/j.gca.2009.11.005

S.N. Raymond, A. Izidoro, Origin of water in the inner Solar System: planetesimals scattered inward during Jupiter and Saturn's rapid gas accretion. Icarus **297**, 134–148 (2017). https://doi.org/10.1016/j.icarus.2017.06.030

F. Rehnmark, K. Zacny, J. Hall, E. Cloninger, C. Hyman, K. Kriechbaum, J. Melko, J. Rabinovitch, B. Wilcox, J. Lambert, N. Traeden, J. Bailey, Sample acquisition drill for Venus in situ explorer (VISE), in *48th Lunar and Planetary Science* (2017). Abstract number 1367. https://www.hou.usra.edu/meetings/lpsc2017/eposter/1367.pdf

M.S. Robinson, J.B. Plescia, B.L. Jolliff, S.J. Lawrence, Soviet lunar sample return missions: landing site identification and geologic context. Planet. Space Sci. **69**, 76–88 (2012). https://doi.org/10.1016/j.pss.2012.03.013

D. Rodgers, M. Gilmore, T. Sweetser, J. Cameron, G.-S. Chen, J. Cutts, R. Gershman, J.L. Hall, V. Kerzhanovich, A. McRonald, E. Nilsen, W. Petrick, C. Sauer, B. Wilcox, A. Yavrouian, W. Zimmerman, Venus sample return. A hot topic, in *Proc. IEEE Aerospace Conf.*, vol. 7 (2000), pp. 473–484. https://trs.jpl.nasa.gov/bitstream/handle/2014/18501/99-1991.pdf?sequence=1

J.J.R. Rosman, P.D.P. Taylor, Isotopic compositions of the elements 1997. Pure Appl. Chem. **70**, 217–236 (1998). https://doi.org/10.1351/pac199870010217

A.E. Rubin, J.T. Wasson, Variations of chondrite properties with heliocentric distance. Meteoritics **30**, 569 (1995) (abstract). http://articles.adsabs.harvard.edu/pdf/1995Metic..30Q.569R

R.Z. Sagdeev, V.M. Linkin, J.E. Blamont, R.A. Preston, The VEGA Venus balloon experiment. Science **231**, 1407–1408 (1986). https://doi.org/10.1126/science.231.4744.1407

N. Sakamoto, Y. Seto, S. Itoh, K. Kuramoto, K. Fujino, K. Nagashima, A.N. Krot, H. Yurimoto, Remnants of the early Solar System water enriched in heavy oxygen isotopes. Science **317**, 231–233 (2007). https://doi.org/10.1126/science.1142021

E.R.D. Scott, A.N. Krot, I.S. Sanders, Isotopic dichotomy among meteorites and its bearing on the protoplanetary disk. Astrophys. J. **854**, 164 (2018). https://doi.org/10.3847/1538-4357/aaa5a5
D.W.G. Sears, A. Sehlke, J.M. Friedrich, M.L. Rivers, D.S. Ebel, X-ray computed tomography of extraterrestrial rocks eradicates their natural radiation record and the information it contains. Meteorit. Planet. Sci. **53**, 2624–2631 (2018). https://doi.org/10.1111/maps.13183
E.V. Shalygin, W.J. Markiewicz, A.T. Basilevsky, D.V. Titov, N.I. Ignatiev, J.W. Head, Active volcanism on Venus in the Ganiki Chasma rift zone. Geophys. Res. Lett. **42**, 4762–4769 (2015). https://doi.org/10.1002/2015GL064088
E. Shibata, Y. Lu, A. Pradeepkumar, J.A. Cutts, S.J. Saika, A Venus atmospheric sample return mission concept: feasibility and technology requirements, in *Planetary Science Vision 2050 Workshop*. LPI Contribution, vol. 1981 (2017). Abstract 8164. https://www.hou.usra.edu/meetings/V2050/pdf/8164.pdf
C-Y. Shih, L.E. Nyquist, H. Wiesmann, Y. Reese, K. Misawa, Rb-Sr and Sm-Nd dating of olivine-phyric shergottite Yamato 980459: petrogenesis of depleted shergottites. Antarct. Meteor. Res. **18**, 46–65 (2005) http://adsabs.harvard.edu/full/2005AMR....18...46S
F.H. Shu, H. Shang, M. Gounelle, A.E. Glassgold, T. Lee, Typhoon. The origin of chondrules and refractory inclusions in chondritic meteorites. Astrophys. J. **548**(2), 1029–1050 (2001). https://doi.org/10.1086/319018
S.E. Smrekar, A. Davaille, C. Sotin, Venus interior structure and dynamics. Space Sci. Rev. **214**, 88 (2018). https://doi.org/10.1007/s11214-018-0518-1
S. Smrekar, S. Hensley, D. Dyar, J. Helbert, VERITAS (Venus Emissivity, Radio science, InSAR, Topography And Spectroscopy): a proposed discovery mission, in *EPSC-DPS Joint Meeting EPSC-DPS2019-1124* (2019). https://meetingorganizer.copernicus.org/EPSC-DPS2019/EPSC-DPS2019-1124-1.pdf
M.L. Spicuzza, J.M.D. Day, L.A. Taylor, J.W. Valley, Oxygen isotope constraints on the origin and differentiation of the Moon. Earth Planet. Sci. Lett. **253**, 254–265 (2007). https://doi.org/10.1016/j.epsl.2006.10.030
N.A. Starkey, I.A. Franchi, M.R. Lee, Isotopic diversity in interplanetary dust particles and preservation of extreme ^{16}O-depletions. Geochim. Cosmochim. Acta **142**, 115–131 (2014). https://doi.org/10.1016/j.gca.2014.07.011
A. Stephant, M. Anand, H.O. Ashcroft, X. Zhao, R.L. Korotev, S. Strekopytove, R.C. Greenwood, E. Humphreys-Williams, Y. Liu, G. Tang, Q. Li, I.A. Franchi, An ancient reservoir of volatiles in the Moon sampled by lunar meteorite Northwest Africa 10989. Geochim. Cosmochim. Acta **266**, 163–183 (2019). https://doi.org/10.1016/j.gca.2019.07.045
D.J. Stevenson, Planetary magnetic fields. Earth Planet. Sci. Lett. **208**, 1–11 (2003). https://doi.org/10.1016/S0012-821X(02)01126-3
D.J. Stevenson, A.N. Halliday, The origin of the Moon. Philos. Trans. R. Soc. A **372**, 20140289 (2014). https://doi.org/10.1098/rsta.2014.0289
F.W. Taylor, *The Scientific Exploration of Venus* (Cambridge University Press, Cambridge, 2014)
F.W. Taylor, J.P. Lebreton, G. Scoon, P. Logonne, P. Masson, H. Waenke, M. Coradini, VSR and MSR study team ESA advanced planetary mission studies: Venus sample return and Mercury sample return, in *American Astronomical Society, DPS Meeting #31, id. 21.06* (1999). http://adsabs.harvard.edu/full/1999BAAS...31.1108T
F.W. Taylor, H. Svedhem, J.W. Head, Venus: the atmosphere, climate, surface, interior and near-space environment of an Earth-like planet. Space Sci. Rev. **214**, 35 (2018). https://doi.org/10.1007/s11214-018-0467-8
M.H. Thiemens, J.E. Heidenreich, The mass-independent fractionation of oxygen: a novel isotope effect and its possible cosmochemical implications. Science **219**, 1073–1075 (1983). https://doi.org/10.1126/science.219.4588.1073
M.H. Thiemens, T.L. Jackson, C.A.M. Brenninkmeijer, Observation of a mass independent oxygen isotopic composition in terrestrial stratospheric CO_2, the link to ozone chemistry, and the possible occurrence in the Martian atmosphere. Geophys. Res. Lett. **22**, 255–257 (1995). https://doi.org/10.1029/94GL02996
E.W. Thommes, M.J. Duncan, H.F. Levison, Oligarchic growth of giant planets. Icarus **161**, 431–455 (2003). https://doi.org/10.1016/S0019-1035(02)00043-X
Z. Tian, H. Chen, B. Fegley, K. Lodders, J.-A. Barrat, J.M.D. Day, K. Wang, Potassium isotopic compositions of howardite-eucrite-diogenite meteorites. Geochim. Cosmochim. Acta **266**, 611–632 (2019). https://doi.org/10.1016/j.gca.2019.08.012
V.T. Titov, N.I. Ignatiev, K. McGouldrick, V. Wilquet, C.F. Wilson, Clouds and hazes of Venus. Space Sci. Rev. **214**, 126–187 (2018). https://doi.org/10.1007/s11214-018-0552-z
G. Torres, F. Fressin, Discovery of the first Earth-sized planets orbiting a star other than our Sun in the Kepler-20 system (2019). eprint arXiv:1905.04309
A.H. Treiman, Venus - bulk and mantle compositions: are Venus and Earth really twins? in *Venus Geochemistry: Progress, Prospects, and New Missions*. LPI Contribution, vol. 1470

(2009), pp. 47–48. Abstract number 2016. https://repository.hou.usra.edu/bitstream/handle/20.500.11753/1177/LPI-001470.pdf?sequence=1

J.M. Trigo-Rodríguez, A. Rimola, S. Tanbakouei, V.C. Soto, M. Lee, Accretion of water in carbonaceous chondrites: current evidence and implications for the delivery of water to early Earth. Space Sci. Rev. **215**(18), 27 (2019). https://doi.org/10.1007/s11214-019-0583-0

A. Trinquier, J-L. Birck, C.J. Allègre, Widespread ^{54}Cr heterogeneity in the inner Solar System. Astrophys. J. **655**, 1179–1185 (2007). https://doi.org/10.1086/510360

A. Trinquier, J.-L. Birck, C.J. Allegre, C. Gopel, D. Ulfbeck, ^{53}Mn–^{53}Cr systematics of the early Solar System revisited. Geochim. Cosmochim. Acta **72**, 5146–5163 (2008). https://doi.org/10.1016/j.gca.2008.03.023

D.L. Turcotte, An episodic hypothesis for venusian tectonics. J. Geophys. Res., Planets **98**, 17061–17068 (1993). https://doi.org/10.1029/93JE01775

A. Udry, G.H. Howarth, C.D.K. Herd, T.J. Lapen, J.M.D. Day, What have meteorites taught us about the Martian interior and surface over the past five years? in *Ninth International Conference on Mars*. LPI Contribution, vol. 2089 (2019). Abstract number 6139. https://www.hou.usra.edu/meetings/ninthmars2019/pdf/6139.pdf

D. Valentian, C. Koppel, P. Mairet, L. Mairet, Venus sample return mission revisited. ESA Vision 2050 White Paper (2019). https://www.cosmos.esa.int/documents/1866264/3219248/ValentianD_Venus+sample+return+mission+revisited_r2.pdf/51e80e8b-8ecd-44a9-8363-5525e6cb35d9?t=1565184752220

E.M.M.E. Van Kooten, D. Wielandt, M. Schiller, K. Nagashima, A. Thomen, K.K. Larsen, M.B. Olsen, Å. Nordlund, A.N. Krot, M. Bizzarro, Isotopic evidence for primordial molecular cloud material in metal-rich carbonaceous chondrites. Proc. Natl. Acad. Sci. **113**, 2011–2016 (2016). https://doi.org/10.1073/pnas.1518183113

Venus Science & Technology Definition Team, Venus Flagship Mission Study, NASA (2009) https://vfm.jpl.nasa.gov/files/Venus+Flagship+Mission+Study+090501-compressed.pdf

P. Voosen, Meteorite divide points to Solar System chaos. Science **359**, 1451 (2018). https://doi.org/10.1126/science.359.6383.1451

K.J. Walsh, H.F. Levison, Planetesimals to terrestrial planets: collisional evolution amidst a dissipating gas disk. Icarus **329**, 88–100 (2019). https://doi.org/10.1016/j.icarus.2019.03.031

K.J. Walsh, A. Morbidelli, S.N. Raymond, D.P. O'Brien, A.M. Mandell, A low mass for Mars from Jupiter's early gas-driven migration. Nature **473**, 206–209 (2011). https://doi.org/10.1038/nature10201

K.J. Walsh, A. Morbidelli, S.N. Raymond, D.P. O'Brien, A.M. Mandell, Populating the asteroid belt from two parent source regions due to the migration of giant planets—"The Grand Tack". Meteorit. Planet. Sci. **47**(12), 1941–1947 (2012). https://doi.org/10.1111/j.1945-5100.2012.01418.x

K. Wang, S.B. Jacobsen, Potassium isotopic evidence for a high-energy giant impact origin of the Moon. Nature **538**, 487–490 (2016). https://doi.org/10.1038/nature19341

P.H. Warren, Stable-isotopic anomalies and the accretionary assemblage of the Earth and Mars: a subordinate role for carbonaceous chondrites. Earth Planet. Sci. Lett. **311**, 93–100 (2011). https://doi.org/10.1016/j.epsl.2011.08.047

S. Watanabe, Y. Tsuda, M. Yoshikawa, S. Tanaka, T. Saiki, S. Nakazawa, Hayabusa2 mission overview. Space Sci. Rev. **208**, 3–16 (2017). https://doi.org/10.1007/s11214-017-0377-1

M.J. Way, A.D. Del Genio, N.Y. Kiang, L.E. Sohl, D.H. Grinspoon, I. Aleinov, M. Kelley, T. Clune, Was Venus the first habitable world of our Solar System? Geophys. Res. Lett. **43**, I8376–8383 (2016). https://doi.org/10.1002/2016GL069790

C.R. Webster, P.R. Mahaffy, G.J. Flesch, P.B. Niles, J.H. Jones, L.A. Leshin, S.K. Atreya, J.C. Stern, L.E. Christensen, T. Owen, H. Franz, R.O. Pepin, A. Steele (The MSL Science Team), Isotope ratios of H, C, and O in CO_2 and H_2O of the Martian atmosphere. Science **341**, 260–263 (2013). https://doi.org/10.1126/science.1237961

S.J. Weidenschilling, Accretion of the asteroids: implications for their thermal evolution. Meteorit. Planet. Sci. **54**, 1115–1132 (2019). https://doi.org/10.1111/maps.13270

M.K. Weisberg, T.J. McCoy, A.N. Krot, in *Systematics and Evaluation of Meteorite Classification. Meteorites and the Early Solar System II*. D.S. Lauretta, H.Y. McSween Jr. (University of Arizona Press, Tucson, 2006), pp. 19–52. https://repository.si.edu/bitstream/handle/10088/20577/min_Weisberg_et_al_2006_In_Meteorites_and_the_Early_Solar_System_II_Univ_of_Arizona_Press_19-52.pdf

L.C. Welzenbach, M.D. Fries, M.M. Grady, R.C. Greenwood, F.M. McCubbin, C.L. Smith, A. Steele, R.A. Zeigler, Potential alteration products of organic materials by X-ray computed tomography of Mars returned samples, in *82nd Annual Meeting of the Meteoritical Society*. LPI Contribution, vol. 2157 (2019). Abstract number 6369. https://www.hou.usra.edu/meetings/metsoc2019/pdf/6369.pdf

G. Whitcomb, J. Lebreton, G. Scoon, Venus sample return mission, in *34th COSPAR Scientific Assembly, the Second World Space Congress* (2002). Abstract 1891. https://ui.adsabs.harvard.edu/abs/2002cosp...34E1891W/abstract

J.L. Whitten, B.A. Campbell, Variations in the radar properties of tesserae across Venus as observed with Magellan data, in *50th Lunar and Planetary Science Conference*. LPI Contribution, vol. 2132 (2019). Abstract 2558. https://www.hou.usra.edu/meetings/lpsc2019/pdf/2558.pdf

U. Wiechert, A.N. Halliday, D-C. Lee, G.A. Snyder, L.A. Taylor, D. Rumble, Oxygen isotopes and the Moon-forming giant impact. Science **294**, 345–348 (2001). https://doi.org/10.1126/science.1063037

D.R. Williams, Venus Fact Sheet. NASA Goddard Space Flight Centre online resource, updated September 2018 (2018a). https://nssdc.gsfc.nasa.gov/planetary/factsheet/venusfact.html

D.R. Williams, Mars Fact Sheet. NASA Goddard Space Flight Centre online resource, updated September 2018 (2018b). https://nssdc.gsfc.nasa.gov/planetary/factsheet/marsfact.html

D.R. Williams, Planetary Fact Sheet, NASA Goddard Space Flight Centre online resource, updated October 2019 (2019a). https://nssdc.gsfc.nasa.gov/planetary/factsheet/

D.R. Williams, Asteroid Fact Sheet. NASA Goddard Space Flight Centre online resource, updated September 2019 (2019b). https://nssdc.gsfc.nasa.gov/planetary/factsheet/asteroidfact.html

D.R. Williams, Chronology of Venus Exploration, NASA Goddard Space Flight Centre online resource, updated February 2020 (2020a). https://nssdc.gsfc.nasa.gov/planetary/chronology_venus.html

D.R. Williams, Chronology of Mars exploration, Exploration, NASA Goddard Space Flight Centre online resource, updated February 2020 (2020a). https://nssdc.gsfc.nasa.gov/planetary/chronology_mars.html

C.F. Wilson et al., Venus: Key to understanding the evolution of terrestrial planets (2019). https://ui.adsabs.harvard.edu/#abs/2019arXiv190804269W/abstract

E.T. Wolf, O.B. Toon, Delayed onset of runaway and moist greenhouse climates for Earth. Geophys. Res. Lett. **41**(1), 167–172 (2014). https://doi.org/10.1002/2013GL058376

A.N. Youdin, J. Goodman, Streaming instabilities in protoplanetary discs. Astrophys. J. **620**, 459–469 (2005). https://doi.org/10.1086/426895

E.D. Young, S.S. Russell, Oxygen reservoirs in the early solar nebula inferred from an Allende CAI. Science **282**, 452–455 (1998). https://doi.org/10.1126/science.282.5388.452

E.D. Young, I.E. Kohl, I.P.H. Warren, D.C. Rubie, S.A. Jacobson, A. Morbidelli, Oxygen isotopic evidence for vigorous mixing during the Moon-forming giant impact. Science **351**, 493–496 (2016). https://doi.org/10.1126/science.aad0525

H. Yurimoto, K. Kuramoto, Molecular cloud origin for the oxygen isotope heterogeneity in the Solar System. Science **305**, 1763–1766 (2004). https://doi.org/10.1126/science.1100989

H. Yurimoto, K. Kuramoto, A.N. Krot, E.R.D. Scott, J.N. Cuzzi, M.H. Thiemens, J.R. Lyons, Origin and evolution of oxygen isotopic compositions of the Solar System, in *Protostars and Planets*, vol. 951, ed. by V.V.B. Reipurth, D. Jewitt, K. Keil (University of Arizona Press, Tucson, 2007), pp. 849–862. https://www.lpi.usra.edu/books/PPV/8057.pdf

C.Z. Zhang, K. Zhang, On the internal structure and magnetic fields of Venus. Earth Moon Planets **69**, 237–247 (1995). https://doi.org/10.1007/BF00643786

Space Sci Rev (2019) 215:54
https://doi.org/10.1007/s11214-019-0622-x

Constraining the Evolutionary History of the Moon and the Inner Solar System: A Case for New Returned Lunar Samples

Romain Tartèse[1] · **Mahesh Anand**[2,3] · **Jérôme Gattacceca**[4] · **Katherine H. Joy**[1] · **James I. Mortimer**[2] · **John F. Pernet-Fisher**[1] · **Sara Russell**[3] · **Joshua F. Snape**[5] · **Benjamin P. Weiss**[6]

Received: 23 August 2019 / Accepted: 25 November 2019 / Published online: 2 December 2019

Abstract The Moon is the only planetary body other than the Earth for which samples have been collected *in situ* by humans and robotic missions and returned to Earth. Scientific investigations of the first lunar samples returned by the Apollo 11 astronauts 50 years ago transformed the way we think most planetary bodies form and evolve. Identification of anorthositic clasts in Apollo 11 samples led to the formulation of the magma ocean concept, and by extension the idea that the Moon experienced large-scale melting and differentiation. This concept of magma oceans would soon be applied to other terrestrial planets and large asteroidal bodies. Dating of basaltic fragments returned from the Moon also showed that a relatively small planetary body could sustain volcanic activity for more than a billion years after its formation. Finally, studies of the lunar regolith showed that in addition to containing a treasure trove of the Moon's history, it also provided us with a rich archive of the past 4.5 billion years of evolution of the inner Solar System. Further investigations of samples returned from the Moon over the past five decades led to many additional discoveries, but also raised new and fundamental questions that are difficult to address with currently available samples, such as those related to the age of the Moon, duration of lunar volcanism, the

Role of Sample Return in Addressing Major Questions in Planetary Sciences
Edited by Mahesh Anand, Sara Russell, Yangting Lin, Meenakshi Wadhwa, Kuljeet Kaur Marhas and Shogo Tachibana

✉ R. Tartèse
romain.tartese@manchester.ac.uk

1 Department of Earth and Environmental Sciences, The University of Manchester, Manchester, M13 9PL, UK

2 Planetary and Space Sciences, School of Physical Sciences, The Open University, Walton Hall, Milton Keynes, MK7 6AA, UK

3 Department of Earth Sciences, The Natural History Museum, Cromwell Road, London, SW7 5BD, UK

4 CNRS, Aix-Marseille Univ, IRD, Coll France, INRA, CEREGE, Aix-en-Provence, France

5 Faculty of Sciences, Department of Earth Sciences, Vrije Universiteit, Amsterdam, The Netherlands

6 Department of Earth, Atmospheric and Planetary Sciences, Massachusetts Institute of Technology, Cambridge, MA, USA

lunar paleomagnetic field and its intensity, and the record on the Moon of the bombardment history during the first billion years of evolution of the Solar System. In this contribution, we review the information we currently have on some of the key science questions related to the Moon and discuss how future sample-return missions could help address important knowledge gaps.

Keywords Earth-Moon system · Lunar evolution · Sample-return · Solar System

1 Overview of Our Lunar Sample Collection

A total of ~382 kg of rock and soil samples were brought back to the Earth by the Apollo and Luna programmes between 1969 and 1976. This sample suite has advanced our scientific understanding of the Moon in a profound manner. Over the last five decades, there are many fundamental questions that have only been addressed through our existing lunar sample collection. These are questions that not only have implications for our understanding of the Earth-Moon system, but also for our understanding of the wider Solar System, such as how planets are built and the nature of the impact flux in the inner Solar System over geological time (e.g., National Research Council 2007). In many cases, this has been possible as a result of the high precision and accuracy of geochemical measurements that can be made in laboratories on Earth relative to geochemical analyses that can be achieved by rovers on the Moon's surface or by remote sensing instruments from lunar orbit.

Despite the many advances the lunar sample collection has enabled, the major limitation of this collection stems from the limited areal extent of the Moon represented by the Apollo and Luna landing sites. Remote sensing orbital spacecraft over the last two decades have highlighted that our existing sample collections come from a compositionally anomalous area of the Moon (Fig. 1). This was first hinted by orbital gamma-ray measurements collected by the Apollo 15 and 16 service modules (Metzger et al. 1977, see also Plate 10.1 of the Lunar Sourcebook—Heiken et al. 1991), and observed in more detail by the 1998 Lunar Prospector and 1994 Clementine missions (Jolliff et al. 2000). This area, referred to as the Procellarum-KREEP terrane (PKT), is enriched in incompatible elements such as potassium (K), rare earth elements (REE), phosphorus (P), uranium (U), and thorium (Th). It also coincides with the extent of the Imbrium basin continuous ejecta blanket (Fig. 1). Consequently, without future samples from other areas such as the feldspathic highlands terrane, or the farside South Pole-Aitken (SPA) basin, there are a number of key lunar science questions that cannot be fully addressed (Table 1).

Attempts have been made to answer many of these questions by lunar orbiter remote sensing missions, and lander and rovers missions. For instance, remote sensing instruments (such as the Moon Mineralogy Mapper (Pieters et al. 2009a; Green et al. 2011)—M^3—on Chandrayaan-1 and Diviner (Paige et al. 2010a, 2010b) on the Lunar Reconnaissance Orbiter) have geologically mapped the Moon globally in high-spatial and spectral detail. Spectroscopic imaging has identified lithologies that are rare or absent in the Apollo or Luna collections, such as dunites and pink spinel troctolites (e.g., Corley et al. 2018; Pieters et al. 2011). However, M^3 has a 70 m/pixel maximum resolution (Green et al. 2011) and Diviner infrared spectrometer has an average resolution of 250 m/pixel (Paige et al. 2010a). Therefore, each reported lithology and associated chemical features represent an average over the entire pixel, only painting a broad-scale view of these complex geological features.

Complementary to remote-sensing datasets, rovers on the lunar surface have the potential to make detailed, localised, *in situ* geological investigations. For instance, China's rover Yutu-2 (part of the Chang'E-4 mission) recently reported orthopyroxene- and olivine-

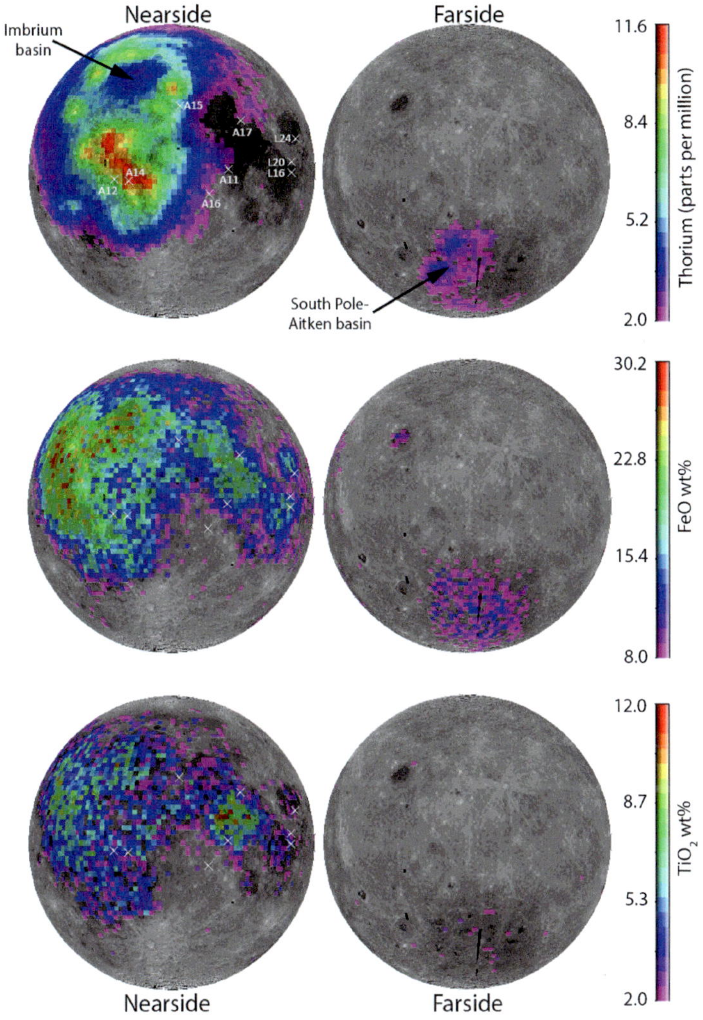

Fig. 1 Nearside and farside of the Moon showing distribution of Th, FeO and TiO_2 abundances obtained by the Lunar Prospector Gamma Ray Spectrometer (2° per pixel data from Prettyman et al. 2006). Data for these elements have been selected to µg g gw abundances of Th > 2µg g^{-1}, FeO > 8 wt.% and TiO_2 > 2 wt.%. These data are overlain on a Clementine albedo map of the Moon in an orthographic projection. Locations of the Apollo (A) and Luna (L) landing sites are from Wagner et al. (2017). The locations of the Imbrium basin and South Pole-Aitken basin are also indicated

bearing lithologies within the Von Kármán crater (situated within the SPA basin; Li et al. 2019). Such a mineral assemblage, together with the geological context of this crater (i.e. material on the Von Kármán crater floor is thought to have been excavated from 72 km below the SPA floor; Huang et al. 2018), led Li et al. (2019) to a suggestion that these lithologies could represent excavated mantle material. This is significant because no mantle rocks have conclusively been identified in the Apollo or Luna collections. However, without laboratory-based geochemical measurements of returned samples, the true provenance of these lithologies and what they tell us about the Moon (for instance, testing predicted lunar magma ocean crystallisation sequences) cannot be unambiguously established.

Table 1 Major outstanding science questions that lunar sample-return missions could address. Adapted from Pernet-Fisher et al. (2019b), based on National Research Council (2007) and Crawford and Joy (2014)

Science questions	Opportunities offered by sample-return	Sampling location
Age of the Moon	Multi-system high precision geochronology; Detailed major and trace element geochemistry; Noble gas geochemistry	Primary crust exposures (ideally pure anorthosite—PAN), non PKT
Primordial crust of the Moon: homogeneous vs. heterogeneous ferroan anorthosite (FAN); Relationship between anorthosites and ancient magmatic (i.e., Mg-suite) rocks	Examine petrogenesis of lunar crustal rocks; Multi-system geochronology; Detailed major and trace element geochemistry	Magnesian anorthosites (MAN) outcrops; Farside PAN
Bombardment history of the inner Solar System	Multi-system high precision geochronology; Detailed major and trace element geochemistry; Search for impactor fragments and/or chemical signatures	Impact-melt material from major basin-forming impacts (SPA, Nectaris, Crisium, etc.); Key stratigraphic marker craters (Copernicus, Tycho); Young lava surfaces
Lunar volcanism: timing, nature, duration, volume, emplacement mechanism	Examine source regions; Multi-system geochronology; High precision stable isotope measurements (e.g., O); Detailed major and trace element geochemistry	Young lava flows; Cryptomare lava flows (access ancient volcanism); Mare basalt bedrock (especially within stacked lava flows that would be accessible in lava pit walls); KREEP-rich basalt flows; Pyroclastic glass bead sites
Structure and composition of the lunar interior	Nearside vs. farside crustal thickness dichotomy; Examine mare basalt source regions; Petrology; Multi-system geochronology; Detailed major and trace element geochemistry; Volatile inventory of the lunar mantle	Crustal rocks from the farside; Deep crustal rocks and possible mantle xenoliths from deep basins that breached the crust
Lunar polar volatiles: composition, physical state, form, distribution, context	Volatile release studies; Stable isotope measurements (H, C, N, O, etc.); Detailed major and trace element geochemistry	Volatile-bearing materials in pristine condition from multiple polar locations
Flux and composition of ancient solar wind composition recorded in lunar regolith	Volatile release studies; Stable isotope measurements (H, C, N, O, etc.); Noble gas geochemistry	Ancient paleoregolith sites; Present day regolith; Regolith samples from swirl regions; Core samples
Galactic record archived in lunar surface samples	Analysis of cosmogenic nuclides (^{3}He, ^{21}Ne, ^{38}Ar, ^{83}Kr, ^{126}Xe, ^{10}Be, ^{36}Cl); analysis of supernova-related isotopes (e.g., ^{60}Fe)	Ancient paleoregolith sites; Present day regolith at various sites; Core samples
Lunar magnetic field and anomalies	High-precision magnetic measurements; Multi-system geochronology	Ancient magmatic bedrock (samples collected with known orientations and from multiple stratigraphic layers); Samples from swirl regions

Undoubtedly, remote sensing and rover missions have made invaluable contributions to our understanding of the Moon by conducting a wide range of science investigations such as identifying new lithologies, globally mapping the chemical heterogeneity of the lunar crust, estimating gravity anomalies, and measuring surface temperature. However, there are some important questions that can only be addressed by detailed geochemical and physical experiments on rock samples in labs on Earth. In this regard, our current best tools for understanding the geology of the Moon outside of the nearside PKT comes from lunar meteorites. Even though lunar meteorites are limited by the fact that they lack exact geographical provenance, precise geological context, and have unknown original orientation, they have uniquely advanced our understanding of the Moon (e.g., Korotev 2005; Joy and Arai 2013). For instance, the lunar meteorite collection contains fragments of the youngest and oldest mare basalts reported to date (e.g., Borg et al. 2009; Snape et al. 2018a; Curran et al. 2019). However, for this example, without known geographic and geological context of where these samples originated on the Moon, it is hard to fully interpret how these mare basalts fit into the wider picture of lunar volcanism.

In this chapter, we detail some of the major outstanding lunar science questions that can only be resolved by the collection of new lunar samples from a geographically diverse set of locations for investigation in labs on Earth (see summary in Table 1).

2 When Did the Moon Form?

2.1 Overview

The formation of the Moon is commonly linked with a catastrophic impact between the proto-Earth and a large (potentially Mars-sized or larger) impactor (e.g., Hartmann and Davis 1975). Given this unique relationship between the Earth and the Moon, the timing of this event represents not just a fundamental starting point for lunar evolution, but a key stage in the early evolution of the Earth. Since the return of the first Apollo samples, there have been numerous efforts to determine the age of the Moon (e.g., Tatsumoto 1970; Tera and Wasserburg 1974; Tilton and Chen 1979; Carlson and Lugmair 1988), and significant differences exist among recently published age estimates (e.g., Borg et al. 2011; Gaffney and Borg 2014; McLeod et al. 2014; Avice and Marty 2014; Connelly and Bizzarro 2016; Barboni et al. 2017; Kruijer and Kleine 2017; Thiemens et al. 2019). These age estimates will often be divided into those proposing either an "old Moon" forming at around 4.5 Ga or earlier, and those advocating a "young Moon" age of <4.5 Ga.

2.2 Estimates for the Time of Lunar Formation

Because the Moon is not a primitive planetary body, and has had a complex geological history of differentiation, impact modification and extensive volcanic activity, there is no single rock whose age can be measured in order to obtain a direct determination for the time of lunar formation. Nonetheless, perhaps the most straightforward approach employed for constraining the time of the Moon's formation has been to determine the ages of the oldest available lunar samples. For example, Borg et al. (2011) determined an age of 4.360 ± 0.003 Ga for the FAN sample 60025, using three independent chronometers (Pb-Pb, ^{147}Sm-^{143}Nd and ^{146}Sm-^{142}Nd). The FAN rocks are believed to represent the remnants of the ancient lunar crust, formed as flotation cumulates in the lunar magma ocean (LMO). As such, the age reported by Borg et al. (2011) can be interpreted as the time at which the LMO crystallised, and by inference suggests a relatively young age for the Moon.

An alternative approach has been to calculate model ages for the mantle sources of a variety of lunar rock types, using different long-lived isotope systems (e.g. U-Pb, Rb-Sr, Sm-Nd and Lu-Hf). While the rocks used in such studies (e.g., basalts formed between ~3.1–4.0 Ga) will often have relatively young crystallisation ages compared to the FAN sample studied by Borg et al. (2011), they will retain an initial isotopic composition from the time at which they crystallised. If these isotopic fingerprints can be determined, it is then possible to calculate the amount of time necessary for radiogenic ingrowth in the mantle sources to generate the initial isotopic composition. Notably, many of these model ages for different lunar lithologies appear to cluster around an age range similar to that of the 60025 FAN sample, as well as that of a number of other ancient lunar crustal rocks (~4350–4400 Ga; e.g., Lugmair and Carlson 1978; Nyquist et al. 1995; Rankenburg et al. 2006; Brandon et al. 2009; Gaffney and Borg 2014; McLeod et al. 2014; Carlson et al. 2014; Snape et al. 2016a). Once again, these model ages have been interpreted as marking the final stages of LMO crystallisation, and potentially suggestive of a relatively young Moon. By contrast, evidence for an older Moon (~4.5 Ga) has been provided by a date of 4.417 ± 0.006 Ga obtained on a zircon grain in Apollo 17 breccia 72215 (Nemchin et al. 2009) and by model ages calculated for zircon grains in lunar breccias based on their initial Hf isotopic compositions and $^{207}Pb/^{206}Pb$ crystallisation ages (Taylor et al. 2009; Barboni et al. 2017).

Frustratingly, even studies looking at the same isotope system have produced different estimates for the age of the Moon. A good example, is the ^{182}Hf-^{182}W system, which can potentially provide unique insights due to the short half-life of ^{182}Hf (*ca.* 9 Ma) and the fact that Hf (lithophile) and W (siderophile) are fractioned during planetary core formation. Therefore, if the formation of a planet's core has occurred during the lifetime of ^{182}Hf, the remaining silicate fraction of the planet will develop an excess of ^{182}W. Analyses of lunar samples have identified a relative excess of ^{182}W (Touboul et al. 2015; Kruijer et al. 2015) and higher Hf/W ratios (Thiemens et al. 2019) compared to terrestrial values. These observations can be explained by radiogenic ingrowth of ^{182}W in the bulk silicate Moon (BSM), provided lunar differentiation occurred no later than ~4.51 Ga (Lee et al. 1997; Halliday and Lee 1999; Righter and Shearer 2003; Kleine et al. 2005; Thiemens et al. 2019). Conversely, it has also been suggested that the Moon could have formed later than 4.5 Ga, with the ^{182}W excess in lunar samples being explained via a disproportionate accretion of meteoritic material to the Earth-Moon system during the so-called "late veneer" (Touboul et al. 2015; Kruijer et al. 2015; Kruijer and Kleine 2017).

A number of studies looking at terrestrial rocks have also provided constraints for the timing of the Moon's formation. For example, Avice and Marty (2014) used I-Pu-Xe isotope systematics to determine an age for the Earth's atmosphere of 40^{+20}_{-10} Ma after Solar System formation, which they suggest might provide a lower limit for the timing of the last giant impact event that could have formed the Moon (i.e. the giant impact occurred prior to *ca.* 40 Ma after Solar System formation). Meanwhile, a much younger estimate for the Moon's age was derived by Connelly and Bizzarro (2016). Based on the assumption that the bulk silicate Earth (BSE) originally had a much lower $^{238}U/^{204}Pb$ ratio (μ-value) than it does today and that the subsequent increase was the result of Pb loss during the giant impact, the authors calculated that this event must have occurred between 4.426–4.417 Ga, in order to explain modern day BSE Pb isotopic compositions.

2.3 How Can New Lunar Samples Help Address This Issue?

The lack of agreement with regard to the time of lunar formation is primarily due to different assumptions that have been made in calculating various ages (e.g., the duration of different stages of planetary differentiation, the occurrence and effect of a late veneer), and how these ages are interpreted (e.g., are they representative of global processes and can they be tied to the LMO model?). It is quite possible that new samples will not completely resolve these uncertainties. Furthermore, they will almost certainly raise new questions. However, a wider range of lunar samples has the potential to address a number of key questions that have arisen from previous studies.

Determining ages for the oldest fragments of the lunar crust is particularly challenging, partly due to the age of these samples and the fact that they have been subjected to more than 4 billion years of impact modification. Furthermore, the Apollo samples were all collected from locations on the nearside of the Moon and near to, and in some cases within, the geochemically anomalous PKT region (Fig. 1). Even the Apollo 16 landing site, which provided many of the known FAN samples (such as 60025), is likely not representative of "typical" lunar highlands as it has a fairly elevated KREEP component (e.g., Korotev et al. 2003). Therefore, returning samples from more representative regions of the lunar highlands, particularly on the lunar farside well away from the PKT, will provide a better chance to determine ages for the most ancient crustal lithologies on the Moon.

One of the most striking features about the model ages derived from the Apollo samples is the preponderance of ~4.35–4.4 Ga ages. If similar model ages are derived from samples collected from a wider range of locations (particularly on the lunar farside), it would provide much stronger support for the age representing final crystallisation stages of a global LMO. This would suggest either that the Moon is significantly younger than 4.5 Ga, or that the LMO cooled more slowly than most models predict (Elkins-Tanton et al. 2011). If, on the other hand, samples returned from the lunar farside indicate that the ~4.35–4.4 Ga model ages are an artefact of the lunar nearside or the PKT, it would support the suggestion that previous model ages represent a later large-scale (but not global) resetting of isotope systems. A potential cause of this event could be a particularly large impact event, such as the one previously suggested to have generated the hypothetical "Procellarum basin" (Cadogan 1974). Remote sensing data have failed to detect the typical gravity anomaly associated with impact basins around the Oceanus Procellarum (Andrews-Hanna et al. 2014). Nevertheless, such a large impact event would not only provide a way to explain observed asymmetries between the lunar nearside and farside (Zhu et al. 2019), but could also provide explanation for the ~4.35–4.4 Ga model ages.

A key goal for future sample-return missions is to investigate predicted exposures of material derived from the lunar mantle, particularly in the SPA basin (Yamamoto et al. 2010; Melosh et al. 2017). Our current understanding of the lunar mantle and its composition is based on inferences from samples such as the lunar mare basalts. While it is generally agreed that these rocks have their origins in partial melts of mantle cumulates, they are not a direct representation of the lunar mantle. Having more direct determinations of lunar mantle compositions, particularly regarding the isotopic systematics of mantle material, could address a number of uncertainties associated with models discussed above, which rely on assumptions regarding the nature of the lunar mantle and the BSM.

In case of the Sm-Nd and Lu-Hf model ages described above, an important constraint is whether or not the mantle sources of the Apollo samples evolved from a BSM with a chondritic composition. The ^{182}Hf-^{182}W studies used to constrain the Moon's age rely on estimates for the ratio of Hf/W in the early BSM (Kruijer and Kleine 2017; Thiemens et al. 2019). Similarly, although recent studies of lunar basalts have made it possible to construct a multistage model of Pb isotopic evolution in the Moon (Snape et al. 2016a), using this model to determine the age of the Moon would require a better understanding of the BSM ^{238}U/^{204}Pb ratio (known as the μ-value) during the Moon's early differentiation. If lunar mantle material indicated that the BSM had very high μ-values (>1000), it would argue strongly in favour of a young Moon (<4.5 Ga), while evidence for lower BSM μ-values more typical of those determined for many mare basalt sources (<500), would imply that the Moon is old (~4.5 Ga) (Snape et al. 2016a).

3 Petrology and Formation History of the Lunar Anorthositic Crust

As a geologically ancient surface, the Moon can inform us about both its own early history and more generally about the expected evolution of the terrestrial planets. The first reports of anorthite-rich lithologies in samples returned from the Apollo missions led to a very simple view of the evolution of the Moon crystallising from a LMO (Wood et al. 1970). In this model, the first minerals to solidify, olivine followed by pyroxene, sank downwards to make a mafic Mg-rich mantle. After solidification of ~75–80% of the LMO, plagioclase began to crystallise (e.g., Snyder et al. 1992). Being less dense than the melt, it rose to form a crust on the lunar surface, bringing with it co-crystallising Fe-bearing mafic minerals. The relative homogeneity of the anorthite composition (An_{94-98}) in Apollo FAN samples supports the LMO model (see discussion in Gross and Joy 2016, and references therein).

Evidence for formation of the different silicate layers in planetary bodies from a magma ocean is most obvious for the Moon, but a fully molten state may be a ubiquitous stage in the evolution of planetary sized bodies and learning about the LMO has implications for planetary evolution more generally. The howardite, eucrite and diogenite (HED) meteorites, thought to originate from asteroid 4-Vesta (e.g., McCord et al. 1970; McSween et al. 2013), have genetic similarities to each other that may be explained by formation from a magma ocean (Righter and Drake 1997). This is likely to be replicated in many large rocky asteroids (McCoy et al. 2006). Mars may also have had a magma ocean, although the wide variety of mineral compositions observed in martian meteorites requires a post-magma ocean period of considerable overturn, mixing and reprocessing (Elkins-Tanton et al. 2003).

The Earth's structure and composition is also consistent with it having a magma ocean stage (Righter et al. 1997). A recent model suggests that the Earth may have been partially molten when it was impacted by a large impactor to produce the Moon. Its molten state may have enabled most of the lunar material to have originated from the Earth rather than from the impactor (Hosono et al. 2019). Thus, magma oceans can have critical implications for how bodies evolve and indeed, how the Moon itself formed.

The study of lunar meteorites, which originate from random locations on the lunar surface, and the comparison of these meteorites to returned Apollo samples, have increased our understanding of crustal heterogeneities and, therefore, of the evolution of the LMO. Most lunar anorthositic meteorites have compositions clearly different from those of the majority of returned Apollo samples in both major and trace element compositions, leading to recent questions about the lunar magma ocean evolution. In terms of major element composition, around 80% of lunar meteorites are significantly more magnesian (and named MAN for magnesian anorthosites) than Apollo FAN samples (Gross et al. 2014). Gross et al. (2014) interpreted this as an indication that much of the lunar crust cannot be the product of a primary magma ocean, in which more ferroan compositions are expected, but likely formed from complex multiple melting events. Trace element analyses of lunar meteorite minerals show a similar picture of complexity. Anorthite in lunar meteorites differs from that in Apollo samples in REE abundances that cannot be explained by crystallisation from a single magma composition (Russell et al. 2014). More recently, FAN clasts in Apollo 15 and 16 samples have been demonstrated to exhibit wide variations in REE abundances that overlap

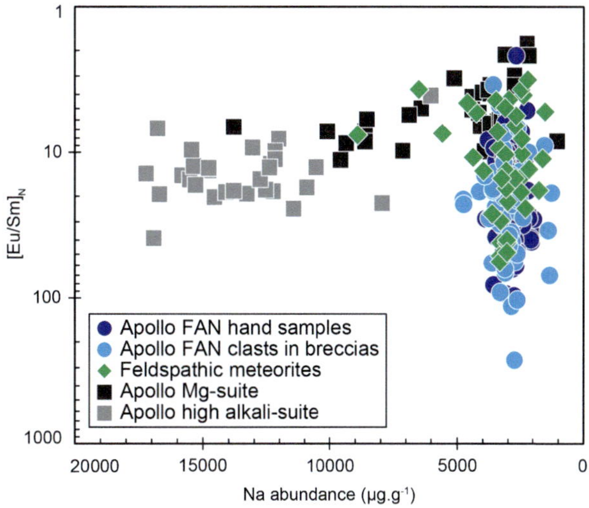

Fig. 2 CI chondrite-normalised (CI chondrite REE abundances from Barrat et al. 2012) Eu/Sm vs. Na abundance in lunar plagioclase from Apollo 15 and 16 hand samples and clasts in breccias (after Joy 2013; Pernet-Fisher et al. 2019a), feldspathic lunar meteorites (from Cahill et al. 2004; Treiman et al. 2010; Joy et al. 2011a; Russell et al. 2014), and Apollo Mg-suite and high alkali-suite samples (from Papike et al. 1996; Shervais and McGee 1998, 1999)

those observed in lunar meteorites, and which cannot be explained by fractional crystallisation from a single source (Pernet-Fisher et al. 2019a) (Fig. 2). The Nd isotope compositions of lunar meteorites and FAN samples likewise suggest that their parent magmas vary in REE abundances, sometimes being enriched in HREE and sometimes in LREE, whereas according to the LMO model the residual melt should always trend towards LREE enrichment, as observed in KREEP samples (Nyquist et al. 2010).

All these recent results acquired on lunar meteorites and clasts within Apollo suite samples require a more complex geological history than that provided by simple LMO solidification models. While the Moon is indeed likely to have experienced a magma ocean stage, it seems unlikely that the majority of the lunar surface is comprised mostly of unmodified primordial crust formed about 4.5 billion years ago. Sample-return missions at selected locations could help answer some outstanding questions about the degree of remelting of the lunar crust post-LMO. In order to understand this very early history of the Moon, we require more samples from a variety of lunar locations. The far side of the Moon may be more magnesium-rich than the nearside FAN lithologies (Arai et al. 2008; Ohtake et al. 2012) and, therefore, some anorthosite outcrops from these terrains (Ohtake et al. 2009; Donaldson Hanna et al. 2014; Yamamoto et al. 2015; Lemelin et al. 2015) may be a good focus for a sample-return landing site. Returning samples from anorthosite areas from regions remote from Apollo and Luna landing sites will enable investigating these outstanding issues related to the LMO formation, crystallisation, and the subsequent lunar igneous evolution.

4 The Bombardment History of the Inner Solar System

Impact bombardment is one of the main geological processes affecting all bodies throughout the Solar System. Because it is not affected by geological reprocessing and hydrospheric/atmospheric weathering, the Moon has preserved a near complete record of the inner Solar System impact history since its formation (e.g., Joy et al. 2016 and references therein) around 4.5 billion years ago (see Sect. 2: When did the Moon form?). The traces left on the Moon by impact bombardment range in size from several thousands of kilometres across (e.g., $\sim 2000 \times 2400$ km for the SPA basin; Garrick-Bethell and Zuber 2009) down to micron-sized crater pits observed on pyroclastic glass beads (e.g., McKay et al. 1991). Deciphering the lunar bombardment record can provide crucial information on (i) the secular evolution of bombardment in the inner Solar System, and (ii) the source of material involved and possible changes of impactor types through time.

4.1 Timing of Lunar Bombardment

Constraining the temporal evolution of impactor delivery to the Moon is one of the top priorities of research in Solar System science (e.g., National Research Council 2007). This is because the lunar cratering chronology anchors the cratering chronology of all other inner Solar System bodies, and is used to derive crater count model ages for geological units at their surface. Surface observations and gravity data enable deciphering the relative ages of the different lunar basins and craters compared to one another (e.g., Shoemaker and Hackman 1962; Hartmann and Wood 1971; Wilhelms 1987; Stöffler et al. 2006; Fassett et al. 2012; Neumann et al. 2015). However, only samples can provide absolute age constraints derived from precise analysis of parent-daughter radiometric chronometers (e.g., Turner 1970a, 1970b; Tera et al. 1974; Cohen et al. 2000; Norman et al. 2006; Fernandes et al. 2013; Snape et al. 2016a).

The temporal evolution of the flux of impactors to the Moon for its first billion years of evolution is still highly controversial (see Bottke and Norman 2017; Morbidelli et al. 2018 for recent reviews). It is generally believed that immediately after the formation of the Moon the impact flux was high and dominated by leftovers from the planetary accretion epoch, a time period commonly referred to as 'late accretion' during which impactors may have imprinted their chemical signatures into the crystallising LMO (e.g., Walker 2009; Day and Walker 2015; Barnes et al. 2016). Once the crust was formed and the LMO fully solidified, it is unlikely that any extraneous sources could have been added and efficiently mixed into the lunar mantle. The Moon then suffered an intense period of bombardment, as recorded by *ca.* 30–40 impact basins larger than 300 km in diameter scattered at its surface (Fig. 3 and Table 2) (e.g., Fassett et al. 2012; Neumann et al. 2015). However, the formation ages of most of these basins are not precisely known. The SPA basin is probably the oldest—crater counting studies and $^{40}Ar/^{39}Ar$ data obtained on samples that are postulated to have originated from SPA, based on their formation depth for example, have tentatively dated formation of SPA at *ca.* 4.25–4.3 Ga (e.g., Hiesinger et al. 2012; Fernandes et al. 2013; Garrick-Bethell and Miljković 2018). Orientale is probably the youngest basin but its age is indirectly constrained at *ca.* 3.9–3.7 Ga (e.g., Stöffler et al. 2006; Whitten et al. 2011). The Imbrium basin is one of the youngest basins, and its age has recently been proposed to be 3.926 ± 0.002 Ga based on U-Pb dating of zircon in an Apollo 12 impact melt breccia and in lunar meteorite Sayh al Uhaymir 169, and of phosphates in four Apollo 14 impact melt breccias (e.g., Snape et al. 2016b). About 30 impact basins larger than 300 km in diameter are proposed to have formed over a time window of $\sim$350–400 million years between the formation of SPA and Imbrium (see Table 2).

Whether impact events during this basin-forming period largely occurred as a cataclysmic spike at *ca.* 3.9 Ga ago (e.g., Tera et al. 1974; Ryder 1990; Cohen et al. 2000; Marchi et al. 2012) or decayed monotonically after the end of planetary accretion from $\sim$4.3–4.2 Ga until $\sim$3.9–3.8 Ga ago (e.g., Hartmann 1970; Neukum et al. 2001; Morbidelli et al. 2012; Werner 2014; Morbidelli et al. 2018), or even after (e.g., Fernandes et al. 2013; Zellner 2017), remains highly debated. One of the explanations for the apparent concentration of chronometric dates at *ca.* 3.9 Ga links it to a bias in largely sampling Imbrium

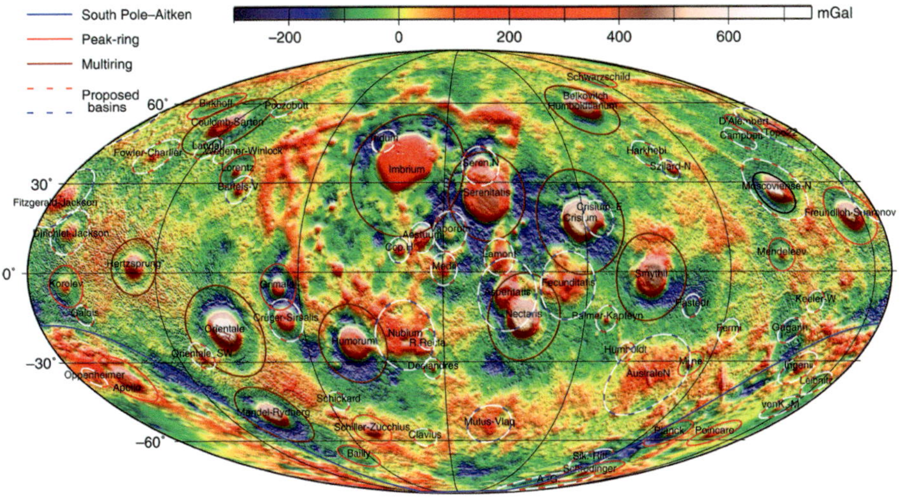

Fig. 3 Bouguer anomaly map of the Moon showing the location of large mass concentrations under impact basins (from Neumann et al. 2015). Bouguer-corrected GRAIL gravity anomaly in Mollweide equal-area projection centred on the nearside at 7°E longitude

ejecta in the Apollo sample collection (Fig. 1) (e.g., Haskin 1998; Spudis et al. 2011; Fernandes et al. 2013; Zellner 2017).

4.2 Sources of Impactors

The lunar record contains direct and indirect evidence that projectiles have hit the Moon throughout its history (see Joy et al. 2016 for a recent review). During the 'late accretion' window, a time period of up to around 200 million years between the segregation of the lunar core and the end of the LMO crystallisation (e.g., Elkins-Tanton et al. 2011), impactors may have been efficiently mixed into the lunar mantle and modified its chemical composition. We only have indirect, chemical and isotopic, evidence to estimate the flux and sources of impactors during this early period. The highly siderophile element (HSE) abundances and Os isotope signatures of mare basalts suggest that *ca.* 0.02 wt.% of the mass of the Moon has been added by chondritic impactors during late accretion (e.g., Day et al. 2007; Day and Walker 2015). These chondritic projectiles may have also delivered the bulk of the indigenous lunar H and N budgets (Barnes et al. 2016).

Projectiles continued hitting the Moon after crystallisation of the LMO and formation of a thick anorthosite crust. Volatile elements (H, C, N, S) and HSE abundances and isotopic compositions of lunar regolith and impact-melt samples provide indirect evidence for the types of impactors involved. Analysis of Apollo 14, 15, 16 and 17 samples show a broadly linear correlation between HSE/Ir and $^{187}Os/^{188}Os$ signatures (Puchtel et al. 2008; Fischer-Gödde and Becker 2012; Sharp et al. 2014; Liu et al. 2015; Gleißner and Becker 2017), reflecting the addition of chondritic and non-chondritic components to the lunar crust, the non-chondritic component HSE composition resembling that of iron meteorites, and contributing up to ~30 wt.% HSE in the most fractionated Apollo 16 impact-melt samples (e.g., Gleißner and Becker 2017). Siderophile element abundances in mature regolith samples that have been exposed at the lunar surface for hundreds of millions of years suggest the addition of *ca.* 1–3 wt.% of a CI chondrite-like micrometeorite component (e.g., Keays et al. 1970;

Table 2 Sequence of lunar basins with diameters greater than 300 km (after Fassett et al. 2012). Basins in brackets are those whose diameters and/or stratigraphic relationships are uncertain because of observational challenges. Main ring and Bouguer anomaly diameters are from Neumann et al. (2015); ring diameters italicised are probable/possible

Name	Period	Main ring diameter (km)	Bouguer anomaly diameter (km)	Superposition relationships
South Pole-Aitken	Pre-Nectarian	2400		no basin older
Coulomb-Sarton	Pre-Nectarian	*672*	330 ± 18	> Birkhoff
Dirichlet-Jackson	Pre-Nectarian	*452*	220	> Korolev
Cruger-Sirsalis	Pre-Nectarian	*430*	268	
Smythii	Pre-Nectarian	878	438 ± 62	> Crisium
[Schiller-Zucchius]	Pre-Nectarian	361	210	
[Amundsen-Ganswindt]	Pre-Nectarian	378	170	> Schrödinger
Nubium	Pre-Nectarian	*690*	416	> Humorum
[Poincaré]	Pre-Nectarian	312	188	
Lorentz	Pre-Nectarian	351	156	
[Fitzgerald-Jackson]	Pre-Nectarian	*600*	334	> Freundlich-Sharonov
[Birkhoff]	Pre-Nectarian	334	130	
[Ingenii]	Pre-Nectarian	342	154	
[Serenitatis]	Pre-Nectarian?	*923*	556 ± 64	> Nectaris (?)
Apollo	Pre-Nectarian/Nectarian	492	264	> Korolev, Hertzsprung
Freundlich-Sharonov	Pre-Nectarian/Nectarian	582	318	> Moscoviense
Nectaris	Start of Nectarian	885	440 ± 61	
Korolev	Nectarian/Pre-Nectarian	417	202	> Hertzsprung (?)
[Mendeleev]	Nectarian/Pre-Nectarian	331	156	
Hertzsprung	Nectarian/Pre-Nectarian	571	254 ± 38	
[Grimaldi]	Nectarian/Pre-Nectarian	460	220	> Mendel-Rydberg
Mendel-Rydberg	Nectarian/Pre-Nectarian	650	328 ± 26	
[Planck]	Nectarian/Pre-Nectarian	321	128	> Schrödinger
Moscoviense	Nectarian	421		
Crisium	Nectarian	1076	498 ± 31	> Humboldtianum
Humorum	Nectarian	816	360 ± 21	
Humboldtianum	Nectarian	618	312 ± 27	
Imbrium	Start of Imbrian	1321	684 ± 45	
Schrödinger	Imbrian	326	154	
Orientale (no buff.)	Imbrian	937	436 ± 20	

Morgan et al. 1972; Korotev et al. 2003; Wolf et al. 2009) often present in the form of micrometre-sized metallic particles, with, on average, about 70% of the total metallic iron content of lunar soils being derived from micrometeorite impactors (Morris 1980).

The lunar regolith also contains relict fragments of various types of projectiles. These are dominated by asteroidal materials, ranging from enstatite chondrite-like to carbonaceous chondrites and iron meteorites (see Joy et al. 2016 for a full list of recovered meteoritic debris). To date no cometary fragments have been positively identified in lunar regolith samples, although there are suggestions that some lunar regolith samples have volatile element signatures consistent with an added cometary component (Gibson and Moore 1973). Finally, a recent investigation of a *ca.* 4.0 billion year old clast from Apollo 14 breccia 14321 suggests that it formed at oxygen fugacity and pressure conditions incompatible with a lunar crustal environment, potentially representing the first positive identification of a terrestrial meteorite in the lunar regolith (Bellucci et al. 2019).

4.3 Outstanding Questions and the Potential of Future Sample-Return

4.3.1 Determining the Magnitude and Duration of the Basin-Forming Period

To test the magnitude and duration of the postulated late heavy bombardment, we need samples to be collected and returned to Earth for analysis in our laboratories collected from one, or ideally several, of the older, pre-Nectarian lunar basins. Accessing samples will require careful assessment of site geology and sampling of specific targets to provide well-understood geological context (Norman 2009; Pernet-Fisher et al. 2019b). Site selection ideally needs to consider (i) accessing basins of different ages to test different key stratigraphic markers throughout the early bombardment interval, and (ii) where possible collecting material directly from impact melt deposits, or reworked impact melt deposits to be able to access material formed within the basin event rather than re-processed by later superimposed impact events. Key periods, with targets providing opportunities for sampling multi-ring basin targets, include:

- *Pre-Nectarian age:* The SPA basin, on the lunar farside, which is the largest (>2000 km diameter) and oldest preserved basin on the Moon (Wilhelms 1987). The formation age of SPA is unknown from direct geological sampling. However, it is inferred from crater size-frequency distribution (CSFD) age models to be >4 Ga, with estimates ranging from ~4.25 Ga to ≤4.33 Ga (Hiesinger et al. 2012; Morbidelli et al. 2012; Fernandes et al. 2013; Marchi et al. 2013; Garrick-Bethell and Miljković 2018). Anchoring the age of SPA will constrain the onset of bombardment (Jolliff et al. 2017) after the lunar crust formed and had the material strength to be able to retain a large basin structure (Conrad et al. 2018).
- *Pre-Nectarian or Nectarian age:* The Serenitatis basin (~900 km diameter), for which there is significant debate whether it is an older Pre-Nectarian basin, or a Nectarian basin younger than Nectaris and only just older than Imbrium at *ca.* 3.93 Ga (Spudis et al. 2011; Fassett et al. 2012). This debate is important as the lunar sample community is currently challenged about how to interpret the ages of samples collected from the North and South Massif at the Apollo 17 landing site, and notably if these samples represent Imbrium, Nectaris or Serenitatis ejecta (e.g., Hurwitz and Kring 2016; Thiessen et al. 2017).
- *Nectarian age:* The Nectaris basin (~900 km diameter) is a key stratigraphic marker that gives its name to the Nectarian period (Stuart-Alexander and Wilhelms 1975). It has been postulated that Nectarian ejecta was sampled at the Apollo 16 landing site at North Ray Crater, with basin age estimates ranging from 3.9 Ga to 4.2 Ga (e.g., Fischer-Gödde and Becker 2011, 2012). Collecting various impact melt samples from this nearside highlands site (e.g., Norman 2009; Cohen et al. 2016) is of key interest for potentially determining the timing of formation of about ten basins formed between Nectaris and Imbrium (Table 2). This would provide crucial constraints on whether the putative late heavy bombardment was a true cataclysm (Ryder 1990), followed a longer duration, less intense, 'sawtooth' model (e.g., Turner 1979; Morbidelli et al. 2012), or consisted in a monotonic decline over time (e.g., Hartmann 1970; Neukum et al. 2001; Morbidelli et al. 2018).

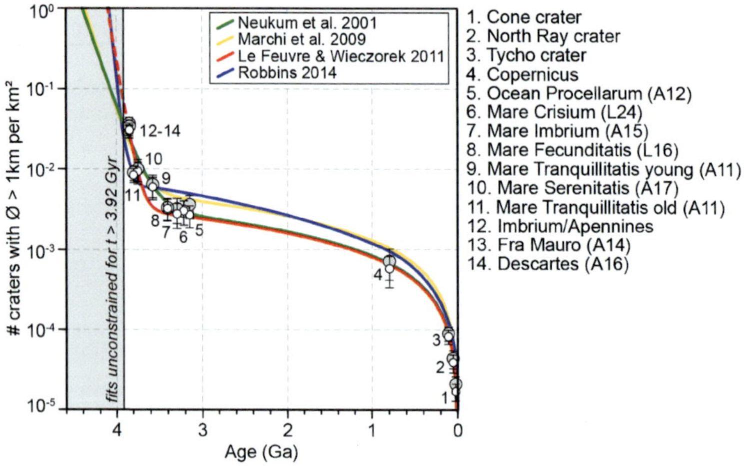

Fig. 4 Examples of different lunar crater chronology calibration curves and the Apollo sample age calibration points used (redrawn after Le Feuvre and Wieczorek 2011). Note that (4) Copernicus and (3) Tycho are from indirect sampling at the Apollo 12 and Apollo 17 landing sites, respectively

4.3.2 Anchoring the Lunar Cratering Chronology

Samples collected from known locations on the Moon, and in particular the mare basalts, have made it possible with CSFDs to quantify the relationship between the age of a geologic unit on a planetary surface and the density of impact craters on that surface (Fig. 4) (Neukum et al. 2001; Stöffler and Ryder 2001; Robbins 2014; Fassett 2016). Returning and dating any new samples that can be confidently linked with particular locations and geologic units on the Moon would provide valuable new data-points for future cratering chronology studies. The importance of this development extends beyond just understanding the impact history of the Moon, since the lunar cratering chronology model is commonly extrapolated to studies of other terrestrial planets (Neukum et al. 2001; Marchi et al. 2009).

The current plans for the China National Space Administration (CNSA) Chang'E-5 mission, planned for late 2019—early 2020, involve returning samples from the Mons Rümker region in northern Oceanus Procellarum (Qian et al. 2018; Zhao et al. 2017), with potential target landing sites close to the young (∼1.2–1.5 Ga) basalt flows identified in previous remote sensing studies (Hiesinger et al. 2003; Stadermann et al. 2018). Likewise there are NASA mission proposals to sample *ca.* 1 Ga-old lava flows adjacent to the Aristarchus region of the nearside of the Moon (Draper et al. 2019). Given that current lunar crater chronology studies still rely predominantly on samples with ages between 3.0–4.0 Ga (Fig. 4), samples from these young basalt flows would be immensely valuable, helping to calibrate parts of the lunar cratering curve more relevant to Mars and Venus surface geological processes. Similarly, as meteorite bombardment is thought to have been so much more intense during the first billion years of solar system history, returning samples from any exposed basaltic flows older than 4.0 Ga would place better constraints at the other end of the timescale.

4.3.3 Secular Evolution of Types of Projectiles Hitting the Moon

A thorough investigation of potential changes in the origin(s) of projectiles hitting the Moon requires precise age constraints on (i) impact-melt lithologies in which chemical and isotopic signatures of impactors are measured, and of (ii) regolith and soil samples from which projectile fragments are recovered. Projectiles in regolith samples >3.5 Ga-old could, for example, contain ancient terrestrial and martian crustal fragments that would be key for tackling questions linked with habitability of terrestrial planets and the origin of life on Earth (Armstrong et al. 2002; Crawford et al. 2008; Armstrong 2010; Bellucci et al. 2019).

There are different proxies for determining the duration of exposure of regolith samples to the space environment (see Joy et al. 2016), including (i) the ferromagnetic resonance maturity index (I_s/FeO—this corresponds to the intensity of the ferromagnetic resonance caused by non-interacting fine-grained metal particles ratioed to the FeO abundance, since the amount of fine-grained metal is proportional to both the duration of surface exposure and the amount of FeO available for reduction; Morris 1976, 1978) used to classify surface exposure of regolith samples into immature ($I_s/FeO \leq 29$), submature ($I_s/FeO = 30$–59) and mature ($I_s/FeO \geq 60$) (Morris 1976, 1978), (ii) the amount of 'trapped' solar wind-derived ^{36}Ar (e.g., Wieler and Heber 2003), or (iii) the agglutinate content (McKay et al. 1972). Quantitative exposure ages can be derived from measurement of the abundance of cosmogenic noble gases (^{3}He, ^{21}Ne, ^{38}Ar, ^{81}Kr, ^{126}Xe) produced during regolith interaction with cosmic rays in the top few meters of regolith (e.g., Eugster et al. 2001). However, none of these proxies allow estimating when the samples were exposed at the lunar surface, and exposure ages could even be influence by several discrete episodes of exposure due to the constant gardening of the lunar regolith. A semi-quantitative method of estimating regolith closure ages can be found using the $^{40}Ar/^{36}Ar$ antiquity indicator, which relates the amount of solar wind-derived ^{36}Ar to the amount of 'parentless' ^{40}Ar derived from the lunar exosphere (e.g., McKay et al. 1986; Eugster et al. 2001; Joy et al. 2011b; Fagan et al. 2014; Wieler 2016). The model closure ages for 74 Apollo, Luna and lunar meteorite regolith breccias range from *ca.* 0.1 Ga for some Apollo 15 samples up to *ca.* 3.8–3.9 Ga for some Apollo 14 samples, displaying a rough bimodal distribution with broad peaks between ~3.4–3.8 Ga and 1–2 Ga (Fig. 5), indicating that most regolith breccia samples in our collections formed around 1–2 Ga ago and 3.5–3.8 Ga ago, and that none is older than *ca.* 4 Ga.

Sample-return missions targeting regolith breccias with lithification ages older than *ca.* 3.8–3.9 Ga would greatly complement the record present in our collections, potentially providing crucial information on the types of projectiles hitting the Earth-Moon system during the basin-forming epoch before 3.9 Ga ago. The most scientifically important sample sites would be 'palaeoregolith' horizons, which are ancient regolith layers or lenses trapped between datable geological units, such as lava flows or impact melt flows (Crawford et al. 2007, 2010; Fagents et al. 2010; Crawford and Joy 2014). Whilst deep drilling might be necessary to access deeply seated palaeoregolith horizons, some natural access might be provided through layers exposed in the walls of lava tube skylights or in layered boulders in which palaeoregolith horizons may be preserved between volcanic layers (Fig. 6). For instance, Lunar Reconnaissance Orbiter Camera (LROC) Narrow Angle Camera (NAC) images of a Mare Tranquillitatis pit show that several tens of meters of layered units are present below the surface (Robinson et al. 2012a). The basaltic units at the surface where the pit is located are ~3.6 Ga old (Hiesinger et al. 2000), so the deepest palaeoregolith layers in the Mare Tranquillitatis pit may well be over 4 Ga old, considering typical regolith formation rates on the order of 1–2 mm/Ma (Hörz et al. 1991).

5 Nature, Timing, and Duration of Lunar Mare Volcanism

5.1 Overview

The most obvious expressions of volcanic activity on the Moon are the mare basalt infillings of impact basins, particularly on the lunar nearside. The majority of these maria

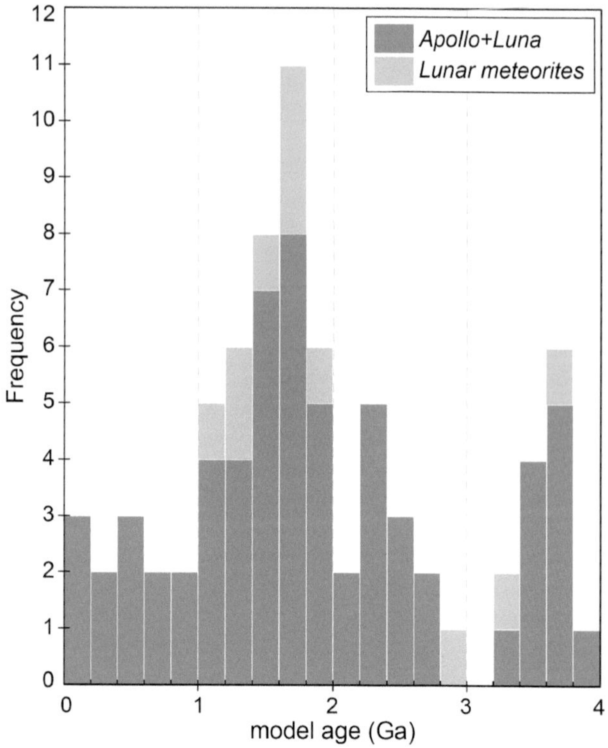

Fig. 5 Histogram of regolith breccias model closure ages (data and sources can be found in Fagan et al. 2014)

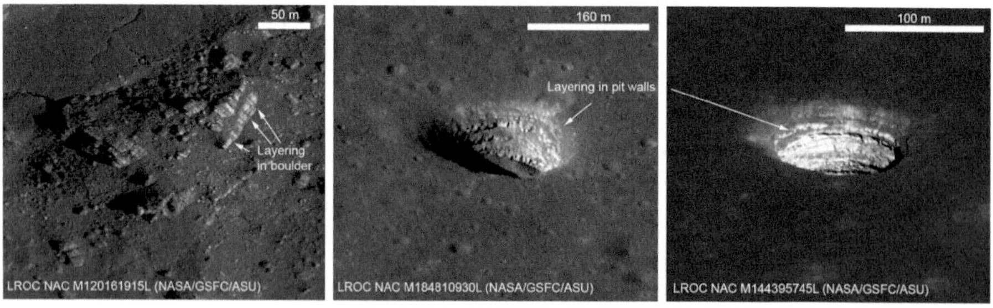

Fig. 6 Oblique LROC NAC view of layered boulders found on the lunar surface within Aristarchus crater (left), and of layered pits in Mare Ingenii (centre) and Mare Tranquillitatis (right). Scale bars in all cases have been estimated from the pixel resolution of NAC images (images: NASA/GSFC/Arizona State University, modified by K.H. Joy)

appear to have been emplaced between approximately 3.8–3.1 Ga (Nyquist and Shih 1992; Hiesinger et al. 2003, 2010). Older rocks identified in both the Apollo collection and in lunar meteorites, indicate that some form of basaltic volcanism was occurring on the Moon as early as approximately 4.35 Ga (Terada et al. 2007a; Snape et al. 2018a; Curran et al. 2019). It is not clear to what extent this early volcanism resembled the large scale eruptions of the younger mare basalts, or if these samples resulted from more localised volcanic activity. Meanwhile, the presence of less heavily cratered mare basalt flows indicates that mare volcanism continued until approximately 1.0 Ga in restricted geographical locations (Hiesinger et al. 2003, 2011; Stadermann et al. 2018), but examples of such young lunar basalts have yet to be identified in the sample collection. In this section, we review the current understanding of the Moon's volcanic history based on the Apollo, Luna and meteoritic samples, and highlight outstanding issues that could be addressed by future sample-return missions.

5.2 The Nature and Origin of Lunar Volcanism

Discussions of volcanism on the Moon have typically divided samples into those formed as a result of mare and non-mare (or pre-mare) volcanism (Nyquist and Shih 1992). The focus of this paper is primarily on the geochemical nature of samples generated by volcanic processes. The readers interested in recent studies on magma ascent from the lunar mantle and on the possible eruption mechanisms that produced these samples and associated volcanic structures are referred to Head and Wilson (2017) and Wilson and Head (2017, 2018). In addition to basaltic rocks, the mare volcanic samples also include glass beads, produced during pyroclastic "fire fountain" eruptions (Rutherford et al. 2017). Moving beyond these top-level classifications, petrologic and geochemical analyses of samples returned by the Apollo and Luna missions have revealed a compositional diversity in the Moon's basaltic rocks that necessitates more specific classification schemes (Papike et al. 1976; Neal and Taylor 1992). Using the classification proposed by (Neal and Taylor 1992), the lunar basalts are defined first by their bulk TiO_2 content (where: >6 wt% = high-Ti; 1–6 wt% = low-Ti; <1 wt% = very low-Ti [VLT]), then by Al_2O_3 content (>11 wt% = high-Al; <11 wt% = low-Al) and finally by K content (>2000 ppm = high-K; <2000 ppm = low-K). At a more detailed level, even within the mare basalt samples collected from individual landing sites, it is possible to define multiple different basaltic suites based on petrologic texture, relative abundance of minerals, chemical composition and isotope systematics (Rhodes et al. 1976, 1977; Beaty and Albee 1978; Warner et al. 1979; Paces et al. 1991; Jerde et al. 1994; Neal et al. 1994; Snyder et al. 1994). One notable group of samples that are not described by the above classification scheme are the KREEP-rich basalts, which are best represented by samples collected during the Apollo 15 mission (Nyquist et al. 1975; Carlson and Lugmair 1979; Nyquist and Shih 1992), and are characterised by elevated trace element compositions similar to the KREEP signature identified in other Apollo samples (Warren and Wasson 1979).

All of these volcanic samples are generally accepted to have been derived from 5–10% partial melting in the Moon's interior (e.g. Shearer et al. 2006 and references therein), with a common starting point for most petrogenetic models being mafic cumulates formed during the LMO crystallisation. Simply varying the proportions of these different cumulates cannot adequately explain the variability observed in lunar basalts, and so a variety of mechanisms have been proposed to address this issue. This includes the presence of more evolved residual LMO melt enriched in incompatible trace elements (ITE) in the lunar mantle (Snyder et al. 1997; Borg et al. 2004; Hallis et al. 2014), the introduction of ilmenite-bearing material during density-driven overturn of the lunar mantle (Snyder et al. 1997), and the assimilation of KREEP and more evolved plutonic rocks (e.g. granite) by the magmas (Neal and Kramer 2006). Additionally, it is not clear what the heat source for the melting of this material would have been. Trapped ITE-rich material in the lunar mantle provides one potential heat source, but external heat sources, such as meteoritic bombardment and tidal heating, have also been proposed (Borg et al. 2004; Elkins-Tanton et al. 2004; Garrick-Bethell et al. 2006; Shearer et al. 2006).

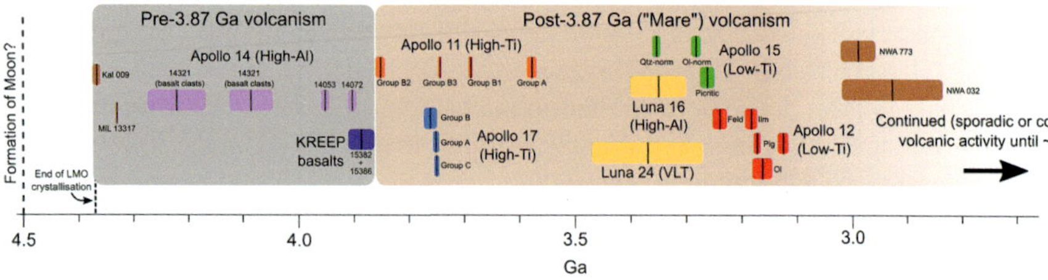

Fig. 7 Compilation of crystallisation ages determined for lunar basaltic samples. The boxes are coloured to correspond to different groups of basaltic samples. The widths of the boxes represent 2 standard deviation uncertainties of either individual sample ages (in the case of the four lunar meteorites; Borg et al. 2004, 2009; Terada et al. 2007a; Sokol et al. 2008; Snape et al. 2018a), or weighted averages of multiple age determinations (including data from: Turner 1970a, 1970b; Papanastassiou and Wasserburg 1971a, 1971b; Compston et al. 1972; Stettler et al. 1973; Turner et al. 1973; Nyquist et al. 1975, 1976; Murthy and Coscio 1976; Guggisberg et al. 1979; Carlson and Lugmair 1979; Paces et al. 1991; Snyder et al. 1994, 1997; Cohen et al. 2001; Fernandes and Burgess 2005; Snape et al. 2016a, 2018b, 2019)

5.3 Timing of Lunar Volcanism

Currently, evidence for the oldest basaltic volcanism on the Moon comes from two lunar meteorites: the basaltic breccia Kalahari (Kal) 009 (4.369 ± 0.007 Ga; Terada et al. 2007a; Sokol et al. 2008; Snape et al. 2018a) and the basalt-bearing regolith breccia, Miller Range (MIL) 13317 (4.332 ± 0.002 Ga; Snape et al. 2018a) (Fig. 7). Prior to the discovery of these meteorites, the oldest samples of lunar basalt were the high-Al basalts collected during the Apollo 14 mission. Early studies of the volcanic rocks returned from the Apollo 14 landing site yielded ages of approximately 3.9–4.1 Ga, based on internal Rb-Sr isochrons (Compston et al. 1971; Papanastassiou and Wasserburg 1971a; Compston et al. 1972), but further investigations of basaltic clasts in several Apollo 14 breccias determined ages of up to 4.3 Ga (Taylor et al. 1983; Dasch et al. 1987) (Fig. 7). The KREEP basalts collected during the Apollo 15 and 17 missions appear to have formed soon after (and potentially overlapping with) the eruption of the last high-Al basalts, with the largest of these samples (15382 and 15386) having crystallised at *ca.* 3.9 Ga (Stettler et al. 1973; Turner et al. 1973; Nyquist et al. 1975; Carlson and Lugmair 1979; Snape et al. 2016a) (Fig. 7).

The formation of the high-Al and KREEP basalts was followed by the emplacement of the mare basalts. Based on the samples currently available, this period of lunar volcanism appears to have begun with the eruption of high-Ti basalts. Samples collected at the Apollo 11 landing site have ages varying from ∼3.6–3.85 Ga (Turner 1970a, 1970b; Papanastassiou and Wasserburg 1971b; Guggisberg et al. 1979; Snyder et al. 1994; Snape et al. 2016a, 2019), while those sampled by the Apollo 17 mission have a more narrow range of ages, between ∼3.75–3.8 Ga (Murthy and Coscio 1976; Nyquist et al. 1976; Paces et al. 1991; Snape et al. 2019) (Fig. 7). The oldest low-Ti mare basalts (∼3.28–3.35 Ga; Snape et al. 2019) collected during the Apollo missions were returned from the Hadley-Appennine province by Apollo 15 (Papanastassiou and Wasserburg 1973; Husain 1974; Snape et al. 2019) (Fig. 7). By comparison, the low-Ti basalts collected during the Apollo 12 mission have ages between ∼3.24–3.13 Ga (Snyder et al. 1997; Snape et al. 2018b), making them the youngest lunar basalts to have been returned by human sampling (Fig. 7).

Sample-based evidence for more recent magmatic activity on the Moon has been identified in the meteorites Northwest Africa (NWA) 773 (Borg et al. 2004) and NWA 032 (Borg et al. 2009). Both of these samples have ages of ∼2.9 Ga (determined by internal Sm-Nd and

Rb-Sr isochrons) (Fig. 7). In the case of NWA 773, it is notable that the sample appears to be strongly enriched in ITEs, and exhibits a KREEP-rich geochemical signature, despite it being significantly younger than any of the KREEP-rich samples returned by Apollo, while NWA 032, by contrast, has relatively low ITE concentrations. While these are currently the youngest identified volcanic samples from the Moon, crater size-frequency distribution studies have identified basaltic flows that appear to have been emplaced at $\sim$1.0 Ga (Hiesinger et al. 2003; Stadermann et al. 2018), and potentially some even younger "irregular mare patches" for which ages of <0.1 Ga have been proposed (Braden et al. 2014).

5.4 Outstanding Questions and the Potential of Future Sample-Return

5.4.1 Duration and Flux of Lunar Volcanism

Identifying and sampling the youngest lunar basalts would provide new insight into how long small terrestrial planetary bodies can sustain volcanic activity. This issue is connected to a more general issue of the changing flux of lunar volcanism with time (e.g., Hiesinger et al. 2011; Robinson et al. 2012b). Based on the ages of the Apollo, Luna and meteorite basaltic samples, together with crater size-density frequency studies of exposed mare basalt units, the majority of the lunar maria appear to have been emplaced between 3.8–3.1 Ga (Nyquist and Shih 1992; Hiesinger et al. 2003, 2010). Upcoming missions such as the CNSA Chang'E-5 or the Inner SOlar System CHRONology (ISOCHRON) missions could constitute a first step in this direction, since it may involve returning samples from very young basalt flows ($\sim$1.2–1.5 Ga) identified in previous remote sensing studies (Hiesinger et al. 2003; Qian et al. 2018; Stadermann et al. 2018; Draper et al. 2019).

Assessing the amount of recent volcanic activity should be a relatively simple issue of identifying and mapping exposed volcanic units. By contrast, determining how common ancient (>4.0 Ga) lunar volcanism was, is likely to be more challenging due to the potential for older units to have been obscured by subsequent eruptions and impact processes. In addition to the ancient volcanic lunar samples discussed previously, remote sensing studies have identified evidence for ancient (>4.0 Ga) mare volcanism in the form of "dark-haloed" craters. These were interpreted as instances where basaltic flows had been buried by the ejecta deposits from large impact craters and basins, and then subsequently re-exposed by smaller impacts (Schultz and Spudis 1979, 1983). Buried basaltic flows of this kind have since been designated by the name "cryptomare" (Head and Wilson 1992; Whitten and Head 2015a, 2015b). Terada et al. (2007a) proposed that the Kal 009 meteorite represented a fragment from such a deposit, and it is true that the bulk-rock composition and mineral chemistry data from the sample are consistent with mare basalts collected during the Apollo and Luna missions (Sokol et al. 2008). Nonetheless, it remains unclear just how much basaltic material could have been buried by impact ejecta.

Sample-return missions to identified dark-haloed craters would provide a valuable first step in testing the cryptomare theory. By performing detailed petrologic, geochemical and isotopic studies of potential cryptomare basalts it would be possible to determine how similar they are to previously studied lunar basalts, as well as when they crystallised. However, individual sample-return missions will almost certainly struggle to provide a definitive answer to the larger question of how common such deposits are. This will likely require a more prolonged human presence on the Moon, enabling protracted and focused geologic sorties with the potential to obtain core samples and map the extent of cryptomare deposits.

5.4.2 Petrogenesis of Lunar Basalts

Returning samples of young and old lunar basalts, together with an increased knowledge of the locations and extent of cryptomare deposits, will be vital to gaining a more complete understanding of mantle sources for lunar volcanism, and addressing questions related to the petrogenesis of lunar basalts. One of the most striking aspects about the lunar mare basalts is their assymetric distribution between the lunar nearside and farside, as well as the spatial correlation between the exposed mare units, the nearside lunar basins and the PKT (Jolliff et al. 2000) (see Fig. 1). Determining the significance of these correlations appears to be fundamental to understanding the sources of lunar basalts and the mechanism(s) that led to partial melting in the lunar mantle.

One suggested explanation for the asymmetric distribution of mare basalts is a propencity for more extrusive volcanism on the thinner nearside crust compared with the thicker crust of the lunar farside (Head and Wilson 2017). Given the location of so many mare basalt flows within large impact basins, there is also a temptation to try and link the period of apparently more intense volcanic activity on the Moon (from $\sim$3.8–3.1 Ga) with the end of the basin-forming period of lunar history and the proposed cataclysm (at $\sim$3.9 Ga). One suggestion is that the formation of the large impact basins on the nearside of the Moon could have led to adiabatic melting in the lunar mantle (e.g., Solomon and Head 1980; Elkins-Tanton et al. 2004). For example, in the model proposed by Elkins-Tanton et al. (2004), partial melting occurs in two stages: (i) *in situ* melting first occurs in response to the decompression associated with basin excavation; (ii) a second stage of adiabatic melting then occurs in response to convection at the lithosphere-astenosphere boundary, triggered by the isostatic rebound of the lithosphere after basin formation. In the case of the Orientale basin, a significant (60–100 Ma) time gap has been estimated between the formation of the basin ($\sim$3.68 Ga) and the subsequent eruption of the Mare Orientale units ($\sim$3.58 Ga), which has been cited as evidence against such impact-induced melting (Whitten et al. 2011). However, melts produced during the second stage of adiabatic melting of the Elkins-Tanton et al. (2004) model could form as late as *ca.* 350 million years after the impact event (Elkins-Tanton et al. 2004). Identifying the distribution of ancient (>4.0 Ga) volcanic units, and determining the times at which they were emplaced by dating returned samples, would be a significant step towards testing the link between lunar volcanism and impact events.

The connection between impact basins and basalts does not, in itself, explain the correlation between the mare basalts and the PKT or the extent to which lunar volcanism is connected with KREEP. Returning basaltic samples further away from the PKT than the Apollo basalts, would help address this issue, and a hint of what might be expected from such samples may be seen in several basaltic lunar meteorites (Yamato-793169, Asuka-881757, Meteorite Hills 01210 and MIL 05035). These samples have lower concentrations of ITEs than most Apollo mare basalts and appear to have been derived from mantle sources with low ratios of U/Pb and Rb/Sr, and high Sm/Nd ratios (Misawa et al. 1993; Terada et al. 2007b; Joy et al. 2008; Liu et al. 2009; Arai et al. 2010). These features have been interpreted as evidence that these rocks may be derived from partial melts of more primitive lunar mantle cumulates than those from which the majority of Apollo basalts originated (Joy et al. 2008; Liu et al. 2009; Arai et al. 2010). The low-ITE concentrations and lack of a KREEP signature in these samples also indicate that their parental magmas did not assimilate significant amounts of KREEP-rich material. Furthermore, while the presence of ITE-rich material in the lunar mantle may have triggered the partial melting that eventually produced many of the Apollo basalts, these samples indicate that the same mechanism may not be sufficient to explain the petrogenesis of all lunar basalts. If these characteristics were found to be common in basalt samples collected away from the PKT, it could provide important insights into the petrogenesis of lunar basalts, the distribution of the proposed urKREEP layer, and the early magmatic evolution and differentiation of the Moon.

6 Structure and Composition of the Lunar Interior

6.1 Geophysical Insights

The Apollo Lunar Surface Experiments Packages (ALSEP) deployed across the Moon's surface by astronauts on the Apollo 12, 14, 15, 16 and 17 missions are one of the key legacies of the Apollo programme. These included active and passive seismic experiments, gravimeters, magnetometers, and heat flow probes at the Apollo 15, 16 and 17 landing sites, for example (e.g., Eichelman and Lauderdale 1974; Wieczorek et al. 2006; Jaumann et al. 2012; Weber 2014). This lunar network of seismometers recorded *ca.* 13000 catalogued events between 1969 and 1977 (e.g., Nakamura et al. 2008), including non-natural events related to impacts of booster rockets and lunar modules from Apollo spacecraft. Natural seismic sources on the Moon have been divided into three categories: meteoroid impacts, shallow moonquakes and deep moonquakes. Deep moonquakes are more frequent than shallow moonquakes. About 7,000 deep moonquakes have been recorded, originating from distinct source regions *ca.* 700–1200 km deep (e.g., Jaumann et al. 2012; Weber 2014). The precise location and periodicity of deep moonquakes are not well understood, but it seems that most deep moonquakes occur at monthly intervals, suggesting that they could be triggered with tidal stress (e.g., Latham et al. 1971). Shallow moonquakes are less frequent than deep moonquakes—28 events have been recorded between 1969 and 1977 (Nakamura 1980)—and more powerful, with magnitudes of *ca.* 1.5 to 5 on a Richter-equivalent scale (e.g., Lammlein 1977; Nakamura 1980). Estimates for the depth of their epicentres range from $\sim$10 km down to $\sim$100 km deep (e.g., Gillet et al. 2017), the shallowest ones being potentially linked with recent thrusting activity on fault scarps suggesting that the Moon is still tectonically active (Watters et al. 2019). Seismic data also point to possible compositional stratification within the lunar mantle—modelling results suggest that there may be a seismic discontinuity *ca.* 500–600 km deep, below which P- and S-wave velocity seems to increase with depth (Nakamura et al. 1982; Khan et al. 2000). The presence of a zone of mineralogical phase transition and change in mantle composition toward more magnesian and aluminous compositions below $\sim$500 km may be interpreted as indicating (i) the base of the LMO, if the whole Moon was not fully melted, (ii) the limit between olivine-dominated early cumulates and orthopyroxene-rich later cumulates in a whole Moon melted scenario, or (iii) the maximum depth of partial melting of the basaltic melt source regions (see Wieczorek et al. 2006 for further discussion).

One of the key findings that arose from interpretation of lunar seismic data is that the Moon is a differentiated body, consisting of a crust, a mantle and a *ca.* 400–500 km radius metallic core comprising a solid inner core and a fluid outer core (e.g., Weber et al. 2011). Gravity data also provide valuable constraints on the Moon's internal structure. The gravity field of the Moon has been mapped with increasing levels of details by successive orbiting spacecraft (NASA Lunar Prospector in the 1990s, JAXA SELENE in the 2000s and NASA GRAIL in the 2010s), the latest maps produced by the GRAIL mission being actually the highest resolution gravity maps for any Solar System body including the Earth (e.g., Zuber et al. 2013). Combined with constraints from Apollo seismic experiments, GRAIL data suggest that the lunar crust has an average thickness of 34–43 km, with the farside crust being *ca.* 10–15 km thicker compared to that of the nearside (Wieczorek et al. 2013). The other main feature revealed by gravity data is the presence of anomalous mass concentrations, known as 'mascons' (Fig. 3). These are associated with large impact basins and thought to originate from the excavation and collapse of impact basins, followed by isostatic adjustment and cooling and contraction of voluminous melt pools (e.g., Melosh et al. 2013).

6.2 Sample-Based Insights

Samples of the lunar mantle have yet to be found in our Apollo, Luna and lunar meteorite collections. Our knowledge of the composition of the lunar mantle thus mostly relies on petrological and geochemical studies of mare basalts and pyroclastic glasses produced via mantle partial melting. Importantly, their compositions only reflect that of their mantle source regions at the time of their extraction from the mantle *ca.* 4.3 to 3 billion years ago (see Sect. 5: Nature, timing, and duration of lunar mare volcanism). Numerous primitive lunar basaltic melts, sampled by pyroclastic glasses, are rich in FeO and TiO_2 (e.g., Delano 1986). Experimental work suggests that most of these primitive melts formed at pressures of *ca.* 1.5 to 3 GPa, roughly corresponding to depths of ~300–600 km (e.g., Delano and Livi 1981; Longhi 1992; Grove and Krawczynski 2009; Brown and Grove 2015), while crystalline mare basalts seem to originate from partial melting at depth shallower than ~350 km (e.g., Wieczorek et al. 2006). This thus provides evidence for the presence of Fe- and Ti-rich sources deep in the lunar mantle, which is not predicted by LMO crystallisation models (e.g., Snyder et al. 1992; Charlier et al. 2018; Rapp and Draper 2018). To account for this observation, it has been proposed that Rayleigh-Taylor instability causes dense ilmenite-rich cumulates and underlying Fe-rich upper mantle layers formed at the end of the LMO crystallisation to sink and mix in with earlier-formed Mg-rich lower mantle (e.g., Hess and Parmentier 1995). This mantle overturn mechanism also provides a way to bring radioactive incompatible elements such as K, U and Th, enriched in late stage LMO liquids, deep into the mantle. However, it remains unclear whether heating via radioactive decay of these ITEs is a viable mechanism for sustaining mantle partial melting during the main period of mare basalt volcanism ~3.8 to 3.0 Ga ago or if the influx of hotter deep mantle material was required to trigger partial melting in cold mare basalt and pyroclastic glasses source regions (e.g., Spohn et al. 2001; Grove and Krawczynski 2009).

Compositionally, the mare basalt source regions are characterised by (i) low alkali and siderophile element abundances, (ii) Sm-Nd and Lu-Hf isotope compositions fractionated compared to chondritic compositions, and (iii) very low oxygen fugacities, as indicated by the presence of Fe metal and reduced valence state of Fe, Ti and Cr in mare basalts (see Wieczorek et al. 2006 et references therein). For decades after the return of Apollo samples, it has been assumed that the lunar mantle was extremely dry, with estimates of lunar mantle water content lower than 1 part per billion (Taylor et al. 2006). However, with technological advances in lab instrumentation detection limits for species such as water, carbon or halogens such as fluorine and chlorine greatly improved, which permitted the successful detection in 2008 of indigenous water in lunar pyroclastic glasses (Saal et al. 2008). This was followed by a wealth of studies reporting the abundance of water and other volatile species (e.g., C, N, F, S, and Cl) in trapped melt inclusions and mineral phases such as apatite in basaltic lava products (see recent reviews by Anand et al. 2014; McCubbin et al. 2015; Hauri et al. 2017). Reconstructing mantle volatile abundances from those measured in volcanic products is not trivial, notably because volatile elements can be modified by numerous processes such as magmatic degassing, assimilation of solar wind-enriched soils, spallogenic reactions at the surface, or contamination during sample preparation on Earth. Therefore, current estimates for the water abundance in the mantle source regions of mare basalts and pyroclastic glasses range between *ca.* 1 and 100 parts per million (e.g., McCubbin et al. 2015; Hauri et al. 2017). It remains unclear whether this fairly large range of estimates reflects uncertainties in back-calculating water contents for mare basalt source regions, heterogeneous distribution of water in the lunar mantle, or a combination of both.

6.3 Outstanding Questions and the Potential of Future Sample-Return

The location of Apollo landing sites within or just around the anomalous PKT area (Fig. 1) suggests that Apollo samples provide us with a biased view of the Moon (see Sect. 5: Nature, timing, and duration of lunar mare volcanism). This is also where ALSEP packages were deployed by Apollo astronauts, suggesting that inferences about the stratification of the lunar interior based on seismic data might also be biased. A global geophysical network (e.g., Neal 2011; Smith et al. 2012; Neal et al. 2019) with stations distributed all over the lunar surface would provide invaluable information for better constraining the structure of the Moon's interior, including the possible discontinuity at ~500 km depth, the presence of partially molten layers deep in the lower mantle, and the size and structure of the core. Global geophysical data such as the gravity data obtained by GRAIL support the crustal thickness dichotomy, the farside crust being *ca.* 10–15 km thicker compared to the nearside crust (Wieczorek et al. 2013). Several scenarios have been proposed to explain such dichotomy, including (i) asymmetric convection during LMO crystallisation and crustal growth (Loper and Werner 2002; Ohtake et al. 2012), (ii) asymmetric impact cratering (Wood 1973), (iii) ejecta deposition from SPA (Zuber et al. 1994), (iv) inhomogeneous early tidal heating (Garrick-Bethell et al. 2010), (v) accretion of a companion moon (Jutzi and Asphaug 2011), or (vi) giant impact of a ~500–800 km diameter impactor onto the lunar nearside soon after the Moon's formation (e.g., Zhu et al. 2019). Returning samples from the farside crust would certainly help in disentangling between these different scenarios, in addition to providing opportunities to evaluate the possible chemical differences between farside magnesian anorthosites and nearside ferroan anorthosites (see Sect. 3: Petrology and formation history of the lunar anorthitic crust).

Sampling pieces of the lunar mantle would prove crucial to our understanding of its mineralogical and chemical nature. In particular, this would allow direct measurement of its volatile and water inventory, which has key implications for investigations of how the Moon formed (e.g., Hauri et al. 2017), and of LMO crystallisation and thermal evolution of the Moon's interior (e.g., Evans et al. 2014; Khan et al. 2014). Impact cratering processes may have breached the crust periodically and could have brought up the pieces of lunar mantle at the surface; SELENE/Kaguya spectral data indeed suggest that numerous olivine-rich (mantle) exposures are found around large impact basins characterised by thin crustal thickness, including Mare Moscoviense, Crisium, Imbrium, Humorum, Nectaris, Serenitatis and basins within SPA such as Schrödinger (Yamamoto et al. 2010; see location of basins on Fig. 3). The Yutu-2 rover, which landed successfully in Von Kármán crater within the SPA basin on the lunar farside (bottom right corner on Fig. 3) on-board the CNSA Chang'e 4 lander, may indeed have identified, remotely through near infrared spectroscopy, samples excavated from the lunar mantle (Li et al. 2019). Such impact basins with very thin crust thus represent key targets for sample-return missions of pieces of the Moon's mantle.

7 Icy Deposits at the Lunar Poles

7.1 Evidence for the Presence of Volatiles at the Moon's Poles

The occurrence of water ice at the lunar poles has been postulated for decades (e.g., Watson et al. 1961; Arnold 1979) because of the presence of deep topographic lows at the

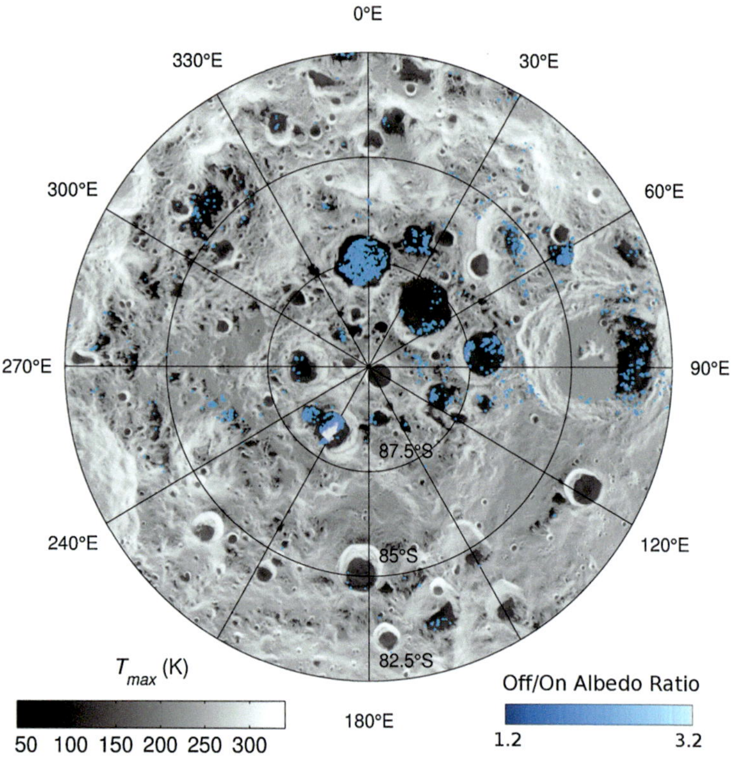

Fig. 8 Locations of anomalous UV albedo consistent with water ice (figure from Hayne et al. 2015). Colours indicate points with off/on-band albedo ratio values >1.2 and Lyman-α albedo <0.03, increasing from deep blue (1.2) to white (>3.2). The average Moon outside of cold traps has a ratio of 0.9. Ratio values in the range 1.2–4.0 are consistent with water ice concentrations of 0.1–2.0 wt.%. If patchy exposures of pure water ice are mixed with dry regolith, the abundance could be up to 10 wt.%

poles and the low obliquity of the Moon. In the 1990s, possible detection of water ice in polar regions of Mercury using Earth-based radar observations (Harmon and Slade 1992) refuelled the speculation that the lunar poles might also act as a trap for water ice and other volatiles. At the same time, data returned by instruments on board the two NASA lunar orbiters Clementine and Lunar Prospector supported the presence of water ice in permanently shadowed regions (PSR) at the lunar poles (Nozette et al. 1996, 1997; Feldman et al. 1998, 2000), although subsequent Earth-based radar investigations challenged the existence of lunar polar water ice, which became a topic of strong scientific controversy (see Campbell et al. 2006, and references therein). Further remote observations carried out by a broad range of instruments on-board India's Chandrayaan-1 and NASA's Deep Impact, Cassini, and Lunar Reconnaissance Orbiter (LRO) missions, coupled with the direct detection of *ca.* 5 wt.% water and other volatile species in the plume formed by the LCROSS impact-experiment in the Cabeus crater, strongly suggested that there is indeed OH/H_2O molecules and water ice at the surface or sub-surface of lunar polar regions (Clark 2009; Pieters et al. 2009b; Sunshine et al. 2009; Colaprete et al. 2010; Gladstone et al. 2010; Hayne et al. 2010, 2015; Mitrofanov et al. 2010; Paige et al. 2010b; Schultz et al. 2010; Spudis et al. 2013; Fisher et al. 2017; Li and Milliken 2017; Li et al. 2018) (Fig. 8). However, we have little insights on how ice occurs (i.e. solid lumps of ice vs.

ice mixed in with lunar soils), on its vertical and lateral distributions, and on its origin(s). For example, the production of OH/H_2O species at the surface could result from interaction of solar-wind protons with the regolith (e.g., Farrell et al. 2015). Cometary and volatile-rich asteroidal objects that have impacted the Moon for billions of years are obvious sources for accumulation of ice at the lunar poles. The migration of water molecules from lower latitudes to polar cold traps in permanently shadowed craters is also possible (e.g., Schorghofer and Taylor 2007), implying that magmatic water degassed onto the lunar surface during volcanic eruptions could have made its way to polar traps.

7.2 Prospecting Missions to the Poles

At this stage only future prospecting missions to the lunar surface can provide firmer constraints on the abundance, form, distribution, and origin(s) of water ice and other lunar polar volatiles. Landers with drilling capabilities should be able to provide crucial information on the composition, abundance and vertical distribution of polar volatiles at a given location. In addition to providing important scientific constraints on the source and processing of these polar volatiles, each single point measurement will represent ground truth for orbital measurements. PROSPECT (Package for Resource Observation and in-Situ Prospecting for Exploration Commercial Exploitation and Transportation) is a payload currently developed by the European Space Agency (ESA) to carry out these types of activities (Carpenter et al. 2016). Broadly, PROSPECT can be deployed on a mobile platform or a fixed lander, and its technical capabilities can be tailored to different objectives. Currently, the first deployment of PROSPECT on the lunar surface is planned on the Russian Luna-27 mission that should land in a lunar south polar region. PROSPECT is comprised of a drill (ProSEED) that will extract regolith samples from up to *ca.* 1.2 m depth. The samples will be delivered to a miniature chemical laboratory (ProSPA) where volatile species (e.g., H_2, H_2O, CO, CO_2, N_2, noble gases) will be degassed in order to measure their abundances and stable isotope compositions (Barber et al. 2017). Thanks to their mobility, rovers could provide complementary information on the composition, abundance and lateral distribution of subsurface polar volatiles across traverses up to a few kilometres long, which would also reveal key scientific information and ground truth for orbital measurements. One such example is the NASA Volatiles Investigating Polar Exploration Rover (VIPER) mission concept (Colaprete et al. 2019). To characterise the budget of polar volatiles, VIPER would rove across the lunar surface and locate areas of high hydrogen concentration using its neutron spectrometer. Once the presence of subsurface volatiles is confirmed, regolith samples would be drilled and delivered to a miniature chemical laboratory in which volatile species (e.g., H_2, H_2O, CO, CO_2, N_2, noble gases) would be degassed and their abundances quantified (Colaprete et al. 2019).

If the presence of polar volatiles at the lunar subsurface is confirmed by prospecting missions, the next logical step would then be to collect and return samples to the Earth for further characterisation of polar volatiles. Only sample-return missions allow fully comprehensive scientific investigations using present day state-of-the-art analytical equipment. In the future, technological improvements should allow addressing new questions provided the returned samples are curated properly. The paradigm shift seen over the last decade on the issue of indigenous lunar water, through re-investigation of Apollo samples using 21st century lab equipment, perfectly illustrates the scientific importance of collection and curation of samples (e.g., Anand et al. 2014; McCubbin et al. 2015). In addition to providing ground truth for orbital measurements, a sample-return mission from a polar site where cold trapped volatiles occur would establish a comprehensive inventory of what species are present. Analysis of the isotopic composition of species such as H, C, N, O, S, and noble gases would

Fig. 9 Astronaut Buzz Aldrin deploying the Solar-Wind Experiment about 5 meters north-northwest of the Apollo 11 lunar module. Credit NASA photograph AS11-40-5872

provide key constraints on the source(s) of these volatiles and the processes that may have helped to concentrate them. To this end great care would be needed during sample collection and handling on the lunar surface, in the returned capsule to Earth, and on Earth. Temperature increase during sample collection through drilling for example could lead to ice and gas loss through sublimation. Therefore, the thermal and mechanical history of the samples collected will need to be precisely recorded.

8 The Lunar Regolith as a Recorder of the Sun's Composition, Activity and Galactic Environment

For half a century, the lunar regolith record has been used to decipher the elemental and isotopic composition of the solar wind (SW). In fact the first experiment deployed on the Moon by Apollo astronauts was designed to directly collect SW (Fig. 9). A 30 cm × 140 cm, 15-micron thick, aluminium foil sheet was exposed to solar rays at the lunar surface for 77 minutes in order to determine the SW He, Ne and Ar abundances and isotopic compositions (Geiss et al. 1969).

The SW is dominated by H ($\sim$95%) and He ($\sim$4%), with heavier elements such as C, N, O, Ne, Mg, Si and Fe comprising <1% (e.g., Anders and Grevesse 1989; Ogilvie and Coplan 1995). If the outermost layers of recently irradiated (i.e. in the last 100 Ma or so) lunar soil grains preserve an accurate record of the present-day composition of the SW, it has been proposed that systematic variations of isotopic ratios of some elements such as He, Ne, Ar, N or C, in soils irradiated billions of years ago could reflect changes of the SW composition through time (e.g., Becker and Clayton 1975; Kerridge 1975; Becker 1980; Clayton and Thiemens 1980; Becker and Pepin 1989; Geiss and Bochsler 1982, 1991; Kerridge et al. 1991; Benkert et al. 1993). Several hypotheses have been proposed to account for such changes, including (i) variations in acceleration conditions in the solar corona for light species such as He, (ii) contamination of the composition of the solar convective zone by interstellar material with exotic compositions, or (iii) change of spallation reaction rates and/or depth of occurrence in the photosphere with time (e.g., Kerridge 1980). However, coordinated analyses of multiple species in ilmenite from Apollo 12, 14 and 17 soil and regolith breccia samples via progressive etching (e.g., Heber et al. 2003), combined with high precision measurement of the noble gas and nitrogen isotope compositions of the SW directly collected in ultrapure targets at the L1 Lagrangian point for more than 2 years by the NASA Genesis spacecraft, seem to indicate that no such secular variations of the SW isotopic composition are recorded in lunar regolith samples (see Wieler 2016 for an extensive discussion on this topic).

8.1 Solar Wind Isotope Compositions Estimated from Lunar Sample Analysis

Estimates of the solar wind isotope compositions for the noble gases He, Ne and Ar, and for H, C, N, and O, obtained through analysis of lunar samples are summarised in Table 3.

8.1.1 Hydrogen, Carbon, Nitrogen and Oxygen

Both bulk and secondary ion mass spectrometry (SIMS) analysis of H isotope ratios in lunar soils, regolith breccia samples, soil minerals, and glassy agglutinates show that SW is an important source of H in the lunar regolith, as indicated by very low δD_{VSMOW} signatures[1] down to *ca.* −850‰ to −970‰ (e.g., Epstein and Taylor 1970; Friedman et al. 1970; Hashizume et al. 2000; Liu et al. 2012). The lowest δD_{VSMOW} values measured in lunar samples are slightly higher than the SW δD_{VSMOW} value of *ca.* −999‰ estimated for a Genesis Si collector (Huss et al. 2012a), indicating that lunar surface soils may contain some non-solar deuterium that may have formed via spallation reactions for example. The SW D/H ratio does not seem to have evolved much, if any, over time since sample 79035, which has a $^{40}Ar/^{36}Ar$-derived antiquity age of around 2 Ga (e.g., Wieler et al. 1996; Heber et al. 2003), yielded $(\delta D_{VSMOW})_{SW} < -950‰$ (Hashizume et al. 2000).

Depth profile analyses in grains from soil 79035, which contains surface H dominated by SW sources (Hashizume et al. 2000), have yielded very low $\delta^{13}C_{VPDB}$ values[2] of *ca.* −105 ±

[1]This δD notation corresponds to the per mil deviation of a measured D/H ratio from that of the Vienna Standard Mean Ocean Water ($VSMOW - D/H = 1.5576 \times 10^{-4}$), given by $[(D/H)_{sample}/(D/H)_{VSMOW} - 1] \times 1000$.

[2]This $\delta^{13}C$ notation corresponds to the per mil deviation of a measured $^{13}C/^{12}C$ ratio from that of the Vienna Pee Dee Belemnite ($VPDB - {}^{13}C/{}^{12}C = 0.0112$), given by $[({}^{13}C/{}^{12}C)_{sample}/({}^{13}C/{}^{12}C)_{VPDB} - 1] \times 1000$.

Table 3 Summary of some lunar sample-based estimates for solar wind isotope compositions and compositions measured on Genesis samples or estimated from absorption lines in the solar photosphere

Ratio	Value	±	Analysed target	References
$^{3}He/^{4}He$ ($\times 10^4$)	4.64	0.18	Genesis DoS	1
	4.26	0.22	Apollo SWE	2
	4.57	0.16	Ilmenite in "young" soil 71501	3
	4.47	0.26	Average of ilmenite from soils and breccias	4
$^{20}Ne/^{22}Ne$	13.78	0.06	Genesis DoS	1
	13.97	0.05	Genesis AloS	5
	14.00	0.04	Genesis Au foil, AuoS and Al collectors	6
	13.70	0.30	Apollo SWE	2
	13.80	0.20	Ilmenite in "young" soil 71501	3
$^{36}Ar/^{38}Ar$	5.47	0.02	Genesis DoS	1
	5.50	0.02	Genesis Si collectors	7
	5.50	0.01	Genesis AloS	5
	5.50	0.01	Genesis Au foil, AuoS and Al collectors	6
	5.40	0.30	Apollo SWE	2
	5.48	0.10	Ilmenite in "young" soil 71501	3
	5.53	0.06	From surface-correlated Ar from Apollo 15 soil feldspar	8
δD_{VSMOW} (‰)	< -999		Genesis Si collector	9
	< -970		Bulk Apollo 11 breccia samples	10
	< -950		Depth profile in soil grains in sample 79035	11
	< -840		Lowest δD in Apollo 11 and 17 agglutinates	12
$\delta^{13}C_{VPDB}$ (‰)	?		Genesis collectors	–
	−105	40	Depth profile in soil grains in sample 79035	13
	−85	25	SW estimates from solar photosphere $\delta^{13}C$ value obtained by analysing CO absorption lines	14
$\delta^{15}N_{AIR}$ (‰)	−390	180	Genesis Au cross arms	15
	−407	7	Genesis concentrator	16
	−417	100	Genesis DoS	17
	< -240		Depth profile in soil grains in sample 79035	11
$\delta^{18}O_{VSMOW}$ (‰)	< -40	15	Depth profile in metal grains in sample 79035	18, 19
	−102.3	6.6	Genesis SiC collector	20

Abbreviations: DoS—diamond-like carbon on silicate substrate; AloS—aluminium deposited on sapphire; AuoS—gold deposited on sapphire. References: 1—Heber et al. (2009); 2—Geiss et al. (2004); 3—Benkert et al. (1993); 4—Heber et al. (2003); 5—Meshik et al. (2007); 6—Pepin et al. (2012); 7—Vogel et al. (2011); 8—Frick et al. (1975); 9—Huss et al. (2012a); 10—Friedman et al. (1970); 11—Hashizume et al. (2000); 12—Liu et al. (2012); 13—Hashizume et al. (2004); 14—Lyons et al. (2018); 15—Marty et al. (2010); 16—Marty et al. (2011); 17—Huss et al. (2012b); 18—Hashizume and Chaussidon (2005); 19—Hashizume and Chaussidon (2009); 20—McKeegan et al. (2011)

40‰ at a depth of 15–30 nm (Hashizume et al. 2004). Such values cannot be compared to Genesis data since the $(^{13}C/^{12}C)_{SW}$ has not yet been determined. However, it matches fairly well with recent estimates for $(\delta^{13}C_{VPDB})_{SW}$ of *ca.* -75 ± 15 to -90 ± 15‰ derived from a new estimate for the $\delta^{13}C_{VPDB}$ of the Sun's photosphere of $-48 \pm 15‰$ and modelling of C isotope fractionation during transport from the photosphere to the corona via Coulomb drag (Lyons et al. 2018). Overall this suggests that the SW is depleted in ^{13}C compared to most planetary and asteroidal materials that are characterised by $\delta^{13}C_{VPDB}$ ratios around $0 \pm 30‰$ (e.g., Lyons et al. 2018 and references therein).

Analyses of nitrogen isotopes in Apollo and Luna regolith samples have yielded $\delta^{15}N_{AIR}$[3] values ranging from *ca.* $-200‰$ up to around $+100‰$ (e.g., Becker and Clayton 1975; Kerridge 1975, 1993; Thiemens and Clayton 1980; Frick et al. 1988; Füri et al. 2012; Mortimer et al. 2016). Based on the apparent correlation of increasing $\delta^{15}N_{AIR}$ with decreasing $^{40}Ar/^{36}Ar$ antiquity ratios, it has been suggested that the SW nitrogen isotope composition evolved through time to present-day values around $+100‰$ (e.g., Becker and Clayton 1975; Kerridge 1975; Thiemens and Clayton 1980). However, based on depth profile analyses of individual mineral grains from soil and regolith samples by SIMS, Hashizume et al. (2000) proposed that the present-day SW is depleted in ^{15}N, with $\delta^{15}N_{AIR}$ values $\leq -240‰$. A ^{15}N-depleted solar nitrogen was eventually confirmed by analysis of Genesis targets that yielded a best $\delta^{15}N_{AIR}$ estimate of $-407 \pm 7‰$ for SW nitrogen (Marty et al. 2011). Recent high-resolution step-heating analyses of Apollo 12, 14, 15, 16 and 17 surface soil samples and Luna 24 drill core samples are consistent with such a ^{15}N-depleted SW nitrogen (Füri et al. 2012; Mortimer et al. 2016).

Based on depth profile O isotope analyses in metal grains isolated from lunar soils, Hashizume and Chaussidon (2005, 2009) have argued that solar gas is characterised by $\delta^{18}O_{VSMOW}$[4] $<$ *ca.* $-40 \pm 15‰$. This is consistent with analysis of Genesis SiC collectors that yielded a ($\delta^{18}O_{VSMOW}$) value of $-102 \pm 7‰$ (McKeegan et al. 2011). However, the fact that such ^{16}O-rich values have not yet been found in lunar samples suggests that either some processes have fractionated the SW oxygen isotope composition during or after its implantation in lunar materials, or that samples with unfractionated SW have not been analysed so far.

8.1.2 The Noble Gases He, Ne and Ar

Heber et al. (2003) carried out high resolution acid etching experiments on ilmenite separates from soil and regolith breccia samples with a wide range of antiquity ages from *ca.* 0.1 to 3.7 Ga (see Sect. 4.3.3 for a discussion of how antiquity ages are estimated). The $(^3He/^4He)_{SW}$ ratios they measured are well correlated with antiquity ages of the samples, suggesting an apparent increase of $(^3He/^4He)_{SW}$ by $\sim$6% per billion years. Similarly, the $(^{20}Ne/^{22}Ne)_{SW}$ ratios measured by Heber et al. (2003) on the samples were also correlated with antiquity ages, with $(^{20}Ne/^{22}Ne)_{SW}$ apparently increasing with time by $\sim$2% per billion years. However, Heber et al. (2003) observed that the apparent temporal changes of $(^3He/^4He)_{SW}$ and $(^{20}Ne/^{22}Ne)_{SW}$ ratios were correlated with each other, mimicking the correlation displayed by single etch steps and indicating progressive etching of deeper sited isotopically heavier solar energetic particles. This led Heber et al. (2003) to conclude that the apparent temporal variations of the $(^3He/^4He)_{SW}$ and $(^{20}Ne/^{22}Ne)_{SW}$ ratios resulted from secondary processes on the lunar surface causing partial removal of surface-correlated SW.

[3]This $\delta^{15}N$ notation corresponds to the per mil deviation of a measured $^{15}N/^{14}N$ ratio from that of the terrestrial air ($AIR - {}^{15}N/^{14}N = 0.003676$), given by $[(^{15}N/^{14}N)_{sample}/(^{15}N/^{14}N)_{AIR} - 1] \times 1000$.

[4]This $\delta^{18}O$ notation corresponds to the per mil deviation of a measured $^{18}O/^{16}O$ ratio from that of the Vienna Standard Mean Ocean Water ($VSMOW - {}^{18}O/^{16}O = 2.005 \times 10^{-3}$), given by $[(^{18}O/^{16}O)_{sample}/(^{18}O/^{16}O)_{VSMOW} - 1] \times 1000$.

The average corrected $(^3He/^4He)_{SW}$ ratio of 4.47 (±0.26) × 10^{-4} obtained on ilmenite from six samples with antiquity ages from *ca.* 0.1 up to ~3.7 Ga by Heber et al. (2003) is indeed identical to the $(^3He/^4He)_{SW}$ ratio of 4.57 (±0.16) × 10^{-4} estimated for recent SW implanted in the last 100 Ma on ilmenite grains in lunar soil 71501 (Benkert et al. 1993). The He isotope composition estimated for SW from lunar regolith samples matches very well with the average $(^3He/^4He)_{SW}$ ratio of 4.64 (±0.18) × 10^{-4} obtained for SW collected for ~2.5 years by the Genesis spacecraft (Heber et al. 2009). The $(^3He/^4He)_{SW}$ ratio derived from lunar soil and Genesis analyses appear to be slightly higher than the average $(^3He/^4He)_{SW}$ ratio of 4.26 (±0.22) × 10^{-4} for SW collected by the Apollo 11, 12, 14, 15 and 16 collectors (Geiss et al. 2004). However, this average value may be slightly uncertain considering that Apollo 11 and 12 experiments yielded 5.38 (±0.40) × 10^{-4} and 4.08 (±0.17) × 10^{-4}, respectively, with $(^3He/^4He)_{SW}$ ratios for Apollo 14 (4.48 ± 0.28 × 10^{-4}), 15 (4.33 ± 0.22 × 10^{-4}) and 16 (4.42 ± 0.20 × 10^{-4}) being consistent with the $(^3He/^4He)_{SW}$ ratio derived from lunar soil and Genesis analyses (Benkert et al. 1993; Heber et al. 2003, 2009; Geiss et al. 2004).

As discussed above the apparent increase of $(^{20}Ne/^{22}Ne)_{SW}$ with antiquity measured in Apollo samples by Heber et al. (2009) has been interpreted as an artefact related to secondary processes while samples resided on the lunar surface. The $(^{20}Ne/^{22}Ne)_{SW}$ of 13.80 ± 0.20 measured in ilmenite in sample 71051 (Benkert et al. 1993) is consistent with the $^{20}Ne/^{22}Ne$ isotope ratio of 13.70 ± 0.30 measured on Apollo collectors (Geiss et al. 2004), both measurements being consistent within error with the $(^{20}Ne/^{22}Ne)_{SW}$ ratios of 13.78±0.06 (Heber et al. 2009), 13.97 ± 0.05 (Meshik et al. 2007) and 14.00 ± 0.04 (Pepin et al. 2012) measured in different types of Genesis collectors. Note that the $(^{20}Ne/^{22}Ne)_{SW}$ ratio calculated by Heber et al. (2009) is not consistent with those obtained by Meshik et al. (2007) and Pepin et al. (2012) thought, a discrepancy that is still not yet understood (see Pepin et al. 2012 for further discussion on this). At present, our best estimate for the $(^{20}Ne/^{22}Ne)_{SW}$ ratio is thus *ca.* 13.9 ± 0.1.

Lunar sample studies yielded estimates for the $(^{36}Ar/^{38}Ar)_{SW}$ ratio of *ca.* 5.50 (e.g., Frick et al. 1975; Benkert et al. 1993), which are consistent with both the relatively imprecise $(^{36}Ar/^{38}Ar)_{SW}$ ratio of 5.40 ± 0.30 determined on Apollo collectors (Geiss et al. 2004) and with the much more precise $(^{36}Ar/^{38}Ar)_{SW}$ ratios of 5.50 ± 0.05 obtained on Genesis collectors (Meshik et al. 2007; Heber et al. 2009; Vogel et al. 2011; Pepin et al. 2012). Isotopic analyses of young lunar regolith materials also provided very good estimates for the SW Kr and Xe isotope compositions (Pepin et al. 1995), with Kr and Xe isotope ratios within 1% of current best estimates for SW compositions measured on Genesis samples (e.g., Crowther and Gilmour 2013; Meshik et al. 2014).

8.2 Investigating Past Solar Activity and Changes of Galactic Environments

Characterising the evolution of the Sun's activity over the past 4.6 billion years is key to investigating the evolution of inner Solar System planetary surfaces and atmospheres, and by extension the habitability of terrestrial planetary bodies (e.g., Airapetian et al. 2016). Investigation of the meteorite record suggests enhanced solar activity during the early Solar System history (e.g., McKeegan et al. 2000; Aléon et al. 2005; Sossi et al. 2017; Kööp et al. 2018), which is consistent with greater flare activity associated with a shorter rotational period for the young Sun (e.g., Güdel 2007). The lunar regolith can also help characterise the rotational rate of the Sun during its first billion years of evolution since solar flares and associated coronal mass ejection (CME) directly affect the surface of airless bodies and could have driven loss of volatile species such as Na and K from the Moon's surface (e.g., Killen et al. 2012; Saxena et al. 2019). Based on the abundance of Na and K in the lunar crust, and on modelling of the effect of CME on Na and K loss, Saxena et al. (2019) argued that the Sun likely rotated relatively slowly during its first billion years of evolution compared to other young stars.

The lunar regolith also preserves a record of changes of the Solar System galactic environment as a whole (Spudis 1996). Over the past 4.5 billion years the Solar System has orbited the centre of the Milky Way approximately 20 times (e.g., Gies and Helsel 2005) and passed through spiral arms 2 to 6 times (Gillman et al. 2019). While moving around the Galaxy, the lunar regolith could have recorded changes in the flux and types of impactors delivered to the inner Solar System (see Sect. 4: The bombardment history of the inner Solar System). Over this period of time, the Solar System would have also been exposed to changing galactic environment conditions—passing through the Milky Way spiral arms would have enhanced the probability of passing close to supernovae, which could be recorded in the lunar regolith via supernova ejecta implanted in regolith materials and/or as an enhanced galactic cosmic ray (GCR) flux (e.g., Crawford 2017 and references therein). When GCRs interact with geological materials, a wide array of cosmogenic nuclides are produced in response to neutron capture and spallation reactions, including noble gases such as ^{3}He, ^{21}Ne, ^{38}Ar, ^{83}Kr and ^{126}Xe and other isotopes such as ^{10}Be and ^{36}Cl. However, our current Apollo and Luna collections are not ideally suited for investigating possible variations of the GCR flux through time. All these samples were collected at the surface or within the top few meters of the lunar regolith and have undergone constant regolith gardening over the past 3–4 billion years. As a result their exposure history records an integrated GCR flux since they were first exposed, which is generally complicated and poorly constrained (e.g., Levine et al. 2007; Füri et al. 2018). In addition to an increased GCR flux, passing by a supernova could also be recorded by deposition of ejecta enriched in radioactive isotopes such as ^{60}Fe, for which spikes have been identified in $\sim$2–3 and 6–9 Ma-old terrestrial ocean sediments and linked to a supernova origin (see Wallner et al. 2016 and references therein). Spikes of ^{60}Fe have also been measured in Apollo samples and have been linked with deposition of supernova ejecta $\sim$2 Ma ago (Cook et al. 2009; Fimiani et al. 2016). This bears promise for investigating galactic changes through time; however, similarly to examining variations of the GCR flux, detecting older supernova events in the Apollo and Luna collections is complicated by regolith gardening, which would average supernova-related spikes over the period of exposure of studied samples (e.g., Crawford 2017).

8.3 Opportunities Offered by Future Sample Return Missions

Study of soil and regolith samples have provided us with solid estimates for the noble gas isotopic compositions of the SW that were later confirmed by samples returned by the Genesis mission. The composition of the SW for other important elements such as Si, Mg, Fe, S, Cl has not yet been measured in Genesis samples (Burnett et al. 2019), and could be estimated based on targeted analyses on constituents of lunar soil and regolith samples already present in our collections. However, thoroughly tackling questions related to possible changes in solar activity, solar wind composition, and galactic environment through time will require accessing regolith samples that have been exposed to the space environment at a known point in time and for a well constrained duration. Such samples could be returned from palaeoregolith horizons (see Sect. 4.3.3: Secular evolution of types of projectiles hitting the Moon) trapped between datable lava flows or impact melt flows, some of them being potentially accessible to robotic or human explorers through layers exposed in the walls of lava skylights or in layered boulders (Fig. 6) (Crawford et al. 2007, 2010; Fagents et al. 2010; Crawford and Joy 2014).

9 The Lunar Magnetic Field

9.1 State of the Art

The search for evidence of past or present lunar magnetism was one of the objectives of the very first missions to the Moon. The second robotic mission to the Moon, Luna 2 in 1959, carried a magnetometer that found no sign of a detectable global magnetic field (Dolginov et al. 1962). After this was confirmed by Explorer 35 (Sonett et al. 1967), the Moon was generally thought to be magnetically uninteresting, in agreement with the preconception that the Moon was too small to have sustained a dynamo field. However, one of the bigger surprises of the early Apollo missions was the discovery of past lunar magnetism. Examination of Apollo 11 samples revealed strong natural remanent magnetizations (Doell and Grommé 1970) indicative of ancient surface fields with intensities comparable to that of the present-day Earth. Subsequent measurements with the Apollo 12 and subsequent surface magnetometers revealed significant present-day remanent magnetic fields (tens to hundreds of nanotesla—nT) on the surface of the Moon indicating near-surface sources of crustal remanent magnetization (Dyal et al. 1970). Paleomagnetic measurements of lunar samples started right after the return of Apollo 11 and were very active over the whole Apollo sample-return period with about 70 samples studied (see Fuller and Cisowski 1987 for a review). After a period of relative disinterest for more than 20 years, this field of research was reactivated with a series of detailed studies on key samples that benefited from improved analytical techniques and methods for paleomagnetism, as well as better geo- and thermochronology, and an overall more comprehensive understanding of the geological evolution of the Moon (see Weiss and Tikoo 2014 for a review; Tikoo et al. 2017; Lepaulard et al. 2019).

It is now established that the Moon generated a magnetic field through advection in its liquid core for a large portion of its history (Fig. 10). This magnetic field was highest between about 4.25 Ga and 3.5 Ga (during the so called "high-field epoch"), with an average paleointensity of 55 μT, much higher than the maximum value of a few tenths of a μT estimated from dynamo scaling laws (e.g., Laneuville et al. 2014, Evans et al. 2014, 2018). The field then decayed but a dynamo was still operating with surface fields in the several μT range around 1.5–2.0 Ga (Tikoo et al. 2017). Today, lunar surface magnetic fields are mostly below 10 nT, with localised higher values (and a maximum measured surface field of 327 nT at the Apollo 12 landing site; Dyal and Daily 1978). This spatial variability of the lunar surface field reflects lithological heterogeneities and/or different magnetisation histories of the underlying crustal rocks.

9.2 Outstanding Questions Requiring Sample-Return Missions (See Summary in Table 4)

9.2.1 Intensity of the Dynamo Field with Time

The paleomagnetic data (Fig. 10) have notable temporal gaps. In particular, the absence of paleomagnetic data between about 3.0 Ga and 1.5 Ga prevents the understanding of the transition between the high and low field epochs. The apparent sharp transition at the precipitous collapse of the dynamo around 3.5 Ga (Tikoo et al. 2014) has been recently reevaluated with suggestions for paleointensities in the 10–50 μT range between 3.5 Ga and 3.0 Ga (Lepaulard et al. 2019). The extent of the weak field period and the behaviour of the magnetic field during this period are also largely unconstrained with only one published

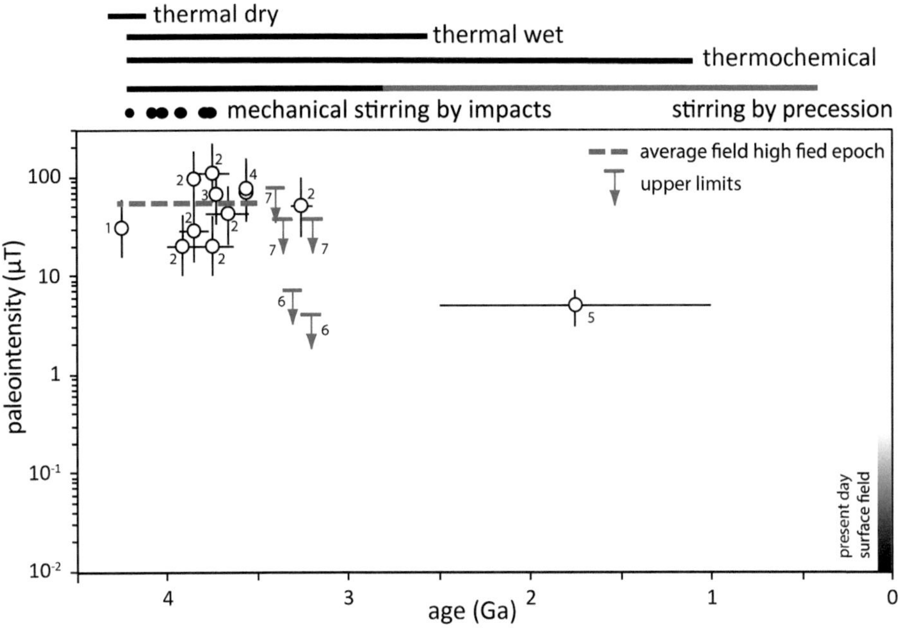

Fig. 10 Intensity of lunar surface field versus time. Note the lack of data between 3.0 and 2.0 Ga and the scarcity of data between 2.0 and 1.0 Ga. Bars and dots above the diagram represent the time intervals over which the various dynamo generation mechanisms listed may have operated. References: 1—Garrick-Bethell et al. (2017), 2—Cournède et al. (2012), 3—Shea et al. (2012), 4—Suavet et al. (2013), 5—Tikoo et al. (2017), 6—Tikoo et al. (2012), 7—Tikoo et al. (2014)

Table 4 Science questions and sample requirements for lunar paleomagnetic studies

Sample requirements	Science questions
Regolith breccias younger than 3 Ga	Constrain weak field epoch intensity and duration
Regolith breccias with age between 3 Ga and 2 Ga	Transition between high and low field epochs
Regolith breccias older than 3 Ga	Confirm the high paleointensity during the high field epoch using thermal paleointensity experiments
One oriented sample from one location	Testing the dipole geometry (if location of the paleopole is assumed), locating the paleopole if a dipolar geometry is assumed
Several oriented samples from one location	Testing the occurrence of reversals, constraining the field geometry, determining the paleopole with implications for true polar wander
Oriented samples from several locations	Testing the dipole geometry and defining the paleopole with implications for true polar wander

paleointensity estimate over the last 2 Ga of the Moon history (Tikoo et al. 2017). Clearly more paleomagnetic studies of samples with magnetisation ages younger than 3.0 Ga are required. Regolith breccias are excellent material for lunar paleomagnetism because their glassy matrices contain very fine metal grains that are favourable for recording and maintaining a record of the paleomagnetic field. However, the age at which a regolith breccia was assembled and, therefore, the age of its magnetisation is not easy to determine notably because materials it contains may have been affected by multiple impact events (e.g., Mercer et al. 2015). Basalt samples younger than 3.0 Ga would be a prime target for lunar paleomagnetism. Providing that their formation ages can be accurately determined through high resolution *in situ* geochronological analyses, regolith breccias with formation ages in the interval 3.0–2.0 Ga, and older than 3 Ga, if they exist, are other important targets.

9.2.2 Geometry of the Lunar Surface Field

All lunar paleomagnetic studies were performed on samples collected in the regolith. Therefore, only the paleointensity of the lunar field can be retrieved, but no paleodirectional information. As with terrestrial paleomagnetism, such directional information would be invaluable. It would allow testing the hypothesis that the lunar field was dipolar. Such a geometry is expected by analogy with the dynamo observed today on Earth, Mercury and other bodies. Substantial non-dipolar geometry would imply rapid decay of the field away from the core such that the high surface paleointensities inferred from Apollo samples would indicate even more unrealistic core fields in view of the small size of the lunar core relative to the size of the Moon. Ideally, testing the dipolar hypothesis would require oriented sampling of in-place bedrock samples at different locations on the Moon, preferably at different latitudes. A few outcrops offer the opportunity to sample rocks that have never been transported by impact processes (i.e. they are at the same place than where they formed) at the surface of the Moon. For example, lava pits may offer the possibility to sample such samples through sequences of basaltic lava flows that still have their original orientations (Fig. 6).

A single oriented sample from one location can only test the dipole hypothesis if one assumes the location of the paleo spin pole. However, a sequence of oriented samples collected at a single location would enable testing whether the field reversed. This occurrence, with evidence for two antipodal sets of paleopoles would strengthen the dipole hypothesis and help indentify the location of the paleopole. If a paleopole is found away from the location of the present-day rotation axis of the Moon, this may indicate that the Moon experienced true polar wander, i.e. large differential rotation of the mantle with respect to the core (e.g., Takahashi et al. 2014)

9.2.3 Mechanisms Behind Dynamo Generation

Generating a lunar dynamo requires advection in its liquid core. A number of mechanisms have been proposed that can sustain convection in the Moon: thermal convection, thermochemical convection, and mechanical stirring by mantle precession or impacts. It is noteworthy that these mechanisms are expected to have been restricted to specific, although sometimes overlapping, periods in lunar history (Weiss and Tikoo 2014). For instance, pure thermal convection with a dry mantle is very unlikely to have operated after 4 Ga, impact-generated mechanical stirring is limited to before 3.7 Ga (age of Orientale basin, the youngest impact basin on the Moon) (Suavet et al. 2013), whereas thermochemical and precession generated mechanical stirring could operate almost up to the present-day. Therefore, a better knowledge of the evolution of the field intensity with time will also shed light on which mechanisms lie behind the lunar core convection, with implications for the dynamo theory and modelling community.

9.3 Limitation of Current Lunar Rocks

No basalts younger than *ca.* 3.0 Ga are available in the Apollo collection, and regolith breccia samples older than 3.0–3.5 Ga are rare or absent (e.g., Fagan et al. 2014) because these rocks are being modified continuously at the surface of the Moon.

Although the orientation of some samples in the regolith was documented *in situ* or can be determined *a posteriori*, the original orientation of these rocks on the Moon is unknown. The paleohorizontal of some rocks may be assessed using for instance the crystal preferred orientation (e.g., Cournède et al. 2012), or vesicles orientation. Even so, the up and down remain unresolvable, and the constraints on the field geometry remain poor.

Meteorites are another prime source of lunar rocks and, with 378 lunar meteorites for 408.6 kg of material as of August 2019, based on samples classified in the Meteoritical Bulletin Database, they now outweigh the total mass of samples collected by Apollo and Luna missions. Moreover, meteorites sample the lunar surface in a more global way than sample-return missions, and, therefore, are likely to provide more representative sampling of the lunar surface, both spatially and temporally. However, meteorites pose challenges for paleomagnetism. All lunar meteorites are "finds", meteorites found at the Earth surface long after their fall, typically several kyr and up to several hundred kyr for Antarctic meteorites. The magnetic carriers of lunar rocks are Ni-poor iron-nickel alloys that are prone to weathering on Earth. Weathering of magnetic carriers leads to the loss of the primary magnetization, and, worse, possible acquisition of magnetization (chemical remanent magnetisation) in the Earth field. The long stay of these meteorites on Earth also means that they have been in the geomagnetic field for long periods, leading to the acquisition of viscous remanent magnetisation. Although these overprints can be identified and partly erased in some cases, the existence of viscous and chemical remanent magnetisations add to the complexity of lunar paleomagnetism studies. Moreover, 90% of lunar meteorites come from hot deserts, in particular from the Sahara, where they are systematically tested with magnets by private collectors and dealers. Exposure to strong magnets usually totally obliterates the primary magnetisation of the meteorites, which renders them useless for paleomagnetic studies. Among these three magnetic contaminations (viscous by long term exposure to the geomagnetic field, isothermal by exposure to strong fields, and chemical by crystallisation of weathering phases), it is noteworthy that the first two can also affect return samples. Apollo samples are curated in the ambient geomagnetic field and have all been contaminated by viscous remanent magnetisations. Storage in a weak magnetic field would be a simple and efficient solution to implement for future returned samples. Exposure to strong magnetic field can also occur during sampling on the Moon, during return from the Moon, and during sample curation and preparation. The use of non-magnetic tools during curation and the absence of strongly magnetic parts in the sampling and return material are therefore highly recommended for future sample-return missions.

10 Summary

Despite a large body of research over the past five decades, many new discoveries are being made through continued laboratory investigations of lunar meteorites and samples returned by Apollo and Luna missions. Insights from recent orbital and landed lunar missions have allowed integration of laboratory data acquired on returned samples to address some of the outstanding questions related to the Moon's origin and evolution. At the same time, new fundamental questions have arisen, which can only be addressed by future sample-return missions from areas of the Moon identified as being more representative than was the case for Apollo and Luna missions. As we enter into a new era of sample-return missions from various Solar System bodies, and with an increasing interest in lunar exploration among the worldwide community, new samples returned from the Moon will have immense potential to address key science questions not just related to the Moon but also to the inner Solar System.

Acknowledgements We thank the organisers of the "Role of Sample Return in Addressing Major Outstanding Questions in Planetary Sciences" workshop that took place at the International Space Science Institute in Bern in February 2018. Jim Head and two anonymous reviewers are thanked for their thorough and insightful evaluation of this manuscript. For financial support, we are grateful to the UK Science and Technology Facilities Council (STFC) (grants ST/P005225/1 to RT, ST/M001253/1 to KHJ, and ST/P000657/1 to MA), the Royal Society (grant RS/UF140190 to KHJ), the European Commission Horizon 2020 Research and Innovation programme (Marie Skłodowska-Curie Actions Fellowship grant 794287 to JFS), the Agence Nationale de la Recherche (project Maglune ANR-14-CE33-0012 to JG), the NASA Solar System Workings program (grant NNX15AL62G to BPW) and the NASA Solar System Exploration Virtual Institute (grant NNA14AB01A to BPW).

References

V.S. Airapetian, A. Glocer, G. Gronoff, E. Hébrard, W. Danchi, Prebiotic chemistry and atmospheric warming of early Earth by an active young Sun. Nat. Geosci. **9**, 452–455 (2016)

J. Aléon, F. Robert, J. Duprat, S. Derenne, Extreme oxygen isotope ratios in the early Solar System. Nature **437**, 385–388 (2005)

M. Anand, R. Tartèse, J.J. Barnes, Understanding the origin and evolution of water in the Moon through lunar sample studies. Philos. Trans. R. Soc. A **372**, 20130254 (2014). https://doi.org/10.1098/rsta.2013.0254

E. Anders, N. Grevesse, Abundances of the elements: meteoritic and solar. Geochim. Cosmochim. Acta **53**, 197–214 (1989)

J.C. Andrews-Hanna, J. Besserer, J.W. Head, C.J.A. Howett, W.S. Kiefer, P.J. Lucey, P.J. McGovern, H.J. Melosh, G.A. Neumann, R.J. Phillips, P.M. Schenk, D.E. Smith, S.C. Solomon, M.T. Zuber, Structure and evolution of the lunar Procellarum region as revealed by GRAIL gravity data. Nature **514**, 68–71 (2014)

T. Arai, H. Takeda, A. Yamaguchi, M. Ohtake, A new model of lunar crust: asymmetry in crustal composition and evolution. Earth Planets Space **60**, 433–444 (2008)

T. Arai, B.R. Hawke, T.A. Giguere, K. Misawa, M. Miyamoto, H. Kojima, Antarctic lunar meteorites Yamato-793169, Asuka-881757, MIL 05035, and MET 01210 (YAMM): launch pairing and possible cryptomare origin. Geochim. Cosmochim. Acta **74**, 2231–2248 (2010)

J.C. Armstrong, Distribution of impact locations and velocities of Earth meteorites on the Moon. Earth Moon Planets **107**, 43–54 (2010)

J.C. Armstrong, L.E. Wells, G. Gonzalez, Rummaging through Earth's Attic for remains of ancient life. Icarus **160**, 183–196 (2002)

J.R. Arnold, Ice in the lunar polar regions. J. Geophys. Res. **84**, 5659–5668 (1979)

G. Avice, B. Marty, The iodine-plutonium-xenon age of the Moon-Earth system revisited. Philos. Trans. R. Soc. A **372**, 20130260 (2014). https://doi.org/10.1098/rsta.2013.0260

S.J. Barber, P.H. Smith, I.P. Wright, F. Abernethy, M. Anand, K.R. Dewar, M. Hodges, P. Landsberg, M.R. Leese, G.H. Morgan, A.D. Morse, J. Mortimer, H.M. Sargeant, I. Sheard, S. Sheridan, A. Verchovsky, F. Goesmann, C. Howe, T. Morse, N. Lillywhite, A. Quinn, N. Missaglia, M. Pedrali, P. Reiss, F. Rizzi, A. Rusconi, M. Savoia, A. Zamboni, J.A. Merrifield, E.K. Gibson Jr., J. Carpenter, R. Fisackerly, B. Houdou, ProSPA: the science laboratory for the processing and analysis of lunar polar volatiles within PROSPECT, in *Lunar Planet. Sci. Conf.* (2017), abstract 2171

M. Barboni, P. Boehnke, B. Keller, I.E. Kohl, B. Schoene, E.D. Young, K.D. McKeegan, Early formation of the Moon 4.51 billion years ago. Sci. Adv. **3**, e1602365 (2017). https://doi.org/10.1126/sciadv.1602365

J.J. Barnes, D.A. Kring, R. Tartèse, I.A. Franchi, M. Anand, S.S. Russell, An asteroidal origin for water in the Moon. Nat. Commun. **7**, 11684 (2016). https://doi.org/10.1038/ncomms11684

J.A. Barrat, B. Zanda, F. Moynier, C. Bollinger, C. Liorzou, G. Bayon, Geochemistry of CI chondrites: major and trace elements, and Cu and Zn isotopes. Geochim. Cosmochim. Acta **83**, 79–92 (2012)

D.W. Beaty, A.L. Albee, Comparative petrology and possible genetic relations among the Apollo 11 basalts, in *Lunar Planet. Sci. Conf. Proc.* (1978), pp. 359–463

R.H. Becker, Evidence for a secular variation in the $^{13}C/^{12}C$ ratio of carbon implanted in lunar soils. Earth Planet. Sci. Lett. **50**, 189–196 (1980)

R.H. Becker, R.N. Clayton, Nitrogen abundances and isotopic compositions in lunar samples, in *Lunar Sci. Conf. Proc.* (1975), pp. 2131–2149

R.H. Becker, R.O. Pepin, Long-term changes in solar wind elemental and isotopic ratios: a comparison of two lunar ilmenites of different antiquities. Geochim. Cosmochim. Acta **53**, 1135–1146 (1989)

J.J. Bellucci, A.A. Nemchin, M. Grange, K.L. Robinson, G. Collins, M.J. Whitehouse, J.F. Snape, M.D. Norman, D.A. Kring, Terrestrial-like zircon in a clast from an Apollo 14 breccia. Earth Planet. Sci. Lett. **510**, 173–185 (2019)

J.-P. Benkert, H. Baur, P. Signer, R. Wieler, He, Ne, and Ar from the solar wind and solar energetic particles in lunar ilmenites and pyroxenes. J. Geophys. Res., Planets **98**, 13147–13162 (1993)

L.E. Borg, C.K. Shearer, Y. Asmerom, J.J. Papike, Prolonged KREEP magmatism on the Moon indicated by the youngest dated lunar igneous rock. Nature **432**, 209–211 (2004)

L.E. Borg, A.M. Gaffney, C.K. Shearer, D.J. DePaolo, I.D. Hutcheon, T.L. Owens, E. Ramon, G. Brennecka, Mechanisms for incompatible-element enrichment on the Moon deduced from the lunar basaltic meteorite Northwest Africa 032. Geochim. Cosmochim. Acta **73**, 3963–3980 (2009)

L.E. Borg, J.N. Connelly, M. Boyet, R.W. Carlson, Chronological evidence that the Moon is either young or did not have a global magma ocean. Nature **477**, 70–72 (2011)

W.F. Bottke, M.D. Norman, The late heavy bombardment. Annu. Rev. Earth Planet. Sci. **45**, 619–647 (2017)

S.E. Braden, J.D. Stopar, M.S. Robinson, S.J. Lawrence, C.H. van der Bogert, H. Hiesinger, Evidence for basaltic volcanism on the Moon within the past 100 million years. Nat. Geosci. **7**, 787–791 (2014)

A.D. Brandon, T.J. Lapen, V. Debaille, B.L. Beard, K. Rankenburg, C.R. Neal, Re-evaluating $^{142}Nd/^{144}Nd$ in lunar mare basalts with implications for the early evolution and bulk Sm/Nd of the Moon. Geochim. Cosmochim. Acta **73**, 6421–6445 (2009)

S.M. Brown, T.L. Grove, Origin of the Apollo 14, 15, and 17 yellow ultramafic glasses by mixing of deep cumulate remelts. Geochim. Cosmochim. Acta **171**, 201–215 (2015)

D.S. Burnett, A.J.G. Jurewicz, D.S. Woolum, The future of Genesis science. Meteorit. Planet. Sci. **54**, 1092–1114 (2019)

P.H. Cadogan, Oldest and largest lunar basin? Nature **250**, 315–316 (1974)

J.T. Cahill, C. Floss, M. Anand, L.A. Taylor, M.A. Nazarov, B.A. Cohen, Petrogenesis of lunar highlands meteorites: Dhofar 025, Dhofar 081, Dar al Gani 262, and Dar al Gani 400. Meteorit. Planet. Sci. **39**, 503–529 (2004)

D.B. Campbell, B.A. Campbell, L.M. Carter, J.L. Margot, N.J.S. Stacy, No evidence for thick deposits of ice at the lunar South pole. Nature **443**, 835–837 (2006)

R.W. Carlson, G.W. Lugmair, Sm-Nd constraints on early lunar differentiation and the evolution of KREEP. Earth Planet. Sci. Lett. **45**, 123–132 (1979)

R.W. Carlson, G.W. Lugmair, The age of ferroan anorthosite 60025: oldest crust on a young Moon? Earth Planet. Sci. Lett. **90**, 119–130 (1988)

R.W. Carlson, L.E. Borg, A.M. Gaffney, M. Boyet, Rb-Sr, Sm-Nd and Lu-Hf isotope systematics of the lunar Mg-suite: the age of the lunar crust and its relation to the time of Moon formation. Philos. Trans. R. Soc. A **372**, 20130246 (2014). https://doi.org/10.1098/rsta.2013.0246

J. Carpenter, R. Fisackerly, B. Houdou, Establishing lunar resource viability. Space Policy **37**, 52–57 (2016)

B. Charlier, T.L. Grove, O. Namur, F. Holtz, Crystallization of the lunar magma ocean and the primordial mantle-crust differentiation of the Moon. Geochim. Cosmochim. Acta **234**, 50–69 (2018)

R.N. Clark, Detection of adsorbed water and hydroxyl on the Moon. Science **326**, 562–564 (2009)

R.N. Clayton, M.H. Thiemens, Lunar nitrogen: evidence for secular change in the solar wind, in *Proceedings of the Conference on the Ancient Sun*, ed. by R.O. Pepin, J.A. Eddy, R.B. Merrill (Pergamon, New York, 1980), pp. 463–473

B.A. Cohen, T.D. Swindle, D.A. Kring, Support for the lunar cataclysm hypothesis from lunar meteorite impact melt ages. Science **290**, 1754–1756 (2000)

B.A. Cohen, G.A. Snyder, C.M. Hall, L.A. Taylor, M.A. Nazarov, Argon-40-argon-39 chronology and petrogenesis along the eastern limb of the Moon from Luna 16, 20 and 24 samples. Meteorit. Planet. Sci. **36**, 1345–1366 (2001)

B.A. Cohen, S.J. Lawrence, N.E. Petro, G.D. Bart, R.N. Clegg-Watkins, B.W. Denevi, R.R. Ghent, R.L. Klima, G.A. Morgan, P.D. Spudis, J.D. Stopar, Identifying and characterizing impact melt outcrops in the Nectaris basin, in *Lunar Planet. Sci. Conf.* (2016), p. abstract 1389

A. Colaprete, P. Schultz, J. Heldmann, D. Wooden, M. Shirley, K. Ennico, B. Hermalyn, W. Marshall, A. Ricco, R.C. Elphic, D. Goldstein, D. Summy, G.D. Bart, E. Asphaug, D. Korycansky, D. Landis, L. Sollitt, Detection of water in the LCROSS ejecta plume. Science **330**, 463–468 (2010)

A. Colaprete, D. Andrews, W. Bluethmann, R.C. Elphic, B. Bussey, J. Trimble, K. Zacny, J.E. Captain, An overview of the Volatiles Investigating Polar Exploration Rover (VIPER) mission, in *American Geophysical Union Fall Meeting* (2019), abstract P34B-03

W. Compston, M.J. Vernon, H. Berry, R. Rudowski, The age of the Fra Mauro formation: a radiometric older limit. Earth Planet. Sci. Lett. **12**, 55–58 (1971)
W. Compston, M.J. Vernon, H. Berry, R. Rudowski, C.M. Gray, N. Ware, Apollo 14 mineral ages and the thermal history of the Fra Mauro formation, in *Lunar Planet. Sci. Conf. Proc.*, ed. by A.E. Metzger, J.I. Trombka, L.E. Peterson, R.C. Reedy, J.R. Arnold (1972), pp. 1487–1501
J.N. Connelly, M. Bizzarro, Lead isotope evidence for a young formation age of the Earth–Moon system. Earth Planet. Sci. Lett. **452**, 36–43 (2016)
J.W. Conrad, F. Nimmo, C.I. Fassett, S. Kamata, Lunar impact history constrained by GRAIL-derived basin relaxation measurements. Icarus **314**, 50–63 (2018)
D.L. Cook, E. Berger, T. Faestermann, G.F. Herzog, K. Knie, G. Korschinek, M. Poutivtsev, G. Ruge, F. Serefiddin, ^{60}Fe, ^{10}Be, and ^{26}Al in lunar cores 12025/8 and 60006/7: search for a nearby supernova, in *Lunar Planet. Sci. Conf. Proc.* (2009), p. abstract 1129
L.M. Corley, P.J. McGovern, G.Y. Kramer, M. Lemelin, D. Trang, J.J. Gillis-Davis, G.J. Taylor, K.E. Powell, W.S. Kiefer, M. Wieczorek, M.T. Zuber, Olivine-bearing lithologies on the Moon: constraints on origins and transport mechanisms from M^3 spectroscopy, radiative transfer modeling, and GRAIL crustal thickness. Icarus **300**, 287–304 (2018)
C. Cournède, J. Gattacceca, P. Rochette, Magnetic study of large Apollo samples: possible evidence for an ancient centered dipolar field on the Moon. Earth Planet. Sci. Lett. **33**, 31–42 (2012)
I.A. Crawford, The Moon as a recorder of nearby supernovae, in *Handbook of Supernovae*, ed. by A.W. Alsabti, P. Murdin (Springer, Berlin, 2017), pp. 2507–2522
I.A. Crawford, K.H. Joy, Lunar exploration: opening a window into the history and evolution of the inner Solar System. Philos. Trans. R. Soc. A **372**, 20130315 (2014). https://doi.org/10.1098/rsta.2013.0315
I.A. Crawford, S.A. Fagents, K.H. Joy, Full Moon exploration. Astron. Geophys. **48**, 3.18–3.21 (2007)
I.A. Crawford, E.C. Baldwin, E.A. Taylor, J.A. Bailey, K. Tsembelis, On the survivability and detectability of terrestrial meteorites on the Moon. Astrobiology **8**, 242–252 (2008)
I.A. Crawford, K.H. Joy, S.A. Fagents, M.E. Rumpf, Lunar palaeoregolith deposits as recorders of the galactic environment of the Solar System and implications for astrobiology. Earth Moon Planets **107**, 75–85 (2010)
S.A. Crowther, J.D. Gilmour, The Genesis solar wind composition and its relationship to planetary xenon signatures. Geochim. Cosmochim. Acta **123**, 17–34 (2013)
N.M. Curran, K.H. Joy, J.F. Snape, J.F. Pernet-Fisher, J.D. Gilmour, A.A. Nemchin, M.J. Whitehouse, R. Burgess, The early geological history of the Moon inferred from ancient lunar meteorite Miller Range 13317. Meteorit. Planet. Sci. **54**, 1401–1430 (2019)
E.J. Dasch, C.Y. Shih, B.M. Bansal, H. Wiesmann, L.E. Nyquist, Isotopic analysis of basaltic fragments from lunar breccia 14321: Chronology and petrogenesis of pre-Imbrium mare volcanism. Geochim. Cosmochim. Acta **51**, 3241–3254 (1987)
J.M.D. Day, R.J. Walker, Highly siderophile element depletion in the Moon. Earth Planet. Sci. Lett. **423**, 114–124 (2015)
J.M.D. Day, D.G. Pearson, L.A. Taylor, Highly siderophile element constraints on accretion and differentiation of the Earth-Moon system. Science **315**, 217–219 (2007)
J.W. Delano, Pristine lunar glasses: criteria, data, and implications. J. Geophys. Res. **91**, 201–213 (1986)
J.W. Delano, K. Livi, Lunar volcanic glasses and their constraints on mare petrogenesis. Geochim. Cosmochim. Acta **45**, 2137–2149 (1981)
E. Doell, C. Grommé, Survey of magnetic properties of Apollo 11 samples at the Lunar Receiving Laboratory, in *Apollo 11 Lunar Sci. Conf. Proc.* (1970), pp. 2093–2096
S.S. Dolginov, E.U. Eroshenko, L.I. Zhuzgov, N.V. Pushkov, A study of the magnetic field of the Moon. The Moon, in *IAU Symposium*, vol. 14 (1962), pp. 45–52
K.L. Donaldson Hanna, L.C. Cheek, C.M. Pieters, J.F. Mustard, B.T. Greenhagen, I.R. Thomas, N.E. Bowles, Global assessment of pure crystalline plagioclase across the Moon and implications for the evolution of the primary crust. J. Geophys. Res., Planets **119**, 1516–1545 (2014)
D.S. Draper, R.L. Klima, S.J. Lawrence, B.W. Denevi, The Inner SOlar system CHRONology (ISOCHRON) Discovery mission: Returning samples of the youngest lunar mare basalts, in *Lunar Planet. Sci. Conf.* (2019), p. abstract 1110
P. Dyal, W.D. Daily, Crustal magnetic scale sizes measured at Apollo landing sites, in *Lunar Planet. Sci. Conf. Proc.* (1978), pp. 273–275
P. Dyal, C. Parkin, C. Sonett, Apollo 12 Magnetometer: measurement of a steady magnetic field on the surface of the Moon. Science **169**, 762–764 (1970)
W.F. Eichelman, W.W. Lauderdale, Apollo scientific experiments data handbook. NASA Technical Memorendum X-58131 (1974)
L.T. Elkins-Tanton, E.M. Parmentier, P.C. Hess, Magma ocean fractional crystallization and cumulate overturn in terrestrial planets: implications for Mars. Meteorit. Planet. Sci. **38**, 1753–1771 (2003)

L.T. Elkins-Tanton, B.H. Hager, T.L. Grove, Magmatic effects of the lunar late heavy bombardment. Earth Planet. Sci. Lett. **222**, 17–27 (2004)

L.T. Elkins-Tanton, S. Burgess, Q.Z. Yin, The lunar magma ocean: reconciling the solidification process with lunar petrology and geochronology. Earth Planet. Sci. Lett. **304**, 326–336 (2011)

S. Epstein, H.P. Taylor, $^{18}O/^{16}O$, $^{30}Si/^{28}Si$, D/H, and $^{13}C/^{12}C$ studies of lunar rock and minerals. Science **167**, 533–535 (1970)

O. Eugster, D. Terribilini, E. Polnau, J. Kramers, The antiquity indicator argon-40/argon-36 for lunar surface samples calibrated by uranium-235-xenon-136 dating. Meteorit. Planet. Sci. **36**, 1097–1115 (2001)

A.J. Evans, M.T. Zuber, B.P. Weiss, S.M. Tikoo, A wet,heterogeneous lunar interior: lower mantle and core dynamo evolution. J. Geophys. Res., Planets **119**, 1061–1077 (2014)

A.J. Evans, S.M. Tikoo, J.C. Andrews-Hanna, The case against an early lunar dynamo powered by core convection. Geophys. Res. Lett. **45**, 98–107 (2018)

A.L. Fagan, K.H. Joy, D.D. Bogard, D.A. Kring, Ages of globally distributed lunar paleoregoliths and soils from 3.9 Ga to the present. Earth Moon Planets **112**, 59–71 (2014)

S.A. Fagents, M.E. Rumpf, I.A. Crawford, K.H. Joy, Preservation potential of implanted solar wind volatiles in lunar paleoregolith deposits buried by lava flows. Icarus **207**, 595–604 (2010)

W.M. Farrell, D.M. Hurley, M.I. Zimmerman, Solar wind implantation into lunar regolith: hydrogen retention in a surface with defects. Icarus **255**, 116–126 (2015)

C.I. Fassett, Analysis of impact crater populations and the geochronology of planetary surfaces in the inner Solar System. J. Geophys. Res., Planets **121**, 1900–1926 (2016)

C.I. Fassett, J.W. Head, S.J. Kadish, E. Mazarico, G.A. Neumann, D.E. Smith, M.T. Zuber, Lunar impact basins: stratigraphy, sequence and ages from superposed impact crater populations measured from Lunar Orbiter Laser Altimeter (LOLA) data. J. Geophys. Res., Planets **117**, E00H06 (2012). https://doi.org/10.1029/2011JE003951

W.C. Feldman, S. Maurice, A.B. Binder, B.L. Barraclough, R.C. Elphic, D.J. Lawrence, Fluxes of fast and epithermal neutrons from Lunar Prospector: Evidence for water ice at the lunar poles. Science **281**, 1496–1500 (1998)

W.C. Feldman, D.J. Lawrence, R.C. Elphic, B.L. Barraclough, S. Maurice, I. Genetay, A.B. Binder, Polar hydrogen deposits on the Moon. J. Geophys. Res. **105**, 4175–4195 (2000)

V.A. Fernandes, R. Burgess, Volcanism in Mare Fecunditatis and Mare Crisium: Ar-Ar age studies. Geochim. Cosmochim. Acta **69**, 4919–4934 (2005)

V.A. Fernandes, J. Fritz, B.P. Weiss, I. Garrick-Bethell, D.L. Shuster, The bombardment history of the Moon as recorded by ^{40}Ar-^{39}Ar chronology. Meteorit. Planet. Sci. **48**, 241–269 (2013)

L. Fimiani, D.L. Cook, T. Faestermann, J.M. Gómez-Guzmán, K. Hain, G. Herzog, K. Knie, G. Korschinek, P. Ludwig, J. Park, R.C. Reedy, G. Rugel, Interstellar ^{60}Fe on the surface of the Moon. Phys. Rev. Lett. **116**, 151104 (2016). https://doi.org/10.1103/PhysRevLett.116.151104

M. Fischer-Gödde, H. Becker, What is the age of the Nectaris basin? New Re-Os constraints for a pre-4.0 Ga bombardment history of the Moon, in *Lunar Planet. Sci. Conf.* (2011), p. abstract 1414

M. Fischer-Gödde, H. Becker, Osmium isotope and highly siderophile element constraints on ages and nature of meteoritic components in ancient lunar impact rocks. Geochim. Cosmochim. Acta **77**, 135–156 (2012)

E.A. Fisher, P.G. Lucey, M. Lemelin, B.T. Greenhagen, M.A. Siegler, E. Mazarico, O. Aharonson, J.P. Williams, P.O. Hayne, G.A. Neumann, D.A. Paige, D.E. Smith, M.T. Zuber, Evidence for surface water ice in the lunar polar regions using reflectance measurements from the Lunar Orbiter Laser Altimeter and temperature measurements from the Diviner Lunar Radiometer Experiment. Icarus **292**, 74–85 (2017)

U. Frick, H. Baur, H. Ducati, H. Funk, D. Phinney, P. Signer, On the origin of helium, neon, and argon isotopes in sieved mineral separates from an Apollo 15 soil, in *Lunar Sci. Conf. Proc.* (1975), pp. 2097–2129

U. Frick, R.H. Becker, R.O. Pepin, Solar wind record in the lunar regolith: nitrogen and noble gases, in *Lunar Planet. Sci. Conf. Proc.* (1988), pp. 87–120

I. Friedman, J.R. O'Neill, L.H. Adami, J.D. Gleason, K. Hardcastle, Water, hydrogen, deuterium, carbon, carbon-13, and oxygen-18 content of selected lunar material. Science **167**, 533–535 (1970)

M. Fuller, S.M. Cisowski, Lunar paleomagnetism. Geomagnetism **2**, 307–455 (1987)

E. Füri, B. Marty, S.S. Assonov, Constraints on the flux of meteoritic and cometary water on the Moon from volatile element (N–Ar) analyses of single lunar soil grains, Luna 24 core. Icarus **218**, 220–229 (2012)

E. Füri, L. Zimmermann, A.E. Saal, Apollo 15 green glass He-Ne-Ar signatures—In search for indigenous lunar noble gases. Geochem. Perspect. Lett. **8**, 1–5 (2018)

A.M. Gaffney, L.E. Borg, A young solidification age for the lunar magma ocean. Geochim. Cosmochim. Acta **140**, 227–240 (2014)

I. Garrick-Bethell, K. Miljković, Age of the lunar South pole-Aitken basin, in *Lunar Planet. Sci. Conf.* (2018), p. abstract 2633
I. Garrick-Bethell, M.T. Zuber, Elliptical structure of the lunar South pole-Aitken basin. Icarus **204**, 399–408 (2009)
I. Garrick-Bethell, J. Wisdom, M.T. Zuber, Evidence for a past high-eccentricity lunar orbit. Science **313**, 652–655 (2006)
I. Garrick-Bethell, F. Nimmo, M.A. Wieczorek, Structure and formation of the lunar farside highlands. Science **330**, 949–951 (2010)
I. Garrick-Bethell, B.P. Weiss, D.L. Shuster, S.M. Tikoo, M.M. Tremblay, Further evidence for early lunar magnetism from troctolite 76535. J. Geophys. Res., Planets **122**, 76–93 (2017)
J. Geiss, P. Bochsler, Nitrogen isotopes in the Solar System. Geochim. Cosmochim. Acta **46**, 529–548 (1982)
J. Geiss, P. Bochsler, Long time variations in solar wind properties: possible causes versus observations, in *The Sun in Time*, ed. by C.P. Sonett, M.S. Giampapa, M.S. Matthews (University of Arizona Press, Tucson, 1991), pp. 98–117
J. Geiss, F. Bühler, H. Cerutti, P. Eberhardt, C.H. Filleux, J. Meister, P. Signer, The Apollo SWC experiment: results, conclusions, consequences. Space Sci. Rev. **110**, 307–335 (2004)
J. Geiss, P. Eberhardt, P. Signer, F. Buehler, J. Meister, The Solar-Wind Composition experiment, in *Apollo 11 Preliminary Science Report* (NASA SP-214, Washington, DC, 1969), pp. 183–186
E.K. Gibson, G.W. Moore, Volatile-rich lunar soil: evidence of possible cometary impact. Science **179**, 69–71 (1973)
D.R. Gies, J.W. Helsel, Ice age epochs and the Sun's path through the Galaxy. Astrophys. J. **626**, 844–848 (2005)
K. Gillet, L. Margerin, M. Calvet, M. Monnereau, Scattering attenuation profile of the Moon: implications for shallow moonquakes and the structure of the megaregolith. Phys. Earth Planet. Inter. **262**, 28–40 (2017)
M.P. Gillman, H.E. Erenler, P.J. Sutton, Mapping the location of terrestrial impacts and extinctions onto the spiral arm structure of the Milky Way. Int. J. Astrobiol. **18**, 323–328 (2019)
G.R. Gladstone, D.M. Hurley, K.D. Retherford, P.D. Feldman, W.R. Pryor, J.Y. Chaufray, M. Versteeg, T.K. Greathouse, A.J. Steffl, H. Throop, J.W. Parker, D.E. Kaufmann, A.F. Egan, M.W. Davis, D.C. Slater, J. Mukherjee, P.F. Miles, A.R. Hendrix, A. Colaprete, A. Stern, LRO-LAMP observations of the LCROSS impact plume. Science **330**, 472–476 (2010)
P. Gleißner, H. Becker, Formation of Apollo 16 impactites and the composition of late accreted material: constraints from Os isotopes, highly siderophile elements and sulfur abundances. Geochim. Cosmochim. Acta **200**, 1–24 (2017)
R.O. Green, C. Pieters, P. Mouroulis, M. Eastwood, J. Boardman, T. Glavich, P. Isaacson, M. Annadurai, S. Besse, D. Barr, B. Buratti, D. Cate, A. Chatterjee, R. Clark, L. Cheek, J. Combe, D. Dhingra, V. Essandoh, S. Geier, J.N. Goswami, R. Green, V. Haemmerle, J. Head, L. Hovland, S. Hyman, R. Klima, T. Koch, G. Kramer, A.S.K. Kumar, K. Lee, S. Lundeen, E. Malaret, T. McCord, S. McLaughlin, J. Mustard, J. Nettles, N. Petro, K. Plourde, C. Racho, J. Rodriquez, C. Runyon, G. Sellar, C. Smith, H. Sobel, M. Staid, J. Sunshine, L. Taylor, K. Thaisen, S. Tompkins, H. Tseng, G. Vane, P. Varanasi, M. White, D. Wilson, The Moon Mineralogy Mapper (M^3) imaging spectrometer for lunar science: instrument description, calibration, on-orbit measurements, science data calibration and on-orbit validation. J. Geophys. Res., Planets **116**, E00G19 (2011). https://doi.org/10.1029/2011JE003797
J. Gross, K.H. Joy, Evolution, lunar: from magma ocean to crust formation, in *Encyclopedia of Lunar Science*, ed. by B. Cudnik (Springer, Berlin, 2016). https://doi.org/10.1007/978-3-319-05546-6_39-1
J. Gross, A.H. Treiman, C.N. Mercer, Lunar feldspathic meteorites: constraints on the geology of the lunar highlands, and the origin of the lunar crust. Earth Planet. Sci. Lett. **388**, 318–328 (2014)
T.L. Grove, M.J. Krawczynski, Lunar mare volcanism: where did the magmas come from? Elements **5**, 29–34 (2009)
M. Güdel, The Sun in time: activity and environment. Living Rev. Sol. Phys. **4**, 3 (2007). https://doi.org/10.12942/lrsp-2007-3
S. Guggisberg, P. Eberhardt, J. Geiss, N. Groegler, A. Stettler, G.M. Brown, A. Peckett, Classification of the Apollo-11 mare basalts according to Ar^{39}-Ar^{40} ages and petrological properties, in *Lunar Planet. Sci. Conf. Proc.*, ed. by N.W. Hinners (1979), pp. 1–39
A.N. Halliday, D.-C. Lee, Tungsten isotopes and the early development of the Earth and Moon. Geochim. Cosmochim. Acta **63**, 4157–4179 (1999)
L.J. Hallis, M. Anand, S. Strekopytov, Trace-element modelling of mare basalt parental melts: implications for a heterogeneous lunar mantle. Geochim. Cosmochim. Acta **134**, 289–316 (2014)
J.K. Harmon, M.A. Slade, Radar mapping of Mercury: full-disk images and polar anomalies. Science **258**, 640–643 (1992)
W.K. Hartmann, Lunar cratering chronology. Icarus **13**, 299–301 (1970)

W.K. Hartmann, D.R. Davis, Satellite-sized planetesimals and lunar origin. Icarus **24**, 504–515 (1975)
W.K. Hartmann, C.A. Wood, Moon: origin and evolution of multi-ring basins. Moon **3**, 3–78 (1971)
K. Hashizume, M. Chaussidon, A non-terrestrial ^{16}O-rich isotopic composition for the protosolar nebula. Nature **434**, 619–622 (2005)
K. Hashizume, M. Chaussidon, Two oxygen isotopic components with extra-selenial origins observed among lunar metallic grains—In search for the solar wind component. Geochim. Cosmochim. Acta **73**, 3038–3054 (2009)
K. Hashizume, M. Chaussidon, B. Marty, F. Robert, Solar wind record on the Moon: deciphering presolar from planetary nitrogen. Science **290**, 1142–1145 (2000)
K. Hashizume, M. Chaussidon, B. Marty, K. Terada, Protosolar carbon isotopic composition: implications for the origin of meteoritic organics. Astrophys. J. **600**, 480–484 (2004)
L.A. Haskin, The Imbrium impact event and the thorium distribution at the lunar highlands surface. J. Geophys. Res., Planets **103**, 1679–1689 (1998)
E.H. Hauri, A.E. Saal, M. Nakajima, M. Anand, M.J. Rutherford, J.A. Van Orman, M. Le Voyer, Origin and evolution of water in the Moon's interior. Annu. Rev. Earth Planet. Sci. **45**, 89–111 (2017)
P.O. Hayne, B.T. Greenhagen, M.C. Foote, M.A. Siegler, A.R. Vasavada, D.A. Paige, Diviner lunar radiometer observations of the LCROSS impact. Science **330**, 477–479 (2010)
P.O. Hayne, A. Hendrix, E. Sefton-Nash, M.A. Siegler, P.G. Lucey, K.D. Retherford, J.P. Williams, B.T. Greenhagen, D.A. Paige, Evidence for exposed water ice in the Moon's south polar regions from Lunar Reconnaissance Orbiter ultraviolet albedo and temperature measurements. Icarus **255**, 58–69 (2015)
J.W. Head III, L. Wilson, Generation, ascent and eruption of magma on the Moon: new insights into source depths, magma supply, intrusions and effusive/explosive eruptions (part 2: observations). Icarus **283**, 176–223 (2017)
J.W. Head, L. Wilson, Lunar mare volcanism: stratigraphy, eruption conditions, and the evolution of secondary crusts. Geochim. Cosmochim. Acta **56**, 2155–2175 (1992)
V.S. Heber, H. Baur, R. Wieler, Helium in lunar samples analyzed by high-resolution stepwise etching: implications for the temporal constancy of solar wind isotopic composition. Astrophys. J. **597**, 602–614 (2003)
V.S. Heber, R. Wieler, H. Baur, C. Olinger, T.A. Friedmann, D.S. Burnett, Noble gas composition of the solar wind as collected by the Genesis mission. Geochim. Cosmochim. Acta **73**, 7414–7432 (2009)
G.H. Heiken, D.T. Vaniman, B.M. French, *Lunar Sourcebook* (Cambridge University Press, Cambridge, 1991)
P.C. Hess, E.M. Parmentier, A model for the thermal and chemical evolution of the Moon's interior: implications for the onset of mare volcanism. Earth Planet. Sci. Lett. **134**, 501–514 (1995)
H. Hiesinger, R. Jaumann, G. Neukum, J.W. Head, Ages of mare basalts on the lunar nearside. J. Geophys. Res. **105**, 29239–29275 (2000)
H. Hiesinger, J.W. Head, U. Wolf, R. Jaumann, G. Neukum, Ages and stratigraphy of mare basalts in Oceanus Procellarum, Mare Nubium, Mare Cognitum, and Mare Insularum. J. Geophys. Res., Planets **108**, 5065 (2003). https://doi.org/10.1029/2002JE001985
H. Hiesinger, J.W. Head, U. Wolf, R. Jaumann, G. Neukum, Ages and stratigraphy of lunar mare basalts in Mare Frigoris and other nearside maria based on crater size-frequency distribution measurements. J. Geophys. Res., Planets **115**, E03003 (2010). https://doi.org/10.1029/2009JE003380
H. Hiesinger, J.W. Head, U. Wolf, R. Jaumann, G. Neukum, Ages and stratigraphy of lunar mare basalts: a synthesis, in *Recent Advances and Current Research Issues in Lunar Stratigraphy*, ed. by W.A. Ambrose, D.A. Williams. Geol. Soc. of America Special Paper, vol. 477 (2011), pp. 1–51. https://doi.org/10.1130/2011.2477(01)
H. Hiesinger, C.H. van der Bogert, J.H. Pasckert, N. Schmedemann, M.S. Robinson, B. Jolliff, N. Petro, New crater size-frequency distribution measurements of the South pole-Aitken basin, in *Lunar Planet. Sci. Conf.* (2012), p. abstract 2863
F. Hörz, R. Grieve, G. Heiken, P. Spudis, A. Binder, Lunar surface processes, in *Lunar Sourcebook*, ed. by G. Heiken, D. Vaniman, B.M. French (Cambridge University Press, Cambridge, 1991), pp. 61–120
N. Hosono, S. Karato, J. Makino, T.R. Saitoh, Terrestrial magma ocean origin of the Moon. Nat. Geosci. **12**, 418–423 (2019)
J. Huang, Z. Xiao, J. Flahaut, M. Martinot, J. Head, X. Xiao, M. Xie, L. Xiao, Geological characteristics of Von Kármán crater, northwestern South pole-Aitken basin: Chang'E-4 landing site region. J. Geophys. Res., Planets **123**, 1684–1700 (2018)
D. Hurwitz, D.A. Kring, Identifying the geologic context of Apollo 17 impact melt breccias. Earth Planet. Sci. Lett. **436**, 64–70 (2016)
L. Husain, ^{40}Ar-^{39}Ar chronology and cosmic ray exposure ages of the Apollo 15 samples. J. Geophys. Res. **79**, 2588–2606 (1974)

G.R. Huss, K. Nagashima, D.S. Burnett, A.J.G. Jurewicz, C.T. Olinger, A new upper limit on the D/H ratio in the solar wind, in *Lunar Planet. Sci. Conf. Proc.* (2012a), abstract 1709

G.R. Huss, K. Nagashima, A.J.G. Jurewicz, D.S. Burnett, C.T. Olinger, The isotopic composition and fluence of solar-wind nitrogen in a Genesis B/C array collector. Meteorit. Planet. Sci. **47**, 1436–1448 (2012b)

R. Jaumann, H. Hiesinger, M. Anand, I.A. Crawford, R. Wagner, F. Sohl, B.L. Jolliff, F. Scholten, M. Knapmeyer, H. Hoffmann, H. Hussmann, M. Grott, S. Hempel, U. Köhler, K. Krohn, N. Schmitz, J. Carpenter, M. Wieczorek, T. Spohn, M.S. Robinson, J. Oberst, Geology, geochemistry, and geophysics of the Moon: status of current understanding. Planet. Space Sci. **74**, 15–41 (2012)

E.A. Jerde, G.A. Snyder, L.A. Taylor, Y.G. Liu, R.A. Schmitt, The origin and evolution of lunar high-Ti basalts: periodic melting of a single source at Mare Tranquillitatis. Geochim. Cosmochim. Acta **58**, 515–527 (1994)

B.L. Jolliff, J.J. Gillis, L.A. Haskin, R.L. Korotev, M.A. Wieczorek, Major lunar crustal terranes: surface expressions and crust-mantle origins. J. Geophys. Res. **105**, 4197–4216 (2000)

B.L. Jolliff, R.C. Watkins, N.E. Petro, D.P. Moriarty, S.J. Lawrence, J.W. Head, C.M. Pieters, J.J. Hagerty, R.L. Fergason, T.M. Hare, L.R. Gaddis, P.O. Hayne, Selecting and certifying a landing site for moonrise in South pole-Aitken basin, in *Lunar Planet. Sci. Conf.* (2017), p. abstract 1326

K.H. Joy, Trace elements in lunar plagioclase as indicators of source lithology, in *Lunar Planet. Sci. Conf.* (2013), p. abstract 1033

K.H. Joy, T. Arai, Lunar meteorites: new insights into the geological history of the Moon. Astron. Geophys. **54**, 4.28–4.32 (2013). https://doi.org/10.1093/astrogeo/att121

K.H. Joy, I.A. Crawford, M. Anand, R.C. Greenwood, I.A. Franchi, S.S. Russell, The petrology and geochemistry of Miller range 05035: a new lunar gabbroic meteorite. Geochim. Cosmochim. Acta **72**, 3822–3844 (2008)

K.H. Joy, R. Burgess, R. Hinton, V.A. Fernandes, I.A. Crawford, A.T. Kearsley, A.J. Irving, EIMF, Petrogenesis and chronology of lunar meteorite Northwest Africa 4472: a KREEPy regolith breccia from the Moon. Geochim. Cosmochim. Acta **75**, 2420–2452 (2011a)

K.H. Joy, D.A. Kring, D.D. Bogard, D.S. McKay, M.E. Zolensky, Re-examination of the formation ages of the Apollo 16 regolith breccias. Geochim. Cosmochim. Acta **75**, 7208–7225 (2011b)

K.H. Joy, I.A. Crawford, N.M. Curran, M. Zolensky, A.F. Fagan, D.A. Kring, The Moon: an archive of small body migration in the Solar System. Earth Moon Planets **118**, 133–158 (2016)

M. Jutzi, E. Asphaug, Forming the lunar farside highlands by accretion of a companion Moon. Nature **476**, 69–72 (2011)

R.R. Keays, R. Ganapathy, J.C. Laul, E. Anders, G.F. Herzog, P.M. Jeffery, Trace elements and radioactivity in lunar rocks: implications for meteorite infall, solar-wind flux, and formation conditions of Moon. Science **167**, 490–493 (1970)

J.F. Kerridge, Solar nitrogen: evidence for a secular increase in the ratio of nitrogen-15 to nitrogen-14. Science **188**, 162–164 (1975)

J.F. Kerridge, Secular variations in composition of the solar wind: evidence and causes, in *Proc. Conf. Ancient Sun* (1980), pp. 475–489

J.F. Kerridge, Long-term compositional variation in solar corpuscular radiation: Evidence from nitrogen isotopes in the lunar regolith. Rev. Geophys. **31**, 423–437 (1993)

J.F. Kerridge, P. Signer, R. Wieler, R.H. Becker, R.O. Pepin, Long-term changes in composition of solar particles implanted in extraterrestrial materials, in *The Sun in Time*, ed. by C.P. Sonett, M.S. Giampapa, M.S. Matthews (University of Arizona Press, Tucson, 1991), pp. 389–412

A. Khan, K. Mosegaard, K.L. Rasmussen, A new seismic velocity model for the Moon from a Monte Carlo inversion of the Apollo lunar seismic data. Geophys. Res. Lett. **27**, 1591–1594 (2000)

A. Khan, J.A.D. Connolly, A. Pommier, J. Noir, Geophysical evidence for melt in the deep lunar interior and implications for lunar evolution. J. Geophys. Res., Planets **119**, 2197–2221 (2014)

R.M. Killen, D.M. Hurley, W.M. Farrell, The effect on the lunar exosphere of a coronal mass ejection passage. J. Geophys. Res. **117**, E00K02 (2012). https://doi.org/10.1029/2011JE00401

T. Kleine, H. Palme, K. Mezger, A.N. Halliday, Hf-W chronometry of lunar metals and the age and early differentiation of the Moon. Science **310**, 1671–1674 (2005)

L. Kööp, P.R. Heck, H. Busemann, A.M. Davis, J. Greer, C. Maden, M.M.M. Meier, R. Wieler, High early solar activity inferred from helium and neon excesses in the oldest meteorite inclusions. Nat. Astron. **2**, 709–713 (2018)

R.L. Korotev, Lunar geochemistry as told by lunar meteorites. Chem. Erde Geochem. **65**, 297–346 (2005)

R.L. Korotev, B.L. Jolliff, R.A. Zeigler, J.J. Gillis, L.A. Haskin, Feldspathic lunar meteorites and their implications for compositional remote sensing of the lunar surface and the composition of the lunar crust. Geochim. Cosmochim. Acta **67**, 4895–4923 (2003)

T.S. Kruijer, T. Kleine, Tungsten isotopes and the origin of the Moon. Earth Planet. Sci. Lett. **475**, 15–24 (2017)

T.S. Kruijer, T. Kleine, M. Fischer-Godde, P. Sprung, Lunar tungsten isotopic evidence for the late veneer. Nature **520**, 534–537 (2015)

D.R. Lammlein, Lunar seismicity, structure, and tectonic. Philos. Trans. R. Soc. Lond. A **285**, 451–461 (1977)

M. Laneuville, M.A. Wieczorek, D. Breuer, J. Albert, G. Morard, T. Rückriemen, A long-lived lunar dynamo powered by core crystallization. Earth Planet. Sci. Lett. **40**, 251–260 (2014)

G. Latham, M. Ewing, J. Dorman, D. Lammlein, F. Press, N. Toksoz, G. Sutton, F. Duennebier, Y. Nakamura, Moonquakes. Science **174**, 687–692 (1971)

M. Le Feuvre, M.A. Wieczorek, Nonuniform cratering of the Moon and a revised crater chronology of the inner Solar System. Icarus **214**, 1–20 (2011)

D.C. Lee, A.N. Halliday, G.A. Snyder, L.A. Taylor, Age and origin of the Moon. Science **278**, 1098–1103 (1997)

M. Lemelin, P.G. Lucey, E. Song, G.J. Taylor, Lunar central peak mineralogy and iron content using the Kaguya Multiband Imager: reassessment of the compositional structure of the lunar crust. J. Geophys. Res., Planets **120**, 869–887 (2015)

C. Lepaulard, J. Gattacceca, M. Uehara, P. Rochette, Y. Quesnel, R.J. Macke, S.J.W. Kiefer, A survey of the natural remanent magnetization and magnetic susceptibility of Apollo whole rocks. Phys. Earth Planet. Inter. **290**, 36–43 (2019)

J. Levine, P.R. Renne, R.A. Muller, Solar and cosmogenic argon in dated lunar impact spherules. Geochim. Cosmochim. Acta **71**, 1624–1635 (2007)

S. Li, R.E. Milliken, Water on the surface of the Moon as seen by the Moon Mineralogy Mapper: distribution, abundance, and origins. Sci. Adv. **3**, e1701471 (2017). https://doi.org/10.1126/sciadv.1701471

S. Li, P.G. Lucey, R.E. Milliken, P.O. Hayne, E. Fisher, J.P. Williams, D.M. Hurley, R.C. Elphic, Direct evidence of surface exposed water ice in the lunar polar regions. Proc. Natl. Acad. Sci. USA **115**, 8907–8912 (2018)

C. Li, D. Liu, B. Liu, X. Ren, J. Liu, Z. He, W. Zuo, X. Zeng, R. Xu, X. Tan, X. Zhang, W. Chen, R. Shu, W. Wen, Y. Su, H. Zhang, Z. Ouyang, Chang'E-4 initial spectroscopic identification of lunar far-side mantle-derived materials. Nature **569**, 378–382 (2019)

Y. Liu, C. Floss, J.M.D. Day, E. Hill, L.A. Taylor, Petrogenesis of lunar mare basalt meteorite Miller Range 05035. Meteorit. Planet. Sci. **44**, 261–284 (2009)

Y. Liu, Y. Guan, Y. Zhang, G.R. Rossman, J.M. Eiler, L.A. Taylor, Direct measurement of hydroxyl in the lunar regolith and the origin of lunar surface water. Nat. Geosci. **5**, 779–782 (2012)

J. Liu, M. Sharp, R.D. Ash, D.A. Kring, R.J. Walker, Diverse impactors in Apollo 15 and 16 impact rocks: evidence from osmium isotopes and highly siderophile elements. Geochim. Cosmochim. Acta **155**, 122–153 (2015)

J. Longhi, Experimental petrology and petrogenesis of mare volcanics. Geochim. Cosmochim. Acta **56**, 2235–2251 (1992)

D.E. Loper, C.L. Werner, On lunar asymmetries 1. Tilted convection and crustal asymmetry. J. Geophys. Res. **107**, 5046 (2002)

G.W. Lugmair, R.W. Carlson, The Sm-Nd history of KREEP, in *Lunar Planet. Sci. Conf. Proc.* (1978), pp. 689–704

J.R. Lyons, E. Gharib-Nezhad, T.R. Ayres, A light carbon isotope composition for the Sun. Nat. Commun. **9**, 908 (2018). https://doi.org/10.1038/s41467-018-03093-3

S. Marchi, S. Mottola, G. Cremonese, M. Massironi, E. Martellato, A new chronology for the Moon and Mercury. Astron. J. **137**, 4936–4948 (2009)

S. Marchi, W.F. Bottke, D.A. Kring, A. Morbidelli, The onset of the lunar cataclysm as recorded in its ancient crater populations. Earth Planet. Sci. Lett. **325**, 27–38 (2012)

S. Marchi, W.F. Bottke, B.A. Cohen, K. Wünnemann, D.A. Kring, H.Y. McSween, M.C. De Sanctis, D.P. O'Brien, P. Schenk, C.A. Raymond, C.T. Russell, High-velocity collisions from the lunar cataclysm recorded in asteroidal meteorites. Nat. Geosci. **6**, 303–307 (2013)

B. Marty, L. Zimmermann, P.G. Burnard, R. Wieler, V.S. Heber, D.L. Burnett, R.C. Wiens, P. Bochsler, Nitrogen isotopes in the recent solar wind from the analysis of Genesis targets: evidence for large scale isotope heterogeneity in the early Solar System. Geochim. Cosmochim. Acta **74**, 340–355 (2010)

B. Marty, M. Chaussidon, R.C. Wiens, A.J.G. Jurewicz, D.S. Burnett, A ^{15}N-poor isotopic composition for the Solar System as shown by Genesis solar wind samples. Science **332**, 1533–1536 (2011)

T.B. McCord, J.B. Adams, T.V. Johnson, Asteroid Vesta: spectral reflectivity and compositional implications. Science **168**, 1445–1447 (1970)

T.J. McCoy, D.W. Middlefehldt, L. Wilson, Asteroid differentiation, in *Meteorites and the Early Solar System II*, ed. by D.S. Lauretta, H.Y. McSween Jr. (University of Arizona Press, Tucson, 2006), pp. 733–745

F.M. McCubbin, K.E. Vander Kaaden, R. Tartèse, R.L. Klima, Y. Liu, J. Mortimer, J.J. Barnes, C.K. Shearer, A.H. Treiman, D.J. Lawrence, S.M. Elardo, D.M. Hurley, J.W. Boyce, M. Anand, Magmatic volatiles

(H, C, N, F, S, Cl) in the lunar mantle, crust, and regolith: abundances, distributions, processes, and reservoirs. Am. Mineral. **100**, 1668–1707 (2015)
D.S. McKay, G.A. Heiken, G.M. Taylor, U.S. Clanton, D.A. Morrison, G.H. Ladle, Apollo 14 soils: size distribution and particle types, in *Lunar Sci. Conf. Proc.* (1972), pp. 983–994
D.S. McKay, D.D. Bogard, R.V. Morris, R.L. Korotev, P. Johnson, S.J. Wentworth, Apollo 16 regolith breccias: characterization and evidence for early formation in the mega-regolith. J. Geophys. Res. **91**, D277–D303 (1986)
D.S. McKay, G. Heiken, A. Basu, G. Blanford, S. Simon, R. Reedy, B.M. French, J.J. Papike, The lunar regolith, in *Lunar Sourcebook*, ed. by G. Heiken, D. Vaniman, B.M. French (Cambridge University Press, Cambridge, 1991), pp. 286–356
K.D. McKeegan, M. Chaussidon, F. Robert, Incorporation of short-lived ^{10}Be in a calcium–aluminum-rich inclusion from the Allende meteorite. Science **289**, 1334–1337 (2000)
K.D. McKeegan, A.P.A. Kallio, V.S. Heber, G. Jarzebinski, P.H. Mao, C.D. Coath, T. Kunihiro, R.C. Wiens, J.E. Nordholt, R.W. Moses Jr., D.B. Reisenfeld, A.J.G. Jurewicz, D.S. Burnett, The oxygen isotopic composition of the Sun inferred from captured solar wind. Science **332**, 1528–1532 (2011)
C.L. McLeod, A.D. Brandon, R.M.G. Armytage, Constraints on the formation age and evolution of the Moon from ^{142}Nd-^{143}Nd systematics of Apollo 12 basalts. Earth Planet. Sci. Lett. **396**, 179–189 (2014)
H.Y. McSween Jr., R.P. Binzel, M.C. De Sanctis, E. Ammannito, T.H. Prettyman, A.W. Beck, V. Reddy, L. Le Corre, M. Gaffey, T.B. McCord, C.A. Raymond, C.T. Russell, Dawn, the Vesta–HED connection, and the geologic context for eucrites, diogenites, and howardites. Meteorit. Planet. Sci. **48**, 2090–2104 (2013)
H.J. Melosh, A.M. Freed, B.C. Johnson, D.M. Blair, J.C. Andrews-Hanna, G.A. Neumann, R.J. Phillips, D.E. Smith, S.C. Solomon, M.A. Wieczorek, M.T. Zuber, The origin of lunar Mascon basins. Science **340**, 1552–1555 (2013)
H.J. Melosh, J. Kendall, B. Horgan, B.C. Johnson, T. Bowling, P.G. Lucey, G.J. Taylor, South pole-Aitken basin ejecta reveal the Moon's upper mantle. Geology **45**, 1063–1066 (2017)
C.M. Mercer, K.E. Young, J.R. Weirich, K.V. Hodges, B.L. Jolliff, J.A. Wartho, M.C. van Soest, Refining lunar impact chronology through high spatial resolution ^{40}Ar/^{39}Ar dating of impact melts. Sci. Adv. **1**, e1400050 (2015). https://doi.org/10.1126/sciadv.1400050
A. Meshik, J. Mabry, C. Hohenberg, Y. Marrocchi, O. Pravdivtseva, D. Burnett, C. Olinger, R. Wiens, D. Reisenfeld, J. Allton, K. McNamara, E. Stansbery, A.J.G. Jurewicz, Constraints on neon and argon isotopic fractionation in solar wind. Science **318**, 433–435 (2007)
A. Meshik, C. Hohenberg, O. Pravdivtseva, D. Burnett, Heavy noble gases in solar wind delivered by Genesis mission. Geochim. Cosmochim. Acta **127**, 326–347 (2014)
A.E. Metzger, E.L. Haines, R.E. Parker, R.G. Radocinski, Thorium concentrations in the lunar surface. I—Regional values and crustal content, in *Lunar Planet. Sci. Conf. Proc.* (1977), pp. 949–999
K. Misawa, M. Tatsumoto, G.B. Dalrymple, K. Yanai, An extremely low U/Pb source in the Moon: U-Th-Pb, Sm-Nd, Rb-Sr, and ^{40}Ar/^{39}Ar isotopic systematics and age of lunar meteorite Asuka 881757. Geochim. Cosmochim. Acta **57**, 4687–4702 (1993)
I.G. Mitrofanov, A.B. Sanin, W.V. Boynton, G. Chin, J.B. Garvin, D. Golovin, L.G. Evans, K. Harshman, A.S. Kozyrev, M.L. Litvak, A. Malakhov, E. Mazarico, T. McClanahan, G. Milikh, M. Mokrousov, G. Nandikotkur, G.A. Neumann, I. Nuzhdin, R. Sagdeev, V. Shevchenko, V. Shvetsov, D.E. Smith, R. Starr, V.I. Tretyakov, J. Trombka, D. Usikov, A. Varenikov, A. Vostrukhin, M.T. Zuber, Hydrogen mapping of the Lunar South Pole using the LRO neutron detector experiment LEND. Science **330**, 483–486 (2010)
A. Morbidelli, S. Marchi, W.F. Bottke, D.A. Kring, A sawtooth-like timeline for the first billion years of lunar bombardment. Earth Planet. Sci. Lett. **355–356**, 144–151 (2012)
A. Morbidelli, D. Nesvorny, V. Laurenz, S. Marchi, D.C. Rubie, L. Elkins-Tanton, M. Wieczorek, S. Jacobson, The timeline of the lunar bombardment: revisited. Icarus **305**, 262–276 (2018)
J.W. Morgan, U. Krähenbühl, R. Ganapathy, E. Anders, Trace elements in Apollo 15 samples: implications for meteorite influx and volatile depletion on the Moon, in *Lunar Sci. Conf. Proc.* (1972), pp. 1361–1376
R.V. Morris, Surface exposure indices of lunar soils—A comparative FMR study, in *Lunar Planet. Sci. Conf. Proc.* (1976), pp. 315–335
R.V. Morris, The surface exposure (maturity) of lunar soils—Some concepts and I_S/FeO compilation, in *Lunar Planet. Sci. Conf. Proc.* (1978), pp. 2287–2297
R.V. Morris, Origins and size distribution of metallic iron particles in the lunar regolith, in *Lunar Planet. Sci. Conf. Proc.* (1980), pp. 1697–1712
J. Mortimer, A.B. Verchovsky, M. Anand, Predominantly non-solar origin of nitrogen in lunar soils. Geochim. Cosmochim. Acta **193**, 36–53 (2016)
V.R. Murthy, M.R. Coscio Jr., Rb-Sr ages and isotopic systematics of some Serenitatis mare basalts, in *Lunar Planet. Sci. Conf. Proc.*, ed. by R.B. Merrill (1976), pp. 1529–1544

Y. Nakamura, Shallow moonquakes: how they compare with earthquakes, in *Lunar Planet. Sci. Conf. Proc.* (1980), pp. 1847–1853

Y. Nakamura, G.V. Latham, H.J. Dorman, Apollo lunar seismic experiment final summary, in *Lunar Planet. Sci. Conf. Proc.* (1982), pp. A117–A123

Y. Nakamura, G.V. Latham, H.J. Dorman, J.E. Harris, Passive Seismic Experiment, Long Period Event Catalog, Final Version (1969 Day 202–1977 Day 273, ALSEP Stations 11, 12, 13, 14, 15, and 16). Univ. Texas Institute Geophys. Tech. Report No. 18 (originally published 19 June 1981, revised 2 October 2008) (2008)

National Research Council, The Scientific Context for the Exploration of the Moon (The National Academies Press, Washington, DC, 2007). https://www.nap.edu/catalog/11954/the-scientific-context-for-exploration-of-the-Moon. Accessed 12 June 2019

C.R. Neal, Establishing a lunar geophysical network for exploration and Solar System science, in *Annual Meeting Lunar Exploration Analysis Group* (2011), abstract 2052

C.R. Neal, G.Y. Kramer, The petrogenesis of the Apollo 14 high-Al mare basalts. Am. Mineral. **91**, 1521–1535 (2006)

C.R. Neal, L.A. Taylor, Petrogenesis of mare basalts: a record of lunar volcanism. Geochim. Cosmochim. Acta **56**, 2177–2212 (1992)

C.R. Neal, M.D. Hacker, G.A. Snyder, L.A. Taylor, Y.G. Liu, R.A. Schmitt, Basalt generation at the Apollo 12 site, part 1: new data, classification, and re-evaluation. Meteoritics **29**, 334–348 (1994)

C.R. Neal, W.B. Banerdt, C. Beghein, P. Chi, D. Currie, S. Del'Agnello, I. Garrick-Bethell, R. Grimm, M. Grott, H. Haviland, S. Kedar, S. Nagihara, M. Panning, N. Petro, N. Schmerr, M. Siegler, R. Weber, M. Wieczorek, K. Zacny, The lunar geophysical network mission, in *Lunar Planet. Sci. Conf.* (2019), abstract 2455

A. Nemchin, N. Timms, R. Pidgeon, T. Geisler, S. Reddy, C. Meyer, Timing of crystallization of the lunar magma ocean constrained by the oldest zircon. Nat. Geosci. **2**, 133–136 (2009)

G. Neukum, B. Ivanov, W.K. Hartmann, Cratering records in the inner Solar System in relation to the lunar reference system. Space Sci. Rev. **96**, 55–86 (2001)

G.A. Neumann, M.T. Zuber, M.A. Wieczorek, J.W. Head, D.M.H. Baker, S.C. Solomon, D.E. Smith, F.G. Lemoine, E. Mazarico, T.J. Sabaka, S.J. Goossens, H.J. Melosh, R.J. Phillips, S.W. Asmar, A.S. Konopliv, J.G. Williams, M.M. Sori, J.M. Soderblom, K. Miljković, J.C. Andrews-Hanna, F.W.S. Nimmo, W. Kiefer, Lunar impact basins revealed by gravity recovery and interior laboratory measurements. Sci. Adv. **1**, e1500852 (2015). https://doi.org/10.1126/sciadv.1500852

M.D. Norman, The lunar cataclysm: reality or "mythconception"? Elements **5**, 23–28 (2009)

M.D. Norman, R.A. Duncan, J.J. Huard, Identifying impact events within the lunar cataclysm from ^{40}Ar-^{39}Ar ages and compositions of Apollo 16 impact melt rocks. Geochim. Cosmochim. Acta **70**, 6032–6049 (2006)

S. Nozette, C.L. Lichtenberg, P.D. Spudis, R. Bonner, W. Ort, E. Malaret, M. Robinson, E.M. Shoemaker, The Clementine bistatic radar experiment. Science **274**, 1495–1498 (1996)

S. Nozette, E.M. Shoemaker, P.D. Spudis, C.L. Lichtenberg, The possibility of ice on the Moon. Science **278**, 144–145 (1997)

L.E. Nyquist, C.-Y. Shih, The isotopic record of lunar volcanism. Geochim. Cosmochim. Acta **56**, 2213–2234 (1992)

L.E. Nyquist, B.M. Bansal, H. Wiesmann, Rb-Sr ages and initial ^{87}Sr/^{86}Sr for Apollo 17 basalts and KREEP basalt 15386, in *Lunar Planet. Sci. Conf. Proc.* (1975), pp. 1445–1465

L.E. Nyquist, B.M. Bansal, H. Wiesmann, Sr-isotopic constraints on the petrogenesis of Apollo 17 mare basalts, in *Lunar Planet. Sci. Conf. Proc.*, ed. by R.B. Merrill (1976), pp. 1507–1528

L.E. Nyquist, H. Wiesmann, B. Bansal, C.Y. Shih, J.E. Keith, C.L. Harper, ^{146}Sm-^{142}Nd formation interval for the lunar mantle. Geochim. Cosmochim. Acta **59**, 2817–2837 (1995)

L.E. Nyquist, C.Y. Shih, D. Reese, J. Park, D.D. Bogard, D.H. Garrison, A. Yamaguchi, Lunar crystal history recorded in lunar anorthosites, in *Lunar Planet. Sci. Conf.* (2010), p. abstract 1383

K.W. Ogilvie, M.A. Coplan, Solar wind composition. Rev. Geophys. **33**, 615–622 (1995)

M. Ohtake, T. Matsunaga, J. Haruyama, Y. Yokota, T. Morota, C. Honda, Y. Ogawa, M. Torii, H. Miyamoto, T. Arai, N. Hirata, A. Iwasaki, R. Nakamura, T. Hiroi, T. Sugihara, H. Takeda, H. Otake, C.M. Pieters, K. Saiki, K. Kitazato, M. Abe, N. Asada, H. Demura, Y. Yamaguchi, S. Sasaki, S. Kodama, J. Terazono, M. Shirao, A. Yamaji, S. Minami, H. Akiyama, J.L. Josset, The global distribution of pure anorthosite on the Moon. Nature **461**, 236–240 (2009)

M. Ohtake, H. Takeda, T. Matsunaga, Y. Yokota, J. Haruyama, T. Morota, S. Yamamoto, Y. Ogawa, T. Hiroi, Y. Karouji, K. Saiki, P.G. Lucey, Asymmetric crustal growth on the Moon indicated by primitive farside high materials. Nat. Geosci. **5**, 384–388 (2012)

J.B. Paces, C.R. Neal, L.A. Taylor, S. Nakai, A.N. Halliday, D.C. Lee, A strontium and neodymium isotopic study of Apollo 17 high-Ti mare basalts: resolution of ages, evolution of magmas, and origins of source heterogeneities. Geochim. Cosmochim. Acta **55**, 2025–2043 (1991)

D.A. Paige, M.C. Foote, B.T. Greenhagen, J.T. Schofield, S. Calcutt, A.R. Vasavada, D.J. Preston, F.W. Taylor, C.C. Allen, K.J. Snook, B.M. Jakosky, B.C. Murray, L.A. Soderblom, B. Jau, S. Loring, J. Bulharowski, N.E. Bowles, I.R. Thomas, M.T. Sullivan, C. Avis, E.M. De Jong, W. Hartford, D.J. McCleese, The Lunar Reconnaissance Orbiter Diviner lunar radiometer experiment. Space Sci. Rev. **150**, 125–160 (2010a)

D.A. Paige, M.A. Siegler, J.A. Zhang, P.O. Hayne, E.J. Foote, K.A. Bennett, A.R. Vasavada, B.T. Greenhagen, J.T. Schofield, D.J. McCleese, M.C. Foote, E. DeJong, B.G. Bills, W. Hartford, B.C. Murray, C.C. Allen, K. Snook, L.A. Soderblom, S. Calcutt, F.W. Taylor, N.E. Bowles, J.L. Bandfield, R. Elphic, R. Ghent, T.D. Glotch, M.B. Wyatt, P.G. Lucey, Diviner Lunar radiometer observations of cold traps in the Moon's South Polar Region. Science **330**, 479–482 (2010b)

D.A. Papanastassiou, G.J. Wasserburg, Rb-Sr ages of igneous rocks from the Apollo 14 mission and the age of the Fra Mauro formation. Earth Planet. Sci. Lett. **12**, 36–48 (1971a)

D.A. Papanastassiou, G.J. Wasserburg, Lunar chronology and evolution from Rb-Sr studies of Apollo 11 and 12 samples. Earth Planet. Sci. Lett. **11**, 37–62 (1971b)

D.A. Papanastassiou, G.J. Wasserburg, Rb-Sr ages and initial strontium in basalts from Apollo 15. Earth Planet. Sci. Lett. **17**, 324–337 (1973)

J.J. Papike, F.N. Hodges, A.E. Bence, M. Cameron, J.M. Rhodes, Mare basalts: crystal chemistry, mineralogy, and petrology. Rev. Geophys. **14**, 475–540 (1976)

J.J. Papike, G.W. Fowler, C.K. Shearer, G.D. Layne, Ion microprobe investigation of plagioclase and orthopyroxene from lunar Mg-suite norites: implications for calculating parental melt REE concentrations and for assessing postcrystallization REE redistribution. Geochim. Cosmochim. Acta **60**, 3967–3978 (1996)

R.O. Pepin, R.H. Becker, P.E. Rider, Xenon and Krypton in extraterrestrial regolith soil and in the solar wind. Geochim. Cosmochim. Acta **59**, 4997–5022 (1995)

R.O. Pepin, D.J. Schlutter, R.H. Becker, D.B. Reisenfeld, Helium, neon, and argon composition of the solar wind as recorded in gold and other Genesis collector materials. Geochim. Cosmochim. Acta **89**, 62–80 (2012)

J.F. Pernet-Fisher, E. Deloule, K.H. Joy, Evidence of chemical heterogeneity within lunar anorthosite parental magmas. Geochim. Cosmochim. Acta (2019a). https://doi.org/10.1016/j.gca.2019.03.033

J.F. Pernet-Fisher, F.E. McDonald, R.A. Zeigler, K.H. Joy, 50 years on: legacies of the Apollo programme. Astron. Geophys. **60**, 4.22–4.28 (2019b)

C.M. Pieters, J. Boardman, B. Buratti, A. Chatterjee, R. Clark, T. Glavich, R. Green, J. Head III, P. Isaacson, E. Malaret, T. McCord, J. Mustard, N. Petro, C. Runyon, M. Staid, J. Sunshine, L. Taylor, S. Tompkins, P. Varanasi, M. White, The Moon Mineralogy Mapper (M3) on Chandrayaan-1. Current. Science **96**, 500–505 (2009a)

C.M. Pieters, J.N. Goswami, R.N. Clark, M. Annadurai, J. Boardman, B. Buratti, J.P. Combe, M.D. Dyar, R. Green, J.W. Head, C. Hibbitts, M. Hicks, P. Isaacson, R. Klima, G. Kramer, S. Kumar, E. Livo, S. Lundeen, E. Malaret, T. McCord, J. Mustard, J. Nettles, N. Petro, C. Runyon, M. Staid, J. Sunshine, L.A. Taylor, S. Tompkins, P. Varanasi, Character and spatial distribution of OH/H_2O on the surface of the Moon seen by M^3 on Chandrayaan-1. Science **326**, 568–572 (2009b)

C.M. Pieters, S. Besse, J. Boardman, B. Buratti, L. Cheek, R.N. Clark, J.P. Combe, D. Dhingra, J.N. Goswami, R.O. Green, J.W. Head, P. Isaacson, R. Klima, G. Kramer, S. Lundeen, E. Malaret, T. McCord, J. Mustard, J. Nettles, N. Petro, C. Runyon, M. Staid, J. Sunshine, L.A. Taylor, K. Thaisen, S. Tompkins, J. Whitten, Mg-spinel lithology: a new rock type on the lunar farside. J. Geophys. Res., Planets **116**, E00G08 (2011). https://doi.org/10.1029/2010JE003727

T.H. Prettyman, J.J. Hagerty, R.C. Elphic, W.C. Feldman, D.J. Lawrence, G.W. McKinney, D.T. Vaniman, Elemental composition of the lunar surface: analysis of gamma ray spectroscopy data from lunar prospector. J. Geophys. Res., Planets **111**, E12007 (2006)

I.S. Puchtel, R.J. Walker, O.B. James, D.A. Kring, Osmium isotope and highly siderophile element systematic of lunar impact melt breccias: implications for the late accretion history of the Moon and Earth. Geochim. Cosmochim. Acta **72**, 3022–3042 (2008)

Y.Q. Qian, L. Xiao, S.Y. Zhao, J.N. Zhao, J. Huang, J. Flahaut, M. Martinot, J.W. Head, H. Hiesinger, G.X. Wang, Geology and scientific significance of the Rümker region in Northern Oceanus Procellarum: China's Chang'E-5 landing region. J. Geophys. Res., Planets **123**, 1407–1430 (2018)

K. Rankenburg, A.D. Brandon, C.R. Neal, Neodymium isotope evidence for a chondritic composition of the Moon. Science **312**, 1369–1372 (2006)

J.F. Rapp, D.S. Draper, Fractional crystallization of the lunar magma ocean: updating the dominant paradigm. Meteorit. Planet. Sci. **53**, 1432–1455 (2018)

J.M. Rhodes, N.J. Hubbard, H. Wiesmann, K.V. Rodgers, J.C. Brannon, B.M. Bansal, Chemistry, classification and petrogenesis of Apollo 17 mare basalts, in *Lunar Planet. Sci. Conf.*, ed. by D.C. Kinsler (1976), pp. 1467–1489

J.M. Rhodes, D.P. Blanchard, M.A. Dungan, J.C. Brannon, K.V. Rodgers, Chemistry of Apollo 12 mare basalts: magma types and fractionation processes, in *Lunar Planet. Sci. Conf.*, ed. by R.B. Merril (1977), pp. 1305–1338

K. Righter, M.J. Drake, A magma ocean on Vesta: core formation and petrogenesis of eucrites and diogenites. Meteorit. Planet. Sci. **32**, 929–944 (1997)

K. Righter, C.K. Shearer, Magmatic fractionation of Hf and W: constraints on the timing of core formation and differentiation in the Moon and Mars. Geochim. Cosmochim. Acta **67**, 2497–2507 (2003)

K. Righter, M.J. Drake, G. Yaxley, Prediction of siderophile element metal-silicate partition coefficients to 20 GPa and 2800°C: the effects of pressure, temperature, oxygen fugacity, and silicate and metallic melt compositions. Phys. Earth Planet. Inter. **100**, 115–134 (1997)

S.J. Robbins, New crater calibrations for the lunar crater-age chronology. Earth Planet. Sci. Lett. **403**, 188–198 (2014)

M.S. Robinson, J.W. Ashley, A.K. Boyd, R.V. Wagner, E.J. Speyerer, B. Ray Hawke, H. Hiesinger, C.H. van der Bogert, Confirmation of sublunarean voids and thin layering in mare deposits. Planet. Space Sci. **69**, 18–27 (2012a)

K.L. Robinson, A.H. Treiman, K.H. Joy, Basaltic fragments in lunar feldspathic meteorites: connecting sample analyses to orbital remote sensing. Meteorit. Planet. Sci. **47**, 387–399 (2012b)

S.S. Russell, K.H. Joy, T.E. Jeffries, G.J. Consolmagno, A.T. Kearsley, Heterogeneity in lunar anorthosite meteorites: implications for the lunar magma ocean model. Philos. Trans. R. Soc. A **372**, 20130241 (2014). https://doi.org/10.1098/rsta.2013.0241

M.J. Rutherford, J.W. Head III, A.E. Saal, E.H. Hauri, L. Wilson, Model for the origin, ascent and eruption of lunar picritic magmas. Am. Mineral. **102**, 2045–2053 (2017)

G. Ryder, Lunar samples, lunar accretion, and the early bombardment history of the Moon. Eos **71**, 322–323 (1990)

A.E. Saal, E.H. Hauri, M.L. Cascio, J.A. Van Orman, M.C. Rutherford, R.F. Cooper, Volatile content of lunar volcanic glasses and the presence of water in the Moon's interior. Nature **454**, 192–195 (2008)

P. Saxena, R.M. Killen, V. Airapetian, N.E. Petro, N.M. Curran, A.M. Mandell, Was the Sun a Slow Rotator? Sodium and Potassium Constraints from the Lunar Regolith. Astrophys. J. Lett. **876**, 1 (2019). https://doi.org/10.3847/2041-8213/ab18fb

N. Schorghofer, G.J. Taylor, Subsurface migration of H_2O at lunar cold traps. J. Geophys. Res., Planets **112**, E02010 (2007). https://doi.org/10.1029/2006JE002779

P.H. Schultz, P.D. Spudis, Evidence for ancient mare volcanism, in *Lunar Planet. Sci. Conf.*, ed. by N.W. Hinners (1979), pp. 2899–2918

P.H. Schultz, P.D. Spudis, Beginning and end of lunar mare volcanism. Nature **302**, 233–236 (1983)

P.H. Schultz, B. Hermalyn, A. Colaprete, K. Ennico, M. Shirley, W.S. Marshall, The LCROSS cratering experiment. Science **330**, 468–472 (2010)

M. Sharp, I. Gerasimenko, L.C. Loudin, J. Liu, O.B. James, I.S. Puchtel, R.J. Walker, Characterization of the dominant impactor signature for Apollo 17 impact melt rocks. Geochim. Cosmochim. Acta **131**, 62–80 (2014)

E.K. Shea, B.P. Weiss, W.S. Cassata, D.L. Shuster, S.M. Tikoo, J. Gattacceca, T.L. Grove, M.D. Fuller, A long-lived lunar core dynamo. Science **335**, 453–456 (2012)

C.K. Shearer, P.C. Hess, M.A. Wieczorek, M.E. Pritchard, E.M. Parmentier, L.E. Borg, J. Longhi, L.T. Elkins-Tanton, C.R. Neal, I. Antonenko, R.M. Canup, A.N. Halliday, T.L. Grove, B.H. Hager, D.C. Lee, U. Wiechert, Thermal and magmatic evolution of the Moon. Rev. Mineral. Geochem. **60**, 365–518 (2006)

J.W. Shervais, J.J. McGee, Ion and electron microprobe study of troctolites, norite, and anorthosites from Apollo 14: evidence for urKREEP assimilation during petrogenesis of Apollo 14 Mg-suite rocks. Geochim. Cosmochim. Acta **62**, 3009–3023 (1998)

J.W. Shervais, J.J. McGee, KREEP cumulates in the western lunar highlands: ion and electron microprobe study of alkali-suite anorthosites and norites from Apollo 12 and 14. Am. Mineral. **84**, 806–820 (1999)

E.M. Shoemaker, R.J. Hackman, Stratigraphic basis for a lunar time scale, in *The Moon*, ed. by Z. Kopal, Z.K. Mikhailov (Academic, San Diego, 1962), pp. 289–300

A. Smith, I.A. Crawford, R.A. Gowen, R. Ambrosi, M. Anand, B. Banerdt, N. Bannister, N. Bowles, C. Braithwaite, P. Brown, J. Chela-Flores, T. Cholinser, P. Church, A.J. Coates, T. Colaprete, G. Collins, G. Collinson, T. Cook, R. Elphic, G. Fraser, Y. Gao, E. Gibson, T. Glotch, M. Grande, A. Griffiths, J. Grygorczuk, M. Gudipati, A. Hagermann, J. Heldmann, L.L. Hood, A.P. Jones, K.H. Joy, O.B. Khavroshkin, G. Klingelhoefer, M. Knapmeyer, G. Kramer, D. Lawrence, W. Marczewski, S. McKenna-Lawlor, K. Miljkovic, S. Narendranath, E. Palomba, A. Phipps, W.T. Pike, D. Pullan, J. Rask, D.T. Richard, K. Seweryn, S. Sheridan, M. Sims, M. Sweeting, T. Swindle, D. Talboys, L. Taylor, N. Teanby, V. Tong, S. Ulamec, R. Wawrzaszek, M. Wieczorek, L. Wilson, I. Wright, Lunar Net—a proposal in response to an ESA M3 call in 2010 for a medium sized mission. Exp. Astron. **33**, 587–644 (2012)

J.F. Snape, A.A. Nemchin, J.J. Bellucci, M.J. Whitehouse, R. Tartèse, J.J. Barnes, M. Anand, I.A. Crawford, K.H. Joy, Lunar basalt chronology, mantle differentiation and implications for determining the age of the Moon. Earth Planet. Sci. Lett. **451**, 149–158 (2016a)
J.F. Snape, A.A. Nemchin, M.L. Grange, J.J. Bellucci, F. Thiessen, M.J. Whitehouse, Phosphate ages in Apollo 14 breccias: resolving multiple impact events with high precision U–Pb SIMS analyses. Geochim. Cosmochim. Acta **174**, 13–29 (2016b)
J.F. Snape, N.M. Curran, M.J. Whitehouse, A.A. Nemchin, K.H. Joy, T. Hopkinson, M. Anand, J.J. Bellucci, G.G. Kenny, Ancient volcanism on the Moon: insights from Pb isotopes in the MIL 13317 and Kalahari 009 lunar meteorites. Earth Planet. Sci. Lett. **502**, 84–95 (2018a)
J.F. Snape, B. Davids, A.A. Nemchin, M.J. Whitehouse, J.J. Bellucci, Constraining the timing and sources of volcanism at the Apollo 12 landing site using new Pb isotopic compositions and crystallisation ages. Chem. Geol. **482**, 101–112 (2018b)
J.F. Snape, A.A. Nemchin, M.J. Whitehouse, R.E. Merle, T. Hopkinson, M. Anand, The timing of basaltic volcanism at the Apollo landing sites. Geochim. Cosmochim. Acta (2019). https://doi.org/10.1016/j.gca.2019.07.042
G.A. Snyder, L.A. Taylor, C.R. Neal, A chemical model for generating the sources of mare basalts—combined equilibrium and fractional crystallization of the lunar magmasphere. Geochim. Cosmochim. Acta **56**, 3809–3823 (1992)
G.A. Snyder, D.C. Lee, L.A. Taylor, A.N. Halliday, E.A. Jerde, Evolution of the upper mantle of the Earth's Moon: neodymium and strontium isotopic constraints from high-Ti mare basalts. Geochim. Cosmochim. Acta **58**, 4795–4808 (1994)
G.A. Snyder, C.R. Neal, L.A. Taylor, A.N. Halliday, Anatexis of lunar cumulate mantle in time and space: clues from trace-element, strontium, and neodymium isotopic chemistry of parental Apollo 12 basalts. Geochim. Cosmochim. Acta **61**, 2731–2747 (1997)
A.K. Sokol, V.A. Fernandes, T. Schulz, A. Bischoff, R. Burgess, R.N. Clayton, C. Münker, K. Nishiizumi, H. Palme, L. Schultz, G. Weckwerth, K. Mezger, M. Horstmann, Geochemistry, petrology and ages of the lunar meteorites Kalahari 008 and 009: new constraints on early lunar evolution. Geochim. Cosmochim. Acta **72**, 4845–4873 (2008)
S.C. Solomon, J.W. Head, Lunar Mascon Basins: lava filling, tectonics, and evolution of the lithosphere. Rev. Geophys. **18**, 107–141 (1980)
C.P. Sonett, D.S. Colburn, R.G. Currie, J.D. Mihalov, The geomagnetic tail topology, reconnection and interaction with the Moon, in *Physics of the Magnetosphere*, ed. by R.L. Carovillano, J.F. McClay, H.F. Radoski (Reidel, Dordrecht, 1967), pp. 461–484
P.A. Sossi, F. Moynier, M. Chaussidon, J. Villeneuve, C. Kato, M. Gounelle, Early Solar System irradiation quantified by linked vanadium and beryllium isotope variations in meteorites. Nat. Astron. **1**, 0055 (2017). https://doi.org/10.1038/s41550-017-0055
T. Spohn, W. Konrad, D. Breuer, R. Ziethe, The longevity of lunar volcanism: implications of thermal evolution calculations with 2D and 3D mantle convection models. Icarus **149**, 54–65 (2001)
P.D. Spudis, *The Once and Future Moon* (Smithsonian Inst. Press, Washington, 1996)
P.D. Spudis, D.E. Wilhelms, M.S. Robinson, The sculptured hills of the taurus highlands: implications for the relative age of serenitatis, basin chronologies and the cratering history of the Moon. J. Geophys. Res., Planets **116**, E00H03 (2011). https://doi.org/10.1029/2011JE003903
P.D. Spudis, D.B.J. Bussey, S.M. Baloga, J.T.S. Cahill, L.S. Glaze, G.W. Patterson, R.K. Raney, T.W. Thompson, B.J. Thomson, E.A. Ustinov, Evidence for water ice on the Moon: results for anomalous polar craters from the LRO mini-RF imaging radar. J. Geophys. Res., Planets **118**, 2016–2029 (2013)
A.C. Stadermann, M.R. Zanetti, B.L. Jolliff, H. Hiesinger, C.H. van der Bogert, C.W. Hamilton, The age of lunar mare basalts south of the Aristarchus Plateau and effects of secondary craters formed by the Aristarchus event. Icarus **309**, 45–60 (2018)
A. Stettler, P. Eberhardt, J. Geiss, N. Grögler, P. Maurer, Ar^{39}-Ar^{40} ages and Ar^{37}-Ar^{38} exposure ages of lunar rocks, in *Lunar Planet. Sci. Conf.* (1973), pp. 1865–1888
D. Stöffler, G. Ryder, Stratigraphy and isotope ages of lunar geologic units: chronological standard for the inner Solar System. Space Sci. Rev. **96**, 9–54 (2001)
D. Stöffler, G. Ryder, B.A. Ivanov, N.A. Artemieva, M.J. Cintala, R.A.F. Grieve, Cratering history and lunar chronology. Rev. Mineral. Geochem. **60**, 519–596 (2006)
D.E. Stuart-Alexander, D.E. Wilhelms, The Nectarian system, a new lunar time-stratigraphic unit. J. Res. U.S. Geol. Surv. **3**, 53–58 (1975)
C. Suavet, B.P. Weiss, W.S. Cassata, D.L. Shuster, J. Gattacceca, L. Chan, I. Garrick-Bethell, J.W. Head, T.L. Grove, M.D. Fuller, Persistence and origin of the lunar core dynamo. Proc. Natl. Acad. Sci. USA **110**, 8453–8456 (2013)
J.M. Sunshine, T.L. Farnham, L.M. Feaga, O. Groussin, F. Merlin, R.E. Milliken, M.F. A'Hearn, Temporal and spatial variability of Lunar hydration as observed by the Deep Impact spacecraft. Science **326**, 565–568 (2009)

F. Takahashi, H. Tsunakawa, H. Shimizu, H. Shibuya, M. Matsushima, Reorientation of the early lunar pole. Nat. Geosci. **7**, 409–412 (2014)

M. Tatsumoto, Age of the Moon: an isotopic study of U-Th-Pb systematics of Apollo 11 lunar samples—II, in *Apollo 11 Lunar Sci. Conf. Proc.* (1970), pp. 1595–1612

L.A. Taylor, J.W. Shervais, R.H. Hunter, C.-Y. Shih, B.M. Bansal, J. Wooden, L.E. Nyquist, L.C. Laul, Pre-4.2 AE mare-basalt volcanism in the lunar highlands. Earth Planet. Sci. Lett. **66**, 33–47 (1983)

S.R. Taylor, C.M. Pieters, G.J. MacPherson, Earth-Moon system, planetary science, and lessons learned. Rev. Mineral. Geochem. **60**, 657–704 (2006)

D.J. Taylor, K.D. McKeegan, T.M. Harrison, Lu-Hf zircon evidence for rapid lunar differentiation. Earth Planet. Sci. Lett. **279**, 157–164 (2009)

F. Tera, G.J. Wasserburg, U-Th-Pb systematics on lunar rocks and inferences about lunar evolution and the age of the Moon, in *Lunar Sci. Conf. Proc.* (1974), pp. 1571–1599

F. Tera, D.A. Papanastassiou, G.J. Wasserburg, Isotopic evidence for a terminal lunar cataclysm. Earth Planet. Sci. Lett. **22**, 1–21 (1974)

K. Terada, M. Anand, A.K. Sokol, A. Bischoff, Y. Sano, Cryptomare magmatism 4.35 Gyr ago recorded in lunar meteorite Kalahari 009. Nature **450**, 849–853 (2007a)

K. Terada, Y. Sasaki, M. Anand, K.H. Joy, Y. Sano, Uranium-lead systematics of phosphates in lunar basaltic regolith breccia, Meteorite Hills 01210. Earth Planet. Sci. Lett. **259**, 77–84 (2007b)

M.H. Thiemens, R.N. Clayton, Ancient solar wind in lunar microbreccias. Earth Planet. Sci. Lett. **47**, 34–42 (1980)

M.M. Thiemens, P. Sprung, R.O.C. Fonseca, F.P. Leitzke, C. Münker, Early Moon formation inferred from hafnium-tungsten systematics. Nat. Geosci. (2019). https://doi.org/10.1038/s41561-019-0398-3

F. Thiessen, A.A. Nemchin, J.F. Snape, M.J. Whitehouse, J.J. Bellucci, Impact history of the Apollo 17 landing site revealed by U-Pb SIMS ages. Meteorit. Planet. Sci. **52**, 584–611 (2017)

S.M. Tikoo, B.P. Weiss, J. Buz, E.A. Lima, E.K. Shea, G. Melo, T.L. Grove, Magnetic fidelity of lunar samples and implications for an ancient core dynamo. Earth Planet. Sci. Lett. **337–338**, 93–103 (2012)

S.M. Tikoo, B.P. Weiss, W.S. Cassata, D.L. Shuster, J. Gattacceca, C. Suavet, F. Nimmo, M.D. Fuller, Decline of the lunar core dynamo. Earth Planet. Sci. Lett. **404**, 89–97 (2014)

S.M. Tikoo, B.P. Weiss, D.L. Shuster, C. Suavet, H. Wang, T.L. Grove, A two-billion-year history for the lunar dynamo. Sci. Adv. **3**, e1700207 (2017). https://doi.org/10.1126/sciadv.1700207

G.R. Tilton, J.H. Chen, Lead isotope systematics of three Apollo 17 mare basalts, in *Lunar Planet. Sci. Conf. Proc.* (1979), pp. 259–274

M. Touboul, I.S. Puchtel, R.J. Walker, Tungsten isotopic evidence for disproportional late accretion to the Earth and Moon. Nature **520**, 530–533 (2015)

A.H. Treiman, A.K. Maloy, C.K. Shearer Jr., J. Gross, Magnesian anorthositic granulites in lunar meteorites Allan Hills A81005 and Dhofar 309: geochemistry and global significance. Meteorit. Planet. Sci. **45**, 163–180 (2010)

G. Turner, Argon-40/Argon-39 dating of lunar rock samples. Science **167**, 466–468 (1970a)

G. Turner, Argon-40/argon-39 dating of lunar rock samples, in *Lunar Sci. Conf. Proc.* (1970b), pp. 1665–1684

G. Turner, A Monte Carlo fragmentation model for the production of meteorites: implications for gas retention ages, in *Lunar Planet. Sci. Conf. Proc.* (1979), pp. 1917–1941

G. Turner, P.H. Cadogan, C.J. Yonge, Argon selenochronology, in *Lunar Sci. Conf. Proc.* (1973), pp. 1889–1914

N. Vogel, V.S. Heber, H. Baur, D.S. Burnett, R. Wieler, Argon, krypton, and xenon in the bulk solar wind as collected by the Genesis mission. Geochim. Cosmochim. Acta **75**, 3057–3071 (2011)

R.V. Wagner, D.M. Nelson, J.B. Plescia, M.S. Robinson, E.J. Speyerer, E. Mazarico, Coordinates of anthropogenic features on the Moon. Icarus **283**, 92–103 (2017)

R.J. Walker, Highly siderophile elements in the Earth, Moon and Mars: update and implications for planetary accretion and differentiation. Chem. Erde Geochem. **69**, 101–125 (2009)

A. Wallner, J. Feige, N. Kinoshita, M. Paul, L.K. Fifield, R. Golser, M. Honda, U. Linnemann, H. Matsuzaki, S. Merchel, G. Rugel, S.G. Tims, P. Steier, T. Yamagata, S.R. Winkler, Recent near-Earth supernovae probed by global deposition of interstellar radioactive ^{60}Fe. Nature **532**, 69–72 (2016)

R.D. Warner, G.J. Taylor, G.H. Conrad, H.R. Northrop, S. Barker, K. Keil, M.-S. Ma, R. Schmitt, Apollo 17 high-Ti mare basalts: new bulk compositional data, magma types, and petrogenesis, in *Lunar Planet. Sci. Conf. Proc.*, ed. by N.W. Hinners (1979), pp. 225–247

P.H. Warren, J.T. Wasson, The origin of KREEP. Rev. Geophys. **17**, 73–88 (1979)

K. Watson, B.C. Murray, H. Brown, The behavior of volatiles on the lunar surface. J. Geophys. Res. **66**, 3033–3045 (1961)

T.R. Watters, R.C. Weber, G.C. Collins, I.J. Howley, N.C. Schmerr, C.L. Johnson, Shallow seismic activity and young thrust faults on the Moon. Nat. Geosci. **12**, 411–417 (2019)

R.C. Weber, Interior of the Moon, in *Encyclopedia of the Solar System* (2014), pp. 539–554. https://doi.org/10.1016/B978-0-12-415845-0.00024-4. Chap. 24

R.C. Weber, P.Y. Lin, E.J. Garnero, Q. Williams, P. Lognonné, Seismic detection of the lunar core. Science **331**, 309–312 (2011)

B.P. Weiss, S.M. Tikoo, The lunar dynamo. Science **346**, 1246753-1–1246753-10 (2014). https://doi.org/10.1126/science.1246753

S.C. Werner, Moon, Mars, Mercury: basin formation ages and implications for the maximum surface age and the migration of gaseous planets. Earth Planet. Sci. Lett. **400**, 54–65 (2014)

J.L. Whitten, J.W. Head III, Lunar cryptomaria: physical characteristics, distribution, and implications for ancient volcanism. Icarus **247**, 150–171 (2015a)

J.L. Whitten, J.W. Head III, Lunar cryptomaria: mineralogy and composition of ancient volcanic deposits. Planet. Space Sci. **106**, 67–81 (2015b)

J. Whitten, J.W. Head III, M.I. Staid, C.M. Pieters, J.F. Mustard, R. Clark, J.W. Nettles, R.L. Klima, L.A. Taylor, Lunar mare deposits associated with the Orientale impact basin: New insights into mineralogy, history, mode of emplacement, and relation to Orientale Basin evolution from Moon Mineralogy Mapper (M3) data from Chandrayaan-1. J. Geophys. Res. **116**, E00G09 (2011). https://doi.org/10.1029/2010JE003736

M.A. Wieczorek, B.L. Jolliff, A. Khan, M.E. Pritchard, B.P. Weiss, J.G. Williams, L.L. Hood, K. Righter, C.R. Neal, C.K. Shearer, I.S. McCallum, S. Tompkins, B.R. Hawke, C. Peterson, J.J. Gillis, B. Bussey, The constitution and structure of the lunar interior. Rev. Mineral. Geochem. **60**, 221–364 (2006)

M.A. Wieczorek, G.A. Neumann, F. Nimmo, W.S. Kiefer, G.J. Taylor, H.J. Melosh, R.J. Phillips, S.C. Solomon, J.C. Andrews-Hanna, S.W. Asmar, A.S. Konopliv, F.G. Lemoine, D.E. Smith, M.M. Watkins, J.G. Williams, M.T. Zuber, The crust of the Moon as seen by GRAIL. Science **339**, 671–675 (2013)

R. Wieler, Do lunar and meteoritic archives record temporal variations in the composition of solar wind noble gases and nitrogen? A reassessment in the light of Genesis data. Chem. Erde **76**, 463–480 (2016)

R. Wieler, V.S. Heber, Noble gas isotopes on the Moon. Space Sci. Rev. **106**, 197–210 (2003)

R. Wieler, K. Kehm, A.P. Meshik, C.M. Hohenberg, Secular changes in the xenon and krypton abundances in the solar wind recorded in single lunar grains. Nature **384**, 46–49 (1996)

D.E. Wilhelms, The geologic history of the Moon. USGS Prof. Pap. **1348** (1987)

L. Wilson, J.W. Head III, Generation, ascent and eruption of magma on the Moon: new insights into source depths, magma supply, intrusions and effusive/explosive eruptions (part 1: theory). Icarus **283**, 146–175 (2017)

L. Wilson, J.W. Head III, Controls on lunar basaltic volcanic erupton structure and morphology: gas release patterns in sequential eruption phases. Geophys. Res. Lett. **45**, 5852–5859 (2018)

S.F. Wolf, M.S. Wang, M.E. Lipschutz, Labile trace elements in basaltic achondrites: can they distinguish between meteorites from the Moon, Mars, and V-type asteroids? Meteorit. Planet. Sci. **44**, 891–903 (2009)

J.A. Wood, Bombardment as a cause of the lunar asymmetry. Moon **8**, 73–103 (1973)

J.A. Wood, J.S. Dickey, U.B. Marvin, B.N. Powell, Lunar anorthosites and a geophysical model of the Moon, in *Lunar Sci. Conf. Proc.* (1970), pp. 965–988

S. Yamamoto, R. Nakamura, T. Matsunaga, Y. Ogawa, Y. Ishihara, T. Morota, N. Hirata, M. Ohtake, T. Hiroi, Y. Yokota, J. Haruyama, Possible mantle origin of olivine around lunar impact basins detected by SELENE. Nat. Geosci. **3**, 533–536 (2010)

S. Yamamoto, R. Nakamura, T. Matsunaga, Y. Ogawa, Y. Ishihara, T. Morota, N. Hirata, M. Ohtake, T. Hiroi, Y. Yokota, J. Haruyama, Featureless spectra on the Moon as evidence of residual lunar primordial crust. J. Geophys. Res., Planets **120**, 2190–2205 (2015)

N.E.B. Zellner, Cataclysm no more: new views on the timing and delivery of lunar impactors. Orig. Life Evol. Biosph. **47**, 261–280 (2017)

J. Zhao, L. Xiao, L. Qiao, T.D. Glotch, Q. Huang, The Mons Rümker volcanic complex of the Moon: a candidate landing site forthe Chang'e-5 mission. J. Geophys. Res., Planets **122**, 1419–1442 (2017)

M.H. Zhu, K. Wünnemann, R.W.K. Potter, T. Kleine, A. Morbidelli, Are the Moon's nearside-farside asymmetries the result of a giant impact? J. Geophys. Res., Planets (2019). https://doi.org/10.1029/2018JE005826

M.T. Zuber, D.E. Smith, F.G. Lemoine, G.A. Neumann, The shape and internal structure of the Moon from the Clementine mission. Science **266**, 1839–1843 (1994)

M.T. Zuber, D.E. Smith, M.M. Watkins, S.W. Asmar, A.S. Konopliv, F.G. Lemoine, H.J. Melosh, G.A. Neumann, R.J. Phillips, S.C. Solomon, M.A. Wieczorek, J.G. Williams, S.J. Goossens, G. Kruizinga, E. Mazarico, R.S. Park, D.N. Yuan, Gravity Field of the Moon from the Gravity Recovery and Interior Laboratory (GRAIL) Mission. Science **339**, 668–671 (2013)

Space Sci Rev (2020) 216:55
https://doi.org/10.1007/s11214-020-00675-w

The Non-carbonaceous–Carbonaceous Meteorite Dichotomy

T. Kleine[1] · G. Budde[1] · C. Burkhardt[1] · T.S. Kruijer[2] · E.A. Worsham[1] · A. Morbidelli[3] · F. Nimmo[4]

Received: 10 July 2019 / Accepted: 17 April 2020 / Published online: 12 May 2020

Abstract The isotopic dichotomy between non-carbonaceous (NC) and carbonaceous (CC) meteorites indicates that meteorite parent bodies derive from two genetically distinct reservoirs, which presumably were located inside (NC) and outside (CC) the orbit of Jupiter and remained isolated from each other for the first few million years of the solar system. Here we review the discovery of the NC–CC dichotomy and its implications for understanding the early history of the solar system, including the formation of Jupiter, the dynamics of terrestrial planet formation, and the origin and nature of Earth's building blocks. The isotopic difference between the NC and CC reservoirs is probably inherited from the solar system's parental molecular cloud and has been maintained through the rapid formation of Jupiter that prevented significant exchange of material from inside (NC) and outside (CC) its orbit. The growth and/or migration of Jupiter resulted in inward scattering of CC bodies, which accounts for the co-occurrence of NC and CC bodies in the present-day asteroid belt and the delivery of presumably volatile-rich CC bodies to the growing terrestrial planets. Earth's primitive mantle, at least for siderophile elements like Mo, has a mixed NC–CC composition, indicating that Earth accreted CC bodies during the final stages of its growth, perhaps through the Moon-forming giant impactor. The late-stage accretion of CC bodies to Earth is sufficient to account for the entire budget of Earth's water and highly volatile species.

Keywords Isotope anomalies · NC–CC dichotomy · Meteorites · Jupiter · Earth's accretion · Solar system dynamics

Role of Sample Return in Addressing Major Questions in Planetary Sciences
Edited by Mahesh Anand, Sara Russell, Yangting Lin, Meenakshi Wadhwa, Kuljeet Kaur Marhas and Shogo Tachibana

✉ T. Kleine
thorsten.kleine@uni-muenster.de

[1] Institut für Planetologie, University of Münster, Wilhelm-Klemm-Str. 10, 48149 Münster, Germany

[2] Lawrence Livermore National Laboratory, Nuclear and Chemical Sciences Division, 7000 East Avenue, Livermore, CA 94550, USA

[3] Observatoire de la Cote d'Azur, CS 34229, 06304 Nice Cedex 4, France

[4] Department of Earth and Planetary Sciences, University of California Santa Cruz, 1156 High Street, Santa Cruz, CA 95064, USA

1 Introduction

Most meteorites are fragments of asteroids located between Mars and Jupiter. It has traditionally been thought that these bodies formed close to their present-day location, but this view was challenged recently by the discovery of an isotopic dichotomy between non-carbonaceous (NC) and carbonaceous (CC) meteorites (Budde et al. 2016; Warren 2011). These two suites of meteorites derive from distinct reservoirs within the solar protoplanetary disk that remained spatially separated for several million years (Ma), most likely through the presence of Jupiter in between them (Kruijer et al. 2017). Independent of the discovery of the isotope dichotomy, dynamical models aimed at explaining the small mass of Mars suggested that S-type and C-type asteroids represent bodies from inside and outside Jupiter's orbit, respectively, and that these bodies were implanted into the present-day asteroid belt through the inward-then-outward migration of Jupiter and Saturn (the so-called 'Grand Tack' model) (Walsh et al. 2011). Later work showed that planetesimals from beyond Jupiter may have also been implanted into the asteroid belt during the growth of the giant planets, without the need to invoke a specific planetary migration pattern (Raymond and Izidoro 2017). These models are remarkably consistent with the isotopic dichotomy observed among meteorites, and collectively these studies indicate that S-type asteroids and NC meteorites represent inner solar system material, whereas C-type asteroids and CC meteorites derive from the outer solar system.

The purpose of this paper is to review the meteorite dichotomy and its origin, and discuss its implications for understanding the early evolution of the solar accretion disk, the role of Jupiter in the early solar system, the formation of planetesimals, and the accretion history and building blocks of the terrestrial planets.

2 Nucleosynthetic Isotope Anomalies in Meteorites

The distinction between non-carbonaceous (NC) and carbonaceous (CC) meteorites is based on nucleosynthetic isotope anomalies. We will therefore start with briefly reviewing the definition, notation and origin of these isotope anomalies, with a particular focus on those that are fundamental to the definition of the isotope dichotomy. Nucleosynthetic isotope anomalies are characterized by enrichments or depletions in nuclides formed by particular stellar nucleosynthetic processes. As such, these anomalies are distinct from mass-dependent isotope fractionations, which result from physical or chemical processes that separate the isotopes according to their mass. Isotope fractionations can also be non-mass-dependent (sometimes also called mass-independent) when they depart from the accepted mass fractionation law. For instance, O isotope variations among meteorites reveal both mass-dependent and non-mass-dependent isotope fractionations, where the latter are typically expressed as $\Delta^{17}O$ as the deviation from the terrestrial mass fractionation line.

Nucleosynthetic isotope anomalies are not the result of isotope fractionation, but instead arise from the heterogeneous distribution of presolar matter derived from multiple nucleosynthetic sources. Some of this material has been identified in the matrix of primitive chondrites as presolar grains, whose isotopic compositions deviate strongly from the solar composition and reflect the particular stellar environment in which these grains formed (Zinner 2014). Isotope anomalies in other meteoritic materials are much smaller than those found in presolar grains, attesting to the near complete homogenization of presolar matter in the parental molecular cloud or the solar nebula itself. Whereas the isotopic composition of presolar grains is commonly reported as absolute isotope ratio, the much smaller isotope anomalies in meteoritic and planetary materials are commonly reported as variations in isotope ratios that are normalized to another ratio of the same element, for which a fixed, terrestrial value is assumed. This normalization is necessary to correct for the effects of natural and instrumental mass fractionation, and it is typically done using the exponential mass fractionation law. Further, as the isotope variations in meteoritic and planetary samples are typically small, they are commonly reported in the ε-unit notation as parts-per-ten-thousand deviations from terrestrial standard values. For instance, Mo isotope anomalies are reported as variations in the $^{i}Mo/^{96}Mo$ ratio, which are normalized to the terrestrial $^{98}Mo/^{96}Mo$ and are given as $\varepsilon^{i}Mo$ values ($i = 92, 94, 95, 97, 100$). Sometimes the anomalies are also given as μ-values ($1\varepsilon = 100\mu$).

For most elements nucleosynthetic anomalies are most prominent in Ca-Al-rich inclusions (CAIs), the solar system's oldest dated solids (Amelin et al. 2010; Connelly et al. 2012). Of these, FUN (Fractionated and Unknown Nuclear effects) CAIs, platy hibonite crystals (PLACs) and spinel hibonite inclusions (SHIBs) typically have the largest nucleosynthetic anomalies, whereas the more common 'normal' CAIs display smaller anomalies (for an overview see Dauphas and Schauble 2016). Most of the initial work on nucleosynthetic anomalies in meteoritic materials was focused on CAIs, because only for these could anomalies be clearly resolved (e.g., Birck and Lugmair 1988; Lee et al. 1978; McCulloch and Wasserburg 1978; Niederer et al. 1981; Niemeyer and Lugmair 1981; Papanastassiou 1986). Nevertheless, some of these earlier studies already provided hints for nucleosynthetic isotope anomalies in other components of primitive chondrites (e.g., chondrules) and also showed that bulk carbonaceous chondrites are characterized by excesses in the neutron-rich isotopes of ^{50}Ti and ^{54}Cr (Niederer et al. 1985; Niemeyer 1988; Niemeyer and Lugmair 1984; Rotaru et al. 1992). With improvements in analytical techniques over the past two decades, nucleosynthetic isotope anomalies have now been identified for many elements, and it has been shown that they exist not only in bulk carbonaceous chondrites, but also in other groups of meteorites (e.g., Burkhardt et al. 2011; Dauphas et al. 2002; Leya et al. 2008; Regelous et al. 2008; Trinquier et al. 2007, 2009). Because nucleosynthetic isotope anomalies are widespread, they are powerful tracers for inferring genetic relationships between and among meteorites and planets (Burkhardt et al. 2016, 2011; Dauphas 2017; Render et al. 2017; Warren 2011), and for assessing transport and mixing processes in the solar protoplanetary disk (Ebert et al. 2018; Gerber et al. 2017; Olsen et al. 2016; Render et al. 2019; Van Kooten et al. 2016). As such, nucleosynthetic anomalies constitute a powerful tool to constrain models of planet formation.

The processes that led to a heterogeneous distribution of presolar matter in the solar accretion disk are not fully understood and several mechanisms have been proposed. These include inherited heterogeneity from the solar system's parental molecular cloud (Burkhardt et al. 2019; Dauphas et al. 2002; Nanne et al. 2019), grain type- or size-sorting in the solar protoplanetary disk (Dauphas et al. 2008; Regelous et al. 2008), thermal processing of dust in the disk (Burkhardt et al. 2012; Poole et al. 2017; Trinquier et al. 2009), and late injection of diverse nucleosynthetic components (Brennecka et al. 2013). These processes are not mutually exclusive and may all have played a role.

Although nucleosynthetic anomalies in bulk meteorites and meteorite components have been reported for a considerable number of elements, the anomalies in Ti, Cr, Ni, and Mo are particularly important, because they have been instrumental for identifying the isotope dichotomy of meteorites, and for constraining its origin. The original identification of the NC–CC dichotomy also included O isotopes (Fig. 1b), but we note that for $\Delta^{17}O$ there is substantial overlap between NC and CC meteorites, and so the dichotomy is essentially defined by isotope variation in the other elements. This paper will therefore largely be focused

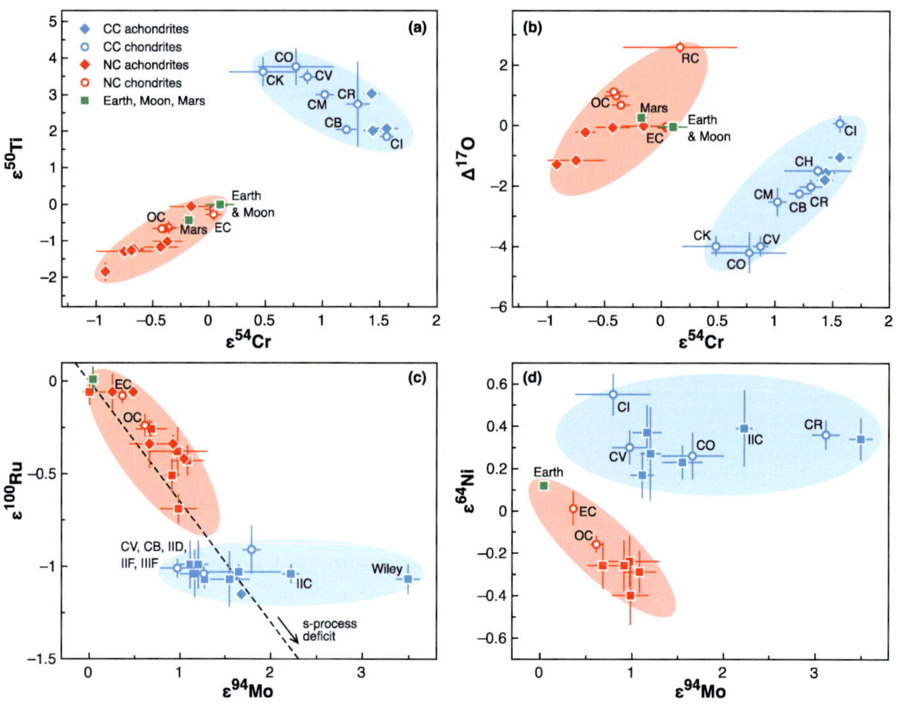

Fig. 1 (**a**) ε^{54}Cr versus ε^{50}Ti and (**b**) ε^{54}Cr versus Δ^{17}O. These two plots were instrumental for the original definition of the NC–CC dichotomy. Note that the O isotope anomalies are *not* thought to be nucleosynthetic in origin, but that they are nevertheless indicative of spatial or temporal changes within the accretion disk (McKeegan et al. 2011). Data taken from the compilations in Burkhardt et al. (2017) and Dauphas and Schauble (2016). (**c**) ε^{94}Mo versus ε^{100}Ru and (**d**) ε^{94}Mo versus ε^{64}Ni. These two plots for siderophile elements are useful for illustrating that the NC–CC dichotomy for iron meteorites extends to several elements. Note that the Ni isotope anomalies most likely reside on ^{58}Ni, but they are shown as apparent variations in ε^{64}Ni, because for this normalization the apparent anomalies are larger and more precisely measured. Data taken from the compilations in Nanne et al. (2019), Worsham et al. (2019), and Hopp et al. (2020). Importantly, in all plots the NC and CC fields are consistently defined by the same meteorite groups. Note the wide gap between the NC (red) and CC (blue) fields

on isotope anomalies for Ti, Cr, Ni, and Mo. The isotope anomalies in Ti and Cr predominantly reflect variations in the neutron-rich isotopes of these elements (i.e., ^{50}Ti, ^{54}Cr), both of which are thought to be produced in supernovae (Leya et al. 2008; Trinquier et al. 2007, 2009; Zhang et al. 2012). For Cr, one potential carrier of these anomalies has been identified as ^{54}Cr-rich nanospinels that probably condensed in the outflow of supernovae (Dauphas et al. 2010; Qin et al. 2011). Such grains may also be strongly enriched in ^{50}Ti (Nittler et al. 2018). Anomalies in Ni appear to have distinct origins in bulk meteorites and CAIs. Whereas the Ni isotope anomalies in CAIs, similar to Cr and Ti, appear to reflect enrichments in the neutron-rich isotopes ^{62}Ni and ^{64}Ni (Birck and Lugmair 1988; Render et al. 2018), those in bulk meteorites largely seem to reflect variations in ^{58}Ni (Nanne et al. 2019; Steele et al. 2012; Tang and Dauphas 2012).

A key element with respect to the NC–CC dichotomy is Mo. The seven stable Mo isotopes are produced by three distinct nucleosynthetic processes, namely the proton (p-) process, the slow neutron capture (s-) process and the rapid neutron capture (r-) process. While the p- and r-processes are thought to occur in neutron-rich stellar environments such as

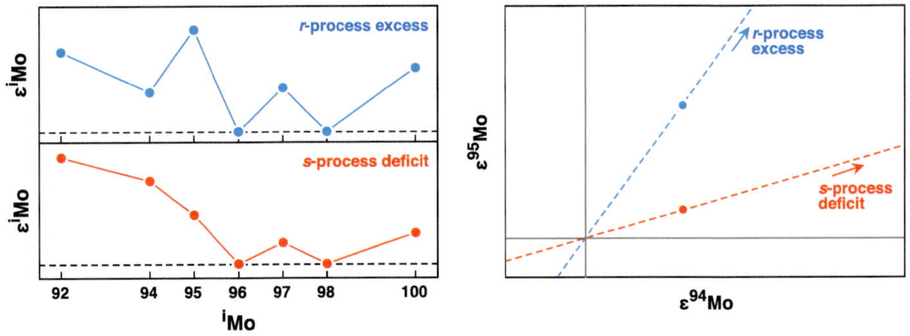

Fig. 2 *Left*: Different Mo isotope patterns for s-process deficit and r-process excess. The differences are most pronounced at ε^{95}Mo and ε^{94}Mo. *Right*: In a plot of ε^{95}Mo versus ε^{94}Mo, s- and r-process variations plot along lines of different slope. Thus, for a given ε^{94}Mo, an r-process excess results in a much larger ε^{95}Mo than does an s-process deficit

supernovae or neutron star mergers, the s-process takes place in AGB stars. Of the stable Mo nuclides, ^{92}Mo is exclusively produced in the p-process, ^{96}Mo is only produced in the s-process, and ^{100}Mo is an almost pure r-process nuclide. Each of the remaining isotopes is produced by two different nucleosynthetic processes: ^{94}Mo (p, s); ^{95}Mo, ^{97}Mo, and ^{98}Mo (s, r). An important feature of Mo isotopes is that the p-process isotopes ^{92}Mo and ^{94}Mo are sufficiently abundant to allow for precise measurement. This in turn makes it possible to distinguish between the heterogeneous distribution of s- and r-process carriers (Burkhardt et al. 2011), which results in distinctly different Mo isotope patterns (Fig. 2). This feature has been essential for identifying the dichotomy between NC and CC meteorites (Budde et al. 2016).

3 Isotopic Dichotomy Between Non-carbonaceous and Carbonaceous Meteorites

3.1 Discovery and Characteristic of the NC–CC Dichotomy

The first hint of what now is recognized as the NC–CC dichotomy was observed by Trinquier et al. (2007). These authors showed that in Δ^{17}O–ε^{54}Cr space, differentiated meteorites and carbonaceous chondrites plot in two distinct fields and on this basis argued that two distinct reservoirs existed in the early solar system. By compiling available isotopic data for Cr, Ti, Ni, and O, Warren (2011) then showed that for these elements meteorites always define two distinct isotopic clusters (Fig. 1a, b). Warren (2011) also introduced the terms non-carbonaceous (NC) and carbonaceous (CC) meteorites to distinguish between these two groups of materials. Importantly, the NC and CC compositional fields are separated by a wide gap, indicating that NC and CC meteorites represent fundamentally distinct materials with little to no mixing between them. On this basis, Warren (2011) suggested that the NC–CC dichotomy may either reflect a temporal change in solar nebula composition (and the later formation of CC bodies), or derivation of NC and CC materials from the inner and outer solar system, respectively.

Budde et al. (2016) subsequently demonstrated that the NC–CC dichotomy also exists for Mo. The dichotomy is most obvious in a plot of ε^{95}Mo versus ε^{94}Mo, in which the difference between s- and r-process variability is most clearly resolved (see above and Fig. 2).

Budde et al. (2016) showed that in this plot, NC and CC meteorites plot on two distinct and approximately parallel lines (termed the NC- and CC-lines). As will be discussed in more detail below (Sect. 3.2), the isotopic variations along both lines reflect the heterogeneous distribution of s-process material, while the offset between these two lines reflects an excess in r-process (and possibly p-process) Mo in the CC over the NC reservoir (Fig. 3). As a siderophile element, and unlike Cr and Ti, Mo can be analyzed in essentially all meteorites, including chondrites and iron meteorites, which derive from bodies that accreted over a time span of ~ 4 million years (Ma) (see below). As such, the discovery of the Mo isotopic dichotomy made it possible to investigate how the compositions of the NC and CC reservoir evolved over time with unprecedented temporal and spatial resolution. The Mo isotope data have, therefore, been key for recognizing that the NC–CC dichotomy reflects the spatial separation of two genetically distinct disk reservoirs, and not merely a temporal change in disk composition (see Sect. 3.4 below). Several subsequent studies have reported high-precision Mo isotopic data for a large set of diverse meteorites, and these studies collectively demonstrate that all known asteroidal meteorites belong to either the NC or the CC group (Bermingham et al. 2018; Budde et al. 2019; Kruijer et al. 2017; Poole et al. 2017; Worsham et al. 2017, 2019; Yokoyama et al. 2019).

Subsequent work demonstrated that the NC–CC dichotomy extends to several other siderophile elements, including W, Ru, and Ni. For instance, CC meteorites are characterized by ^{58}Ni (Nanne et al. 2019; Steele et al. 2012) and ^{183}W excesses (Budde et al. 2018; Kruijer et al. 2017; Worsham et al. 2019), and they typically also have larger ^{100}Ru deficits than NC meteorites (Bermingham et al. 2018; Fischer-Gödde et al. 2015; Fischer-Gödde and Kleine 2017; Hopp et al. 2020; Worsham et al. 2019). When the isotope anomalies for these elements are plotted against each other, meteorites can always be subdivided into two clusters (Fig. 1c, d). Importantly, for all these elements a given meteorite always belongs to the same cluster, indicating that the NC–CC dichotomy is consistently present for a wide range of elements (Fig. 1). For some element pairs, the isotope anomalies are correlated in one reservoir but not the other. For instance, Ru and Mo anomalies are correlated in the NC but not the CC reservoir (Fig. 1c), whereas Mo and W anomalies are correlated in the CC but not the NC reservoir (Worsham et al. 2019). The disparate correlations of isotope anomalies most likely reflect different redox and thermal conditions under which presolar carriers have been processed in the two reservoirs, providing further evidence that NC and CC meteorites represent material from fundamentally different areas of the solar protoplanetary disk (Worsham et al. 2019).

In summary, the dichotomy between NC and CC meteorites universally holds for all known meteorite classes. It exists for a large number of elements of different geochemical and cosmochemical behaviour, including lithophile (Cr, Ti, O) and siderophile elements (Ni, Mo, Ru, W), as well as refractory (Ti, Mo, Ru, W) and non-refractory (Cr, Ni) elements. The NC–CC dichotomy, therefore, is a ubiquitous and fundamental characteristic of early-formed planetary objects in the solar system. Compared to the NC reservoir, the CC reservoir exhibits excesses in ^{50}Ti, ^{54}Cr, ^{58}Ni, and r-process Mo. The CC reservoir, therefore, is characterized by an enrichment in nuclides produced in neutron-rich stellar environments. Finally, for several element pairs, the isotopic compositions of NC and CC meteorites define two clusters that are separated by a wide gap, indicating that the NC and CC reservoirs represent fundamentally different materials with no or very limited mixing between them (Fig. 1). In contrast to most elements where the NC–CC dichotomy is revealed by distinct anomalies in a single isotope ratio, Mo isotopes are uniquely important, because for them the NC–CC dichotomy is clearly resolved in multiple isotope ratios of a single element (Fig. 3).

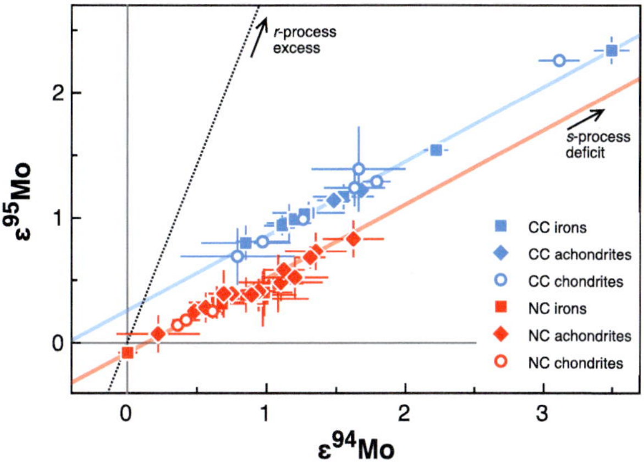

Fig. 3 ε^{95}Mo versus ε^{94}Mo for bulk meteorites. In this plot the difference between s- and r-process isotope variations in most clearly resolved, and so the NC–CC dichotomy for Mo is best illustrated in this plot. Note that the Mo isotope variability along the NC- and CC-lines is predominantly governed by s-process variations, whereas the offset between the two lines reflects an approximately constant r-process excess in the CC over the NC reservoir. Data are from various sources (Table 1)

3.2 Mo Isotopic Characteristics of the NC and CC Reservoirs

Given the importance of Mo isotopes for the NC–CC dichotomy, we will describe the characteristic Mo isotope systematics of NC and CC meteorites in some more detail. As noted above, the isotopic variations along the NC- and CC-lines are predominantly governed by s-process heterogeneity. This conclusion follows from the observation that the slopes of both lines are consistent with the expected slope as derived from isotopic data for presolar mainstream SiC grains. The latter are thought to have condensed in AGB stars, the dominant site of the s-process.

The slope of s-process mixing lines in ε^iMo–ε^{94}Mo plots can be determined from presolar grain data using the formalism of Dauphas et al. (2004):

$$m = \frac{\rho^i\text{Mo} - \rho^{98}\text{Mo} \cdot \mu^i\text{Mo}}{\rho^{94}\text{Mo} - \rho^{98}\text{Mo} \cdot \mu^{94}\text{Mo}}$$

where the parameter ρ^iMo represents the s-process composition relative to the terrestrial Mo isotope composition:

$$\rho^i\text{Mo} = \frac{(^i\text{Mo}/^{96}\text{Mo})_s}{(^i\text{Mo}/^{96}\text{Mo})} - 1$$

and μ^iMo is the mass difference of the isotope ratio relative to that of the normalizing pair used for correction of instrumental mass fractionation:

$$\mu^i\text{Mo} = \frac{i - 96}{98 - 96}$$

Information about the s-process Mo isotope composition is provided by data for mainstream circumstellar SiC grains, which correspond to ρ^iMo values of 1, 0.964, 0.615, 0.532, 0.256, and 0.980 for $i = 92, 94, 95, 97, 98$, and 100 (Stephan et al. 2019).

Budde et al. (2019) showed that the slopes of the NC- and CC-lines are indistinguishable from the predicted slopes of s-process mixing lines. These authors used a comprehensive set of Mo isotopic data for meteorites, including bulk meteorites, meteorite components from carbonaceous chondrites (i.e., chondrules, matrix, and metal samples from primitive chondrites), and acid leachates of primitive carbonaceous and ordinary chondrites. Linear

Table 1 Summary of Mo isotope data for major groups of meteorites[a]

Sample	ε^{92}Mo ($\pm 2\sigma$)	ε^{94}Mo ($\pm 2\sigma$)	ε^{95}Mo ($\pm 2\sigma$)	ε^{97}Mo ($\pm 2\sigma$)	ε^{100}Mo ($\pm 2\sigma$)	Δ^{95}Mo ($\pm 2\sigma$)	t_{accr} (Ma) ($\pm 2\sigma$)[b]	References[c]
Carbonaceous (CC) meteorites								
Chondrites								
CI	1.12 ± 0.59	0.79 ± 0.41	0.69 ± 0.23	0.26 ± 0.19	0.44 ± 0.34	22 ± 34	3.6 ± 0.5	(1)
CM	6.44 ± 0.39	4.82 ± 0.20	3.17 ± 0.16	1.66 ± 0.14	2.28 ± 0.22	30 ± 20	3.6 ± 0.6	(1)
CO	2.41 ± 0.15	1.66 ± 0.34	1.39 ± 0.34	0.71 ± 0.28	0.97 ± 0.11	40 ± 40	2.7 ± 0.2	(2)
CV	1.41 ± 0.27	0.97 ± 0.19	0.81 ± 0.05	0.40 ± 0.08	0.44 ± 0.12	23 ± 12	2.7 ± 0.3	(3)
CR[d]	4.14 ± 0.12	3.11 ± 0.15	2.26 ± 0.04	1.18 ± 0.04	1.40 ± 0.17	23 ± 4	3.6 ± 0.6	(4)
CH	2.32 ± 0.11	1.79 ± 0.10	1.29 ± 0.04	0.62 ± 0.09	0.66 ± 0.11	22 ± 7	3.2 ± 0.3	(5)
CB	1.53 ± 0.08	1.26 ± 0.04	0.99 ± 0.04	0.51 ± 0.04	0.45 ± 0.04	24 ± 5	4.5 ± 0.5	(5)
Achondrites								
Tafassasset (ungr.)	2.24 ± 0.15	1.68 ± 0.06	1.22 ± 0.06	0.62 ± 0.07	0.60 ± 0.03	22 ± 7	1.9 ± 0.2	(5)
Iron meteorites								
IIC	2.87 ± 0.10	2.25 ± 0.10	1.56 ± 0.07	0.83 ± 0.07	0.93 ± 0.12	22 ± 9	1.0 ± 0.2	(6)
IID	1.67 ± 0.11	1.18 ± 0.06	0.99 ± 0.08	0.51 ± 0.01	0.63 ± 0.11	29 ± 9	1.0 ± 0.2	(6)
IIF	1.52 ± 0.06	1.10 ± 0.03	0.96 ± 0.04	0.51 ± 0.03	0.62 ± 0.03	30 ± 5	1.0 ± 0.2	(6)
IIIF	1.62 ± 0.29	1.21 ± 0.05	0.97 ± 0.08	0.50 ± 0.19	0.60 ± 0.03	24 ± 9	1.0 ± 0.2	(6)
IVB	2.27 ± 0.28	1.54 ± 0.10	1.16 ± 0.05	0.55 ± 0.03	0.70 ± 0.09	24 ± 8	1.0 ± 0.2	(7)
Non-carbonaceous (NC) meteorites								
Chondrites								
EH	0.67 ± 0.63	0.50 ± 0.16	0.20 ± 0.10	0.10 ± 0.02	0.27 ± 0.27	-10 ± 14	1.8 ± 0.1	(8, 9)
EL	0.34 ± 0.12	0.33 ± 0.10	0.13 ± 0.06	0.08 ± 0.03	0.12 ± 0.04	-7 ± 9	1.8 ± 0.1	(8, 9)
OC (H, L, LL)	0.78 ± 0.22	0.63 ± 0.08	0.24 ± 0.04	0.12 ± 0.03	0.23 ± 0.11	-14 ± 7	2.1 ± 0.1	(7, 8, 9)
RC	0.45 ± 0.15	0.42 ± 0.10	0.18 ± 0.05	0.08 ± 0.05	0.11 ± 0.12	-7 ± 8	2.1 ± 0.1	(5)

Table 1 (*continued*)

Sample	ε^{92}Mo ($\pm 2\sigma$)	ε^{94}Mo ($\pm 2\sigma$)	ε^{95}Mo ($\pm 2\sigma$)	ε^{97}Mo ($\pm 2\sigma$)	ε^{100}Mo ($\pm 2\sigma$)	Δ^{95}Mo ($\pm 2\sigma$)	t_{accr} (Ma) ($\pm 2\sigma$)[b]	References[c]
Achondrites								
Acapulcoites-lodranites	1.05 ± 0.11	0.92 ± 0.07	0.48 ± 0.03	0.23 ± 0.02	0.24 ± 0.07	-7 ± 5	1.3 ± 0.3	(10)
Winonaites	0.36 ± 0.31	0.25 ± 0.15	0.09 ± 0.09	0.02 ± 0.02	0.03 ± 0.13	-6 ± 13	~ 1.8	(9)
Brachinites	1.33 ± 0.15	1.12 ± 0.15	0.58 ± 0.08	0.33 ± 0.04	0.32 ± 0.09	-9 ± 12		(5)
Ureilites	0.98 ± 0.12	0.89 ± 0.09	0.38 ± 0.04	0.21 ± 0.03	0.15 ± 0.07	-15 ± 7	1.0 ± 0.3	(5)
Angrites	0.78 ± 0.28	0.75 ± 0.11	0.39 ± 0.06	0.26 ± 0.10	0.07 ± 0.12	-6 ± 9	0.5 ± 0.4	(5)
Aubrites	0.56 ± 0.13	0.48 ± 0.05	0.25 ± 0.06	0.19 ± 0.03	0.16 ± 0.10	-4 ± 7	1.5 ± 0.1	(5)
Mesosiderites	1.21 ± 0.14	1.04 ± 0.08	0.46 ± 0.05	0.25 ± 0.03	0.19 ± 0.04	-16 ± 7	0.9 ± 0.3	(10)
Iron meteorites								
IAB (MG, sL)	-0.03 ± 0.30	0.04 ± 0.10	-0.07 ± 0.05	0.00 ± 0.02	0.02 ± 0.03	-9 ± 8	~ 1.4	(11)
IC	0.88 ± 0.23	0.83 ± 0.08	0.36 ± 0.06	0.23 ± 0.08	0.22 ± 0.16	-13 ± 8	0.2 ± 0.2	(6)
IIAB	1.37 ± 0.34	1.15 ± 0.11	0.51 ± 0.06	0.26 ± 0.05	0.30 ± 0.12	-18 ± 9	0.2 ± 0.2	(7)
IIIAB	1.32 ± 0.25	1.06 ± 0.09	0.45 ± 0.06	0.21 ± 0.04	0.41 ± 0.09	-18 ± 8	0.2 ± 0.2	(11, 12)
IIIE	1.05 ± 0.05	0.94 ± 0.02	0.44 ± 0.07	0.29 ± 0.05	0.28 ± 0.11	-12 ± 7	0.2 ± 0.2	(6)
IVA	1.08 ± 0.18	0.79 ± 0.10	0.36 ± 0.05	0.17 ± 0.04	0.26 ± 0.12	-11 ± 7	0.2 ± 0.2	(11, 12)

[a]Data for ungrouped meteorites are not shown (except for Tafassasset) because accretion ages for these samples are usually unknown

[b]Accretion ages are from Sugiura and Fujiya (2014), Desch et al. (2018). For the iron meteorites they are from Kruijer et al. (2014) and Kruijer et al. (2017), except for the IAB irons, which are from Hunt et al. (2018)

[c]References: (1) Burkhardt et al. (2011); (2) Burkhardt et al. (2014); (3) Budde et al. (2016); (4) Budde et al. (2018); (5) Budde et al. (2019); (6) Worsham et al. (2019); (7) Yokoyama et al. (2019); (8) Render et al. (2017); (9) Worsham et al. (2017); (10) Hopp et al. (2020); (11) Bermingham et al. (2018); (12) Poole et al. (2017)

[d]Δ^{95}Mo values calculated from mean of data for CR components (metal, chondrules, silicates), which results in a much more precise value than the single bulk analyses

regression of these data provides ε^{95}Mo–ε^{94}Mo slopes of 0.596 ± 0.008 and 0.596 ± 0.006 for the NC- and CC-lines, respectively. These slopes are in good agreement with a slope of 0.609 ± 0.008 as calculated from the presolar SiC grain data, demonstrating that the isotopic variations predominantly reflect s-process variations.

As noted above, the offset between the NC- and CC-lines is consistent with an excess of r-process material in the CC reservoir. This is because for a given ε^{94}Mo, a r-process excess results in a more positive ε^{95}Mo than an s-process deficit (Fig. 2). Further, as the NC- and CC-lines are parallel, all CC meteorites have an approximately constant r-process excess compared to NC meteorites, because otherwise the NC- and CC-lines would not be parallel. To quantify the offset between the NC and CC reservoirs it is useful to introduce the Δ^{95}Mo notation as follows (Budde et al. 2019):

$$\Delta^{95}\text{Mo} = \left(\varepsilon^{95}\text{Mo} - 0.596 \times \varepsilon^{94}\text{Mo}\right) \times 100$$

where the slope of 0.596 is the slope of the s-process mixing line as obtained from NC and CC meteorites (using the 0.609 slope obtained from SiC grain data would change the Δ^{95}Mo values by less than ~ 1 ppm). As such, Δ^{95}Mo provides the parts-per-million deviation of a sample from an s-process mixing line that passes through the origin. In other words, Δ^{95}Mo 'removes' the s-process variations among samples from a given reservoir, and therefore is a measure of any deviation from pure s-process variations relative to the composition of the standard. The characteristic values for the NC and CC reservoir are $\Delta^{95}\text{Mo}_{\text{NC}} = -9 \pm 2$ and $\Delta^{95}\text{Mo}_{\text{CC}} = +26 \pm 2$ (Budde et al. 2019). These values are unique isotope signatures of the NC and CC reservoirs, irrespective of the absolute anomaly of a given sample.

The Mo isotopic data allow for small r-process (and possibly p-process) heterogeneities within the NC- and CC-reservoirs (Budde et al. 2019; Poole et al. 2017; Worsham et al. 2017; Yokoyama et al. 2019). For instance, when only the data for bulk NC meteorites are considered, the slope of the NC-line seems to be slightly shallower (Budde et al. 2019; Yokoyama et al. 2019), suggesting that there are small additional r- or p-process variations among these samples. However, these heterogeneities are small compared to the r-process excess of the CC over the NC reservoir (Fig. 3). Thus, in spite of the heterogeneous distribution of an s-process Mo carrier in both reservoirs, the presolar material responsible for the r-process offset between the NC and CC reservoirs is fairly homogeneously distributed within both reservoirs. This is not only true for bulk meteorites, but also for most components of primitive chondrites (with the exception of CAI). For instance, chondrule and matrix separates from the CV3 chondrite Allende (Budde et al. 2016), as well as metal and silicate fractions from CR chondrites (Budde et al. 2018) exhibit large s-process Mo isotope variations, yet they all plot on the CC-line (not shown in Fig. 3). Collectively, these observations suggest that the excess r-process material in the CC reservoir is not hosted in individual presolar carriers, but is characteristic for the bulk isotopic composition of this reservoir (Budde et al. 2019; Burkhardt et al. 2019; Nanne et al. 2019).

3.3 Chronology of NC and CC Meteorites

The dichotomy between NC and CC meteorites may either reflect formation in two spatially separated regions of the solar nebula or distinct formation times of NC and CC meteorites combined with a temporal change of the solar nebula's isotopic composition (Warren 2011). For instance, because carbonaceous chondrites formed relatively late, up to ~ 4 Ma after CAI formation (Budde et al. 2018; Schrader et al. 2017), their distinct Ti–Cr–O isotopic composition (Fig. 1a, b) could in principle reflect a change in the solar nebula's isotopic composition prior to their formation, but after accretion of NC meteorite parent bodies (Warren 2011). Distinguishing between a spatial and temporal origin of the NC–CC dichotomy, therefore, requires knowledge of the timescales of meteorite parent body accretion. Recent summaries of accretion ages can be found in Sugiura and Fujiya (2014), Desch et al. (2018), and Kruijer et al. (2020). Here we will not provide a detailed review of these accretion ages, but focus on those that are most important for understanding the formation of the NC and CC reservoirs, namely the accretion ages of the parent bodies of iron meteorites and chondrites.

The Hf–W chronology of 'magmatic' iron meteorites [i.e., iron meteorites from the metallic cores of differentiated bodies (Scott and Wasson 1975)] indicates that core formation in their parent bodies occurred at ~ 0.5–1.5 Ma (NC irons) and ~ 2–3 Ma (CC irons) after CAI formation (Kruijer et al. 2017, 2014). Core formation within a few Ma after CAI formation is consistent with decay of ^{26}Al as the dominant heat source facilitating melting and differentiation (Hevey and Sanders 2006), and thermal modeling of planetesimals heated internally by ^{26}Al decay to obtain inferred core formation ages indicates that the NC iron parent bodies accreted within ~ 0.5 Ma, whereas the CC iron parent bodies accreted slightly later, at ~ 1 Ma after CAI formation (Kruijer et al. 2017, 2014). Thus, early-formed iron meteorite parent bodies exist in both the NC and CC reservoirs, indicating that the NC–CC dichotomy was already present when the iron meteorite parent bodies formed at < 1 Ma after CAI formation.

The accretion ages for chondrite parent bodies can be inferred from the chronology of chondrules, small igneous spherules that are the dominant component of most chondrites and formed prior to or during the accretion of their host chondrite (Alexander et al. 2008). The Al–Mg chronology of chondrules from ordinary, CV, and CO chondrites provide a narrow interval of formation ages of ~ 2–3 Ma after CAI formation (Kita and Ushikubo 2012; Kunihiro et al. 2004; Kurahashi et al. 2008; Nagashima et al. 2017; Pape et al. 2019; Rudraswami and Goswami 2007; Rudraswami et al. 2008; Villeneuve et al. 2009), indicating that chondrite parent body accretion in the NC and CC reservoirs occurred mostly within ~ 2–3 Ma. However, Pb–Pb ages for chondrules from ordinary, CV, and CR chondrites are more variable and include chondrules that appear to be as old as CAIs (Bollard et al. 2017; Connelly et al. 2012). The discrepancy between Al–Mg and Pb–Pb ages may reflect local variations of ^{26}Al/^{27}Al in the disk, and, more specifically, a reduced initial ^{26}Al/^{27}Al in the inner disk (Bollard et al. 2019). This interpretation is difficult to reconcile, however, with the good agreement of Hf–W and Al–Mg ages for several meteorites and meteorite components, including samples from the inner and outer disk, and including samples with formation ages varying by almost 5 Ma (Budde et al. 2018; Kleine et al. 2018; Kruijer et al. 2020). Moreover, the Pb–Pb ages for pooled chondrule separates from CV and CR chondrites are also in good agreement with the Al–Mg and Hf–W ages of CV and CR chondrules, in contrast to the Pb–Pb ages for single chondrules (see summary in Kruijer et al. 2020). Altogether, these observations indicate that the discrepancy between Al–Mg and Pb–Pb ages for single chondrules is not due to spatial heterogeneities of ^{26}Al/^{27}Al. As such, the ancient Pb–Pb ages reported for some chondrules do not seem to record the time of chondrule formation. One possibility is that the Pb–Pb ages of some chondrules appear older because of the loss of intermediate decay products within the U–Pb decay chains (Pape et al. 2019), but more work is needed to understand the significance of Pb–Pb ages for single chondrules.

The accretion timescales of chondrite parent bodies can also be inferred by combining ages for processes on the parent body (e.g., aqueous alteration, thermal metamorphism) with thermal modeling of meteorite parent bodies heated internally by ^{26}Al decay. For instance,

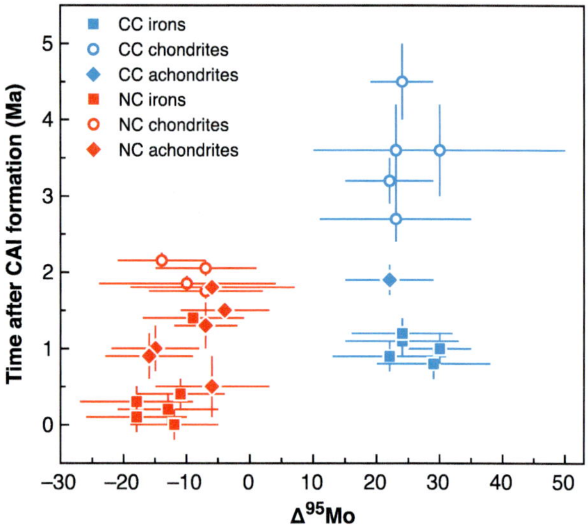

Fig. 4 Plot of meteorite accretion ages versus Δ^{95}Mo. Only meteorites for which Δ^{95}Mo is known to within better than ± 20 ppm are shown. The NC–CC dichotomy combined with the accretion ages demonstrates that NC and CC meteorites derive from two spatially separated and contemporaneous reservoirs. The two reservoirs were most plausibly separated by Jupiter. For data sources see Table 1

the cooling history of equilibrated (i.e., thermally metamorphosed) ordinary chondrites as deduced from Pb–Pb and Hf–W chronometry combined with thermal modeling consistently returns an accretion age of ∼ 2 Ma after CAI formation (Blackburn et al. 2017; Hellmann et al. 2019; Henke et al. 2012). Likewise, ^{53}Mn–^{53}Cr ages of aqueously formed secondary fayalites, combined with their formation conditions (i.e., temperature, water/rock ratio) and with a thermal model of their parent body provides accretion ages of ∼ 2 Ma (ordinary chondrites), ∼ 2.2 Ma (CO chondrites), and ∼ 2.5 Ma (CV chondrites), post CAI formation (Doyle et al. 2015; Jogo et al. 2017). Together, these accretion ages are in excellent agreement with the Al–Mg ages for chondrules from these chondrites. This not only provides further evidence that the Al–Mg chondrule ages are chronologically meaningful, but also indicates that the chondrule ages closely approximate the time of chondrite accretion.

In summary, the chronological data for meteorites and meteorite components indicate that chondrite parent bodies accreted later than the iron meteorite parent bodies, starting at ∼ 2 Ma in the NC reservoir (i.e., the ordinary chondrites) and continuing until at least ∼ 4 Ma after CAI formation in the CC reservoir (i.e., the CR chondrites). Importantly, for both the NC and CC reservoirs, iron meteorites and chondrites plot on single *s*-process mixing lines (i.e., the NC- and CC-lines), demonstrating that CC irons and chondrites have the same characteristic *r*-process excess over the NC irons and chondrites (Fig. 3). This is illustrated in a plot of accretion age versus Δ^{95}Mo, which shows that the early- and late-formed NC and CC meteorites have the same, characteristic Δ^{95}Mo of their respective reservoir (Fig. 4). Thus, there was only limited influx of CC material into the NC reservoir at least for the period between ∼ 0.5 Ma up until ∼ 2 Ma (i.e., the accretion period of NC irons and chondrites), and there was no significant influx of NC material into the CC reservoir at least from ∼ 1 Ma to at least ∼ 4 Ma (i.e., the accretion period of CC irons and chondrites) (Fig. 4). These observations indicate that the NC and CC reservoirs represent two spatially separated reservoirs of the solar protoplanetary disk that co-existed for several Ma (Kruijer et al. 2017).

3.4 The Role of Jupiter

The prolonged spatial separation of the NC and CC reservoirs implies that there was a physical barrier between them, because otherwise the diffusive mixing and radial drift of solids in the disk would have rapidly erased the isotopic difference between the two reservoirs. A plausible mechanism to efficiently separate two disk reservoirs for an extended period of time is the accretion of a giant planet in between them, generating a gap within the disk and inhibiting the inward drift of dust grains (e.g., Morbidelli et al. 2016). Being the largest and nearest gas-giant planet, Jupiter is therefore the most likely candidate for separating the NC and CC reservoirs (Kruijer et al. 2017). Jupiter may have also helped preserving the NC–CC dichotomy by accreting planetesimals with intermediate isotopic compositions (Jacquet et al. 2019). Recently, Brasser and Mojzsis (2020) proposed that the initial separation of the NC and CC reservoirs was not caused by Jupiter itself, but resulted from a pressure maximum in the disk, near the location at which Jupiter later formed. Thus, while these different models differ in the efficiency by which Jupiter itself was responsible for the isolation of the NC and CC reservoirs, they all agree in that the NC and CC reservoirs represent material from inside and outside Jupiter's orbit and, therefore, the inner and outer solar system, respectively.

The efficiency of the Jupiter barrier depends on the size of the dust grains drifting inwards. In general, the barrier is more efficient for large grains, whereas smaller grains that are well coupled to the gas may pass through the barrier as part of the gas flow. For instance, Weber et al. (2018) showed that the flux of mm-size grains is diminished by seven orders of magnitude by the presence of a gap, whereas there is almost no reduction in the flux of 100 μm-size grains. For comparison, the flux of mm-size grains is reduced by a factor ~ 300 if the gap is opened by a growing gas giant that has reached half the mass of Jupiter. The filtering is even stronger at small viscosity (deeper gap), reaching a factor of ~ 100 even for 100 μm-size dust in the case of a gap opened by a Jupiter-mass planet. Thus, while the Jupiter barrier provides an efficient means for preserving an isotopic difference between the NC and CC reservoirs, it likely was permeable for small dust grains, especially when Jupiter had not grown to its full size yet. On this basis, Schiller et al. (2018) argued that CC dust from the outer solar system passed through the Jupiter barrier and continuously changed the isotopic composition of the inner solar system. These authors also speculated that this CC dust was only accreted by larger bodies—such as Vesta, Mars, and Earth—but not by smaller bodies such as the enstatite and ordinary chondrite parent bodies. The preservation of the NC–CC dichotomy may then not only reflect the spatial separation of the two reservoirs, but also the different efficiencies with which CC dust is accreted by planetary bodies. However, in this model the close isotopic link between enstatite chondrites and Earth would be mere coincidence, which seems unlikely. More recently, Spitzer et al. (2020) argued that the isotopic composition of the inner disk may have changed continuously through the addition of CC dust, and that this process affected all meteorite parent bodies, including those of enstatite and ordinary chondrites. Clearly, understanding the efficiency of the Jupiter barrier, and the effect of inward drifting CC dust on the isotopic composition of the inner disk, will be important tasks for future studies.

Regardless of whether Jupiter is responsible for the initial separation of the NC and CC reservoirs, its growth and migration provides a mechanism for the later scattering of CC bodies into the inner solar system to their present-day location in the main asteroid belt between Mars and Jupiter. This inward scattering of CC bodies resulted either from the inward-then-outward migration of Jupiter as proposed in the Grand Tack model (Walsh et al. 2011), or from the runaway growth of Jupiter on a fixed orbit (Raymond and Izidoro 2017). Consequently, the formation of Jupiter can account for both the prolonged spatial separation of the NC and CC reservoirs, and the co-occurrence of NC and CC bodies in the present-day asteroid belt.

4 Early Evolution of the Solar Accretion Disk Inferred from the NC–CC Dichotomy

4.1 Origin of the Dichotomy

The NC–CC dichotomy exists for refractory and non-refractory elements and is consistently present for nuclides produced in neutron-rich stellar environments (i.e., *r*-process Mo, ^{58}Ni, ^{50}Ti, ^{54}Cr). As such it is unlikely that the dichotomy has been produced by thermal processing of presolar carriers within the disk, because this would require the efficient removal of presolar components that were likely hosted in different carriers over a wide range of volatility (Burkhardt et al. 2019; Jacquet et al. 2019; Nanne et al. 2019). Moreover, the efficiency of thermal processing depends on the thermal stability of a carrier and not the particular stellar environment from which this carrier originated, and so there is no reason why thermal processing would preferentially affect presolar carriers from neutron-rich stellar sources.

Many of the isotopic characteristics of the CC reservoir are also found, in more extreme form, in CAIs (Burkhardt et al. 2019; Nanne et al. 2019). This includes enrichments in *r*-process Mo (Brennecka et al. 2013; Burkhardt et al. 2011), ^{54}Cr (Papanastassiou 1986), and coupled ^{46}Ti–^{50}Ti excesses (Davis et al. 2018; Trinquier et al. 2009). Specifically, the isotopic composition of the CC reservoir is intermediate between those of the NC reservoir and CAIs, implying that compared to the NC reservoir, the CC reservoir contains a higher proportion of material with a CAI-like isotopic composition. Nevertheless, the addition of CAIs themselves to the CC reservoir cannot be the sole cause for the NC–CC dichotomy, because this dichotomy exists for non-refractory elements (i.e., Cr, Ni), which are not enriched in CAIs (Burkhardt et al. 2019; Nanne et al. 2019). To affect refractory and non-refractory elements in a similar manner, the formation of the NC–CC dichotomy must, therefore, have involved the addition of isotopically CAI-like material whose bulk chemical composition was broadly chondritic (Burkhardt et al. 2019; Nanne et al. 2019). In this case, CAIs represent only the refractory part of an early disk reservoir, which was enriched in those nuclides that define the NC–CC dichotomy (Burkhardt et al. 2019; Jacquet et al. 2019). Of note, this scenario also accounts for the ^{46}Ti–^{50}Ti correlation observed among bulk meteorites. Because ^{46}Ti and ^{50}Ti are produced in distinct stellar sites, the ^{46}Ti–^{50}Ti correlation was interpreted as evidence for thermal processing during which Ti from distinct presolar carriers evaporated and then re-condensed together into a common Ti-bearing phase, which subsequently was heterogeneously distributed at the bulk meteorite and planetary scale (Trinquier et al. 2009). However, in light of the NC–CC dichotomy the coupled ^{46}Ti–^{50}Ti excesses of CAIs simply reflect the composition of an early disk reservoir, and the ^{46}Ti–^{50}Ti correlation is easily understood as the result of mixing of this material to a disk reservoir characterized by ^{46}Ti–^{50}Ti deficits.

All these observations combined suggest that the NC–CC dichotomy reflects a change in the isotopic composition of the material that is accreted to the disk during infall from the Sun's collapsing protostellar envelope. The isotopic composition of the infalling material may have changed because material infalling at different times derives from different parts of the parental molecular cloud. Thus, an isotopic heterogeneity within the cloud would have, at least in part, been imparted on the isotopic composition of the disk (Jacquet et al. 2019; Visser et al. 2009). The distance where mass is added to the disk increases with time outwards to several AU (e.g., Hueso and Guillot 2005; Yang and Ciesla 2012), and so different regions of the disk received varying proportions of material infalling at different times. This scenario can readily account for the observation that the dichotomy is consistently present for nuclides produced in neutron-rich stellar environments, exactly because this is

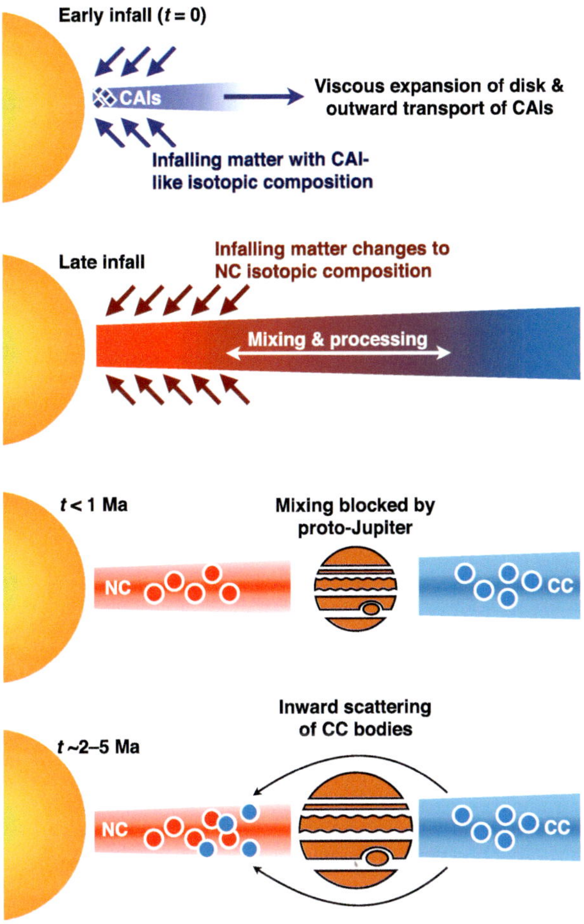

Fig. 5 Formation of the NC and CC reservoirs by late infall of isotopically distinct material (not to scale). Rapid expansion of early infalling material forms an initial disk, whose isotopic composition is represented by CAIs. Later infalling material dominates the composition of the inner disk and has a distinct, NC-like isotopic composition, which compared to the earlier infalling material is depleted in nuclides from neutron-rich stellar environments. Mixing within the disk reduced the isotopic difference between the inner and outer disk, but the rapid formation of Jupiter prevented completed homogenization. Jupiter grew large enough after ~ 2 Ma but before ~ 4–5 Ma after CAIs to induce scattering of CC bodies into the inner solar system. Figure adapted from Nanne et al. (2019) and Burkhardt et al. (2019)

the material that is heterogeneously distributed between the two reservoirs (Nanne et al. 2019).

Figure 5, adopted from Nanne et al. (2019) and Burkhardt et al. (2019), is a cartoon illustrating this model for the formation of the NC–CC dichotomy during infall. During the earliest stage of infall, CAIs formed close to the Sun and were transported outwards by rapid radial expansion of the early infalling material, forming an initial disk (Desch et al. 2018; Jacquet et al. 2011, 2019; Pignatale et al. 2018). The isotopic composition of this initial disk is given by that of CAIs, meaning that this disk was enriched in nuclides produced in neutron-rich stellar environments, such as r-process Mo, ^{50}Ti, and ^{54}Cr. Yet, the later infalling material was depleted in these nuclides and, because most of the material in the inner disk derives from this infall, this material likely had a NC-like composition. In the outer disk, beyond the radius at which later material is infalling, the composition of the initial disk is preserved, in diluted form, as the composition of the CC reservoir. Thus, in this model the isotopic composition of the CC reservoir reflects a mixture between that of the earliest disk, which is preserved in CAIs, and later infalling material with an NC-like isotopic composition.

After the infall of material stopped, the proto-Sun continued to accrete material from its disk, as a result of inward transport of mass and outward movement of angular momentum (Yang and Ciesla 2012). Thus, to maintain an isotopic difference between the CC and NC reservoirs, the inward transport of isotopically distinct outer disk material must have been blocked before planetesimal formation in the NC and CC reservoir had begun (Fig. 4). As argued above, this most likely reflects the presence of a barrier (either Jupiter itself or a pressure bump in the disk where Jupiter later formed) that prohibited significant inward drift of CC material into the inner disk. In brief, "the NC–CC dichotomy reflects an inherited isotopic heterogeneity from the solar system's parental molecular cloud, which was established during infall from the Sun's protostellar envelope and has been preserved through the (...) formation of Jupiter" (Nanne et al. 2019).

4.2 Planetesimal Formation in the Inner Solar System

The presence of a barrier against the inward drift of CC dust into the inner solar system likely resulted in the rapid depletion of dust in the inner disk. This is because the available dust either is rapidly accreted into a first generation of planetesimals or drifts into the Sun. Consequently, once most of the dust in the NC reservoir has been incorporated into early-formed planetesimals, the inner disk remains depleted in dust because it cannot be replenished by the inward drift of CC dust (so as to preserve the NC–CC dichotomy), with the exception of the smallest particles (< 100 μm), which can pass through the Jupiter barrier (Haugbølle et al. 2019; Weber et al. 2018) but are too small to be efficiently accreted. However, the chronology of NC meteorites indicates that there is more than one generation of planetesimals in the inner disk and that planetesimal formation occurred over a period of at least ~ 2 Ma (Fig. 4). This then raises the question of how there was sufficient dust in the inner disk for this prolonged period of planetesimal formation.

One possibility is that the dust for later-formed planetesimals was produced by collisions among pre-existing planetesimals. As the youngest sampled planetesimals from the NC reservoir are the ordinary and enstatite chondrites (see Fig. 4), this would imply that these chondrites predominantly consist of secondary dust from an earlier generation of planetesimals. Both chondrites contain abundant metal and, in some cases (H and EH chondrites) have approximately solar Fe/Si ratios (Palme et al. 2014). Thus, if these chondrites formed from secondary dust, then this dust must have contained metal and silicate grains in approximately solar proportions. This secondary dust, therefore, probably did not derive primarily from collisions among differentiated planetesimals, because these underwent large-scale metal-silicate separation and core formation. Instead, it is more likely that most secondary dust was produced by collisions among small planetesimals that remained either undifferentiated or partially differentiated, so that no or only minor metal-silicate melt segregation occurred. Alternatively, the secondary dust may also derive from the primitive, undifferentiated crust of larger planetesimals (Elkins-Tanton et al. 2011).

The collisional production of dust would naturally occur in the late stages of the protoplanetary disk, when the damping of the planetesimals' velocity dispersion exerted by the gas becomes weaker and protoplanets become more massive and can scatter the planetesimals more effectively (Gerbig et al. 2019). The low amount of remaining gas favors the triggering of the streaming instability by the generated dust, because the dust/gas ratio can be large even if the total amount of dust is not so high (Carrera et al. 2017). This would produce in a natural way an asteroid population with a small total mass, as advocated in some models (Hansen 2009; Izidoro et al. 2016).

The NC chondrites, like CC chondrites, contain presolar grains (Huss and Lewis 1995). Although some of these grains may derive from the colliding primitive planetesimals, they may also derive from inward drifting, fine-grained CC dust that was able to pass the Jupiter barrier and mixed into the secondary dust. The amount of this CC dust must have been low, because otherwise the isotopic composition of late- and early-formed NC bodies would be different, but it would likely be a sufficient source for presolar grains in NC chondrites. This would be consistent with the observation that all chondrites, including CC and NC, sampled the same solar-system-wide reservoir of presolar grains (Huss and Lewis 1995).

The formation of NC chondrites from collisionally produced dust does not necessarily imply that these collisions also resulted in the melting that produced NC chondrules. The chondrule-melting events may have occurred at a later time, and they may have been unrelated to the collisions that produced their precursor dust. Nevertheless, it also possible that the same collisions that produced the secondary dust also resulted in chondrule formation (e.g., Stewart et al. 2019).

Clearly, more work is needed to reconcile the early formation of the NC–CC dichotomy with the 'late' formation of NC chondrites. The formation of secondary dust by collisions among an earlier generation of planetesimals is one possibility, but requires a more thorough assessment of how well the textural, chemical, and isotopic properties of NC chondrites and chondrules can be reproduced in such a model. Regardless, the NC–CC dichotomy combined with the chronology of chondrite accretion provides a fundamental constraint on models of planetesimal formation in the inner disk.

4.3 Growth History of Jupiter

The standard model for the formation of Jupiter (and other gas giant planets in the solar system) is the core accretion model, in which the planet's gaseous envelope is accreted onto a pre-existing solid core of about 10–20 Earth masses ($M_\oplus$) (Pollack et al. 1996). In this model, gas accretion is not a steady process, but proceeded slowly initially until Jupiter reached a mass of $\sim 50\ M_\oplus$, which was then followed by runaway gas accretion to Jupiter's final mass of $\sim 318\ M_\oplus$. Depending on the mass reached at different stages of its growth, Jupiter affected bodies inside and outside its orbit in different ways. Specifically, once Jupiter's core reached a mass of $\sim$ 10–20 $M_\oplus$, it would have substantially hampered exchange of material from inside (NC) and outside (CC) its orbit (Lambrechts et al. 2014). When Jupiter reached a mass of $\sim 50\ M_\oplus$, a gap opened in the disk (Crida et al. 2006), leading to inward migration of Jupiter (Walsh et al. 2011). The migration and/or the further growth of Jupiter results in gravitational scattering of bodies from inside and outside Jupiter's orbit, ultimately leading to dynamical mixing of NC and CC bodies (Raymond and Izidoro 2017). Within this framework, Jupiter's accretion timescale can be estimated from the formation time and longevity of the NC–CC dichotomy (Kruijer et al. 2017).

A key observation from the Mo isotopic data is that the NC iron meteorites together with ordinary and enstatite chondrites plot on the NC-line (Fig. 2). Thus, there was no significant influx of CC material into the NC reservoir since the formation of the earliest NC bodies at ~ 0.5 Ma after CAI formation, because otherwise early (i.e., NC irons) and late formed NC bodies (i.e., ordinary chondrites) would not plot on a single NC line (i.e., have the same characteristic Δ^{95}Mo; Fig. 4). As noted above, the data allow some r-process heterogeneity within the NC reservoir, meaning that there might have been some influx of CC material after accretion of the oldest NC bodies. This potential change is small, however, compared to the difference between the NC and CC reservoirs (Spitzer et al. 2020). Both reservoirs were, therefore, probably already separated from each other when the NC iron meteorite parent bodies accreted. Consequently, under the assumption that the formation of Jupiter's core is responsible for the initial separation of the NC and CC reservoirs, Jupiter reached

~ 10–$20\ M_\oplus$ within ~ 0.5 Ma after CAI formation. This would probably require formation of Jupiter's core by pebble accretion (Lambrechts and Johansen 2014; Levison et al. 2015). Brasser and Mojzsis (2020) argued that owing to the inefficiency of the pebble accretion process, a large number of CC pebbles would nevertheless drift past the growing Jupiter, and would then contaminate the NC reservoir. This predicted large influx of CC pebbles into the inner disk is inconsistent with the early preservation of the NC–CC dichotomy, however. To overcome this problem, Brasser and Mojzsis (2020) proposed that there was a pressure maximum in the disk near the location where Jupiter later formed, which limited the inward drift of CC pebbles. In this case, the initial separation of the NC and CC reservoirs may not have been caused by the growth of Jupiter's core itself, which would make the meteoritic constraints on Jupiter's early growth more uncertain.

Following accretion of iron meteorite parent bodies in the NC and CC reservoirs, planetesimal formation in both reservoirs continued for an extended period of time, until at least ~ 2 Ma in the NC and ~ 4 Ma in the CC reservoir (Fig. 4). As both NC and CC chondrites plot on their respective NC- and CC-lines together with the NC and CC irons, respectively, there was no substantial mixing between the NC and CC reservoirs at least until after accretion of the youngest chondrite parent body in each reservoir. The formation time of the youngest CC bodies does not provide tight constraints on the early growth of Jupiter, because planetesimal formation in the CC reservoir may have still occurred while Jupiter already scattered earlier-formed CC bodies into the inner solar system (Raymond and Izidoro 2017). A more direct constraint may be obtained from the observation that planetesimal formation in the NC reservoir seems to have stopped at ~ 2 Ma, as inferred from the accretion age of the youngest NC meteorites (Fig. 4). For instance, within the framework of the Grand Tack model (Walsh et al. 2011), Jupiter's migration through the asteroid belt would have terminated planetesimal formation there, implying that the migration occurred at ~ 2 Ma and Jupiter reached a mass of $\sim 50\ M_\oplus$ at about this time. However, if Jupiter did not migrate through the asteroid belt, then the termination of NC planetesimal formation may simply reflect the lack of substantial dust in the inner disk (Sect. 4.2) or the almost complete depletion of gas, which would not allow the dust to clump into new planetesimals. Both possibilities are more difficult to relate to a specific stage of Jupiter's growth. Nevertheless, there is so far no observational evidence for the inward scattering of CC bodies during the time of NC planetesimal formation, which may mean that Jupiter did not reach $\sim 50\ M_\oplus$ before ~ 2 Ma (Alibert et al. 2018; Kruijer et al. 2020). This conclusion, however, depends on the assumed formation location of the scattered CC bodies and on whether Jupiter's growth was solely responsible for the inward scattering of CC bodies.

Wang et al. (2017) argued that the final stage of Jupiter's growth to $\sim 318\ M_\oplus$ was completed by ~ 4–5 Ma at the latest. This argument is based on the observation that angrites, a class of basaltic achondrites dated at ~ 4–5 Ma after CAI formation (Amelin 2008; Connelly et al. 2008; Kleine et al. 2012; Spivak-Birndorf et al. 2009), record the absence of a nebular magnetic field, indicating that at this time the nebula gas had dissipated. As Jupiter can only grow to its final size in the presence of nebular gas, its accretion must have been completed by this time. This is consistent with evidence for the addition of CC material to the eucrite and angrite parent bodies (both are NC bodies) at ~ 4–5 Ma after CAI formation (Sarafian et al. 2017, 2014), which may be related to inward scattering of CC bodies by a nearly fully-formed Jupiter (Raymond and Izidoro 2017).

Taking all these observations together reveals that the meteorite data are consistent with the following growth history of Jupiter: after the rapid initial accretion of its ~ 10–$20\ M_\oplus$ core within less than ~ 0.5 Ma, Jupiter reached $\sim 50\ M_\oplus$ after ~ 2 Ma, and its final mass of $\sim 318\ M_\oplus$ before ~ 4–5 Ma (Kruijer et al. 2017, 2020). This growth history and the distinct rate of gas accretion are approximately consistent with the core accretion model for the formation of Jupiter (Alibert et al. 2018). Nevertheless, as outlined above, the growth history deduced from the meteorite data depends on several assumptions on how Jupiter affected the interaction between material inside and outside its orbit, and so other growth histories are possible.

5 Accretion of Earth

Accretion of Earth, and potentially the other terrestrial planets, continued past the inward scattering of CC bodies resulting from the growth of the gas giants, and so Earth likely accreted CC bodies during its growth (O'Brien et al. 2018; Raymond and Izidoro 2017). These objects are the most likely source of Earth's water and highly volatiles species (Alexander et al. 2012; Marty 2012). Thus, understanding when, how, and how much CC material was accreted by the growing Earth is of considerable interest (McCubbin and Barnes 2019; O'Brien et al. 2018). It also provides a potential way of discriminating between different models of Earth's accretion (e.g., Fischer et al. 2018).

Given that NC and CC meteorites have distinct isotopic signatures, the isotopic composition of Earth's primitive mantle (or bulk silicate Earth, BSE) can be used to reconstruct the accretion of CC material to Earth. For instance, Dauphas (2017) argued that Earth accreted $\sim 5\%$ CC material only during the early stages of accretion, and that the later stages consisted entirely of material with an enstatite chondrite-like isotopic composition. This conclusion depends on the assumption that enstatite, ordinary, and carbonaceous chondrites are representative for the building blocks of Earth. If instead a broader range of isotopic compositions is assumed, including those of some achondrites having the lowest ε^{50}Ti and ε^{54}Cr values observed among meteorites (Fig. 1), then the Earth may have incorporated a much larger amount of CC material of up to ~ 40–50% (Schiller et al. 2018; Warren 2011). Thus, the total amount of CC material accreted by Earth is not well known.

Unlike all meteorites analysed to date, the Mo isotopic composition of the bulk silicate Earth (BSE) falls between those of the NC and CC reservoirs ($\Delta^{95}\text{Mo}_{\text{BSE}} = +7 \pm 5$; Fig. 6), indicating that the BSE's Mo derives from both reservoirs (Budde et al. 2019). This observation may seem unsurprising, given that Earth is expected to have accreted some CC bodies. However, it is important to recognize that Mo, as a siderophile element, only records the late stages of Earth's growth, probably the last 10–20% of accretion (Dauphas 2017). This is because the Mo from earlier accreted objects was largely removed into Earth's core. Consequently, the mixed NC–CC Mo isotopic composition of the BSE indicates that Earth accreted CC material in the final 10–20% of its growth, because otherwise the BSE would plot on the NC-line (Fig. 6). Earth may have also accreted CC bodies during earlier stages, but this would not be visible in the BSE's Mo isotopic composition (Budde et al. 2019; Dauphas 2017; Fischer et al. 2018).

The position of the BSE among the NC- and CC-lines also reveals that no combination of known meteorites yields the BSE's Mo isotopic composition, indicating that Earth incorporated material that is distinct from known bulk meteorites (Fig. 6). This material must be enriched in s-process Mo, to balance the s-process depleted composition of known meteorites, relative to the composition of the BSE. The s-process–enriched material may derive from either the NC or the CC reservoir (Fig. 6), but it is noteworthy that for essentially all elements investigated thus far, Earth's isotopic composition is most similar to that of enstatite chondrites. This suggests that Earth predominantly accreted from material with an enstatite chondrite-like isotopic composition (Dauphas 2017; Dauphas et al. 2014), in which case the BSE's Mo isotopic composition most likely reflects a mixture of this material with s-

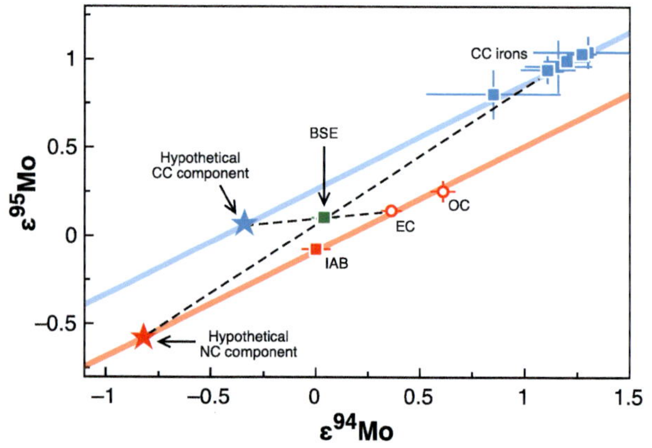

Fig. 6 ε^{95}Mo versus ε^{94}Mo showing the position of the bulk silicate Earth (BSE) relative to the NC-line (red) and CC-line (blue). For clarity, data points for only a few meteorites are shown. Also shown are two hypothetical components, which are characterized by an *s*-process excess relative to the BSE, and are from either the CC or the NC reservoir. At least one *s*-process–enriched component must have been incorporated into the Earth, to accommodate the inclusion of material represented by known meteorites, all of which are characterized by an *s*-process deficit compared to Earth. Figure adapted from Budde et al. (2019)

process–enriched material from the CC reservoir (Fig. 6). In this model, the BSE's isotopic composition would naturally be similar to enstatite chondrites and would deviate only for elements that were added late, such as Mo (Budde et al. 2019).

Because the NC- and CC-lines are parallel, the BSE composition divides any tie line between them into two segments whose ratio to each other is constant (Fig. 6). Consequently, irrespective of the position of the endmembers on the NC- and CC-lines (i.e., of the absolute Mo isotope anomalies), the resulting mixing ratio between NC and CC material remains the same. The Mo isotopic data, therefore, allow quantifying the fraction of CC material in the BSE irrespective of the NC and CC endmember compositions. Using the Δ^{95}Mo values for the BSE and the NC- and CC reservoirs, the CC/(NC + CC) ratio of the BSE is 0.46 ± 0.15, indicating that between $\sim 30\%$ and $\sim 60\%$ of the BSE's Mo derives from the CC reservoir (Budde et al. 2019). This high mass fraction of CC material refers only to the last 10–20% of Earth's accretion. As such, it should not be confused with the total amount of CC material accreted by Earth, which likely was smaller. For instance, if Earth accreted CC material only during the last $\sim$ 10–20% of accretion, then it incorporated a total of $\sim$ 3–6% CC material. Still it seems likely that Earth also accreted some CC material during earlier accretion stages, and so the total amount of CC material is likely higher, and, as noted above, could be as high as $\sim$ 40–50%.

The last 10–20% of Earth's accretion are dominated by the Moon-forming impactor, which is the last large body thought to collide with Earth (Canup and Asphaug 2001), and the late veneer, the material added to Earth's mantle after the giant impact and cessation of core formation (e.g., Walker et al. 2015). As such, the Mo in the present-day BSE can be regarded as a mixture of Mo from proto-Earth's mantle (i.e., the mantle before the giant impact), the Moon-forming impactor, and the late veneer. To quantify the contribution of these components to the observed Mo isotopic composition of the BSE, Budde et al. (2019) calculated the expected Δ^{95}Mo of the Earth's mantle for different assumed compositions of these three components. The results reveal that the CC-derived Mo in the BSE most likely derives from the Moon-forming impactor or the late veneer (Fig. 7). This conclusion holds unless the core of the Moon-forming impactor did not equilibrate even partially with the

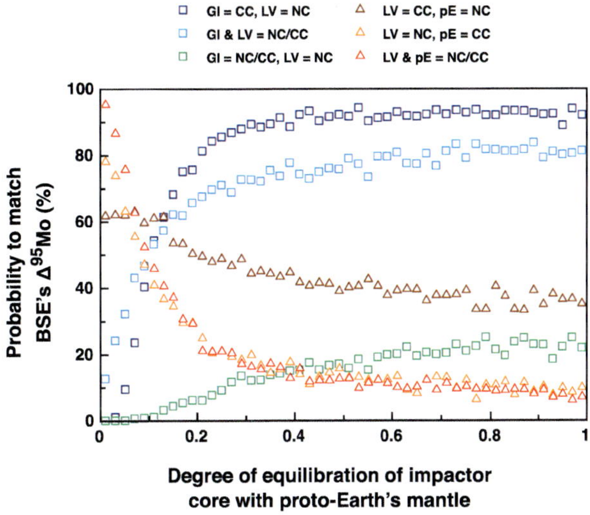

Fig. 7 Probability of matching the BSE's Δ^{95}Mo for different compositions of proto-Earth's mantle (pE), the Moon-forming impactor (GI), and the late veneer (LV) as a function of impactor core equilibration during formation of the Moon. Scenarios in which the Moon-forming impactor contains CC material have a much higher probability than scenarios in which the Moon-forming impactor has a pure NC composition. A pure CC composition for the LV also provides a good match, but this possibility is ruled out based on Ru isotopes (see text). Figure modified from Budde et al. (2019)

Earth's mantle before merging with Earth's core (Fig. 7). This is because if the impactor core did not equilibrate with Earth's mantle, then the impactor would not have contributed significantly to the BSE's Mo isotopic composition. In this case, the CC-derived Mo in the BSE may derive from objects accreted before the Moon-forming impact or from the late veneer (Budde et al. 2019).

One possibility is that Earth accreted CC material in the form of pebbles, which drifted sunwards and were then captured by the growing proto-Earth (Schiller et al. 2018, 2020). This process requires the presence of gas, implying that most of Earth's mass accreted within the lifetime of the solar nebula of ~ 5 Ma. In this model, the Mo isotopic composition of Earth's mantle would be entirely dominated by the CC pebbles, and so proto-Earth's mantle just prior to the Moon-forming impact would have had a CC-like Mo isotopic composition. This is because as a siderophile element, the Mo in the mantle predominantly derives from late-added material (see above). Reproducing the mixed NC–CC Mo isotopic composition of the present-day BSE then requires a very low degree of impactor core re-equilibration during the Moon-forming impact, because otherwise the BSE's Mo isotopic composition would predominantly reflect that of the impactor (Fig. 7). This poses a problem, however, because the rapid accretion of Earth required in the pebble accretion model is compatible with the ^{182}Hf–^{182}W chronology of Earth's core only for a very high degree of core-mantle re-equilibration throughout accretion, and a late (> 50 Ma) Moon-forming impact (Kleine and Walker 2017; Rudge et al. 2010). This particularly applies to the pebble accretion model, because the rapid initial accretion of Earth inevitably results in a very radiogenic ^{182}W signature of Earth's mantle. This radiogenic ^{182}W must then be removed by a late Moon-forming impact, which in turn requires a high degree of impactor core re-equilibration during this event. Thus, in the pebble accretion model, the BSE would be expected to have a NC-like Mo isotopic composition, because otherwise the previously accumulated CC-derived Mo would be entirely removed into Earth's core. This problem can only be overcome if the Moon-forming impactor itself already contained appreciable amounts of CC material.

The mixed NC–CC Mo isotopic composition of the BSE may also reflect the accretion of CC planetesimals (instead of pebbles) prior the Moon-forming impact. Again, this would require a low degree impactor core re-equilibration during this impact to avoid complete removal of the CC-derived Mo into Earth's core (Fig. 7). Unlike for pebble accretion, however, addition of planetesimals to Earth may have occurred over a much longer timescale because it does not require the presence of gas. For this more protracted accretion timescale, the Hf–W systematics of the BSE are compatible with a lower degree of core-mantle re-equilibration (Nimmo et al. 2010; Rudge et al. 2010), and so the accretion of CC planetesimals to Earth would be consistent with the Hf–W chronology of Earth. However, the planetesimal population in the inner disk is dominated by NC planetesimals (Raymond and Izidoro 2017), rendering it unlikely that accretion of planetesimals to Earth prior to the Moon-forming impact produced a CC-like Mo isotopic composition of the proto-Earth's mantle.

Finally, based on Ru isotope anomalies it seems unlikely that Earth accreted CC material solely through the late veneer. As a highly siderophile element, almost all the Ru in the BSE derives from the late veneer, and so the isotopic composition of Ru solely reflects that of the late veneer. The Ru isotopic composition of the BSE is distinct from all known CC meteorites, but overlaps with some NC meteorites (Bermingham et al. 2018; Fischer-Gödde et al. 2015; Fischer-Gödde and Kleine 2017; Hopp et al. 2020; Worsham et al. 2019). As such, the late veneer most likely had either a pure NC or a mixed NC–CC composition (Hopp et al. 2020). Such a late veneer composition cannot solely account for the observed Mo isotopic composition of the BSE, however (Fig. 7).

Taking all these observations together reveals that the CC-derived Mo in the BSE at least partially derives from the Moon-forming impactor. As such, the impactor may have been a CC embryo from the outer solar system, or it may have been a body with a mixed NC–CC composition itself. Some dynamical models predict that due to the orbital evolution of the giant planets, the final large impactor on Earth (i.e., the Moon-forming impactor) tends to be an embryo from beyond ~ 2.5 AU (O'Brien et al. 2018). Thus, one possibility is that the CC impactor was implanted on a stable orbit in the asteroid belt during the early migration and growth of Jupiter (Raymond and Izidoro 2017; Walsh et al. 2011), and then was destabilized again during a later orbital instability of the gas giants, ultimately resulting in its collision with the Earth. Consistent with this, recent studies suggest that a giant planet instability occurred within the first ~ 100 Ma of the solar system (Morbidelli et al. 2018; Nesvorný et al. 2018) and, therefore, during the likely time of the Moon-forming event (Barboni et al. 2017; Borg et al. 2019; Kleine and Walker 2017).

A final implication of a CC (or mixed NC–CC) heritage for the Moon-forming impactor is that this object may have delivered all of Earth's water and highly volatile species, which requires at least $2 \pm 1\%$ CI chondrite-like material (Alexander et al. 2012; Marty 2012). It is not known if the Moon-forming impactor was as volatile-rich as CI chondrites, but given that it probably delivered about 10% of Earth's mass (Canup and Asphaug 2001), it seems likely that it contained the equivalent of $\sim 2\%$ CI material in terms of water and highly volatiles (Budde et al. 2019).

6 Conclusions

The discovery of an isotopic dichotomy between non-carbonaceous (NC) and carbonaceous (CC) meteorites has fundamentally changed our understanding of early solar system evolution, including the origin of meteorites and how they can be used to constrain the dynamical evolution of the early solar system, and the formation of terrestrial planets. The key observations may be summarized as follows:

- Meteorites derive from two genetically distinct reservoirs that probably were located inside (NC) and outside (CC) Jupiter's orbit, and remained isolated from each other for the first few million years of the solar system.

- The isotopic difference between the NC and CC reservoirs probably results from a temporal change in the isotopic composition of infalling matter from the Sun's protostellar envelope, which led to the formation of two early, isotopically distinct disk reservoirs. The outer reservoir had a CAI-like isotopic composition, while the inner reservoir was NC-like. Mixing between these two reservoirs resulted in the formation of the CC reservoir, whose isotopic composition is intermediate between those of the two earlier disk reservoirs.
- The rapid growth of Jupiter prevented significant exchange of material from inside and outside its orbit over several million years, thereby maintaining an isotopic difference between the NC and CC reservoirs. The efficiency of the Jupiter barrier and how it affected the inward drift of CC dust into the inner solar system is subject to further study.
- The chronology of meteorite parent body accretion, combined with the NC–CC dichotomy, may be used to constrain the timescale of Jupiter's growth. One possible scenario is that Jupiter's core accreted within the first ~ 0.5 Ma after CAI formation, whereas runaway gas accretion and migration of Jupiter started no earlier than ~ 2 Ma, and was completed by ~ 4–5 Ma after CAI formation. However, other growth scenarios are possible.
- The growth and/or migration of Jupiter resulted in inward scattering of CC bodies, which accounts for the co-occurrence of NC and CC bodies in the present-day asteroid belt. The compositional diversity of bodies in the asteroid belt, therefore, reflects formation of these objects over a wide range of orbital distances. Nevertheless, Earth, and presumably also the other terrestrial planets, also incorporated material that has yet not been identified among meteorites.
- The inward scattering of CC bodies resulted in the incorporation of these objects into the growing terrestrial planets, resulting in a mixed NC–CC composition for the Earth's mantle, at least for siderophile elements like Mo. The final embryo colliding with Earth and resulting in the formation of the Moon probably had either a CC or a mixed NC–CC composition.
- The late-stage accretion of CC material to Earth is sufficient to account for the entire budget of Earth's water and highly volatile species, although additional, earlier accretion of CC bodies is also possible.

Collectively, these observations bear testimony to the important role of Jupiter in shaping the architecture of the inner solar system, including the compositional mixture of the asteroid belt, the presence of water and volatile elements on Earth, and perhaps even the very existence of the Moon.

Acknowledgements Open Access funding provided by Projekt DEAL. This paper benefited from the comments and suggestions provided by R. Brasser and two anonymous reviewers. We thank M. Anand for editorial handling of this paper, and most of all for this patience. TK, CB, and EAW acknowledge funding by the Deutsche Forschungsgemeinschaft (DFG, German Research Foundation)—Project-ID 263649064—TRR 170. FN acknowledges support from NASA (NNX17AE27G). This study was in part performed under the auspices of the US DOE by Lawrence Livermore National Laboratory under Contract DE-AC52-07NA27344. This is TRR 170 pub. no. 94.

References

C.M.O. Alexander, J.N. Grossman, D.S. Ebel, F.J. Ciesla, Science **320**, 1617–1619 (2008)
C.M.O.D. Alexander, R. Bowden, M.L. Fogel, K.T. Howard, C.D.K. Herd, L.R. Nittler, Science **337**, 721–723 (2012)
Y. Alibert, J. Venturini, R. Helled, S. Ataiee, R. Burn, L. Senecal, W. Benz, L. Mayer, C. Mordasini, S.P. Quanz, M. Schönbächler, Nat. Astron. **2**, 873–877 (2018)
Y. Amelin, Geochim. Cosmochim. Acta **72**, 221–232 (2008)
Y. Amelin, A. Kaltenbach, T. Iizuka, C.H. Stirling, T.R. Ireland, M. Petaev, S.B. Jacobsen, Earth Planet. Sci. Lett. **300**, 343–350 (2010)
M. Barboni, P. Boehnke, B. Keller, I.E. Kohl, B. Schoene, E.D. Young, K.D. McKeegan, Sci. Adv. **3**, e1602365 (2017)
K.R. Bermingham, E.A. Worsham, R.J. Walker, Earth Planet. Sci. Lett. **487**, 221–229 (2018)
J.L. Birck, G.W. Lugmair, Earth Planet. Sci. Lett. **90**, 131–143 (1988)
T. Blackburn, C.M.O.D. Alexander, R. Carlson, L.T. Elkins-Tanton, Geochim. Cosmochim. Acta **200**, 201–217 (2017)
J. Bollard, J.N. Connelly, M.J. Whitehouse, E.A. Pringle, L. Bonal, J.K. Jørgensen, Å. Nordlund, F. Moynier, M. Bizzarro, Sci. Adv. **3**, e1700407 (2017)
J. Bollard, N. Kawasaki, N. Sakamoto, M. Olsen, S. Itoh, K. Larsen, D. Wielandt, M. Schiller, J.N. Connelly, H. Yurimoto, M. Bizzarro, Geochim. Cosmochim. Acta **260**, 62 (2019)
L.E. Borg, A.M. Gaffney, T.S. Kruijer, N.A. Marks, C.K. Sio, J. Wimpenny, Earth Planet. Sci. Lett. **523**, 115706 (2019)
R. Brasser, S.J. Mojzsis, Nature Astronomy (2020)
G.A. Brennecka, L.E. Borg, M. Wadhwa, Proc. Natl. Acad. Sci. **110**, 17241–17246 (2013)
G. Budde, C. Burkhardt, G.A. Brennecka, M. Fischer-Gödde, T.S. Kruijer, T. Kleine, Earth Planet. Sci. Lett. **454**, 293–303 (2016)
G. Budde, T.S. Kruijer, T. Kleine, Geochim. Cosmochim. Acta **222**, 284–304 (2018)
G. Budde, C. Burkhardt, T. Kleine, Nat. Astron. **3**, 736–741 (2019)
C. Burkhardt, T. Kleine, F. Oberli, A. Pack, B. Bourdon, R. Wieler, Earth Planet. Sci. Lett. **312**, 390–400 (2011)
C. Burkhardt, T. Kleine, N. Dauphas, R. Wieler, Earth Planet. Sci. Lett. **357–358**, 298–307 (2012)
C. Burkhardt, R.C. Hin, T. Kleine, B. Bourdon, Earth Planet. Sci. Lett. **391**, 201–211 (2014)
C. Burkhardt, L.E. Borg, G.A. Brennecka, Q.R. Shollenberger, N. Dauphas, T. Kleine, Nature **537**, 394–398 (2016)
C. Burkhardt, N. Dauphas, H. Tang, M. Fischer-Gödde, L. Qin, J.H. Chen, S.S. Rout, A. Pack, P.R. Heck, D.A. Papanastassiou, Meteorit. Planet. Sci. **52**, 807–826 (2017)
C. Burkhardt, N. Dauphas, U. Hans, B. Bourdon, T. Kleine, Geochim. Cosmochim. Acta **261**, 145–170 (2019)
R.M. Canup, E. Asphaug, Nature **412**, 708–712 (2001)
D. Carrera, U. Gorti, A. Johansen, M.B. Davies, Astrophys. J. **839**, 16 (2017)
J. Connelly, M. Bizzarro, K. Thrane, J.A. Baker, Geochim. Cosmochim. Acta **72**, 4813–4824 (2008)
J.N. Connelly, M. Bizzarro, A.N. Krot, A. Nordlund, D. Wielandt, M.A. Ivanova, Science **338**, 651–655 (2012)
A. Crida, A. Morbidelli, F. Masset, Icarus **181**, 587–604 (2006)
N. Dauphas, Nature **541**, 521–524 (2017)
N. Dauphas, E.A. Schauble, Annu. Rev. Earth Planet. Sci. **44**, 709–783 (2016)
N. Dauphas, B. Marty, L. Reisberg, Astrophys. J. **565**, 640–644 (2002)
N. Dauphas, A.M. Davis, B. Marty, L. Reisberg, Earth Planet. Sci. Lett. **226**, 465–475 (2004)
N. Dauphas, D.L. Cook, A. Sacarabany, C. Frohlich, A.M. Davis, M. Wadhwa, A. Pourmand, T. Rauscher, R. Gallino, Astrophys. J. **686**, 560–569 (2008)
N. Dauphas, L. Remusat, J.H. Chen, M. Roskosz, D.A. Papanastassiou, J. Stodolna, Y. Guan, C. Ma, J.M. Eiler, Astrophys. J. **720**, 1577–1591 (2010)
N. Dauphas, J.H. Chen, J. Zhang, D.A. Papanastassiou, A.M. Davis, C. Travaglio, Earth Planet. Sci. Lett. **407**, 96–108 (2014)

A.M. Davis, J. Zhang, N.D. Greber, J. Hu, F.L.H. Tissot, N. Dauphas, Geochim. Cosmochim. Acta **221**, 275–295 (2018)

S.J. Desch, A. Kalyaan, C.M.O'D. Alexander, Astrophys. J. Suppl. Ser. **238**(1), 11 (2018)

P.M. Doyle, K. Jogo, K. Nagashima, A.N. Krot, S. Wakita, F.J. Ciesla, I.D. Hutcheon, Nat. Commun. **6**, 7444 (2015)

S. Ebert, J. Render, G.A. Brennecka, C. Burkhardt, A. Bischoff, S. Gerber, T. Kleine, Earth Planet. Sci. Lett. **498**, 257–265 (2018)

L.T. Elkins-Tanton, B.P. Weiss, M.T. Zuber, Earth Planet. Sci. Lett. **305**, 1–10 (2011)

R.A. Fischer, F. Nimmo, D.P. O'Brien, Earth Planet. Sci. Lett. **482**, 105–114 (2018)

M. Fischer-Gödde, T. Kleine, Nature **541**, 525–527 (2017)

M. Fischer-Gödde, C. Burkhardt, T.S. Kruijer, T. Kleine, Geochim. Cosmochim. Acta **168**, 151–171 (2015)

S. Gerber, C. Burkhardt, G. Budde, K. Metzler, T. Kleine, Astrophys. J. Lett. **841**, L17 (2017)

K. Gerbig, C.T. Lenz, H. Klahr, Astron. Astrophys. **629**, A116 (2019)

B.M.S. Hansen, Astrophys. J. **703**, 1131–1140 (2009)

T. Haugbølle, P. Weber, D.P. Wielandt, P. Benítez-Llambay, M. Bizzarro, O. Gressel, M.E. Pessah, Astron. J. **158**, 55 (2019)

J.L. Hellmann, T.S. Kruijer, J.A. Van Orman, K. Metzler, T. Kleine, Geochim. Cosmochim. Acta **258**, 290–309 (2019)

S. Henke, H.P. Gail, M. Trieloff, W.H. Schwarz, T. Kleine, Astron. Astrophys. **545**, A135 (2012)

P.J. Hevey, I.S. Sanders, Meteorit. Planet. Sci. **41**, 95–106 (2006)

T. Hopp, G. Budde, T. Kleine, Earth Planet. Sci. Lett. **534**, 116065 (2020)

R. Hueso, T. Guillot, Astron. Astrophys. **442**, 703–725 (2005)

A.C. Hunt, D.L. Cook, T. Lichtenberg, P.M. Reger, M. Ek, G.J. Golabek, M. Schönbächler, Earth Planet. Sci. Lett. **482**, 490–500 (2018)

G.R. Huss, R.S. Lewis, Geochim. Cosmochim. Acta **59**, 115–160 (1995)

A. Izidoro, S.N. Raymond, A. Pierens, A. Morbidelli, O.C. Winter, D. Nesvorny, Astrophys. J. **833**, 40 (2016)

E. Jacquet, S. Fromang, M. Gounelle, Astron. Astrophys. **526**, L8 (2011)

E. Jacquet, F.C. Pignatale, M. Chaussidon, S. Charnoz, Astrophys. J. **884**, 32 (2019)

K. Jogo, T. Nakamura, M. Ito, S. Wakita, M.Y. Zolotov, S.R. Messenger, Geochim. Cosmochim. Acta **199**, 58–74 (2017)

N.T. Kita, T. Ushikubo, Meteorit. Planet. Sci. **47**, 1108–1119 (2012)

T. Kleine, R.J. Walker, Annu. Rev. Earth Planet. Sci. **45**, 389–417 (2017)

T. Kleine, U. Hans, A.J. Irving, B. Bourdon, Geochim. Cosmochim. Acta **84**, 186–203 (2012)

T. Kleine, G. Budde, J.L. Hellmann, T.S. Kruijer, C. Burkhardt, in *Chondrules: Records of Protoplanetary Disk Processes*, ed. by A.N. Krot, H.C. Connolly Jr., S.S. Russell (Cambridge University Press, Cambridge, 2018), pp. 276–299

T.S. Kruijer, M. Touboul, M. Fischer-Godde, K.R. Bermingham, R.J. Walker, T. Kleine, Science **344**, 1150–1154 (2014)

T.S. Kruijer, C. Burkhardt, G. Budde, T. Kleine, Proc. Natl. Acad. Sci. **114**, 6712–6716 (2017)

T.S. Kruijer, T. Kleine, L.E. Borg, Nat. Astron. **4**, 32–40 (2020)

T. Kunihiro, A.E. Rubin, K.D. McKeegan, J.T. Wasson, Geochim. Cosmochim. Acta **68**, 2947–2957 (2004)

E. Kurahashi, N.T. Kita, H. Nagahara, Y. Morishita, Geochim. Cosmochim. Acta **72**, 3865–3882 (2008)

M. Lambrechts, A. Johansen, Astron. Astrophys. **572**, A107 (2014)

M. Lambrechts, A. Johansen, A. Morbidelli, Astron. Astrophys. **572**, A35 (2014)

T. Lee, D.A. Papanastassiou, G.J. Wasserburg, Astrophys. J. **220**, L21 (1978)

H.F. Levison, K.A. Kretke, M.J. Duncan, Nature **524**, 322–324 (2015)

I. Leya, M. Schonbachler, U. Wiechert, U. Krahenbuhl, A.N. Halliday, Earth Planet. Sci. Lett. **266**, 233–244 (2008)

B. Marty, Earth Planet. Sci. Lett. **313–314**, 56–66 (2012)

F.M. McCubbin, J.J. Barnes, Earth Planet. Sci. Lett. **526**, 115771 (2019)

M.T. McCulloch, G.J. Wasserburg, Geophys. Res. Lett. **5**, 599–602 (1978)

K.D. McKeegan, A.P.A. Kallio, V.S. Heber, G. Jarzebinski, P.H. Mao, C.D. Coath, T. Kunihiro, R.C. Wiens, J.E. Nordholt, R.W. Moses, D.B. Reisenfeld, A.J.G. Jurewicz, D.S. Burnett, Science **332**, 1528 (2011)

A. Morbidelli, B. Bitsch, A. Crida, M. Gounelle, T. Guillot, S. Jacobson, A. Johansen, M. Lambrechts, E. Lega, Icarus **267**, 368–376 (2016)

A. Morbidelli, D. Nesvorny, V. Laurenz, S. Marchi, D.C. Rubie, L. Elkins-Tanton, M. Wieczorek, S. Jacobson, Icarus **305**, 262–276 (2018)

K. Nagashima, A.N. Krot, M. Komatsu, Geochim. Cosmochim. Acta **201**, 303–319 (2017)

J.A.M. Nanne, F. Nimmo, J.N. Cuzzi, T. Kleine, Earth Planet. Sci. Lett. **511**, 44–54 (2019)

D. Nesvorný, D. Vokrouhlický, W.F. Bottke, H.F. Levison, Nat. Astron. **2**, 878–882 (2018)

F.R. Niederer, D.A. Papanastassiou, G.J. Wasserburg, Geochim. Cosmochim. Acta **45**, 1017–1031 (1981)

F.R. Niederer, D.A. Papanastassiou, G.J. Wasserburg, Geochim. Cosmochim. Acta **49**, 835–851 (1985)
S. Niemeyer, Geochim. Cosmochim. Acta **52**, 309–318 (1988)
S. Niemeyer, G.W. Lugmair, Earth Planet. Sci. Lett. **53**, 211–225 (1981)
S. Niemeyer, G.W. Lugmair, Geochim. Cosmochim. Acta **48**, 1401–1416 (1984)
F. Nimmo, D.P. O'Brien, T. Kleine, Earth Planet. Sci. Lett. **292**, 363–370 (2010)
L.R. Nittler, C.M.O'D. Alexander, N. Liu, J. Wang, Astrophys. J. **856**, L24 (2018)
D.P. O'Brien, A. Izidoro, S.A. Jacobson, S.N. Raymond, D.C. Rubie, Space Sci. Rev. **214**, 47 (2018)
M.B. Olsen, D. Wielandt, M. Schiller, E.M.M.E. Van Kooten, M. Bizzarro, Geochim. Cosmochim. Acta **191**, 118–138 (2016)
H. Palme, K. Lodders, A. Jones, in *Treatise on Geochemistry*, ed. by K.K. Turekian 2nd edn. (Elsevier, Oxford, 2014), pp. 15–36
D.A. Papanastassiou, Astrophys. J. **308**, L27–L30 (1986)
J. Pape, K. Mezger, A.S. Bouvier, L.P. Baumgartner, Geochim. Cosmochim. Acta **244**, 416–436 (2019)
F.C. Pignatale, S. Charnoz, M. Chaussidon, E. Jacquet, Astrophys. J. Lett. **867**, L23 (2018)
J.B. Pollack, O. Hubickyj, P. Bodenheimer, J.J. Lissauer, M. Podolak, Y. Greenzweig, Icarus **124**, 62–85 (1996)
G.M. Poole, M. Rehkämper, B.J. Coles, T. Goldberg, C.L. Smith, Earth Planet. Sci. Lett. **473**, 215–226 (2017)
L. Qin, L.R. Nittler, C.M.O.D. Alexander, J. Wang, F.J. Stadermann, R.W. Carlson, Geochim. Cosmochim. Acta **75**, 629–644 (2011)
S.N. Raymond, A. Izidoro, Icarus **297**, 134–148 (2017)
M. Regelous, T. Elliott, C.D. Coath, Earth Planet. Sci. Lett. **272**, 330–338 (2008)
J. Render, M. Fischer-Gödde, C. Burkhardt, T. Kleine, Geochem. Perspect. Lett. **3**, 170–178 (2017)
J. Render, G.A. Brennecka, S.-J. Wang, L.E. Wasylenki, T. Kleine, Astrophys. J. **862**, 26 (2018)
J. Render, S. Ebert, C. Burkhardt, T. Kleine, G.A. Brennecka, Geochim. Cosmochim. Acta **254**, 40–53 (2019)
M. Rotaru, J.L. Birck, C.J. Allegre, Nature **358**, 465–470 (1992)
J.F. Rudge, T. Kleine, B. Bourdon, Nat. Geosci. **3**, 439–443 (2010)
N.G. Rudraswami, J.N. Goswami, Earth Planet. Sci. Lett. **257**, 231–244 (2007)
N.G. Rudraswami, J.N. Goswami, B. Chattopadhyay, S.K. Sengupta, A.P. Thapliyal, Earth Planet. Sci. Lett. (2008). https://doi.org/10.1016/j.epsl.2008.07.004
A.R. Sarafian, S.G. Nielsen, H.R. Marschall, F.M. McCubbin, B.D. Monteleone, Science **346**, 623 (2014)
A.R. Sarafian, S.G. Nielsen, H.R. Marschall, G.A. Gaetani, E.H. Hauri, K. Righter, E. Sarafian, Geochim. Cosmochim. Acta **212**, 156–166 (2017)
M. Schiller, M. Bizzarro, V.A. Fernandes, Nature **555**, 507 (2018)
M. Schiller, M. Bizzarro, J. Siebert, Sci. Adv. **6**, eaay7604 (2020)
D.L. Schrader, K. Nagashima, A.N. Krot, R.C. Ogliore, Q.-Z. Yin, Y. Amelin, C.H. Stirling, A. Kaltenbach, Geochim. Cosmochim. Acta **201**, 275–302 (2017)
E.R.D. Scott, J.T. Wasson, Rev. Geophys. **13**, 527–546 (1975)
F. Spitzer, C. Burkhardt, G. Budde, T.S. Kruijer, T. Kleine, Lunar Planet. Sci. Conf. **51**, 3040 (2020)
L. Spivak-Birndorf, M. Wadhwa, P.E. Janney, Geochim. Cosmochim. Acta **73**, 5202–5211 (2009)
R.C.J. Steele, C.D. Coath, M. Regelous, S. Russell, T. Elliott, Astrophys. J. **758**, 59 (2012)
T. Stephan, R. Trappitsch, P. Hoppe, A.M. Davis, M.J. Pellin, O.S. Pardo, Astrophys. J. **877**, 101 (2019)
S.T. Stewart, P.J. Carter, E.J. Davies, S.J. Lock, R.G. Kraus, S. Root, M.I. Petaev, S.B. Jacobsen, Lunar Planet. Sci. Conf. **50**, 1251 (2019)
N. Sugiura, W. Fujiya, Meteorit. Planet. Sci. **49**, 772–787 (2014)
H.L. Tang, N. Dauphas, Earth Planet. Sci. Lett. **359**, 248–263 (2012)
A. Trinquier, J.L. Birck, C.J. Allègre, Astrophys. J. **655**, 1179–1185 (2007)
A. Trinquier, T. Elliott, D. Ulfbeck, C. Coath, A.N. Krot, M. Bizzarro, Science **324**, 374–376 (2009)
E.M.M.E. Van Kooten, D. Wielandt, M. Schiller, K. Nagashima, A. Thomen, K.K. Larsen, M.B. Olsen, Å. Nordlund, A.N. Krot, M. Bizzarro, Proc. Natl. Acad. Sci. USA **113**, 2011–2016 (2016)
J. Villeneuve, M. Chaussidon, G. Libourel, Science **325**, 985–988 (2009)
R. Visser, E.F. van Dishoeck, S.D. Doty, C.P. Dullemond, Astron. Astrophys. **495**, 881–897 (2009)
R.J. Walker, K. Bermingham, J. Liu, I.S. Puchtel, M. Touboul, E.A. Worsham, Chem. Geol. **411**, 125–142 (2015)
K.J. Walsh, A. Morbidelli, S.N. Raymond, D.P. O'Brien, A.M. Mandell, Nature **475**, 206–209 (2011)
H. Wang, B.P. Weiss, X.-N. Bai, B.G. Downey, J. Wang, J. Wang, C. Suavet, R.R. Fu, M.E. Zucolotto, Science **355**, 623 (2017)
P.H. Warren, Earth Planet. Sci. Lett. **311**, 93–100 (2011)
P. Weber, P. Benítez-Llambay, O. Gressel, L. Krapp, M.E. Pessah, Astrophys. J. **854**, 153 (2018)
E.A. Worsham, K.R. Bermingham, R.J. Walker, Earth Planet. Sci. Lett. **467**, 157–166 (2017)

E.A. Worsham, C. Burkhardt, G. Budde, M. Fischer-Gödde, T.S. Kruijer, T. Kleine, Earth Planet. Sci. Lett. **521**, 103–112 (2019)
L. Yang, F.J. Ciesla, Meteorit. Planet. Sci. **47**, 99–119 (2012)
T. Yokoyama, Y. Nagai, R. Fukai, T. Hirata, Astrophys. J. **883**, 62 (2019)
J.J. Zhang, N. Dauphas, A.M. Davis, I. Leya, A. Fedkin, Nat. Geosci. **5**, 251–255 (2012)
E. Zinner, in *Treatise on Geochemistry*, ed. by H.D. Holland, K.K. Turekian 2nd edn. (Elsevier, Oxford, 2014), pp. 181–213

Space Sci Rev (2020) 216:59
https://doi.org/10.1007/s11214-020-00671-0

Exploring the Bimodal Solar System via Sample Return from the Main Asteroid Belt: The Case for Revisiting Ceres

Thomas H. Burbine[1] · **Richard C. Greenwood**[2]

Received: 24 July 2019 / Accepted: 7 April 2020 / Published online: 18 May 2020

Abstract Sample return from a main-belt asteroid has not yet been attempted, but appears technologically feasible. While the cost implications are significant, the scientific case for such a mission appears overwhelming. As suggested by the "Grand Tack" model, the structure of the main belt was likely forged during the earliest stages of Solar System evolution in response to migration of the giant planets. Returning samples from the main belt has the potential to test such planet migration models and the related geochemical and isotopic concept of a bimodal Solar System.

Isotopic studies demonstrate distinct compositional differences between samples believed to be derived from the outer Solar System (CC or carbonaceous chondrite group) and those that are thought to be derived from the inner Solar System (NC or non-carbonaceous group). These two groups are separated on relevant isotopic variation diagrams by a clear compositional gap. The interface between these two regions appears to be broadly coincident with the present location of the asteroid belt, which contains material derived from both groups.

The Hayabusa mission to near-Earth asteroid (NEA) (25143) Itokawa has shown what can be learned from a sample-return mission to an asteroid, even with a very small amount of sample. One scenario for main-belt sample return involves a spacecraft launching a projectile that strikes an object and flying through the debris cloud, which would potentially allow multiple bodies to be sampled if a number of projectiles are used on different asteroids. Another scenario is the more traditional method of landing on an asteroid to obtain the sample.

A significant range of main-belt asteroids are available as targets for a sample-return mission and such a mission would represent a first step in mineralogically and isotopically mapping the asteroid belt. We argue that a sample-return mission to the asteroid belt does

Role of Sample Return in Addressing Major Questions in Planetary Sciences
Edited by Mahesh Anand, Sara Russell, Yangting Lin, Meenakshi Wadhwa, Kuljeet Kaur Marhas and Shogo Tachibana

✉ T.H. Burbine
tburbine@mtholyoke.edu

[1] Department of Astronomy, Mount Holyoke College, South Hadley, MA 01075, USA

[2] Planetary and Space Sciences, School of Physical Sciences, The Open University, Walton Hall, Milton Keynes, MK7 6AA, UK

not necessarily have to return material from both the NC and CC groups to viably test the bimodal Solar System paradigm, as material from the NC group is already abundantly available for study. Instead, there is overwhelming evidence that we have a very incomplete suite of CC-related samples.

Based on our analysis, we advocate a dedicated sample-return mission to the dwarf planet (1) Ceres as the best means of further exploring inherent Solar System variation. Ceres is an ice-rich world that may be a displaced trans-Neptunian object. We almost certainly do not have any meteorites that closely resemble material that would be brought back from Ceres. The rich heritage of data acquired by the Dawn mission makes a sample-return mission from Ceres logistically feasible at a realistic cost. No other potential main-belt target is capable of providing as much insight into the early Solar System as Ceres. Such a mission should be given the highest priority by the international scientific community.

Keywords Asteroid · Sample return · Main belt · Ceres · Meteorites · Spacecraft

1 Introduction

1.1 Why Is Sample Return Important?

Thanks to the regular delivery of meteorite samples to Earth, we already have an extensive and diverse range of relatively pristine extraterrestrial materials in our worldwide collections. These falls are augmented by an even more extensive inventory of meteorite finds, mainly recovered from hot (e.g., Sahara) and cold (e.g., Antarctica) desert regions. The result is that there are currently more than 60,000 individual meteorite samples (Meteoritical Bulletin Database 2020) available on Earth for analysis. In addition to meteorites, we also have significant collections of Interplanetary Dust Particles (IDPs) and micrometeorites collected in a diverse range of terrestrial environments, including the stratosphere, deep ocean basins, and polar regions (Brownlee 1985; Bradley 2014; Noguchi et al. 2015).

With such a diverse assemblage of, essentially low-cost, extraterrestrial materials available for detailed characterization studies, it is a valid question to ask why significantly more expensive samples need to be collected by robotic missions.

The problem is that apart from relatively rare fireball trajectory information (Devillepoix et al. 2018), and one-off events, such as the tracking of asteroid 2008 TC_3 prior to atmospheric entry (Shaddad et al. 2010), most Earth-recovered extraterrestrial materials provide little detailed information about which bodies such samples actually originated from. In contrast, collecting material directly from a well characterized source body, as was the case of asteroid (25143) Itokawa (Nakamura et al. 2011), provides a geologic context that is unavailable for Earth-recovered samples. A very important aspect of these sample-return missions is that they link together laboratory (e.g., isotopic analyses, mineralogy) and remote sensing (e.g., images, reflectance spectra) data.

Another problem with the extraterrestrial samples recovered on Earth is that they are likely to be a very unrepresentative and biased sampling of Solar System bodies (Campbell-Brown 2019). Very friable carbonaceous materials are likely under-represented in the meteorite record as they tend not to be able to survive atmospheric entry intact (e.g., Sears 1998). Poorly consolidated meteorites, such as the ungrouped carbonaceous chondrite Tagish Lake, are rare exceptions (Brown et al. 2000).

Sample return is not a new concept and in fact has already made a very significant contribution to our understanding of Solar System evolution. It is not an exaggeration to say that some of the greatest engineering feats of the late 20th and early 21st century have been sample-return missions from planetary bodies. Samples have been returned from the Moon (Apollo and Luna missions) (e.g., Lunar Sample Preliminary Examination Team 1969), a comet [Stardust mission to comet 81P/Wild (Wild 2)] (e.g., Zolensky et al. 2006), and a near-Earth asteroid (NEA) [Hayabusa mission to (25143) Itokawa] (e.g., Nakamura et al. 2011). Solar wind particles have also been collected from space by the Genesis mission (e.g., Grimberg et al. 2006). Two NEA sample-return missions are currently being conducted: the Hayabusa2 spacecraft to (162173) Ryugu (e.g., Wada et al. 2018; Watanabe et al. 2019) and the OSIRIS-REx spacecraft to (101955) Bennu (e.g., Bierhaus et al. 2018; Lauretta et al. 2019). Both asteroids are C-complex bodies. C-complex asteroids tend to have low albedos, absorption features (when present) due to hydrated silicates, and have been typically linked with carbonaceous chondrites. A Japanese sample-return mission (MMX: Martian Moons eXploration) to Phobos is currently being planned (Usui et al. 2020). Except for the Apollo missions, where the sampling was undertaken by humans, all sample-return missions so far have been done robotically.

These sample-return missions have helped to "solve" a number of planetary science questions. Lunar samples have given evidence for an impact origin for the Moon (Hartmann and Davis 1975), the presence of an early magma ocean (Wood et al. 1970), and the likely existence of a relatively high water content in the lunar interior (Saal et al. 2008). Samples from comet Wild 2 have revealed the extent to which mixing of high temperature solids took place throughout the disk during the earliest stages of Solar System evolution (Zolensky et al. 2006; Brownlee 2014; Westphal et al. 2017). Analyses of samples collected by Hayabusa show evidence for space weathering (Noguchi et al. 2011) and confirm that at least one S-complex asteroid has an ordinary chondrite composition (Nakamura et al. 2011). S-complex bodies have absorption features that tend to be due to olivine and/or pyroxene and have been typically linked with ordinary chondrites. Analyses of solar wind samples collected by the Genesis mission have allowed the oxygen isotopic composition of the Sun to be estimated (McKeegan et al. 2011).

1.2 Why do We Need Samples from the Asteroid Belt?

While NEAs have been sampled by the Hayabusa mission (Nakamura et al. 2011) and are currently the targets of Hayabusa2 and OSIRIS-REx missions (Watanabe et al. 2019; Lauretta et al. 2019), what has so far never been attempted is a sample-return mission from a main-belt object, which have semi-major axes (a) between $\sim$2.1 and $\sim$3.3 AU. While the cost implications of such a mission are significant, the scientific case for it is now overwhelming. The reason for this is that potentially the main belt holds the key to a more profound understanding of early Solar System evolution. This is because the structure of the main belt was likely forged during the earliest stages of Solar System history, most probably in response to migration of the giant planets (Walsh et al. 2011, 2012). This migration would have scattered bodies into their present-day locations in the asteroid belt.

Returning samples from a main-belt asteroid has the potential, not only to test current planet migration models (e.g., Walsh et al. 2011, 2012), but also to furnish critical evidence with which to test current geochemical concepts of a bimodal Solar System (e.g., Warren 2011; Kruijer et al. 2017, 2020; Scott et al. 2018; Kleine et al. 2020) by mineralogically and isotopically mapping the asteroid belt. This bimodality is in the form of compositional and isotopic differences between materials that most likely originated in the inner Solar System compared to those that probably formed in the outer Solar System (Warren 2011). As we will discuss, models for the early dynamical evolution of the main belt are inextricably linked to the concept of such a bimodal Solar System (Morbidelli et al. 2015). The central goal of a sample-return mission to the main belt would be to rigorously test models of Solar System formation. The return and subsequent detailed laboratory analyses of samples collected directly from main-belt asteroids represent the most fundamental approach available, both in terms of testing our existing migration models and also obtaining critical data that may lead to the development of radically new ideas about the formation of our Solar System and its early evolution. The latest planetary decadal survey (National Research Council 2011) states that one of the important questions to try to answer in the future is "What were the initial stages, conditions, and processes of Solar System formation and the nature of the interstellar matter that was incorporated?"

The problem with sample return from just NEAs is that these bodies have dynamical lifetimes of only a few million years (Gladman et al. 1997) and are fragments derived from much larger bodies. The probability that a NEA or meteorite with a known orbit originates from a particular source region (e.g., Hungaria region, ν_6 secular resonance, 3:1 mean motion resonance) (e.g., Bottke et al. 2002a; Granvik and Brown 2018) can be calculated; however, it is currently not possible to definitively link any NEA or meteorite with any main-belt object. Therefore, sample return from a NEA does not give any definitive information on the exact location in the main belt where the NEA originated.

In this paper, we will look in detail at why sample return from the main belt is a critical next step in the exploration of the Solar System and has the potential to significantly improve our understanding of its early evolution. We will look first at the present-day structure of the asteroid belt and review dynamical models for its evolution. We will then examine recent geochemical concepts that invoke an essentially bimodal composition for Solar System materials. We will discuss the importance of sample return from main-belt objects and possible scenarios for bringing fragments of these bodies back to Earth. We will examine possible destinations and finally present what we believe is the single most strategic target to "solve" the bimodal Solar System paradigm.

2 Structure of the Main Belt and Its Link to Gas Giant Migration

The asteroid belt is usually defined as the region lying roughly between the orbits of Mars and Jupiter (Fig. 1). One of the most significant features of the main belt is its low mass. It currently contains only ~3% of the mass of the Moon. Estimates for the primordial mass of the asteroid belt range from between three and four orders of magnitude (Clement et al. 2019) to ~3.5 times more massive (Levison et al. 2015) than its current mass. Significant mass loss from the belt likely took place at a very early stage in Solar System evolution, prior to Jupiter attaining its full size (Bottke et al. 2005; Clement et al. 2019).

At the broadest scale, the main belt has a distinct structure with the outer regions (~2.8-3.3 AU) dominated by C-complex and P-type asteroids, whereas its inner regions (~2.1-2.5 AU) contain a high percentage of S-complex asteroids (Gradie and Tedesco 1982; DeMeo and Carry 2014). However, on a more detailed scale, there is significant mixing of different spectral classes throughout the belt (DeMeo and Carry 2014).

Dynamic models have been developed in part to explain the structure of the main belt and in part the small size of Mars. The "Grand Tack" model is based on the premise that prior to dispersion of the gas phase in the solar nebula, Jupiter and Saturn underwent an initial inwards migration with Jupiter reaching ~1.5 AU, roughly the current location of Mars (Walsh et al. 2011, 2012) (Fig. 2) and then both bodies subsequently migrated outwards to their present positions. Prior to migration, Jupiter and Saturn were still in the final

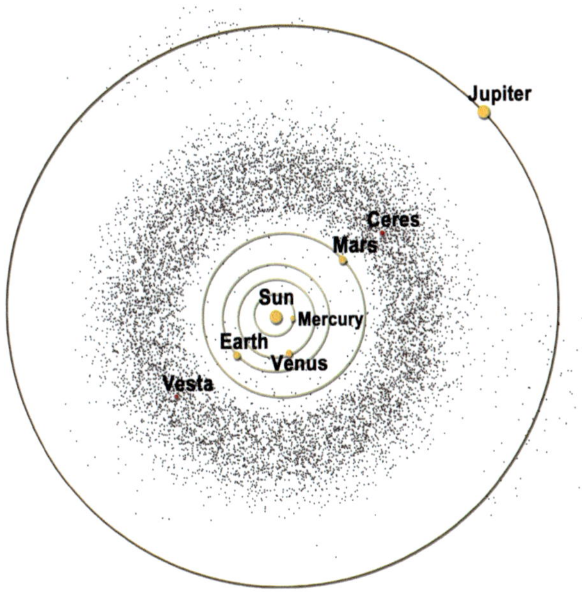

Fig. 1 Artist's depiction of the structure and location of the asteroid belt (2007). The locations of the Sun, Ceres, and Vesta are given. The orbits of Mercury, Venus, Earth, Mars, and Jupiter are also shown. Trojan asteroids, which are found at Jupiter's distance from the Sun, are also displayed. Image credit: NASA/McREL

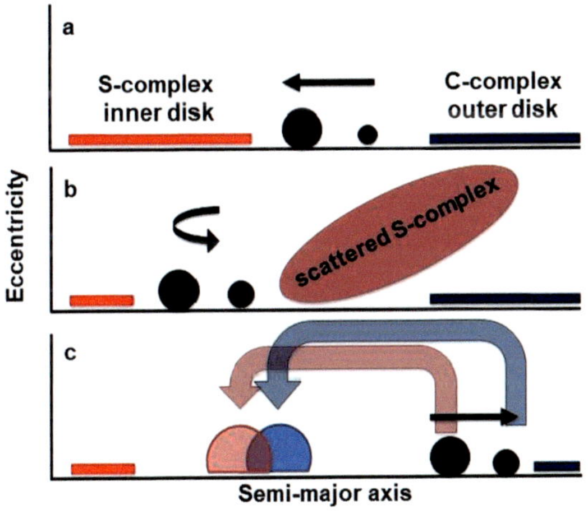

Fig. 2 A cartoon of the "Grand Tack" model for the early evolution of the Solar System. The large dark circle is Jupiter and the smaller dark circle is Saturn. (**a**) Jupiter and Saturn are growing in a region outside the current location of the main belt with distinct populations of planetesimals in the inner and outer parts of the Solar System. (**b**) Jupiter and Saturn undergo gas-driven inward migration with many of the S-complex asteroids from the inner Solar System being scattered outwards. (**c**) Subsequent outwards migration of Jupiter and Saturn populates the inner main-belt region with S-complex asteroids that had originally been scattered outwards and then subsequently populates the outer margins of the main belt with C-complex asteroids, which are scattered inwards during the final stages of the gas giants' migration. Plot used with the permission of Kevin Walsh

stages of accretion and a gas phase was present in the nebula. Migration would therefore have been early, only a few million years after the formation of the earliest nebular solids, calcium-aluminium-rich inclusions (CAIs). CAIs have mineralogies consistent with being early condensates from a hot gas of solar composition (Grossman 1972). The population of planetesimals at this early stage of Solar System evolution is considered to have been bimodal (Fig. 2a). Inwards of the gas giants (Jupiter and Saturn), planetesimals are believed to be essentially anhydrous S-complex bodies and outwards they were hydrated and volatile-rich (C-complex bodies). The initial inward migration of the gas giants to ~1.5 AU scattered the S-complex planetesimals outwards, cleaning out the primordial belt region and truncating the disk (Fig. 2b). The restricted amount of material left at ~1.5 AU is argued to be the reason for the relatively small size of Mars compared to the other terrestrial planets (Walsh et al. 2011). The inward migration of Jupiter and Saturn was then reversed when Saturn became captured in a 2:3 mean motion resonance (Pierens and Raymond 2011). As the outward migration took place, the gas giants would first have encountered the remnants of the S-complex bodies that had been scattered during the initial inward migration (Fig. 2c). These would then have been scattered back into the inner part of the main belt. As migration continued, the C-complex asteroids that would have populated the outer part of the Solar System would then have been encountered and a fraction of them scattered inwards to populate what became the outer portion of the current main belt (Walsh et al. 2011, 2012). The outward migration phase ended once the nebular gas had dissipated (Walsh et al. 2012).

The "Grand Tack" model provides a viable explanation for the origin and structure of the asteroid belt. It offers a coherent mechanism to explain how the main belt lost much of its primordial mass and also why it shows an overall trend from predominantly S-complex bodies at its inner margin to C-complex bodies at its outer margin. The "Grand Tack" model (Fig. 2) is inextricably linked to the concept of a bimodal Solar System in which materials that formed inwards of the "snow line" were distinctly different from those that formed outside of it (Lecar et al. 2006).

However, the degree of scattering of bodies in the "Grand Tack" model has not been tested with the analysis of samples derived from objects with known locations in the asteroid belt. Sample return from main-belt asteroids would allow mineralogical and isotopic compositions to be correlated with present distance from the Sun. Our meteorite collections contain evidence for between ~100-150 parent bodies (e.g., Greenwood et al. 2020), but the "Grand Tack" model groups asteroids into only two types of general compositions (C- and S-complex bodies). Samples from main-belt bodies should allow the "Grand Tack" model to be refined to better duplicate the compositional and isotopic differences in the Solar System.

It is important to note that the "Grand Tack" is not the only recent model put forward to explain the present day structure of the main belt. The results of a dynamical simulation study undertaken by Raymond and Izidoro (2017) are consistent with the main belt forming empty and only acquiring mass as a consequence of scattering and drag processes from the inner terrestrial planet region and outer Solar System. Invoking a pebble accretion model, Kretke et al. (2017) reach similar conclusions.

Irrespective of whether the main belt evolved by significant mass loss or alternatively never had much mass is the first place, the concept of a compositionally bimodal Solar System is supported by significant geochemical evidence. We look in detail at the supporting evidence for a bimodal Solar System concept in the next section.

3 Bimodality in the Early Solar System

Warren (2011), based on data from a range of earlier studies (e.g., Trinquier et al. 2007; Qin et al. 2010a), demonstrated that Solar System materials show a distinct bimodal distribution. with respect to a number of isotopic systems. This variation is clearly seen on a plot of $\Delta^{17}O$ versus $\varepsilon^{54}Cr$ (Fig. 3). Two distinct groupings are present. One cluster contains all

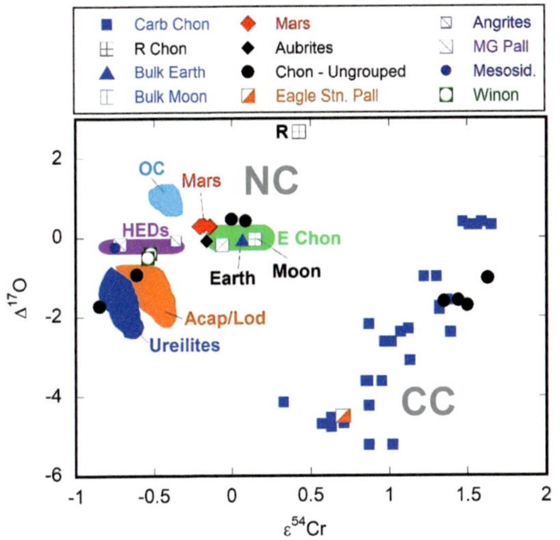

Fig. 3 Plot of Δ^{17}O versus ε^{54}Cr for a range of Solar System materials. This figure is modified from plots found in Warren (2011) and Greenwood et al. (2020). Two distinct clusters of data are present. The cluster in the upper left comprises the non-carbonaceous chondrite (NC) group and includes the ordinary chondrites (OC); enstatite chondrites (E Chon); R chondrites (R); Bulk Earth; Bulk Moon; Mars; howardites, eucrites, and diogenites (HEDs); main-group pallasites (MG Pall); aubrites; angrites; mesosiderites (Mesosid.); winonaites (Winon); and ureilites. The carbonaceous chondrite (CC) cluster in the lower right includes all the carbonaceous chondrites (Carb Chon), Eagle Station pallasites (Eagle Stn. Pall), and various ungrouped chondrites (Chon – Ungrouped). The NC and CC groups are separated by a well-defined gap, which is sometimes referred to as "The Warren Gap" (e.g., Voosen 2018) based on its initial identification by Warren (2011). By definition, the Earth falls at an ε^{54}Cr value of 0 and a Δ^{17}O value of 0

the carbonaceous chondrites and a relatively minor subset of achondrites. The other consists of all other Solar System materials, including planetary-derived samples (Mars, Earth, Moon); ordinary, enstatite, and R chondrites; and a wide range of achondrites (main-group pallasites, howardites, eucrites, diogenites, ureilites, aubrites). Warren (2011) suggested that the carbonaceous chondrite (CC) group may represent material that accreted in the outer Solar System, whereas the non-carbonaceous chondrite (NC) group might be materials derived from the inner Solar System. The gap separating the NC and CC groups is sometimes referred to as "The Warren Gap" (e.g., Voosen 2018).

While the NC-CC dichotomy was originally identified with reference to the isotopes of a relatively limited number of elements (Cr, Ti, O, Ni, Mo, Mg) (Warren 2011; Budde et al. 2016; Van Kooten et al. 2016; Kruijer et al. 2017, 2020), this variation has now been documented for a much wider range of elements (Burkhardt et al. 2019). In addition to silicate-dominated types, iron meteorites also display a distinct bimodal distribution of nucleosynthetic anomalies (e.g., Budde et al. 2016).

With the notable exception of oxygen, isotopic mass-independent variation in extraterrestrial samples, not caused by spallation or radioactive decay, reflects nucleosynthetic processes in the feeder stars to the Solar System (Dauphas and Schauble 2016; Scott et al. 2018; Burkhardt et al. 2019; Nanne et al. 2019). Such anomalies are a reflection, at various scales, of the heterogeneous distribution of the presolar grains that carried these anomalies and which would have been derived from a range of stellar sources (Kruijer et al. 2020). The processes of mixing and homogenization that took place in the parental molecular cloud of the Solar System, and later within the solar protoplanetary disk, were not sufficient to erase these anomalies (Burkhardt et al. 2019; Nanne et al. 2019; Kruijer et al. 2020). In the case of oxygen, mass-independent variation may be the result of selective UV dissociation of CO, either in the presolar giant molecular cloud or the solar nebula (Clayton 2002; Yurimoto and Kuramoto 2004; Lyons and Young 2005). The oxygen isotope anomalies produced by this process may then have become locked into different phases, including water ice, gas, and dust. Preservation of these oxygen isotopic differences may also reflect incomplete homogenization in the protosolar nebula (Ireland et al. 2020).

Recent studies have invoked a change in the composition of the infalling material (Burkhardt et al. 2019; Nanne et al. 2019; Kruijer et al. 2020) to produce this isotopic dichotomy. CAIs are believed to have formed close to the proto-Sun and preserve the isotopic composition of the earliest Solar System solids. This material was enriched in nuclides produced in neutron-rich stellar environments and would have been transported outwards through viscous spreading. Later infalling material, assumed to be depleted in neutron-rich nuclides, would have tended to accumulate in the inner part of the disk and have diluted the isotopic signature within the NC region.

That Solar System materials should show significant variation with respect to a range of isotopic systems is not a novel finding and, particularly with respect to oxygen, has been well documented (Greenwood et al. 2017, 2020). However, the preservation of a distinct compositional gap between inner and outer Solar System-derived materials, as demonstrated by Warren (2011), is unexpected. The early Solar System was likely to have been a highly energetic environment, with considerable mixing taking place between different reservoirs of gas and dust (Misener et al. 2019). One explanation that has been advanced to explain the preservation of the NC-CC dichotomy is that it can be related to the early, rapid accretion of Jupiter, which then acted as a barrier between the inner and outer Solar System regions preventing complete isotopic homogenization within the disk (Kruijer et al. 2017, 2020). This scenario has been disputed by Brasser and Mojzsis (2020) who argue that Jupiter accreted at too slow a rate to have represented a significant barrier to mixing within the disk. Instead, Brasser and Mojzsis (2020) invoke a pressure maximum in the disk close to the present location of Jupiter and suggest that it was this feature, rather than the planet itself, which prevented significant mixing between the inner and outer Solar System regions. Brasser and Mojzsis (2020) speculate that the early Solar System may have had multiple rings, such as those observed by ALMA (Atacama Large Millimeter/submillimeter Array) in the disk around the young star TW Hydrae (Fig. 4) (Andrews et al. 2016). If this interpretation is correct then additional data clusters, over and above the dichotomy so far identified, might be present on isotopic plots such as shown in Fig. 3. The fact that additional clusters have not so far been identified may be a reflection of inadequate sampling of Solar System materials, an insufficient level of analytical resolution to identify such clusters, or alternatively these features do not actually exist.

Bimodality in Solar System materials is not just confined to isotopic anomalies. It has been well known for many decades that carbonaceous chondrites on the one hand and ordinary and enstatite chondrites on the other show distinctive and contrasting characteristics (Weisberg et al. 2006; Krot et al. 2014). Carbonaceous chondrites generally contain abundant CAIs, predominantly plot below the terrestrial fractionation line (TFL) on oxygen three-isotope plots and often show evidence of having accreted water ice into their parent bodies (e.g., Grimm and McSween 1989). In contrast, ordinary and enstatite chondrites generally have a relatively low CAI content, plot on or above the TFL, and show relatively anhydrous characteristics. Thus, a bimodal, inner versus outer Solar System dichotomy has always been evident in meteoritical studies, it is just that the new isotopic evidence has brought this concept into sharper focus (e.g., Scott et al. 2018).

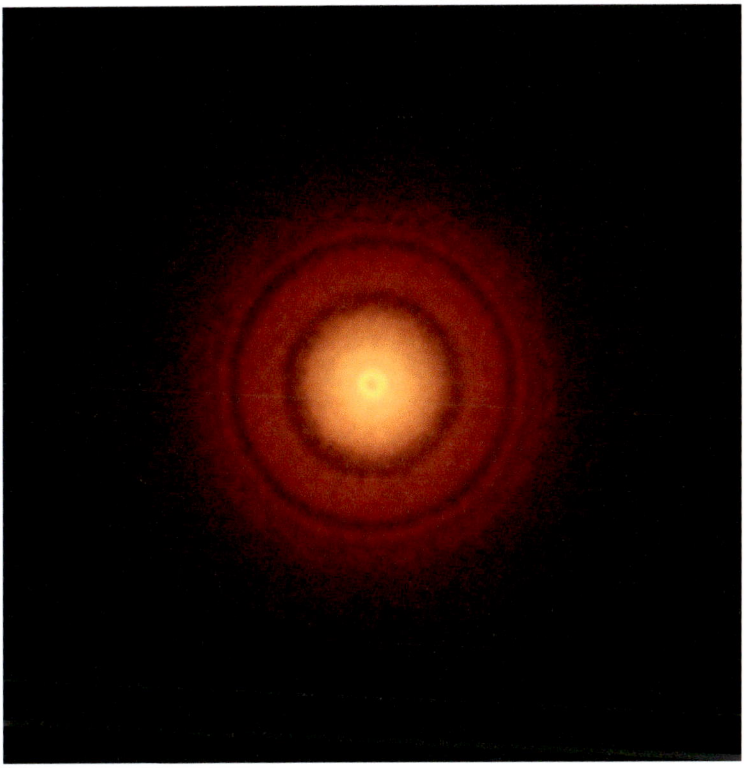

Fig. 4 ALMA (Atacama Large Millimeter/submillimeter Array) image of protoplanetary disk around the T-Tauri star TW Hydrae (Andrews et al. 2016). This object is the closest T-Tauri star to the Solar System (196 light years away). The gaps in the disk are generally considered to be due to planets that are forming. Brasser and Mojzsis (2020) suggest that pressure maxima in the early Solar System protoplanetary disk may have resulted in a similar ring system. These pressure maxima may have prevented radial mixing. Image credit: S. Andrews (Harvard-Smithsonian CfA); B. Saxton (NRAO/AUI/NSF); ALMA (ESO/NAOJ/NRAO)

The concept of bimodality in the meteorite record mirrors the overall structure of the Solar System itself, which is conventionally divided into the inner terrestrial planet region and the outer gas and ice giant region (Morbidelli et al. 2015). What lies between the two is the asteroid belt. Unsurprisingly, one of the most pertinent means of testing the bimodal Solar System concept would be to collect samples from both CC and NC main-belt asteroids. In theory, if enough samples were collected from different bodies with known locations, the asteroid belt could potentially be mapped both mineralogically and isotopically. However, there are clearly significant practical and economic difficulties in adopting such a global sampling approach. In Sects. 6 and 7, we look at various scenarios aimed at addressing the bimodal Solar System concept with sample return.

4 Asteroid Families

The asteroid belt is dominated by groupings of asteroids in which individual members share similar proper orbital elements (Nesvorný et al. 2015). These clusterings are known as asteroidal families, with each family thought to be derived from the breakup of a single larger body or are derived from the debris ejected after a "large" impact on the parent body. With estimated ages that rarely exceed 1 billion years (Spoto et al. 2015), currently recognized families are significantly younger than the age of the Solar System and hence relatively ephemeral features of the asteroid belt, albeit very long-lived ones.

The advantage of an asteroid family member as a target for a sample return mission is that observations and samples collected at a single asteroid are relevant to a much larger number of related bodies. In particular, close-up exploration and subsequent collection of material from one member of a family would provide an independent verification of Earth-based observations (e.g., reflectance spectra) made on the family as a whole. There may be a huge scientific advantage in targeting a well-defined member of a well-constrained asteroid family (e.g., Nesvorný et al. 2015). For the purpose of testing the bimodal Solar System paradigm, amongst some of the most favourable targets could be C-complex and S-complex asteroid families.

5 Meteorites and Their Relationship to Main-Belt Asteroids

Meteorites currently represent our principal source of extraterrestrial materials and so provide our best means of undertaking detailed mineralogical and isotopic studies of rocky planetary and asteroidal bodies. While our collections do contain materials derived from the Moon and Mars, most meteorites are asteroidal in origin. In terms of asteroidal sources, the vast majority of samples are likely derived from main-belt objects, rather than near-Earth asteroids. The cosmic-ray exposure ages of stony meteorites tend to be much longer than the dynamical lifetimes of bodies delivered to near-Earth space by resonances. These long cosmic-ray exposure ages for stony meteorites are consistent with entry into near-Earth space through slow Yarkovsky drift of meter-sized bodies derived from objects located anywhere in the main belt and not derivation directly from near-Earth bodies (e.g., Farinella et al. 1998; Bottke et al. 2002b). A cosmic-ray exposure age is the period of time that a surface (up to approximately one meter in depth) has been exposed to cosmic rays (e.g., Eugster et al. 2006) and is calculated by measuring the abundances of isotopes produced by cosmic-ray exposure. The Yarkovsky effect is the force acting on rotating objects due to the anisotropic emission of photons from blackbody radiation due to differences in temperature across the surface of the body.

There is a significant question mark about how representative the meteorites that arrive on Earth are of the material present in the main belt. In particular, atmospheric entry certainly acts as a filter preferentially destroying weak, friable objects at the expense of tough, strong meteoroids (Campbell-Brown 2019). That the meteorite flux is compositionally distinct from the micrometeorite flux is well established. Most micrometeorites have compositions that resemble CI and CM chondrites (Taylor et al. 2012) whereas these meteorite groups comprise only about 2% of meteorite falls (Burbine 2017). While this discrepancy is certainly not just a result of atmospheric entry processes, it serves to illustrate that major differences exist in the composition and, by implication, the likely origin of the different size fractions of extraterrestrial samples recovered on Earth.

Asteroids linked with "primitive" types of carbonaceous chondritic material predominate in the outer portion of the asteroid belt and so are less likely to arrive on Earth as recoverable samples. The Tagish Lake meteorite may be a notable exception due to its spectral similarity to D-type asteroids (Hiroi et al. 2001), which are abundant among Jupiter Trojans (e.g., Bendjoya et al. 2004). However, Granvik and Brown (2018) find that the orbit of Tagish Lake is most consistent with originating from the inner main-belt and not the outer main-belt. D-type asteroids are known to exist in the inner main-belt (DeMeo et al. 2014). Therefore, the parent body of Tagish Lake may have been scattered from its most likely original formation location in the outer belt to the inner main-belt.

There is an assumption that C-complex bodies primarily represent carbonaceous chondritic material and S-complex bodies primarily represent non-carbonaceous chondritic material such as ordinary chondrites. This assumption was validated for one body by the Hayabusa mission. The returned sample from S-complex NEA (25143) Itokawa has an LL-chondrite composition (Nakamura et al. 2011). However, the ureilite Almahata Sitta, which resides in the NC group (Qin et al. 2010b) of the Warren plot (Fig. 3), is derived from the C-complex NEA 2008 TC_3 (Jenniskens et al. 2009), which was disrupted on impact with the atmosphere over the Sudan. This event illustrates the fact that it cannot be assumed that C-complex bodies will universally equate to the CC group. Targeting C-complex asteroids to recover material related to the CC group clearly needs to be done with care and with the benefit of significant input from remote sensing data (see Sect. 7 for further discussion).

Extraterrestrial samples that survive atmospheric entry will become altered to a variable extent due to interaction with the terrestrial environment. Meteorites that contain a significant metallic iron content are particularly susceptible to terrestrial weathering (Lee and Bland 2004; Greenwood et al. 2012). However, carbonaceous chondrites are also known to be affected by terrestrial alteration processes (e.g., Alexander et al. 2018). The well-studied CI chondrite Orgueil, which fell in France in 1864, has been noted to have been chemically altered (e.g., disappearance of ammonium sulfates) during its residence on Earth (Gounelle and Zolensky 2014). Direct collection and recovery of carbonaceous chondrite samples from the asteroid belt would help mitigate these alteration issues and is also likely to provide material that may not arrive on Earth as meteorites due to atmospheric entry processes.

6 Asteroid Sample Return

6.1 A Historical Perspective: Near-Earth Asteroid Sample Return (Hayabusa Mission)

The Hayabusa mission to Itokawa was the first mission to return samples from a near-Earth asteroid. Hayabusa was launched in May 2003 and rendezvoused with Itokawa in September 2005. In November 2005, Hayabusa landed on Itokawa to collect a sample. During the touchdown, a small pellet was supposed to be fired at the surface to cause the ejection of material into the sampling container; however, the pellet did not discharge. The hope was, as later confirmed, that during the encounter with the surface, small particles would have been ejected upward into the sampling container. In June 2010, the re-entry capsule returned to Earth, landing in the Australian desert. Approximately 700 grains from Itokawa have been catalogued (Okada et al. 2017). Hayabusa confirmed (Nakamura et al. 2011) the postulated compositional (Binzel et al. 2001) and isotopic similarity of Itokawa to LL chondrites.

Bulk and grain-specific mineralogy can be studied in extreme detail using a wide range of techniques, including scanning and analytical electron microscopy, electron probe analysis, electron backscattered diffraction, transmission electron microscopy, and Raman spectroscopy. Mass spectrometry can be used to measure both radiogenic and stable isotopic compositions. Such measurements provide information relevant to the formation region of the samples and the formation age of their parent body.

These studies returned a number of interesting results. Nakamura et al. (2011) determined that the fayalite and forsterite contents matched LL chondrites. Yurimoto et al. (2011) found that the Itokawa grains had an oxygen isotopic composition similar to LL chondrites. Noguchi et al. (2011) found that nanophase iron particles, the proposed cause of space weathering, were present in the grains. Jin and Bose (2019) measured the hydrogen isotopic composition and water content of a number of "anhydrous" minerals found in Itokawa grains and were able to derive the water content of the bulk silicate Itokawa parent body.

Studying returned particles has a number of advantages compared to studying whole meteorites. The particles would not have been exposed to the atmosphere during re-entry and would have a much more limited exposure to the terrestrial atmosphere. Individual grains from an asteroid's surface can also be studied. Different grains may have distinct "personal histories" (a term coined by Tomoki Nakamura) on the asteroid due to experiencing different shock effects, space weathering, and/or thermal heating. The Hayabusa mission clearly demonstrates the benefit that sample return can have in providing extraterrestrial particles that would not otherwise have survived atmospheric entry without experiencing significant modification at best or, more likely, complete destruction.

6.2 The Benefits of Combining Sample Return with Detailed Spacecraft Characterization Studies

Sample return from an asteroid would allow extraterrestrial material to be studied within a geologic context. When a meteorite is recovered on Earth, there is essentially no information available on where the sample originated, apart from some fireball trajectory information for a few relatively limited events (e.g., Spurný et al. 2012). However, when a sample is recovered directly from an asteroid, the location of the obtained material from the particular object is "known." Prior to retrieval of the sample, the spacecraft undertakes detailed analysis and observations of the asteroid's surface. These measurements provide critical geologic context and help to define the optimal sampling location. The terrain of the recovered sample and its relationship to different geologic features (e.g., craters, plains) on the surface can be established. The contextual information provided by the spacecraft's remote sensing measurements, coupled with the subsequent results of detailed laboratory studies once material has been returned to Earth, constitutes a very powerful approach likely to yield important scientific results.

Sample return from main-belt asteroids is also important because it allows extraterrestrial material to be studied that may sample the surfaces of meteorite parent bodies and/or the oldest surviving planetesimals in the Solar System. Morbidelli et al. (2009) has argued that the minimum size of the initial planetesimals in our Solar System was $\sim$100 km, which would mean that $\sim$200 original planetesimals are possibly currently intact in the main belt.

The capabilities of Earth-based laboratory equipment still dwarf the capabilities of spacecraft instruments. Meteoritic material can often be analysed at scales of tens of nanometers (e.g., Kebukawa et al. 2019) to sub-nanometer (e.g., Parman et al. 2019), depending on the technique. Isotopic measurements of presolar grains of one micron or smaller can routinely be made in the laboratory (e.g., Davis 2011). Only bulk elemental compositions and estimated silicate mineralogies have currently been determined remotely for main-belt asteroids using spacecrafts (Burbine 2016). The use of well-calibrated standards allows a wide range of high-precision measurements to be obtained in the laboratory that are simply impossible using spacecraft-based techniques. So while sample-return missions may be expensive, the scientific yield from such missions more than offsets the cost.

6.3 How to Return a Sample from a Main-Belt Asteroid

A number of studies have been done to investigate scenarios for main-belt sample return. Turtle et al. (1999) proposed the AMBASSADOR (A Main-Belt Asteroid Seismic study and Sample Acquisition to Determine meteorite ORigins) mission to study and return a sample from S-complex body (7) Iris (e.g., Gaffey et al. 1993; Noonan et al. 2019). The mission would consist of an orbiter and a lander. Material would be collected by two different methods. One would be a chipping device able to collect regolith using two circular blades that would drive material into a collection basket. Another would involve coring devices that are fired downward into the surface where recoil would return the cored material within the coring device back to the lander.

Sukhanov et al. (2001) proposed a low-cost sample-return mission to a main-belt asteroid. The spacecraft would eject a projectile that would strike the asteroid and then fly through the resulting dust cloud. Fragments would then be picked up by a collector on the spacecraft. This type of mission is a combination of the primary components of two cometary missions: Deep Impact and Stardust. Deep Impact sent an impactor to strike a comet Tempel 1 (9P/Tempel) and study the resulting crater and debris, while Stardust collected fragments of a comet's (Wild 2) coma as it flew through it. Sukhanov et al. (2001) predicts that the collected mass would be ~0.1-1 mg of material for their postulated sample-return mission to M-type (16) Psyche. Psyche is the current target of the Psyche mission (Elkins-Tanton et al. 2020), which will be launched in 2022. The orbiting spacecraft will determine the shape, the geology, the elemental composition, and the magnetic field of Psyche.

Deep Impact's impactor was ~360 kg in mass at impact. The impactor's payload included a copper "cratering mass," an Impactor Targeting Sensor (ITS), thrusters, a high-precision star tracker, and a radio receiver (Henderson and Blume 2015). Impactors proposed for main-belt asteroid sample-return missions are much smaller with masses of 10 kg or less (e.g., Morimoto et al. 2004). Stardust (Brownlee 2014) used a two-sided collector (~0.1 m^2) with containers filled with aerogel to slow down the fragments so they would not be vaporized on impact, plus aluminum foil where the fragments would actually vaporize. Besides fragments from the coma, Stardust also collected interstellar dust as it flew through the interstellar dust stream. However, collected grains may have experienced a short period of intense heating due to their high velocity when entering the aerogel (Roskosz et al. 2008). Approximately 1 mg of material was returned by Stardust. Stardust collected thousands of cometary dust particles plus a few chondrule fragments and CAIs (Westphal et al. 2017).

Morimoto et al. (2004) studied sample-return missions from main-belt asteroid families. Their proposed missions to the primarily S-complex Koronis family (Rivkin et al. 2011) and the multi-taxonomic Nysa-Polana complex (Walsh et al. 2013; Dykhuis and Greenberg 2015), respectively, also uses projectiles to impact each asteroid's surface with the spacecraft flying through the resulting dust cloud. They propose that ~10 kg projectiles would be released. Their target bodies have a wide variety of interpreted mineralogies.

Dachwald et al. (2008) studied whether a sample-return mission to C-complex (19) Fortuna using solar electric propulsion was feasible. The missions proposed by Dachwald et al. (2008) consist either of a lander with a sample-return vehicle powered by chemical propulsion or employ a spacecraft that samples the asteroid itself with a re-entry capsule to return the collected material to Earth by electric propulsion.

Sample-return missions to the C-complex dwarf planet (1) Ceres has also been the subject of a number of studies. Based on the discovery of active volatile-rich ejecta plumes on Ceres by the Herschel Space Observatory, Poncy et al. (2014) proposed a "low-cost" flyby mission to sample the plumes and return material to Earth. Fisher and Graham (2019) pointed out that Ceres represents a prime target for sample return and also advocated a "low-cost" option by utilizing a modified version of the OSIRIS-REx spacecraft configuration. GAUSS (Genesis of Asteroids and evolUtion of the Solar System) is a joint Chinese-European concept sample-return mission to Ceres, and the subject of a white paper (Shi et al. 2020) submitted to the European Space Agency's Voyage 2050 initiative. Their study highlights the importance of returning samples from Ceres for our understanding of early Solar System processes and identifies various sampling sites and spacecraft configurations that might be used to accomplish this task. GAUSS is named after mathematician Johann Carl Friedrich Gauss (1777–1855) who was able to predict the position of Ceres after this body was "lost" (Cunningham 2016). Gauss' technique to determine the preliminary orbit of a body using at least three observations is now called the Gauss method (Marsden 1985).

Not including the launch vehicle, the cost for the Stardust mission (launched in 1999) was approximately 200 million dollars, while the cost for OSIRIS-REx (launched in 2016) was approximately 800 million dollars. The cost of a main-belt sample-return mission is likely to be significantly more expensive than for OSIRIS-REx and so would probably exceed one billion dollars. However, as discussed in earlier sections, the scientific return from such a mission would be extremely high.

6.4 What Objects Should We Sample?

A successful sample-return mission to any main-belt asteroid would represent a major engineering triumph with a huge scientific return. To investigate the isotopic bimodality of the Solar System, a number of bodies, expected to have different isotopic compositions, could be sampled. From our current knowledge, our best guess is that the C-complex bodies are the best analogues for CC material and that S-complex bodies are the best analogues for NC material. But what objects should be prioritized? Here we review potential target asteroids before setting out in Sect. 7 our preferred mission scenario for investigating the bimodal Solar System.

As discussed earlier, observations and samples collected from a member of an asteroid family will be relevant to a much larger number of related bodies. Two possible candidate C-complex families are the Nemesis (Carruba and Barletta 2019) and Adeona families (Carruba et al. 2003). Also of interest would be samples from members of a $\sim$4 billion year old family identified by Delbó et al. (2017) extending across the inner main-belt. Objects in this family tend to have very low albedos. Two possible S-complex family candidates are the Flora (Vernazza et al. 2008) and Koronis families (e.g., Rivkin et al. 2011).

A number of relatively large well-studied bodies could also be possible targets. Dwarf planet Ceres ($a \approx 2.8$ AU), the second target of the Dawn mission (Fig. 5), has an interpreted mineralogy broadly consistent with carbonaceous chondrites (McSween et al. 2018) but no specific carbonaceous chondrite meteorite analogue has been identified. The interpreted presence of ammoniated hydrated silicates (King et al. 1992; Ammannito et al. 2016) may indicate Ceres' formation in the outer Solar System where ammonia would have been stable. Isotopic analyses of material from Ceres would give considerable insight on possible formation locations. Another possible target could be (19) Fortuna ($a \approx 2.4$ AU). Fortuna has been proposed to be a possible parent body (Burbine 1998) for the CM2 chondrites due to spectral similarities in the visible and near-infrared and its location near the 3:1 mean-motion resonance.

Relatively large bodies that are expected to have non-carbonaceous isotopic compositions include (4) Vesta, (6) Hebe, and (16) Psyche. Vesta ($a \approx 2.4$ AU) has been studied in detail by the Dawn mission (Russell et al. 2013). Vesta has been linked with most HED

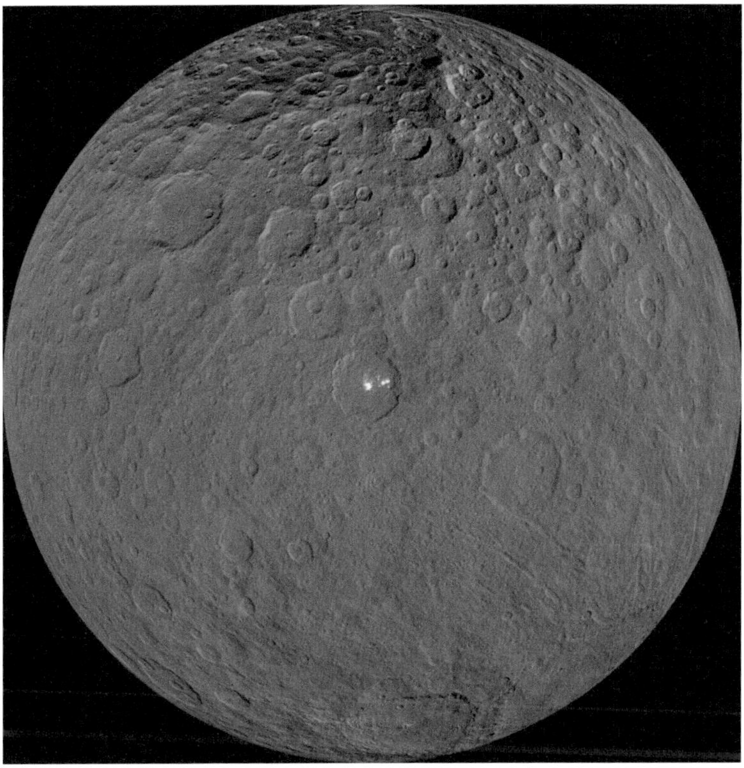

Fig. 5 Orthographic projection of Ceres as seen by NASA's Dawn spacecraft. The bright features in Occator crater are called faculae and appear to be due to hydrated magnesium sulfates (Nathues et al. 2015). Image credit: NASA/JPL-Caltech/UCLA/MPS/DLR/IDA

(howardite, eucrite, diogenite) meteorites (McCord et al. 1970; Consolmagno and Drake 1977; McSween et al. 2013) due to spectral similarities. Vesta is the only large (~525 km in diameter) body with an HED-like spectrum. However, a number of eucrites (e.g., NWA 011) have oxygen isotopic compositions that are anomalous compared to the values for most HEDs (Scott et al. 2009), implying multiple HED parent bodies. Sample return from Vesta could also "confirm" which HEDs are fragments of Vesta and whether it has the expected NC isotopic composition.

Asteroid (6) Hebe ($a \approx 2.4$ AU) would be another interesting target since it has been linked with both the H chondrites and IIE irons (Gaffey and Gilbert 1998) due to spectral similarities and Hebe's location near the meteorite-supplying 3:1 resonance. However, Vernazza et al. (2014) has indicated from spectral studies that a number of bodies with interpreted mineralogies similar to H chondrites exist in the main belt. Sample return from Hebe would help us decipher whether Hebe is the parent body of the H chondrites and IIE irons.

Psyche ($a \approx 2.9$ AU) is commonly thought to be the core of a disrupted differentiated body (e.g., Ostro et al. 1985). Absorption features due to silicates have also been identified on the surface of Psyche (e.g., Hardersen et al. 2005). Fragments of Psyche could test whether Psyche is a core of a fragmented differentiated body and whether it has a NC or CC isotopic composition.

There are also a number of possible targets that have mineralogies that have perplexed ground-based researchers and may provide interesting insights on the history of the early

Solar System. For example, asteroids such as (387) Aquitania ($a \approx 2.7$ AU) and (980) Anacostia ($a \approx 2.7$ AU) have long been known to have unusual spectral properties (Burbine et al. 1992). Their spectra appear consistent with a surface enriched in spinel-rich CAIs (Burbine et al. 1992; Sunshine et al. 2008). CAIs are the oldest dated material in the Solar System with the oldest CAI currently having an age of 4,568.22 ± 0.17 Ma (Bouvier and Wadhwa 2010). However, the implied CAI abundances for their surfaces are much higher than those found in chondrites. Sunshine et al. (2008) argues that these bodies are prime candidates for sample return since these objects might be some of the oldest bodies still existing today since they may have formed before the injection of radiogenic ^{26}Al into the Solar System. The presence of significant concentrations of ^{26}Al in these inclusions would be expected to have resulted in the melting of these bodies. Sample return from such bodies may allow CAIs that are older than currently known refractory inclusions to be studied. The implied high CAI abundances are consistent with a CC isotopic composition for these bodies.

7 Preferred Mission Scenario: Return to Ceres

In the previous section, we looked at a wide range of potential target asteroids that could be sampled to investigate in further detail the bimodal Solar System concept. But as this is likely to be a relatively costly and technically-challenging mission, a coherent and well-defined strategy will be needed. Here we set out our preferred option for a sample-return mission to the main belt.

7.1 Defining the Mission Strategy and Target Body

Based on the experience gained from previous sample-return missions, there are two distinct strategies that could be employed to successfully collect and return samples from the asteroid belt. These we term: (1) flyby and (2) direct sampling strategies. The flyby approach would involve visiting either a single, or preferably multiple asteroids with a spacecraft equipped with Deep Impact-style impactors. The impactors would be used to create a debris cloud that would be sampled by the spacecraft. By keeping the instrumentation to a minimum, costs could be reduced, thus enhancing the financial viability of such an approach. Visiting multiple target asteroids should provide a range of materials that hopefully would be representative of both the NC and CC groups.

However, there are also some downsides to the flyby approach. From past experience, the amount of material collected is likely to be limited and the complexity of visiting multiple targets increases the inherent risks of this type of mission. One particular problem, as exemplified by the Almahata Sitta meteorite (Sect. 5), is that not all C-complex asteroids are members of the CC group. Targeting smaller, less well-characterized C-complex main-belt asteroids leaves open the possibility that a flyby mission might fail to return CC group material.

While it would appear to be advantageous for a main-belt sample return mission investigating Solar System bimodality to return material from both NC- and CC-related bodies this concept may not stand up to more detailed scrutiny during the proposal review process. We already have a very large selection of NC-related materials and we actually live on the largest NC parent body. Rather than obtaining material from the asteroid belt to investigate inner Solar System evolution, we argue that a more fruitful approach and a better use of resources (time and money) would be to sample another large, inner Solar System body, such as Venus (Greenwood and Anand 2020). Venus would be expected to fall within the NC group.

In terms of the bimodal Solar System, we have a much poorer understanding of the CC group than the NC group. The amount of CC-related material delivered to Earth is highly biased and it is likely that friable and unconsolidated lithologies do not survive transit from the asteroid belt, for reasons already discussed. As a consequence, and paradoxically, rather than returning multiple samples from the main belt, the more targeted approach would be to return an unambiguous example of CC-related material. While returning material from the regions of Jupiter and beyond is unlikely to be feasible for the foreseeable future, one candidate body in the asteroid belt stands out as an exemplar of an outer Solar System icy-world. That object is Ceres, the largest body in the main belt. Ceres comprises one-third of the total mass of the main-belt region and is a possible refugee from the outer Solar System. Our preferred approach for investigating the bimodal Solar System would be direct sampling of a well-characterized location on Ceres. In the next section, we look in further detail at why sample return from Ceres would provide a first order scientific return capable of significantly improving our understanding of early Solar System evolution.

7.2 Ceres After Dawn: An Icy World from the Outer Solar System

As a result of the Dawn mission, we now have a significant body of detailed topographic, geologic and compositional data for Ceres (Williams et al. 2018). A sample-return mission to Ceres would be visiting an already well characterized object, rather than a relatively unknown main-belt asteroid requiring detailed orbital observations prior to any attempt to collect material from its surface. The observations already completed by Dawn means that a sample-return mission to Ceres would need to carry a far less comprehensive suite of instruments than required for other main-belt bodies, thus ensuring significant cost savings.

The Dawn mission began orbiting Ceres in March 2015 (Russell et al. 2016) following its earlier phase of activity studying Vesta (Russell et al. 2013). Dawn continued to make orbital observations of Ceres until November 2018 when it ran out of fuel. Dawn's observations resulted in a remarkable amount of information being learnt about the asteroid belt's largest object. The surface was mapped at a resolution of 35 m/pixel during the LAMO (Low-Altitude Mapping Orbit) phase of operations (Williams et al. 2018). Compositional and mineralogical information was obtained with three instruments: the Framing Camera (FC), the Visible and InfraRed mapping spectrometer (VIR), and the Gamma Ray and Neutron Detector (GRaND) (Russell et al. 2006).

The results from the Dawn mission present a relatively coherent picture of Ceres as a carbonaceous chondrite-related, differentiated icy-world that has experienced a protracted history of water/rock interaction and alteration (McCord and Castillo-Rogez 2018). There is clear evidence on its surface in the form of bright deposits indicating geologically recent hydrothermal brine deposition (Scully et al. 2019). The possible presence of ammoniated phyllosilicates on its surface have been interpreted as indicating that Ceres may not have formed in its present location, but possibly further out in the trans-Neptunian disk (De Sanctis et al. 2015). This scenario is supported by the presence of ammonia ice on a number of trans-Neptunian objects, including (134340) Pluto, Pluto's moon Charon, and (90482) Orcus (Brown and Calvin 2000; Barucci et al. 2008; DeMeo et al. 2015; Dalle Ore et al. 2019). De Sanctis et al. (2015) speculates that Ceres may have been implanted into the main belt during the "Grand Tack" migration (Walsh et al. 2011). On the basis of Dawn observations and our present understanding of the Solar System, there is currently very little doubt that Ceres is a bona fide representative of the CC group and as such

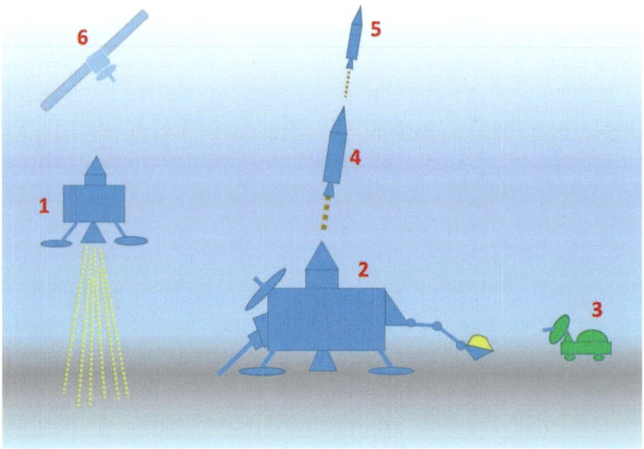

Fig. 6 Schematic of a possible dedicated sample-return mission to Ceres. 1: Spacecraft arrives and is slowed during the landing phase by chemical propulsion. 2: Lander collects sample material. 3: Small reconnaissance rover has the ability to collect small amounts of additional material. 4 and 5: Two-stage rocket leaves base unit with samples once operations are complete. 6: For communication and imaging of the surface, a small orbiter would be required but would take no part in the return of samples to Earth

would represent a target that is clearly relevant to the investigation of the bimodal Solar System concept.

Ceres bears some compositional resemblance to carbonaceous chondrites (CI/CM), but the match is not perfect (Castillo-Rogez et al. 2020). Ceres potentially contains more water and organics than CIs and CMs and the latter do not contain significant ammoniated phyllosilicates. The carbonate mineralogy detected on Ceres is significantly more diverse than found in CIs. In particular, sodium carbonate has never been found in a CI, but appears to be present on Ceres (Castillo-Rogez et al. 2020). The clear implication of these observations is that we do not have samples in our meteorite collections that are good matches to the mineralogy of Ceres. The high water content estimates for Ceres have been disputed by Zolotov (2020) who suggests instead that the body may have a high organic content of between 12 and 29 vol%. Based on the detection of ammoniated phyllosilicates and a potentially high organic content, it has been suggested that Ceres may be derived from further out in the Solar System than either the CIs or CMs (De Sanctis et al. 2015; Zolotov 2020).

7.3 How and Where Would Samples Be Collected on Ceres?

While it is beyond the scope of this study to identify a specific spacecraft configuration for returning a sample from Ceres, we do make some suggestions concerning a possible mission scenario. A flyby mission would not seem appropriate for the reasons discussed earlier. An orbiter/lander rendezvous maneuver following collection of material on the surface would also be a risky option. We would advocate a single lander approach to a well-scoped landing site (Fig. 6). Significant research has been done in developing strategies and equipment for returning both volatiles and non-volatiles from an ice-rich body back to Earth (e.g., Glavin et al. 2019). A small rover for reconnaissance studies could also be employed. Once the samples have been collected, a portion of the lander would then return to Earth without involving any transfer of material to another orbiting spacecraft. However, a small orbiter would be used for communication and imaging of the surface.

In terms of specific locations on Ceres for the recovery of material, potential sampling sites would clearly need to be the subject of detailed evaluation. However, the Occator crater (Fig. 7), which has been the subject of a large number of detailed studies (e.g., Scully et al. 2019), does appear to be a particularly attractive possibility (Shi et al. 2020).

Fig. 7 Close-up view of Occator crater on Ceres. The crater is approximately ~90.5 km in diameter (Nathues et al. 2015) and is located at a latitude of 20° N and a longitude of 239° E. Image credit: NASA/JPL-Caltech/UCLA/MPS/DLR/IDA

7.4 Sample Analysis Strategy Back on Earth

A full suite of mineralogical and isotopic analyses would need to be undertaken on the returned material and consequently a Sample Analysis Plan (SAP) similar to that developed for the OSIRIS-REx mission (Lauretta et al. 2017) would need to be formulated and implemented. Material returned from Ceres would be expected to be hydrated and organic-rich (Raponi et al. 2018; Marchi et al. 2019; Zolotov 2020). Consequently, such samples would have major astrobiological implications (Castillo-Rogez et al. 2020). Although water ice is not expected to be close to the surface at equatorial latitudes, GRaND measurements suggests it is present in significant amounts at mid- to high-latitudes (Prettyman et al. 2017). Organic material appears to be particularly abundant on the surface of Ceres and likely to be of endogenous origin and not the result of meteorite infall, although this remains an issue of current debate (De Sanctis et al. 2019). Hydrogen isotopic analysis of organics and water (derived from hydrated silicates and possibly also ices) would provide additional constraints relevant to the origin of Ceres (Alexander et al. 2012).

In this study, we have focused on the bimodal variation in Solar System materials as exemplified by the range of isotopic systems (Fig. 3). However, there is an additional scenario raised by the study of Brasser and Mojzsis (2020), namely that the early Solar System was not just divided into two regions, but multiple ring-like regions were present (Fig. 4). It is possible that we now view the early Solar System as bimodal because we only receive samples from its two inner zones. If Ceres really did form much further out in the disk than anything we have so far in our meteorite collections, we could be in for an unexpected surprise. There is a real chance that Ceres is formed from material that we have never hitherto been able to sample. It is a tantalizing possibility.

8 Conclusions

Sample return from the main belt is the logical next step in exploring the dichotomy present in Solar System materials. Isotopic studies indicate that a dichotomy existed in the early Solar System, with material from the inner region (NC group) being separated from outer region material (CC group) by a clear compositional gap on relevant variation diagrams. The interface between the NC and CC regions appears to be broadly coincident with the present location of the asteroid belt, which contains material derived from both groups. A significant range of compositionally diverse asteroids are available as targets of such a mission. A flyby mission using "Deep Impact"-style impactors to collect material from multiple targets has merits as a potential collection scenario. However, there are a number of downsides to this approach, not least of which is its potential complexity and consequent failure risks.

We argue that a sample-return mission to the asteroid belt does not necessarily have to return material from both the NC and CC groups to viably test the bimodal Solar System paradigm as the NC group is already abundantly available for study. Instead, there is overwhelming evidence that we have a very incomplete suite of CC-related samples. Based on our analysis, we advocate a dedicated sample-return mission to the dwarf planet Ceres as the best means of further exploring inherent Solar System variation. Ceres is an ice-rich world that may be a displaced trans-Neptunian object. We almost certainly do not have any samples in our current collections that would resemble material that could be brought back from Ceres. The rich heritage of data acquired by the Dawn mission makes a sample-return mission from Ceres logistically feasible at a realistic cost. No other potential target appears capable of providing as much insight into the early Solar System as Ceres. It should be given the highest priority by the international scientific community.

Acknowledgements THB would like to thank the Remote, In Situ, and Synchrotron Studies for Science and Exploration 2 (RISE2) Solar System Exploration Research Virtual Institute (SSERVI) (NASA grant 80NSSC19M0215) for support. THB would also like to thank The Open University for hosting him. Oxygen isotope studies at The Open University are funded by a Consolidated Grant from the Science and Technology Facilities Council (STFC), UK GRANT NUMBER: ST/P000657/1. The authors would like to thank Bill Bottke, Mikael Granvik, and an anonymous reviewer for very helpful comments that greatly improved the manuscript. The authors would also like to thank guest editor Mahesh Anand for inviting us to write this article. The authors would like to thank Kevin Walsh for the use of his "Grand Tack" cartoon.

References

C.M.O'D. Alexander, R. Bowden, M.L. Fogel, K.T. Howard, C.D.K. Herd, L.R. Nittler, The provenances of asteroids, and their contributions to the volatile inventories of the terrestrial planets. Science **337**, 721–723 (2012)

C.M.O'D. Alexander, R.C. Greenwood, R. Bowden, J.M. Gibson, K.T. Howard, I.A. Franchi, A multi-technique search for the most primitive CO chondrites. Geochim. Cosmochim. Acta **221**, 406–420 (2018)

E. Ammannito, M.C. DeSanctis, M. Ciarniello, A. Frigeri, F.G. Carrozzo, J.-Ph. Combe, B.L. Ehlmann, S. Marchi, H.Y. McSween, A. Raponi, M.J. Toplis, F. Tosi, J.C. Castillo-Rogez, F. Capaccioni, M.T. Capria, S. Fonte, M. Giardino, R. Jaumann, A. Longobardo, S.P. Joy, G. Magni, T.B. McCord, L.A. McFadden, E. Palomba, C.M. Pieters, C.A. Polanskey, M.D. Rayman, C.A. Raymond, P.M. Schenk, F. Zambon, C.T. Russell, Distribution of phyllosilicates on the surface of Ceres. Science **353**, aaf4279 (2016)

S.M. Andrews, D.J. Wilner, Z. Zhu, T. Birnstiel, J.M. Carpenter, L.M. Pérez, X.-N. Bai, K.I. Öberg, A.M. Hughes, A. Isella, L. Ricci, Ringed substructure and a gap at 1 au in the nearest protoplanetary disk. Astrophys. J. Lett. **820**, L40 (2016)

M.A. Barucci, F. Merlin, A. Guilbert, C. de Bergh, A. Alvarez-Candal, O. Hainaut, A. Doressoundiram, C. Dumas, T. Owen, A. Coradini, Surface composition and temperature of the TNO Orcus. Astron. Astrophys. **479**, L13–L16 (2008)

P. Bendjoya, A. Cellino, M. Di Martino, L. Saba, Spectroscopic observations of Jupiter Trojans. Icarus **168**, 374–384 (2004)

E.B. Bierhaus, B.C. Clark, J.W. Harris, K.S. Payne, R.D. Dubisher, D.W. Wurts, R.A. Hund, R.M. Kuhns, T.M. Linn, J.L. Wood, A.J. May, J.P. Dworkin, E. Beshore, D.S. Lauretta, The OSIRIS-REx spacecraft and the Touch-and-Go Sample Acquisition Mechanism (TAGSAM). Space Sci. Rev. **214**, 107 (2018)

R.P. Binzel, A.S. Rivkin, S.J. Bus, J.M. Sunshine, T.H. Burbine, MUSES-C target asteroid (25143) 1998 SF36: a reddened ordinary chondrite. Meteorit. Planet. Sci. **36**, 1167–1172 (2001)

W.F. Bottke Jr., A. Morbidelli, R. Jedicke, J.-M. Petit, H.F. Levison, P. Michel, T.S. Metcalfe, Debiased orbital and absolute magnitude distribution of the near-Earth objects. Icarus **156**, 399–433 (2002a)

W.F. Bottke Jr., D. Vokrouhlický, D.P. Rubincam, M. Brož, The effect of Yarkovsky thermal forces on the dynamical evolution of asteroids and meteoroids, in *Asteroids III*, ed. by W.F. Bottke Jr., A. Cellino, P. Paolicchi, R.P. Binzel (University of Arizona Press, Tucson, 2002b), pp. 395–408

W.F. Bottke, D.D. Durda, D. Nesvorný, R. Jedicke, A. Morbidelli, D. Vokrouhlický, H. Levison, The fossilized size distribution of the main asteroid belt. Icarus **175**, 111–140 (2005)

A. Bouvier, M. Wadhwa, The age of the Solar System redefined by the oldest Pb-Pb age of a meteoritic inclusion. Nat. Geosci. **3**, 637–641 (2010)

J.P. Bradley, Interplanetary dust particles, in *Meteorites and Cosmochemical Processes, Treatise on Geochemistry*, ed. by H.D. Holland, K.K. Turekian, 2nd edn. (Elsevier, San Diego, 2014), pp. 689–711

R. Brasser, S.J. Mojzsis, The partitioning of the inner and outer Solar System by a structured protoplanetary disk. Nat. Astron. **4**, 492–499 (2020)

M.E. Brown, W.M. Calvin, Evidence for crystalline water and ammonia ices on Pluto's satellite Charon. Science **287**, 107–109 (2000)

P.G. Brown, A.R. Hildebrand, M.E. Zolensky, M. Grady, R.N. Clayton, T.K. Mayeda, E. Tagliaferri, R. Spalding, N.D. MacRae, E.L. Hoffman, D.W. Mittlefehldt, J.F. Wacker, J.A. Bird, M.D. Campbell, R. Carpenter, H. Gingerich, M. Glatiotis, E. Greiner, M.J. Mazur, P.J.A. McCausland, H. Plotkin, T.R. Mazur, The fall, recovery, orbit, and composition of the Tagish Lake meteorite: a new type of carbonaceous chondrite. Science **290**, 320–325 (2000)

D.E. Brownlee, Cosmic dust—collection and research. Annu. Rev. Earth Planet. Sci. **13**, 147–173 (1985)

D. Brownlee, The Stardust mission: analyzing samples from the edge of the Solar System. Annu. Rev. Earth Planet. Sci. **42**, 179–205 (2014)

G. Budde, C. Burkhardt, G.A. Brennecka, M. Fischer-Gödde, T.S. Kruijer, T. Kleine, Molybdenum isotopic evidence for the origin of chondrules and a distinct genetic heritage of carbonaceous and non-carbonaceous meteorites. Earth Planet. Sci. Lett. **454**, 293–303 (2016)

T.H. Burbine, Could G-class asteroids be the parent bodies of the CM chondrites? Meteorit. Planet. Sci. **33**, 253–258 (1998)

T.H. Burbine, Advances in determining asteroid chemistries and mineralogies. Chem. Erde **76**, 181–195 (2016)

T.H. Burbine, *Asteroids: Astronomical and Geological Bodies* (Cambridge University Press, Cambridge, 2017)

T.H. Burbine, M.J. Gaffey, J.F. Bell, S-asteroids 387 Aquitania and 980 Anacostia: possible fragments of the breakup of a spinel-bearing parent body with CO3/CV3 affinities. Meteoritics **27**, 424–434 (1992)

C. Burkhardt, N. Dauphas, U. Hans, B. Bourdon, T. Kleine, Elemental and isotopic variability in Solar System materials by mixing and processing of primordial disk reservoirs. Geochim. Cosmochim. Acta **261**, 145–170 (2019)
M.D. Campbell-Brown, Meteoroid structure and fragmentation. Planet. Space Sci. **169**, 1–7 (2019)
V. Carruba, W. Barletta, The influence of Ceres on the dynamical evolution of the Zdenekhorsky/Nemesis asteroid family. Planet. Space Sci. **165**, 10–18 (2019)
V. Carruba, J.A. Burns, W. Bottke, D. Nesvorný, Orbital evolution of the Gefion and Adeona asteroid families: close encounters with massive asteroids and the Yarkovsky effect. Icarus **162**, 308–327 (2003)
J.C. Castillo-Rogez, M. Neveu, J.E.C. Scully, C.H. House, L.C. Quick, A. Bouquet, K. Miller, M. Bland, M.C. De Sanctis, A. Ermakov, A.R. Hendrix, T.H. Prettyman, C.A. Raymond, C.T. Russell, B.E. Sherwood, E. Young, Astrobiological target and possible ocean world. Astrobiology **20**, 269–291 (2020)
R.N. Clayton, Solar System: self-shielding in the solar nebula. Nature **415**, 860–861 (2002)
M.S. Clement, S.N. Raymond, N.A. Kaib, Excitation and depletion of the asteroid belt in the early instability scenario. Astron. J. **157**, 38 (2019)
G.J. Consolmagno, M.J. Drake, Composition and evolution of the eucrite parent body: evidence from rare Earth elements. Geochim. Cosmochim. Acta **41**, 1271–1282 (1977)
C.J. Cunningham, *Discovery of the First Asteroid, Ceres* (Springer, Dordrecht, 2016)
B. Dachwald, W. Seboldt, H.W. Loeb, K.H. Schartner, Main-belt asteroid sample return mission using solar electric propulsion. Acta Astronaut. **63**, 91–101 (2008)
C.M. Dalle Ore, D.P. Cruikshank, S. Protopapa, F. Scipioni, W.B. McKinnon, J.C. Cook, W.M. Grundy, B. Schmitt, S.A. Stern, J.M. Moore, A. Verbiscer, A.H. Parker, K.N. Singer, O.M. Umurhan, H.A. Weaver, C.B. Olkin, L.A. Young, K. Ennico, (New Horizons Surface Composition Science Theme Team), Detection of ammonia on Pluto's surface in a region of geologically recent tectonism. Sci. Adv. **5**, eaav5731 (2019)
N. Dauphas, E.A. Schauble, Mass fractionation laws, mass-independent effects, and isotopic anomalies. Annu. Rev. Earth Planet. Sci. **44**, 709–783 (2016)
A.M. Davis, Stardust in meteorites. Proc. Natl. Acad. Sci. **108**, 19142–19146 (2011)
M.C. De Sanctis, E. Ammannito, A. Raponi, S. Marchi, T.B. McCord, H.Y. McSween, F. Capaccioni, M.T. Capria, F.G. Carrozzo, M. Ciarniello, A. Longobardo, F. Tosi, S. Fonte, M. Formisano, A. Frigeri, M. Giardino, G. Magni, E. Palomba, D. Turrini, F. Zambon, J.-P. Combe, W. Feldman, R. Jaumann, L.A. McFadden, C.M. Pieters, T. Prettyman, M. Toplis, C.A. Raymond, C.T. Russell, Ammoniated phyllosilicates with a likely outer Solar System origin on (1) Ceres. Nature **528**, 241–244 (2015)
M.C. De Sanctis, V. Vinogradoff, A. Raponi, E. Ammannito, M. Ciarniello, F.G. Carrozzo, S. De Angelis, C.A. Raymond, C.T. Russell, Characteristics of organic matter on Ceres from VIR/Dawn high spatial resolution spectra. Mon. Not. R. Astron. Soc. **482**, 2407–2421 (2019)
M. Delbó, K. Walsh, B. Bolin, C. Avdellidou, A. Morbidelli, Identification of a primordial asteroid family constrains the original planetesimal population. Science **357**, 1026–1029 (2017)
F.E. DeMeo, B. Carry, Solar System evolution from compositional mapping of the asteroid belt. Nature **505**, 629–634 (2014)
F.E. DeMeo, R.P. Binzel, B. Carry, D. Polishook, N.A. Moskovitz, Unexpected D-type interlopers in the inner main belt. Icarus **229**, 392–399 (2014)
F.E. DeMeo, C. Dumas, J.C. Cook, B. Carry, F. Merlin, A.J. Verbiscer, R.P. Binzel, Spectral variability of Charon's 2.21-μm feature. Icarus **246**, 213–219 (2015)
H.A.R. Devillepoix, E.K. Sansom, P.A. Bland, M.C. Towner, M. Cupák, R.M. Howie, T. Jansen-Sturgeon, M.A. Cox, B.A.D. Hartig, G.K. Benedix, J.P. Paxman, The Dingle Dell meteorite: a Halloween treat from the main belt. Meteorit. Planet. Sci. **53**, 2212–2227 (2018)
M.J. Dykhuis, R. Greenberg, Collisional family structure within the Nysa-Polana complex. Icarus **252**, 199–211 (2015)
L.T. Elkins-Tanton, E. Asphaug, J.F. Bell III, H. Bercovici, B. Bills, R. Binzel, W.F. Bottke, S. Dibb, D.J. Lawrence, S. Marchi, T.J. McCoy, R. Oran, R.S. Park, P.N. Peplowski, C.A. Polanskey, T.H. Prettyman, C.T. Russell, L. Schaefer, B.P. Weiss, M.A. Wieczorek, D.A. Williams, M.T. Zuber, Observations, meteorites, and models: a preflight assessment of the composition and formation of (16) Psyche. J. Geophys. Res.: Planets **125**, e2019JE006296 (2020)
O. Eugster, G.F. Herzog, K. Marti, M.W. Caffee, Irradiation records, cosmic-ray exposure ages, and transfer times of meteorites, in *Meteorites and the Early Solar System II*, ed. by D.S. Lauretta, H.Y. McSween Jr. (University of Arizona Press, Tucson, 2006), pp. 829–851
P. Farinella, D. Vokrouhlický, W.K. Hartmann, Meteorite delivery via Yarkovsky orbital drift. Icarus **132**, 378–387 (1998)
K.R. Fisher, L.D. Graham, Ceres: a prime target for robotic sample return and future human exploration, in *50th Lunar and Planetary Science Conference*, Lunar and Planetary Institute, Houston (2019). Abstract #3062

M.J. Gaffey, S.L. Gilbert, Asteroid 6 Hebe: the probable parent body of the H-type ordinary chondrites and the IIE iron meteorites. Meteorit. Planet. Sci. **33**, 1281–1295 (1998)
M.J. Gaffey, J.F. Bell, R.H. Brown, T.H. Burbine, J. Piatek, K.L. Reed, D.A. Chaky, Mineralogic variations within the S-type asteroid class. Icarus **106**, 573–602 (1993)
B. Gladman, F. Migliorini, A. Morbidelli, V. Zappalà, P. Michel, A. Cellino, Ch. Froeschlé, H. Levison, M. Bailey, M. Duncan, Dynamical lifetimes of objects injected into asteroid belt resonances. Science **277**, 197–201 (1997)
D.P. Glavin, S.W. Squyres, P.C. Chu, P.A. Gerakines, K. Yamada, J.E. Parker, D.C. Wegel, H.L. Peabody, M.B. Houghton, S. Aslam, N. Gorius, G. Quilligan, R.S. Thorpe, J. Spring, K.A. Zacny, S. Gorevan, S. Okazaki, Y. Satoh, Y. Maru, T. Nakao, A. Kukita, T. Shimoda, T. Yamawaki, A.G. Hayes, D.S. Lauretta, S. Messenger, K. Nakamura-Messenger, M.J. Mumma, S.N. Milam, J.I. Lunine, J.L. Mitchell, M.M. Hasegawa, P.M. Takahara, T. Nakamura, J.P. Dworkin, S. Clemett, G.A. Blake, J.M. Soderblom, Y. Furukawa, Y. Kimura, A. Takigawa, (CAESAR Project Team), The CAESAR New Frontiers mission: 3. Sample acquisition and preservation, in *50th Lunar and Planetary Science Conference*, Lunar and Planetary Institute, Houston (2019). Abstract #2541
M. Gounelle, M.E. Zolensky, The Orgueil meteorite: 150 years of history. Meteorit. Planet. Sci. **49**, 1769–1794 (2014)
J. Gradie, E. Tedesco, Compositional structure of the asteroid belt. Science **216**, 1405–1407 (1982)
M. Granvik, P. Brown, Identification of meteorite source regions in the Solar System. Icarus **311**, 271–287 (2018)
R.C. Greenwood, M. Anand, What is the oxygen isotope composition of Venus? The scientific case for sample return from Earth's "sister" planet. Space Sci. Rev. **216**, 52 (2020)
R.C. Greenwood, I.A. Franchi, J.M. Gibson, G.K. Benedix, Oxygen isotope variation in primitive achondrites: the influence of primordial, asteroidal and terrestrial processes. Geochim. Cosmochim. Acta **94**, 146–163 (2012)
R.C. Greenwood, T.H. Burbine, M.F. Miller, I.A. Franchi, Melting and differentiation of early-formed asteroids: the perspective from high precision oxygen isotope studies. Chem. Erde **77**, 1–43 (2017)
R.C. Greenwood, T.H. Burbine, I.A. Franchi, Linking asteroids and meteorites to the primordial planetesimal population. Geochim. Cosmochim. Acta **277**, 377–406 (2020)
A. Grimberg, H. Baur, P. Bochsler, F. Bühler, D.S. Burnett, C.C. Hays, V.S. Heber, A.J.G. Jurewicz, R. Wieler, Solar wind neon from Genesis: implications for the lunar noble gas record. Science **314**, 1133–1135 (2006)
R.E. Grimm, H.Y. McSween Jr., Water and the thermal evolution of carbonaceous chondrite parent bodies. Icarus **82**, 244–280 (1989)
L. Grossman, Condensation in the primitive solar nebula. Geochim. Cosmochim. Acta **36**, 597–619 (1972)
P.S. Hardersen, M.J. Gaffey, P.A. Abell, Near-IR spectral evidence for the presence of iron-poor orthopyroxenes on the surfaces of six M-type asteroids. Icarus **175**, 141–158 (2005)
W.K. Hartmann, D.R. Davis, Satellite-sized planetesimals and lunar origin. Icarus **24**, 50–515 (1975)
M. Henderson, W. Blume, Deep Impact—a review of the world's pioneering hypervelocity impact mission. Procedia Eng. **103**, 165–172 (2015)
T. Hiroi, M.E. Zolensky, C.M. Pieters, The Tagish Lake meteorite: a possible sample from a D-type asteroid. Science **293**, 2234–2236 (2001)
T.R. Ireland, J. Avila, R.C. Greenwood, L.J. Hicks, J.C. Bridges, Oxygen isotopes and sampling of the Solar System. Space Sci. Rev. **216**, 25 (2020)
P. Jenniskens, M.H. Shaddad, D. Numan, S. Elsir, A.M. Kudoda, M.E. Zolensky, L. Le, G.A. Robinson, J.M. Friedrich, D. Rumble, A. Steele, S.R. Chesley, A. Fitzsimmons, S. Duddy, H.H. Hsieh, G. Ramsay, P.G. Brown, W.N. Edwards, E. Tagliaferri, M.B. Boslough, R.E. Spalding, R. Dantowitz, M. Kozubal, P. Pravec, J. Borovicka, Z. Charvat, J. Vaubaillon, J. Kuiper, J. Albers, J.L. Bishop, R.L. Mancinelli, S.A. Sandford, S.N. Milam, M. Nuevo, S.P. Worden, The impact and recovery of asteroid 2008 TC_3. Nature **458**, 485–488 (2009)
Z. Jin, M. Bose, New clues to ancient water on Itokawa. Sci. Adv. **5**, eaav8106 (2019)
Y. Kebukawa, H. Kobayashi, N. Urayama, N. Baden, M. Kondo, M.E. Zolensky, K. Kobayashi, Nanoscale infrared imaging analysis of carbonaceous chondrites to understand organic-mineral interactions during aqueous alteration. Proc. Natl. Acad. Sci. **116**, 753–758 (2019)
T.V.V. King, R.N. Clark, W.M. Calvin, D.M. Sherman, R.H. Brown, Evidence for ammonium-bearing minerals on Ceres. Science **255**, 1551–1553 (1992)
T. Kleine, G. Budde, C. Burkhardt, T.S. Kruijer, E.A. Worsham, A. Morbidelli, F. Nimmo, The non-carbonaceous—carbonaceous meteorite dichotomy. Space Sci. Rev. **216**, 55 (2020)
K.A. Kretke, W.F. Bottke, H.F. Levison, D.A. Kring, Mixing of the asteroid belt due to the formation of the giant planets, in *Accretion: Building New Worlds Conference*, Lunar and Planetary Institute, Houston (2017). Abstract #2027

A.N. Krot, K. Keil, E.R.D. Scott, C.A. Goodrich, M.K. Weisberg, Classification of meteorites and their genetic relationships, in *Meteorites and Cosmochemical Processes, Treatise on Geochemistry*, ed. by H.D. Holland, K.K. Turekian, 2nd edn. (Elsevier, San Diego, 2014), pp. 1–63

T.S. Kruijer, C. Burkhardt, G. Budde, T. Kleine, Age of Jupiter inferred from the distinct genetics and formation times of meteorites. Proc. Natl. Acad. Sci. **114**, 6712–6716 (2017)

T.S. Kruijer, T. Kleine, L.E. Borg, The great isotopic dichotomy of the early Solar System. Nat. Astron. **4**, 32–40 (2020)

D.S. Lauretta, S.S. Balram-Knutson, E. Beshore, W.V. Boynton, C. Drouet d'Aubigny, D.N. DellaGiustina, H.L. Enos, D.R. Golish, C.W. Hergenrother, E.S. Howell, C.A. Bennett, E.T. Morton, M.C. Nolan, B. Rizk, H.L. Roper, A.E. Bartels, B.J. Bos, J.P. Dworkin, D.E. Highsmith, D.A. Lorenz, L.F. Lim, R. Mink, M.C. Moreau, J.A. Nuth, D.C. Reuter, A.A. Simon, E.B. Bierhaus, B.H. Bryan, R. Ballouz, O.S. Barnouin, R.P. Binzel, W.F. Bottke, V.E. Hamilton, K.J. Walsh, S.R. Chesley, P.R. Christensen, B.E. Clark, H.C. Connolly, M.K. Crombie, M.G. Daly, J.P. Emery, T.J. McCoy, J.W. McMahon, D.J. Scheeres, S. Messenger, K. Nakamura-Messenger, K. Righter, S.A. Sandford, OSIRIS-REx: sample return from asteroid (101955) Bennu. Space Sci. Rev. **212**, 925–984 (2017)

D.S. Lauretta, D.N. DellaGiustina, C.A. Bennett, D.R. Golish, K.J. Becker, S.S. Balram-Knutson, O.S. Barnouin, T.L. Becker, W.F. Bottke, W.V. Boynton, H. Campins, B.E. Clark, H.C. Connolly Jr., C.Y. Drouet d'Aubigny, J.P. Dworkin, J.P. Emery, H.L. Enos, V.E. Hamilton, C.W. Hergenrother, E.S. Howell, M.R.M. Izawa, H.H. Kaplan, M.C. Nolan, B. Rizk, H.L. Roper, D.J. Scheeres, P.H. Smith, K.J. Walsh, C.W.V. Wolner, (The OSIRIS-REx Team), The unexpected surface of asteroid (101955) Bennu. Nature **568**, 55–60 (2019)

M. Lecar, M. Podolak, D. Sasselov, E. Chiang, On the location of the snow line in a protoplanetary disk. Astrophys. J. **640**, 1115–1118 (2006)

M.R. Lee, P.A. Bland, Mechanisms of weathering of meteorites recovered from hot and cold deserts and the formation of phyllosilicates. Geochim. Cosmochim. Acta **68**, 893–916 (2004)

H.F. Levison, K.A. Kretke, K.J. Walsh, W.F. Bottke, Growing the terrestrial planets from the gradual accumulation of submeter-sized objects. Proc. Natl. Acad. Sci. **112**, 14180–14185 (2015)

Lunar Sample Preliminary Examination Team, Preliminary examination of lunar samples from Apollo 11. Science **165**, 1211–1227 (1969)

J.R. Lyons, E.D. Young, CO self-shielding as the origin of oxygen isotope anomalies in the early solar nebula. Nature **435**, 317–320 (2005)

S. Marchi, A. Raponi, T.H. Prettyman, M.C. De Sanctis, J. Castillo-Rogez, C.A. Raymond, E. Ammannito, T. Bowling, M. Ciarniello, H. Kaplan, E. Palomba, C.T. Russell, V. Vinogradoff, N. Yamashita, An aqueously altered carbon-rich Ceres. Nat. Astron. **3**, 140–145 (2019)

B.G. Marsden, Initial orbit determination: the pragmatist's point of view. Astron. J. **90**, 1541–1547 (1985)

T.B. McCord, J.C. Castillo-Rogez, Ceres's internal evolution: the view after Dawn. Meteorit. Planet. Sci. **53**, 1778–1792 (2018)

T.B. McCord, J.B. Adams, T.V. Johnson, Asteroid Vesta: spectral reflectivity and compositional implications. Science **168**, 1445–1447 (1970)

K.D. McKeegan, A.P.A. Kallio, V.S. Heber, G. Jarzebinski, P.H. Mao, C.D. Coath, T. Kunihiro, R.C. Wiens, J.E. Nordholt, R.W. Moses, D.B. Reisenfeld, A.J.G. Jurewicz, D.S. Burnett, The oxygen isotopic composition of the Sun inferred from captured solar wind. Science **332**, 1528–1532 (2011)

H.Y. McSween, R.P. Binzel, M.C. de Sanctis, E. Ammannito, T.H. Prettyman, A.W. Beck, V. Reddy, L. Le Corre, M.J. Gaffey, T.B. McCord, C.A. Raymond, C.T. Russell, (The Dawn Science Team), Dawn; the Vesta-HED connection; and the geologic context for eucrites, diogenites, and howardites. Meteorit. Planet. Sci. **48**, 2090–2104 (2013)

H.Y. McSween Jr., J.P. Emery, A.S. Rivkin, M.J. Toplis, J.C. Castillo-Rogez, T.H. Prettyman, M.C. De Sanctis, C.M. Pieters, C.A. Raymond, C.T. Russell, Carbonaceous chondrites as analogs for the composition and alteration of Ceres. Meteorit. Planet. Sci. **53**, 1793–1804 (2018)

Meteoritical Bulletin Database (2020). https://www.lpi.usra.edu/meteor/

W. Misener, S. Krijt, F.J. Ciesla, Tracking dust grains during transport and growth in protoplanetary disks. Astrophys. J. **885**, 118 (2019)

A. Morbidelli, W.F. Bottke, D. Nesvorný, H.F. Levison, Asteroids were born big. Icarus **204**, 558–573 (2009)

A. Morbidelli, M. Lambrechts, S. Jacobson, B. Bitsch, The great dichotomy of the Solar System: small terrestrial embryos and massive giant planet cores. Icarus **258**, 418–429 (2015)

M. Morimoto, H. Yamakawa, M. Yoshikawa, M. Abe, H. Yano, Trajectory design of multiple asteroid sample return missions. Adv. Space Res. **34**, 2281–2285 (2004)

T. Nakamura, T. Noguchi, M. Tanaka, M.E. Zolensky, M. Kimura, A. Tsuchiyama, A. Nakato, T. Ogami, H. Ishida, M. Uesugi, T. Yada, K. Shirai, A. Fujimura, R. Okazaki, S.A. Sandford, Y. Ishibashi, M. Abe, T. Okada, M. Ueno, T. Mukai, M. Yoshikawa, J. Kawaguchi, Itokawa dust particles: a direct link between S-type asteroids and ordinary chondrites. Science **333**, 1113–1116 (2011)

J.A.M. Nanne, F. Nimmo, J.N. Cuzzi, T. Kleine, Origin of the non-carbonaceous–carbonaceous meteorite dichotomy. Earth Planet. Sci. Lett. **511**, 44–54 (2019)

A. Nathues, M. Hoffmann, M. Schaefer, L. Le Corre, V. Reddy, T. Platz, E.A. Cloutis, U. Christensen, T. Kneiss, J.-Y. Li, K. Mengel, N. Schmedemann, T. Schaefer, C.T. Russell, D.M. Applin, D.L. Buczkowski, M.R.M. Izawa, H.U. Keller, D.P. O'Brien, C.M. Pieters, C.A. Raymond, J. Ripken, P.M. Schenk, B.E. Schmidt, H. Sierks, M.V. Sykes, G.S. Thangjam, J.-B. Vincent, Sublimation in bright spots on (1) Ceres. Nature **528**, 237–240 (2015)

National Research Council, *Vision and Voyages for Planetary Science in the Decade 2013–2022* (The National Academies Press, Washington, 2011)

D. Nesvorný, M. Brož, V. Carruba, Identification and dynamical properties of asteroid families, in *Asteroids IV*, ed. by P. Michel, F.E. DeMeo, W.F. Bottke (University of Arizona Press, Tucson, 2015), pp. 297–321

T. Noguchi, T. Nakamura, M. Kimura, M.E. Zolensky, M. Tanaka, T. Hashimoto, M. Konno, A. Nakato, T. Ogami, A. Fujimura, M. Abe, T. Yada, T. Mukai, M. Ueno, T. Okada, K. Shirai, Y. Ishibashi, R. Okazaki, Incipient space weathering observed on the surface of Itokawa dust particles. Science **333**, 1121–1125 (2011)

T. Noguchi, N. Ohashi, S. Tsujimoto, T. Mitsunari, J.P. Bradley, T. Nakamura, S. Toh, T. Stephan, N. Iwata, N. Imae, Cometary dust in Antarctic ice and snow: past and present chondritic porous micrometeorites preserved on the Earth's surface. Earth Planet. Sci. Lett. **410**, 1–11 (2015)

J.W. Noonan, V. Reddy, W.M. Harris, W.F. Bottke, J.A. Sanchez, R. Furfaro, Z. Brown, R. Fernandes, T. Kareta, C. Lejoly, R.T. Nallapu, H. Khan Niazi, L.R. Slick, L. Schatz, B.N.L. Sharkey, A. Springmann, G. Angle, L. Bailey, D.D. Acuna, C. Lewin, K. Marchese, M. Meshel, N. Quintero, K. Tatum, G. Wilburn, Search for the H chondrite parent body among the three largest S-type asteroids: (3) Juno, (7) Iris, and (25) Phocaea. Astron. J. **158**, 213 (2019)

T. Okada, R.P. Binzel, H.C. Connolly Jr., T. Yada, K. Ohtsuki, Special issue "Science of Solar System materials examined from Hayabusa and future missions (II)". Earth Planets Space **69**, 31 (2017)

S.J. Ostro, D.B. Campbell, I.I. Shapiro, Mainbelt asteroids: dual-polarization radar observations. Science **229**, 442–446 (1985)

S. Parman, S. Jacobsen, M. Petaev, A. Akey, Atom probe tomography of opaque assemblage in Allende CAI, in *50th Lunar and Planetary Science Conference*, Lunar and Planetary Institute, Houston (2019). Abstract #2819

A. Pierens, S.N. Raymond, Two phase, inward-then outward migration of Jupiter and Saturn in the gaseous solar nebula. Astron. Astrophys. **533**, A131 (2011)

J. Poncy, J. Fontdecaba, P. Couzin, A robust mission concept for a low-cost Ceres plume sample return, in *EPSC Abstracts Vol. 9*, European Planetary Science Congress, Strasbourg (2014). EPSC2014-637

T.H. Prettyman, N. Yamashita, M.J. Toplis, H.Y. McSween, N. Schörghofer, S. Marchi, W.C. Feldman, J. Castillo-Rogez, O. Forni, D.J. Lawrence, E. Ammannito, B.L. Ehlmann, H.G. Sizemore, S.P. Joy, C.A. Polanskey, M.D. Rayman, C.A. Raymond, C.T. Russell, Extensive water ice within Ceres' aqueously altered regolith: evidence from nuclear spectroscopy. Science **355**, 55–59 (2017)

L. Qin, C.M.O'D. Alexander, R.W. Carlson, M.F. Horan, T. Yokoyama, Contributions to chromium isotope variations of meteorites. Geochim. Cosmochim. Acta **74**, 1122–1145 (2010a)

L. Qin, D. Rumble, C.M.O'D. Alexander, R.W. Carlson, P. Jenniskens, M.H. Shaddad, The chromium isotopic composition of Almahata Sitta. Meteorit. Planet. Sci. **45**, 1771–1777 (2010b)

A. Raponi, M.C. De Sanctis, A. Frigeri, E. Ammannito, M. Ciarniello, M. Formisano, J.-P. Combe, G. Magni, F. Tosi, F.G. Carrozzo, S. Fonte, M. Giardino, S.P. Joy, C.A. Polanskey, M.D. Rayman, F. Capaccioni, M.T. Capria, A. Longobardo, E. Palomba, F. Zambon, C.A. Raymond, C.T. Russell, Variations in the amount of water ice on Ceres' surface suggest a seasonal water cycle. Sci. Adv. **4**, eaao3757 (2018)

S.N. Raymond, A.Z. Izidoro, The empty primordial asteroid belt. Sci. Adv. **3**, e1701138 (2017)

A.S. Rivkin, C.A. Thomas, D.E. Trilling, M.-t. Enga, J.A. Grier, Ordinary chondrite-like colors in small Koronis family members. Icarus **211**, 1294–1297 (2011)

M. Roskosz, H. Leroux, H.C. Watson, Thermal history, partial preservation and sampling bias recorded by Stardust cometary grains during their capture. Earth Planet. Sci. Lett. **273**, 195–202 (2008)

C.T. Russell, F. Capaccioni, A. Coradini, U. Christensen, M.C. De Sanctis, W.C. Feldman, R. Jaumann, H.U. Keller, A. Konopliv, T.B. McCord, L.A. McFadden, H.Y. McSween, S. Mottola, G. Neukum, C.M. Pieters, T.H. Prettyman, C.A. Raymond, D.E. Smith, M.V. Sykes, B. Williams, M.T. Zuber, Dawn discovery mission to Vesta and Ceres: present status. Adv. Space Res. **38**, 2043–2048 (2006)

C.T. Russell, C.A. Raymond, R. Jaumann, H.Y. McSween, M.C. De Sanctis, A. Nathues, T.H. Prettyman, E. Ammannito, V. Reddy, F. Preusker, D.P. O'Brien, S. Marchi, B.W. Denevi, D.L. Buczkowski, C.M. Pieters, T.B. McCord, J.-Y. Li, D.W. Mittlefehldt, J.-P. Combe, D.A. Williams, H. Hiesinger, R.A. Yingst, C.A. Polanskey, S.P. Joy, Dawn completes its mission at 4 Vesta. Meteorit. Planet. Sci. **48**, 2076–2089 (2013)

C.T. Russell, C.A. Raymond, E. Ammannito, D.L. Buczkowski, M.C. De Sanctis, H. Hiesinger, R. Jaumann, A.S. Konopliv, H.Y. McSween, A. Nathues, R.S. Park, C.M. Pieters, T.H. Prettyman, T.B. McCord, L.A. McFadden, S. Mottola, M.T. Zuber, S.P. Joy, C. Polanskey, M.D. Rayman, J.C. Castillo-Rogez, P.J. Chi, J.P. Combe, A. Ermakov, R.R. Fu, M. Hoffmann, Y.D. Jia, S.D. King, D.J. Lawrence, J.-Y. Li, S. Marchi, F. Preusker, T. Roatsch, O. Ruesch, P. Schenk, M.N. Villarrea, N. Yamashita, Dawn arrives at Ceres: exploration of a small, volatile-rich world. Science **353**, 1008–1010 (2016)

A.E. Saal, E.H. Hauri, M.L. Cascio, J.A. Van Orman, M.C. Rutherford, R.F. Cooper, Volatile content of lunar volcanic glasses and the presence of water in the Moon's interior. Nature **454**, 192–195 (2008)

E.R.D. Scott, R.C. Greenwood, I.A. Franchi, I.S. Sanders, Oxygen isotopic constraints on the origin and parent bodies of eucrites, diogenites, and howardites. Geochim. Cosmochim. Acta **73**, 5835–5853 (2009)

E.R.D. Scott, A.N. Krot, I.S. Sanders, Isotopic dischotomy among meteorites and its bearing on the protoplanetary disk. Astrophys. J. **854**, 164 (2018)

J.E.C. Scully, T. Bowling, C. Bu, D.L. Buczkowski, A. Longobardo, A. Nathues, A. Neesemann, E. Palomba, L.C. Quick, A. Raponi, O. Ruesch, P.M. Schenk, N.T. Stein, E.C. Thomas, C.T. Russell, J.C. Castillo-Rogez, C.A. Raymond, R. Jaumann, (Dawn Science Team), The formation and evolution of Ceres' Occator crater. Icarus **320**, 213–225 (2019)

D.W.G. Sears, The case for rarity of chondrules and calcium-aluminum-rich inclusions in the early Solar System and some implications for astrophysical models. Astrophys. J. **498**, 773–778 (1998)

M.H. Shaddad, P. Jenniskens, D. Numan, A.M. Kudoda, S. Elsir, I.F. Riyad, A.E. Ali, M. Alameen, N.M. Alameen, O. Eid, A.T. Osman, M.I. Abubaker, M. Yousif, S.R. Chesley, P.W. Chodas, J. Albers, W.N. Edwards, P.G. Brown, J. Kuiper, J.M. Friedrich, The recovery of asteroid 2008 TC_3. Meteorit. Planet. Sci. **45**, 1557–1589 (2010)

X. Shi, J. Castillo-Rogez, H. Hsieh, H. Hui, W.-H. Ip, H. Lei, J.-Y. Li, F. Tosi, L. Zhou, J. Agarwal, A. Barucci, P. Beck, A. Campo Bagatin, F. Capaccioni, A. Coates, G. Cremonese, R. Duffard, R. Jaumann, G. Jones, M. Grande, E. Kallio, Y. Lin, O. Mousis, A. Nathues, J. Oberst, A. Showman, H. Sierks, S. Ulamec, M. Wang, *GAUSS – A Sample Return Mission to Ceres ESA Voyage 2050 White Paper* (2020). https://arxiv.org/ftp/arxiv/papers/1908/1908.07731.pdf

F. Spoto, A. Milani, Z. Knežević, Asteroid family ages. Icarus **257**, 275–289 (2015)

P. Spurný, P.A. Bland, L. Shrbený, J. Borovička, Z. Ceplecha, A. Singelton, A.W.R. Bevan, D. Vaughan, M.C. Towner, T.P. McClafferty, R. Toumi, G. Deacon, The Bunburra Rockhole meteorite fall in SW Australia: Fireball trajectory, luminosity, dynamics, orbit, and impact position from photographic and photoelectric records. Meteorit. Planet. Sci. **47**, 163–185 (2012)

A.A. Sukhanov, O. Durão, D. Lazzaro, Low-cost main-belt asteroid sample return. J. Spacecr. Rockets **38**, 736–744 (2001)

J.M. Sunshine, H.C. Connolly, T.J. McCoy, S.J. Bus, L.M. La Croix, Ancient asteroids enriched in refractory inclusions. Science **320**, 514–517 (2008)

S. Taylor, G. Matrajt, Y. Guan, Fine-grained precursors dominate the micrometeorite flux. Meteorit. Planet. Sci. **47**, 550–564 (2012)

A. Trinquier, J.-L. Birck, C.J. Allègre, Widespread ^{54}Cr heterogeneity in the inner Solar System. Astrophys. J. **655**, 1179–1185 (2007)

E.P. Turtle, M.E. Minitti, B.A. Cohen, N.L. Chabot, D. Tourbier, C. Bachman, J. Brock, R. Foerstner, G.V. Hoppa, J. Kay, C.A. Lewicki, R.M.E. Mastrapa, J. Patel, N. Sherman, J.N. Spitale, A.S. Rivkin, D.E. Trilling, D. Villegas, C.M. Weitz, (The JPL Advanced Projects Design Team), AMBASSADOR: asteroid sample return mission to 7 Iris. Acta Astronaut. **45**, 415–422 (1999)

T. Usui, K.-i. Bajo, W. Fujiya, Y. Furukawa, M. Koike, Y.N. Miura, H. Sugahara, S. Tachibana, Y. Takano, K. Kuramoto, The importance of Phobos sample return for understanding the Mars-Moon system. Space Sci. Rev. **216**, 49 (2020)

E.M.M.E. Van Kooten, D. Wielandt, M. Schiller, K. Nagashima, A. Thomen, K.K. Larsen, M.B. Olsen, Å. Nordlund, A.N. Krot, M. Bizzarro, Isotopic evidence for primordial molecular cloud material in metal-rich carbonaceous chondrites. Proc. Natl. Acad. Sci. **113**, 2011–2016 (2016)

P. Vernazza, R.P. Binzel, C.A. Thomas, F.E. DeMeo, S.J. Bus, A.S. Rivkin, A.T. Tokunaga, Compositional differences between meteorites and near-Earth asteroids. Nature **454**, 858–860 (2008)

P. Vernazza, B. Zanda, R.P. Binzel, T. Hiroi, F.E. DeMeo, M. Birlan, R. Hewins, L. Ricci, P. Barge, M. Lockhart, Multiple and fast: the accretion of ordinary chondrite parent bodies. Astrophys. J. **791**, 120 (2014)

P. Voosen, Meteorite divide points to solar system chaos. Science **359**, 1451 (2018)

K. Wada, M. Grott, P. Michel, K.J. Walsh, A.M. Barucci, J. Biele, J. Blum, C.M. Ernst, J.T. Grundmann, B. Gundlach, A. Hagermann, M. Hamm, M. Jutzi, M.-J. Kim, E. Kührt, L. Le Corre, G. Libourel, R. Lichtenheldt, A. Maturilli, S.R. Messenger, T. Michikami, H. Miyamoto, S. Mottola, T. Müller, A.M. Nakamura, L.R. Nittler, K. Ogawa, T. Okada, E. Palomba, N. Sakatani, S.E. Schröder, H. Senshu, D. Takir, M.E. Zolensky, (International Regolith Science Group (IRSG) in Hayabusa2 project), Asteroid Ryugu before the Hayabusa2 encounter. Prog. Earth Planet. Sci. **5**, 82 (2018)

K.J. Walsh, A. Morbidelli, S.N. Raymond, D.P. O'Brien, A.M. Mandell, A low mass for Mars from Jupiter's early gas-driven migration. Nature **473**, 206–209 (2011)

K.J. Walsh, A. Morbidelli, S.N. Raymond, D.P. O'Brien, A.M. Mandell, Populating the asteroid belt from two parent source regions due to the migration of giant planets—"The Grand Tack". Meteorit. Planet. Sci. **47**, 1941–1947 (2012)

K.J. Walsh, M. Delbó, W.F. Bottke, D. Vokrouhlický, D.S. Lauretta, Introducing the Eulalia and new Polana asteroid families: re-assessing primitive asteroid families in the inner main-belt. Icarus **225**, 283–297 (2013)

P.H. Warren, Stable-isotopic anomalies and the accretionary assemblage of the Earth and Mars: a subordinate role for carbonaceous chondrites. Earth Planet. Sci. Lett. **311**, 93–100 (2011)

S. Watanabe, M. Hirabayashi, N. Hirata, Na. Hirata, R. Noguchi, Y. Shimaki, H. Ikeda, E. Tatsumi, M. Yoshikawa, S. Kikuchi, H. Yabuta, T. Nakamura, S. Tachibana, Y. Ishihara, T. Morota, K. Kitazato, N. Sakatani, K. Matsumoto, K. Wada, H. Senshu, C. Honda, T. Michikami, H. Takeuchi, T. Kouyama, R. Honda, S. Kameda, T. Fuse, H. Miyamoto, G. Komatsu, S. Sugita, T. Okada, N. Namiki, M. Arakawa, M. Ishiguro, M. Abe, R. Gaskell, E. Palmer, O.S. Barnouin, P. Michel, A.S. French, J.W. McMahon, D.J. Scheeres, P.A. Abell, Y. Yamamoto, S. Tanaka, K. Shirai, M. Matsuoka, M. Yamada, Y. Yokota, H. Suzuki, K. Yoshioka, Y. Cho, S. Tanaka, N. Nishikawa, T. Sugiyama, H. Kikuchi, R. Hemmi, T. Yamaguchi, N. Ogawa, G. Ono, Y. Mimasu, K. Yoshikawa, T. Takahashi, Y. Takei, A. Fujii, C. Hirose, T. Iwata, M. Hayakawa, S. Hosoda, O. Mori, H. Sawada, T. Shimada, S. Soldini, H. Yano, R. Tsukizaki, M. Ozaki, Y. Iijima, K. Ogawa, M. Fujimoto, T.-M. Ho, A. Moussi, R. Jaumann, J.-P. Bibring, C. Krause, F. Terui, T. Saiki, S. Nakazawa, Y. Tsuda, Hayabusa2 arrives at the carbonaceous asteroid 162173 Ryugu—a spinning top–shaped rubble pile. Science **364**, 268–272 (2019)

M.K. Weisberg, T.J. McCoy, A.N. Krot, Systematics and evaluation of meteorite classification, in *Meteorites and the Early Solar System II*, ed. by D.S. Lauretta, H.Y. McSween Jr. (University of Arizona Press, Tucson, 2006), pp. 19–52

A.J. Westphal, J.C. Bridges, D.E. Brownlee, A.L. Butterworth, B.T. De Gregorio, G. Dominguez, G.J. Flynn, Z. Gainsforth, H.A. Ishii, D. Joswiak, L.R. Nittler, R.C. Ogliore, R. Palma, R.O. Pepin, T. Stephan, M.E. Zolensky, The future of Stardust science. Meteorit. Planet. Sci. **52**, 1859–1898 (2017)

D.A. Williams, D.L. Buczkowski, S.C. Mest, J.E.C. Scully, T. Platz, T. Kneissl, Introduction: the geologic mapping of Ceres. Icarus **316**, 1–13 (2018)

J.A. Wood, J.S. Dickey Jr., U.B. Marvin, B.N. Powell, Lunar anorthosites. Science **167**, 602–604 (1970)

H. Yurimoto, K. Kuramoto, Molecular cloud origin for the oxygen isotope heterogeneity in the Solar System. Science **305**, 1763–1766 (2004)

H. Yurimoto, K.-i. Abe, M. Abe, M. Ebihara, A. Fujimura, M. Hashiguchi, K. Hashizume, T.R. Ireland, S. Itoh, J. Katayama, C. Kato, J. Kawaguchi, N. Kawasaki, F. Kitajima, S. Kobayashi, T. Meike, T. Mukai, K. Nagao, T. Nakamura, H. Naraoka, T. Noguchi, R. Okazaki, C. Park, N. Sakamoto, Y. Seto, M. Takei, A. Tsuchiyama, M. Uesugi, S. Wakaki, T. Yada, K. Yamamoto, M. Yoshikawa, M.E. Zolensky, Oxygen isotopic compositions of asteroidal materials returned from Itokawa by the Hayabusa mission. Science **333**, 1116–1119 (2011)

M.E. Zolensky, T.J. Zega, H. Yano, S. Wirick, A.J. Westphal, M.K. Weisberg, I. Weber, J.L. Warren, M.A. Velbel, A. Tsuchiyama, P. Tsou, A. Toppani, N. Tomioka, K. Tomeoka, N. Teslich, M. Taheri, J. Susini, R. Stroud, T. Stephan, F.J. Stadermann, C.J. Snead, S.B. Simon, A. Simionovici, T.H. See, F. Robert, F.J.M. Rietmeijer, W. Rao, M.C. Perronnet, D.A. Papanastassiou, K. Okudaira, K. Ohsumi, I. Ohnishi, K. Nakamura-Messenger, T. Nakamura, S. Mostefaoui, T. Mikouchi, A. Meibom, G. Matrajt, M.A. Marcus, H. Leroux, L. Lemelle, L. Le, A. Lanzirotti, F. Langenhorst, A.N. Krot, L.P. Keller, A.T. Kearsley, D. Joswiak, D. Jacob, H. Ishii, R. Harvey, K. Hagiya, L. Grossman, J.N. Grossman, G.A. Graham, M. Gounelle, P. Gillet, M.J. Genge, G. Flynn, T. Ferroir, S. Fallon, S. Fakra, D.S. Ebel, Z. Rong Dai, P. Cordier, B. Clark, M. Chi, A.L. Butterworth, D.E. Brownlee, J.C. Bridges, S. Brennan, A. Brearley, J.P. Bradley, P. Bleuet, P.A. Bland, R. Bastien, Mineralogy and petrology of comet 81P/Wild 2 nucleus samples. Science **314**, 1735–1739 (2006)

M.Yu. Zolotov, The composition and structure of Ceres' interior. Icarus **335**, 113404 (2020)

Space Sci Rev (2020) 216:45
https://doi.org/10.1007/s11214-020-00670-1

Concepts of the Small Body Sample Return Missions - the 1st 10 Million Year Evolution of the Solar System

Yangting Lin[1] · Yonghe Zhang[2] · Sen Hu[1] · Yuchen Xu[3] · Weijia Zhou[4] · Shijie Li[5] · Wei Yang[1] · Yang Gao[6] · Mingtao Li[3] · Qingzhu Yin[7] · Douglas Lin[8,9] · Wing Ip[10]

Received: 18 July 2019 / Accepted: 7 April 2020 / Published online: 20 April 2020

Abstract Each type of asteroids and comets are important, serving as the unique puzzle pieces of the solar system. The countless number of small bodies spread vastly from the near-Earth orbits to the main belt and beyond Jupiter. Thus, in order to complete the whole puzzle, and hence requires a well-designed roadmap of sample return (SR) missions and international coordination. The main consideration is the accreting locations of various types of asteroids, which may be referred through their redox status and abundances of water and volatile components. C-complex asteroids are water and volatile-rich, likely accreted in the outer solar system. Two C-complex asteroids are being explored by Hayabusa-2 and OSIRIS-REx missions, respectively. In contrast, enstatite chondrite-like asteroids formed under extremely reducing conditions in the inner solar system. The samples returned from enstatite chondrite-like asteroids will clarify the nebular processes in the zone closest to the

Role of Sample Return in Addressing Major Questions in Planetary Sciences
Edited by Mahesh Anand, Sara Russell, Yangting Lin, Meenakshi Wadhwa, Kuljeet Kaur Marhas and Shogo Tachibana

✉ Y. Lin
LinYT@mail.iggcas.ac.cn

1 Key Laboratory of Earth and Planetary Physics, Institute of Geology and Geophysics, Chinese Academy of Sciences, Beijing, China

2 Innovation Academy for Microsatellites Chinese Academy of Sciences, Shanghai, China

3 National Space Science Center, Chinese Academy of Sciences, Beijing, China

4 Shenyang Institute of Automation, Chinese Academy of Sciences, Shenyang, China

5 Institute of Geochemistry, Chinese Academy of Sciences, Guiyang, China

6 Technology and Engineering Center for Space Utilization, Chinese Academy of Sciences, Beijing, China

7 University of California, Davis, USA

8 University of California, Santa Cruz, USA

9 Institute for Advanced Study, Tsinghua University, Beijing, China

10 Institute of Astronomy, National Central University, Taoyuan, Taiwan

Sun, and reveal fractionation of the solar nebula along the radial direction, via comparing with those of the C-complex asteroids. The exploration will also shed light on the bulk compositions of the Earth and terrestrial planets accreted in the same region of the inner solar system.

The SR missions will focus on the first 10 Ma history of the solar system, including the initial condition, the nebular processes, and the accretion of planetesimals. Because the secondary processes that took place in planetesimals, such as thermal metamorphism, aqueous alteration, melting and differentiation, could largely erase the records of the nebular events, it is critically important to choose the primordial asteroid targets. Although C-complex asteroids accreted at low temperature in the outer solar nebula, they, especially those hydrated, could have suffered severe aqueous alteration as observed in CM chondrites. Other preferred targets are L-type asteroids, which probably contain abundant Ca-, Al-rich inclusions, the first solid assemblages of the solar system. Based on the roadmap of SR missions, we propose to return samples first from enstatite chondrite-like or L-type Near-Earth asteroids.

Keywords Solar system evolution · Protoplanetary disk · Asteroids · Comets · Sample return missions

1 Introduction

Small celestial bodies, namely asteroids and comets, are the remained planetesimals and/or their broken debris accreted in the protoplanetary disk. They have recorded the formation and early history of the solar system. In addition, the small bodies are referred to as the 'building blocks" of the Earth and terrestrial planets; hence they are the key to probe the bulk compositions of the latter. Asteroids are classified into various spectral types (Bus and Binzel 2002; Demeo et al. 2009; Vernazza and Beck 2016), suggestive of different compositions and therefore their formation locations in the protoplanetary disk. Each type of asteroids and comets need to be explored and sampled, in order to draw the whole picture of the solar system evolution. Up to date, a number of asteroids and comets have been explored or selected as the targets of future missions by NASA, ESA, and JAXA. However, only one small body mission was so far planned by the China National Space Administration (CNSA), which will return samples from the asteroid 2016 HO3, a quasi-satellite of the Earth, followed by remote sensing of the comet 133P in the main belt. One of the reasons is that CNSA has already planned many other missions before 2035, including the phase 4 of Chang'E program (consisting of 3 landing missions at the lunar south pole), two Mars missions (the 1st one consisting of an orbiter, a lander and rover, and the 2nd one returning samples), and one mission to Jupiter and its satellites.

In order to have more contributions to exploring of the origin and early evolution of the solar system, we need to plan and carry out additional SR missions of asteroids and comets. This is possible, because the cost of small body SR missions can be reduced to a low level within the budget of the Strategic Priority Program on Space Science of the Chinese Academy of Sciences (CAS). For this purpose, the in advance study project of "sample return missions of small bodies" was selected and supported by CAS. The goals of this in advance study are to clarify the key scientific questions addressed by small body sample return missions, outline the exploration roadmap, and demonstrate the key techniques including self-navigation, sampling and sample capsule. The additional small body SR missions will be complementary to the deep space exploration program of CNSA, and should be coordinated with ongoing and future missions via international cooperation.

What is the strategy of the small body SR missions? There are more than one million of asteroids observed plus a large number of comets. The reflectance spectra of asteroids were classified into about 26 types (Bus and Binzel 2002; Demeo et al. 2009). The representative asteroids of the major types have been explored or chosen as the candidate targets for future missions, e.g. Vesta (V-type) by Dawn mission (Coradini et al. 2011), Psyche (M-type, metallic) by Psyche mission, Itokawa (S-type) by Hayabusa mission (Abe et al. 2006), Ryugu (C-type) by Hayabusa-2 (Kitazato et al. 2019), and Bennu (B-type) by OSIRIS-REx (Lauretta et al. 2019). On the other hand, the asteroid targets were also selected based on their orbits, e.g. Eros and Mathilde (Near Earth asteroids, NEA) by NEAR mission (Veverka et al. 2000), Vesta and Ceres (main belt asteroids) by Dawn mission, and Trojan asteroids of Jupiter by Lucy mission. The asteroid 2016 HO_3 (the quasi-satellite of the Earth) will be explored and sampled by China SR mission; whereas Phobos and Deimos (the satellites of Mars) are the targets of MMX mission of JAXA (Galimov 2010). Different from previous missions, we made a proposal of a serial SR missions based on the key questions of the origin and early evolution of the solar system. Furthermore, the proposed SR missions will combine with other ongoing and planned missions, in order to depict the early history of the solar system. After drawing up the potential roadmap of small body exploration, we propose to explore and return samples first from enstatite chondrite-like asteroids, which accreted under extremely reducing conditions in the inner solar system likely closest to the Sun.

2 Key Questions of the 1st 10 Ma Evolution of the Solar System

Asteroids and comets are the "fossil" of the solar system, preserving records of the initial condition and various processes of the solar nebula, accretion of planetesimals, and various secondary events that took place in asteroids (including thermal metamorphism, aqueous alteration, melting and differentiation). In addition, they were the building blocks of the Earth and terrestrial planets. In order to achieve the main scientific goals of SR missions, the key questions are focused on the 1st 10 Ma evolution of the solar system; whereas the secondary processes took place in the planetesimals after accretion will not be discussed in this work.

According to the formation hypothesis of the solar system and the astronomic observations of extrasolar systems, the solar system originated from a piece of collapsed molecular cloud, and the nebular disk evolved into the highly heterogeneous planetary system (Boss and Keiser 2015; Ciesla and Charnley 2006). The 1st 10 Ma evolution and fractionation of the protoplanetary disk largely determined the bulk compositions of the Earth and other terrestrial planets, hence their diverse physical properties. The main episodes of the 1st 10 Ma history can be divided into: (1) The initial solar nebula before condensation of the first solid assemblages (i.e. Ca-, Al-rich inclusions, CAIs hereafter); (2) Various nebular processes, including gas-solid condensation, heating events, solid-gas reactions (alteration), and transportation of gas and solid assemblages; (3) Accretion of planetesimals, to be followed by migration effects of planets. These main episodes are discussed in detail in the following subsections.

2.1 Initial Condition of the Protoplanetary Disk

The original solar nebula was composed of dust and gaseous molecules. A key question is about the dust/gas initial ratio. The nebular gas is expected homogeneous in chemical and isotopic compositions, whereas the dust particles should have preserved their distinct isotopic compositions of various stellar sources. The pre-existing dust particles are therefore referred to as the presolar grains, which have been found in the fine-grained matrix of primordial chondrites, based on their large isotopic anomalies (Nittler and Ciesla 2016; Zinner 1998). Except for the fine-grained matrix, other components in primitive chondrites including CAIs, silicate chondrules and metal-sulfide assemblages have experienced various high temperature processes in the nebula, which should have destroyed any pre-existing presolar grains. The abundances of presolar grains in chondrites were therefore reported as the fine-grained matrix-normalized values, and the maximum values are ~1000 ppm for presolar diamond (Lewis et al. 1987; Zinner 1998), up to ~200 ppm for presolar SiC (Floss and Stadermann 2009; Zhao et al. 2013) and 125-220 ppm for presolar silicates (Floss and Haenecour 2016; Floss and Stadermann 2009; Haenecour et al. 2018; Nittler et al. 2018). Higher abundances of presolar silicates up to an average of ~500 ppm (with a maximum of 1.5%) were reported in primitive interstellar dust particles (IDPs) (Busemann et al. 2009; Floss and Haenecour 2016). However, all the reported abundances of presolar grains are rather low, especially the abundances of presolar silicates in primitive chondrites are at least a factor of 2 orders too low (Hoppe et al. 2017). The very low abundances of presolar grains indicate a very low dust/gas ratio of the initial solar nebula, although it cannot be excluded that the presolar dust grains were highly dominated by a single stellar reservoir. In this scenario, the predominant presolar grains from this stellar reservoir should have the same solar isotopic compositions, hence they were not identified as presolar origins by in situ isotope measurement. Alternatively, most of the solid materials of the planetary disk re-condensed from the homogenized nebular gas, which could have been produced by evaporation of the pre-existing dust via heating events or in the hot zone closest to the Sun.

Another key question is the spatial distribution of presolar grains in the initial solar nebula, homogeneous or heterogeneous? Most studies of presolar grains were focused on carbonaceous chondrites, with a few on non-carbonaceous chondrites. Isotope mapping of the acid-residue from Qingzhen EH3 chondrite revealed the abundance ratios of various presolar grains (e.g. ~1/400 of X SiC grains/mainstream SiC, 1/4 of X-Si_3N_4/X-SiC) (Lin et al. 2002) different from those of Murchison CM2 carbonaceous chondrite (~1% X SiC grains relative to Mainstream, 1/30 of X-Si_3N_4/X-SiC) (Zinner 1998). The relatively lower abundance of X-grains in Qingzhen EH3 chondrite was confirmed by the ion mapping of SiC from Sahara 97166 (EH3), which resulted in 0.36% X SiC relative to Mainstream SiC (Besmehn et al. 2001). Another hint for heterogeneous distribution of presolar grains was the isotopic anomalies of Cr, Ti, Ni and O between carbonaceous chondrites and non-carbonaceous ones (Trinquier et al. 2009; Warren 2011b). The identification of nanoparticles of Cr-oxides with δ^{54}Cr of > 1500‰ supplied with convincing evidence that the ^{54}Cr-enrichment was due to the addition of supernova ejecta (Qin et al. 2011). Is it possible that the initial solar nebula was homogeneous but was later mixed with the supernova ejecta?

Another key question is the temperature and its spatial and temporal variation in the solar nebula. After the formation of the Sun, a temperature gradient in the protoplanetary disk could be expected because of the sunlight and the strong irradiation from the T-Tauri proto-Sun (Shu et al. 1996). The presence of presolar SiC and/or oxides in the primordial enstatite chondrites (Lin et al. 2002), ordinary chondrites (Choi et al. 1998; Nittler et al. 1994) and carbonaceous chondrites (Tang and Anders 1988) suggests that the temperature in the accreting regions of these parent asteroids was low (<300-400 °C), in order to preserve the submicron-sized dust grains. This is consistent with the presence of presolar grains exclusively in primitive chondrites with petrographic type ≤3. The matrix-normalized abundance of presolar SiC (based on noble gas isotopes) seems quite similar among these chemical groups, 21.9 ppm for Qingzhen (EH3), 20.1 ppm for Semarkona (LL3.0) and 28.5 ppm for Orgueil (CI), but dropped immediately in thermal metamorphosed meteorites (Huss and Lewis 1995). The similar abundances of presolar SiC from the extremely reduced EH group (likely closest to the Sun) to the oxidized C-groups (far from the Sun) suggests that the wide nebular regions from the inner to the outer solar system have never been heated high enough to destroy the submicron-sized presolar SiC. However, the low temperature requirement for preserving the submicron-sized presolar grains seems to be in conflict with the high temperature gas-solid condensation at least in the enstatite chondrite-forming regions.

2.2 Condensation Under Various Redox Conditions

Regardless of the initial low temperature of the molecular cloud and the presence of presolar grains that require a maximum temperature of <300-400 °C as discussed above, fine-grained CAIs are high-temperature condensate assemblages of various refractory oxides and silicates found commonly in carbonaceous chondrites (e.g., Krot 2019; MacPherson and Grossman 1984). The coexistence and formation sequence of these refractory minerals of the fine-grained CAIs are well consistent with condensation from the gas with CI composition (Grossman 1972, 2010; Yoneda and Grossman 1995). Furthermore, the mineral assemblages and bulk compositions of the fine-grained CAIs vary from corundum-bearing ultra-refractory, to spinel-melilite-rich, and to spinel-pyroxene-rich, along the condensation trajectory from high to low temperatures (Lin and Kimura 2003). Amoeboid olivine assemblages (AOAs) can be referred to as gas-solid condensation at lower temperatures than typical CAIs, with spinel-/anorthite-bearing AOAs filling the gap between CAIs and AOAs (Aléon et al. 2002; Komatsu et al. 2001; Krot et al. 2004). The presence of fine-grained CAIs suggests gas-solid condensation in the solar nebula. However, it is unknown where the condensation process took place. One possibility is that fine-grained CAIs condensed in the region closest to the Sun. This is supported by the typical ^{16}O-enriched oxygen isotopic compositions of CAIs (Clayton et al. 1973), which has been identified as the composition of the Sun (Hashizume and Chaussidon 2005; McKeegan et al. 2011). Recent oxygen isotope mapping of primitive chondrites found unusually ^{16}O-enriched CAIs with $\delta^{17}O$ and $\delta^{18}O$ of $-77‰$ and $-75‰$, respectively (Sakamoto and Kawasaki 2019), probably formed closest to the Sun and hence recorded its undiluted oxygen isotopic composition.

Another key question of the solar nebular condensation was raised by the formation of enstatite chondrites. These meteorites are characteristic of unique opaque mineral assemblages, consisting of various sulfides of typical lithophile elements (e.g. oldhamite, niningerite, alabandite, djerfisherite, sphalerite, daubréelite), perryite, schreibersite and Si-rich Fe-Ni metal (El Goresy et al. 1988; Keil 1989; Lin and El Goresy 2002). In addition, the silicates are predominant FeO-poor enstatite. All these minerals and their compositions required extremely reducing conditions, which could be achieved by increasing the C/O ratio from the solar value (0.54) up to >0.8 (Grossman et al. 2008; Larimer and Bartholomay 1979), although the mechanism of enhancing C/O ratio of the solar nebula remained unknown. On the other hand, the high-temperature condensation of the predominant constituents of enstatite chondrites seems to be inconsistent with the preservation of presolar grains, because the latter requires a low temperature (<300-400 °C) as discussed above. This is different from the case of fine-grained CAIs, because they are minor objects of chondrites and could be carried from the region closest to the Sun and accreted together with presolar grains in chondrite-forming ones. In contrast, it is difficult to have a mechanism of transporting all these components (sulfides, Si-bearing metal and enstatite, with different densities and other properties) to the same region, where they accreted together with the pre-existing presolar grains. Alternatively, the constituent minerals condensed under extremely reducing conditions in the enstatite chondrite-accreting regions, and the presolar grains were added after the condensation from the outer region. In addition, other questions are the origin of the extremely reducing condition and the spatial and genetic relationship between the condensation regions of CAIs and enstatite chondrites.

Variation in redox along the solar disk was revealed by oxygen fugacity of various groups of chondrites and the redox of terrestrial planets. It becomes more reducing towards the Sun, as indicated by the decreasing of FeO contents of silicates and increasing of the metallic core/silicate mantle ratios in the sequence of Mars, Earth, Venus and Mercury. Meteorites show a wide range of redox, from highly reducing of enstatite chondrites to ordinary chondrites and then the oxidizing CV chondrites. Although the accreting regions of their parental asteroids are unknown, the oxygen fugacity was positively correlated with the abundances of water and volatile components. It is reasonable to assume that the protoplanetary nebula became more reducing towards the Sun. In addition, the oxygen fugacity profile should have been established before the condensation of the fine-grained CAIs and the constituent minerals of enstatite chondrites. Hence, the related key question is how the solar nebula evolved from homogeneous into heterogeneous before condensation?

2.3 Heating Events

Distinct from the condensation of fine-grained CAIs depicted above, there were at least two transient heating events in the protoplanetary disk. One of the heating events corresponds to the formation of coarse-grained CAIs. The coarse-grained CAIs share the similar mineral, chemical and isotopic compositions of the fine-grained ones, except for their igneous textures. It was generally accepted that the coarse-grained CAIs crystallized from Ca-, Al-rich melts. However, the heating mechanism and the precursors of these refractory inclusions are controversial. A common hypothesis is the X-wind model (Shu et al. 1996), which suggests CAIs as relict refractory materials by evaporating loss (up to 90%) of Mg and Si from CI-like precursor at a distance about 0.06 AU to the Sun. Many observations of coarse-grained CAIs can be explained by the X-wind model, e.g. the igneous textures, heavier isotope enrichment of Mg and Si, fast cooling, and identical features of CAIs from different groups of chondrites. However, the X-wind model has also been critically examined, and a number of internal inconsistencies were reported (Desch et al. 2010). In addition, the overlapping of bulk compositions between the fine-grained CAIs and the coarse-grained ones (compact type A and B) along the condensation trajectory suggests that the latter can be produced by melting of the former without a significant evaporating loss (Lin and Kimura 2003). As discussed above, the fine-grained CAIs show convincing evidence for gas-solid condensation, instead of being heated and melted by a transient event. Another discovery is the genetic relationship between type C CAIs and the fine-grained anorthite-spinel-rich inclusions (ASIs). The typical Type A CAIs reacted with the nebular gaseous SiO_2, producing the ASIs; and the latter was finally heated and melted to crystallize Type C CAIs (Lin and Kimura 1998). It was also noticed that a few of type B CAIs had experienced multiple heating events, as indicated by their high Na abundances and the correlation with the åkermanite contents of melilite (Lin and Kimura 2000; MacPherson and Davis 1993). Based on the mineralogy, petrography, oxygen isotope compositions, and trace element abundances, the igneous CAIs formed by incomplete melting of refractory precursors in an ^{16}O-rich gas likely near the protosun (Krot et al. 2018).

The other heating event was related to the formation of chondrules. This heating event was probably later by about 1-3 Ma than the formation of typical CAIs based on Al-Mg and U-Pb radiometric dating (Amelin et al. 2002; Bouvier and Wadhwa 2010; Kita et al.

2005). There are some chondrules, especially Al-rich chondrules, with the initial $^{26}Al/^{27}Al$ ratios plotted between CAIs and chondrules (Bizzarro et al. 2004), they could be due the presence of mini-CAIs or the debris of CAIs captured by the chondrules (Krot et al. 2006). Most chondrules formed via incomplete melting of isotopically diverse solid precursors, including CAIs, AOAs, fragments of pre-existing chondrules, and fine-grained matrix, by local transient heating events in ^{16}O-poor regions (Krot et al. 2018). In addition, the heating events of chondrules are distinct from that of the igneous CAIs, since they show different cooling rates with $10 \sim 1000$ °C/hr for chondrules (Lofgren 1996) whereas < 5 °C/hr for the igneous CAIs (Stolper and Paque 1986). A much higher cooling rate up to 10^5-10^6 K/hr was reported in a phenocryst in a CO chondrite type-II chondrule (Yurimoto and Wasson 2002). The possible mechanisms are the X-wind model (Shu et al. 2001, 1996), lightning heating and shock heating (Connolly et al. 2006). The FeO contents of olivine and pyroxene in chondrules are related with the redox condition of their host chondrites, favoring for local heating of chondrules. This is also distinct from the common reservoir of CAIs from various groups of chondrites discussed in the next subsection.

2.4 Transportation and Fractionation of the Protoplanetary Disk

Fractionation of the protoplanetary disk, from initially homogeneous to highly heterogeneous, requires transportation of materials along the radial direction of the disk. The coexistence of CAIs with presolar grains argues for their different origins and delivering from distinct locations as discussed above. It was noticed that most CAIs from various groups of chondrites share the same petrographic features, mineral chemistry, oxygen isotopic compositions and the initial $^{26}Al/^{27}Al$ ratios (Guan et al. 2000; Lin et al. 2003; McKeegan et al. 1998), suggestive of the same reservoir. The differences of CAIs between the host chondrites are likely secondary, which could be attributed to alteration reactions in the accretion regions and/or the parent bodies of chondrites. This discovery further supports that CAIs in chondrites were not local products but delivered from other regions. According to the X-wind model, all CAIs were produced in the region very close to the proto-Sun, and then ejected to the whole protoplanetary disk (Shu et al. 1996). The discovery of CAIs in comet wild 2 sampled by Stardust mission supplies with robust evidence for the long distance transportation of CAIs (McKeegan et al. 2006; Zolensky et al. 2006). In addition, the recently reported highly C rich clast (~100 μm) in CR2 LaPaz Icefield 02342 is probably a cometary microxenolith (Nittler et al. 2019). This discovery supports the idea of a radially inward transportation of materials from the outer protoplanetary disk into the CR reservoir. Another convincing evidence for the transportation in the nebular disk is the presence of dust-like pyroxene clasts in primitive enstatite chondrites, e.g. Qingzhen (EH3) and Yamato 691 (EH3) (Lin and El Goresy 2002; Rambaldi et al. 1983). The dust-like particles are Ni-poor micron-sized Fe metal, reduced from FeO of the silicates. The high FeO-contents of the silicate clasts suggest formation under oxidizing conditions, and then transported into the reducing regions of enstatite chondrites.

Before the condensation of the fine-grained CAIs and the constituent minerals of enstatite chondrites, the protoplanetary disk had already been evolved into heterogeneous. As discussed above, the fine-grained CAIs condensed from the nebula with the solar C/O ratio of ~0.54, whereas the formation of enstatite chondrites required a much reducing condition with the C/O ratio >0.8. A simplified mechanism of enhancing the C/O ratio is fractionation between H, O and C via condensation and evaporation of water ice and organic materials across the snow line in the protoplanetary disk. The H_2O vapor and volatile organics could diffuse outwards, and re-condensed beyond the snow line; whereas the water ice and solid organics moved inwards due to the drag force of gas. This process would separate H_2O from the refractory organics, and consequently increasing the C/O ratio towards the Sun. On the other hand, water ice would concentrate close to the snow line, increasing the dust/gas ratio there and consequently feeding the formation of Jupiter. This scenario also explains the variation of redox among chondrites, and the correlation of redox with water and volatile abundances. It is expected that the isotopic anomalies of water ice (or preserved as oxides via reactions with metallic Fe-Ni and/or sulfides) and organics should be more significant in comets that accreted far from the snow line.

2.5 Accretion of Planetesimals and the Consequent Processes

CAIs, chondrules, opaque mineral assemblages, mineral fragments, presolar grains and the fine-grained matrix components all deposited together onto the midplane of the protoplanetary disk, forming pebble-sized chunks that grew into km-sized planetesimals probably by gravitational instability and turbulent collision (Chambers 2004). Little knowledge of the growing process of planetesimals was confirmed, since all meteorites are small in size and there were no representative samples of such large planetesimals. On the other hand, the explored asteroids are commonly composed of two parts (contact binary asteroids), which appears to be connected together via a collision of two smaller lobes. Simulations on the formation of planetesimals also showed that a part of the materials were contributed from other regions far away from the accreting site (Chambers 2004). Because of the isotopic heterogeneity of the protoplanetary disk, it is possible to track the distinct reservoirs of asteroids by precise isotope analysis of the samples collected from different sites of the targets.

After accretion, planetesimals usually experienced various degrees of secondary processes, such as thermal metamorphism, aqueous alteration, melting and differentiation. These processes are important, especially for the formation of planets. On the other hand, they would have largely erased the records of the nebular processes. C-complex asteroids are generally considered most primitive, and they are the preferred targets for exploration. However, some C-complex asteroids, especially those highly enriched in water, likely have suffered severely aqueous alteration. An example of aqueous alteration is CM2 chondrites. In Murchison and many other CM2 chondrites, most olivine grains in the fine-grained matrix and even in chondrules were heavily altered into phyllosilicates. Another indicator of aqueous alteration is the abundance of presolar silicates, which can be shown as the abundance ratio of presolar silicate/SiC. As mentioned above, presolar SiC grains are submicron-sized and very sensitive to temperature. Presolar silicates are also submicron-sized (0.2-0.5 μm predominantly), hence sensitive to temperature too. In addition, presolar silicates were readily erased by aqueous alteration even under low temperature. NanoSIMS mapping of presolar grains in the CM Sutter's Mill meteorite reported the presolar silicate/SiC abundance ratio of $<1/32$ (Zhao et al. 2014), much lower than those in CR and other less aqueous altered primitive chondrites (e.g. 33 in GRV 021710) (Zhao et al. 2013).

2.6 Relationship Between the Inner and Outer Solar System

A new classification schematic of meteorites and planetary materials was proposed as carbonaceous and non-carbonaceous groups, based on the bimodal isotopic anomalies of Cr, Ti, and O (Warren 2011a). This bimodality was confirmed by isotope analyses of more elements including Ni (Nanne et al. 2019), Ba (Bermingham et al. 2016), Mo (Budde et al. 2016), Ru and W (Worsham et al. 2019). The carrier of ^{54}Cr-anomaly was successfully separated and identified as nano-particles of Cr-oxides, which was likely ejected by the explosion of a nearby supernova (Qin et al. 2011). Carbonaceous meteorites accreted in the outer solar system, and show enrichment in ^{54}Cr, ^{50}Ti and ^{62}Ni, in comparison with non-carbonaceous materials (including ordinary chondrites, enstatite chondrites, achondrites, the Earth, the Moon, Mars, Vesta) representative of the inner solar system (Qin et al. 2011; Warren 2011a). These discoveries revealed heterogeneous isotopic compositions in the protoplanetary disk, and the spatial isolation between CC and non-CC materials before the accretion of planetesimals. The key questions are how and when this process took place? It is obviously important to explore and sample both the representative CC and non-CC asteroids, in order to address these questions.

2.7 Distribution and Evolution of Organic Matter in the Protoplanetary Disk

Organic materials are common in comets (Capaccioni et al. 2015; Cody et al. 2011; Goesmann et al. 2015; Sandford et al. 2006), IDPs (Busemann et al. 2009) and primitive chondrites (e.g., Alexander et al. 2017; Busemann et al. 2006; Cooper et al. 2001; Engel and Macko 1997; Nakamura-Messenger et al. 2006). They are widespread especially in the outer solar system. Organics in the protoplanetary disk constituted the source materials of the "building blocks" required by life, but their formation mechanisms and the subsequent modification processes are still poorly understood. The insoluble organic matter (IOM) usually shows large and highly heterogeneous D and ^{15}N-enrichments especially in the most primitive carbonaceous chondrites and IDPs (Busemann et al. 2006; Nakamura-Messenger et al. 2006), suggestive of the preservation of the primitive organics of the early solar system. However, the origins of the isotopic anomalies are unresolved, produced in the interstellar medium or the outer solar system (Alexander et al. 2017, 2008). In addition, organic materials can be readily modified by thermal and aqueous alteration reactions in the parent asteroids. The IOM in the Tagish Lake carbonaceous chondrite shows correlation with the parent body aqueous alteration, suggesting that at least some organic compounds formed in the parent asteroid (Herd et al. 2011). Other organic grains from CR chondrites showed association with carbonates and have more processed organic matter in the rims, providing evidence for highly aromatic and N-bearing organic synthesis by the parent body fluid (Vollmer et al. 2014). Hence, the samples returned from comets and the most primitive asteroids (such as D-type) that avoided the thermal and aqueous alteration reactions in the parent bodies are critically important to understand the formation and evolution of organic matter in the interstellar medium and the outer solar nebula.

Other key question is the distribution and modification of organic matter in the inner solar system. Compared with those accreted in the outer solar system, organic matter in ordinary chondrites and enstatite chondrites show distinct isotopic features. The organics in the thermally metamorphosed Abee enstatite chondrite is the most D-depleted of extraterrestrial samples (Remusat et al. 2012), whereas IOM in unequilibrated ordinary chondrites is extremely D-enriched and negatively related with the H/C ratios (Alexander et al. 2010; Remusat et al. 2016). Furthermore, organic carbon could play a key role in the redox state of the inner solar system, although ordinary chondrites and enstatite chondrites contain low organic matter and the majority of the organic materials of the Earth could be delivered from carbonaceous asteroids based on the H and N isotopes (Furi and Marty 2015; Marty et al. 2016). However, any possible genetic link between organics, amorphous and crystalline graphite in the nebula close to the Sun is unknown. The step heating of the Abee enstatite chondrite suggested that the graphite was not derived from a pure thermal solid-state graphitization of the organic matter (Remusat et al. 2012). In addition, the EL3 Almahata Sitta MS-17 and MS-177 fragments contain two petrographic settings of graphite with distinct and heterogeneous ^{13}C-depleted compositions (El Goresy et al. 2017), suggestive of complex formation of graphite. Samples returned from the primitive asteroids accreted close to the Sun is very important to understand the evolution of organic materials in the inner solar system.

3 Concepts of Small Body Sample Return Missions

Distinct from the exploration of planets and their satellites, samples returned from small bodies play the unique roles of understanding the formation and evolution of the solar system, especially during the first 10 Ma. The major key questions can only be addressed by analyzing the returned samples using the state-of-art facilities in laboratories, which requires extremely high spatial resolution and/or very high precision analysis of isotopes and trace elements. The unique advantages of SR missions are below: (1) New samples not yet in the meteorite collections; (2) Linking the spectral types of asteroids with the groups of meteorites; (3) Maximum elimination of terrestrial contamination, especially for water and organics; (4) With full contextual information of the parental asteroids and/or comets; (5) Multi-site sampling of the same planetesimals; (6) Intact surface regolith samples for study of space weathering on small bodies at various orbits.

The major scientific goals will be focused on the 1st 10 Ma history of the solar system and how it evolved into the highly heterogeneous planetary system, including the initial condition of the solar nebula, the major nebular processes, and accretion of planetesimals. In order to achieve these purposes efficiently, it is required to design a roadmap of SR missions based on the key questions. There are other two key considerations of the target selection: (1) primordial asteroids avoided possible thermal metamorphism and aqueous alteration after accretion; (2) various spectral types of asteroids and comets, representative of different accretion locations in the protoplanetary disk. It is critically important to return samples from the inner and outer solar system, respectively. The concepts of small body SR missions are illustrated in Fig. 1.

3.1 Scientific Goals of SR Missions

The scientific goals of SR missions are to address the key questions discussed above, mainly focused on the 1st 10 Ma history of the solar system. This period can be referred to as the birth of the solar system, and it has largely determined the formation and bulk compositions of the Earth and other terrestrial planets. Those secondary processes took place within planetesimals, such as thermal and aqueous metamorphism, melting and differentiation, could erase the nebular records and hence should be avoided in our project. The main scientific goals of small body SR missions include below:

(1) To understand the spatial distribution of presolar dust along the radial direction of the protoplanetary disk, the dust/gas ratio and its possible variation with time (or temperature gradient increasing towards the Sun). It is required to distinguish presolar grains from the solar condensates. These questions can be addressed via surveying of presolar grains in the returned samples from most primitive asteroids and comets, which accreted in various locations in the inner and outer solar system.
(2) To understand the nebular processes and their possible differences between the inner and outer systems. The key processes include high temperature gas-solid condensation, the heating events related to the formation of coarse-grained CAIs and chondrules, and

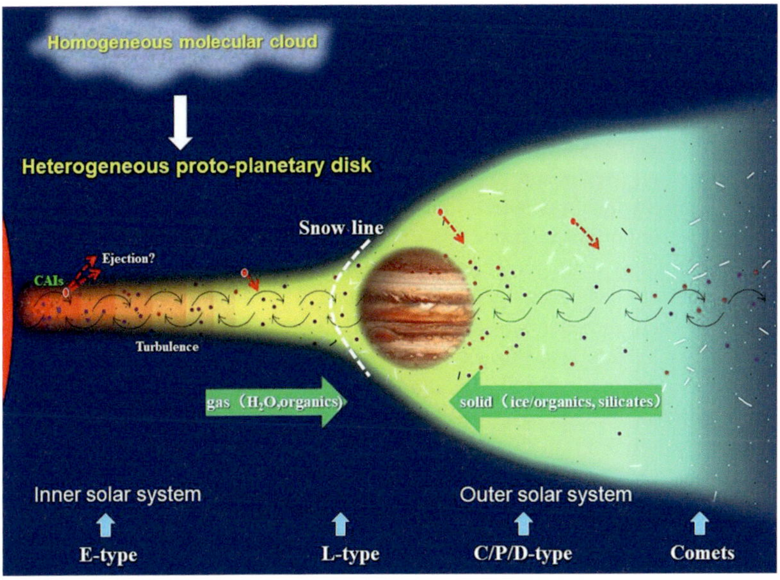

Fig. 1 Concepts of small body sample return missions

solid-gas reactions (alteration). It is required to clarify where and when these processes took place. Is there any genetic relationship between the condensation (cooling process) and the fast heating events?

(3) To understand the chemical fractionation of the protoplanetary disk in temporal and spatial dimensions. The fractionation in spatial dimensions requires samples returned from the representative asteroids and comets accreted in various locations, especially those from the inner and outer solar system, respectively. The key questions are the distributions of oxygen fugacity, water, organics and other volatile components along the radial direction of the protoplanetary disk. In addition, the fractionations were achieved by the transportation of materials, which probably started very early before the condensation of fine-grained CAIs and the unique opaque minerals of enstatite chondrites.

(4) To understand the isotopic anomalies and their correlation between CC and non-CC materials. Do the isotopic anomalies relate with the distribution of presolar dust or were they ejected from a nearby supernova? Besides the representative objects formed in the inner and outer solar system, respectively, candidate asteroids accreted close to or in the transition zone between them will shed light on the origin of the bimodality. A possibility is that the bimodality could be related to the fast formation of Jupiter, which placed a role of a barrier to stop the transportation of materials between the inner and outer solar system because of the strong gravity field of Jupiter.

(5) To understand the accretion of planetesimals, especially the possible reservoirs tracked by high precision isotope analysis. It could reveal how the different reservoirs were delivered to the same region and then accreted into the planetesimals.

(6) To establish the connection between the spectral types of asteroids with the chemical groups of chondrites or the returned samples if they are not in the collection of meteorites.

(7) To understand formation and evolution of organic compounds in the protoplanetary disk.

3.2 Target Selection

Primordial Asteroids After accretion, planetesimals likely experienced various degrees of thermal metamorphism and aqueous alteration. Thermal metamorphism is readily recognized, referred to as the petrographic types of chondrites. However, the effects of aqueous alteration were often overlooked, which could severely erase most nebular records even at low temperatures. In fact, water-rich C-complex asteroids were favorably selected as the targets. They are important to probe the distribution of water and organics in the protoplanetary disk, but may not a good choice in order to reveal the nebular processes. This has been demonstrated by the extensive aqueous alteration of CM2 chondrites, with most olivine and pyroxene (except the phenocrysts in chondrules) altered to serpentine-tochilinite intergrowths (Poorly Characterized Phases, PCP). Carbonates (calcite, dolomite) and magnetite are also common products of aqueous alteration, and their oxygen isotopes are ^{16}O-poor and show significant mass-dependent fractionation (e.g., Benedix et al. 2003; Jenniskens et al. 2012; Jilly-Rehak et al. 2018; Tyra et al. 2012). Another index of the aqueous alteration is the abundance ratios of presolar silicate/SiC and presolar silicate/oxide, since the submicron-sized presolar silicates are very sensitive to aqueous alteration, whereas both SiC and oxides are acid-resistant. Sutter's Mill CM chondrite shows severe aqueous alteration, and the ion mapping resulted in the presolar silicate/SiC abundance ratio of 2/55 (Zhao et al. 2014). The very low abundance of presolar silicates relative to SiC indicates that they have been destroyed by aqueous alteration. In contrast, GRV 021710 CR chondrite shows little aqueous alteration, with the presolar silicate/SiC and silicate/oxide ratios of 170/182 and 33/1 (Zhao et al. 2013), respectively. Hence, it is also important to select intact water-poor carbonaceous asteroids as targets, in order to avoid possible aqueous alteration. Dehydrated carbonaceous asteroids are also water-poor, but they have probably experienced aqueous alteration followed by degassing of thermal metamorphism.

Asteroids Formed in the Inner vs Those in the Outer Solar System In order to characterize the nebular processes at different locations in the protoplanetary disk, and to clarify the chemical fractionation of the solar system, it is essential to return samples from representative asteroids accreted at different locations, especially from both the inner and outer solar system. The formation locations of asteroids may be indicated by the oxygen fugacity and the abundances of water and volatile components. Enstatite chondrites are the most reduced meteorites, formed under a high C/O ratio ≥ 0.8. They can be referred to as the representative materials accreted closest to the Sun. On the other hand, carbonaceous chondrites are water-rich, and distinct from non-carbonaceous chondrites by the isotopic anomalies of O, Cr, Ti and other elements as depicted above. They accreted likely far from the Sun in the outer solar system. There are other types of asteroids dominantly in the outer main belt, including types D and P. Both D- and P-types are considered the most primitive among the asteroid population, probably formed far from the Sun. Their reflectance spectra are comparable to those of Hilda and Jupiter Trojans (DeMeo and Carry 2014; Grav et al. 2011; Vernazza and Beck 2016), respectively. D-type asteroids show a weak link with the Tagish Lake C2 ungrouped carbonaceous chondrite (Barucci et al. 2018; Fujiya et al. 2019; Hiroi et al. 2001), but the latter experienced severe aqueous alteration and contains abundant carbonates (Brown et al. 2000; Hiroi et al. 2001). Recently several small D-type asteroids (<600 m in diameter) were found by NEOShield-2 project, which are accessible to SR missions ($\Delta V < 7$ km/s) (Barucci et al. 2018). Other important types of asteroids are those not related to any meteorites. One of them is L-type, which probably contains the most abundant of CAIs. Spinel is a common mineral in CAIs and shows an absorption at about 2.4 μm, which are similar to the spectra of L-type asteroids (Sunshine et al. 2008).

Near Earth Asteroids The orbits of asteroids are readily modified by the gravity fields of planets, and the Near Earth asteroids likely immigrated from a wide range of spatial location. The spectral types of NEA and their distribution pattern are comparable to those of the asteroid belt (Binzel et al. 2019, 2004c; DeMeo and Carry 2014; Popescu et al. 2019). Hence, the Near Earth asteroids are the preferred targets for asteroid sample return missions. Other considerations are the suitable sizes, low rotation speeds, readily approachable orbits, and the launch windows.

Comets with Different Types of Orbits Comets are the most primitive small bodies, because they have been preserved at very low temperatures and never been melted. They suffered no aqueous alteration, hence preserved most records of the solar nebular processes and events. The orbits of comets can be classified as short period (Jupiter-family, Halley-type) and long period. It is possible that comets formed close to the snow line might have accreted more re-condensed water ice and organics from the vapor diffusing from the inner side of the snow line. In contrast, long period comets probably originated beyond the Kuiper belt, e.g. from the Oort cloud. The long period comets should have preserved the intact interstellar water ice and organics originated from various stellar sources. The samples returned from comets are especially important for understanding the distribution and evolution of organics, water and other highly volatile components in the solar system.

3.3 Heterogeneity and Multi-Site Sampling

According to numerical simulations of planetesimals and planets, the accretion at each zone has contributions from other zones of the protoplanetary disk, although the majority materials are local (Chambers 2004). In addition, the previous exploration of asteroids shows that the contact binary asteroids are common, e.g. asteroid 4179 Toutatis observed by the flyby of Chang'E-2 (Huang et al. 2013). Polymict breccia of chondrites were also reported. The Kaidun meteorite is composed of carbonaceous and enstatite chondrites, including various breccia of EH3-5, EL3, CV3, CR, CI-like, CM1-2 and R chondrites (MacPherson et al. 2009; Zolensky and Ivanov 2003). The Almahata Sitta meteorite consists mainly of ureilite with breccia of EH3, EL3, R and ordinary chondrites (Bischoff et al. 2010). Hence, it is required to obtain and return samples from different locations on the possibly heterogeneous surface of the target. High precision analyses of the isotopes and trace elements of the returned multi-site samples will shed light on the accretion process of planetesimals and transportation of materials in the protoplanetary disk.

3.4 Sampling and Curating Techniques

Besides common techniques of deep space exploration, e.g. energy, propulsion, communication, there are specific concerns on how to preserve the original status of the samples and to eliminate any possible contamination, which could happen during sampling, the return journey to the Earth, and curation in the laboratories. Special attention should be paid to organic materials, water, and other volatile components. In addition, new techniques of extremely low temperature from sampling process to the return journey, curation and analysis should be developed before returning samples from comets.

Sampling on a small body surface could be the most challenge technique and the most risky process of SR missions. Touch-and-go is still the preferable strategy in the first mission. Different from Hayabusa and OSIRIS-REx missions, full mechanical design of the sampling system will be applied, consisting of a robotic arm and a mechanical sampling device. The robotic arm supplies with a flexible touch with the surface regolith of the target, which can maintain the contact at least 3 seconds for sampling operation. Both main parts were connected with the self-adaptive joint, making sure well contact of the sampling device with the rough surface of the asteroid. This design has the advantage that the sampling device can be replaced for different tasks in future missions. Several sampling techniques, i.e. brushing, grabbing and electro-magnetic attracting (efficient only for chondritic samples, but erasing the magnetic information), are under consideration and test. The sampling system is designed to collect the surface regolith from three sites of the target, with the minimum mass of 2 g at each site. The materials used to manufacture the sampling device and the sample capsules will be carefully selected and verified, in order to eliminate possible contamination. The sample capsules will be vacuum-sealed and thermally isolated, with the temperature $< 80\ ^\circ\mathrm{C}$ when entering the atmosphere.

It will be a great challenge to return intact samples from comets. The key technique is how to preserve the comet samples under extremely low temperatures (e.g. $< -170\ ^\circ\mathrm{C}$) during the mission, including the sampling process, the long journey back to the Earth and re-entry of the atmosphere, and finally long-term curation, sample preparing and analysis. Besides water ice, comets also contain abundant high-volatile compounds, such as CH_4, CO, CO_2 and O_2 (Bieler et al. 2015; Capaccioni et al. 2015; Goesmann et al. 2015). In addition, contamination could be a severe issue, especially for organic materials. It is also critically important to keep the comet samples in their original status and forms, in order to preserve the micro-textures and heterogeneity in chemical and isotopic compositions, which recorded the various interstellar origins of these highly volatile components, the low-temperature processes of the solar nebula and formation of comets.

The lessons learnt from curation of Apollo lunar samples, comet dust samples, asteroid Itokawa samples, meteorites and IDPs can be used to design the small body return sample curation facilities. Besides preservation in ultra-pure N_2 atmosphere, a proportion of the asteroid samples could be kept in various conditions, e.g. high vacuum, liquid N_2, liquid He. Nondestructive analytical techniques, especially high resolution CT, can be applied to characterize and allocate the samples. However, new techniques should be developed in order to curate, prepare and analyze the comet samples. All of the procedures must be carried out under extremely low temperatures. In addition, the samples should be completely isolated from the lab atmosphere, in order to avoid possible contamination via condensation of the atmosphere components under such low temperatures.

3.5 Road Map of Sample Return Missions

In order to decode the first 10 Ma history of the solar system, a series of SR missions need to be planned and carried out in sequence. The targets should be primordial and their accretion zones cover the whole protoplanetary disk, from the inner to outer solar system in an order of enstatite chondrite-like asteroids (closest to the Sun), L-type (probably most CAIs-rich), C-, D- and P-types (outer solar system) and comets (most primitive objects).

The order of exploring of each type of the small bodies is scheduled mainly based on the approachability of the targets and the novelty of the samples. Because both asteroids Ryugu (C-type) and Bennu (B-type) originated likely in the outer solar system, hence the asteroids formed in the inner solar system closest to the Sun, i.e. enstatite chondrite-like asteroids, are selected as the top priority targets for the first SR mission of the space science program of CAS. In addition, L-type asteroids are also selected as the additional candidates.

The space technology will be developed accordingly, starting from those with higher Technology Readiness Level (TRL). SR missions of comets require special techniques of operating under extremely low temperatures in order to preserve the samples intact.

4 First Asteroid Sample Return Mission

4.1 Scientific Goals

According to the above discussion and the roadmap, the main scientific goals of the first asteroid SR mission of the space science program of CAS will aim to reveal the initial status and the consequent nebular processes under the extremely reducing condition in the inner solar disk. The discoveries will be compared with those by Hayabusa-2 and OSIRIS-REx missions, to clarify the chemical fractionation between the inner and outer solar system and understand how the solar system evolved into highly heterogeneous. In addition, the returned samples would shed light on the formation and the compositions of the Earth, Mercury and Venus accreted also in the inner solar system.

The specific questions to be addressed are: (1) The original status of the inner solar disk, including the abundance ratios of the dust (presolar grains)/gas and presolar silicate/SiC, and their differences from the outer solar system; (2) Inherent presence of water, organics and volatiles in the inner solar disk, and their fractionation along the radial direction towards the outer solar system; (3) Condensation under highly reducing conditions, and how the oxygen fugacity varied along the protoplanetary disk; (4) Possible heterogeneous accretion of planetesimals in the inner region, via orbit remote sensing and multi-site sampling of the target; (5) Space weathering and physical properties of the parental asteroids of enstatite chondrites; (6) Genetic relationship of the enstatite chondrite-like asteroids and enstatite chondrites, and their spatial distribution in the main melt; (7) Distribution of organic materials in the inner solar system, and their evolution revealed by comparison with those in the outer solar system.

4.2 Candidate Asteroids

According to the concepts and roadmap of small body SR missions, the high priority type of asteroids would be enstatite chondrite-like. The possible types include E, Xc, and Xe (Bus and Binzel 2002; Vernazza and Beck 2016). With consideration of accessibility, two NEA candidates have been tentatively selected, i.e. 1989 ML (10302) and 1982 DB (4660 Nereus). 1989 ML is about 248 m in diameter, with the rotation period of 19 hrs. It was classified as X-type or may be a shock-darkened ordinary chondrite (Binzel et al. 2001), or E-type (Mueller et al. 2007). 1989 ML was also chosen as a back-up target for MUSES-C (Fujiwara et al. 2000). A launch window is in July 2025, and the mission duration is about 4 years. 1982 DB is $500 \times 300 \times 240$ m in size, with the rotation period of 15 hrs (Brozovic et al. 2009). It was classified as Xe-type (Binzel et al. 2004a). The launch window is Feb. 2026, and the mission duration is about 5 years.

There is a large uncertainty whether or not 1989 ML and 1982 DB are really enstatite chondrite-like, because the spectra of enstatite chondrites measured in laboratory are almost featureless in terms of mafic silicate absorption bands. In addition, both 1989 ML and 1982 DB have high albedo, more close to enstatite achondrite (aubrite) than enstatite chondrite. Much more studies are required to compare the spectra of enstatite chondrites with the observations of candidate asteroids. Another helpful feature is that oldhamite shows high albedo at 0.56 μm (63%) and strong absorption band near 0.40 μm and weaker features at 0.546, 0.763, 1.151, and 1.831 μm (Izawa et al. 2013). Although oldhamite is a minor phase, it is the REE-rich Ca-sulfide and condensed rather early. The oldhamite-rich enstatite-chondrite-like asteroid would be very interesting.

L-type asteroids are rare, and not in the meteorite collections. They shows unique ∼2 micron absorptions attributed to FeO-bearing spinel, the most common constituent in CAIs. Based on the fitting of spectra, the abundance of CAIs in L-type asteroids could be up to 30 vol% (Sunshine et al. 2008; Tanga et al. 2015), as high by a factor of 3 as the most CAI-rich Allende CV3 chondrites. Two candidate asteroids of L-type, i.e. 2001 CC21 and 2012 DA14 (Binzel et al. 2004b; Takahashi et al. 2014), were selected. 2001 CC21 has a size of 0.6-1.3 km, with a rotation period of 5 hrs. The next launch window is June 2026. 2012 DA14 is much smaller, ∼30 m in diameter, with a rotation period of 9.5 hrs. The launch window is Feb. 2025.

4.3 Scouting of the Target Asteroid

In order to conduct the sampling safely on the surface of the target and to obtain the geological background, the global mapping of the asteroid will be carried out first. The essential information and key properties of the target include the Digital Elevation Model (DEM), distribution of boulders, mineral compositions and spatial distribution, presence and distribution of water and organics, thermal inertia and surface regolith, internal structure, magnetic field, the mass and density. Based on the density and composition of the surface materials revealed by spectral data, the porosity of the target can be well estimated. In addition, ground-penetrating radar can directly determine the inner structure of the asteroid. The interior structure of the target will be critical to distinguish a rubble-piled up object from the originally accreted one, and has constraints on its collisional evolution.

In order to reduce the budget of the SR mission, only the key scientific payloads will be considered. The tentatively configuration of the remote sensing instruments includes cameras, VNIS and thermal imaging system, radar sounder and receiver, magnetometer, and Rb atom clock and radio transmitter.

5 Concluding Remarks

Exploration of asteroids and comets plays the key role of understanding the origin and evolution of the solar system, and it is a long term task. Besides the CNSA SR mission aiming at the asteroid 2016 HO3, the quasi-satellite of the Earth, a series of small body sample return missions should be well designed and carried out at a lower cost. The main scientific goals will be focused on the first 10 Ma history of the solar system, including the initial condition of the solar nebula, the nebular processes and events, and the accretion of planetesimals. The exploration and study of the returned samples will reveal how the protoplanetary disk evolved into the highly heterogeneous solar system. To achieve these goals, primordial asteroids of various spectral types and comets, which accreted in different locations of the disk, should be explored and sampled. The accretion locations could be inferred from the redox status and the abundances of water, organics and other volatiles of the asteroids. The top priority targets are enstatite chondrite-like, because they formed under extremely reducing conditions likely closest to the Sun in the inner solar system. These asteroids are also the key to understand the formation and their bulk compositions of the Earth and terrestrial planets. They will shed light on the genetic relationship and fractionation between the inner and outer solar system, via comparing with those achieved by Hayabusa-2 and OSIRIS-REx missions both returning samples from the outer solar system. Another high priority targets are L-type asteroids, which probably contain the highest abundance of CAIs.

Two enstatite chondrite-like asteroids, 1989 ML and 1982 DB, and another two L-type asteroids, 2001 CC21, and 2012 DA14, have been tentatively selected as the candidate targets for the first SR mission of the space science program of CAS. All of them are Near Earth Asteroids, and approachable. The topography, mineral compositions, temperature distribution, density and internal structure of the target will be explored first, followed by touch-and-go sampling at two or three sites of the surface. The sampling system consists of a robotic arm and a collecting device that can be replaced with various designs, including brushing, grabbing and electro-magnetic attracting.

Because of the large number of the spectral types of asteroids and the long mission duration, the international coordination and collaboration could play a key role in the SR missions, ranging from scientific research, to selection and astronomical observation of the targets, and to the development of payloads.

Acknowledgements The authors are grateful to three anonymous reviewers for their constructive comments and suggestions, which significantly improved the manuscript. This study was supported by the Strategic Priority Research Program on Space Science, Chinese Academy of Sciences (XDA15020300), and the National Natural Science Foundation of China (41673069).

References

M. Abe, Y. Takagi, K. Kitazato, S. Abe, T. Hiroi, F. Vilas, B.E. Clark, P.A. Abell, S.M. Lederer, K.S. Jarvis et al., Near-infrared spectral results of asteroid Itokawa from the Hayabusa spacecraft. Science **312**, 1334–1338 (2006)

J. Aléon, A.N. Krot, K.D. McKeegan, Calcium-aluminum-rich inclusions and amoeboid olivine aggregates from the CR carbonaceous chondrites. Meteorit. Planet. Sci. **37**, 1729–1755 (2002)

C.M.O.D. Alexander, G.D. Cody, M. Fogel, H. Yabuta, Organics in meteorites-Solar or interstellar? in *Proceedings of the International Astronomical Union*, vol. 251 (2008), p. 293

C.M.O.D. Alexander, S.D. Newsome, M.L. Fogel, L.R. Nittler, H. Busemann, G.D. Cody, Deuterium enrichments in chondritic macromolecular material–Implications for the origin and evolution of organics, water and asteroids. Geochim. Cosmochim. Acta **74**, 4417–4437 (2010)

C.M.O.D. Alexander, G.D. Cody, B.T. De Gregorio, L.R. Nittler, R.M. Stroud, The nature, origin and modification of insoluble organic matter in chondrites, the major source of Earth's C and N. Geochemistry **77**, 227–256 (2017)

Y. Amelin, A.N. Krot, I.D. Hutcheon, A.A. Ulyanov, Lead isotopic ages of chondrules and calcium-aluminum-rich inclusions. Science **297**, 1678–1683 (2002)

M.A. Barucci, D. Perna, M. Popescu, S. Fornasier, A. Doressoundiram, C. Lantz, F. Merlin, M. Fulchignoni, E. Dotto, S. Kanuchova, Small D-type asteroids in the NEO population: new targets for space missions. Mon. Not. R. Astron. Soc. **476**, 4481–4487 (2018)

G.K. Benedix, L.A. Leshin, J. Farquhar, T. Jackson, M.H. Thiemens, Carbonates in CM2 chondrites: constraints on alteration conditions from oxygen isotopic compositions and petrographic observations. Geochim. Cosmochim. Acta **67**, 1577–1588 (2003)

K.R. Bermingham, K. Mezger, E.E. Scherer, M.F. Horan, R.W. Carlson, D. Upadhyay, T. Magna, A. Pack, Barium isotope abundances in meteorites and their implications for early Solar System evolution. Geochim. Cosmochim. Acta **175**, 282–298 (2016)

A. Besmehn, S. Mostefaoui, P. Hoppe, Presolar minerals in the enstatite chondrite Sahara 97166. Meteorit. Planet. Sci. **36**, A20 (2001)

A. Bieler, K. Altwegg, H. Balsiger, A. Bar-Nun, J.J. Berthelier, P. Bochsler, C. Briois, U. Calmonte, M. Combi, J. De Keyser et al., Abundant molecular oxygen in the coma of comet 67P/Churyumov-Gerasimenko. Nature **526**, 678 (2015)

R.P. Binzel, A.W. Harris, S.J. Bus, T.H. Burbine, Spectral properties of near-Earth objects: Palomar and IRTF results for 48 objects including spacecraft targets (9969) Braille and (10302) 1989 ML. Icarus **151**, 139–149 (2001)

R.P. Binzel, M. Birlan, S.J. Bus, A.W. Harris, A.S. Rivkin, S. Fornasier, Spectral observations for near-Earth objects including potential target 4660 Nereus: results from Meudon remote observations at the NASA Infrared Telescope Facility (IRTF). Planet. Space Sci. **52**, 291–296 (2004a)
R.P. Binzel, E. Perozzi, A.S. Rivkin, A. Rossi, A.W. Harris, S.J. Bus, G.B. Valsecchi, S.M. Slivan, Dynamical and compositional assessment of near-Earth object mission targets. Meteorit. Planet. Sci. **39**, 351–366 (2004b)
R.P. Binzel, A.S. Rivkin, J.S. Stuart, A.W. Harris, S.J. Bus, T.H. Burbine, Observed spectral properties of near-Earth objects: results for population distribution, source regions, and space weathering processes. Icarus **170**, 259–294 (2004c)
R.P. Binzel, F.E. DeMeo, E.V. Turtelboom, S.J. Bus, A. Tokunaga, T.H. Burbine, C. Lantz, D. Polishook, B. Carry, A. Morbidelli et al., Compositional distributions and evolutionary processes for the near-Earth object population: results from the MIT-Hawaii Near-Earth Object Spectroscopic Survey (MITHNEOS). Icarus **324**, 41–76 (2019)
A. Bischoff, M. Horstmann, A. Pack, M. Laubenstein, S. Haberer, Asteroid 2008 TC3- Almahata Sitta: a spectacular breccia containing many different ureilitic and chondritic lithologies. Meteorit. Planet. Sci. **45**, 1638–1656 (2010)
M. Bizzarro, J.A. Baker, H. Haack, Mg isotope evidence for contemporaneous formation of chondrules and refractory inclusions. Nature **431**, 275–278 (2004)
A.P. Boss, S.A. Keiser, Triggering collapse of the presolar dense cloud core and injecting short-lived radioisotopes with a shock wave. IV. Effects of rotational axis orientation. Astrophys. J. **809**, 1 (2015)
A. Bouvier, M. Wadhwa, The age of the Solar System redefined by the oldest Pb-Pb age of a meteoritic inclusion. Nat. Geosci. **3**, 637–641 (2010)
P.G. Brown, A.R. Hildebrand, M.E. Zolensky, M. Grady, R.N. Clayton, T.K. Mayeda, E. Tagliaferri, R. Spalding, N.D. MacRae, E.L. Hoffman et al., The fall, recovery, orbit, and composition of the Tagish lake meteorite: a new type of carbonaceous chondrite. Science **290**, 320–325 (2000)
M. Brozovic, S.J. Ostro, L.A.M. Benner, J.D. Giorgini, R.F. Jurgens, R. Rose, M.C. Nolan, A.A. Hine, C. Magri, D.J. Scheeres et al., Radar observations and a physical model of asteroid 4660 Nereus, a prime space mission target. Icarus **201**, 153–166 (2009)
G. Budde, C. Burkhardt, G.A. Brennecka, M. Fischer-Gödde, T.S. Kruijer, T. Kleine, Molybdenum isotopic evidence for the origin of chondrules and a distinct genetic heritage of carbonaceous and non-carbonaceous meteorites. Earth Planet. Sci. Lett. **454**, 293–303 (2016)
S.J. Bus, R.P. Binzel, Phase II of the small main-belt asteroid spectroscopic survey: a feature-based taxonomy. Icarus **158**, 106–145 (2002)
H. Busemann, A.F. Young, C.M.O.D. Alexander, P. Hoppe, S. Mukhopadhyay, L.R. Nittler, Interstellar chemistry recorded in organic matter from primitive meteorites. Science **312**, 727–730 (2006)
H. Busemann, A.N. Nguyen, G.D. Cody, P. Hoppe, A.L.D. Kilcoyne, R.M. Stroud, T.J. Zega, L.R. Nittler, Ultra-primitive interplanetary dust particles from the comet 26P/Grigg–Skjellerup dust stream collection. Earth Planet. Sci. Lett. **288**, 44–57 (2009)
F. Capaccioni, A. Coradini, G. Filacchione, S. Erard, G. Arnold, P. Drossart, M.C. De Sanctis, D. Bockelee-Morvan, M.T. Capria, F. Tosi, The organic-rich surface of comet 67P/Churyumov-Gerasimenko as seen by VIRTIS/Rosetta. Science **347**, aaa0628 (2015)
J.E. Chambers, Planetary accretion in the inner Solar System. Earth Planet. Sci. Lett. **223**, 241–252 (2004)
B.G. Choi, G.R. Huss, G.J. Wasserburg, R. Gallino, Presolar corundum and spinel in ordinary chondrites: origins from AGB stars and a supernova. Science **282**, 1284–1289 (1998)
F.J. Ciesla, S.B. Charnley, The physics and chemistry of nebular evolution, in *Meteorites and the Early Solar System II* (2006), pp. 209–230
R.N. Clayton, L. Grossman, T.K. Mayeda, A component of primitive nuclear composition in carbonaceous meteorites. Science **182**, 485–488 (1973)
G.D. Cody, E. Heying, C.M.O. Alexander, L.R. Nittler, A.L.D. Kilcoyne, S.A. Sandford, R.M. Stroud, Establishing a molecular relationship between chondritic and cometary organic solids. Proc. Natl. Acad. Sci. **108**, 19171–19176 (2011)
H.C. Connolly Jr., S.J. Desch, R.D. Ash, R.H. Jones, Transient heating events in the protoplanetary nebula, in *Meteorites and the Early Solar System II* (2006), pp. 383–397
G. Cooper, N. Kimmich, W. Belisle, J. Sarinana, K. Brabham, L. Garrel, Carbonaceous meteorites as a source of sugar-related organic compounds for the early Earth. Nature **414**, 879–883 (2001)
A. Coradini, D. Turrini, C. Federico, G. Magni, Vesta and Ceres: crossing the history of the solar system. Space Sci. Rev. **163**, 25–40 (2011)
F.E. DeMeo, B. Carry, Solar System evolution from compositional mapping of the asteroid belt. Nature **505**, 629 (2014)
F.E. Demeo, R.P. Binzel, S.M. Slivan, S.J. Bus, An extension of the Bus asteroid taxonomy into the near-infrared. Icarus **202**, 160–180 (2009)

S.J. Desch, M.A. Morris, H.C. Connolly, A.P. Boss, A critical examination of the X-wind model for chondrule and calcium-rich, aluminum-rich inclusion formation and radionuclide production. Astrophys. J. **725**, 692–711 (2010)
A. El Goresy, H. Yabuki, K. Ehlers, D. Woolum, E. Pernicka, Qingzhen and Yamato-691: a tentative alphabet for the EH chondrites. Proc. NIPR Symp. Antarct. Meteor. **1**, 65–101 (1988)
A. El Goresy, Y. Lin, M. Miyahara, A. Gannoun, M. Boyet, E. Ohtani, P. Gillet, M. Trieloff, A. Simionovici, L. Feng et al., Origin of EL3 chondrites: evidence for variable C/O ratios during their course of formation-A state of the art scrutiny. Meteorit. Planet. Sci. **52**, 1–26 (2017)
M.H. Engel, S.A. Macko, Isotopic evidence for extraterrestrial non-racemic amino acids in the Murchison meteorite. Nature **389**, 265–268 (1997)
C. Floss, P. Haenecour, Presolar silicate grains: abundances, isotopic and elemental compositions, and the effects of secondary processing. Geochem. J. **50**, 3–25 (2016)
C. Floss, F. Stadermann, Auger nanoprobe analysis of presolar ferromagnesian silicate grains from primitive CR. Geochim. Cosmochim. Acta **73**, 2415–2440 (2009). chondrites QUE 99177 and MET 00426
A. Fujiwara, T. Mukai, J. Kawaguchi, K.T. Uesugi, Sample return mission to NEA: MUSES-C. Adv. Space Res. **25**, 231–238 (2000)
W. Fujiya, P. Hoppe, T. Ushikubo, K. Fukuda, P. Lindgren, M.R. Lee, M. Koike, K. Shirai, Y. Sano, Migration of D-type asteroids from the outer Solar System inferred from carbonate in meteorites. Nat. Astron. **3**, 910–915 (2019)
E. Furi, B. Marty, Nitrogen isotope variations in the Solar System. Nat. Geosci. **8**, 515–522 (2015)
E. Galimov, Phobos sample return mission: scientific substantiation. Sol. Syst. Res. **44**, 5–14 (2010)
F. Goesmann, H. Rosenbauer, J.H. Bredehöft, M. Cabane, P. Ehrenfreund, T. Gautier, C. Giri, H. Krüger, L. Le Roy, A.J. MacDermott, Organic compounds on comet 67P/Churyumov-Gerasimenko revealed by COSAC mass spectrometry. Science **349**, aab0689 (2015)
T. Grav, A.K. Mainzer, J. Bauer, J. Masiero, T. Spahr, R.S. McMillan, R. Walker, R. Cutri, E. Wright, P.R. Eisenhardt et al., WISE/NEOWISE observations of the Hilda population: preliminary results. Astrophys. J. **744**, 197 (2011)
L. Grossman, Condensation in the primitive solar nebula. Geochim. Cosmochim. Acta **36**, 597–619 (1972)
L. Grossman, Vapor-condensed phase processes in the early solar system. Meteorit. Planet. Sci. **45**, 7–20 (2010)
L. Grossman, J.R. Beckett, A.V. Fedkin, S.B. Simon, F.J. Ciesla, Redox conditions in the solar nebula: observational, experimental, and theoretical constraints. Rev. Mineral. Geochem. **68**, 93–140 (2008)
Y. Guan, G.R. Huss, G.J. MacPherson, G.J. Wasserburg, Calcium-aluminum-rich inclusions from enstatite chondrites: indigenous or foreign? Science **289**, 1330–1333 (2000)
P. Haenecour, C. Floss, T.J. Zega, T.K. Croat, A. Wang, B.L. Jolliff, P. Carpenter, Presolar silicates in the matrix and fine-grained rims around chondrules in primitive CO3.0 chondrites: evidence for pre-accretionary aqueous alteration of the rims in the solar nebula. Geochim. Cosmochim. Acta **221**, 379–405 (2018)
K. Hashizume, M. Chaussidon, A non-terrestrial ^{16}O-rich isotopic composition for the protosolar nebula. Nature **434**, 619–622 (2005)
C.D.K. Herd, A. Blinova, D.N. Simkus, Y. Huang, R. Tarozo, C.M.O. Alexander, F. Gyngard, L.R. Nittler, G.D. Cody, M.L. Fogel et al., Origin and evolution of prebiotic organic matter as inferred from the Tagish lake meteorite. Science **332**, 1304–1307 (2011)
T. Hiroi, M.E. Zolensky, C.M. Pieters, The Tagish lake meteorite: a possible sample from a D-type asteroid. Science **293**, 2234–2236 (2001)
P. Hoppe, J. Leitner, J. Kodolányi, The stardust abundance in the local interstellar cloud at the birth of the Solar System. Nat. Astron. **1**, 617–620 (2017)
J.C. Huang, J.H. Ji, P.J. Ye, X.L. Wang, J. Yan, L.Z. Meng, S. Wang, C.L. Li, Y. Li, D. Qiao et al., The ginger-shaped asteroid 4179 Toutatis: new observations from a successful flyby of Chang'e-2. Sci. Rep. **3**, 3411 (2013)
G. Huss, R.S. Lewis, Presolar diamond, SiC, and graphite in primitive chondrites: abundances as a function of meteorite class and petrologic type. Geochim. Cosmochim. Acta **59**, 115–160 (1995)
M.R.M. Izawa, D.M. Applin, P. Mann, M.A. Craig, E.A. Cloutis, J. Helbert, A. Maturilli, Reflectance spectroscopy (200–2500 nm) of highly-reduced phases under oxygen- and water-free conditions. Icarus **226**, 1612–1617 (2013)
P. Jenniskens, M.D. Fries, Q.-Z. Yin, M. Zolensky, A.N. Krot, S.A. Sandford, D. Sears, R. Beauford, D.S. Ebel, J.M. Friedrich et al., Radar-enabled recovery of the Sutter's Mill meteorite, a carbonaceous chondrite regolith breccia. Science **338**, 1583–1587 (2012)
C.E. Jilly-Rehak, G.R. Huss, K. Nagashima, D.L. Schrader, Low-temperature aqueous alteration on the CR chondrite parent body: implications from in situ oxygen-isotope analyses. Geochim. Cosmochim. Acta **222**, 230–252 (2018)

K. Keil, Enstatite meteorites and their parent bodies. Meteoritics **24**, 195–208 (1989)
N.T. Kita, G.R. Huss, S. Tachibana, Y. Amelin, L.E. Nyquist, I.D. Hutcheon, Constraints on the origin of chondrules and CAIs from short-lived and long-lived radionuclides, in *Chondrites and the Protoplanetary Disk*, vol. 341, ed. by A.N. Krot, E.R.D. Scott, B. Reipurth (Astronomical Society of the Pacific, San Francisco, 2005), pp. 558–587
K. Kitazato, R.E. Milliken, T. Iwata, M. Abe, M. Ohtake, S. Matsuura, T. Arai, Y. Nakauchi, T. Nakamura, M. Matsuoka et al., The surface composition of asteroid 162173 Ryugu from Hayabusa2 near-infrared spectroscopy. Science **364**, eaav7432 (2019)
M. Komatsu, A.N. Krot, M.I. Petaev, A.A. Ulyanov, K. Keil, M. Miyamoto, Mineralogy and petrography of amoeboid olivine aggregates from the reduced CV3 chondrites Efremovka, Leoville and Vigarano: products of nebular condensation, accretion and annealing. Meteorit. Planet. Sci. **36**, 629–641 (2001)
A.N. Krot, Refractory inclusions in carbonaceous chondrites: records of early solar system processes. Meteorit. Planet. Sci. **54**, 1647–1691 (2019)
A.N. Krot, M.I. Petaev, H. Yurimoto, Amoeboid olivine aggregates with low-Ca pyroxenes: a genetic link between refractory inclusions and chondrules? Geochim. Cosmochim. Acta **68**, 1923–1941 (2004)
A.N. Krot, I.D. Hutcheon, A.J. Brearley, O.V. Pravdivtseva, M.I. Petaev, C.M. Hohenberg, Timescales and settings for alteration of chondritic meteorites, in *Meteorites and the Early Solar System II* (2006), pp. 525–553
A.N. Krot, K. Nagashima, G. Libourel, K.E. Miller, Multiple mechanisms of transient heating events in the protoplanetary disk: evidence from precursors of chondrules and igneous Ca, Al-rich inclusions, in *Chondrules and the Protoplanetary Disk*, ed. by S.S. Russell, H.C.J. Connolly, A.N. Krot (Cambridge University Press, Cambridge, 2018), pp. 11–56
J.W. Larimer, M. Bartholomay, The role of carbon and oxygen in cosmic gases: some applications to the chemistry and mineralogy of enstatite chondrites. Geochim. Cosmochim. Acta **43**, 1455–1466 (1979)
D.S. Lauretta, D.N. DellaGiustina, C.A. Bennett, D.R. Golish, K.J. Becker, S.S. Balram-Knutson, O.S. Barnouin, T.L. Becker, W.F. Bottke, W.V. Boynton et al., The unexpected surface of asteroid (101955) Bennu. Nature **568**, 55–60 (2019)
R.S. Lewis, M. Tang, J.F. Wacker, E. Anders, E. Steel, Interstellar diamonds in meteorites. Nature **326**, 160–162 (1987)
Y. Lin, A. El Goresy, A comparative study of opaque phases in Qingzhen (EH3) and MAC 88136 (EL3): representative of EH and EL parent bodies. Meteorit. Planet. Sci. **37**, 577–599 (2002)
Y. Lin, M. Kimura, Anorthite-spinel-rich inclusions in the Ningqiang carbonaceous chondrite: genetic links with type A and C inclusions. Meteorit. Planet. Sci. **33**, 435–446 (1998)
Y. Lin, M. Kimura, Two unusual type B refractory inclusions in the Ningqiang carbonaceous chondrite: evidence for relicts, xenoliths and multi-heating. Geochim. Cosmochim. Acta **64**, 4031–4047 (2000)
Y. Lin, M. Kimura, Ca-Al-rich inclusions from the Ningqiang meteorite: continuous assemblages of the nebular condensates and genetic link to type Bs. Geochim. Cosmochim. Acta **67**, 2251–2267 (2003)
Y. Lin, S. Amari, O. Pravdivtseva, Presolar grains from the Qingzhen (EH3) meteorite. Astrophys. J. **575**, 257–263 (2002)
Y. Lin, M. Kimura, H. Hiyagon, A. Monoi, Unusually abundant refractory inclusions from Sahara 97159 (EH3): a comparative study with other groups of chondrites. Geochim. Cosmochim. Acta **67**, 4935–4948 (2003)
G.E. Lofgren, A dynamic crystallization model for chondrule melts, in *Chondrules and the Protoplanetary Disk*, ed. by R.H. Hewins, R.H. Jones, E.R.D. Scott (Cambridge University Press, Cambridge, 1996), pp. 187–196
G.J. MacPherson, A.M. Davis, A petrologic and ion microprobe study of a Vigarano type B refractory inclusion: evolution by multiple stages of alteration and melting. Geochim. Cosmochim. Acta **57**, 231–243 (1993)
G.J. MacPherson, L. Grossman, "Fluffy" type A Ca-, Al-rich inclusions in the Allende meteorite. Geochim. Cosmochim. Acta **48**, 29–46 (1984)
G.J. MacPherson, D.W. Mittlefehldt, M.E. Lipschutz, R.N. Clayton, E.S. Bullock, A.V. Ivanov, T.K. Mayeda, M.-S. Wang, The Kaidun chondrite breccia: petrology, oxygen isotopes, and trace element abundances. Geochim. Cosmochim. Acta **73**, 5493–5511 (2009)
B. Marty, G. Avice, Y. Sano, K. Altwegg, H. Balsiger, M. Hässig, A. Morbidelli, O. Mousis, M. Rubin, Origins of volatile elements (H, C, N, noble gases) on Earth and Mars in light of recent results from the ROSETTA cometary mission. Earth Planet. Sci. Lett. **441**, 91–102 (2016)
K.D. McKeegan, L.A. Leshin, S.S. Russell, G.J. MacPherson, Oxygen isotopic abundances in calcium-aluminum-rich inclusions from ordinary chondrites: implications for nebular heterogeneity. Science **280**, 414–418 (1998)

K.D. McKeegan, J. Aleon, J. Bradley, D. Brownlee, H. Busemann, A. Butterworth, M. Chaussidon, S. Fallon, C. Floss, J. Gilmour et al., Isotopic compositions of cometary matter returned by stardust. Science **314**, 1724–1728 (2006)
K.D. McKeegan, A.P.A. Kallio, V.S. Heber, G. Jarzebinski, P.H. Mao, C.D. Coath, T. Kunihiro, R.C. Wiens, J.E. Nordholt, R.W. Moses et al., The oxygen isotopic composition of the Sun inferred from captured solar wind. Science **332**, 1528–1532 (2011)
M. Mueller, A.W. Harris, A. Fitzsimmons, Size, albedo, and taxonomic type of potential spacecraft target asteroid (10302) 1989 ML. Icarus **187**, 611–615 (2007)
K. Nakamura-Messenger, S. Messenger, L.P. Keller, S.J. Clemett, M.E. Zolensky, Organic globules in the Tagish Lake Meteorite: remnants of the protosolar disk. Science **314**, 1439–1442 (2006)
J.A.M. Nanne, F. Nimmo, J.N. Cuzzi, T. Kleine, Origin of the non-carbonaceous–carbonaceous meteorite dichotomy. Earth Planet. Sci. Lett. **511**, 44–54 (2019)
L.R. Nittler, F. Ciesla, Astrophysics with extraterrestrial materials, in *Annual Review of Astronomy and Astrophysics*, vol. 54, ed. by S.M. Faber, E. VanDishoeck (Annual Reviews, Palo Alto, 2016), pp. 53–93
L.R. Nittler, C.M.O. Alexander, X. Gao, R.M. Walker, E.K. Zinner, Interstellar oxide grains from the Tieschitz ordinary chondrite. Nature **370**, 443 (1994)
L.R. Nittler, C.M.O.D. Alexander, J. Davidson, M.E.I. Riebe, R.M. Stroud, J. Wang, High abundances of presolar grains and ^{15}N-rich organic matter in CO3.0 chondrite Dominion Range 08006. Geochim. Cosmochim. Acta **226**, 107–131 (2018)
L.R. Nittler, R.M. Stroud, J.M. Trigo-Rodríguez, B.T. De Gregorio, C.M.O.D. Alexander, J. Davidson, C.E. Moyano-Cambero, S. Tanbakouei, A cometary building block in a primitive asteroidal meteorite. Nat. Astron. **3**, 659–666 (2019)
M. Popescu, O. Vaduvescu, J. de León, R.M. Gherase, J. Licandro, I.L. Boacă, A.B. Şonka, R.P. Ashley, T. Močnik, D. Morate et al., Near-Earth asteroids spectroscopic survey at Isaac Newton Telescope. Astron. Astrophys. **627**, A124 (2019)
L. Qin, L.R. Nittler, C.M.O. Alexander, J. Wang, F.J. Stadermann, R.W. Carlson, Extreme ^{54}Cr-rich nano-oxides in the CI chondrite Orgueil -implication for a late supernova injection into the solar system. Geochim. Cosmochim. Acta **75**, 629–644 (2011)
E.R. Rambaldi, R.S. Rajan, D. Wang, R.M. Housley, Evidence for relict grains in chondrules of Qingzhen, an E3 type enstatite chondrite. Earth Planet. Sci. Lett. **66**, 11–24 (1983)
L. Remusat, J.N. Rouzaud, E. Charon, C. Le Guillou, Y. Guan, J. Eiler, D-depleted organic matter and graphite in the Abee enstatite chondrite. Geochim. Cosmochim. Acta **96**, 319–335 (2012)
L. Remusat, L. Piani, S. Bernard, Thermal recalcitrance of the organic D-rich component of ordinary chondrites. Earth Planet. Sci. Lett. **435**, 36–44 (2016)
N. Sakamoto, N. Kawasaki, Extreme ^{16}O-rich refractory inclusions in the Isheyevo chondrite. Meteorit. Planet. Sci. **54**, #6069 (2019)
S.A. Sandford, J. Aleon, C.M.O.D. Alexander, T. Araki, S. Bajt, G.A. Baratta, J. Borg, J.P. Bradley, D.E. Brownlee, J.R. Brucato et al., Organics captured from comet 81P/Wild 2 by the stardust spacecraft. Science **314**, 1720–1724 (2006)
F.H. Shu, H. Shang, T. Lee, Toward an astrophysical theory of chondrites. Science **271**, 1545–1552 (1996)
F.H. Shu, H. Shang, M. Gounelle, A.E. Glassgold, T. Lee, The origin of chondrules and refractory inclusions in chondritic meteorites. Astrophys. J. **548**, 1029–1050 (2001)
E. Stolper, J.M. Paque, Crystallization sequences of Ca-Al-rich inclusions from Allende: the effects of cooling rate and maximum temperature. Geochim. Cosmochim. Acta **50**, 1785–1806 (1986)
J.M. Sunshine, H.C. Connolly Jr., T.J. McCoy, S.J. Bus, L.M. La Croix, Ancient asteroids enriched in refractory inclusions. Science **320**, 514–517 (2008)
J. Takahashi, S. Urakawa, T. Terai, H. Hanayama, A. Arai, S. Honda, Y. Takagi, Y. Itoh, T. Zenno, M. Ishiguro, Near-infrared colors of asteroid 2012DA14 at its closest approach to Earth: observations with the Nishiharima Infrared Camera (NIC). Publ. Astron. Soc. Jpn. **66**, 53 (2014)
M. Tang, E. Anders, Isotopic anomalies of Ne, Xe, and C in meteorites. II. Interstellar diamond and SiC: carriers of exotic noble gases. Geochim. Cosmochim. Acta **52**, 1235–1244 (1988)
P. Tanga, B. Carry, F. Colas, M. Delbo, A. Matter, J. Hanuš, V. Alí Lagoa, A.H. Andrei, M. Assafin, M. Audejean et al., The non-convex shape of (234) Barbara, the first Barbarian. Mon. Not. R. Astron. Soc. **448**, 3382–3390 (2015)
A. Trinquier, T. Elliott, D. Ulfbeck, C. Coath, A.N. Krot, M. Bizzarro, Origin of nucleosynthetic isotope heterogeneity in the solar protoplanetary disk. Science **324**, 374–376 (2009)
M.A. Tyra, J. Farquhar, Y. Guan, L.A. Leshin, An oxygen isotope dichotomy in CM2 chondritic carbonates—a SIMS approach. Geochim. Cosmochim. Acta **77**, 383–395 (2012)
P. Vernazza, P. Beck, Composition of Solar System small bodies, in *Planetesimals: Early Differentiation and Consequences for Planets*, ed. by L.T. Elkins-Tanton, B.P. Weiss (Cambridge University Press, Cambridge, 2016)

J. Veverka, M. Robinson, P. Thomas, S. Murchie, J.F. Bell, N. Izenberg, C. Chapman, A. Harch, M. Bell, B. Carcich et al., NEAR at Eros: imaging and spectral results. Science **289**, 2088–2097 (2000)
C. Vollmer, D. Kepaptsoglou, J. Leitner, H. Busemann, N.H. Spring, Q.M. Ramasse, P. Hoppe, L.R. Nittler, Fluid-induced organic synthesis in the solar nebula recorded in extraterrestrial dust from meteorites. Proc. Natl. Acad. Sci. **111**, 15338–15343 (2014)
P.H. Warren, Stable-isotopic anomalies and the accretionary assemblage of the Earth and Mars: a subordinate role for carbonaceous chondrites. Earth Planet. Sci. Lett. **311**, 93–100 (2011a)
P.H. Warren, Stable isotopes and the noncarbonaceous derivation of ureilites, in common with nearly all differentiated planetary materials. Geochim. Cosmochim. Acta **75**, 6912–6926 (2011b)
E.A. Worsham, C. Burkhardt, G. Budde, M. Fischer-Gödde, T.S. Kruijer, T. Kleine, Distinct evolution of the carbonaceous and non-carbonaceous reservoirs: Insights from Ru, Mo, and W isotopes. Earth Planet. Sci. Lett. **521**, 103–112 (2019)
S. Yoneda, L. Grossman, Condensation of CaO-MgO-Al_2O_3-SiO_2 liquids from cosmic gases. Geochim. Cosmochim. Acta **59**, 3413–3444 (1995)
H. Yurimoto, J.T. Wasson, Extremely rapid cooling of a carbonaceous-chondrite chondrule containing very ^{16}O-rich olivine and a ^{26}Mg-excess. Geochim. Cosmochim. Acta **66**, 4355–4363 (2002)
X. Zhao, C. Floss, Y. Lin, M. Bose, Stardust investigation into the CR chondrite Grove Mountain 021710. Astrophys. J. **769**, 49 (2013)
X. Zhao, Y. Lin, Q.-Z. Yin, J. Zhang, J. Hao, M. Zolensky, P. Jenniskens, Presolar grains in the CM2 chondrite Sutter's Mill. Meteorit. Planet. Sci. **49**, 2038–2046 (2014)
E. Zinner, Leonard award address-trends in the study of presolar dust grains from primitive meteorites. Meteorit. Planet. Sci. **33**, 549–564 (1998)
M. Zolensky, A. Ivanov, The Kaidun microbreccia meteorite: a harvest from the inner and outer asteroid belt. Chem. Erde, Geochem. **63**, 185–246 (2003)
M.E. Zolensky, T.J. Zega, H. Yano, S. Wirick, A.J. Westphal, M.K. Weisberg, I. Weber, J.L. Warren, M.A. Velbel, A. Tsuchiyama et al., Report - mineralogy and petrology of comet 81P/Wild 2 nucleus samples. Science **314**, 1735–1739 (2006)

Space Sci Rev (2020) 216:54
https://doi.org/10.1007/s11214-020-00679-6

Organic Matter in the Solar System—Implications for Future on-Site and Sample Return Missions

Zita Martins[1] · **Queenie Hoi Shan Chan**[2,3] · **Lydie Bonal**[4] · **Ashley King**[2,5] · **Hikaru Yabuta**[6]

Received: 9 July 2019 / Accepted: 23 April 2020 / Published online: 12 May 2020

Abstract Solar system bodies like comets, asteroids, meteorites and dust particles contain organic matter with different abundances, structures and chemical composition. This chapter compares the similarities and differences of the organic composition in these planetary bodies. Furthermore, these links are explored in the context of detecting the most pristine organic material, either by on-site analysis or sample return missions. Finally, we discuss the targets of potential future sample return missions, as well as the contamination controls that should be in place in order to successfully study pristine organic matter.

Keywords Organic matter · Solar system · Comets · Asteroids · Extraterrestrial samples · Sample return mission

Role of Sample Return in Addressing Major Questions in Planetary Sciences
Edited by Mahesh Anand, Sara Russell, Yangting Lin, Meenakshi Wadhwa, Kuljeet Kaur Marhas and Shogo Tachibana

✉ Z. Martins
zita.martins@tecnico.ulisboa.pt

1 Centro de Química Estrutural, Departamento de Engenharia Química, Instituto Superior Técnico, Universidade de Lisboa, Avenida Rovisco Pais 1, 1049-001 Lisboa, Portugal

2 Planetary and Space Sciences, School of Physical Sciences, The Open University, Walton Hall, Milton Keynes MK7 6AA, UK

3 Present address: Department of Earth Sciences, Royal Holloway University of London, Egham, Surrey TW20 0EX, UK

4 Institut de Planétologie et d'Astrophysique de Grenoble, Univ. Grenoble Alpes, CNRS, CNES, 38000 Grenoble, France

5 The Natural History Museum, Cromwell Road, London SW7 5BD, UK

6 Department of Earth and Planetary Systems Science, Hiroshima University, 1-3-1 Kagamiyama, Hiroshima 739-8526, Japan

1 Introduction

Organic matter is widespread throughout the solar system, and the access to its primitive form is achieved by analysing extraterrestrial samples from comets (Bockelée-Morvan et al. 2019), asteroids, meteorites and dust particles. Comets contain a wide variety of extraterrestrial molecules, including glycine and a significant number of carbon-rich species (Altwegg et al. 2016; Bockelée-Morvan et al. 2004; Crovisier and Bockelée-Morvan 1999; Crovisier et al. 2009; Elsila et al. 2009; Goesmann et al. 2015; Mumma and Charnley 2011; Wright et al. 2015). Dust particles, such as micrometeorites, interplanetary dust particles (IDPs), and Ultra-carbonaceous Antarctic Micrometeorites (UCAMMs) have also been analysed for organic molecules. Meteorites have been thoroughly analysed for organic matter in the last few decades. However, the origin of the containing organic material (indigenous *versus* terrestrial contamination) is not certain (Brinton et al. 1998; Clemett et al. 1993, 1998; Flynn et al. 2003, 2004; Glavin et al. 2004; Keller et al. 2004; Matrajt et al. 2004, 2005). Furthermore, alteration that has occurred in the meteorite parent body(ies) (i.e., thermal metamorphism, aqueous alteration, impact shock and brecciation) may have influenced the chemical composition of chondrites (Browning et al. 1996; Palmer and Lauretta 2011; Vinogradoff et al. 2017), which are meteorites that have not been melted since formation, and constitute the majority (more than 80%) of meteorites that reach the Earth surface. Carbonaceous chondrites are some of the most primitive chondrites as their bulk chemical composition is very similar to that of the sun (Lodders 2003). They consist of coarse (>10's μm) silicate components such as chondrules and calcium-aluminium-rich inclusions (CAIs) set within a fine-grained (<1 μm) matrix with dark appearance. Carbonaceous chondrites have high bulk meteoritic carbon content (up to ∼3.5 weight total (wt)%) (Alexander et al. 2007, 2017) present as silicon carbide, graphite, diamonds, carbonate and organic matter (Amari et al. 1990, 1993; Anders and Zinner 1993; Benedix et al. 2003; Fredriksson and Kerridge 1988; Grady et al. 1988; Hoppe et al. 1996; Lewis et al. 1987; Smith and Kaplan 1970; Zinner et al. 1995). Their organic content may be divided into two fractions: the solvent-soluble (SOM) one, which is obtained by extraction using different solvents, and the acid-insoluble (IOM) (Cody and Alexander 2005; Cody et al. 2002; Gardinier et al. 2000), which is obtained after demineralization (Robert and Epstein 1982).

The IOM is a dark, macromolecular organic solid which accounts for more than half of total organic carbon. Although the intact molecular structure of IOM is still unknown due to its complex configuration, determination of the macromolecular composition has been addressed by destructive techniques (e.g., pyrolysis, chemical degradation, and thermochemolysis) and non-destructive techniques (e.g. solid state-nuclear magnetic resonance (NMR), Fourier-transform infrared spectroscopy (FTIR), X-ray absorption near-edge structure (XANES), and electron paramagnetic resonance (EPR)) (See reviews by Sephton 2002; Pizzarello et al. 2006; Alexander et al. 2017; Glavin et al. 2018). Elemental compositions of IOM in the primitive meteorites are estimated as $C_{100}H_{58.8}N_{3.3}O_{18.3}$ (Alexander et al. 2007). Pyrolysis of IOM yielded one to six ring-polycyclic aromatic hydrocarbons (PAHs) and O-, N-, S- containing aromatic compounds (e.g., Komiya et al. 1993; Sephton et al. 2004). Dichromic acid ($H_2Cr_2O_7$) oxidation (Hayatsu et al. 1977), alkaline copper oxide (CuO) oxidation (Hayatsu et al. 1980), tetramethylammonium hydroxide (TMAH) thermochemolysis (Remusat et al. 2005), and ruthenium tetroxide (RuO_4) (Remusat et al. 2005) demonstrated that IOM contain short alkyl chains (C_1-C_9). It has been suggested from solid state ^{13}C NMR studies that IOM in primitive carbonaceous chondrites is mainly constituted of polycyclic aromatic rings with aliphatic substituents (Cody et al.

2002; Cronin et al. 1987; Gardinier et al. 2000). Cody et al. (2002) further concluded that IOM is composed of a small ring ($<C_{20}$) aromatic network ($\sim$60%) crosslinking with short-branched aliphatic chains ($\sim$20%) and various oxygen ($\sim$20%) functional groups (e.g., ether, carbonyl groups). Variable enrichments in D and ^{15}N from the IOM in the primitive carbonaceous chondrites have been reported (Alexander et al. 2007; Kerridge et al. 1987; Robert and Epstein 1982; Yang and Epstein 1983, 1984). Using nanoscale secondary ion mass spectrometry (NanoSIMS), it has been reported that D and ^{15}N enrichments of IOM are located in submicron areas (Busemann et al. 2006) and they are often associated with nanoglobule textures (Nakamura-Messenger et al. 2006). Hydrous heating experiments of IOM indicated the presence of two fractions in the IOM, labile (enriched in deuterium D, ^{13}C and ^{15}N), and refractory (depleted in ^{13}C and ^{15}N) (Sephton et al. 2003, 2004).

There have been several hypotheses about the origin and synthetic mechanisms of IOM in the early solar system. It has been suggested that IOM is originated in extremely cold environments such as interstellar or solar nebula environments, which can explain enrichments of D and ^{15}N in IOM in meteorites and IDPs (Alexander et al. 2007, 2010). On the other hand, contributions of Fischer–Tropsch (FTT) synthesis (e.g. Hill and Nuth 2003) and irradiation reaction of nebular gas (Kuga et al. 2015) have been also suggested for the formation of IOM. There is also another suggestion that the first IOM might have been formed via formose-type reaction of formaldehyde and ammonia on the meteorite parent body (Cody et al. 2011; Kebukawa et al. 2013).

The SOM content of meteorites has been extensively analysed and reviewed in the literature (e.g., Martins and Sephton 2009; Martins 2019), showing a high molecular diversity with tens of thousands of different molecular compositions (Schmitt-Kopplin et al. 2010). For more details (abundances, distribution and isotopic composition) regarding the specific SOM present in carbonaceous chondrites, including carboxylic acids, amino acids, diamino acids, dipeptides, diketopiperazines, sulfonic and phosphonic acids, purines, pyrimidines, sugars and sugar-related compounds, hydrocarbons, alcohols, amines, amides, aldehydes and ketones, please see Martins (2019). Their origin has been discussed against four main points: i) by detecting molecules that are not synthesised by terrestrial organisms; ii) by comparison to the organic content of samples collected from the terrestrial fall-site; iii) enantiomeric ratios for chiral molecules; and when that is not possible, and/or we have access to large quantities of sample, iv) measurement of the compound specific carbon/hydrogen/nitrogen isotopic compositions. This last item also points not only towards the source (terrestrial *versus* interstellar/parent body processing), but also towards the synthetic pathway of the organic compounds present in the SOM (Aléon and Robert 2004; Elsila et al. 2012; Millar et al. 1989; Robert 2003; Sandford et al. 2001; Terzieva and Herbst 2000; Tielens 1983; Yang and Epstein 1983). Indigenous soluble organic meteoritic compounds have been found to be enriched in D, ^{13}C and ^{15}N (e.g. Cronin et al. 1993; Cooper et al. 1997; Elsila et al. 2012; Krishnamurthy et al. 1992; Yuen et al. 1984), with the enrichment in D and ^{15}N of the SOM suggesting the synthesis of their precursors at cold temperatures, either in the interstellar medium or in the outer protoplanetary disk (Aléon and Robert 2004; Millar et al. 1989; Robert 2003; Sandford et al. 2001; Terzieva and Herbst 2000; Tielens 1983; Yang and Epstein 1983). However, one cannot exclude the possibility of extraterrestrial soluble organic compounds having low D, ^{13}C and ^{15}N. Details on the stable isotope measurements of each specific soluble organic compound present in carbonaceous chondrites have also been presented in detail in the literature (Aponte et al. 2011, 2014a, 2014b, 2015, 2016; Cooper et al. 1997, 2001; Cronin et al. 1993; Engel and Macko 1997; Ehrenfreund et al. 2001; Epstein et al. 1987; Gilmour and Pillinger 1994; Huang et al. 2005, 2015; Krishnamurthy et al. 1992; Martins

2018; Martins et al. 2007b, 2008; Naraoka et al. 2000; Pizzarello and Huang 2002, 2005; Pizzarello and Yarnes 2016; Pizzarello et al. 1991, 2001, 2003, 2004, 2010; Yuen et al. 1984; for a review see Martins 2019).

2 The Organic Matter Link Between Meteorites and Asteroids

Ground- and space-based remote sensing measurements have linked the carbonaceous chondrites to asteroids with dark surfaces (i.e. low albedo). An increased awareness of this relationship first occurred in 1971 with the discovery that asteroid Bamberga has a very low albedo (∼0.03) similar to the carbonaceous chondrites (Cruikshank and Kerridge 1992). It has since been established that dark, carbonaceous (C-type) asteroids comprise ∼70% of all known asteroids, in particular dominating the population in the cold, outer (>3 AU) regions of the main asteroid belt (e.g. DeMeo and Carry 2015). The relationship between some carbonaceous chondrites and C-type asteroids has been further strengthened by directly comparing telescope and spacecraft observations of asteroids with laboratory analyses of meteorites, and through a better understanding of space and terrestrial weathering effects (e.g. Cloutis et al. 2011a, 2011b; Cloutis et al. 2012a).

Most meteorite-asteroid comparisons focus on the visible to near infrared (VNIR, ∼0.5–4 μm) spectral region. Carbonaceous chondrites and C-type asteroids both have low overall reflectance and either neutral, slightly red (increasing reflectance with wavelength) or blue spectral slopes (decreasing reflectance with wavelength). Absorption features in this region are usually weak, with the most significant feature typically occurring at ∼0.7 μm due to Fe^{2+}–Fe^{3+} charge transfer, at ∼1 μm from anhydrous silicates, and at ∼3 μm from –OH/H_2O bound in hydrated minerals. The 3.3–3.6 μm region is diagnostic for organic materials as it contains absorptions from chemical groups such as CH and NH. These features are regularly identified in laboratory spectra of bulk carbonaceous chondrites and extracted IOM and SOM, although again the absorptions are often weak due to the high opacity of the fine-grained samples, while terrestrial contamination of meteorites is always an issue (e.g. Kebukawa et al. 2009; Takir et al. 2013; Orthous-Daunay et al. 2013; Trigo-Rodríguez et al. 2014). It should also be noted that the surface-measured spectra of asteroids are influenced by the effects of space weathering, whereas with meteorites we are able to analyse sub-surface pieces, i.e., either the meteorite was fractured during atmospheric entry or was intentionally freshly fractured during its curation before being analysed, plus some meteoroids could be inner parts of asteroids and therefore not spatially weathered.

The spectral signatures of organic materials have only been observed for a small number of C-type asteroids, plus comets such as 67P/Churyumov-Gerasimenko (Capaccioni et al. 2015). Absorptions due to organics, which are present at only low levels in carbonaceous chondrites, are expected to be very weak in diffuse reflectance spectra from asteroids and therefore require high signal-to-noise remote measurements for their detection. The first report of organics on the surface of an asteroid was by Cruikshank and Brown (1987), who tentatively identified a feature at 3.4 μm, which they attributed to C-H, in the spectra of asteroid Elektra. From independent observations Campins et al. (2010) and Rivkin and Emery (2010) both suggested that the position and shape of features at ∼3.3 μm and ∼3.4 μm in spectra of asteroid Themis were consistent with the presence of aromatic and aliphatic CH_2 and CH_3 functional groups on the surface, respectively. Similarly, Licandro et al. (2011) reported that the reflectance spectra of asteroid Cybele are comparable to Themis and inferred the presence of complex organic materials from absorption features between 3.2–3.6 μm.

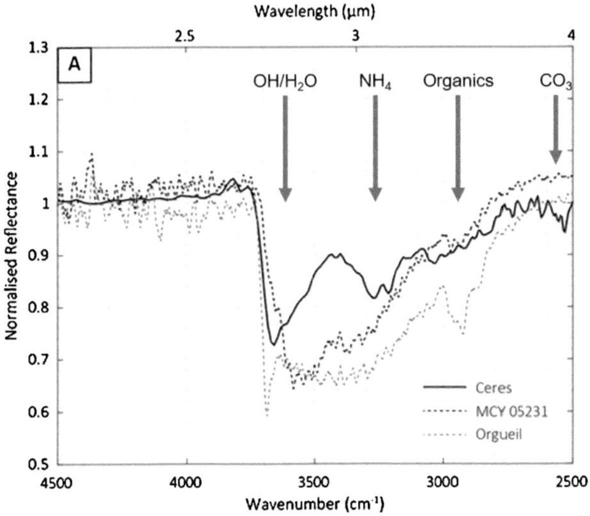

Fig. 1 The "3 μm" spectral region for Ceres compared to the hydrated carbonaceous chondrites Orgueil (CI1) and MacKay Glacier (MCY) 05231 (CM1). The Ceres spectrum is from De Sanctis et al. (2017) and the figure is taken from Bates et al. (2020)

Ground-based observations have failed to unambiguously detect organic species on the surface of the dwarf planet Ceres, which is spectrally similar to the hydrated carbonaceous chondrites. However, in 2015 NASA's Dawn mission entered orbit around Ceres and characterised the surface in detail using a VNIR mapping spectrometer. Reflectance spectra collected by the Dawn spacecraft confirmed the presence of aliphatic organic compounds from C–H vibrational absorptions at ~3.4 μm (De Sanctis et al. 2017; Fig. 1). Although it cannot be ruled out that the organics are widespread on the surface at low abundances, they are typically found concentrated in small, fresh (<50 Ma) impact craters (De Sanctis et al. 2017; Pieters et al. 2017). The depth of the ~3.4 μm band in these localised areas are significantly stronger than those found for meteoritic IOM, hinting at the presence of IOM-like organics at an abundance higher than that present in carbonaceous chondrites (De Sanctis et al. 2017; Kaplan et al. 2018). An important but currently unresolved question is whether the organics on Ceres come from an exogenous source (i.e. delivered by an organic-rich impactor) or result from an internal or near-surface endogenous process. Distinguishing between these two possibilities will have important implications for understanding the geophysical and geochemical history of Ceres and other carbonaceous bodies.

The Dawn mission ended in 2018 but there are currently two active space missions, JAXA's Hayabusa2 and NASA's OSIRIS-REx, characterising the carbonaceous near-Earth asteroids Ryugu and Bennu, which were selected due to their spectral links to the volatile- and organic-rich carbonaceous chondrites. Preliminary results suggest that the surface of asteroid Ryugu contains a significant fraction of dehydrated minerals (Kitazato et al. 2019). Organic matter has not yet been detected on Ryugu but its surface is very dark (Watanabe et al. 2019; Sugita et al. 2019), possibly indicating a high abundance of carbon. Furthermore, a small lander of Hayabusa2 spacecraft, the Mobile Asteroid Surface Scout (MASCOT) observed a number of inclusions from the dark boulders on Ryugu, which are similar textures to those of less altered carbonaceous meteorites (Jaumann et al. 2019). MASCOT even revealed that Ryugu has low thermal conductivity and high porosity unlike any chondritic meteorites, while these thermal properties are similar to comets (Grott et al. 2019). In contrast, VNIR reflectance spectra from the surface of asteroid Bennu indicate an affinity with the hydrated carbonaceous chondrites (Hamilton et al. 2019). Again, organic materials have not been found but future higher-spatial-resolution spectral observations are expected to confirm their presence or absence on the surface of Bennu. A major aim for both Hayabusa2 and OSIRIS-REx is to collect samples from the surface of asteroids Ryugu and Bennu and return them to Earth (in 2020 for Hayabusa2, 2023 for OSIRIS-REx). Laboratory analyses of the returned samples will provide new insights into the formation and evolution of extra-terrestrial organics and ground-truth remote sensing observations of solar system bodies.

3 Organic Matter Comparison Between Meteorites and Dust Particles

Several thousands of tons of extra-terrestrial matter fall on Earth, with meteorites representing the largest samples in terms of individual size. The major fraction in terms of total mass is nevertheless related to dust—among which, micrometeorites are recovered on the Earth's surface, and IDPs in the Earth's stratosphere. Chondrites are commonly considered as pieces of non-differentiated asteroids, although a cometary origin is discussed for some of them (Gounelle et al. 2006, 2008). The specific asteroidal class which chondrites are sampled from is not definitively established for each chondritic group. Ordinary chondrites are related to S-type asteroids; this was confirmed through the Hayabusa mission (Ebihara et al. 2011; Nagao et al. 2011; Nakamura et al. 2011; Noguchi et al. 2011; Tsuchiyama et al. 2011; Yurimoto et al. 2011). Moreover, type 1 and 2 carbonaceous chondrites are considered to be derived from C-type asteroids (e.g., Cloutis et al. 2011a, 2011b; Vernazza et al. 2015a). Nevertheless, the relative proportions of ordinary and carbonaceous chondrites is not similar to the relative abundances of the presumed asteroidal parent bodies (e.g., DeMeo and Carry 2013). Moreover, the limited number of chondrite groups does not reflect the large spectral variability observed among asteroids. On the other hand, dust can potentially be of asteroidal or cometary origin. There is a growing idea that some particles of dust might be samples from asteroidal parent bodies, which are not the parent bodies of meteorites. For example, some IDPs might be related to D- and P-type asteroids, most likely unrepresented among chondrites (Levison et al. 2009; Vernazza et al. 2015b).

Primitive meteorites and dust particles, which are preserved from severe heating resulting from terrestrial atmospheric entry, all contain a substantial fraction of organic compounds. Several reviews were published on organic matter in meteorites (e.g., Alexander et al. 2017; Gilmour 2014; Glavin et al. 2018). Some comparison of the organic matter in dust particles (IDPs and micrometeorites mostly collected in Antarctica) were previously published (Pizzarello et al. 2006; Quirico and Bonal 2019). The objective of this section is to summarise the key-points of the comparison on organic matter between chondrites and dust particles: mostly abundances, elemental and chemical compositions, and structure, as already stated in the literature. One of the big-picture questions the scientific community wants to address is related to the existence (or not) of several organic reservoir(s) in the solar protoplanetary disk potentially sampled by distinct parent bodies. Isotopic compositions of the organic matter appear to be key constraints. In this context, some new isotopic data acquired on Antarctic Micro-Meteorites (AMMs), completed by a review of the existing literature on IDPs and chondrites will also be discussed.

The chemical demineralization process, commonly applied to chondrites to isolate SOM and IOM, is highly challenging to apply on individual dust. So, most studies on dust are dealing with bulk analysis and not SOM or IOM analysis as for chondrites. Moreover, due to the small size and mass of individual dust, the analytical techniques that can be successfully applied are restricted in comparison to chondrites. Dust particles appear to be enriched in elemental carbon in comparison to chondrites. While carbonaceous chondrites contain only a few wt% of carbon, the bulk elemental abundance of carbon in IDPs can be as high as 45 wt% with a mean value around 15 wt% (Thomas et al. 1993), and in UCAMMs it can be up to the extreme abundance of 90 wt% (Dartois et al. 2013). At the micron scale, the C/H ratios in bulk individual IDPs vary between 1 and 3 (Aléon et al. 2001), and between 2 and 6 in UCAMMs (Duprat et al. 2010). Emblematic chondrites are characterised in bulk by C/H of 2.2–2.5, 4.8, and 1.9 in Orgueil (CI), Semarkona (UOC) and Murchison (CM), respectively (Alexander et al. 2012). The N/C ratios range from 0.03 to 0.24 in comet 81P/Wild dust particles (Cody et al. 2008; De Gregorio et al. 2011), from 0.05 to 0.15 in UCAMMs (Dartois et al. 2013; Yabuta et al. 2017), and from 0.03 to 0.04 in IOM primitive carbonaceous chondrites (Alexander et al. 2007). The O/C ratios range from 0.11 to 0.67 in comet 81P/Wild dust particles (Cody et al. 2008) while they range from 0.10 to 0.22 in IOM primitive carbonaceous chondrites (Alexander et al. 2007). Nevertheless, it should be noted that distinct analytical techniques were used (e.g., NanoSIMS and SIMS *vs.* Elemental Analyzer). The ionization probability of a certain species being matrix dependent, the relationship between secondary ion yield of that species with its concentrations is not direct (e.g., Havette and Slodzian 1980). The determination of elemental ratios (potentially different matrix effects for distinct elements) is thus challenging by secondary ionization mass spectrometry.

Infrared micro-spectrometry was performed on IDPs (e.g., Flynn et al. 2003; Keller et al. 2004; Matrajt et al. 2005; Merouane et al. 2003; Muñoz Caro et al. 2006), AMMs (Battandier et al. 2018), UCAMMs (Dartois et al. 2013, 2018), and chondritic IOMs (e.g., Kebukawa et al. 2011; Orthous-Daunay et al. 2013). With the exception of UCAMMs, all these samples exhibit the spectral bands related to the CH_2 and CH_3 functions. The CH_2/CH_3 tend to be higher in dust particles in comparison to chondritic IOM (Quirico and Bonal 2019). However, (*i*) the spectral acquisition and reduction were not always led in the same conditions (e.g., under vacuum or not); (*ii*) the CH_2/CH_3 ratio observed in dust could be modified through flash heating during atmospheric entry (e.g., discussed in Battandier et al. 2018); (*iii*) bulk measurements on dust might not be directly comparable to those on chondritic IOMs.

The structure of the polyaromatic matter present in the distinct families of primitive cosmomaterials was characterised by Raman micro-spectrometry by different research groups (chondrites: e.g., Bonal et al. 2006, 2007, 2016; Busemann et al. 2007; Quirico et al. 2003. AMMs: Dobrica et al. 2011; Battandier et al. 2018. IDPs: e.g., Wopenka 1988; Davidson et al. 2012; Starkey et al. 2013. Stardust particles: Sandford et al. 2006; Rotundi et al. 2008; Chan et al. 2019a). The experimental and analytical conditions are not always strictly similar, but the comparison stays informative. Dust collected in the stratosphere of the Earth or at the surface of the Earth obviously survived their atmospheric entry, but they have been heated to some degree, and thus preserve the pristine information to different degrees (Chan et al. 2019a). The consequences of atmospheric entry is particularly visible on the texture of MM and is used as a criterium of classification (Genge et al. 2008). Nevertheless, it was recently shown that even though their textures do not necessarily reveal heating through atmospheric entry, the organic matter of some fine-grained AMMs (Fg-AMMs) was in fact slightly modified (Battandier et al. 2018). Indeed, Fg-AMMs can either be made of hydrated or non-hydrated minerals, and non-hydrated ones tend to be characterised by a distinct structure of the polyaromatic organic (shift of some Raman spectral parameters) and a lower abundance of aliphatic organic matter in comparison to the hydrated ones. This was interpreted as resulting from heating through the atmospheric entry (Battandier et al. 2018). There is not an exact similitude in the spectral parameters describing the structure of the polyaromatic matter present in pristine dusts—the ones having best escaped atmospheric heating (Busemann et al. 2009; Dobrică et al. 2011; Davidson et al. 2012;

Starkey et al. 2013; Battandier et al. 2018) and primitive chondrites (e.g., Bonal et al. 2016; Busemann et al. 2007; Quirico et al. 2018). This tends to underline the sampling of potentially distinct organic reservoirs, but dedicated experimental simulations are required to definitely rule out the effect of heating/oxidation and irradiation in space (Quirico and Bonal 2019).

The origin of isotopic enrichments observed in extra-terrestrial organics is still debated in the literature (solar and/or interstellar, see the review by Alexander et al. 2017). The precursor of organics in primitive cosmomaterials is considered by some authors as resulting from the assemblage of several components of interstellar and protoplanetary disk origins (e.g., Aléon 2010). The isotopic variability observed among cosmomaterials is then considered as reflecting variable abundance and preservation of the distinct components. On the other hand, large variations of C, H, and N isotopic compositions observed among chondritic IOMs have been interpreted by other authors as largely derived from secondary processes of a common precursor (e.g., Alexander et al. 1998, 2007). Isotopic anomalies may also imply that only a very small fraction of volatile interstellar matter survived the events of solar system formation, as an interstellar component with highly fractionated H and unfractionated N isotopes has been detected in ordinary chondrites (Aléon 2010). Isotopic heterogeneities at the micron- and submicron-scales are also observed within chondritic IOM (e.g., Busemann et al. 2006). Similar variations are observed between and within IDPs (Messenger 2000; Aléon et al. 2001; Messenger et al. 2003; Aléon et al. 2003; Keller et al. 2004; Floss et al. 2006; Busemann et al. 2009; Davidson et al. 2012; Starkey et al. 2014; Chan et al. 2019a). With the exception of UCAMMs (Duprat et al. 2010; Dartois et al. 2013; Yabuta et al. 2017), isotopic measurements in MM are very scarce, but bulk C and N isotopic measurements were recently performed in a series of Fg-AMMs (Battandier 2018) (Fig. 2). The C and N isotopic compositions of these Fg-AMMs are is highly variable in contrast to bulk compositions within individual group of chondrites (Kerridge 1985; Alexander et al. 2012; Floss et al. 2014). Moreover, there is no perfect match between individual Fg-AMMs and chondrite groups in terms of C and N isotopic compositions and there is no systematic variation in the bulk C or N isotopic compositions with the degree of hydration of the individual AMMs. Lastly, bulk chondrites are systematically ^{13}C-poor, while dusts are roughly characterised by $-60 < \delta^{13}C$ (‰) $< +50$ (Fig. 2).

One could easily be tempted to interpret these differences between chondrites and dust in terms of distinct parent bodies, having sampled distinct organic reservoirs. However, some analytical details are worth being taken into consideration. It is interesting to note that among the measurements obtained by Battandier (2018) there seems to be a systematic difference between the measurements performed with two distinct instruments: Fg-AMMs characterised at Washington University (WU) appear to be systematically enriched in ^{13}C in comparison to those characterised at the Open University (OU), although no spectroscopic information allows to distinguish the considered samples in two groups. Interestingly IDPs characterised by Messenger (2000) are typically depleted in ^{13}C (only 2 out of 29 IDPs are ^{13}C-rich), similar to those characterised by Starkey et al. (2014) (1 out of 10). In contrast, IDPs characterised by Floss et al. (2006) and Davidson et al. (2012) are ^{13}C-enriched (only 3 out 29 and 1 out 10 are ^{13}C-poor, respectively). Are these systematic ^{13}C-enrichment or depletion related to specific series of samples, or to a potential uncontrolled analytical bias during individual measurement sessions? The measurements were performed following the established analytical methods in each of the laboratories. The standards were different: a mixture of synthetic SiC and Si_3N_4 ($^{14}N/^{15}N = 272$) at WU *vs.* a HOBt polymer (hydroxybenzotriazole; $C_6H_5N_3O$) at OU, and although the e-Gun has never been used at WU, it was used for some measurements at OU. The limited size and mass of individual dust

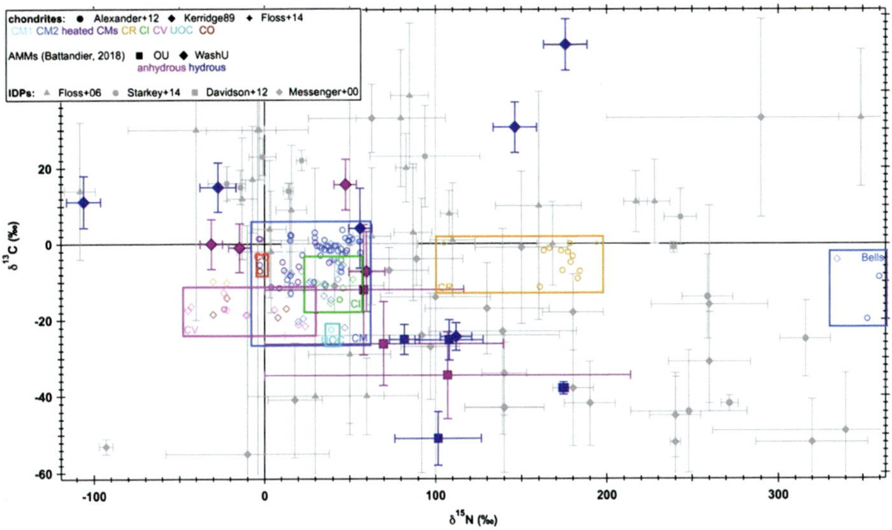

Fig. 2 Bulk C and N isotopic composition of Fg-AMMs, in comparison to bulk chondrites (open-colored symbols) and IDPs (grey symbols). Measurements on AMMs were performed with NanoSIMS at the Open University (square symbol) and at Washington University (diamond symbols) (Battandier 2018). Characterization by IR-spectroscopy of the Fg-AMMs: hydrated (in blue) or non-hydrated (in red) (Battandier et al. 2018). Data for chondrites and IDPs are from the literature, as cited in the text

clearly prevents repeat measurements using destructive techniques or using the same analytical technique as for chondrites (isotope-ratio mass spectrometry). Comparing the isotopic composition of dusts and chondrites, in particular C, and interpreting the difference in terms of potential variability of organic reservoirs in the solar system might require performing systematic and multiple NanoSIMS measurements on chondrites. This analytical approach would allow (*i*) comparing the isotopic compositions obtained independently by two distinct instruments; (*ii*) confirming the absence of analytical biases; (*iii*) confirming (or not) the large variability of the C-isotopic composition of dusts in contrast to chondrites.

4 In Which Type of Body We Expect to Find the Most Pristine Organic Material: On-Site Versus Sample Return

Extensive worldwide meteorite collection has expanded year after year. Decades of meteoritic studies have improved our understanding of the organic distribution in our solar system. However, our meteorite record is biased as porous carbonaceous materials are under-represented owing to their fragile nature that is prone to destruction during the atmospheric entry process. Also, most meteorite samples are *finds* with a long residence time on Earth (terrestrial ages for Antarctica finds rarely exceed 150 ka (Welten et al. 2006), terrestrial ages in hot desert rarely exceed 50 ka (Jull 2006), but in the Atacama desert the average terrestrial age is 710 ka (Drouard et al. 2019)), and are therefore more likely to be contaminated by the fall site surroundings that can overprint any intrinsic organic signatures (Bada et al. 1998; Glavin et al. 1999; Martins et al. 2007a). Therefore, the elucidation of the main scientific problems related to the nature, distribution and evolution of organics necessitates the return of the most pristine extra-terrestrial samples to Earth. In addition, uncertainties remained in the source and geological information for the chemical evolution, which are not clearly recorded in the fallen extraterrestrial materials. In order to link the cosmochemical features of meteorites, IDPs, and AMMs with the geology of their original parent bodies, the exploration of small bodies is essential (Yabuta 2019).

4.1 On-Site Versus Sample Return

On-site space exploration can be conducted at reasonable costs and has provided us with insights into the organic inventory of our planetary and cometary neighbours (e.g., Freissinet et al. 2015; Goesmann et al. 2015; Klein 1978, 1979; Wright et al. 2015), demonstrating that organic compounds are ubiquitous in the solar system. In contrast, sample return missions (with the exception of Stardust so far) are often limited to closer inner solar system bodies (e.g. OSIRIS-REx, Hayabusa2, Hayabusa, Genesis). Unlike sample return missions, of which the returned samples are analysed in a terrestrial environment, on-site analysis of extra-terrestrial samples has a relatively lower risk of contamination. However, the samples can still be subjected to potential sources of contaminants that have bypassed the methodically controlled pre-flight, flight, and post-flight procedures, such as module engine exhaust, solvent, and chemical agents involved in wet chemistry experiments like organic extraction and derivatisation (Elsila et al. 2016; Freissinet et al. 2015) (further details regarding potential sources of terrestrial contamination is given in Chan et al. 2020, this issue). On-site exploration can consist of analytical instruments onboard spacecrafts or mobile vehicles, such as rovers that offer mobility and enables large-scale comparison of the organic properties to different surface features across the whole body.

However, despite our recent advancements in science and technology, the instrumentation required for high-precision and high-sensitivity characterisation of the organic composition cannot be easily miniaturised into flight compatible size. This downside of on-site exploration underscores the merit of a sample return mission. Any sample returned to Earth can be analysed by state-of-the-art instrumentation available in laboratories beyond the realm of on-site measurements by robotic spacecraft and processed by wet chemistry too complicated to be conducted without a dedicated laboratory facility. These instrumentations allow high-resolution imaging of the sample (e.g. synchrotron FTIR micro-spectroscopy, XANES spectroscopy, HRTEM and atomic force microscope infrared-spectroscopy (AFM-IR)) and isotopic analyses (e.g. NanoSIMS, gas chromatography combustion isotope ratio mass spectrometry (GC-C-IRMS)), both of which are crucial in order to elucidate the origins of the organics detected.

4.2 Potential Future Sample Return Mission Target

In order to detect some of the key compounds that play an important role in the origin of life on Earth and evolution of volatile components in the early solar system, it is imperative that we identify and return to Earth samples from a body that hosts *abundant and pristine* organic material, to be studied by the most precise and sensitive instrumentation available today, in order to answer more complex scientific questions such as where did organic symmetry breaking (i.e., prebiotic molecules with a slight enantiomeric excess) take place, and how did that influence the onset of life on Earth?

The significant proportion of hydrous minerals in carbonaceous chondrites suggests that their asteroid parent bodies have been processed by warm liquid water after accretion (Brearley 2006; Weisberg et al. 2006). Carbonaceous chondrites showing signs of aqueous alteration were also found to contain abundant organic material that include simple soluble organic precursors as well as complex and structurally diverse reaction products (Elsila et al. 2016; Glavin et al. 2010; Martins 2019; Martins et al. 2007b, 2015), therefore their parent bodies – C-, D- and P-type asteroids – are amongst the most scientifically exciting targets of sample return missions. In addition, L-enantiomeric excesses (L*ee*) of amino acids (e.g. the abiotic amino acid isovaline) were observed in meteorites that were aqueously altered, which indicates aqueous alteration as a plausible symmetry breaking mechanism (Glavin and Dworkin 2009; Pizzarello et al. 2003). Such non-racemic meteoritic amino acid composition is one of a few examples of molecular asymmetry present in the soluble fraction of carbonaceous meteorites, which offers important insights into the origin of terrestrial homochirality. Other examples include L-lactic acid in a diversity of carbonaceous meteorites (Pizzarello et al. 2010), sugar acids in several carbonaceous meteorites that contain large excesses of the D-enantiomer (Cooper and Rios 2016), and even IOM (Kawasaki et al. 2006). Different degrees of aqueous alteration processes are commonly associated with amino acid synthetic processes such as Strecker synthesis and Michael addition (Cronin and Chang 1993; Kebukawa et al. 2017; Peltzer and Bada 1978), although other more complex processes to synthesize amino acids may exist. It should be noted meteorites that have experienced a prolonged aqueous alteration history, such as the petrologic type 1 chondrites, have lower abundances of amino acids than meteorites that are mildly altered by water (Ehrenfreund et al. 2001; Martins et al. 2015). In fact, extensive aqueous alteration on the meteorite parent body may result in the decomposition of α-amino acids (Botta et al. 2007; Martins et al. 2007b), as increasing low-temperature chemical oxidation removes the aliphatic portions of the soluble and macromolecular organic material (Cody and Alexander 2005; Martins et al. 2006). Racemization of amino acids is particularly rapid in the presence of liquid (Bada and McDonald 1995; Bada et al. 1994; Cohen and Chyba 2000) that could erase signs of molecular asymmetry. Aqueously altered meteorites that were subsequently dehydrated, commonly referred to as thermally metamorphosed carbonaceous chondrites (TMCCs), have yet lower contents of soluble organic molecules (Shimoyama et al. 1989, 1990; Burton et al. 2014; Chan et al. 2016, 2018). Therefore, we should target a body which shows only a minor to mild degree of aqueous alteration due to its involvement in synthesis of organic material, the presence of ancient liquid water, and the role of aqueous processing in the origin of homochirality.

One of the most effective methods commonly used to identify such a primitive body is telescopic remote observation based on the spectrophotometric evidence of alteration minerals in their reflectance spectra in the wavelength range, as aqueous alteration in asteroids is evident by the absorption features in the 2.7–3.0 μm region attributed to the presence of structural hydroxyl and adsorbed water, and thermal alteration evident by the absorption features indicative of high temperature anhydrous minerals (e.g. the absorption feature of olivine in the 0.8–1.3 μm region) (Rivkin et al. 2002; Hiroi et al. 1996; Cloutis et al. 2012b). Hence, asteroids that show absorption features in both regions and have been altered aqueously and thermally would be less ideal. It is because the combined effect of both alteration regimes would have a profound destruction effect on the organics, and the complicated alteration history must be deciphered in order to appropriately address its influences on the organics. Indeed, the alteration event should be very mild and at low temperatures (e.g. aqueous alteration occurred at <150 °C on CM parent body (Zolensky et al. 1989; Verdier-Paoletti et al. 2016)). For instance, in the CM Paris meteorite, one of the most primitive carbonaceous chondrites, isovaline was found to be in racemic ratio (equal abundance between L- and D-isovaline), which indicates that the meteorite was only subjected to limited aqueous alteration, and therefore it likely reflects interstellar and/or presolar cloud organic inventory accreted prior to parent body processing (Martins et al. 2015).

Although meteorites with strong evidence of a high-temperature (>300 °C) thermal alteration history do not necessarily lack organic components (Bonal et al. 2006, 2007, 2016; Burton et al. 2012, 2014; Chan et al. 2012, 2016; Kitajima et al. 2002; Naraoka et al. 2004; Yabuta et al. 2005, 2010), the organic abundances are significantly lower in these meteorites. Their organic compounds are shown to be more structurally altered, and thus any pre-accretion information would have been modified alongside the heating processes. Hence, future sample return missions that aim at investigating primitive organics should avoid objects that show signs of significant thermal metamorphism. In addition, the aliphatic moieties are reduced by decarboxylation or conversion into aromatic C upon pyrolysis (e.g. Martins et al. 2006; Alexander et al. 2007; Chan et al. 2019a, 2019b). Amino acids can also be decomposed at temperatures as low as 100 °C through processes such as decarboxylation, deamination and chain homolyses (e.g., Ratcliff et al. 1974; Bada and McDonald 1995; Pietrucci et al. 2018). Therefore, the amino acid abundances measured in meteorites that had experienced parent body metamorphism, including petrologic type 3 to 6 chondrites, are at least an order of magnitude lower than those in the mildly aqueously altered carbonaceous chondrites (Burton et al. 2012; Burton et al. 2014; Chan et al. 2012, 2016; Martins et al. 2007a). Parent body metamorphism may also modify the organic content by altering the relative abundances of different organic compounds, thus the detected organic composition does not represent the intrinsic population formed upon the asteroid accretion. For example, proteic amino acids are typically more thermodynamically unstable than non-proteic amino acids (e.g. β-alanine, γ-aminobutyric acid (γ-ABA)) (Kitadai 2016), while hydrothermal alteration can produce simple volatile organic compounds such as amines, carboxylic acids and hydrocarbons at the expense of amino acids (Pietrucci et al. 2018). Therefore, the original organic compounds present in parent bodies that have experienced severe thermal metamorphism have been destroyed, i.e. their pre-accretion signature would be lost.

Data suggest that amino acids could be formed at low temperatures on the parent bodies of meteorites (0 to −10 °C) (Pizzarello and Groy 2011), which opens up new horizons for potential sample return missions from solar system objects that involve low temperature processes. Organic compounds may also be formed by solid-phase reactions on interstellar ice grains by energetic processing (e.g. cosmic rays and UV irradiation) (Augé et al. 2016, 2019). Laboratory analogues of the physical and chemical conditions of these interstellar ices (or pre-cometary icy grains) results in the formation of a residue that may contain glycerol, urea, glycolic acid (Nuevo et al. 2010), nucleobases (Nuevo et al. 2009, 2017), hydantoin (de Marcellus et al. 2011a), glyceraldehyde (de Marcellus et al. 2015), aldopentoses and sugar-related compounds (Meinert et al. 2016), and amino acids (Briggs et al. 1992; Kasamatsu et al. 1997; Bernstein et al. 2002; Muñoz Caro et al. 2002; Meinert and Meierhenrich 2012; de Marcellus et al. 2011b; Modica et al. 2014, 2018). Comets are also interesting targets as they are preserved cold – well below the typical liquid water phase temperatures – and thus have lower effects of thermal and aqueous processing, allowing the preservation of information on primitive grain aggregation and the initial inventory of pristine organics. Studies of comet 81P/Wild 2 samples show that the organics are more labile than the materials seen in meteorites, and enriched in heavy isotopes inherited from the interstellar medium (Brownlee et al. 2006; Sandford et al. 2006), therefore cometary organics offer a unique window into understanding the interstellar/protostellar organic inventory. Organic matter has also been detected in comet 67P/Churyumov-Gerasimenko. The Rosetta Orbiter Spectrometer for Ion and Neutral Analysis (ROSINA) and the Reflectron-type Time-Of-Flight mass spectrometer (RTOF) of the Rosetta mission determined the relative densities of the four major volatile species in the coma of the comet (H_2O, CO_2, CO and O_2) and their seasonal evolution, showing that CO_2 and CO were located on the surface of the southern hemisphere as a result of the strong erosion during the previous perihelion (Hoang et al. 2019; Combi et al. 2020). ROSINA also detected several hydrocarbons such as methane, ethane, ethylene, benzene and toluene, as well as CHO-bearing molecules (alcohols up to pentanol, aldehydes and acetone) (Altwegg et al. 2017). Several other classes of organic compounds were detected, e.g. CHN compounds (amines, hydrogen cyanide and acetonitrile), and CHNO (amides, isocyanic acid and glycine) (Altwegg et al. 2019). The presence of both glycine and methylamine in comet 67P suggests a possible synthesis pathway for amino acids without liquid water by the photochemistry of interstellar ices (methylamine and CO_2) (Bossa et al. 2010).

Despite the availability of cometary samples such as the comet 81P/Wild 2 samples, UCAMMs and chondritic porous (CP) IDPs, they are of very limited total mass with only low amino acid and amines abundances (Elsila et al. 2009; Glavin et al. 2008). For example, the concentration of the amino acids in the Stardust-returned aerogel cells backing foil samples was very low and only glycine and ε-amino-n-caproic acid (EACA) were present at abundances barely sufficient for compound-specific carbon isotope measurements (Elsila et al. 2009). Although this implies the presence of amino acids derived from impacts of large cometary grains, their low abundances require large amounts of sample for their successful detection, which is rather difficult for sample return mission as samples are only available at a finite amount. Other than comets, icy objects such as the Saturnian and Jovian moons have also caught the attention as they were suggested to involve large-scale hydrovolcanism and have the potential of hosting a large water ocean beneath the icy crust (Carr et al. 1998; Postberg et al. 2008, 2009). The existence of organic molecules in the atmosphere of the Saturnian moon Enceladus account for a favourable environment suited for the formation of life's precursors and thus possibly host information on the origin of life (e.g., Khawaja et al. 2019; Postberg et al. 2008, 2009). As sample return missions aim to deliver the most pristine, unaltered samples from another celestial object to Earth, proper contamination control is imperative so as not to compromise the pristine nature of the extra-terrestrial sample (Chan et al. (2020) and Ferriere et al. (2020) – this issue).

5 Conclusions

The meteorite collections available all over the world have been a precious asset in the study of primitive organic matter. However, a precise conclusion is jeopardised if the samples are not carefully handled and curated. Alternatively, on-site and sample return space missions from asteroids and/or comets, while time-consuming, costly and complicated are fundamental towards obtaining pristine organic matter. While deciding between those two types of space missions (*on-site* versus sample return), one must consider the risk of contamination, and the available equipment. In both space missions, the indigenous organic content may encounter terrestrial organic molecules; even with excellent terrestrial curation facilities, returned samples will always be in contact with the terrestrial environment and therefore organic contaminants. In the case of on-site samples, the contamination may come from the module engine exhaust and chemical reagents present in wet chemistry experiments. Regarding the available equipment, sample return missions will certainly have access to more diverse and state-of-the-art instruments on Earth. Finally, the degree of alteration of the target body is a crucial factor to consider as it is necessary to have access to pristine and primitive samples. Target bodies should have minor to mild degrees of aqueous alteration, and no thermal metamorphism in order to increase the chances of acquiring data directly related to the accreted interstellar and/or presolar cloud organic inventory. Alternatively, the target bodies should have very low temperatures (below 0 °C) in order to avoid alteration (aqueous and thermal) processes, e.g. space missions to comets. In summary, when planning future space missions, it is necessary to pay attention to all the above points. These considerations are fundamental to obtain extraterrestrial samples that will inform the planetary scientific community.

Acknowledgements Zita Martins was financed by FEDER—Fundo Europeu de Desenvolvimento Regional funds through the COMPETE 2020—Operational Programme for Competitiveness and Internationalisation (POCI), and by Portuguese funds through FCT—Fundação para a Ciência e a Tecnologia in the framework of the project POCI-01-0145-FEDER-029932 (PTDC/FIS-AST/29932/2017). Centro de Química Estrutural acknowledges the financial support of Fundação para a Ciência e Tecnologia (UIDB/00100/2020). This research is framed within the College on Polar and Extreme Environments (Polar2E) of the University of Lisbon. Queenie Chan and Ashley King were supported by Science and Technology Facilities Council (STFC) grants #ST/P000657/1 and #ST/R000727/1, respectively. Lydie Bonal was supported by CNES through its "Système Solaire" program. Hikaru Yabuta was supported by JSPS KAKENHI Grant Numbers JP19H01954, JP18H04461 and by the Astrobiology Center Program of National Institutes of Natural Sciences (NINS) (Grant Number AB312007). The NanoSIMS measurements at OU have received funding from the European Union's Horizon 2020 research and innovation programme under grant agreement No 654208. Dr I. Franchi (OU) and Dr. F. Gyngard (WU) are thanked for their help with NanoSIMS measurements. We acknowledge the constructive comments from the Guest Editor and the anonymous Reviewers that improved the quality of this manuscript.

References

J. Aléon, Multiple origins of nitrogen isotopic anomalies in meteorites and comets. Astrophys. J. **722**, 1342–1351 (2010)

J. Aléon, F. Robert, Interstellar chemistry recorded by nitrogen isotopes in Solar System organic matter. Icarus **167**, 424–430 (2004)

J. Aléon, C. Engrand, F. Robert et al., Clues to the origin of interplanetary dust particles from the isotopic study of their hydrogen-bearing phases. Geochim. Cosmochim. Acta **65**, 4399–4412 (2001)

J. Aléon, F. Robert, M. Chaussidon, B. Marty, Nitrogen isotopic composition of macromolecular organic matter in interplanetary dust particles. Geochim. Cosmochim. Acta **67**, 3773–3783 (2003)

C.M.O'D. Alexander, S.S. Russell, J.W. Arden et al., The origin of chondritic macromolecular organic matter: a carbon and nitrogen isotope study. Meteorit. Planet. Sci. **33**, 603–622 (1998)

C.M.OD. Alexander, M. Fogel, H. Yabuta, G.D. Cody, The origin and evolution of chondrites recorded in the elemental and isotopic compositions of their macromolecular organic matter. Geochim. Cosmochim. Acta **71**, 4380–4403 (2007)

C.M.O'D.Alexander, S.D. Newsome, M.L. Fogel, L.R. Nittler, H. Busemann, G.D. Cody, Deuterium enrichments in chondritic macromolecular material—implications for the origin and evolution of organics, water and asteroids. Geochim. Cosmochim. Acta **74**, 4417–4437 (2010)

C.M.O'D. Alexander, R. Bowden, M.L. Fogel et al., The provenances of asteroids, and their contributions to the volatile inventories of the terrestrial planets. Science **337**, 721–723 (2012)

C.M.O'D. Alexander, G.D. Cody, B.T. De Gregorio et al., The nature, origin and modification of insoluble organic matter in chondrites, the major source of Earth's C and N. Chemie der Erde. Geochemistry **77**, 227–256 (2017)

K. Altwegg, H. Balsiger, A. Bar-Nun et al., Prebiotic chemicals—amino acid and phosphorus—in the coma of comet 67P/Churyumov-Gerasimenko. Sci. Adv. **2**, e1600285–e1600285 (2016)

K. Altwegg, H. Balsiger, J.J. Berthelier et al., Organics in comet 67P – a first comparative analysis of mass spectra from ROSINA–DFMS, COSAC and Ptolemy. Mon. Not. R. Astron. Soc. **469**, S130 (2017)

K. Altwegg, H. Balsiger, S.A. Fuselier, Cometary chemistry and the origin of icy solar system bodies: the view after Rosetta. Annu. Rev. Astron. Astrophys. **57**, 113–155 (2019)

S. Amari, A. Anders, A. Virag et al., Interstellar graphite in meteorites. Nature **345**, 238–240 (1990)

S. Amari, P. Hoppe, E. Zinner et al., The isotopic compositions and stellar sources of meteoritic graphite grains. Nature **365**, 806–809 (1993)

E. Anders, E. Zinner, Interstellar grains in primitive meteorites—diamond, silicon carbide, and graphite. Meteoritics **28**, 490–514 (1993)

J.C. Aponte, M.R. Alexandre, Y. Wang et al., Effects of secondary alteration on the composition of free and IOM-derived monocarboxylic acids in carbonaceous chondrites. Geochim. Cosmochim. Acta **75**, 2309–2323 (2011)

J.C. Aponte, R. Tarozo, M.R. Alexandre et al., Chirality of meteoritic free and IOM-derived monocarboxylic acids and implications for prebiotic organic synthesis. Geochim. Cosmochim. Acta **131**, 1–12 (2014a)

J.C. Aponte, J.P. Dworkin, J.E. Elsila, Assessing the origins of aliphatic amines in the Murchison meteorite from their compound-specific carbon isotopic ratios and enantiomeric composition. Geochim. Cosmochim. Acta **141**, 331–345 (2014b)

J.C. Aponte, J.P. Dworkin, J.E. Elsila, Indigenous aliphatic amines in the aqueously altered Orgueil meteorite. Meteorit. Planet. Sci. **50**, 1733–1749 (2015)

J.C. Aponte, H.L. McLain, J.P. Dworkin et al., Aliphatic amines in Antarctic CR2, CM2, and CM1/2 carbonaceous chondrites. Geochim. Cosmochim. Acta **189**, 296–311 (2016)

B. Augé, E. Dartois, C. Engrand et al., Irradiation of nitrogen-rich ices by Swift heavy ions. Clues for the formation of ultracarbonaceous micrometeorites. Astron. Astrophys. **592**, A99 (2016)

B. Augé, E. Dartois, J. Duprat et al., Hydrogen isotopic anomalies in extraterrestrial organic matter: role of cosmic ray irradiation and implications for UCAMMs. Astron. Astrophys. **627**, A122 (2019)

J.L. Bada, G.D. McDonald, Amino acid racemization on Mars: implications for the preservation of biomolecules from an extinct martian biota. Icarus **114**, 139–143 (1995)

J.L. Bada, X.S. Wang, H.N. Poinar et al., Amino acid racemization in amber-entombed insects: implications for DNA preservation. Geochim. Cosmochim. Acta **58**, 3131–3135 (1994)

J.L. Bada, D.P. Glavin, G.D. McDonald, L. Becker, A search for endogenous amino acids in martian meteorite ALH84001. Science **279**, 362–365 (1998)

H.C. Bates, A.J. King, K.L. Donaldson Hanna et al., Linking mineralogy and spectroscopy of highly aqueously altered CM and CI carbonaceous chondrites in preparation for primitive asteroid sample return. Meteorit. Planet. Sci. **55**, 77–101 (2020)

M. Battandier, Etude d'une série de micrométéorites antarctiques: caractérisation multi analytique et comparaison à des chondrites carbonées. Doctoral dissertation from, Université Grenoble Alpes (2018)

M. Battandier, L. Bonal, E. Quirico, P. Beck, C. Engrand, J. Duprat, E. Dartois, Characterization of the organic matter and hydration state of Antarctic micrometeorites: a reservoir distinct from carbonaceous chondrites. Icarus **306**, 74–93 (2018)

G.K. Benedix, L.A. Leshin, J. Farquhar et al., Carbonates in CM2 chondrites: constraints on alteration conditions from oxygen isotopic compositions and petrographic observations. Geochim. Cosmochim. Acta **67**, 1577–1588 (2003)

M.P. Bernstein, J.P. Dworkin, S.A. Sandford et al., Racemic amino acids from the ultraviolet photolysis of interstellar ice analogues. Nature **416**, 401–403 (2002)

D. Bockelée-Morvan et al., AMBITION—Comet Nucleus Cryogenic Sample Return. A white paper for ESA Voyage 2050 long-term plan (2019). https://arxiv.org/pdf/1907.11081.pdf

D. Bockelée-Morvan, J. Crovisier, M.J. Mumma et al., The composition of cometary volatiles, in *Comets II*, ed. by M. Festou, H.U. Keller, H.A. Weaver (University of Arizona Press, Tucson, 2004), pp. 391–423

L. Bonal, E. Quirico, M. Bourot-Denise et al., Determination of the petrologic type of CV3 chondrites by Raman spectroscopy of included organic matter. Geochim. Cosmochim. Acta **70**, 1849–1863 (2006)

L. Bonal, M. Bourot-Denise, E. Quirico et al., Organic matter and metamorphic history of CO chondrites. Geochim. Cosmochim. Acta **71**, 1605–1623 (2007)

L. Bonal, E. Quirico, L. Flandinet et al., Thermal history of type 3 chondrites from the Antarctic meteorite collection determined by Raman spectroscopy of their polyaromatic carbonaceous matter. Geochim. Cosmochim. Acta **189**, 312–337 (2016)

J.B. Bossa, F. Borget, F. Duvernay et al., How a usual carbamate can become an unusual intermediate: a new chemical pathway to form glycinate in the interstellar medium. J. Phys. Org. Chem. **23**, 333–339 (2010)

O. Botta, Z. Martins, P. Ehrenfreund, Amino acids in Antarctic CM1 meteorites and their relationship to other carbonaceous chondrites. Meteorit. Planet. Sci. **42**, 81–92 (2007)

A.J. Brearley, *The Action of Water* (The University of Arizona Press, Tucson, 2006), pp. 584–624

R. Briggs, G. Ertem, J.P. Ferris et al., Comet Halley as an aggregate of interstellar dust and further evidence for the photochemical formation of organics in the interstellar medium. Orig. Life Evol. Biosph. **22**, 287–307 (1992)

K.L.F. Brinton, C. Engrand, D.P. Glavin et al., A search for extraterrestrial amino acids in carbonaceous Antarctic micrometeorites. Orig. Life Evol. Biosph. **28**, 413–424 (1998)

L.B. Browning, H.Y. McSween, M.E. Zolensky, Correlated alteration effects in CM carbonaceous chondrites. Geochim. Cosmochim. Acta **60**, 2621–2633 (1996)

D. Brownlee, P. Tsou, J. Aléon et al., Comet 81P/Wild 2 under a microscope. Science **314**, 1711–1716 (2006)

A.S. Burton, J.E. Elsila, M.P. Callahan et al., A propensity for n-ω-amino acids in thermally altered Antarctic meteorites. Meteorit. Planet. Sci. **47**, 374–386 (2012)

A.S. Burton, S. Grunsfeld, J.E. Elsila et al., The effects of parent-body hydrothermal heating on amino acid abundances in CI-like chondrites. Polar Sci. **8**, 255–263 (2014)

H. Busemann, A.F. Young, C.M.O'D. Alexander et al., Interstellar chemistry recorded in organic matter from primitive meteorites. Science **312**, 727–730 (2006)

H. Busemann, C.M.OD. Alexander, L.R. Nittler, Characterization of insoluble organic matter in primitive meteorites by micro-Raman spectroscopy. Meteorit. Planet. Sci. **42**, 1387–1416 (2007)

H. Busemann, A.N. Nguyen, G.D. Cody et al., Ultra-primitive interplanetary dust particles from the comet 26P/Grigg-Skjellerup dust stream collection. Earth Planet. Sci. Lett. **288**, 44–57 (2009)

H. Campins, K. Hargrove, N. Pinilla-Alonso et al., Water ice and organics on the surface of the asteroid 24 Themis. Nature **464**, 1320–1321 (2010)

F. Capaccioni et al., The organic-rich surface of comet 67P/Churyumov-Gerasimenko as seen by VIRTIS/Rosetta. Science **347**, aaa0628 (2015)

M.H. Carr, M.J.S. Belton, C.R. Chapman, Evidence for a subsurface ocean on Europa. Nature **391**, 363 (1998)

H.-S. Chan, Z. Martins, M.A. Sephton, Amino acid analyses of type 3 chondrites Colony, Ornans, Chainpur, and Bishunpur. Meteorit. Planet. Sci. **47**, 1502–1516 (2012)

Q.H.S. Chan, Y. Chikaraishi, Y. Takano et al., Amino acid compositions in heated carbonaceous chondrites and their compound-specific nitrogen isotopic ratios. Earth Planets Space **68**, 7 (2016)

Q.H.S. Chan, M. Zolensky, Y. Kebukawa et al., Organic matter in extraterrestrial water-bearing salt crystals. Sci. Adv. **4**, eaao3521 (2018)

Q.H.S. Chan et al., Organics preserved in anhydrous interplanetary dust particles: pristine or not? Meteorit. Planet. Sci. (2019a). https://doi.org/10.1111/maps.13414.

Q.H.S. Chan, A. Nakato, Y. Kebukawa et al., Heating experiments of the Tagish Lake meteorite: investigation of the effects of short-term heating on chondritic organics. Meteorit. Planet. Sci. **54**, 104–125 (2019b)

Q.H.S. Chan, R. Stroud, Z. Martins, H. Yabuta, Concerns of organic contamination for sample return space missions. Space Sci. Rev. (2020). https://doi.org/10.1007/s11214-020-00678-7

S.J. Clemett, C.R. Maechling, R.N. Zare et al., Identification of complex aromatic molecules in individual interplanatery dust particles. Science **262**, 721–725 (1993)

S.J. Clemett, X.D.F. Chillier, S. Gillette et al., Observation of indigenous polycyclic aromatic hydrocarbons in 'giant' carbonaceous Antarctic micrometeorites. Orig. Life Evol. Biosph. **28**, 425–448 (1998)

E.A. Cloutis, T. Hiroi, M.J. Gaffey et al., Spectral reflectance properties of carbonaceous chondrites: 1. CI chondrites. Icarus **212**, 180–209 (2011a)

E.A. Cloutis, P. Hudon, T. Hiroi et al., Spectral reflectance properties of carbonaceous chondrites: 2. CM chondrites. Icarus **216**, 309–346 (2011b)

E.A. Cloutis, P. Hudon, T. Hiroi et al., Spectral reflectance properties of carbonaceous chondrites: 6. CV chondrites. Icarus **221**, 328–358 (2012a)

E.A. Cloutis et al., Spectral reflectance properties of carbonaceous chondrites 4: aqueously altered and thermally metamorphosed meteorites. Icarus **220**, 586–617 (2012b)

G.D. Cody, C.M.OD. Alexander, NMR studies of chemical structural variation of insoluble organic matter from different carbonaceous chondrites groups. Geochim. Cosmochim. Acta **69**, 1085–1097 (2005)

G.D. Cody, C.M.OD. Alexander, F. Tera, Solid-state (1H and 13C) nuclear magnetic resonance spectroscopy of insoluble organic residue in the Murchison meteorite: a self-consistent quantitative analysis. Geochim. Cosmochim. Acta **66**, 1851–1865 (2002)

G.D. Cody, H. Ade, C.M.OD. Alexander et al., Quantitative organic and light-element analysis of comet 81P/Wild 2 particles using C-, N-, and O-μ-XANES. Meteorit. Planet. Sci. **43**, 353–365 (2008)

G.D. Cody, E. Heying, C.M.OD. Alexander et al., Establishing a molecular relationship between chondritic and cometary organic solids. Proc. Natl. Acad. Sci. USA **108**, 19171–19176 (2011)
B.A. Cohen, C.F. Chyba, Racemization of meteoritic amino acids. Icarus **145**, 272–281 (2000)
M. Combi, Y. Shou, N. Fougere et al., The surface distributions of the production of the major volatile species, H2O, CO2, CO and O2, from the nucleus of comet 67P/Churyumov-Gerasimenko throughout the Rosetta Mission as measured by the ROSINA double focusing mass spectrometer. Icarus **335**, 113421 (2020)
G. Cooper, A. Rios, Enantiomer excesses of rare and common sugar derivatives in carbonaceous meteorites. Proc. Natl. Acad. Sci. USA **113**, E3322–E3331 (2016)
G.W. Cooper, M.H. Thiemens, T.L. Jackson et al., Sulfur and hydrogen isotope anomalies in meteorite sulfonic acids. Science **277**, 1072–1074 (1997)
G. Cooper, N. Kimmich, W. Belisle et al., Carbonaceous meteorites as a source of sugar-related organic compounds for the early Earth. Nature **414**, 879–883 (2001)
J.R. Cronin, S. Chang, Organic matter in meteorites: molecular and isotopic analyses of the Murchison meteorite, in *The Chemistry of Life's Origins*, ed. by J.M. Greenberg, C.X. Mendoza-Gómez, V. Pirronello. NATO ASI Series, vol. 416 (Kluwer Academic Publishers, Norwell, 1993), pp. 209–258
J.R. Cronin, S. Pizzarello, J.S. Frye, ^{13}C NMR spectroscopy of the insoluble carbon of carbonaceous chondrites. Geochim. Cosmochim. Acta **51**, 299–303 (1987)
J.R. Cronin, S. Pizzarello, S. Epstein et al., Molecular and isotopic analyses of the hydroxy acids, dicarboxylic acids, and hydroxydicarboxylic acids of the Murchison meteorite. Geochim. Cosmochim. Acta **57**, 4745–4752 (1993)
J. Crovisier, D. Bockelée-Morvan, Remote observations of the composition of cometary volatiles. Space Sci. Rev. **90**, 19–32 (1999)
J. Crovisier, N. Biver, D. Bockelee-Morvan et al., The chemical diversity of comets. Earth Moon Planets **105**, 267–272 (2009)
D.P. Cruikshank, R.H. Brown, Organic matter on asteroid 130 Elektra. Science **238**, 183–184 (1987)
D.P. Cruikshank, J.F. Kerridge, Organic material: asteroids, meteorites, and planetary satellites, in *Exobiology in Solar System Exploration* (1992), pp. 159–176
E. Dartois, C. Engrand, R. Brunetto et al., UltraCarbonaceous Antarctic micrometeorites, probing the Solar System beyond the nitrogen snow-line. Icarus **224**, 243–252 (2013)
E. Dartois, C. Engrand, J. Duprat et al., Dome C ultracarbonaceous Antarctic micrometeorites. Infrared and Raman fingerprints. Astron. Astrophys. **609**, A65 (2018)
J. Davidson, H. Busemann, I.A. Franchi, A NanoSIMS and Raman spectroscopic comparison of interplanetary dust particles from comet Grigg-Skjellerup and non-Grigg Skjellerup collections. Meteorit. Planet. Sci. **47**, 1748–1771 (2012)
B.T. De Gregorio, R.M. Stroud, G.D. Cody et al., Correlated microanalysis of cometary organic grains returned by Stardust. Meteorit. Planet. Sci. **46**, 1376–1396 (2011)
P. de Marcellus, M. Bertrand, M. Nuevo et al., Prebiotic significance of extraterrestrial ice photochemistry: detection of hydantoin in organic residues. Astrobiology **11**, 847–854 (2011a)
P. de Marcellus, C. Meinert, M. Nuevo et al., Non-racemic amino acid production by ultraviolet irradiation of achiral interstellar ice analogs with circularly polarized light. Astrophys. J. Lett. **727**, 1–6 (2011b)
P. de Marcellus, C. Meinert, I. Myrgorodska et al., Aldehydes and sugars from evolved precometary ice analogs: importance of ices in astrochemical and prebiotic evolution. Proc. Natl. Acad. Sci. USA **112**, 965–970 (2015)
M.C. De Sanctis, E. Ammannito, H.Y. McSween et al., Localized aliphatic organic material on the surface of Ceres. Science **355**, 719–722 (2017)
F.E. DeMeo, B. Carry, The taxonomic distribution of asteroids from multi-filter all-sky photometric surveys. Icarus **226**, 723–741 (2013)
F.E. DeMeo, B. Carry, Solar System evolution from compositional mapping of the asteroid belt. Nature **505**, 629–634 (2015)
E. Dobrică, C. Engrand, E. Quirico et al., Raman characterization of carbonaceous matter in CONCORDIA Antarctic micrometeorites. Meteorit. Planet. Sci. **46**, 1363–1375 (2011)
A. Drouard, J. Gattacceca, A. Hutzler, P. Rochette, R. Braucher, D. Bourlès, M. Gounelle, A. Morbidelli, V. Debaille, M. Van Ginneken, M. Valenzuela, Y. Quesnel, R. Martinez, The meteorite flux of the past 2 m.y. recorded in the Atacama Desert. Geology **47**, 673–676 (2019)
J. Duprat, E. Dobrica, C. Engrand et al., Extreme deuterium excesses in ultracarbonaceous micrometeorites from Central Antarctic snow. Science **328**, 742–745 (2010)
M. Ebihara, S. Sekimoto, N. Shirai et al., Neutron activation analysis of a particle returned from asteroid Itokawa. Science **333**, 1119–1121 (2011)
P. Ehrenfreund, D.P. Glavin, O. Botta et al., Extraterrestrial amino acids in Orgueil and Ivuna: tracing the parent body of CI type carbonaceous chondrites. Proc. Natl. Acad. Sci. USA **98**, 2138–2141 (2001)

J.E. Elsila, D.P. Glavin, J.P. Dworkin, Cometary glycine detected in samples returned by Stardust. Meteorit. Planet. Sci. **44**, 1323–1330 (2009)
J.E. Elsila, S.B. Charnley, A.S. Burton et al., Compound-specific carbon, nitrogen, and hydrogen isotopic ratios for amino acids in CM and CR chondrites and their use in evaluating potential formation pathways. Meteorit. Planet. Sci. **47**, 1517–1536 (2012)
J.E. Elsila, J.C. Aponte, D.G. Blackmond et al., Meteoritic amino acids: diversity in compositions reflects parent body histories. ACS Central Science **2**, 370–379 (2016)
M.H. Engel, S.A. Macko, Isotopic evidence for extraterrestrial non-racemic amino acids in the Murchison meteorite. Nature **389**, 265–268 (1997)
S. Epstein, R.V. Krishnamurthy, J.R. Cronin et al., Unusual stable isotope ratios in amino acid and carboxylic acid extracts from the Murchison meteorite. Nature **326**, 477–479 (1987)
C. Floss, F.J. Stadermann, J.P. Bradley et al., Identification of isotopically primitive interplanetary dust particles: a NanoSIMS isotopic imaging study. Geochim. Cosmochim. Acta **70**, 2371–2399 (2006)
C. Floss, C. Le Guillou, A. Brearley, Coordinated NanoSIMS and FIB-TEM analyses of organic matter and associated matrix materials in CR3 chondrites. Geochim. Cosmochim. Acta **139**, 1–25 (2014)
G.J. Flynn, L.P. Keller, M. Feser et al., The origin of organic matter in the solar system: evidence from the interplanetary dust particles. Geochim. Cosmochim. Acta **67**, 4791–4806 (2003)
G.J. Flynn, L.P. Keller, C. Jacobsen et al., An assessment of the amount and types of organic matter contributed to the Earth by interplanetary dust. Adv. Space Res. **33**, 57–66 (2004)
K. Fredriksson, J.F. Kerridge, Carbonates and sulfates in CI chondrites—formation by aqueous activity on the parent body. Meteoritics **23**, 35–44 (1988)
C. Freissinet, D.P. Glavin, P.R. Mahaffy et al., Organic molecules in the Sheepbed Mudstone, Gale Crater, Mars. J. Geophys. Res., Planets **120**, 495–514 (2015)
A. Gardinier, S. Derenne, F. Robert et al., Solid state CP/MAS 13C NMR of the insoluble organic matter of the Orgueil and Murchison meteorites: quantitative study. Earth Planet. Sci. Lett. **184**, 9–21 (2000)
M.J. Genge, C. Engrand, M. Gounelle et al., The classification of micrometeorites. Meteorit. Planet. Sci. **43**, 497–515 (2008)
I. Gilmour, Structural and isotopic analysis of organic matter in carbonaceous chondrites, in *Reference Module in Earth Systems and Environmental Sciences—Treatise on Geochemistry 1* (2014), pp. 215–233
I. Gilmour, C.T. Pillinger, Isotopic compositions of individual polycyclic aromatic hydrocarbons from the Murchison meteorite. Mon. Not. R. Astron. Soc. **269**, 235–240 (1994)
D.P. Glavin, J. Dworkin, Enrichment of the amino acid L-isovaline by aqueous alteration on CI and CM meteorite parent bodies. Proc. Natl. Acad. Sci. USA **106**, 5487–5492 (2009)
D.P. Glavin, J.L. Bada, K.L.F. Brinton et al., Amino acids in the Martian meteorite Nakhla. Proc. Natl. Acad. Sci. **96**, 8835–8838 (1999)
D.P. Glavin, G. Matrajt, J.L. Bada, Re-examination of amino acids in Antarctic micrometeorites. Adv. Space Res. **33**, 106–113 (2004)
D.P. Glavin, J.P. Dworkin, S.A. Sandford, Detection of cometary amines in samples returned by Stardust. Meteorit. Planet. Sci. **43**, 399–413 (2008)
D.P. Glavin, M.P. Callahan, J.P. Dworkin et al., The effects of parent body processes on amino acids in carbonaceous chondrites. Meteorit. Planet. Sci. **45**, 1948–1972 (2010)
D.P. Glavin, C.M.O'D. Alexander, J.C. Aponte, J.P. Dworkin, J.E. Elsila, H. Yabuta, The origin and evolution of organic matter in carbonaceous chondrites and links to their parent bodies, in *Primitive Meteorites and Asteroids Physical, Chemical and Spectroscopic Observations Paving the Way to Exploration* (2018), pp. 205–271
F. Goesmann, H. Rosenbauer, J.H. Bredehöft et al., Organic compounds on comet 67P/Churyumov-Gerasimenko revealed by COSAC mass spectrometry. Science **349**, aab0689 (2015)
M.A. Gounelle, P. Spurný, P.A. Bland, The orbit and atmospheric trajectory of the Orgueil meteorite from historical records. Meteorit. Planet. Sci. **41**, 135–150 (2006)
M. Gounelle, A. Morbidelli, P.A. Bland et al., Meteorites from the outer solar system?, in *The Solar System Beyond Neptune*, ed. by M.A. Barucci, H. Boehnhardt, D.P. Cruikshank, A. Morbidelli (University of Arizona Press, Tucson, 2008), pp. 525–541
M.M. Grady, I.P. Wright, P.K. Swart et al., The carbon and oxygen isotopic composition of meteoritic carbonates. Geochim. Cosmochim. Acta **52**, 2855–2866 (1988)
M. Grott, J. Knollenberg, M. Hamm, Low thermal conductivity boulder with high porosity identified on C-type asteroid (162173) Ryugu. Nat. Astron. **3**, 971–976 (2019)
V.E. Hamilton, A.A. Simon, P.R. Christensen et al., Evidence for widespread hydrated minerals on asteroid (101955) Bennu. Nat. Astron. **3**, 332–340 (2019)
A. Havette, G. Slodzian, Matrix effects in secondary ion emission: quantitative analysis of silicates. J. Phys. Lett. **41**, L247–L250 (1980)

R. Hayatsu, S. Matsuoka, E. Anders et al., Origin of organic matter in the early solar system. VII. The organic polymer in carbonaceous chondrites. Geochim. Cosmochim. Acta **41**, 1325–1339 (1977)

R. Hayatsu, R.E. Winans, R.G. Scott et al., Phenolic ethers in the organic polymer of the Murchison meteorite. Science **207**, 1202–1204 (1980)

H.G.M. Hill, J.A. Nuth, The catalytic potential of cosmic dust: implications for prebiotic chemistry in the solar nebula and other protoplanetary systems. Astrobiology **3**, 291–304 (2003)

T. Hiroi et al., Thermal metamorphism of the C, G, B, and F asteroids seen from the 0.7 μm, 3 μm, and UV absorption strengths in comparison with carbonaceous chondrites. Meteorit. Planet. Sci. **31**, 321–327 (1996)

M. Hoang, P. Garnier, H. Gourlaouen et al., Two years with comet 67P/Churyumov-Gerasimenko: H2O, CO2, and CO as seen by the ROSINA/RTOF instrument of Rosetta. Astron. Astrophys. **630**, A33 (2019)

P. Hoppe, R. Strebel, P. Eberhardt et al., Type II supernova matter in a silicon carbide grain from the Murchison meteorite. Science **272**, 1314–1316 (1996)

Y. Huang, Y. Wang, M.R. Alexandre et al., Molecular and compound-specific isotopic characterization of monocarboxylic acids in carbonaceous meteorites. Geochim. Cosmochim. Acta **69**, 1073–1084 (2005)

Y. Huang, J.C. Aponte, J. Zhao et al., Hydrogen and carbon isotopic ratios of polycyclic aromatic compounds in two CM2 carbonaceous chondrites and implications for prebiotic organic synthesis. Earth Planet. Sci. Lett. **426**, 101–108 (2015)

R. Jaumann, N. Schmitz, T.-M. Ho et al., Images from the surface of asteroid Ryugu show rocks similar to carbonaceous chondrite meteorites. Science **365**, 817–820 (2019)

A.J.T. Jull, Terrestrial Ages of Meteorites, in *Meteorites and the Early Solar System II*, ed. by D.S. Lauretta, H.Y. McSween Jr. (University of Arizona Press, Tucson, 2006), pp. 889–905. 943 pp.

H.H. Kaplan, R.E. Milliken, C.M.OD. Alexander, New constraints on the abundance and composition of organic matter on Ceres. Geophys. Res. Lett. **45**, 5274–5282 (2018)

T. Kasamatsu, T. Kaneko, T. Saito et al., Formation of organic compounds in simulated interstellar media with high energy particles. Bull. Chem. Soc. Jpn. **70**, 1021 (1997)

T. Kawasaki, K. Hatase, Y. Fujii et al., The distribution of chiral asymmetry in meteorites: an investigation using asymmetric autocatalytic chiral sensors. Geochim. Cosmochim. Acta **70**, 5395–5402 (2006)

Y. Kebukawa, S. Nakashima, T. Otsuka et al., Rapid contamination during storage of carbonaceous chondrites prepared for micro FTIR measurements. Meteorit. Planet. Sci. **44**, 545–557 (2009)

Y. Kebukawa, C.M.OD. Alexander, G.D. Cody, Compositional diversity in insoluble organic matter in type 1, 2 and 3 chondrites as detected by infrared spectroscopy. Geochim. Cosmochim. Acta **75**, 3530–3541 (2011)

Y. Kebukawa, A.L.D. Kilcoyne, G.D. Cody, Exploring the potential formation of organic solids in chondrites and comets through polymerization of interstellar formaldehyde. Astrophys. J. **771**, 19 (2013)

Y. Kebukawa, Q.H.S. Chan, S. Tachibana, One-pot synthesis of amino acid precursors with insoluble organic matter in planetesimals with aqueous activity. Sci. Adv. **3**, e1602093 (2017)

L.P. Keller, S. Messenger, G.J. Flynn et al., The nature of molecular cloud material in interplanetary dust. Geochim. Cosmochim. Acta **68**, 2577–2589 (2004)

J.F. Kerridge, Carbon, hydrogen and nitrogen in carbonaceous chondrites: abundances and isotopic compositions in bulk samples. Geochim. Cosmochim. Acta **49**, 1707–1714 (1985)

J.F. Kerridge, S. Chang, R. Shipp, Isotopic characterisation of kerogen-like material in the Murchison carbonaceous chondrite. Geochim. Cosmochim. Acta **51**, 2527–2540 (1987)

N. Khawaja, F. Postberg, J. Hillier et al., Low-mass nitrogen-, oxygen-bearing, and aromatic compounds in Enceladean ice grains. Mon. Not. R. Astron. Soc. **489**, 5231–5243 (2019)

N. Kitadai, Predicting thermodynamic behaviors of non-protein amino acids as a function of temperature and pH. Orig. Life Evol. Biosph. **46**, 3–18 (2016)

F. Kitajima, T. Nakamura, N. Takaoka et al., Evaluating the thermal metamorphism of CM chondrites by using the pyrolytic behavior of carbonaceous macromolecular matter. Geochim. Cosmochim. Acta **66**, 163–172 (2002)

K. Kitazato, R.E. Milliken, T. Iwata et al., The surface composition of asteroid 162173 Ryugu from Hayabusa2 near-infrared spectroscopy. Science **364**, 272–275 (2019)

H.P. Klein, The Viking biological experiments on Mars. Icarus **34**, 666–674 (1978)

H.P. Klein, The Viking mission and the search for life on Mars. Rev. Geophys. Space Phys. **17**, 1655–1662 (1979)

M. Komiya, A. Shimoyama, K. Harada, Examination of organic compounds from insoluble organic matter isolated from some Antarctic carbonaceous chondrites by heating experiments. Geochim. Cosmochim. Acta **57**, 907–914 (1993)

R.V. Krishnamurthy, S. Epstein, J.R. Cronin et al., Isotopic and molecular analyses of hydrocarbons and monocarboxylic acids of the Murchison meteorite. Geochim. Cosmochim. Acta **56**, 4045–4058 (1992)

M. Kuga, B. Marty, Y. Marrocchi, L. Tissandier, Synthesis of refractory organic matter in the ionized gas phase of the solar nebula. Proc. Natl. Acad. Sci. **112**, 7129–7134 (2015)

H.F. Levison, W.F. Bottke, M. Gounelle et al., Contamination of the asteroid belt by primordial trans-Neptunian objects. Nature **460**, 364–366 (2009)

R.S. Lewis, T. Ming, J.F. Wacker et al., Interstellar diamonds in meteorites. Nature **326**, 160–162 (1987)

J. Licandro, H. Campins, M. Kelley et al., (65) Cybele: detection of small silicate grains, water-ice, and organics. Astron. Astrophys. **525**, A34 (2011)

K. Lodders, Solar system abundances and condensation temperatures of the elements. Astrophys. J. **591**, 1220–1247 (2003)

Z. Martins, The nitrogen heterocycle content of meteorites and their significance for the origin of life. Life **8**, 28–43 (2018)

Z. Martins, Organic molecules in meteorites and their astrobiological significance, in *Handbook of Astrobiology*, ed. by V.M. Kolb (CRC Press, Boca Raton, 2019), pp. 177–194. ISBN 9781138065123, CAT# K33392.

Z. Martins, M.A. Sephton, Extraterrestrial amino acids, in *Amino Acids, Peptides and Proteins in Organic Chemistry*, ed. by A.B. Hughes (Wiley, New York, 2009), pp. 3–42

Z. Martins, J.S. Watson, M.A. Sephton et al., Free dicarboxylic and aromatic acids in the carbonaceous chondrites Murchison and Orgueil. Meteorit. Planet. Sci. **41**, 1073–1080 (2006)

Z. Martins, B.A. Hofmann, E. Gnos et al., Amino acid composition, petrology, geochemistry, ^{14}C terrestrial age and oxygen isotopes of the Shisr 033 CR chondrite. Meteorit. Planet. Sci. **42**, 1581–1595 (2007a)

Z. Martins, C.M.OD. Alexander, G.E. Orzechowska et al., Indigenous amino acids in primitive CR meteorites. Meteorit. Planet. Sci. **42**, 2125–2136 (2007b)

Z. Martins, O. Botta, M.L. Fogel et al., Extraterrestrial nucleobases in the Murchison meteorite. Earth Planet. Sci. Lett. **270**, 130–136 (2008)

Z. Martins, P. Modica, Z. Zanda et al., The amino acid and hydrocarbon contents of the Paris meteorite: insights into the most primitive CM chondrite. Meteorit. Planet. Sci. **50**, 926–943 (2015)

G. Matrajt, S. Pizzarello, S. Taylor et al., Concentration and variability of the AIB amino acid in polar micrometeorites: implications for the exogenous delivery of amino acids to the primitive Earth. Meteorit. Planet. Sci. **39**, 1849–1858 (2004)

G. Matrajt, G.M. Muñoz Caro, E. Dartois et al., FTIR analysis of the organics in IDPs: comparison with the IR spectra of the diffuse interstellar medium. Astron. Astrophys. **433**, 979–995 (2005)

C. Meinert, U.J. Meierhenrich, A new dimension in separation science: comprehensive two-dimensional gas chromatography. Angew. Chem., Int. Ed. Engl. **51**, 10460–10470 (2012)

C. Meinert, I. Myrgorodska, P. de Marcellus et al., Ribose and related sugars from ultraviolet irradiation of interstellar ice analogs. Science **352**, 208–212 (2016)

S. Merouane, Z. Djouadi, L. D'Hendecourt, Relationship Between Organics and Silicates in Interplanetary Dust Particles, in *44th Lunar and Planetary Science Conference*. LPI Contribution, No. 1719 (2003), p. 1981

S. Messenger, Identification of molecular-cloud material in interplanetary dust particles. Nature **404**, 968–971 (2000)

S. Messenger, L.P. Keller, F.J. Stadermann et al., Samples of stars beyond the solar system: silicate grains in interplanetary dust. Science **300**, 105–108 (2003)

T.J. Millar, A. Bennett, E. Herbst, Deuterium fractionation in dense interstellar clouds. Astrophys. J. **340**, 906–920 (1989)

P. Modica, C. Meinert, P. de Marcellus et al., Enantiomeric excesses induced in amino acids by ultraviolet circularly polarized light irradiation of extraterrestrial ice analogs: a possible source of asymmetry for prebiotic chemistry. Astrophys. J. **788**, 1–11 (2014)

P. Modica, Z. Martins, C. Meinert, B. Zanda, L.L.S. d'Hendecourt, The amino acid distribution in laboratory analogs of extraterrestrial organic matter: a comparison to CM chondrites. Astrophys. J. **865**, 1–11 (2018)

M.J. Mumma, S.B. Charnley, The chemical composition of comets—emerging taxonomies and natal heritage. Annu. Rev. Astron. Astrophys. **49**, 471–524 (2011)

G.M. Muñoz Caro, U.J. Meierhenrich, W.A. Schutte et al., Amino acids from ultraviolet irradiation of interstellar ice analogues. Nature **416**, 403–406 (2002)

G.M. Muñoz Caro, G. Matrajt, E. Dartois et al., Nature and evolution of the dominant carbonaceous matter in interplanetary dust particles: effects of irradiation and identification with a type of amorphous carbon. Astron. Astrophys. **459**, 147–159 (2006)

K. Nagao, R. Okazaki, T. Nakamura et al., Irradiation history of itokawa regolith material deduced from noble gases in the Hayabusa samples. Science **333**, 1128–1131 (2011)

T. Nakamura, T. Noguchi, M. Tanaka et al., Itokawa dust particles: a direct link between S-type asteroids and ordinary chondrites. Science **333**, 1113–1116 (2011)

K. Nakamura-Messenger, S. Messenger, L.P. Keller et al., Organic globules in the Tagish Lake meteorite: remnants of the protosolar disk. Science **314**, 1439–1442 (2006)
H. Naraoka, A. Shimoyama, K. Harada, Isotopic evidence from an Antarctica carbonaceous chondrite for two reaction pathways of extraterrestrial PAH formation. Earth Planet. Sci. Lett. **184**, 1–7 (2000)
H. Naraoka, H. Mita, M. Komiya et al., A chemical sequence of macromolecular organic matter in the CM chondrites. Meteorit. Planet. Sci. **39**, 401–406 (2004)
T. Noguchi, T. Nakamura, M. Kimura et al., Incipient space weathering observed on the surface of Itokawa dust particles. Science **333**, 1121–1125 (2011)
M. Nuevo, S.N. Milam, S.A. Sandford et al., Formation of uracil from the ultraviolet photo-irradiation of pyrimidine in pure H2O ices. Astrobiology **9**, 683–695 (2009)
M. Nuevo, J.H. Bredehöft, U.J. Meierhenrich et al., Urea, glycolic acid, and glycerol in an organic residue produced by ultraviolet irradiation of interstellar/pre-cometary ice analogs. Astrobiology **10**, 245–256 (2010)
M. Nuevo, S.A. Sandford, G. Cooper, Sugars and sugar derivatives in residues produced from the UV irradiation of astrophysical ice analogs. LPI Contrib. **48**, 2496 (2017)
F.-R. Orthous-Daunay, E. Quirico, P. Beck et al., Mid-infrared study of the molecular structure variability of insoluble organic matter from primitive chondrites. Icarus **223**, 534–543 (2013)
E.E. Palmer, D.S. Lauretta, Aqueous alteration of kamacite in CM chondrites. Meteorit. Planet. Sci. **46**, 1587–1607 (2011)
E.T. Peltzer, J.L. Bada, α-Hydroxycarboxylic acids in the Murchison meteorite. Nature **272**, 443–444 (1978)
C.M. Pieters, A. Nathues, G. Thangjam et al., Geologic constraints on the origin of red organic-rich material on Ceres. Meteorit. Planet. Sci. **53**, 1983–1998 (2017)
F. Pietrucci, J.C. Aponte, R. Starr et al., Hydrothermal decomposition of amino acids and origins of prebiotic meteoritic organic compounds. ACS Earth Space Chem. **2**, 588–598 (2018)
S. Pizzarello, T.L. Groy, Molecular asymmetry in extraterrestrial organic chemistry: an analytical perspective. Geochim. Cosmochim. Acta **75**, 645–656 (2011)
S. Pizzarello, Y. Huang, Molecular and isotopic analyses of Tagish Lake alkyl dicarboxylic acids. Meteorit. Planet. Sci. **37**, 687–696 (2002)
S. Pizzarello, Y. Huang, The deuterium enrichment of individual amino acids in carbonaceous meteorites: a case for the presolar distribution of biomolecule precursors. Geochim. Cosmochim. Acta **69**, 599–605 (2005)
S. Pizzarello, C.T. Yarnes, Enantiomeric excesses of chiral amines in ammonia-rich carbonaceous meteorites. Earth Planet. Sci. Lett. **443**, 176–184 (2016)
S. Pizzarello, R.V. Krishnamurthy, S. Epstein et al., Isotopic analyses of amino acids from the Murchison meteorite. Geochim. Cosmochim. Acta **55**, 905–910 (1991)
S. Pizzarello, Y. Huang, L. Becker et al., The organic content of the Tagish Lake meteorite. Science **293**, 2236–2239 (2001)
S. Pizzarello, M. Zolensky, K.A. Turk, Nonracemic isovaline in the Murchison meteorite: chiral distribution and mineral association. Geochim. Cosmochim. Acta **67**, 1589–1595 (2003)
S. Pizzarello, Y. Huang, M. Fuller, The carbon isotopic distribution of Murchison amino acids. Geochim. Cosmochim. Acta **68**, 4963–4969 (2004)
S. Pizzarello, G.W. Cooper, G.J. Flynn, The nature and distribution of the organic material in carbonaceous chondrites and interplanetary dust particles, in *Meteorites and the Early Solar System II*, ed. by D.S. Lauretta, H.Y. McSween Jr. (University of Arizona Press, Tucson, 2006), pp. 625–651
S. Pizzarello, Y. Wang, G.M. Chaban, A comparative study of the hydroxy acids from the Murchison, GRA 95229 and LAP 02342 meteorites. Geochim. Cosmochim. Acta **74**, 6206–6217 (2010)
F. Postberg, S. Kempf, J.K. Hillier et al., The E-ring in the vicinity of Enceladus: II. Probing the moon's interior—the composition of E-ring particles. Icarus **193**, 438–454 (2008)
F. Postberg, S. Kempf, J. Schmidt et al., Sodium salts in E-ring ice grains from an ocean below the surface of Enceladus. Nature **459**, 1098 (2009)
E. Quirico, L. Bonal, Organic matter in interplanetary dusts and meteorites, in *Biosignatures for Astrobiology. Advances in Astrobiology and Biogeophysics*, ed. by B. Cavalazzi, F. Westall (Springer, Cham, 2019), pp. 23–50
E. Quirico, P.-I. Raynal, M. Bourot-Denise, Metamorphic grade of organic matter in six unequilibrated ordinary chondrites. Meteorit. Planet. Sci. **38**, 795–811 (2003)
E. Quirico, L. Bonal, P. Beck et al., Prevalence and nature of heating processes in CM and C2-ungrouped chondrites as revealed by insoluble organic matter. Geochim. Cosmochim. Acta **241**, 17–37 (2018)
M.A. Ratcliff, E.E. Medley, P.G. Simmonds, Pyrolysis of amino acids. Mechanistic considerations. J. Org. Chem. **39**, 1481–1490 (1974)
L. Remusat, S. Derenne, F. Robert, New insight on aliphatic linkages in the macromolecular organic fraction of Orgueil and Murchison meteorites through ruthenium tetroxide oxidation. Geochim. Cosmochim. Acta **69**, 4377–4386 (2005)

A.S. Rivkin, J.P. Emery, Detection of ice and organics on an asteroidal surface. Nature **464**, 1322–1323 (2010)
A.S. Rivkin et al., Hydrated minerals on asteroids: the astronomical record. Asteroids III **1**, 235–253 (2002)
F. Robert, The D/H ratio in chondrites. Space Sci. Rev. **106**, 87–101 (2003)
F. Robert, S. Epstein, The concentration and isotopic composition of hydrogen, carbon and nitrogen in carbonaceous meteorites. Geochim. Cosmochim. Acta **46**, 81–95 (1982)
A. Rotundi, G.A. Baratta, J. Borg et al., Combined micro-Raman, micro-infrared, and field emission scanning electron microscope analyses of comet 81P/Wild 2 particles collected by Stardust. Meteorit. Planet. Sci. **43**, 367–397 (2008)
S.A. Sandford, M.P. Bernstein, J.P. Dworkin, Assessment of the interstellar processes leading to deuterium enrichment in meteoritic organics. Meteorit. Planet. Sci. **36**, 1117–1133 (2001)
S.A. Sandford et al., Organics captured from Comet 81p/Wild 2 by the stardust spacecraft. Science **314**, 1720–1724 (2006)
P. Schmitt-Kopplin, Z. Gabelica, R.D. Gougeon et al., High molecular diversity of extraterrestrial organic matter in Murchison meteorite revealed 40 years after its fall. Proc. Natl. Acad. Sci. USA **107**, 2763–2768 (2010)
M.A. Sephton, Organic compounds in carbonaceous meteorites. Natl. Prod. Rep. **19**, 292–311 (2002)
M.A. Sephton, A.B. Verchovsky, P.A. Bland et al., Investigating the variations in carbon and nitrogen isotopes in carbonaceous chondrites. Geochim. Cosmochim. Acta **67**, 2093–2108 (2003)
M.A. Sephton, G.D. Love, J.S. Watson et al., Hydropyrolysis of insoluble carbonaceous matter in the Murchison meteorite: new insights into its macromolecular structure. Geochim. Cosmochim. Acta **68**, 1385–1393 (2004)
A. Shimoyama, H. Naraoka, N. Komiya et al., Analyses of carboxylic acids and hydrocarbons in Antarctic carbonaceous chondrites, Yamato-74662 and Yamato-793321. Geochem. J. **23**, 181–193 (1989)
A. Shimoyama, S. Hagishita, K. Harada et al., Search for nucleic acid bases in carbonaceous chondrites from Antarctica. Geochem. J. **24**, 343–348 (1990)
J.W. Smith, I.R. Kaplan, Endogenous carbon in carbonaceous meteorites. Science **167**, 1367–1370 (1970)
N.A. Starkey, I.A. Franchi, C.M.OD. Alexander, A Raman spectroscopic study of organic matter in interplanetary dust particles and meteorites using multiple wavelength laser excitation. Meteorit. Planet. Sci. **48**, 1800–1822 (2013)
N.A. Starkey, I.A. Franchi, M.R. Lee, Isotopic diversity in interplanetary dust particles and preservation of extreme ^{16}O-depletion. Geochim. Cosmochim. Acta **142**, 115–131 (2014)
S. Sugita et al., The geomorphology, color, and thermal properties of Ryugu: implications for parent-body processes. Science **364**, 252 (2019)
D. Takir, J.P. Emery, H.Y. McSween et al., Nature and degree of aqueous alteration in CM and CI carbonaceous chondrites. Meteorit. Planet. Sci. **48**, 1618–1637 (2013)
R. Terzieva, E. Herbst, The possibility of nitrogen isotopic fractionation in interstellar clouds. Mon. Not. R. Astron. Soc. **317**, 563–568 (2000)
K.L. Thomas, G.E. Blanford, L.P. Keller et al., Carbon abundance and silicate mineralogy of anhydrous interplanetary dust particles. Geochim. Cosmochim. Acta **57**, 1551–1566 (1993)
A.G.G.M. Tielens, Surface chemistry of deuterated molecules. Astron. Astrophys. **119**, 177–184 (1983)
J.M. Trigo-Rodríguez, C.E. Moyano-Cambero, J. Llorca et al., UV to far-IR reflectance spectra of carbonaceous chondrites—I. implications for remote characterization of dark primitive asteroids targeted by sample-return missions. Mon. Not. R. Astron. Soc. **437**, 227–240 (2014)
A. Tsuchiyama, M. Uesugi, T. Matsushima et al., Three-dimensional structure of Hayabusa samples: origin and evolution of Itokawa regolith. Science **333**, 1125–1128 (2011)
M.J. Verdier-Paoletti et al., Oxygen isotope constraints on the alteration temperatures of CM chondrites. Earth Planet. Sci. Lett. **458**, 273 (2016)
P. Vernazza, B. Zanda, T. Nakamura et al., The formation and evolution of ordinary chondrite parent bodies, in *Asteroids IV*, ed. by P. Michel, F.E. DeMeo, W.F. Bottke (University of Arizona Press, Tucson, 2015a), pp. 617–634
P. Vernazza, M. Marsset, P. Beck et al., Interplanetary dust particles as samples of icy asteroids. Astrophys. J. **806**, 204–2013 (2015b)
V. Vinogradoff, C. Le Guillou, S. Bernard et al., Paris vs. Murchison: impact of hydrothermal alteration on organic matter in CM chondrites. Geochim. Cosmochim. Acta **212**, 234–252 (2017)
S. Watanabe, M. Hirabayashi, N. Hirata et al., Hayabusa2 arrives at the carbonaceous asteroid 162173 Ryugu – a spinning top-shaped rubble pile. Science **364**, 268–272 (2019)
M.K. Weisberg, T.J. McCoy, A.N. Krot, Systematics and evaluation of meteorite classification, in *Meteorites and the Early Solar System II*, ed. by D.S. Lauretta, H.Y. McSween (University of Arizona Press, Tucson, 2006), pp. 19–52

K.C. Welten, K. Nishiizumi, M.W. Caffee, D.J. Hillegonds, J.A. Johnson, A.J.T. Jull, R. Wieler, L. Folco, Terrestrial ages, pairing, and concentration mechanism of Antarctic chondrites from Frontier Mountain, Northern Victoria Land. Meteorit. Planet. Sci. **41**, 1081–1094 (2006)

B. Wopenka, Raman observations on individual interplanetary dust particles. Earth Planet. Sci. Lett. **88**, 221–231 (1988)

I.P. Wright, S. Sheridan, S.J. Barber et al., CHO-bearing organic compounds at the surface of 67P/Churyumov-Gerasimenko revealed by Ptolemy. Science **349**, aab0673 (2015)

H. Yabuta, Solar system exploration: small bodies and their chemical and physical conditions, in *Handbook of Astrobiology*, ed. by V.M. Kolb (CRC Press, Boca Raton, 2019), pp. 23–42

H. Yabuta, H. Naraoka, K. Sakanishi et al., Solid-state ^{13}C NMR characterization of insoluble organic matter from Antarctic CM2 chondrites: evaluation of the meteoritic alteration level. Meteorit. Planet. Sci. **40**, 779–787 (2005)

H. Yabuta, C.M.OD. Alexander, M.L. Fogel et al., A molecular and isotopic study of the macromolecular organic matter of the ungrouped C2 WIS 91600 and its relationship to Tagish Lake and PCA 91008. Meteorit. Planet. Sci. **45**, 1446–1460 (2010)

H. Yabuta, T. Noguchi, S. Itoh et al., Formation of an ultracarbonaceous Antarctic micrometeorite through minimal aqueous alteration in a small porous icy body. Geochim. Cosmochim. Acta **214**, 172–190 (2017)

J. Yang, S. Epstein, Interstellar organic matter in meteorites. Geochim. Cosmochim. Acta **47**, 2199–2216 (1983)

J. Yang, S. Epstein, Relic interstellar grains in Murchison meteorite. Nature **311**, 544–547 (1984)

G. Yuen, N. Blair, D.J. Des Marais et al., Carbon isotope composition of low molecular weight hydrocarbons and monocarboxylic acids from Murchison meteorite. Nature **307**, 252–254 (1984)

H. Yurimoto, K. Abe, M. Abe et al., Oxygen isotopic compositions of asteroidal materials returned from Itokawa by the Hayabusa mission. Science **333**, 1116–1118 (2011)

E. Zinner, S. Amari, B. Wopenka et al., Interstellar graphite in meteorites: isotopic compositions and structural properties of single graphite grains from Murchison. Meteoritics **30**, 209 (1995)

M.E. Zolensky, W.L. Bourcier, J.L. Gooding, Aqueous alteration on the hydrous asteroids: results of Eq. 3/6 computer simulations. Icarus **78**, 411–425 (1989)

Space Sci Rev (2020) 216:49
https://doi.org/10.1007/s11214-020-00668-9

The Importance of Phobos Sample Return for Understanding the Mars-Moon System

Tomohiro Usui[1,2] · Ken-ichi Bajo[3] · Wataru Fujiya[4] · Yoshihiro Furukawa[5] · Mizuho Koike[1] · Yayoi N. Miura[6] · Haruna Sugahara[1] · Shogo Tachibana[1,7] · Yoshinori Takano[8] · Kiyoshi Kuramoto[1,3]

Received: 16 September 2019 / Accepted: 3 April 2020 / Published online: 28 April 2020

Abstract Phobos and Deimos occupy unique positions both scientifically and programmatically on the road to the exploration of the solar system. Japan Aerospace Exploration Agency (JAXA) plans a Phobos sample return mission (MMX: Martian Moons eXploration). The MMX spacecraft is scheduled to be launched in 2024, orbit both Phobos and Deimos (multiple flybys), and retrieve and return >10 g of Phobos regolith back to Earth in 2029. The Phobos regolith represents a mixture of endogenous Phobos building blocks and exogenous materials that contain solar system projectiles (e.g., interplanetary dust particles and coarser materials) and ejecta from Mars and Deimos. Under the condition that the representativeness of the sampling site(s) is guaranteed by remote sensing observations in the geologic context of Phobos, laboratory analysis (e.g., mineralogy, bulk composition, O-Cr-Ti isotopic systematics, and radiometric dating) of the returned sample will provide crucial information

Role of Sample Return in Addressing Major Questions in Planetary Sciences
Edited by Mahesh Anand, Sara Russell, Yangting Lin, Meenakshi Wadhwa, Kuljeet Kaur Marhas and Shogo Tachibana

✉ T. Usui
usui.tomohiro@jaxa.jp

1 Institute of Space and Astronautical Science, JAXA, 3-1-1 Yoshinodai, Sagamihara, Kanagawa 252-5210, Japan

2 Earth-Life Science Institute, Tokyo Institute of Technology, 2-12-1 Ookayama, Meguro, Tokyo 152-8550, Japan

3 Department of Earth and Planetary Sciences, Faculty of Science, Hokkaido University, N10W8 Kita-ku, Sapporo 060-0810, Japan

4 Ibaraki University, 2-1-1 Bunkyo, Mito, Ibaraki 310-8512, Japan

5 Department of Earth Science, Tohoku University, 6-3 Aza-aoba, Aramaki, Aoba-ku, Sendai 980-8578, Japan

6 Earthquake Research Institute, University of Tokyo, 1-1-1 Yayoi, Bunkyo-ku, Tokyo 113-0032, Japan

7 UTOPS, University of Tokyo, 7-3-1 Hongo, Bunkyo-ku, Tokyo 113-0033, Japan

8 Biogeochemistry Research Center, Japan Agency for Marine-Earth Science and Technology, 2-15 Natsushima, Yokosuka 237-0061, Japan

about the moon's origin: capture of an asteroid or in-situ formation by a giant impact. If Phobos proves to be a captured object, isotopic compositions of volatile elements (e.g., D/H, $^{13}C/^{12}C$, $^{15}N/^{14}N$) in inorganic and organic materials will shed light on both organic-mineral-water/ice interactions in a primitive rocky body originally formed in the outer solar system and the delivery process of water and organics into the inner rocky planets.

Keywords Phobos sample return · Origin of Martian moons · Martian Moons eXploration · Endogenous Phobos material · Mars-originating materials

1 Introduction

The Mars-moon system is a most promising target for exploration in the broad field of planetary science: e.g., geology, geochemistry, astrobiology, and comparative planetology. The international Mars science community has set a long-term goal toward achieving Mars sample return (MSR) and human exploration (Banfield 2018). MSR will address questions on the origin (e.g., source of building blocks), interior evolution (e.g., crust-mantle differentiation and mantle dynamics), the potential biological history (e.g., nature and extent of habitability and preservation of biosignature), and geologic history of Mars with an emphasis on the role of water (Beaty et al. 2019). Previous Mars exploration missions have provided compelling evidence for the presence of liquid water during the earliest geologic era (Noachian: $>\sim 3.9$ Ga) of Mars. Observations of dense valley networks, evaporites (e.g., gypsum) and hydrous minerals (e.g., clays) that are commonly formed by aqueous processes imply that, at some time in the past, Mars had an Earth-like active hydrological cycle with lakes, oceans, possibly also aquifers and groundwater, and had the potential for life (Carr 2006; Ehlmann et al. 2016; Usui 2019).

Along with the geological similarities, Earth and Mars are the only terrestrial planets in the inner solar system that have moons. The moons are witness plates for planet formation processes. Lunar exploration and studies of lunar samples (Apollo and meteorites) in the lab provided significant knowledge of the formation and co-evolution of the Earth-Moon system: e.g., the giant impact and magma ocean hypotheses for terrestrial planet formation, and the late veneer hypothesis for the delivery of water into the inner solar system (e.g., Taylor et al. 2006; Day et al. 2016). These hypotheses may also be tested for the Mars system by virtue of studying Phobos and Deimos. However, our understanding of Phobos and Deimos is sparse mainly due to the complete lack of successful missions dedicated to the exploration of Phobos/Deimos as well as the lack of meteorites from these bodies. The Soviet Union's Phobos 2 spacecraft acquired television images and infrared spectra at $\sim$700 m/pixel resolution (Avanesov et al. 1989; Bibring et al. 1990). Subsequent observations have been made by Earth-based telescopes and Mars orbiters/landers reporting a surface spectral similarity to D-type asteroid (e.g.,Murchie and Erard 1996; Simonelli et al. 1998; Cantor et al. 1999; Thomas et al. 2011; Fraeman et al. 2012; Fraeman et al. 2014). Therefore, Phobos and Deimos occupy unique positions both scientifically and programmatically on the road to the exploration of the solar system (Murchie et al. 2014).

The Japan Aerospace Exploration Agency (JAXA) plans a Martian moons' sample return mission (MMX: Martian Moons eXploration) (Kuramoto et al. 2018; Usui et al. 2018); MMX is the 3rd Japanese sample return mission, following Hayabusa (Fujiwara et al. 2006) and Hayabusa2 (Watanabe et al. 2019). The origin(s) of Phobos and Deimos is still a matter of significant debate, with the most likely candidates being: 1) the capture of asteroids (e.g.,

Table 1 Expected characteristics of endogenous returned samples

	Moon origin			
	capture of asteroid		In-situ formation	
	Outer solar system body	Inner solar system body	Co-accretion	Giant impact
Petrology	Analogous to carbonaceous chondrite, IDP, or cometary material	Analogous to ordinary chondrite	?	Glassy or recrystallized igneous texture
Mineralogy	Rich in oxidized and hydrous alteration phases (e.g., phyllosilicate, carbonates), amorphous silicate	Reduced and mostly anhydrous phases (e.g., pyroxene, olivine, metal, sulfides)	Un-equilibrated mixture of chondritic minerals?	High-T igneous phases (e.g., pyroxene, olivine), Martian crustal (evolved igneous) & mantle (high-P) phases
Bulk chemistry	Chondritic, volatile rich (e.g., high C and high H)	Chondritic, volatile poor	Chondritic (= ∼ bulk Mars?) with nebula-derived volatile?	Mixture of Martian crustal (mafic) and mantle-like (ultramafic) composition possibly with impactor material (high HSE?). Degree of volatile depletion varies due to impact regime
Isotopes	Carbonaceous chondrite signature (e.g., $\Delta^{17}O$, $\varepsilon^{54}Cr$, $\varepsilon^{50}Ti$, εMo, noble gases), primitive solar-system volatile signature (e.g., D/H, $^{15}N/^{14}N$)	Non-carbonaceous chondrite signature (e.g., $\Delta^{17}O$, $\varepsilon^{54}Cr$, $\varepsilon^{50}Ti$, εMo, noble gases), primitive (e.g., chondritic D/H, $^{15}N/^{14}N$)?	Bulk-Mars (?) signature (e.g., $\Delta^{17}O$, $\varepsilon^{54}Cr$, $\varepsilon^{50}Ti$, εMo), planetary volatile (e.g., intermediate D/H, low $^{15}N/^{14}N$)?	Mixture of Martian and impactor (carbonaceous or non-carbonaceous) composition, highly mass fractionated planetary volatile (e.g., low D/H, low $^{15}N/^{14}N$)?
Organics	Primitive organic matter, volatile & semi-volatile organics, soluble organics?	Non-carbonaceous signature ?	?	?

Hartmann 1990; Higuchi et al. 2017) or 2) in-situ formation by either co-accretion (Safronov et al. 1986) or by a giant impact (e.g., Rosenblatt et al. 2016; Hesselbrock and Minton 2017; Hyodo et al. 2017) on Mars (Table 1). In either case, samples from a Martian moon returned by MMX will provide the necessary ground truth to test these theories and to offer an opportunity to directly explore the building blocks or juvenile crust/mantle components of Mars. This new knowledge of Phobos/Deimos and Mars will be further leveraged by constraining the initial condition of the Mars-moon system and has potential for offering vital insights regarding the sources and delivery process of volatiles including water and organics to the inner rocky planets. This paper describes the science goals, mission profile and science payload of the MMX mission (Sect. 2), summarizes the expected characteristics of the returned samples and the prospective scientific outcome from the laboratory analysis (Sects. 3-5), and describes an assessment of contamination control issues (Sect. 6).

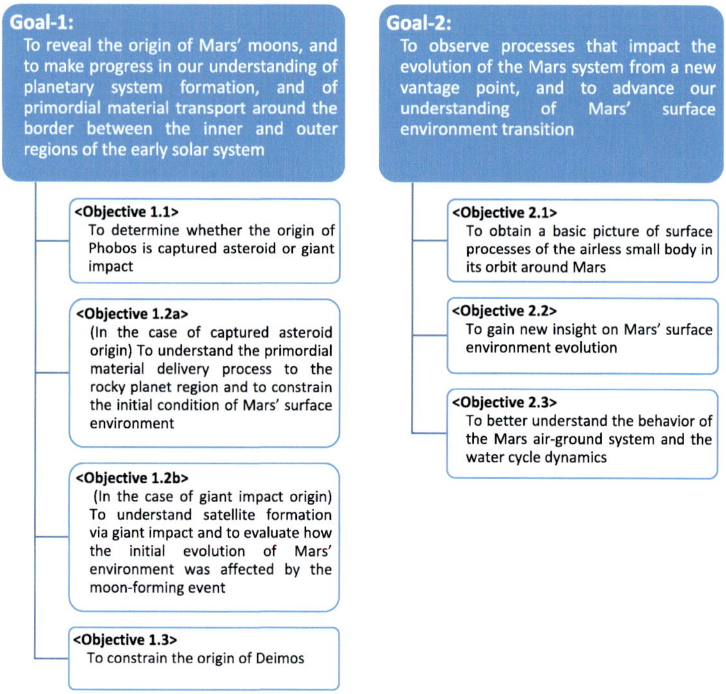

Fig. 1 Mission goals and objectives of MMX

2 Mission Overview

2.1 Mission Goals and Objectives

MMX has two main science goals: Goal-1) to reveal the origin of the Martian moons and make progress in the understanding of planetary system formation and material transport in the solar system, and Goal-2) to observe processes that impact the circumplanetary and surface environments of Mars (Fig. 1). These mission goals lead to the five mission objectives (MO): MO-1.1) to determine whether the origin of Phobos was by capture of an asteroid or by in-situ formation following a giant impact on Mars; MO-1.2) to understand material transport within the Mars-system during the moon formation stage; MO-1.3) to constrain the origin of Deimos; MO-2.1) to obtain a basic picture of surface processes acting on airless small bodies in orbit around Mars; MO-2.2) to gain critical new insight on Mars surface environment evolution; and MO2.3) to better understand the behavior of the Mars atmosphere-surface system and water cycle dynamics. To accomplish the science goals and objectives, the MMX team has defined a mission profile, has selected a comprehensive suite of remote-sensing and in-situ instruments, designed the Phobos regolith sampling system, and will conduct laboratory analyses of returned samples. Analyses of endogenous Phobos constituent materials and exogeneous late accreted materials would specifically achieve the mission goals-1 and -2, respectively (Sect. 3).

2.2 Mission Profile and Science Payload

The MMX spacecraft is scheduled to be launched in 2024, orbit Phobos and Deimos, perform multiple flybys, and retrieve and return >10 g of Phobos regolith back to Earth in 2029 (Kawakatsu et al. 2017). The spacecraft will consist of propulsion, exploration, and return modules (total launch mass = ~4,000 kg). The chemical propulsion system is utilized for Mars orbit injection and an escape maneuver. The outward interplanetary flights will take ~1 year by the most efficient Hohmann-like transfer. The spacecraft will remain in circum-Mars orbits for approximately 3 years for survey and exploration, followed by the ~1 year homeward interplanetary journey to Earth. The Phobos exploration phase will include multiple landing/sampling operations, each taking ~2.5 hours. To prevent sample contamination from thruster plume products and regolith particles that could be ejected from the surface during a thrust descent, the spacecraft will employ ballistic descent to reach a point directly above a landing site before a final free-fall descent without a thruster jet (Kawakatsu et al. 2017; Kawakatsu et al. 2019).

Seven nominal science instruments comprise several payloads for the remote sensing observations (Kuramoto et al. 2018): 1) a narrow-angle camera (TElescopic Nadir imager for GeOmOrphology, TENGOO) (JAXA) (Kameda et al. 2019), 2) a wide-angle multi-spectral camera (Optical RadiOmeter composed of CHromatic Imagers, OROCHI) (JAXA) (Kameda et al. 2019), 3) a near-infrared spectrometer (0.9-3.6 μm) MMX Infrared Spectrometer, MIRS (CNES), 4) a light detection and ranging instrument (LIDAR) (JAXA), 5) a dust instrument for particle sizes >10 μm (Circum-Martian Dust Monitor, CMDM) (JAXA) (Kobayashi et al. 2018), 6) a gamma-ray and neutron instrument (Mars-moon Exploration with GAmma rays and Neutrons, MEGANE) (NASA) (Lawrence et al. 2019), and 7) an ion mass spectrometer (Mass Spectrum Analyzer, MSA) (JAXA). Along with these seven instruments, the MMX spacecraft will carry and deploy a Rover (CNES/DLR) for in-situ surface exploration (Ulamec et al. 2019). The rover will carry several instruments for in-situ observations, including a Raman spectrometer, radiometer as well as navigation and wheel cameras.

2.3 Sampling System

Samples collected by MMX are expected to represent a mixture of endogenous and exogenous materials in Phobos' regolith. The former represents Phobos' building blocks that record information of the moon's origin, while the latter is expected to contain solar system projectiles and ejecta derived from Mars and Deimos (Ramsley and Head 2013; Nayak et al. 2016; Hyodo et al. 2019). Although the depth profile of Phobos regolith regarding material distribution is unknown, the ratio of [exogenous /endogenous] abundances is expected to be highest at the top-most regolith layer, which is where sampling will occur.

MMX plans to employ a double sampling approach: (C) coring and (P) pneumatic. The C-sampler has more than two core soil tubes deployed by a robotic arm; each core has a volume large enough to collect >10 g regolith. The cores would provide access to material beneath the surface (>2 cm) but would collect a mixture of surface and sub-surface materials. The P-sampler, on the other hand, selectively samples the surface veneer and provides a reference of surface components for the C-sampler. The P-sampler will also increase the chance of retrieving invaluable Martian and Deimos materials. Each core sample retrieved by the C-sampler and surface materials collected by the P-sampler will be separately stored in different canisters and returned to Earth. Thus, the C- and P-samplers would constitute complementary approaches to addressing the MMX mission goals.

The double sampling system not only enhances the scientific merits of MMX, but also reduces risks associated with the coring system. The nominal landing operation will execute both C- and P-sampling at each landing site. However, lacking knowledge of physical and chemical properties and conditions of the surface of Phobos (e.g., compositions, temperature gradient/variation, porosity, grain size distribution), we will prepare for scenarios in which the C-sampler cannot penetrate deep enough into a thin regolith layer covering a rigid basement and/or that it cannot be extracted once it has penetrated. Under any conceivable surface conditions, the P-sampler will work effectively and independently to collect fine-grained regolith particles.

3 Expected Characteristics of Returned Samples

3.1 Phobos' Original Building Blocks

3.1.1 A Case of the Captured Asteroid Hypothesis

The characteristics of the returned endogenous samples depend on Phobos' origin (Table 1). In the case of captured asteroid origin (e.g., Hartmann 1990; Higuchi et al. 2017), the returned samples would be analogous to a certain type of chondrites, IDPs (interplanetary dust particles), or even comets, depending on where these moons originally formed in the early solar system. If they formed in the outer solar system (beyond the snow line), they could potentially contain abundant hydrous secondary phases and organic molecules. Such phases could have formed by water-rock-organic interaction under low-temperature conditions in the parent asteroid of Phobos, represented by H-, C-, and N-rich bulk chemistry (Alexander et al. 2012; Krot et al. 2015). Outer solar system formation could be also indicated by crystalline/amorphous silicate dust mixtures with a solar nebular origin, as found with the comet Wild-2 samples returned by the Stardust spacecraft (Brownlee et al. 2006). On the other hand, if the Martian moons formed in the inner solar system (inside of the snow line), they probably consist mostly of anhydrous phases with lower bulk volatile contents and characteristic isotopic differences. These two extreme cases for the captured model may be tested on the basis of the heliocentric gradients of volatile isotopes and abundances (e.g., CO_2/H_2O, D/H, $^{15}N/^{14}N$ and noble gases) and the isotopes of rock-forming elements of O, Cr, and Ti in the solar system (Figs. 2, 3, and 4) (Trinquier et al. 2007; Trinquier et al. 2009; Qin et al. 2010; Warren 2011; Marty 2012; Fujiya et al. 2019).

3.1.2 A Case of In-Situ Formation

If Phobos and Deimos formed in-situ by a giant impact (like the Earth's moon) (e.g., Rosenblatt et al. 2016; Hesselbrock and Minton 2017; Hyodo et al. 2017), the returned samples would be characterized by high-temperature and possibly high-pressure shocked phases and/or glassy/recrystallized phases. Due to the high-temperature impact process (e.g., $\sim$2000 K, Hyodo et al. 2017), endogenous organic materials would be unlikely to be present to a significant degree in the regolith. The bulk chemistry could range from a mafic to ultramafic composition with high abundances of highly siderophile elements, possibly representing a mixture of Martian silicate portions (crust/mantle) and the impactor (Hyodo et al. 2017). The bulk-silicate Mars is characterized by elevated volatile (e.g., Na and K) and siderophile (e.g., Mn, Cr, and W) elements and depletions in chalcophile elements (e.g., Cu), relative to the bulk silicate Earth (Dreibus and Wanke 1985;

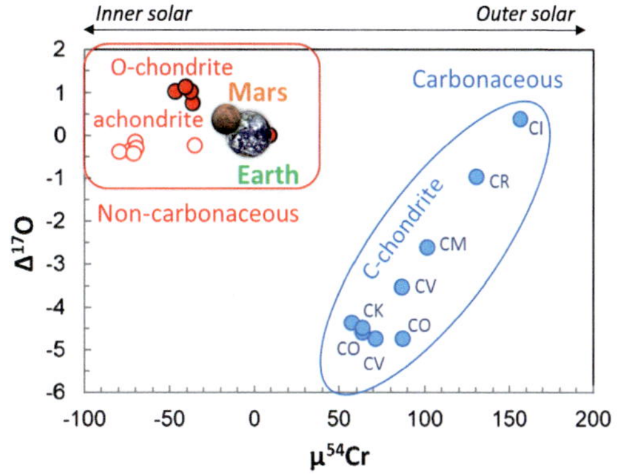

Fig. 2 $\mu^{54}Cr$ vs. $\Delta^{17}O$ isotope diagram of planetary materials. Data compilation by Ryota Fukai

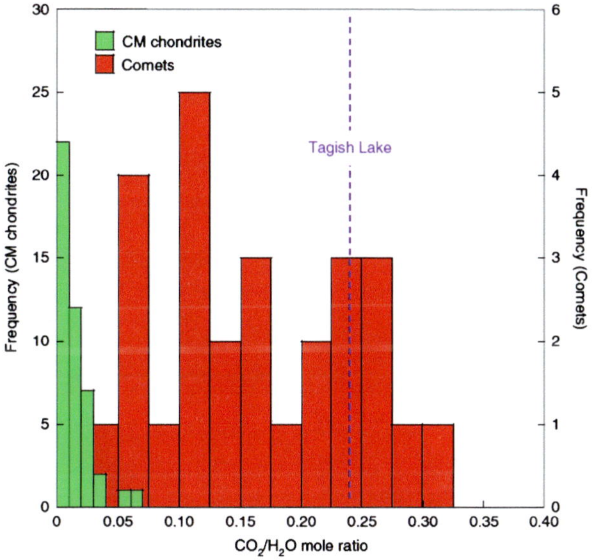

Fig. 3 Histogram of the CO_2/H_2O mole ratios of ice in CM chondrites and comets. The CO_2/H_2O ratios of ice in CM chondrites are calculated from carbonate abundances and hydrogen contents. Modified after Fujiya et al. (2019)

Taylor 2013). Such a volatile-rich nature relative to the Earth and Moon is evident in ratios of K/Th and K/U (volatile/refractory incompatible element) (Fig. 5) and the K/Th ratio of Phobos surface will be measured by MEGANE (Lawrence et al. 2019). On the other hand, in-situ formation by co-accretion with Mars would provide a bulk chemistry represented by the chondritic bulk Mars.

3.2 Late Accreted Materials

The regolith of Phobos is likely to contain ejecta derived from Mars and Deimos due to Phobos' location in Mars' orbit (Ramsley and Head 2013; Nayak et al. 2016; Hyodo et al. 2019). Though the exact nature of Deimos-originating material is unknown, the spectral similarity between Deimos and the dominant red unit of Phobos suggests that Deimos-originating materials resemble the endogenous Phobos materials (Fraeman et al. 2012); note that Phobos' surface shows two distinct units named "blue unit" and "red unit" after the relative slopes of their reflectance spectra (Murchie and Erard 1996). We may have diffi-

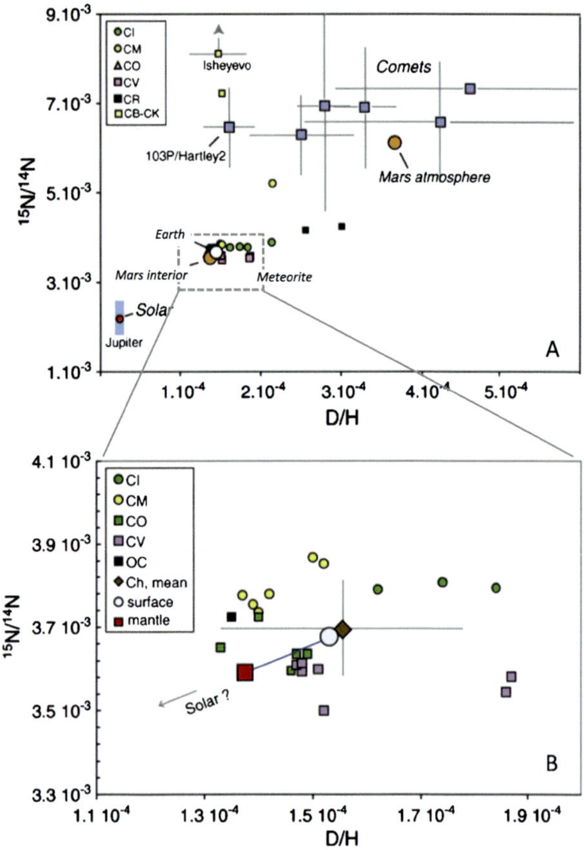

Fig. 4 D/H and $^{15}N/^{14}N$ ratios among solar system reservoirs and objects. The hatched area in A is enlarged B. Modified after Marty (2012)

culty in discriminating Deimos-originating materials from the Phobos regolith unless detailed remote sensing observations by MMX clarify the distinction between the two moons. On the other hand, Mars-originating materials would differ from the endogenous Phobos materials even if Phobos was formed by a giant impact. The endogenous Phobos material formed by a giant impact should have sampled both crust and mantle reservoirs of Mars (Hyodo et al. 2017), whereas the later Mars-originating materials would have mostly derived from the near-surface of Mars (Hyodo et al. 2019). The Martian near-surface materials have oxygen ($\Delta^{17}O$), hydrogen (D/H), and carbon ($\delta^{13}C$) isotopic compositions that are distinct from those of the Martian crust and mantle reservoirs (Leshin et al. 1996; Farquhar and Thiemens 2000; Usui et al. 2012; Agee et al. 2013; Webster et al. 2013; Usui et al. 2015).

The surface of Mars is dominated by a basaltic crust mainly composed of olivine, pyroxene, and plagioclase (McSween et al. 2009) (Fig. 6). Although these igneous phases are common in chondrites and achondrites, their chemistries are distinct (e.g., Fe/Mg and Fe/Mn ratios, and anorthite component) (Karner et al. 2003; Karner et al. 2006). The Martian crust also contains a variety of sedimentary rocks, including both clastic rocks (sandstone and siltstone, shale and mudstone, and conglomerate) and chemically deposited rocks (mostly evaporitic sulfate, and some carbonate) (McSween 2015). Sedimentary rocks in ancient terranes potentially contain organic compounds (Eigenbrode et al. 2018). Global mapping of sedimentary rocks suggests the secular desiccation and acidification of the near-surface en-

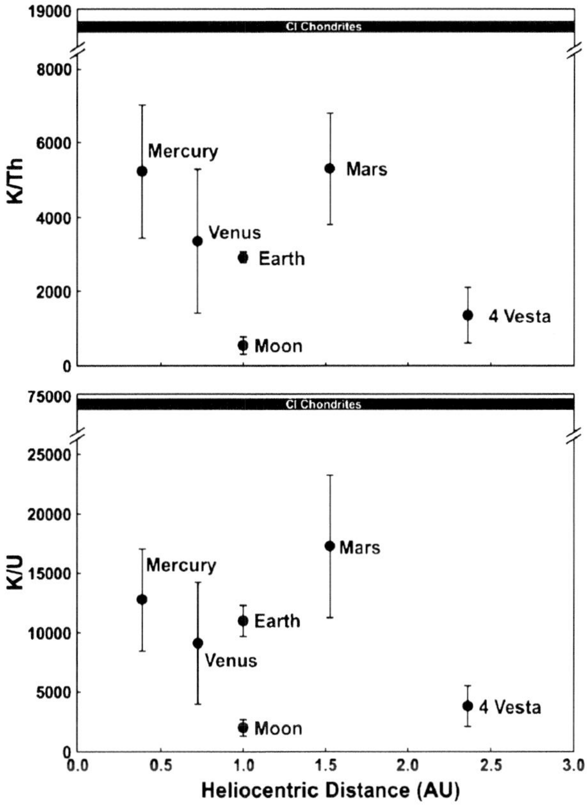

Fig. 5 K/Th and K/U ratios of terrestrial bodies within the inner Solar System plotted as a function of heliocentric distance (McCubbin et al. 2012)

vironment (Bibring et al. 2006; Ehlmann and Edwards 2014) (Fig. 7). The surface mineralogy of early Noachian terranes is characterized by clay minerals such as Fe/Mg smectites, suggesting water-rich and near-neutral fluid conditions. The occurrence and distribution of carbonates and sulfates in the late Noachian to early Hesperian suggest a transition of fluid chemistry to more acidic conditions in this period. The young Amazonian terranes are dominated by anhydrous ferric oxides such as hematite, suggestive of acidic and drier conditions. Meteorite impacts would have randomly sampled these Martian surficial materials, some of which would have subsequently been deposited on the surface of Phobos as late accreting materials in the regolith (Hyodo et al. 2019).

4 Key Analysis of Phobos Sample Return Mission

Compelling evidence for Phobos' origin will be provided by high precision isotopic analyses of lithophile elements. Stable isotopic systematics of O, Cr, Ti, and Mo clearly differentiate the carbonaceous and non-carbonaceous reservoirs (Fig. 2), which are proposed to be spatially distinguished either side of Jupiter's orbit (Trinquier et al. 2007; Trinquier et al. 2009; Qin et al. 2010; Warren 2011; Kruijer et al. 2017). A suite of these isotope analyses, except for Mo, can be carried out using a <100 mg fraction of the returned samples (Table 2). Secondary Ion Mass Spectrometry (SIMS) techniques further permit grain-by-grain characterization by high-spatial-resolution and high-precision O isotope analysis with a primary

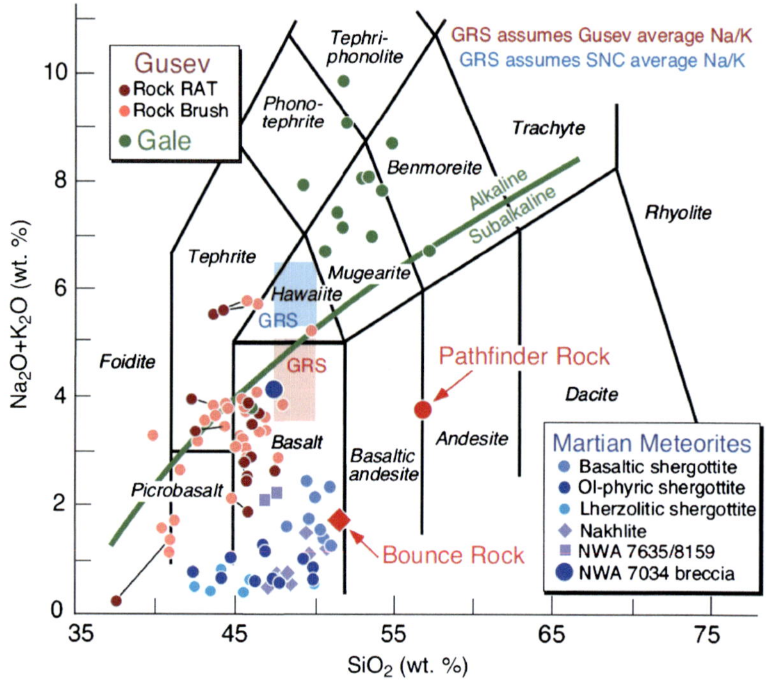

Fig. 6 Compositions of Martian igneous rocks (McSween 2015)

ion beam of ∼15 μm in size and analytical uncertainties of ∼0.5‰ for $\Delta^{17}O$ values (Kita et al. 2010; Yurimoto et al. 2011), potentially providing individual histories of each collected grain. Isotopic data will also be carefully examined alongisde petrographic and mineralogical observations to help discriminate the exogenous materials. For example, major element mineral chemistry of olivine and pyroxene, phases that are common in planetary materials, is distinct between asteroids (chondrite/achondrite parent bodies) and Mars (Brearley 1998; McSween and Treiman 1998; Mittlefehldt et al. 1998). Other lines of evidence for Phobos' origin would also come from the presence/lack of refractory inclusions, amorphous silicates, presolar grains, or organic materials with anomalous H, C, and N isotopic compositions.

Radiometric dating of the returned samples would provide chronological constraints on the origin of Phobos. Mineral isochrons of ^{87}Rb-^{87}Sr, ^{147,146}Sm-^{143,142}Nd, ^{176}Lu-^{176}Hf, and ^{238,235}U-^{206,207}Pb systems obtained by Thermal Ionization Mass Spectrometry (TIMS) and Inductively Coupled Plasma Mass Spectrometry (ICP-MS) could yield formation ages (e.g., timing of crystallization) of individual fragments in the returned regolith sample (e.g., Amelin et al. 2002; Borg et al. 2005; Lapen et al. 2010). Bulk isochron defined by co-genetic fragments in the regolith sample could yield a formation age of the source (e.g., timing of the giant impact). Chronological study of short-lived nuclides (e.g., ^{146}Sm-^{142}Nd, ^{182}Hf-^{182}W, ^{53}Mn-^{53}Cr) would provide more precise knowledge of formation ages (e.g., Kleine et al. 2002; Trinquier et al. 2008; Kruijer et al. 2014). The ^{40}K(^{39}Ar)-^{40}Ar age could represent the timing of a secondary impact on Phobos. Should a statistically meaningful number of ^{40}K(^{39}Ar)-^{40}Ar ages be obtained, they would provide the chronometric ground truth for the crater density distribution obtained by MMX's remote sensing observations. Furthermore, in-situ dating of secondary phases (e.g., ^{53}Mn-^{53}Cr age of carbonate) would provide the

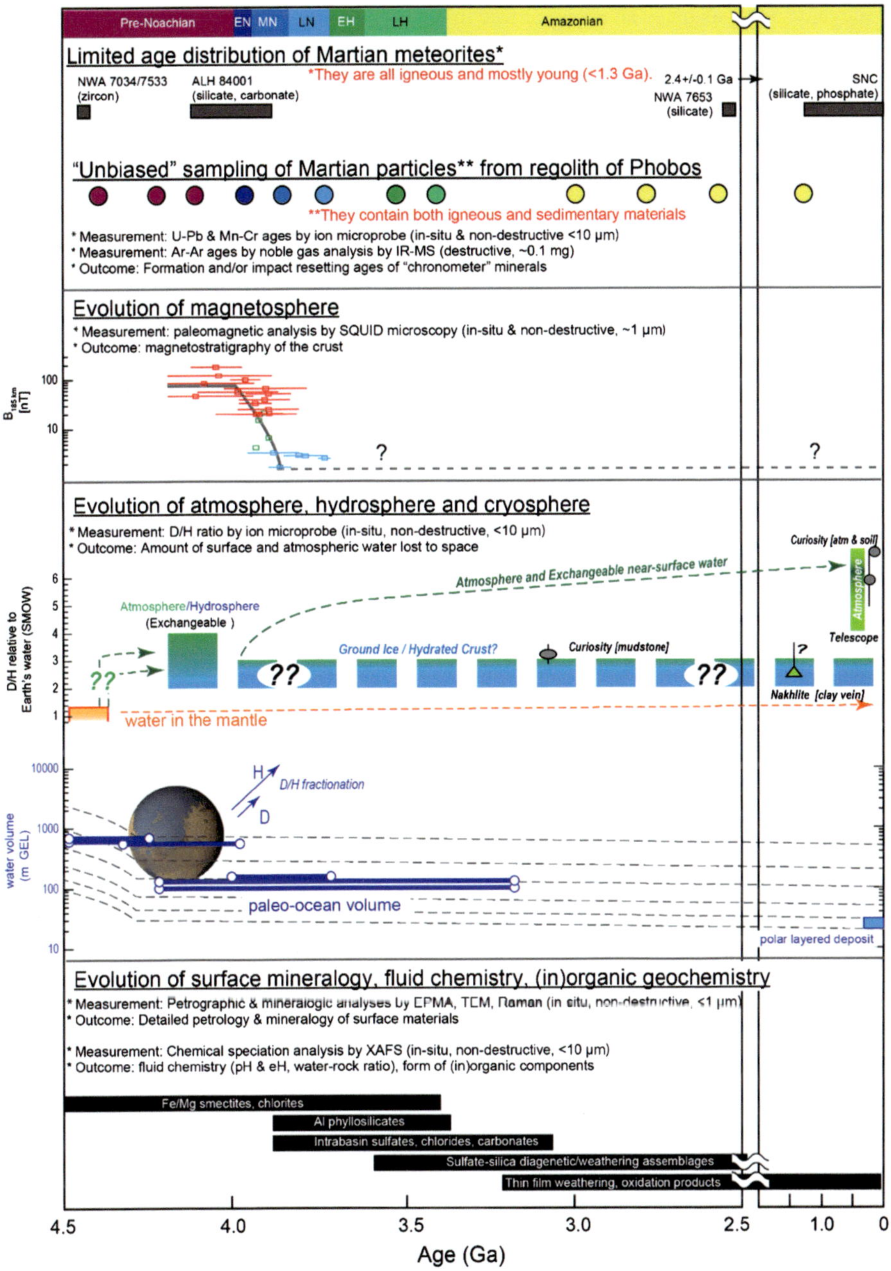

Fig. 7 Timeline of the evolution of magnetosphere, atmosphere, hydrosphere, cryosphere, and surface mineralogy of Mars. Age distribution of Martian meteorites are also plotted (Hyodo et al. 2019)

timing of secondary aqueous alteration events (Fujiya et al. 2012). Temperature of aqueous alteration would be estimated by C-O clumped isotope thermometry (e.g., Halevy et al. 2011).

Organic analysis plays an essential role in the Phobos sample return mission because the regolith of Phobos is expected to contain mature organics derived from the surface of Mars (Eigenbrode et al. 2018; Hyodo et al. 2019). If Phobos is a captured asteroid that formed in the outer solar system then the regolith could also contain a mixture of primitive compounds originating from its parent body (Table 1). In such a case, the MMX sample will expand knowledge about the primitive asteroids Ryugu and Bennu, the target bodies of Hayabusa 2 (Watanabe et al. 2019) and OSIRIS-REx (Lauretta et al. 2019) sample return missions, respectively. The MMX sample would be further important for organic astrochemistry, because Phobos has stayed in Mars' orbit ($\sim$1.5 AU), maintaining average surface temperatures lower than those of near-Earth asteroids Ryugu and Bennu.

Considering our experience with these sample return missions, Phobos sample return requires a comprehensive framework to maximize the scientific harvest through a seamless pathway from non-destructive to semi-destructive and destructive analysis (Table 2). In this workflow, bulk chemical analysis and molecular-based organic analysis will contribute to the careful documentation of volatile organic gas, soluble organic matter, insoluble organic matter, and remaining organic chemistry. For example, compound-specific isotope analyses of hydrogen (D/H), carbon ($^{13}C/^{12}C$), and nitrogen ($^{15}N/^{14}N$) by destructive (e.g., gas chromatography/isotope ratio mass spectrometry) and non-destructive (e.g., cavity ring-down spectroscopy) techniques would provide quantitative estimates of the low-temperature reactions between inorganic phases and organic molecules, many of which occur during aqueous alteration of anhydrous phases (Pizzarello et al. 2001; Kminek et al. 2002). The behavior of elements, isotopes, and molecules in micron-scale regions of interest can be measured by two dimensional-liquid chromatography/fluorescence detector system (2D-LC/FLD) and gas chromatography mass spectrometry (2D-GC/MS) in order to monitor these small-scale organic-mineral-water/ice interactions. These mass spectrometry analyses will complement spectroscopic analytical techniques of scanning transmission x-ray microscopy (STXM), and micro-Raman and infrared (IR) spectroscopy (e.g., Yabuta et al. 2014; Kitajima et al. 2015).

5 Sample Representativeness

To reveal the origin of the Martian moons (Goal-1, Fig. 1), one should evaluate the representativeness of the returned sample in the geologic context of Phobos understood from remote sensing observations; otherwise, the origin of individual grains analyzed by detailed laboratory analyses cannot directly reveal the moon's origin. The surface of Phobos is geologically diverse, which is evident in the spectral difference between the blue unit located around Stickney crater and the red unit found on the remainder of Phobos' surface (Murchie and Erard 1996). Basilevsky et al. (2014) suggest that the red and blue materials might have formed from large blocks originating in the interior of Phobos, although the actual cause of the spectral difference between the red and blue units is still unclear. A comprehensive and high spatial resolution analysis of reflectance spectra obtained by OROCHI and MIRS will document the distribution and mutual relations of the red and blue units in detail. Furthermore, subsurface (<1 m) chemical information of Phobos' surface obtained by MEGANE and in-situ surface observations to be provided by the rover in the vicinity of the landing site will establish the representativeness of the returned sample.

Table 2 List of key analysis for returned sample

Method	Data type	sample amount	Notes
Inorganic			
SIMS	O, H, (C, S, etc.) isotopes	~10 μm dia.	in-situ, destructive but limited to meas. spot
IRMS	O (C, S, N) isotopes	<1 mg	Bulk, destructive
TIMS, ICP-MS	Cr, Ti isotope	<10 mg	Bulk, destructive
TIMS, ICP-MS	Rb-Sr, Sm-Nd, Pb-Pb age	~10 mg	Bulk, destructive
SIMS	Mn-Cr age	~10 μm dia.	in-situ, destructive but limited to meas. spot
NG/IRMS	Ar-Ar age	<0.1 mg	Bulk, destructive, neutron irradiated
NG/IRMS	Noble gas isotopes	<1 mg	Bulk, destructive
PGA/NAA	Bulk chemistry	~1 g	Bulk, non-destructive
ICP-MS	Bulk chemistry	~10 mg	Bulk, destructive
EPMA/TEM/ Raman	Mineral chemistry, crystallography	<3 μm dia.	in-situ, non-destructive
XAFS	Chemical speciation	~1-10 μm dia.	in-situ, non-destructive
SQUID spectroscopy	Magnetic field	~1 μm dia.	in-situ, non-destructive
Organic			
CRDS	Individual gas molecules with H, C, N, O, (S) isotopes	~ 1 ppb	Molecular level, non-destructive
HS-GC/MS	Individual volatile molecules	< 1 mg	Molecular level, destructive
LC/MS	Individual molecules	< 10 mg	Molecular level, destructive
(2D)-LC/FLD	Individual amino acids with enantiomer ratio	< 1 mg	Molecular level (including D/L ratio), destructive
(2D)-GC/MS	Individual molecules	< 10 mg	Molecular level, destructive
GC/IRMS	Individual molecules with H, C, N, (S) isotopes	< 10 mg	Molecular level, destructive
DESI/MS	Molecular imaging	~5 μm dia.	in-situ, destructive but limited to meas. spot
MALDI/TOF-MS	Molecular imaging with accurate mass	~5 μm dia.	in-situ, destructive but limited to meas. spot
EA/IRMS	Bulk composition with H, C, N, O, (S) isotopes	< 1 mg	Bulk, destructive
XAFS	Chemical speciation of organic molecules	~1-10 μm dia.	in-situ, non-destructive
NMR	Molecular structure, chemical speciation	~ 1 mg	Bulk & Molecular level, non-destructive
FT-ICRMS	Comprehensive individual molecules	< 10 mg	Molecular level, destructive

Abbreviations: SIMS, Secondary Ion Mass Spectrometry; IRMS, Isotope Ratio Mass Spectrometry; TIMS, Thermal Ionization Mass Spectrometry; ICP-MS, Inductively Coupled Plasma Mass Spectrometry; NG, Noble Gass; EPMA, Electron Probe Micro Analyzer; TEM, Transmission Electron Microprobe; XAFS, X-ray Absorption Fine Structure; SQUID, Superconducting Quantum Interference Device; CRDS, Cavity Ring-Down spectroscopy; HS-GCMS, Head Space Gas Chromatography Mass Spectrometry; LC/MS, Liquid Chromatography Mass Spectrometry; (2D)-LC/FLD, 2 dimensional LC florescence detector; (2D)-GC/MS, 2 dimensional GC/MS; GC/IRMS, GC isotope ratio MS; DESI/MS, Desorption electrospray ionization MS; MALDI/TOF-MS, matrix-assisted laser desorption/ionization/ time-of-flight MS; EA/IRMS, Elemental analyzer/IRMS; NMR, Nuclear Magnetic Resonance. FT-ICRMS, Fourier transform ion cyclotron resonance mass spectrometry

6 Sample Assessment for Martian Moon Sample Return

Sample return missions offer sample analysis potential that is not otherwise achievable in-situ via the present generation of robotic planetary exploration missions. Nonetheless, preserving the integrity of returned samples to produce meaningful results requires that terrestrial contamination is understood, controlled, and minimized. Planetary protection regulations also serve to prevent the returned samples from contaminating the Earth. Thus, any sample return mission must take into account contamination assessment through design, manufacture, construction, testing, launch, flight and recovery.

MMX will be the first mission to return Phobos regolith to Earth. Sample return of Phobos regolith was classified as an unrestricted Earth-return mission by the Committee on Space Research (COSPAR) (Fujita et al. 2019; Kurosawa et al. 2019). Although MMX is unrestricted, the Phobos regolith is expected to contain a variety of volatile-bearing phases and organics, similar to the C-class asteroid target for the Hayabusa2 mission (Table 1). Thus, MMX has inherited the contamination assessment, control procedures, and practical knowledge of Hayabusa2 (Tachibana et al. 2014; Okazaki et al. 2017; Sawada et al. 2017).

Hayabusa2 is an unrestricted Earth-return mission from the carbonaceous asteroid 162173 Ryugu (Watanabe et al. 2019). The sample catcher and container of Hayabusa2 are designed to avoid terrestrial contamination of volatile and non-volatile organic and inorganic materials into the returned sample container (Okazaki et al. 2017; Sawada et al. 2017). To maintain the interior pressure and the chemical information of the initial gas components originating from the Ryugu sample, the maximum leak rate permitted to mitigate the atmospheric contamination is <1 Pa for 100 hours at one atmosphere. These conditions will preserve the isotope analyses of highly volatile elements such as H, N, and noble gases. For the assembly, test, and launch operations (ATLO) for Hayabusa2, spacecraft cleanliness was carefully monitored. Immediately upon return to Earth, the Hayabusa2 sample container will be brought to the curation facility at ISAS/JAXA for the recovery and initial classification of the sample (Tachibana et al. 2014). The assessment of the organic terrestrial contamination in the curation facility was conducted by analyzing amino acids as an indicator molecule collected on witness coupons ($\sim$ below the order of pico mole cm^{-2} day^{-1}) (Sugahara et al. 2018).

Each sample return mission has a different contamination control policy, as defined by the mission requirements. For MMX, accurate analyses of stable isotope ratios of O, Cr, and Ti in bulk samples are given the highest priority because these are the most important indicators of sample origin (Mission Goal-1) (Fig. 2). We must reduce possible contamination of these elements throughout the mission by <1/1000 of the expected amounts in the samples so that we can distinguish isotopic signatures characterized by individual meteorite groups. Moreover, Mars-originating exogenous materials in the Phobos regolith are expected to contain sedimentary phases that would be rich in volatile elements. Their pristine geochemical information, such as water contents of clays and isotopic compositions of volatiles (H, C, N, O, and S), would also provide important clues for understanding the co-evolution of circum-Mars environments (Mission Goal-2) (Fig. 7).

We will leverage previously acquired knowledge from Hayabusa2 to optimize MMX's ATLO and curation protocols for the Phobos sample return. A critical difference in the sample assessment between MMX and Hayabusa2 is the returned sample mass; the expected minimum sample amounts are 0.1 g for Hayabusa2 and 10 g for MMX. The larger sample amount for MMX should reduce the relative degree of terrestrial contamination because the "cleanliness" is defined by a relative fraction of foreign material to the returned sample in an amount that degrades the ability to analyze the chemistry and mineralogy of the sample. For example, NASA's OSIRIS-REx asteroid sample return mission set the contamination allowance level of the organic carbon contaminants to ±30% precision and accuracy on measurements based on the National Research Council's recommendation (Dworkin et al. 2018).

7 Conclusions

MMX has two mission goals: 1) to reveal the origin of the Martian moons and 2) to observe processes that impact on the evolution of the circum-Mars environment (Fig. 1). These mission goals will be achieved by comprehensive remote sensing observations of Phobos and Deimos and by laboratory analyses of a regolith sample returned from Phobos. Operation of C- and P-samplers will provide access to both Phobos' subsurface building blocks (> 2 cm deep) and the top-most regolith layer, resulting in a total of >10 g of Phobos regolith returned to Earth. The regolith of Phobos likely represents a mixture of endogenous Phobos materials and exogenous materials that accumulated after the moon's formation. Laboratory analyses of these two kinds of materials would allow successful achievement of the critical mission goals.

The characteristics of the endogenous Phobos materials depend on the moon's origin. In the case of the captured asteroid scenario, they would be analogous to a certain type of chondrites. On the other hand, if Phobos formed in-situ by a giant impact, the endogenous Phobos materials should contain high-temperature phases with the bulk chemistry possibly representing a mixture of Martian silicate reservoirs (crust and mantle) and the impactor. In addition to such contrasting characteristics of the endogenous Phobos materials, stable isotopic systematics of O, Cr, and Ti will further differentiate the source of Phobos' building blocks. The exogenous late accreted materials are expected to contain Mars-originating materials. Because the Mars-originating materials could be derived from several geologic units on the surface of Mars by random samplings through impacts, which could be potentially corroborated by radiometric dating of the returned samples, they would provide valuable information about the change in surface environments through the history of Mars.

Acknowledgements We are grateful to B. Marty, H.Y. McSween, and an anonymous reviewer for constructive reviews and M. Anand for editorial handling. We thank the MMX Science Board and the MMX study team for insightful discussions for this project. We also thank M. Zolensky, P. Mitchel, and N. Chabot for comments on the early version of this manuscript, and R. Fukai for compiling meteorite data used in Fig. 2. Our work was supported by ISAS/JAXA as a part of Phase-A activity of MMX and by a Grant-in-Aid from the Japan Society for the Promotion of Science to TU (17H06459, 15KK0153). TU gives special thanks to Europlanet for the opportunity to present our work at the ISSI workshop "Role of Sample Return in Addressing Major Outstanding Questions in Planetary Sciences".

References

C.B. Agee et al., Unique meteorite from early Amazonian Mars: water-rich basaltic breccia Northwest Africa 7034. Science **339**, 780–785 (2013)

C.M.O.D. Alexander et al., The provenances of asteroids, and their contributions to the volatile inventories of the terrestrial planets. Science **337**, 721–723 (2012)

Y. Amelin et al., Lead isotopic ages of chondrules and calcium-aluminium-rich inclusions. Science **297**, 1678–1683 (2002)

G.A. Avanesov et al., Television observations of Phobos. Nature **341**, 585–587 (1989)

D. Banfield, Mars Science Goals, Objectives, Investigations, and Priorities: 2018 Version, Mars Exploration Program Analysis Group (MEPAG) (2018). https://mepag.jpl.nasa.gov/reports/MEPAG%20Goals_Document_2018.pdf

A. Basilevsky et al., The surface geology and geomorphology of Phobos. Planet. Space Sci. **102**, 95–118 (2014)

D.W. Beaty et al., The potential science and engineering value of samples delivered to Earth by Mars sample return: international MSR Objectives and Samples Team (iMOST). Meteorit. Planet. Sci. **54**, S3–S152 (2019)

J.-P. Bibring et al., ISM observations of Mars and PHOBOS-first results, in *Lunar and Planetary Science Conference Proceedings*, vol. 20 (1990), pp. 461–471

J.P. Bibring et al., Global mineralogical and aqueous mars history derived from OMEGA/Mars express data. Science **312**, 400–404 (2006)

L.E. Borg et al., Constraints on the U-Pb isotopic systematics of Mars inferred from a combined U-Pb, Rb-Sr, and Sm-Nd isotopic study of the Martian meteorite Zagami. Geochim. Cosmochim. Acta **69**, 5819–5830 (2005)

A.J. Brearley, Chondritic meteorites, in *Planetary Materials* (1998)

D. Brownlee et al., Comet 81P/Wild 2 under a microscope. Science **314**, 1711–1716 (2006)

B.A. Cantor et al., Phobos disk-integrated photometry: 1994–1997 HST observations. Icarus **142**, 414–420 (1999)

M.H. Carr, *The Surface of Mars* (Cambridge University Press, Cambridge, 2006)

J.M. Day et al., Highly siderophile elements in Earth, Mars, the Moon, and asteroids. Rev. Mineral. Geochem. **81**, 161–238 (2016)

G. Dreibus, H. Wanke, Mars, a volatile-rich planet. Meteoritics **20**, 367–381 (1985)

J. Dworkin et al., OSIRIS-REx contamination control strategy and implementation. Space Sci. Rev. **214**, 19 (2018)

B.L. Ehlmann, C.S. Edwards, Mineralogy of the Martian surface. Annu. Rev. Earth Planet. Sci. **42**, 291–315 (2014)

B. Ehlmann et al., The sustainability of habitability on terrestrial planets: insights, questions, and needed measurements from Mars for understanding the evolution of Earth-like worlds. J. Geophys. Res., Planets **121**, 1927–1961 (2016)

J.L. Eigenbrode et al., Organic matter preserved in 3-billion-year-old mudstones at Gale crater, Mars. Science **360**, 1096–1101 (2018)

J. Farquhar, M.H. Thiemens, Oxygen cycle of the Martian atmosphere-regolith system: $\Delta^{17}O$ of secondary phases in Nakhla and Lafayette. J. Geophys. Res., Planets **105**, 11991–11997 (2000)

A. Fraeman et al., Analysis of disk-resolved OMEGA and CRISM spectral observations of Phobos and Deimos. J. Geophys. Res., Planets **117**, E00J15 (2012)

A. Fraeman et al., Spectral absorptions on Phobos and Deimos in the visible/near infrared wavelengths and their compositional constraints. Icarus **229**, 196–205 (2014)

K. Fujita et al., Assessment of the probability of microbial contamination for sample return from Martian moons I: departure of microbes from Martian surface. Life Sci. Space Res. **23**, 73–84 (2019)

A. Fujiwara et al., The rubble-pile asteroid Itokawa as observed by Hayabusa. Science **312**, 1330–1334 (2006)

W. Fujiya et al., Evidence for the late formation of hydrous asteroids from young meteoritic carbonates. Nat. Commun. **3**, 627 (2012)

W. Fujiya et al., Migration of D-type asteroids from the outer Solar System inferred from carbonate in meteorites. Nat. Astron. **3**, 910–915 (2019). https://doi.org/10.1038/s41550-019-0801-4

I. Halevy et al., Carbonates in the Martian meteorite Allan Hills 84001 formed at 18 ± 4 C in a near-surface aqueous environment. Proc. Natl. Acad. Sci. USA **108**, 16895–16899 (2011)

W.K. Hartmann, Additional evidence about an early intense flux of C asteroids and the origin of Phobos. Icarus **87**, 236–240 (1990)

A.J. Hesselbrock, D.A. Minton, An ongoing satellite–ring cycle of Mars and the origins of Phobos and Deimos. Nat. Geosci. **10**, 266 (2017)

A. Higuchi et al., Temporary capture of asteroids by an eccentric planet. Astrophys. J. **153**, 155 (2017)
R. Hyodo et al., On the impact origin of Phobos and Deimos. I. Thermodynamic and physical aspects. Astrophys. J. **845**, 125 (2017)
R. Hyodo et al., Transport of impact ejecta from Mars to its moons as a means to reveal Martian history. Sci. Rep. **9**, 19833 (2019)
S. Kameda et al., Telescopic Camera (TENGOO) and Wide-Angle Multiband Camera (OROCHI) Onboard Martian Moons eXploration (MMX) Spacecraft, in *The 50th Lunar and Planetary Science Conference* (2019). Abstract #2292
J. Karner et al., Olivine from planetary basalts: chemical signatures that indicate planetary parentage and those that record igneous setting and process. Am. Mineral. **88**, 806–816 (2003)
J. Karner et al., Comparative planetary mineralogy: pyroxene major- and minor-element chemistry and partitioning of vanadium between pyroxene and melt in planetary basalts. Am. Mineral. **91**, 1574–1582 (2006)
Y. Kawakatsu et al., Mission Concept of Martian moons exploration (MMX), in *Proceedings in the 68th International Astronautical Congress. IAC-17-A3.3A.5* (2017). http://www.iafastro.org/events/iac/iac-2017/
Y. Kawakatsu et al., Mission design of Martian Moons eXploration (MMX), in *70th International Astronautical Congress. IAC-18-A3.3A.8* (2019). http://www.iafastro.org/events/iac/iac-2019/
N.T. Kita et al., High precision SIMS oxygen three isotope study of chondrules in LL3 chondrites: role of ambient gas during chondrule formation. Geochim. Cosmochim. Acta **74**, 6610–6635 (2010)
F. Kitajima et al., A micro-Raman and infrared study of several Hayabusa category 3 (organic) particles. Earth Planets Space **67**, 20 (2015)
T. Kleine et al., Rapid accretion and early core formation on asteroids and the terrestrial planets from Hf-W chronometry. Nature **418**, 952–955 (2002)
G. Kminek et al., Amino acids in the Tagish Lake meteorite. Meteorit. Planet. Sci. **37**, 697–701 (2002)
M. Kobayashi et al., In situ observations of dust particles in Martian dust belts using a large-sensitive-area dust sensor. Planet. Space Sci. **156**, 41–46 (2018)
A. Krot et al., Sources of water and aqueous activity on the chondrite parent asteroids, in *Asteroids IV* (2015), pp. 635–660
T. Kruijer et al., Protracted core formation and rapid accretion of protoplanets. Science **344**, 1150–1154 (2014)
T.S. Kruijer et al., Age of Jupiter inferred from the distinct genetics and formation times of meteorites. Proc. Natl. Acad. Sci. **114**, 6712–6716 (2017)
K. Kuramoto et al., Martian Moons Exploration (MMX) conceptual study update, in *The 49th Lunar and Planetary Science Conference* (2018). Abstract #2143
K. Kurosawa et al., Assessment of the probability of microbial contamination for sample return from Martian moons II: the fate of microbes on Martian moons. Life Sci. Space Res. **23**, 85–100 (2019)
T.J. Lapen et al., A younger age for ALH84001 and its geochemical link to shergottite sources in Mars. Science **328**, 347–351 (2010)
D.S. Lauretta et al., The unexpected surface of asteroid (101955) Bennu. Nature **568**, 55–60 (2019)
D.J. Lawrence et al., Measuring the elemental composition of Phobos: the Mars-moon exploration with GAmma rays and NEutrons investigation for the Martian moons eXploration mission. Earth Space Sci (2019). https://doi.org/10.1029/2019EA000811
L.A. Leshin et al., Hydrogen isotope geochemistry of SNC meteorites. Geochim. Cosmochim. Acta **60**, 2635–2650 (1996)
B. Marty, The origins and concentrations of water, carbon, nitrogen and noble gases on Earth. Earth Planet. Sci. Lett. **313–314**, 56–66 (2012)
F.M. McCubbin et al., Is Mercury a volatile-rich planet? Geophys. Res. Lett. **39**, L09202 (2012)
H.Y. McSween, Petrology on Mars. Am. Mineral. **100**, 2380–2395 (2015)
H.Y. McSween, A.H. Treiman, Martian meteorites, in *Planetary Materials*, vol. 36, ed. by J.J. Papike (Mineralogical Society of America, Washington, 1998), p. 53
H.Y. McSween et al., Elemental composition of the Martian crust. Science **324**, 736–739 (2009)
D.W. Mittlefehldt et al., Non-chondritic meteorites from asteroidal bodies, in *Planetary Materials*, vol. 36, ed. by J.J. Papike (Mineralogical Society of America, Washington, 1998), pp. 1–195
S. Murchie, S. Erard, Spectral properties and heterogeneity of Phobos from measurements by Phobos 2. Icarus **123**, 63–86 (1996)
S.L. Murchie et al., The value of Phobos sample return. Planet. Space Sci. **102**, 176–182 (2014)
M. Nayak et al., Effects of mass transfer between Martian satellites on surface geology. Icarus **267**, 220–231 (2016)
R. Okazaki et al., Hayabusa2 sample catcher and container: metal-seal system for vacuum encapsulation of returned samples with volatiles and organic compounds recovered from C-type asteroid Ryugu. Space Sci. Rev. **208**, 107–124 (2017)

S. Pizzarello et al., The organic content of the Tagish Lake meteorite. Science **293**, 2236–2239 (2001)

L. Qin et al., Contributors to chromium isotope variation of meteorites. Geochim. Cosmochim. Acta **74**, 1122–1145 (2010)

K.R. Ramsley, J.W. Head, Mars impact ejecta in the regolith of Phobos: bulk concentration and distribution. Planet. Space Sci. **87**, 115–129 (2013)

P. Rosenblatt et al., Accretion of Phobos and Deimos in an extended debris disc stirred by transient moons. Nat. Geosci. **9**, 581–583 (2016)

V. Safronov et al., Protosatellite swarms, in *Satellites* (University of Arizona Press, Tucson, 1986), pp. 89–116

H. Sawada et al., Hayabusa2 sampler: collection of asteroidal surface material. Space Sci. Rev. **208**, 81–106 (2017)

D.P. Simonelli et al., Photometric properties of Phobos surface materials from Viking images. Icarus **131**, 52–77 (1998)

H. Sugahara et al., Amino acids on witness coupons collected from the ISAS/JAXA curation facility for the assessment and quality control of the Hayabusa2 sampling procedure. Earth Planets Space **70**, 194 (2018)

S. Tachibana et al., Hayabusa2: scientific importance of samples returned from C-type near-Earth asteroid (162173) 1999 JU3. Geochem. J. **48**, 571–587 (2014)

G.J. Taylor, The bulk composition of Mars. Chem. Erde **73**, 401–420 (2013)

S.R. Taylor et al., Earth-moon system, planetary science, and lessons learned. Rev. Mineral. Geochem. **60**, 657–704 (2006)

N. Thomas et al., Spectral heterogeneity on Phobos and Deimos: HiRISE observations and comparisons to Mars Pathfinder results. Planet. Space Sci. **59**, 1281–1292 (2011)

A. Trinquier et al., Widespread 54Cr heterogeneity in the inner solar system. Astrophys. J. **655**, 1179 (2007)

A. Trinquier et al., Mn-53-Cr-53 systematics of the early Solar System revisited. Geochim. Cosmochim. Acta **72**, 5146–5163 (2008)

A. Trinquier et al., Origin of nucleosynthetic isotope heterogeneity in the solar protoplanetary disk. Science **324**, 374–376 (2009)

S. Ulamec et al., A rover for the JAXA MMX Mission to Phobos, in *70th International Astronautical Congress. IAC-19-A3.4.B8* (2019). http://www.iafastro.org/events/iac/iac-2019/

T. Usui, What geology and mineralogy tell us about water on Mars, in *Astrobiology* (Springer, Berlin, 2019), pp. 345–352

T. Usui et al., Origin of water and mantle–crust interactions on Mars inferred from hydrogen isotopes and volatile element abundances of olivine-hosted melt inclusions of primitive shergottites. Earth Planet. Sci. Lett. **357–358**, 119–129 (2012)

T. Usui et al., Meteoritic evidence for a previously unrecognized hydrogen reservoir on Mars. Earth Planet. Sci. Lett. **410**, 140–151 (2015)

T. Usui et al., Martian Moons eXploration (MMX): Japanese Phobos Sample Return Mission, in *42nd COSPAR Scientific Assembly*, Pasadena, California, USA (2018). http://cospar2018.org/

P.H. Warren, Stable-isotopic anomalies and the accretionary assemblage of the Earth and Mars: a subordinate role for carbonaceous chondrites. Earth Planet. Sci. Lett. **311**, 93–100 (2011)

S. Watanabe et al., Hayabusa2 arrives at the carbonaceous asteroid 162173 Ryugu—a spinning top–shaped rubble pile. Science **364**, 268–272 (2019)

C.R. Webster et al., Low upper limit to methane abundance on Mars. Science **342**, 355–357 (2013)

H. Yabuta et al., X-ray absorption near edge structure spectroscopic study of Hayabusa category 3 carbonaceous particles. Earth Planets Space **66**, 156 (2014)

H. Yurimoto et al., Oxygen isotopic compositions of asteroidal materials returned from Itokawa by the Hayabusa mission. Science **333**, 1116–1119 (2011)

Space Sci Rev (2020) 216:51
https://doi.org/10.1007/s11214-020-00676-9

Exploring Mars with Returned Samples

Monica M. Grady[1]

Received: 5 November 2019 / Accepted: 17 April 2020 / Published online: 6 May 2020

Abstract The international Mars Exploration community has been planning to return samples from Mars for many years; the next decade should see the plans becoming a reality. Mars Sample Return (MSR) requires a series of missions, first to collect the samples, then to return them to Earth, whilst preventing the contamination of both Earth and Mars. The first mission in the campaign, Mars 2020, will land at Jezero Crater in early 2021; samples should return to Earth sometime after 2032. The information to be derived from analysis of martian samples in terrestrial laboratories equipped with state-of-the-art instrumentation is more than recompense for the difficulties of the MSR campaign. Results from analyses of returned samples will enable increased understanding of martian geological (and possibly biological) evolution. They will facilitate preparations for human exploration of Mars and by providing a second set of absolute ages for a planetary surface will validate (or otherwise) application of the lunar crater-age scale throughout the Solar System.

Keywords Mars-sample-return · MSR · Mars · Jezero

1 Introduction

Ever since the Mariner 9 mission of 1971-2 returned images of the martian landscape showing networks of craters, dried-up river valleys and towering (but extinct) volcanoes (Mutch and Saunders 1976), it has been known that Mars experienced impact, fluvial, volcanic and aeolian processes – and a potential for martian life to develop. Knowledge of the extent and complexity of these processes has increased in detail with each succeeding space mission. We now have global scale coverage of the planet at visible wavelengths and almost

Role of Sample Return in Addressing Major Questions in Planetary Sciences
Edited by Mahesh Anand, Sara Russell, Yangting Lin, Meenakshi Wadhwa, Kuljeet Kaur Marhas and Shogo Tachibana

✉ M.M. Grady
monica.grady@open.ac.uk

[1] School of Physical Sciences, The Open University, Milton Keynes MK7 6AA, UK

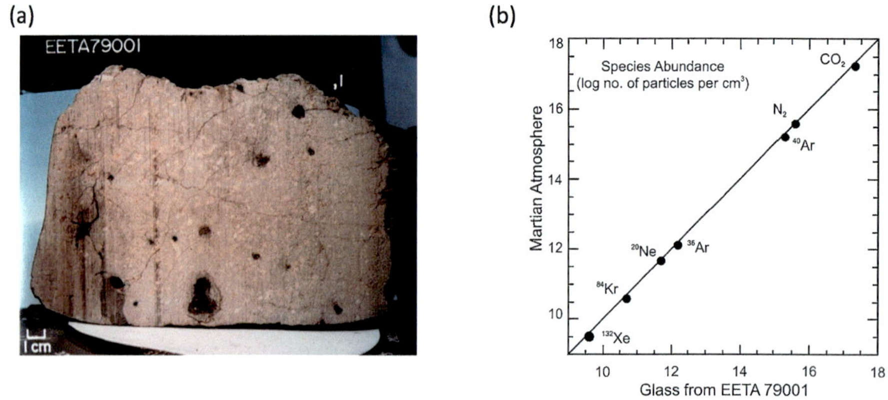

Fig. 1 (**a**) The EETA 79001 meteorite, showing areas of black glass on its cut-surface (image credit NASA-JSC); (**b**) Compositional evidence that gas trapped inside EETA 79001 has the same composition as Mars' atmosphere (reproduced from Pepin 1985)

total coverage in the infrared. We know the distribution of craters across the planet's surface, the location of the main volcanic regions, the existence of a complex network of fluvial features and the composition and dynamic properties of the atmosphere. This has given us a broad idea of Mars' evolutionary history based on a relative chronology that ties together the different processes (Tanaka 1986). Changes in mineralogy brought about by aqueous alteration have been observed from orbit and at the landing sites of the Spirit, Opportunity and Curiosity rovers (e.g., Ehlmann and Edwards 2014; Ruff and Farmer 2016; Bedford et al. 2019). These changes, when related to a cratering chronology, can be modelled to yield information about the timing, extent and duration of fluid flow, as well as constraining the temperature and composition of the fluid (e.g., Griffith and Shock 1997; Ehlmann et al. 2011; Bridges et al. 2015; Zolotov and Mironenko 2016).

Complementary information about Mars is obtained from martian meteorites, a group that has been recognised since the Antarctic meteorite Elephant Moraine (EETA) 79001 was found to contain pockets of clear black glass in which gas with the composition of Mars' atmosphere was trapped (Bogard and Johnson 1983; Becker and Pepin 1984; Pepin 1985 and Fig. 1).

It was inferred that EETA 79001 had been ejected from Mars during an impact cratering event. Once this suggestion was accepted, analysis of oxygen isotopic composition indicated that other igneous meteorites with young crystallisation ages had the same source (Clayton and Mayeda 1983; Franchi et al. 1999). Since then, the number of samples of Mars has grown to over 240 - although the number of individual meteorites is unknown because of the difficulty of matching samples that fragmented prior to collection. The Meteorite Database at https://www.lpi.usra.edu/meteor/metbull.php provides regular updates of this figure. Differences in mineralogy and mineral chemistry imply that they represent at least nine different rock types coming from a number of discrete impact events from different regions of the planet (Righter 2017). Unfortunately, whilst comparisons can be made between the compositions of the Martian meteorites (as measured in a laboratory) and composition of Mars (as measured by orbiting and landed craft), it is not possible to determine the specific region of Mars from which any of the meteorites has been derived (Werner et al. 2014). Thus, their utility in addressing questions regarding the origin and evolution of Mars has limitations. In addition, the meteorites are all igneous rocks (apart from the breccia NWA 7034 and its pairs) so their ability to record evidence of martian biological activity in a hydrous sedimentary context is limited. Admittedly, several of the samples (particularly the nakhlites and the orthopyroxenite ALH 84001) contain assemblages of secondary minerals (combinations of phyllosilicates, carbonates, halite, etc.) produced by water on Mars (e.g., Bridges et al. 2001; Melwani Daswani et al. 2016; Lee et al. 2018) infiltrating fractures and which have the potential to harbour traces of life, but to date, no such traces have been definitively recognised (McKay et al. 1996; Steele et al. 2007).

There are, then, three highly complementary and detailed sets of data from which information about Mars may be inferred. Images and spectral information from orbiting spacecraft have provided global coverage of landforms and associated mineralogical variations. This has allowed a relative chronology for fluvial and volcanic activity to be established. Information obtained by instruments on rovers and landers have enabled much more detailed investigation of smaller, more specific areas on the planet. Finally, the absolute ages of meteorites ejected from Mars during impacts has helped to constrain the timing of periods of volcanic activity and aqueous alteration. Measurement of the elemental and isotopic composition of secondary minerals have informed models of fluid flow and water-rock ratios to illustrate how different generations of alteration relate to each other (Melwani Daswani et al. 2016).

Despite all this information and the insight it has given to Mars' evolutionary history, each dataset is lacking in some way, such that we cannot yet construct a complete (absolute) timeline for Mars. Data from orbiting spacecraft define a relative chronology and give context to data from rovers and landers. Data from rovers and landers provide more detailed chemical and mineralogical compositions to refine global stratigraphy whilst martian meteorites provide absolute ages and detailed compositional data for specific phases but without any contextual information. The most logical next step to increase knowledge about Mars is to bring samples from known locations back to Earth for analysis.

Return of samples from the surface of Mars has been a goal of the international Mars exploration community for many years. There has been much discussion of the profile of a sample return mission, including comparison with the return of the Apollo 11 samples from the Moon. Almost as soon as Armstrong set foot on the lunar surface, he collected a 'contingency sample'. This comprised a few scoops of lunar regolith from close to the lander (Kramer et al. 1977). The idea behind the collection was that if things became non-nominal (i.e., went wrong) and the astronauts had to leave the lunar surface rapidly, they would at least have some material to bring back. Following that example, the utility of a 'grab and go' mission to Mars has been considered. However, as understanding of the interplay between different reservoirs (atmosphere, hydrosphere-cryosphere and lithosphere) has increased and realisation of the potential habitability of Mars has grown, it has become clear that return of a 'grab and go' (i.e., randomly-collected) sample would be insufficient to answer the major questions of interest about Mars, *viz* the absolute age of the bedrocks and the potential for life in the subsurface (Beaty et al. 2019). The most recent iteration of a Mars Sample Return mission is described below.

2 The Pros and Cons of Ex Situ Versus in Situ Analysis of Martian Samples

There are (at least) four advantages to analysing a sample back on Earth (*ex situ*) compared with making analogous measurements on Mars (*in situ*). There are also several disadvantages to removing a sample for measurement elsewhere. We will consider the advantages of sample return first. In doing so, we make the assumption, just as is recommended in the iMOST report (Beaty et al. 2019), that suites of samples are collected from a locality rather than single specimens. It is also assumed that appropriate instrumentation is available and operational.

2.1 Advantages of Ex Situ Analysis

(i) *Sample preparation techniques*: a major advantage of analysis of returned samples compared with analysis of a sample *in situ* on Mars' surface is that more sophisticated sample preparation techniques can be employed to obtain high quality and high precision information. One of the most widely-applied techniques in geological investigations is to make a polished section from a rock sample. A single chip of rock with dimensions of less than $2 \times 2 \times 2$ mm, once embedded in resin, can be sliced to produce several sections or mounts. An entire panoply of measurements can be made on a single polished mount at high spatial resolution revealing details that are invisible at lower magnifications. A polished mount is suitable for analysis by electron microscopy to determine the major and minor element compositions of specific minerals, the texture, grain size and luminescence properties of the rock, its alteration by fluids and its structure and shock history. The mount would also be available for examination by other microbeam techniques, including electron microprobe (for trace elements) and ion probe (for elemental and isotopic compositions). Similarly, spectroscopy (across infrared, visible and ultraviolet wavelengths and Raman) could be performed on a polished mount to obtain structural and organic information. Other important preparation techniques are possible in a terrestrial laboratory that cannot (as yet) be performed on Mars' surface. For example, the picking of individual, specific mineral grains from a specimen, followed by dissolution of the grains in a series of acids for age-dating. Or processing of solid materials through sequential solvent and acid extractions and derivatisation for the analysis of organic molecules. Although the SAM instrument suite on Curiosity is able to perform some wet chemistry experiments, including derivatisation (Freissinet et al. 2015), it is only carrying a very limited number and amount of chemicals. This precludes the more sophisticated analyses carried out in organic chemistry laboratories on Earth, where different 'recipes' for extractions can be substituted if appropriate.

(ii) *Range of analytical techniques and instruments*: another of the major advantages of having samples returned to Earth is the range of analytical techniques that can be applied to them. All missions are restricted in terms of payload: factors such as cost, mass, volume, power requirements and data transfer (bandwidth) limits constrain choices of instruments and place limits on the ensemble of measurements that can be made by any specific mission. Additionally, although instruments onboard rovers are now capable of determining the mineralogy, mineral chemistry, volatile elemental and isotopic compositions of rocks and soil on the surface of Mars, there are still gaps where information is missing because analytical techniques have not yet been adapted or developed for deployment on spacecraft. For example, a CT scanner is a large piece of equipment that is becoming almost a 'workhorse' instrument to produce 3-D images of rocks, showing texture, porosity and composition. If attached to a synchrotron source, the measurements can be at very fine-scales on samples as small as individual grains. It is unlikely, in the foreseeable future, that such an instrument could be miniaturised for deployment on Mars.

(iii) *Responsive investigation pathways*: A further advantage of laboratory analysis is that investigation pathways are responsive to each discovery. Measurements using different techniques on a single aliquot of material (which might be chip, powder, polished mount, separated minerals, etc.) can run in sequence and be designed so that one technique does not interfere with results expected from a technique to be applied subsequently. Measurements can be repeated (by the same or different analysts) for internal checking and calibration. Whilst planning of an analytical sequence is essential to get the best possible results from a limited amount of material, because the measurements are taking place in 'real time' (i.e., not autonomously on a planetary surface where instructions must be uploaded ahead of time), changes to a sequence can be made if unusual results are produced from one technique that might require verification from a different point on the analysis pathway. Finally, the repeatability of critical measurements is readily achievable by inter-laboratory comparisons on Earth, but is impractical on Mars.

(iv) *Spatial and spectral resolution*: Instruments on orbiting spacecraft are generally returning data acquired at low magnification or relatively low spectral and spatial resolution from large areas of the martian surface. The instruments are constrained by limited power availability, mapping geometry, data storage capacity and bandwidth and communication timelines. Although a vast library of information has resulted from orbiting missions, there are gaps where the footprint of the instrument is on a larger-scale than the features being mapped. For instance, one of the outstanding questions about Mars' changing climate is the extent and duration of fluid flow on the surface. The distribution of clay minerals (and other alteration products) is of critical importance in trying to establish the temperature and timing of different fluvial events (e.g., Bishop et al. 2008; Carter et al. 2013), yet differentiation between phyllosilicate species is not always possible from orbit (Ehlmann and Edwards 2014). Similarly, taking a closer view of process-related and diagnostic textural associations using rover-based instruments may show discrepancies between orbital and surface results if there is a much greater variety of minerals present at a site than can be detected from orbit (e.g., Mangold et al. 2017; Rampe et al. 2018). There are definitive analyses (e.g., mineral crystal structure determination by X-ray diffraction) that could be performed on the surface, but appropriate instrumentation has rarely been deployed, thus limiting the intercomparability of results among multiple surface-mission operation areas (Velbel 2018). Laboratory-based equipment can be employed to make measurements at finer spatial scales and get more accurate results for the individual proportions and compositions of minerals present in a mixture of components. As well as gaining information on the source regions for the different minerals, leading to understanding of the environment in which the rock formed, results can be used to calibrate orbital measurements so that the occurrence of the same compositional mixture can be mapped across the surface with greater confidence.

2.2 Disadvantages of *Ex Situ* Analysis

The disadvantages of removing a sample from its environment prior to analysis revolve around changes that might occur because the sample is no longer in thermal or redox equilibrium with its surroundings. Before the material arrives back on Earth, it will have spent at least 10 years encapsulated within sample tubes. For part of that time, the tubes will be on Mars' surface experiencing diurnal and annual thermal cycling that might differ from the cycles the samples experienced when *in situ*. This has the potential to alter the samples, which

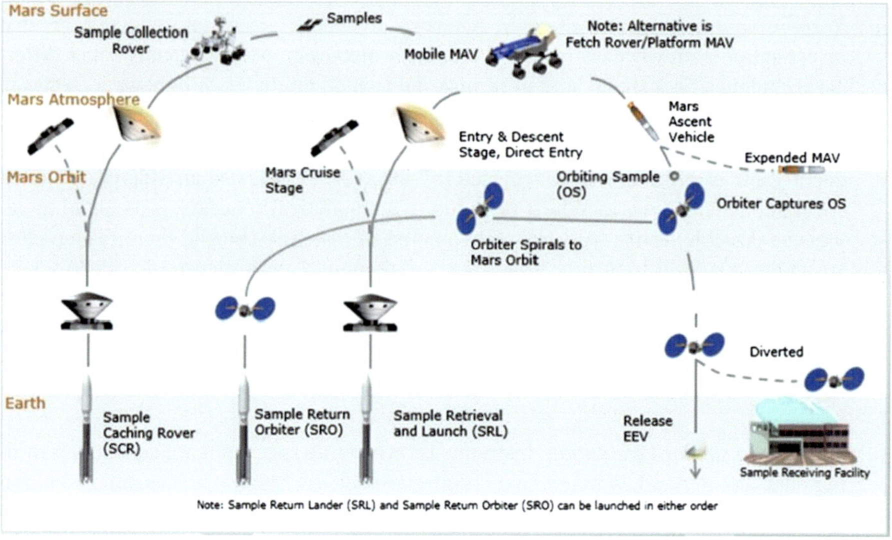

Fig. 2 Possible architecture for a Mars Sample Return Campaign (from Haltigin et al. 2018)

up to the point of collection would have been in chemical equilibrium or steady state with the surrounding strata. Following that, the material will be in orbit around Mars or Earth, or in interplanetary space, each time experiencing a different set of thermal (and gravitational) conditions. However, when the sample tubes are sealed, the material inside will represent the conditions at the time of their collection. The environment of collection will be known and recorded, so enabling back calculation of the initial conditions of the material. This is important if we wish to infer water activity, redox potential, etc.

There are specific mineral species that are sensitive to changing environmental conditions, including phyllosilicates (clay minerals) and hydrated salts. Experiments have shown that reactions between anhydrous minerals and water (especially as brine) occur very rapidly, on a timescale of days, even at low temperature (Phillips-Lander et al. 2019). Hence reactions between mineral assemblages in the MSR collection tubes with water of crystallisation or inter-layer water released from clay minerals could change the chemical balance of the system and not be a true reflection of the hydration state of the original material collected. There could even be movement of cations during the reactions – for example, removal of aluminium from plagioclase would alter fluid composition and change the products of serpentinization of olivine (Pens et al. 2016). The conversion of Fe- and Mg-bearing phyllosilicates to Al-bearing silicates is a key marker for changing environmental conditions on Mars (Bishop et al. 2018). It is essential, if we are to be certain that the samples returned to Earth can be interpreted as representative of the material that was collected on Mars, that analyses are performed of the material whilst it is *in situ* on Mars.

3 MSR Mission Design

As currently planned, Mars Sample Return is not a mission, but a campaign of several missions; one possible architecture for the campaign, with the different component missions is shown in Fig. 2.

(a) (b)

Fig. 3 The Mars2020 landing site, Jezero Crater. (**a**) Context map: Syrtis Major quadrangle MC13 centred on 15°N 292.5°W; (**b**) Close-up of a delta on the NW edge of Jezero Crater. Image Credits: NASA/JPL-Caltech/MSSS/JHU-APL

In this scenario, the first mission is NASA's Mars-2020 rover, due to launch in July 2020. The rover has the same chassis as the highly-successful Curiosity rover, with a different complement of on-board instrumentation. An important part of the equipment is a drill which will produce small cores, about 5 - 6 cm long and 1 cm in diameter. The cores will be stored in sample tubes that will be cached on Mars' surface for retrieval by a subsequent mission. The Mars-2020 rover will land at Jezero Crater (Fig. 3); the prime mission aims to explore part of the delta that once debouched into the crater. The rover will collect and cache about 20 samples from different localities on the crater floor. If an extended campaign is approved, the rover would traverse the crater wall and navigate towards the Noachian deposits of NE Syrtis, caching a further set of samples ($\sim$20).

The return part of the MSR mission is co-funded by ESA and NASA. It comprises four separate elements carried on two spacecraft, both of which are currently scheduled for launch in 2026. The first launch, the NASA-led Sample Retrieval Lander, will carry three of the elements: NASA's Mars Ascent Vehicle (MAV), ESA's Fetch Rover and the Orbiting Sample canister (OS). The rover will collect the samples cached by Mars-2020 and return them to the MAV where they will be loaded into the OS canister; either the rover or the OS could carry tubes designed to collect atmospheric samples. The MAV will carry the OS canister for release into low Mars orbit.

The second launch, the ESA-led Earth Return Orbiter acts as a communications relay for the lander. Once the OS is launched into Mars orbit, the orbiter will pick it up, encapsulating it in the Earth Entry Vehicle and return it to Earth orbit. The planning schedule has landing of the Earth Entry Vehicle in 2032. The mission needs this level of complexity with different vehicles launching from Mars and returning to Earth in order to 'break the chain' of contact between the two planets. This is to conform with planetary protection requirements designed to ensure no contamination of the Earth by material returned directly from Mars (Rummel et al. 2002).

4 Jezero Crater

Jezero Crater (77.5°E 18.4°N) was selected as the landing site for Mars 2020 in October 2018. It is a 45 km diameter crater on the west margin of Isidis Basin and just north of

NE Syrtis (Fig. 3a). In the Late Noachian-Early Hesperian, there is thought to have been an open-basin paleolake up to 250 m deep within the crater (Goudge et al. 2012) into which a river debouched leaving behind a series of deltaic sediments (Goudge et al. 2017, 2018). The remains of a fan-shaped delta can clearly be seen in Fig. 3b. The lake drained a vast watershed to the South and West of Jezero (Fassett and Head 2005); sediments transported into the lake should be a mix of the erosion products of the differing rock types of the source region. The Mars 2020 rover will land on the crater floor, probably on to the series of lacustrine deposits that include olivine- and magnesium carbonate-bearing rocks (Ehlmann et al. 2008). Above these pale-coloured sediments that cover the original basement floor is a series of deltaic deposits displaying a range of alteration species (Goudge et al. 2017). The whole is capped by a unit that may be a lava flow or volcanic sediments derived from a flow (Goudge et al. 2015; Ruff 2017).

In sum, Jezero Crater is a sequence of lacustrine and fluvio-deltaic deposits, sampling of which during the prime mission would respond to the iMOST objectives for exploration of a sedimentary system (Objective 1.1; Beaty et al. 2019). If the mission is extended, then the (current) plan is to traverse the crater rim to collect material from the rim and of the Noachian-age mega-breccia related to the volcanic units of NE Syrtis. It is hoped that samples returned from Jezero would include material from the basement unit of the crater floor (if there are any exposures) plus samples of lacustrine sediments and deltaic material.

4.1 What Might Be Learnt from the Jezero Crater Samples?

As stated above, samples returned from Jezero Crater will be appropriate to address Objective 1.1 of the iMOST objectives: *Interpret the Primary Geologic Processes and History that Formed the Martian Geologic Record, with an Emphasis on the Role of Water* (Beaty et al. 2019). In other words, returned samples will be used to understand the processes, timing, geochemistry, and biological potential of a sedimentary system. Lacustrine basins like that of Jezero Crater keep a good record of paleoenvironmental and paleoclimatic conditions because they often have a continuous stratigraphy (e.g., Smoot and Lowenstein 1991; Renaut and Last 1994), independent of whether or not the basin has been open (i.e., has outflowing rivers or streams or has over-flowed a crater margin) or closed. Knowing already that Jezero Crater contains rocks produced by fluvial, lacustrine and aeolian processes (Goudge et al. 2012, 2015, 2017, 2018), then acquisition of samples from the different strata will be essential to disentangle the history of the Jezero palaeolake and its surrounding catchment area. As well as the provenance of the material that formed the sediments, specific processes that operated on the sediments would be studied, including diagenesis, weathering, transport and deposition. Once the history of surface and near-surface water and its interactions with the sediments and the atmosphere is developed, a record of the ancient climate of Mars preserved in the resultant rocks should be revealed, constraining models for the range of past climate variations. The results will enable construction of an evolutionary timeline for the basin, providing absolute ages for sedimentary deposition under different environments (fluvial, lacustrine, aeolian). If the current understanding of the geology of Jezero Crater is correct, then samples obtained from the crater floor should be volcanic, further elucidating the history of the crater, presumably placing an upper limit on its formation age.

A significant parallel set of investigations to the chronological question is the search for preservation of organic material and assessment of the biological potential of the material. One of the reasons that Jezero crater was selected as the landing site for Mars 2020 is that palaeolakes are an ideal location for the deposition and preservation of organic material. Samples returned from the crater may contain evidence of martian life, even if at the biomarker level rather than as fossils (Summons et al. 2011).

An additional benefit of studying a suite of samples from Jezero crater is that the results will allow reconstruction of an entire martian sedimentary system, from which it should be possible to apply the results to other sedimentary regions observed from orbit.

4.2 Using Returned Samples for Evaluation of Potential Hazards to Human Exploration

The return of samples to Earth from Mars is also an important part of any Human Exploration programme (NASA 2009). At the most basic of levels, the mission would demonstrate that a spacecraft could lift off from a body with significant gravity and an atmosphere and return to Earth safely. As well as this most fundamental aspect of the relevance of MSR, the iMOST report established three separate areas in which returned samples would contribute: planetary protection, engineered systems safety and hazards to human health. Planetary protection has been discussed earlier and there are no specific additional concerns for human space exploration. The safety, efficiency and effectiveness of engineered systems, such as breathing apparatus, transport, power generation, etc., are likely to be reduced following exposure to martian dust. There are many examples from the Apollo missions (e.g., Gaier 2005; Kobrick and Agui 2019) where dust abraded space suits, infiltrating joints and seals, and the likelihood of a similar effect on Mars is high. The presence of a martian atmosphere means that wind can lift and transport dust, turning the dust into a very efficient agent of abrasion; the long duration of a martian mission would cause exposed surfaces to be scoured and seals to degrade.

Whilst there is a need to test the resistance of spacesuits and other systems to the effects of martian dust, it is unlikely that samples returned from Mars would be available for such experiments, even if they were suitable. The MSR mission currently being planned will return limited amounts of sample, mainly rocks. It is not scheduled to collect airfall dust, which is the material required in relatively large quantities for testing. However, the returned tubes, which will have been exposed on the martian surface for around 10 years, will almost certainly be covered in dust - and it is possible that this material might be suitable for the abrasion testing. What is likely to be more useful, though, is that collection and characterisation of the airfall dust from the exterior surfaces of the sample tubes will help in production of a high-quality dust simulant. The grain size, shape, angularity, composition and density of the airfall dust will be replicated and large quantities synthesised, enabling large-scale testing of engineering systems to be undertaken.

Infiltration of dust through joints and seals is a hazard to astronaut health as well as to engineering systems. Analysis of the returned samples, especially the airfall dust, would take place as part of the planetary protection procedures to determine whether there were any biological or geochemical hazards in the samples that might affect humans. "Breaking the chain of contact" when leaving Mars is currently a requirement for containment of all material returning from Mars including any mission items that have been exposed to the martian environment. Whilst this can be put into practice for autonomous missions, it is certainly not practicable for missions involving astronauts – no matter how well-designed future generations of space hardware (including spacesuits) are, it is highly likely that astronauts will come into contact with martian dust, as was the case for Apollo astronauts and lunar dust (NRC 2002; Gaier 2005; NASA 2009; COSPAR 2011). One of the most important goals of sample return will be to determine the toxicity of martian material, and whether or not it comprises a biological hazard.

Table 1 Science Objectives for Mars Sample Return (Beaty et al. 2019)

	Shorthand	Full Statement
Objective 1:	*Geological environment(s)*	Interpret the primary geologic processes and history that formed the martian geologic record, with an emphasis on the role of water
1.1	*Sedimentary System*	Characterize the essential stratigraphic, sedimentologic, and facies variation of a sequence of martian sedimentary rocks
1.2	*Hydrothermal*	Understand an ancient martian hydrothermal system through study of its mineralization products and morphological expression
1.3	*Deep subsurface groundwater*	Understand the rocks and minerals representative of a deep subsurface groundwater environment
1.4	*Subaerial*	Understand water/rock/atmosphere interactions at the martian surface and how they have changed with time
1.5	*Igneous terrane*	Determine the petrogenesis of martian igneous rocks in time and space
Objective 2:	*Life*	Assess and interpret the potential biological history of Mars, including assaying returned samples for the evidence of life
2.1	*Carbon chemistry*	Assess and characterize carbon, including possible organic and pre-biotic chemistry
2.2	*Biosignatures-ancient*	Assay for the presence of biosignatures of past life at sites that hosted habitable environments and could have preserved any biosignatures
2.3	*Biosignatures-modern*	Assess the possibility that any life forms detected are still alive, or were recently alive
Objective 3:	*Geochronology*	Determine the evolutionary timeline of Mars
Objective 4:	*Volatiles*	Constrain the inventory of martian volatiles as a function of geologic time and determine the ways in which these volatiles have interacted with Mars as a geologic system
Objective 5:	*Planetary-scale geology*	Reconstruct the history of Mars as a planet, elucidating those processes that have affected the origin and modification of the crust, mantle and core
Objective 6:	*Environmental hazards*	Understand and quantify the potential martian environmental hazards to future human exploration and the terrestrial biosphere
Objective 7:	*ISRU*	Evaluate the type and distribution of *in situ* resources to support potential future Mars Exploration

5 Science Goals

There are many questions that remain unresolved about Mars. The iMOST report (Beaty et al. 2019) organised the questions into seven objectives (Table 1). Five of the objectives related to understanding the geological evolution of Mars, possible interactions between the geosphere and potential biosphere and signatures of life. The remaining two objectives were in support of human space exploration.

The first objective was specified to *'Interpret the primary geologic processes and history that formed the martian geologic record, with an emphasis on the role of water'*. It had

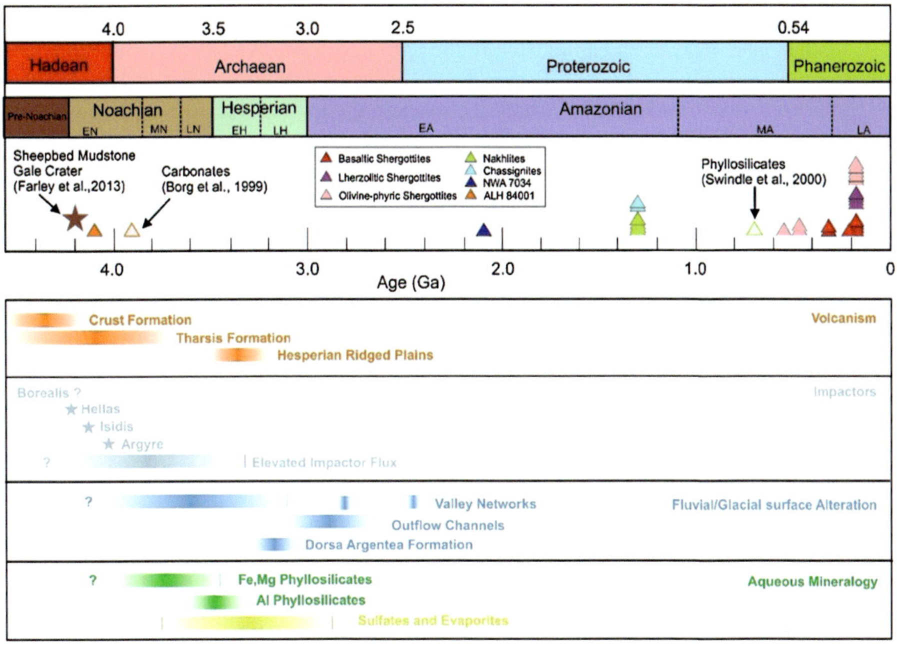

Fig. 4 Timeline of major events in Mars' history, with the geologic aeons of Earth. Question marks indicate cases where processes could also have occurred earlier but the geologic record is obscured by subsequent events. Figure and caption adapted from Wordsworth (2016). Based on data from Ehlmann et al. (2011), Fassett and Head (2011), Head and Pratt (2001), Werner (2019) and Werner and Tanaka (2011). Martian meteorite data from Borg et al. (1999), Swindle et al. (2000), Nyquist et al. (2001), Agee et al. (2013). Gale Crater data from Farley et al. (2013)

five sub-objectives, related to different geological environments (sedimentary, hydrothermal, deep sub-surface groundwater, sub-aerial and igneous). Since selection of Jezero Crater as the landing site for the Mars 2020 rover, the first sub-objective now relates directly to a sedimentary environment. We highlight here two headline issues that demonstrate the importance of sample return in order to address the science goals.

5.1 An Absolute Age for the Martian Surface

One of the most significant pieces of information missing from our understanding of Mars is its age. We know that it aggregated from the same presolar cloud as the rest of the Solar System, so its fundamental age is about 4567 Ma. We also know that there has been significant differentiation of the planet and that it has a crust, mantle and core. The core is almost certainly solid, as is the mantle. Surface morphology and specific features (impact craters, volcanoes and river and valley networks) give relative chronologies (Fig. 4); combined with almost global coverage of high-resolution imagery, a detailed timeline for the physical evolution of Mars can be constructed (Tanaka et al. 2014). Although the extent of spectroscopic data (for chemical and mineralogical compositions) is less complete than that of imagery, it has also been possible to construct a timeline for the chemical evolution of the planet that can be matched with the physical timeline. There are, however, no absolute ages of either bedrock or regolith that can definitively tie down the chronology (e.g., Bibring et al. 2006; Murchie et al. 2009; Ehlmann et al. 2016). We can place some specific anchor points on the timeline from the absolute ages of martian meteorites. The different compositional types have distinct ages which can be linked with the cratering and compositional chronologies - but cannot definitively assign dates to specific events or epochs because we lack knowledge of the specific sites on Mars from which the martian meteorites were ejected. So we have a situation with two end-members: martian meteorites give us an absolute age, but no geographical context. Remote sensing data acquired by orbiting spacecraft have yielded relative ages for physical and chemical events but have no way of anchoring the events to absolute ages.

Instruments on-board the Curiosity rover measured the K-Ar of the Sheepbed Mudstones within Gale Crater as 4.21 ± 0.35 Ga (Farley et al. 2014). A second determination, of the Windjana sandstone failed (Vasconcelos et al. 2016). A third age measurement, of the Mojave 2 mudstone was undertaken in two temperature increments, to distinguish secondary alteration products from primary material (Martin et al. 2017). The results showed that detrital plagioclase (presumably from older igneous rocks) had an age of 4.07 ± 0.63 Ga whilst jarosite produced by subsequent weathering had an age of 2.12 ± 0.36 Ga. These ages are a start in construction of a timeline for the sedimentary history of Gale Crater and the chronology of martian secondary processes but are not able to address issues of the nature and inferred history (primary differentiation and crystallisation) of the mantle source of the magmas from which the sedimentary rocks were derived.

The samples collected at Jezero Crater will mainly be sedimentary in nature and so, by definition, are not primary igneous materials that record a formation age. It is hoped that at least some of the crater floor will be exposed for collection of (potentially) late Noachian volcanic flows. Nonetheless, some of the most ancient ages for terrestrial rocks have come from analyses of detrital minerals in sedimentary-derived materials (Cavosie et al. 2004). U-Pb ages of individual grains of zircon and monazite from terrestrial impact craters have retained their original formation age despite having experienced subsequent hydrothermal processing (Moser 1997). The sediments from Jezero crater may provide analogous materials, as the sediments that fill the crater have been drawn from a catchment area that encompasses primary igneous rocks. Sample return is definitely essential for age-dating studies of this nature, because of the necessity to separate individual grains that have not been altered by subsequent processing.

5.2 Organics and Life?

The driving force for almost all martian exploration has been the prospect of finding life on the planet – if not extant life, then signs that life had been present at times in Mars' past. The first attempt to find life on Mars came with the Viking mission of 1977 (Klein 1978). The two landers each carried an experiment package to test the martian soil for evidence of metabolic activity. The results were almost uniformly negative – and the sole positive result was entirely explicable in terms of a chemical, not biological, reaction (Navarro-González et al. 2010). Since the Viking experiments, we have learnt an enormous amount about the martian surface, especially about the range of highly-oxidising salts (especially perchlorates) present in the soil (Lasne et al. 2016). The salts, together with the ever-present solar UV radiation, combine to ensure that the martian surface is probably free of the types of organic molecules that might be diagnostic of living cells. However, there may well be a more benign environment for survival of organic material below the surface. In July 2022, ESA's ExoMars 2022 mission will be launched; one of its main goals is to search for evidence of organic compounds. To do this, it will deploy a 2m-length drill to penetrate below Mars' surface in order to search for organic (and biological) material below any likely oxidised horizons. This will be the first time that rocks that have not been exposed to surface radiation will be analysed – and it is hoped that the presence of biologically significant molecules will be detected.

One tenet that underlies the search for life on Mars is that it is likely to be carbon-based. There are good physical and chemical explanations for why this is an acceptable assumption, especially given that Mars and Earth were formed from the same starting materials at the same time and for the earliest part of their histories experienced the same processes (impact bombardment, differentiation, etc.). The presence of reduced carbon is taken to be a significant marker for the potential discovery of evidence of life or prebiotic chemistry in a sample (Sephton and Botta 2005). The sediments of Gale Crater have been shown to be relatively rich in organic carbon (Freissinet et al. 2015; Sutter et al. 2017), but as yet have not yielded any evidence for definitive biological signatures.

It is hoped that the samples returned by the current MSR campaign will be collected from below horizons affected by solar irradiation. The second objective of the iMOST report (Beaty et al. 2019) concerned the range of investigations that should be undertaken to investigate the likelihood of life being present in the returned material. The objective was sub-divided into three: firstly, to investigate carbon – its occurrence, sources and characteristics – to determine its biotic or abiotic nature. The second part is designed to look for signs of ancient (fossilised) life by comparing the structure of any carbon-bearing entities found in the martian samples with ancient terrestrial biosignatures. Finally, there would be a search for signs of current life, again using the presence of biosignatures as a diagnostic tool.

Although organics have been measured *in situ* by the Curiosity rover, the series of chemical derivatisations intended to characterise specific organic compounds in the sediments have not yet been successfully completed. The reaction steps required to determine the organic inventory of a sample, and the associated range of chemicals and temperature regimes required, make for an extremely precise and complex undertaking. Such a sequence of steps is only possible on returned samples – and as outlined in Sect. 5, almost all the samples that will be collected from Jezero Crater have the potential to address the question of life on Mars.

6 Planetary Protection (PP)

The United Nations, through COSPAR (Committee on Space Research), has responsibility for Planetary Protection (PP), both the prevention of contamination of the Earth from potentially harmful agents (backward PP) and prevention of contamination of bodies beyond Earth by terrestrial agents (forward PP). For a sample return mission, both aspects of PP must be considered: forward PP on the collected samples and backward PP during transport, curation and analysis.

There are five planetary protection categories defined by COSPAR; all Earth return missions are Category V and a Mars Sample Return mission is Category Vb (Restricted Earth Return), the requirements for which are given as follows:

> The absolute prohibition of destructive impact upon return, the need for containment throughout the return phase of all returned hardware which directly contacted the target body or unsterilized material from the body, and the need for containment of any unsterilized sample collected and returned to Earth. Post-mission, there is a need to conduct timely analyses of any unsterilized sample collected and returned to Earth, under strict containment, and using the most sensitive techniques. If any sign of the

existence of a non-terrestrial replicating entity is found, the returned sample must remain contained unless treated by an effective sterilizing procedure (COSPAR 2011; Kminek et al. 2017).

Design studies for an MSR Receiving and Curation Facility have examined facilities in which biohazardous material is stored on Earth. The most recent reports are Euro-CARES (2017) and MSPG (2019a, 2019b). The most strictly-controlled of such facilities is BSL-4, where all possible precautions are taken to prevent material escaping (backward PP). Less attention is given to material coming into the laboratory (forward PP). Currently, there is no existing facility that maintains high level containment in both the backwards and forwards direction, hence the need for a specially-designed facility for returned martian samples.

It has sometimes been the case in previous design studies that the requirements for PP have been seen to be in conflict with those of the science goals. It is now recognised that many of the analyses designed to respond to PP questions about the potentially hazardous nature of martian materials are the same as those that would be applied to answering science questions about the presence (or absence) of biological matter and its characteristics. The benefit of the change in approach to both PP and Science goals is that there is likely to be a more efficient and stream-lined analytical sequence for the preliminary investigations, possibly with a more rapid outcome to decision-making about release of samples from containment (MSPG 2019a, 2019b).

7 Sample Receiving and Curation Facility(ies)

A hugely important part of an MSR campaign is the facility (or facilities) that will support the processing, basic characterization, and eventual allocation of the material on its return to Earth. At present, although final decisions about the return part of the mission have not yet been taken, it is almost certain that the designated landing site of the cannister containing the samples will be in the U.S. This will require a sample retrieval facility to be based in the landing area, as was the case for the Genesis and Stardust missions (McCubbin et al. 2019). Processing and opening of the cannister and all subsequent activities are governed by planetary protection concerns; the chain of what happens and when and where it happens is under current consideration.

Once the samples are returned from Mars, there will be a systematic process to evaluate the type of material that has been recovered. In order to ensure fair access to samples by the community and also to carry out and evaluate the required series of tests for planetary protection purposes, it is likely that there will be an international board established to oversee a preliminary examination of returned Martian samples as well as its subsequent distribution. The structure and governance of the body that will oversee the MSR sample curation, analysis and distribution is under consideration (MSPG 2019c).

The samples will be moved from the landing site to a Sample Receiving Facility (SRF). This will be combined with (indeed, act as) a Sample Curation Facility (SCF). There will only be a single SRF – and if, as is assumed, the sample cannister lands in the U.S., the SRF will be in the U.S. However, there almost certainly will be more than one SCF: one will be part of the SRF and at least one will be in Europe (MSPG 2019a, 2019b). For simplicity, in the rest of this section, we refer simply to a Sample Curation Facility (SCF).

7.1 Preliminary Analyses

Staff at the SCF would be responsible for sample curation and eventual allocation of samples to the international community; allocation would (presumably) be decided by an international allocation committee operating under the auspices of an oversight board. There have been many studies outlining how such a facility might operate, what instruments should be installed and the analyses that should be performed. Planetary protection protocols for different areas and processes within the SCF are under active discussion, under the auspices of COSPAR. There are many issues that have to be considered prior to analysis of a sample. The most recent study that has deliberated on potential operations models for an SRF is the MSPG joint ESA-NASA Working Group (MSPG 2019a, 2019b). In its reports, the MSPG recognised two phases of examination that the samples would undergo in the first instance: Basic Characterisation (BC) and Preliminary Examination (PE). Both phases would take place under full BSL-4 conditions.

An outline of a possible sequence of events is as follows. BC is non-invasive and non-destructive – it comprises, as a minimum, photography and weighing of the sample tubes. All sample tubes would go through BC on arrival at the SCF. The next phase of investigation, PE, is more detailed, minimally invasive and non-destructive. It is likely to be carried out on a single tube at a time.

There are still huge unresolved issues of how BC and PE would take place: one particular concern is the dust that will cover the tubes – it is likely to permeate the cannister and will have to be collected at different stages of the cannister and sample tube opening procedures. Other issues include: (i) how the headspace gases above the tubes should be extracted and what, if anything, the tubes should be re-pressurised with; (ii) the desirability (or otherwise) of CT scanning of the tubes prior to removal of the sample; (iii) if, when and how material should be sterilised prior to distribution and (iv) how much sample will be assigned for PP testing and how will it be selected, etc. It is clear that whatever instruments and techniques are selected to perform the BC and PE analyses, they must generate sufficient relevant information to characterise the samples, enabling decision-making during the subsequent allocation process. Most of these considerations are detailed in MSPG 2019a, 2019b, and so will not be covered here.

7.2 Instrumentation

The international community has made several different attempts to determine the type of instrumentation and analysis that will be required for complete characterisation of the return samples. The required analyses can be divided up into the following categories: (i) morphology, structure and texture; (ii) mineralogy and mineral chemistry; (iii) organic components and (iv) isotopic composition. As yet undetermined is the split between which instruments and analyses would be carried out within the SCF as part of BC and PE, which analyses have to be undertaken quickly because of time-dependent considerations (e.g., disequilibrium reactions involving water activity once a tube is opened) and which analyses might be carried out on sterilised samples. MSPG (2019a) summarizes recent community consideration of this topic.

The following table is a compilation of the instruments regarded by the Euro-CARES consortium as being appropriate for installation within the SCF. The Table is divided on the basis of the examination stage that a sample is undergoing. It is probably redundant to repeat that no decisions have yet been reached as to which equipment or instruments would be required. It is also almost certain that samples would be processed within the SCF for allocation outside the SCF to individual investigators or groups with specialist instruments that will acquire data beyond that acquired within the SRF by the instruments listed in Table 2. Again, the issue of analytical instrumentation is discussed further in the iMOST and MPSG reports (Beaty et al. 2019; MSPG 2019a, 2019b, 2019c).

Table 2 Range of instruments and equipment identified as required in a Sample Curation Facility (EURO-CARES, 2018)

Sample Preparation Equipment	Comment
Sputter coaters	For SEM analyses
Microtome	For small samples
Ion Micromills	For small samples
Micromanipulators	Electrostatic tips, micro-tweezers for different sizes
Integrated preparation systems	High precision cut, grind and polish system
High precision saws	Diamond wheel (band saw for large samples)
Grind and polish systems	Automated high precision for polished sections
Instruments for Basic Characterisation	Comment
Low magnification microscopes	Multiple microscopes required
High magnification microscopes	Petrographic and Materials microscopes
3D imaging/shape profiler	Requires technology development
High precision balances	Several required for a wide mass range
Scanning Near Field Optical Microscope (SNOM)	Only for some samples (< few micron particle size)
Instruments for Preliminary Examination	Comment
FTIR Microscope	Spectral imaging detectors for rapid high resolution
Laser Raman Microscope	UV resonance Raman useful for organic analyses
X-Ray CT (sample)	Separate X-Ray CT for sample tubes
Micro-XRD	May be superseded by spectral X-Ray CT advances
Analytical SEM	Multi-detector environmental SEM for insulators
Focussed Ion Beam-SEM	Micron-scale selection of areas for TEM study
Analytical TEM	Nanometre- to micrometre-scale analyses
Optical Microscopy	High resolution

8 Summary

Decades of observation of Mars by fly-by, orbiting, landed and roving spacecraft, complemented by data from ground- and space-based telescopes and martian meteorites, have resulted in an enormous dataset of information. This has enabled a detailed picture of the planet to be built up. We have been able to infer relative chronologies for Mars' formation and its atmospheric, fluvial and volcanic histories. We have evaluated the likelihood (or otherwise) of the evolution and survivability of life and the suitability of Mars as a target for human exploration. But despite these advances, we do not know the age of Mars' crust or mantle, or when water flowed across the surface, or when spectacular volcanic eruptions took place. We can only obtain such information by returning samples from Mars to laboratories on Earth. Only then will detailed studies, carried out on carefully-prepared, highly-specific components of the martian samples complemented by high resolution, high magnification imagery, deliver the absolute chronology of Mars.

The return of samples from Mars is an essential next step in understanding the origin and evolution of our neighbouring planet. It is also a critical requirement for the safe return of future martian astronauts.

Table 2 (*Continued*)

Instruments for Planetary Protection Analyses	Type of Biosignature
Optical Microscopy	Morphological
Electron Microscopy SEM	Morphological
Gas Chromatography-Mass Spectrometry (GC-MS)	Chemical
Liquid Chromatography-Mass Spectrometry (LC-MS)	Chemical
Matrix-Assisted Laser Desorption/Ionisation (MALDI) Time of Flight (TOF) mass spectrometry	Chemical
Fluorescence microscope	Chemical
Raman spectrometer	Chemical
High Performance Liquid Chromatography (HPLC)	Chemical
Polymerase Chain Reaction (PCR) system	Biochemical
Enzyme-linked immune-absorbent assay system (ELISA)	Biochemical
Fluorescent In Situ Hybridisation (FISH) system	Biochemical
Sequencing	Biochemical
Chromatography	Biochemical
Protein micro-array (marker chip)	Biochemical
Secondary Ion Mass Spectrometry (SIMS)	Isotopic
Isotope Ratio Mass Spectrometry (IRMS)	Isotopic
^{13}C-Nuclear Magnetic Resonance spectroscopy (NMR)	Isotopic
SEM-EDX	Mineralogical
X-Ray Fluorescence (XRF)	Mineralogical
X-Ray CT	Mineralogical
X-Ray Diffraction (XRD)	Mineralogical

Note: Many of the instruments are duplicated because they are used for several purposes. The list does not imply (unless specifically recognised) that physical duplicates of the instruments are required

Acknowledgements The author has drawn heavily on the iMOST report to produce this paper, and would like to acknowledge the iMOST team, particularly Dave Beaty and Brandi Carrier, for the work that went into producing the document. Gratitude is expressed to Michael Velbel for a thorough and helpful review. Funding from the European Space Agency and the UK Space Agency is gratefully acknowledged.

References

C.B. Agee, N.V. Wilson, F.M. McCubbin, K. Ziegler, V.J. Polyak, Z.D. Sharp, Y. Asmerom, M.H. Nunn, R. Shaheen, M.H. Thiemens, A. Steele, M.L. Fogel, R. Bowden, M. Glamoclija, Z. Zhang, S.M. Elardo,

Unique meteorite from early Amazonian Mars: water-rich basaltic breccia Northwest Africa 7034. Science **339**, 780–785 (2013). https://doi.org/10.1126/science.1228858

D.W. Beaty, M.M. Grady, H.Y. McSween, E. Sefton-Nash, B.L. Carrier (the iMOST Team), Meteorit. Planet. Sci. **54**, S3–S152 (2019). https://doi.org/10.1111/maps.13242

R.H. Becker, R.O. Pepin, The case for a Martian origin of the Shergottites: Nitrogen and Noble gases in EETA79001. Earth Planet. Sci. Lett. **69**, 225–242 (1984)

C.C. Bedford, J.C. Bridges, S.P. Schwenzer, R.C. Wiens, E.B. Rampe, J. Frydenvang, P.J. Gasda, Alteration trends and geochemical source region characteristics preserved in the fluviolacustrine sedimentary record of Gale crater, Mars. Geochim. Cosmochim. Acta **246**, 234–266 (2019). https://doi.org/10.1016/j.gca.2018.11.031

J.-P. Bibring et al., Global mineralogical and aqueous Mars history derived from OMEGA/Mars express data. Science **312**, 400–404 (2006). https://doi.org/10.1126/science.1122659

J.L. Bishop, E.Z.N. Dobrea, N.K. McKeown, M. Parente, B.L. Ehlmann, J.R. Michalski, R.E. Milliken, F. Poulet, G.A. Swayze, J.F. Mustard, S.L. Murchie, J.-P. Bibring, Phyllosilicate diversity and past aqueous activity revealed at Mawrth Vallis, Mars. Science **321**, 830–833 (2008). https://doi.org/10.1126/science.1159699

J.L. Bishop, A.G. Fairén, J.R. Michalski, L. Gago-Duport, L.L. Baker, M.A. Velbel, C. Gross, E.B. Rampe, Surface clay formation during short-term warmer and wetter conditions on a largely cold ancient Mars. Nat. Astron. **2**, 206–213 (2018). https://doi.org/10.1038/s41550-017-0377-9

D.D. Bogard, P. Johnson, Martian gases in an Antarctic meteorite? Science **221**, 651–654 (1983)

L.E. Borg, J.N. Connelly, L.E. Nyquist, C-Y. Shih, H. Wiesmann, Y. Reese, The age of the carbonates in the Martian meteorite ALH84001. Science **286**, 90–94 (1999)

J.C. Bridges, D.C. Catling, J.M. Saxton, T.D. Swindle, I.C. Lyon, M.M. Grady, Alteration assemblages in Martian meteorites: implications for near-surface processes. Space Sci. Rev. **96**, 365–392 (2001)

J.C. Bridges, S.P. Schwenzer, R. Leveille, F. Westall, R.C. Wiens, N. Mangold, T. Bristow, P. Edwards, G. Berger, Diagenesis and clay mineral formation at Gale Crater, Mars. J. Geophys. Res., Planets **120**, 1–19 (2015)

J. Carter, F. Poulet, J.-P. Bibring, N. Mangold, S. Murchie, Hydrous minerals on Mars as seen by the CRISM and OMEGA imaging spectrometers: updated global view. J. Geophys. Res., Planets **118**, 831–858 (2013)

A.J. Cavosie, S.A. Wilde, D. Liu, P.W. Weiblen, J.W. Valley, Internal zoning and U-Th-Pb chemistry of Jack Hills detrital zircons: a mineral record of early Archean to Mesoproterozoic (4348-1576Ma) magmatism. Precambrian Res. **135**, 251–279 (2004). https://doi.org/10.1016/j.precamres.2004.09.001

R.N. Clayton, T.K. Mayeda, Oxygen isotopes in eucrites, shergottites, nakhlites and chassignites. Earth Planet. Sci. Lett. **62**, 1–6 (1983)

COSPAR Planetary Protection Policy, (2011). COSPAR/IAU Workshop on Planetary Protection Retrieved from https://cosparhq.cnes.fr/sites/default/files/pppolicy.pdf

B.L. Ehlmann, C.S. Edwards, Mineralogy of the Martian surface. Annu. Rev. Earth Planet. Sci. **42**, 291–315 (2014)

B.L. Ehlmann, J.F. Mustard, C.I. Fassett, S.C. Schon, J.W. Head III., D.J. Des Marais, J.A. Grant, S.L. Murchie, Clay minerals in delta deposits and organic preservation potential on Mars. Nat. Geosci. **1**, 355–358 (2008). https://doi.org/10.1038/ngeo207

B.L. Ehlmann, J.F. Mustard, R.N. Clark, G.A. Swayze, S.L. Murchie, Evidence for low-grade metamorphism, hydrothermal alteration, and diagenesis on Mars from phyllosilicate mineral assemblages. Clays Clay Miner. **59**, 359–377 (2011)

B.L. Ehlmann, G.A. Swayze, R.E. Milliken, J.F. Mustard, R.N. Clark, S.L. Murchie, G.N. Breit, J.J. Wray, B. Gondet, F. Poulet, J. Carter, W.M. Calvin, W.M. Benzel, K.D. Seelos, Discovery of alunite in Cross crater, Terra Sirenum, Mars: evidence for acidic, sulfurous waters. Am. Mineral. **101**, 1527–1542 (2016). https://doi.org/10.2138/am-2016-5574

Euro-CARES, European curation of astromaterials returned from exploration of space. D7.2 Final Technical Report (2017). http://www.euro-cares.eu/reports

K.A. Farley, J.A. Hurowitz, P.D. Asimow, N.S. Jacobson, J.A. Cartwright, A double-spike method for K–Ar measurement: A technique for high precision in situ dating on Mars and other planetary surfaces. Geochim. Cosmochim. Acta **110**, 1–12 (2013). https://doi.org/10.1016/j.gca.2013.02.010

K.A. Farley, C. Malespin, P. Mahaffy, J.P. Grotzinger, P.M. Vasconcelos, R.E. Milliken, M. Malin, K.S. Edgett, A.A. Pavlov, J.A. Hurowitz, J.A. Grant, H.B. Miller, R. Arvidson, L. Beegle, F. Calef, P.G. Conrad, W.E. Dietrich, J. Eigenbrode, R. Gellert, S. Gupta, V. Hamilton, D.M. Hassler, K.W. Lewis, S.M. McLennan, D. Ming, R. Navarro-González, S.P. Schwenzer, A. Steele, E.M. Stolper, D.Y. Sumner, D. Vaniman, A. Vasavada, K. Williford, R.F. Wimmer-Schweingruber (the MSL Science Team), In situ radiometric and exposure age dating of the martian surface. Science **343** (2014). https://doi.org/10.1126/science.1247166

C.I. Fassett, J.W. Head, Fluvial sedimentary deposits on Mars: ancient deltas in a crater lake in the Nili Fossae region. Geophys. Res. Lett. **32**, L14201 (2005). https://doi.org/10.1029/2005GL023456

C.I. Fassett, J.W. Head, Sequence and timing of conditions on early Mars. Icarus **211**, 1204–1214 (2011). https://doi.org/10.1016/j.icarus.2010.11.014

I.A. Franchi, I.P. Wright, A.S. Sexton, C.T. Pillinger, The oxygen-isotopic composition of Earth and Mars. Meteorit. Planet. Sci. **34**, 657–661 (1999)

C. Freissinet et al., Organic molecules in the Sheepbed Mudstone, Gale Crater, Mars. J. Geophys. Res., Planets **120**, 495–514 (2015)

J. Gaier, The effects of lunar dust on EVA systems during the Apollo missions. NASA/TM—213610 (2005). https://ntrs.nasa.gov/archive/nasa/casi.ntrs.nasa.gov/20050160460.pdf

T.A. Goudge, J.W. Head, J.F. Mustard, C.I. Fassett, An analysis of open-basin lake deposits on Mars: evidence for the nature of associated lacustrine deposits and post-lacustrine modification processes. Icarus **219**, 211–229 (2012). https://doi.org/10.1016/J.ICARUS.2012.02.027

T.A. Goudge, J.F. Mustard, J.W. Head, C.I. Fassett, S.M. Wiseman, Assessing the mineralogy of the watershed and fan deposits of the Jezero crater paleolake system, Mars. J. Geophys. Res., Planets **120**, 775–808 (2015). https://doi.org/10.1002/2014JE004782

T.A. Goudge, R.E. Milliken, J.W. Head, J.F. Mustard, C.I. Fassett, Sedimentological evidence for a deltaic origin of the western fan deposit in Jezero crater, Mars and implications for future exploration. Earth Planet. Sci. Lett. **458**, 357–365 (2017). https://doi.org/10.1016/J.EPSL.2016.10.056

T.A. Goudge, D. Mohrig, B.T. Cardenas, C.M. Hughes, C.I. Fassett, Stratigraphy and paleohydrology of delta channel deposits, Jezero crater, Mars. Icarus **301**, 58–75 (2018). https://doi.org/10.1016/J.ICARUS.2017.09.034

L.L. Griffith, E.L. Shock, Hydrothermal hydration of Martian crust: illustration via geochemical model calculations. J. Geophys. Res. **102**(E4), 9135–9143 (1997)

T. Haltigin, C. Lange, R. Mugnuolo, C. Smith (the iMARS Working Group), iMARS Phase 2: a draft mission architecture and science management plan for the return of samples from Mars. Astrobiology **18**, 1–131 (2018)

J.W. Head, S. Pratt, Extensive Hesperian-aged south polar ice sheet on Mars: evidence for massive melting and retreat, lateral flow and ponding of meltwater. J. Geophys. Res. **106**(E6), 12275–12300 (2001). https://doi.org/10.1029/2000JE001359

H.P. Klein, The Viking biological experiments on Mars. Icarus **34**, 666–674 (1978). https://doi.org/10.1016/0019-1035(78)90053-2

G. Kminek, C. Conley, V. Hipkin, H. Yano, COSPAR's planetary protection policy. Space Res. Today **200**, 12–25 (2017). https://doi.org/10.1016/J.SRT.2017.11.010

R.L. Kobrick, J. Agui, Preparing for planetary surface exploration by measuring habitat dust intrusion with filter tests during an analogue Mars mission. Acta Astron. **160**, 297–307 (2019). https://doi.org/10.1016/j.actaastro.2019.04.040

F.E. Kramer, D.B. Twedell, W.J.A. Walton Jr., Apollo-11 lunar sample information catalogue (revised) (1977). https://www.hq.nasa.gov/alsj/a11/A11SampleCat.pdf

J. Lasne, A. Noblet, C. Szopa, R. Navarro-González, M. Cabane, O. Poch, F. Stalport, P. François, S.K. Atreya, P. Coll, Oxidants at the surface of Mars: a review in light of recent exploration results. Astrobiology **16**, 977–996 (2016). https://doi.org/10.1089/ast.2016.1502

M.R. Lee, L. Daly, B.E. Cohen, L.J. Hallis, S. Griffin, P. Trimby, A. Boyce, D.F. Mark, Aqueous alteration of the Martian meteorite Northwest Africa 817: Probing fluid–rock interaction at the nakhlite launch site. Meteorit. Planet. Sci. **53**, 2395–2412 (2018). https://doi.org/10.1111/maps.13136

N. Mangold, M.E. Schmidt, M.R. Fisk, O. Forni, S.M. McLennan, D.W. Ming, V. Sautterg, D. Sumner, A.J. Williams, S.M. Clegg, A. Cousin, O. Gasnault, R. Gellert, J.P. Grotzinger, R.C. Wiens, Classification scheme for sedimentary and igneous rocks in Gale crater, Mars. Icarus **284**, 1–17 (2017)

P.E. Martin, K.A. Farley, M.B. Baker, C.A. Malespin, S.P. Schwenzer, B.A. Cohen, P.R. Mahaffey, A.C. McAdam, D.W. Ming, P.M. Vasconcelos, R. Navarro-González, A two-step K-Ar experiment on Mars: dating the diagenetic formation of jarosite from Amazonian groundwaters. J. Geophys. Res., Planets **122**, 2803–2818 (2017). https://doi.org/10.1002/2017JE005445

F.M. McCubbin, C.D.K. Herd, T. Yada et al., Advanced curation of astromaterials for planetary science. Space Sci. Rev. **215**, 48 (2019). https://doi.org/10.1007/s11214-019-0615-9

D.S. McKay, E.K. Gibson, K.L. Thomas-Keprta, H. Vali, C.S. Romanek, S.J. Clemett, X.D.F. Chillier, C.R. Maechling, R.N. Zare, Search for life on Mars: possible relic biogenic activity in Martian meteorite ALH84001. Science **273**, 924–930 (1996)

M. Melwani Daswani, S.P. Schwenzer, M.H. Reed, I.P. Wright, M.M. Grady, Alteration minerals, fluids, and gases on early Mars: predictions from 1-D flow geochemical modeling of mineral assemblages in meteorite ALH 84001. Meteorit. Planet. Sci. **51**, 2154–2174 (2016)

MSPG (MSR Science Planning Group: co-chairs M. Meyer, E. Sefton-Nash, f.D.W. Beaty, B.L. Carrier, D. Bass, F. Gaubert, T. Haltigin, A.D. Harrington, M.M. Grady, Y. Liu, D. Martin, B. Marty, R. Mattingly, S. Siljestrom, E. Stansbery, K. Tait, M. Wadhwa, L. White), C.C. Allen, H. Busemann, M. Calaway, M. Chaussidon, C.M. Corrigan, N. Dauphas, V. Debaille, D.P. Glavin, S.M. McLennan, K. Olsson-Francis, R. Shaheen, C.L. Smith, J. Thieme, T. Usui, M.A. Velbel, S.C. Werner, The relationship of MSR science and containment (2019a). Unpublished workshop report, posted 04/01/19 at https://mepag.jpl.nasa.gov/reports/Science%20in%20Containment%20Report%20Final.pdf

MSPG (MSR Science Planning Group: co-chairs M. Meyer, E. Sefton-Nash, f.D.W. Beaty, B.L. Carrier, D. Bass, F. Gaubert, M.M. Grady, T. Haltigin, A.D. Harrington, Y. Liu, D. Martin, B. Marty, R. Mattingly, S. Siljeström, E. Stansbery, K. Tait, M. Wadhwa, L. White), A.M.B. Anesio, L. Bonal, A. Bouvier, J.C. Bridges, J.R. Brucato, K.L. French, U. Gommel, H.V. Graham, J.M.C. Holt, G. Kreck, R. Mackelprang, F.M. McCubbin, K. Olsson-Francis, A.B. Regberg, A. Saverino, M.A. Sephton, C.K. Sio, Science-driven contamination control issues associated with the receiving and initial processing of the MSR samples (2019b). Unpublished workshop report, posted 09/20/19 at https://mepag.jpl.nasa.gov/reports.cfm

MSPG (MSR Science Planning Group: co-chairs M. Meyer, E. Sefton-Nash, f.D.W. Beaty, B.L. Carrier, D. Bass, F. Gaubert, M.M. Grady, T. Haltigin, Y. Liu, D. Martin, B. Marty, R. Mattingly, S. Siljeström, E. Stansbery, K. Tait, M. Wadhwa, L. White), A framework for Mars returned sample science management (2019c). Unpublished white paper, posted 12/11/19 at https://mepag.jpl.nasa.gov/reports.cfm?expand=mspg

D.E. Moser, Dating the shock wave and thermal imprint of the giant Vredefort impact, South Africa. Geology **25**, 7–10 (1997). https://doi.org/10.1130/0091-7613(1997)025<0007:DTSWAT>2.3.CO;2

S.L. Murchie, J.F. Mustard, B.L. Ehlmann et al., A synthesis of Martian aqueous mineralogy after 1 Mars year of observations from the Mars Reconnaissance Orbiter. J. Geophys. Res., Planets **114** (2009). https://doi.org/10.1029/2009JE003342

T.A. Mutch, S. Saunders, The geologic development of Mars: a review. Space Sci. Rev. **19**, 3–57 (1976)

National Aeronautics and Space Administration, Human exploration of Mars: design reference architecture 5.0. NASA Pub. No. NASA-SP-2009-566 (2009). http://www.nasa.gov/pdf/373665main_NASA-SP-2009-566.pdf

R. Navarro-González, E. Vargas, J. de la Rosa, A.C. Raga, C.P. McKay, Reanalysis of the Viking results suggests perchlorate and organics at midlatitudes on Mars. J. Geophys. Res. **115**, E12010 (2010). https://doi.org/10.1029/2010JE003599

NRC (National Research Council), *Safe on Mars: Precursor Measurements Necessary to Support Human Operations on the Martian Surface* (National Academies Press, Washington, 2002). https://doi.org/10.17226/10360

L.E. Nyquist, D.D. Bogard, C-Y. Shih, A. Greshake, D. Stoffler, O. Eugster, Ages and geologic histories of Martian meteorites, in *Chronology and Evolution of Mars (ISSI)*, vol. 96 (Kluwer Academic Publishers, The Netherlands, 2001), pp. 105–164

M. Pens, M. Andreani, I. Daniel, J.-P. Perillat, C. Herve, Contrasted effect of aluminum on the serpentinization rate of olivine and orthopyroxene under hydrothermal conditions. Chem. Geol. **441**, 256–264 (2016)

R.O. Pepin, Evidence of Martian origins. Nature **317**, 473–475 (1985)

C.M. Phillips-Lander, A.S. Elwood Madden, E.M. Hausrath, M.E. Elwood Madden, Aqueous alteration of pyroxene in sulfate, chloride and perchlorate brines: implications for post-Noachian aqueous alteration on Mars. Geochim. Cosmochim. Acta **257**, 336–353 (2019)

E.B. Rampe, M.G.A. Lapotre, T.F. Bristow, R.E. Arvidson, R.V. Morris, C.N. Achilles, C. Weitz, D.F. Blake, D.W. Ming, S.M. Morrison, D.T. Vaniman, S.J. Chipera, R.T. Downs, J.P. Grotzinger, R.M. Hazen, T.S. Peretyazhko, B. Sutter, V. Tu, A.S. Yen, B. Horgan, N. Castle, P.I. Craig, D.J. Des Marais, J. Farmer, R. Gellert, A.C. McAdam, J.M. Morookian, P.C. Sarrazin, A.H. Treiman, Sand mineralogy within the Bagnold Dunes, Gale crater, as observed in situ and from orbit. Geophys. Res. Lett. **45**, 9488–9497 (2018)

R.W. Renaut, W.M. Last (eds.), *Sedimentology and Geochemistry of Modern and Ancient Saline Lakes* (SEPM (Society for Sedimentary Geology) 1994). https://doi.org/10.2110/pec.94.50

K. Righter, Introduction to the Martian Meteorite Compendium (2017). https://curator.jsc.nasa.gov/antmet/mmc/introduction2017update.cfm

S.W. Ruff, Investigating the floor of paleolake Jezero by way of Gusev crater, in *4th Int. Conf. on Early Mars: Geologic, Hydrologic, and Climatic Evolution and the Implications for Life. Abst. #3076* (2017)

S.W. Ruff, J. Farmer, Silica deposits on Mars with features resembling hot spring biosignatures at El Tatio in Chile. Nat. Commun. **7**, 13554 (2016). https://doi.org/10.1038/ncomms13554

J.D. Rummel, M.S. Race, D.L. DeVincenzi, P.J. Schad, P.D. Stabekis, M. Viso, S.E. Acevedo, A draft test protocol for detecting possible biohazards in Martian samples returned to Earth, Washington, DC (2002). https://planetaryprotection.nasa.gov/summary/DraftTestProtocol

M.A. Sephton, O. Botta, Recognizing life in the Solar System: guidance from meteoritic organic matter. Int. J. Astrobiol. **4**, 269–276 (2005). https://doi.org/10.1017/S1473550405002806

J.P. Smoot, T.K. Lowenstein, Depositional environments of non-marine evaporites, in *Developments in Sedimentology*, vol. 50 (1991), pp. 189–347. Ch. 3. https://doi.org/10.1016/S0070-4571(08)70261-9

A. Steele, M.D. Fries, H.E.F. Amundsen, B.O. Mysen, M.L. Fogel, M. Schweizer, N.Z. Boctor, Comprehensive imaging and Raman spectroscopy of carbonate globules from Martian meteorite ALH 84001 and a terrestrial analogue from Svalbard. Meteorit. Planet. Sci. **42**, 1549–1566 (2007)

R.E. Summons, J.P. Amend, D. Bish, R. Buick, G.D. Cody, D.J. Des Marais, G. Dromart, J.L. Eigenbrode, A.H. Knoll, D.Y. Sumner, Preservation of Martian organic and environmental records: final report of the Mars biosignature working group. Astrobiology **11**, 157–181 (2011)

B. Sutter, A.C. McAdam, P.R. Mahaffy, D.W. Ming, K.S. Edgett, E.B. Rampe, J.L. Eigenbrode, H.B. Franz, C. Freissinet, J.P. Grotzinger, A. Steele, C.H. House, P.D. Archer, C.A. Malespin, R. Navarro-González, J.C. Stern, J.F. Bell, F.J. Calef, R. Gellert, D.P. Glavin, L.M. Thompson, A.S. Yen, Evolved gas analyses of sedimentary rocks and eolian sediment in Gale Crater, Mars: results of the Curiosity rover's sample analysis at Mars instrument from Yellowknife Bay to the Namib Dune. J. Geophys. Res., Planets **122**, 2574–2609 (2017). https://doi.org/10.1002/2016JE005225

T.D. Swindle, A.H. Treiman, D.J. Lindstrom, M.K. Burkland, B.A. Cohen, J.A. Grier, B. Li, E.K. Olsen, Noble gases in iddingsite from the Lafayette meteorite: evidence for liquid water on Mars in the last few hundred million years. Meteorit. Planet. Sci. **35**, 107–115 (2000)

K.L. Tanaka, The stratigraphy of Mars. J. Geophys. Res. **91**, E139–E158 (1986)

K.L. Tanaka, J.A. Skinner Jr., J.M. Dohm, R.P. Irwin III, E.J. Kolb, C.M. Fortezzo, T. Platz, G.G. Michael, T.M. Hare, Geologic map of Mars. U.S.G.S. Scientific Investigations Map, #3292 (2014). https://doi.org/10.3133/sim3292

P.M. Vasconcelos, K.A. Farley, C.A. Malespin, P. Mahaffy, D. Ming, S.M. McLennan, J.A. Hurowitz, M.S. Rice, Discordant K-Ar and young exposure dates for the Windjana sandstone, Kimberley, Gale Crater, Mars. J. Geophys. Res., Planets **121**, 2176–2192 (2016). https://doi.org/10.1002/2016JE005017

M.A. Velbel, Crystallography on Mars – Curiosity's Bragging right. Am. Mineral. **103**, 837–838 (2018). https://doi.org/10.2138/am-2018-6468

S.C. Werner, In situ calibration of the Martian cratering chronology. Meteorit. Planet. Sci. **54**, 1182–1193 (2019). https://doi.org/10.1111/maps.13263

S.C. Werner, K.L. Tanaka, Redefinition of the crater-density and absolute-age boundaries for the chronostratigraphic system of Mars. Icarus **215**, 603–607 (2011)

S.C. Werner, A. Ody, F. Poulet, The source crater of martian shergottite meteorites. Science **343**, 1343–1346 (2014). https://doi.org/10.1126/science.1247282

R. Wordsworth, The climate of early Mars. Annu. Rev. Earth Planet. Sci. **44**, 381–408 (2016)

M.Y. Zolotov, M.V. Mironenko, Chemical models for martian weathering profiles: insights into formation of layered phyllosilicate and sulfate deposits. Icarus **275**, 203–220 (2016)

Space Sci Rev (2019) 215:44
https://doi.org/10.1007/s11214-019-0609-7

Sample Collection and Return from Mars: Optimising Sample Collection Based on the Microbial Ecology of Terrestrial Volcanic Environments

Charles S. Cockell[1] · Sean McMahon[1] · Darlene S.S. Lim[2] · John Rummel[3] · Adam Stevens[1] · Scott S. Hughes[4] · Shannon E. Kobs Nawotniak[4] · Allyson L. Brady[5] · Viggo Marteinsson[6,7] · Javier Martin-Torres[8,9,1] · Maria-Paz Zorzano[10,8] · Jesse Harrison[11]

Received: 5 November 2018 / Accepted: 10 September 2019 / Published online: 9 October 2019

Abstract With no large-scale granitic continental crust, all environments on Mars are fundamentally derived from basaltic sources or, in the case of environments such as ices, evaporitic, and sedimentary deposits, influenced by the composition of the volcanic crust. Therefore, the selection of samples on Mars by robots and humans for investigating habitability or testing for the presence of life should be guided by our understanding of the microbial ecology of volcanic terrains on the Earth. In this paper, we discuss the microbial ecology of volcanic rocks and hydrothermal systems on the Earth. We draw on microbiological investigations of volcanic environments accomplished both by microbiology-focused studies and Mars analog studies such as the NASA BASALT project. A synthesis of these data emphasises a number of common patterns that include: (1) the heterogeneous distribution of

Role of Sample Return in Addressing Major Questions in Planetary Sciences
Edited by Mahesh Anand, Sara Russell, Yangting Lin, Meenakshi Wadhwa, Kuljeet Kaur Marhas and Shogo Tachibana

✉ C.S. Cockell
c.s.cockell@ed.ac.uk

1 UK Centre for Astrobiology, School of Physics and Astronomy, University of Edinburgh, James Clerk Maxwell Building, King's Buildings, Edinburgh, EH9 3FD, UK

2 NASA Ames Research Center, Moffett Field, CA, USA

3 SETI Institute, P.O. Box 2445, Friday Harbor, WA 98250, USA

4 Dept. of Geosciences, Idaho State University, 921 S. 8th Ave, Pocatello, ID 83209, USA

5 School of Geography and Earth Sciences, McMaster University, Hamilton, Ontario, Canada

6 MATIS, Research & Innovation, Vínlandsleið 12, 113 Reykjavik, Iceland

7 Faculty of Food Science and Nutrition, University of Iceland, Reykjavik, Iceland

8 Group of Atmospheric Science, Department of Computer Science, Electrical and Space Engineering, Luleå University of Technology, Luleå, Sweden

9 Instituto Andaluz de Ciencias de la Tierra (CSIC-UGR), Armilla, Granada, Spain

10 Centro de Astrobiología (CSIC-INTA), Torrejon de Ardoz, 28850 Madrid, Spain

11 Turku Centre for Biotechnology, University of Turku and Åbo Akademi University, Turku, Finland

biomass and diversity in all studied materials, (2) physical, chemical, and biological factors that can cause heterogeneous microbial biomass and diversity from sub-millimetre scales to kilometre scales, (3) the difficulty of *a priori* prediction of which organisms will colonise given materials, and (4) the potential for samples that are habitable, but contain no evidence of a biota. From these observations, we suggest an idealised strategy for sample collection. It includes: (1) collection of multiple samples in any given material type (~9 or more samples), (2) collection of a coherent sample of sufficient size (~10 cm^3) that takes into account observed heterogeneities in microbial distribution in these materials on Earth, and (3) collection of multiple sample suites in the same material across large spatial scales. We suggest that a microbial ecology-driven strategy for investigating the habitability and presence of life on Mars is likely to yield the most promising sample set of the greatest use to the largest number of astrobiologists and planetary scientists.

Keywords Mars · Sample collection · Microbial ecology · Basaltic · Fumaroles

1 Introduction

The search for habitable conditions or life on any planetary body must necessarily be guided by what we know about life on Earth. A very large number of 'analog' environments on the Earth have been investigated to understand the factors that influence their microbial inhabitants and determine both the biomass and diversity of life they can support. These environments range from deserts to lakes with physico-chemical conditions, such as extreme aridity, acidity or salinity, that correspond to extant or ancient conditions on Mars (e.g., Dickinson and Rosen 2003; Navarro-González et al. 2003; Cabrol et al. 2007; Preston and Dartnell 2014; Hays et al. 2017; Lim et al. 2019).

Over the last two decades our view of Mars as a homogeneous 'red planet' has given way to a highly nuanced understanding of the great heterogeneity of past and present environments at kilometre scales (Cabrol and Grin 2001; Bibring et al. 2006; Ehlmann et al. 2009, 2011) down to millimetre scales, where individual mineral grains have been investigated (Weitz et al. 2006). Some geochemical records indicate paleoenvironmental conditions permissive for life (Grotzinger et al. 2013; Hurowitz et al. 2017). Despite this impressive expansion in our understanding of Mars, the fact remains that Mars is fundamentally a volcanic planet, with a meagre rock cycle (no plate tectonics) and no extensive surface biosphere, both of which vastly increase the range of viable habitats on the Earth. Even in the realm of volcanic lithologies, Mars is limited in its geological diversity compared to the Earth. For example the lack of a highly productive biosphere to generate large abundances of organic carbon (Field et al. 1998) limits the diversity and extent of volcanic soils, called 'andosols', that are a substantial habitat for microbial life on Earth. Although the Earth's basaltic oceanic crust offers a number of lessons about microbial distribution (Edwards et al. 2003; Lysnes et al. 2004; Santelli et al. 2008), the specific chemical composition of Earth's sea water and its continuous flow through the basaltic crust over million-year time scales makes even deep-ocean basaltic microbial habitats of potentially limited direct applicability to Mars. On Mars, the composition of hydrothermal fluids in the subsurface is poorly constrained, the former presence and extent of sustained ocean(s) to generate analogous deep ocean habitats is controversial (Head et al. 1999) and long-term hydrological activity appears to have been curtailed by the loss of most surface water in the late Noachian and Hesperian (Lasue et al. 2013).

There are, of course, environments on the Earth and Mars that are often considered to be independent of the geology of the crust and on Earth host life or preserve biosignatures. For example, ice-sheets and glaciers on Earth host life (Carpenter et al. 2000; Liu et al. 2015; Anesio et al. 2017), as do evaporitic salt deposits (Maturrano et al. 2006; Oren 2008; Robinson et al. 2015). These environments are known to exist on Mars (Head et al. 2003; Lasue et al. 2013; Osterloo et al. 2008, 2010). However, the aqueous and particulate geochemistry of these environments are influenced by the geological substratum and thus the nutrients and energy supplies in such settings on Mars would still be linked to a volcanic geological context. Nevertheless, this paper will not discuss the biological potential and sampling of ices and salts. We draw lessons on sample collection by focusing on primary volcanic rocks and derived weathering materials since they constitute the majority of materials and potentially habitable substrates that would likely be sampled by robotic and human explorers. Although we do not review the literature here, we note that the conclusions we will draw on the heterogeneity of microbial biomass and diversity in primary volcanic samples also apply to ices and salt deposits in which heterogeneous biomass and diversity distribution has been observed (e.g., Liu et al. 2015; Robinson et al. 2015).

The study of the microbial ecology of volcanic terrains relevant to Mars is surprisingly underdeveloped. For example, the microbial ecology of environments with sustained interactions between freshwater and volcanic rocks, including crystalline rocks and glasses, is poorly known. Indeed, limited knowledge exists regarding the overall thickness and extent of volcanic soils and other deposits derived by chemical weathering on Mars. However, a sufficient number of studies have been accomplished to extract knowledge that directly bears on what our most optimistic expectations for microbial ecology on Mars could be (assuming that there is or was life) and thus how one should go about collecting samples. In this paper, we do not attempt an exhaustive review of microbiology studies in volcanic environments. Instead, we provide examples used to draw out general principles. The selection of studies used in this paper are shown in Table 1.

In this paper, we have two objectives. First, we provide a review of microbial life in volcanic rocks to inform the reader about the typical distribution of life in a diversity of materials. This review provides insight into the types of samples of interest to a biologist. One might regard this as an introductory primer on microbial life in volcanic materials. Second, using these examples of the microbial ecology of volcanic materials, we draw out common patterns that allow us to suggest microbial ecology-driven sample collection approaches for investigating habitability and testing the hypothesis of life on Mars.

The conclusions we draw in this paper are applicable to the search for both extant and extinct life. The distribution of biomarkers such as biomolecules (e.g., hopanes) and isotopic signatures in volcanic paleohabitats will depend on the original distribution of biomass as well as its potential to be preserved in those environments (McMahon et al. 2018). Thus, the search for extinct life must also take into account taphonomic considerations relevant to the microbial habitats discussed here.

2 A Review of Igneous Petrology from an Astrobiological Perspective

Igneous petrology is a vast area of research, but from an astrobiological perspective, several points can be made about volcanism that are essential to the later discussion in this paper. Volcanic rocks (i.e. igneous rocks derived from the extrusion of lava at the Earth's surface rather than cooling of magma in the subsurface) vary widely in composition. They

Table 1 Studies of the microbial ecology of volcanic environments relevant to Mars used in this paper to suggest a sampling strategy

Type of environment/material	Location	Comments	Reference
Basaltic/andesitic crystalline rock	Italy	Lichen colonisation of lava flows on Mount Vesuvius	Adamo and Violante (1991), Adamo et al. (1993)
	Iceland	Studies on bacterial inhabitants of crystalline rocks (lava)	Kelly et al. (2011), Cockell et al. (2013)
	Iceland	Bacterial pioneer colonists of crystalline rocks (lava)	Kelly et al. (2014)
	Iceland	Inhabitants of lava flows of differing ages	Byloos et al. (2018)
	USA	Inhabitants of transects from fumarolic to non-fumarolic sites including unaltered and weathered basalts (Hawai'i)	Cockell et al. (2019)
Rhyolitic crystalline rock	Iceland	Bacterial inhabitants of crystalline rocks (lava)	Kelly et al. (2011)
Basaltic glass	Iceland	Bacterial inhabitants of weathered basaltic glass	Cockell et al. (2009)
	Iceland	Bacterial inhabitants of weathered basaltic glass	Kelly et al. (2010)
Obsidian (silica-rich glass)	Iceland	Bacterial inhabitants of obsidian	Herrera et al. (2008, 2009)
	Iceland	Bacterial inhabitants of weathered glasses	Kelly et al. (2010)
Volcanic fumaroles/active geothermal substrates/geothermal lakes	Italy	Endolithic inhabitants of rocks and soils in active geothermal area	Gross et al. (1998)
	USA	Endolithic inhabitants of siliceous geothermal deposits	Walker et al. (2005)
	New Zealand	Endolithic inhabitants of siliceous deposits in active geothermal springs	Gaylarde et al. (2006)
	Chile	Inhabitants of fumaroles and surrounding soils	Costello et al. (2009)
	Antarctica	Inhabitants of fumarolic soils	Soo et al. (2009)
	USA	Inhabitants of nonsulphur, sulphur and iron geothermal steam vents	Benson et al. (2011)
	USA	Inhabitants of fumaroles	Wall et al. (2015)
	Iceland	Geothermal spring microbial and isotopic gradients	Cousins et al. (2018)
	Costa Rica	Microbial diversity in Laguna Caliente lake	Hynek et al. (2018)
	Chile	Inhabitants of fumarolic soils	Solon et al. (2018)
	USA	Inhabitants of transects from fumarolic to non-fumarolic sites (Hawai'i)	Cockell et al. (2019)

Table 1 (*Continued*)

Type of environment/material	Location	Comments	Reference
Mud volcanoes	Romania	Microbial diversity of mud volcanoes	Dahigaonkar and Chavan (2018)
	China	Microbial diversity of mud volcanoes	Huang et al. (2016)
Ash/ash soils/tephra	Iceland	Colonisation of Surtsey Island	Schwabe (1970), Schwabe and Behre (1972); Kristinsson (1970, 1974); Brock (1973); Henriksson and Henriksson (1982); Stefansdottir et al. (2014)
	Iceland	Colonisation of Heimaey Island	Englund (1976)
	USA	Cinder cone colonisation (Hawai'i)	Carson and Brown (1978)
	USA	Soil microbial communities (Hawai'i)	Nüsslein and Tiedje (1998)
	USA	Recent colonisation of ash and their lithotrophic metabolisms (Hawai'i)	King (2003)
	USA	Recent colonisation of ash and their lithotrophic metabolisms (Hawai'i)	Nanba et al. (2004)
	USA	Inhabitants of 42–300 year old Hawaiian volcanic deposits (Hawai'i)	Dunfield and King (2005)
	USA	Inhabitants of numerous soil and ash deposits of different ages (Hawai'i)	Gomez-Alvarez et al. (2006)
	Antarctica	Inhabitants of deposits on geothermally active volcano	Fermani et al. (2007)
	Italy	Inhabitants of different aged soils	Hopkins et al. (2007)
	USA	Inhabitants of different aged soils	Ibekwe et al. (2007)
	Iceland	Biomass and microbial community in fresh volcanic tephra	Amador et al. (2015), Gentry et al. (2017)

are classified into fifteen different compositional groups depending upon their silica and alkali content (in this context alkali means the concentration of Na and K) following convention established by the International Union of Geological Sciences (IUGS) (Le Bas et al. 1992). Ultimately, these categories largely reflect the source composition and thermal evolution of the magma, with more mafic (iron- and magnesium-rich) minerals crystallising from the melt at higher temperatures. Mafic and ultramafic rocks (also referred to as 'basic') are formed between ~1000–1500 °C and contain elevated concentrations of pyroxenes, olivines and plagioclase (predominantly Mg, Fe and Ca-containing minerals). The precipitation of these dense minerals within a cooling magma chamber drives the composition of the residual magma towards higher silica content; thus high-silica volcanic rocks (also referred to as 'acidic' or 'felsic' rocks) generally represent lower temperatures (typically about 700 °C). They also have a higher content of orthoclase feldspar (a K-containing silicate mineral) and quartz. Although rare felsic lavas have been detected on Mars, the martian crust is dominated by the mafic rock, basalt. In this paper, we will not consider samples spanning the entire ultramafic-felsic continuum, but we will provide examples of the microbial ecology of materials, focusing on basaltic rocks. We also provide comparisons to high-silica obsidian where this is useful to illustrate factors that influence colonisation.

A second factor of great importance for a biota is the glassiness or crystallinity of the material. Volcanic melt that comes into contact with ice or water is rapidly quenched before crystals can form, resulting in a homogeneous glass, in contrast to other rocks that are formed subaerially and cool slowly, forming crystalline materials. This is significant in the Martian context because widespread glaciers and ground ice in contact with volcanically active regions (Cousins and Crawford 2011) may imply that past episodes of volcanism have generated substantial quantities of glass and their weathering products (Horgan and Bell 2012).

Why is distinguishing crystalline rock and glass important? First, glasses present a material with a more homogenised composition to a potential biota, whereas in crystalline materials, bioessential elements are generally localised to particular crystals, meaning that fewer bioessential elements can be found in one location at the micron-scale. Second, basaltic glass, in contrast to crystalline rocks, has a distinctive capacity for weathering to palagonite, a soft alteration product (Thorseth et al. 1991; Stroncik and Schminke 2002) that tends to form within basaltic vesicles and fractures and may be a more bio-accessible source of nutrients than the crystalline matrix. Third, in crystalline high-silica rocks, the silica is primarily bound up in quartz. However, in glass, this silica (which has no biological use) is distributed throughout and may retard the release of more biologically useful elements such as Fe (Wolff-Boenisch et al. 2004). *A priori* therefore, in parent rocks on Mars we would expect the concentration of silica and the glassiness of the material to be important considerations in their potential as habitable substrates and how microorganisms would be theoretically distributed within them. These factors are explored in later sections.

The weathering of these primary rock types (and of volcanic ash deposits) in aqueous environments leads to the formation of a variety of cation-rich aluminosilicate clays (or palagonite in the case of basaltic glass). The types of clays formed depend on the thermal regime and aqueous chemistry. On Mars, Fe- and Mg-rich smectites, serpentine, and chlorite, resulting from the alteration of basaltic precursor minerals, are particularly widespread (e.g., Poulet et al. 2005; Milliken and Bish 2010; Ehlmann et al. 2011; Bristow et al. 2018). Interactions with clay particles, including smectite, can profoundly influence the fate and residence time of organic matter in the environment and inhibit or promote the growth of different microbial groups. However, little is currently known about the correlation of microbial diversity and biomass to different clay types formed in volcanic environments, particularly in environments where organic input is low. Most analyses of volcanic clays have occurred in the critical zone (soil-rock interface) in regions of the world where the formation of economically important volcanic soils (andosols) is associated with generally high organic/humic substance concentrations (Kimble et al. 2000). Nevertheless, the studies that have been undertaken in volcanic terrains are sufficient to draw the conclusions we make here.

We recognise that present-day surface conditions on Mars are very different from the Earth. The desiccated surface, high-ultraviolet and ionising radiation and the presence of perchlorates (Cockell et al. 2000; Dartnell et al. 2007; Hecht et al. 2009; Wadsworth and Cockell 2017), among other factors, make the surface inimical to life and may also impact the preservation of ancient biosignatures (particularly organic biomarkers). In this synthesis, we are assuming that martian samples will be collected from environments that experienced sustained contact with water in the past and have either been drilled from the subsurface or were only exposed to martian atmosphere relatively recently, for example by wind erosion (e.g., Farley et al. 2014; Hurowitz et al. 2017). We also note that more extreme conditions found on Mars, particularly in later stages of its hydrological evolution, if they were to theoretically reduce the potential biomass and diversity of life, would merely strengthen the conclusions we draw about low biomass environments and the need for large sample numbers required to test for life or affirm its absence.

A further complication is the aeolian transport of material that becomes entrained in rocks. Microorganisms are known to be transported in atmospheric dust on Earth (Gat et al. 2017). As well as providing a microbial inoculum to new habitats, the aerially transported dust can potentially introduce extraneous geological material into samples, confounding geological analysis. On Mars, aeolian dust transport and mixing over the planetary scale is ubiquitous (Lasue et al. 2018). Aerial transport of particulate matter is also widespread in many volcanic environments on the Earth, including those in which microbial ecological analysis described in this paper have been undertaken, such as environments in Iceland (Arnalds et al. 2016). This transport does not compromise geological analysis of the origin of samples or the conclusions about the heterogeneous distribution of biomass and diversity within the sample that is linked to the primary sample type. This is likely to be the case for sample selection on Mars, particularly if samples are cleaned of dust or collected from the near-surface or interior of outcrops.

3 Limits of 'Analogue' Materials

Not all volcanic environments and materials on Earth are useful analogues to Mars. Two examples are andosols and marine-influenced volcanic rocks. We discuss these here to emphasise caution in drawing lessons about Mars from some volcanic materials on Earth.

Volcanic rocks ultimately weather to soils called andosols, which form an important component of terrestrial volcanic environments (Kimble et al. 2000). Andosols exist in a variety of states such as extreme pH in soils surrounding geothermal pools, hot and cold desert soils, as well as humid andosols (Oskarsson et al. 2004; Parfitt and Kimble 1989; Arnalds 2004; Belobrov and Ovechkin 2005; Hopkins et al. 2007). However, apart from regions free of vegetation and with low precipitation (deserts), most andosols are fertile because of many years of microbial and floral amelioration. They are generally high in Fe and Al. These metals, in addition to high concentrations of humic acids, make andosols highly effective at immobilising phosphate (Lukito et al. 1998). This retention of biologically available phosphorus combined with the precipitation of phosphorus, impedes microbial and floral growth on younger andosols (Kimble et al. 2000). It is not currently known what the initial microbial colonists are for many andosols as it depends on the chemical composition of each individual soil. Richardson (2001) showed that microorganisms were key to the release of organic and inorganic phosphorus. Nevertheless, even young andosols, although relatively hostile, are capable of sustaining highly diverse microbial communities. Nüsslein and Tiedje (1998) investigated a 200 year-old andosol in Hawai'i with high diversity. They showed that the soil biomass was dominated by a distinct microbial diversity of *Pseudomonas*, *Rhizobium-Agrobacterium* and *Rhodospirillum* species. On many andosols, plants strongly influence microbial colonisation. Some of the first floral species to inhabit Mount St. Helens soils following the 1980 eruption were *Lupinus* spp., which provide support for nitrogen-fixing cyanobacteria (Halvorson et al. 1992). Although these data are interesting, they nevertheless illustrate that many terrestrial andosols are likely poor analogues to past or present Martian volcanic environments because of the extent of microbial and plant colonisation and the relatively rich organic carbon cycle established within them.

An extensive body of literature exists on the microbial ecology of the basaltic oceanic crust. Although superficially this carbon- and energy-poor environment might appear to be a good analogue for the Martian deep subsurface, there is uncertainty as to how much the communities associated with ocean crust, their biomass and diversity, are influenced by ocean chemistry and factors linked with the Earth's unusual plate tectonic regime. During the rifting that occurs at plate boundaries in the oceans, basaltic glass is produced. The rate of crustal production may be well over a billion tonnes a year (Bach and Edwards 2003). Insofar as basaltic glass contains many major cations that are biologically useful, including Ca, Mg, K, Na and Fe, then it would be expected to be a source of material for microbial growth (Staudigel et al. 1995, 1998; Thorseth et al. 2001; Santelli et al. 2008).

The extent to which oceanic basaltic glass is used by a biota *in-situ* is unclear (Staudigel et al. 1998, 2006, Torsvik et al. 1998). There is evidence for the deep-ocean biological alteration of glass. In investigations of 0–30 Myr old basalts from the region of mid-ocean ridge known as the Australian–Antarctic Discordance, Thorseth et al. (2003) recorded the presence of Mn-rich encrustations which they suggested could be linked to biological Mn-yielding reactions. Partially fossilised endolithic microorganisms were associated with rims of altered glasses. Basaltic glass from the Knipovich Ridge, Arctic shows pervasive colonisation of fractures, including stalked, coccoid, rod and filamentous forms. Organisms were found to be associated with the alteration products in the fractures and they were often covered in Fe- and Al-rich precipitates (Thorseth et al. 2001).

The organisms that might be involved in weathering glass have been the subject of several investigations. Templeton et al. (2005) showed the presence of Mn-oxidising bacteria in basalts from Loihi Seamount, although autotrophic Mn-oxidisers were not recovered. This suggests Mn oxidation is a secondary process occurring in heterotrophic organisms (Tebo et al. 2005) that might be using organic carbon compounds available in seawater. Diverse neutrophilic Fe-oxidising bacteria have been isolated from oceanic environments (Edwards et al. 2003), although it is not clear to what extent they might be accessing iron from basaltic glass or from the surrounding sea water. The Fe oxide minerals found associated with deep-ocean basalts have been shown to be similar to minerals produced by iron-oxidising bacteria in culture (Daughney et al. 2004).

In a study of the microbial community of the basaltic glasses of the Knipovich Ridge, Arctic, Thorseth et al. (2001) identified heterotrophs and some chemolithotrophs including phylotypes closely matching with S-oxidisers. Fe-reducing organisms were cultured from Arctic Ridge seafloor basaltic glasses by Lysnes et al. (2004) and these authors suggested that this is evidence for an Fe-cycle within seafloor basalts. Phylogenetic analysis of seafloor basaltic glasses around Hawai'i show high diversity (Santelli et al. 2008; Orcutt et al. 2009) which is hypothesised to be linked to the chemolithotrophic use of basaltic glass alteration products (Santelli et al. 2008).

In summary, there appears to be good evidence for microbial use of elements from basaltic glass directly, particularly Fe, that may suggest that these rocks provide an analogue for habitable substrates on Mars. However, the extent to which resources in seawater are used, such as organics and other ions, and the extent to which the biomass and diversity of these habitats is established by continuous water flow in the oceanic crust together with the high production rate of fresh substrate through plate tectonics is not known. Plate tectonics, if it existed on Mars at all, ceased early (Yin 2012). Although there is controversial evidence for a northern ocean (Head et al. 1999), even if it did exist, its chemical composition and the degree to which it might have sustained crustal hydrothermal cycling to generate habitats analogous to the terrestrial oceanic crust is not known.

For these reasons, in this paper we focus on volcanic materials in contact with freshwater, primarily land-based crystalline and glassy volcanic rocks, including materials in land-based volcanic hydrothermal systems. However, heterogeneous distribution of biomass and diversity is observed in ocean crust materials (Santelli et al. 2008), so although we do not consider these to be necessarily the best analogue materials to examine here, the consideration of their microbiology would not change our conclusions.

4 Microorganisms Are Heterogeneously Distributed—the Need for Appropriate Sample Size

Observations of microorganisms in contiguous cryptoendoliths biofilms (e.g., Bell 1993; de la Torre et al. 2003) might give the impression that organisms are generally uniformly distributed in rocks. Although these distinctive endolithic communities can be observed in volcanic geothermal materials (Gaylarde et al. 2006; Gross et al. 1998; Walker et al. 2005) and cooled volcanic rocks (Herrera et al. 2009), organisms within volcanic materials are frequently highly heterogeneously distributed at millimetre and submillimetre scales, with implications for sample collection (Fig. 1). As a case study, it is worth considering the colonisation and distribution of microbes in basaltic glass and its felsic counterpart, obsidian. In both materials, microbes are observed to be distributed heterogeneously, but apparently caused by different factors. The distribution of microbes in obsidian, although not directly relevant to Mars as this material has not be detected, provides insights into the factors that influence microbial distribution in surface and near-surface volcanic materials more generally.

Basaltic glass weathers to palagonite, a relatively soft weathering product, which forms rinds on the glass (Stroncik and Schminke 2001, 2002). Although weathering of obsidian occurs, it is much slower than the weathering of basaltic glass (Wolff-Boenisch et al. 2004, 2006). The higher silica concentration in obsidian impedes the degradation of the material since the Si-O bonds result in a more extensive silica network that is thought to retard leaching of cations, including bioessential cations, from the rock.

These different weathering processes are consistent with observations of varying microbial abundance. Cockell et al. (2009) observed high cell numbers in basalt glass samples with bulk cell numbers of $\sim 10^7$ cells/gram. In this material, most of the observed cells were associated with the weathered palagonite surfaces and potentially a population beneath the palagonite. Very few cells were associated with the unweathered glass surface. In obsidian, which does not so readily form weathered rinds, Herrera et al. (2008), using FISH

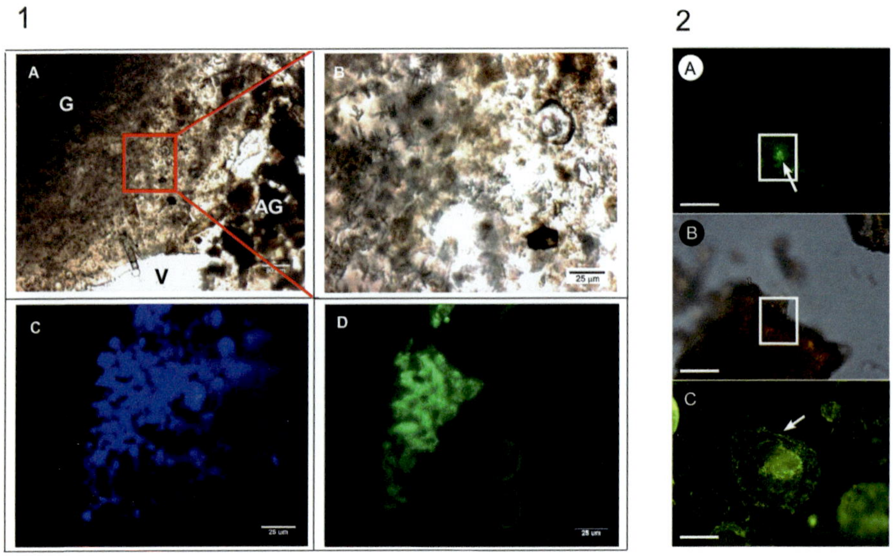

Fig. 1 Cells are distributed heterogeneously in materials for different reasons. Examples of heterogeneity in microbial distribution in volcanic glasses. Image **1** shows cells in obsidian. In (**A**), unaltered glass (G) and altered glass (AG) at the edge of a vesicle (V) are shown. Subimage **B** shows the region located in the square in image **A**. Cells were detected after DAPI staining under UV light (subimage **C**) and after fluorescence in situ hybridisation (FISH) with bacterial Cy3-EUB338I-III primers targeting most of bacteria (subimage **D**). Observations B, C and D correspond to an identical area on the thin section (data from Herrera et al. 2008, 2009). Image **2** shown cells in basaltic glass. Subimage **A** shows SYBR Green I (DNA) staining of microorganisms on a fragment of palagonite also shows in bright field (scale bar 20 µm) (subimage **B**). Subimage **C** shows a SYBR Green I stained section of a vesicle in weathered basaltic glass, arrow indicates DNA staining at the rock–palagonite interface with staining also apparent on the surface of the palagonite (scale bar 40 µm) (data from Cockell et al. 2009)

(Fluorescence In-Situ Hybridisation), found that cells were localised to altered regions in the obsidian glass (Fig. 1). The cells were associated with iron oxide crystals. Subsequent cell enumeration found total bulk cell numbers less than 10^4 cells/gram. Because the cells were so localised within the obsidian, whole rock cell numbers were unrepresentative. Significantly, entire chunks to cm^3 scales, contained no detectable cells. The localised biofilms associated with alteration zones were often only found in exposed rocks or weathered regions where water had penetrated into the material.

Similarly heterogeneous distribution of biomass has been observed in other volcanic environments. The NASA project BASALT (Biologic Analog Science Associated with Lava Terrains; Lim et al. 2019) carried out a study of microbial diversity and biomass in basaltic fumaroles, weathered rocks and unweathered rocks in Hawai'i. Consistent with other studies cited here, all materials were found to host a low diversity, and biomass was highly heterogeneous, ranging from 6.25×10^5 to 4.06×10^7 cells/gram dry weight based on DNA concentration (means shown in Fig. 3). Lipid biomarker-based estimates of biomass also demonstrated heterogeneous distribution within BASALT samples (Brady et al. 2017). The heterogeneous distribution of biomass in basaltic terrains has been noted in both fresh lava (Kelly et al. 2014) and in dispersed volcanic tephra of varying ages (Gentry et al. 2017).

Factors that are likely to account for the observed heterogeneity in microbial biomass in volcanic materials include physical influences such as variations in the availability of fracture surfaces and pore/vesicle space for growth, differences in permeability and thus fluid flow through fractures, albedo and therefore tendency to absorb sunlight and heat; chemical factors such as the proportion of glass, heterogeneity in leaching of elements from the composite minerals or glasses available for redox couples or as nutrients, variations in the chemical weathering reactions mediated by constituent minerals and elements, differences in the kinetic release rates of elements from rocks (Cockell et al. 2011); and biological factors, e.g. priority effects, whereby first colonisers influence subsequent colonists (Lee et al. 2013). Some of these factors are known to be important determinants in the colonisation of soils (Garcia-Pichel et al. 2003; Vos et al. 2013; O'Brien et al. 2016).

These observations across a variety of volcanic materials emphasise the need for appropriate sample size. It is not possible to state an exact size, since the optimal sample size depends on the material. However, the presence of organisms inhabiting millimetre-scale vesicles, crystals and fractures in many volcanic materials shows that samples with integrity at centimetre scales are required. These observations of cm^3 scale samples or segments of samples devoid of life even in materials on Earth, such as obsidian, shows that typical minimum sample sizes of volcanic rock collected in the field for microbiological analysis should be of size $\sim 10\ cm^3$ and this sample size applies to Mars exploration.

5 Every Sample Is a 'Microbial Island'—the Need for Multiple Samples from the Same Material Type

Like biomass, the microbial diversity of volcanic materials varies enormously as a function of physical, chemical and biological factors that relate to the functional requirements of life in that environment and/or the specific organisms that first successfully colonise the material.

Microbial diversity is significant to the search for life on Mars independent of biomass. For example higher-diversity environments could include more organisms with particularly good preservation potential compared with lower-diversity environments simply because such environments contain a larger cross-section of microbial taxa. For an extant biota, microbial diversity provides critical information about the range of functional capabilities that an environment can support and so diversity can be used as a measure of the habitability of any given environment.

Studies have been undertaken on the diversity of volcanic rocks to investigate the role of elemental composition in their colonisation. For example, statistically significant differences were observed in the microbial diversity between basaltic and more silica-rich obsidian glass in Iceland and between basaltic and more silica-rich rhyolitic rocks in juxtaposed locations where environmental factors were unlikely to influence the results (Kelly et al. 2010, 2011).

It is now well established that gradients in both physical parameters such as temperature (Soo et al. 2009) and chemical parameters such as pH and sulphur content (Benson et al. 2011) cause small-scale variations in microbial diversity in volcanic environments. These spatial variations in microbial diversity are observed in all volcanic environments investigated (Table 1) ranging from geothermal fumaroles (Soo et al. 2009; Benson et al. 2011; Wall et al. 2015) to weathered exposed crystalline and glassy volcanic rocks (Herrera et al. 2009; Kelly et al. 2010, 2011), ash and soils (Gomez-Alvarez et al. 2006; Ibekwe et al. 2007), and mud volcanoes (Huang et al. 2016; Dahigaonkar and Chavan 2018). Variations in the microbial ecology caused by temperature and geochemical spatial gradients were examined in detail in a geothermal setting in Iceland, where differences in microbial phylogeny, function and their associated isotope fractionation values were found in single geothermal outflows (Cousins et al. 2018).

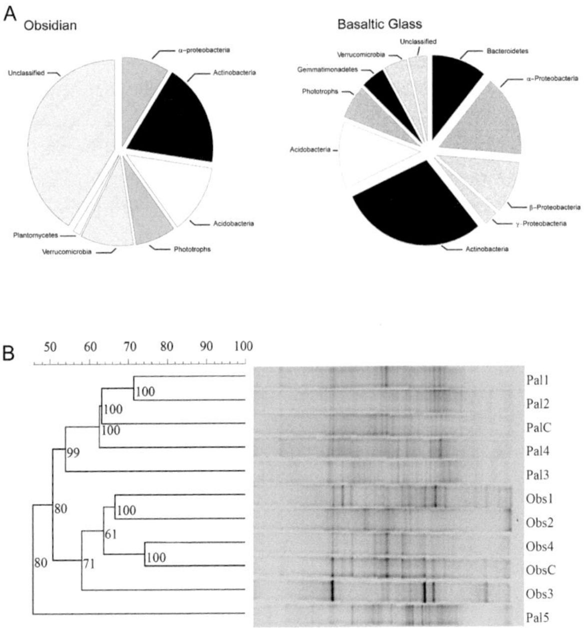

Fig. 2 Diversity varies across materials, but also between samples of the same material. (**A**) Microbial diversity of basaltic glass and obsidian. (**B**) Each sample is a 'microbial island.' DGGE (Denaturing Gradient Gel Electrophoresis) shows that each rock hosts a distinct community. Although particular volcanic materials host distinct types of community, variations occur within those communities at small scales. Thus, multiple samples are required to investigate microbial diversity. In image **B**, cluster analysis of glass microbial communities was based on DGGE (30–70% denaturant) community fingerprints. The Jaccard UPGMA dendrogram was generated from the presence-absence matrix in GelComparII. Values at nodes represent cophenetic correlations. Pal represents basalt glass samples from Valafell, Obs represents rhyolitic glass/obsidian samples from Dómadalshraun. Samples ending in C are composites for the given site (data from Kelly et al. 2010, 2011). Note that obsidian and basalt glass samples tend to cluster into two separate groups, but within each group, the microbial community of each individual sample is distinctive

These published findings suggest that certain factors reliably influence microbial diversity, but that combinations of local environmental factors, such as specific mineralogy and age, uniquely determine the microbial community in any given environment making the *a priori* prediction of which organisms will colonise a given type of environment difficult to achieve. These factors are compounded by biogeographical stochasticity caused by the variations in local microbial populations and their dispersal (Whitaker et al. 2003; Fierer and Jackson 2006).

With respect to sample collection, a notable observation is that in all these studies individual samples of the same material exhibit variations in microbial diversity (Fig. 2). In that sense, we can regard each rock or sample as a microbial island. As sample size becomes smaller, especially in outcrops where rock heterogeneities are on a smaller scale compared to sample size, one might expect that diversity differences could increase as smaller patches of colonies are targeted. At large scales, biogeographical factors (such as environmental differences) will come into play. Thus, within any given sample size, physical, chemical and biological factors are likely to make it extremely unlikely that two samples will exhibit exactly the same microbial diversity.

As with biomass, much of the diversity of surface volcanic environments on Earth is either directly influenced by the oxygenated environment on Earth (e.g., containing aerobic microorganisms) or indirectly (e.g., the presence of electron donors such as organic matter or electron acceptors such as sulfate or oxidised iron). Some of these factors are irrelevant for Mars (such as widespread oxic conditions), while others are relevant; for example, although Mars is generally anoxic, it does host very abundant sulfate and iron oxides (Gendrin et al. 2005; McSween et al. 2009; Grindrod et al. 2012). Extrapolating a theoretical pattern of colonisation of Martian volcanic materials is not possible, since Martian volcanic materials are in some sense uniquely Martian. However, the general lessons we suggest about these environments on the Earth (such as heterogeneous microbial biomass and diversity) would be applicable since physical, chemical and biological factors will influence microbial colonisation of rocks at sub-millimetre to kilometre scales, even if the specific concatenation of those factors was uniquely Martian.

From the point of view of sample collection, these observations emphasise the need for multiple samples to gain statistically meaningful insights into microbial diversity and its variation. These conclusions also apply to the search for relict biomarkers. Indeed, the variation in microbial diversity even at single sample level suggests the conjectural possibility of a rare organism producing a preservable biomarker that was present in one sample, but not another. In Hawai'ian basalts, lipid biomarkers have been shown to vary between samples, illustrating how heterogeneity in microbial diversity also translates into heterogeneity in preservable biomarkers (Brady et al. 2017). Different molecular biomarkers have also been shown to vary in distribution even within the same samples over centimetre scales (Gentry et al. 2017).

6 More Hydrological Activity Does Not Imply Greater Microbial Biomass or Diversity—Select Samples in a Non-discriminatory Way

An implicit assumption in Mars sample selection is the notion that where there is more water or more sustained water, there is likely to be more life. This paradigm has been encouraged by the maxim, 'follow the water' (Hubbard et al. 2002). Although the notion that longer-lived hydrothermal systems or lakes are more likely to sustain long-lived habitable conditions and therefore higher biomass and more diverse microbial communities makes intuitive sense, countervailing results can be observed in many volcanic environments (Fig. 3).

The results of the NASA BASALT study in Hawai'i showed several significant results with respect to Mars sample collection. The hypothesis that active fumaroles (locations where meteoric or magmatic water circulates through active geothermal regions) would be sites of enhanced microbial diversity (since high water availability and the presence of diverse elements in solution would be conducive to life) was not supported (Cockell et al. 2019). Of all sample types, active fumaroles had the lowest diversity. One hypothesis to explain the relatively low diversity of bacterial communities within active fumaroles is that their high temperatures might prohibit colonisation by certain taxa. High temperatures in geothermal settings have previously been found to restrict microbial diversity compared to more mesophilic environments. Sharp et al. (2014) found that in geothermal environments, peak diversity occurred at 24 °C. Thus, merely picking sites that appeared to have hosted active hydrothermal systems may not *a priori* guarantee samples with the highest biological potential. Similarly low diversities are observed in other volcanic Mars-analogue sites where physical and chemical extremes, for example in volcanic lakes, limit life (Hynek et al.

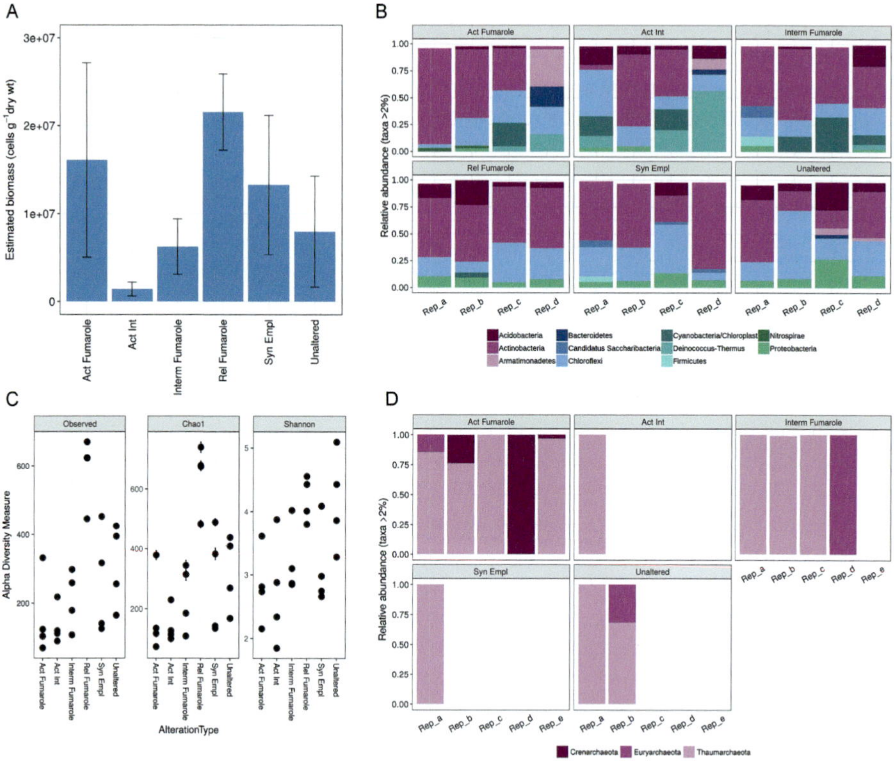

Fig. 3 Both the abundance and diversity of microorganisms do not necessarily correlate directly with the hydrological cycle. Investigation of different volcanic materials in Hawai'i by the NASA BASALT project. The materials are: Active fumaroles (material exposed to active venting within a fumarole), active intermediate (material from the periphery of one of the active fumaroles), intermediate fumarole (material associated with fumaroles showing low levels of activity and/or transient activity, with temperatures between 30–70 °C), relict fumarole (material associated with an inactive fumarole), unaltered (material associated with unweathered basalt), syn-emplacement (material associated with basalt altered during its emplacement primarily by magmatically associated water and gas or meteoric water associated with the heated environment around the lava). The data in (**A**) show that microbial abundance is highly variable and not correlated with material type and its exposure to water. Image **B** shows microbial diversity in the different materials at phylum level. Image **C** shows the microbial diversity of the materials as three different measures of diversity (Observed, Chao 1 and Shannon diversity). Note that relict fumaroles have higher overall bacterial diversity than active fumaroles, but this trend was not obvious for archaeal communities, with one phylum, Crenarchaeota, only occurring in the active fumaroles (**D**). Data from Cockell et al. (2019)

2018). The NASA BASALT study found that as fumaroles transition to a relict state, average alpha diversity increases. This could be explained by the development of less extreme temperature conditions, even though the potential flux of bioavailable nutrients in fumarolic water is reduced.

Despite these findings, fumarolic environments can harbour a diversity of extremophiles in extreme deserts where water flow is enhanced compared to persistently dry environments that surround them (Costello et al. 2009; Soo et al. 2009; Wall et al. 2015) and although extreme physical conditions can limit diversity, fumaroles are nevertheless high priority sites for sample collection and for biological material that may yield insights into deep-branching thermophilic lineages.

The complexity of sample selection is exacerbated by the fact that the most diverse biota may not always be the most likely to produce long-lasting biosignatures. For example, although active fumaroles studied by NASA BASALT had the lowest overall bacterial diversity, they contained the most diverse populations of archaea (Fig. 3; Cockell et al. 2019). Ether-bound intact polar lipids found in archaea are believed to be more stable than ester-bound bacterial lipids and thus expected to survive longer over geological timescales (e.g. Harvey et al. 1986; Logemann et al. 2011). A preferential survival of archaeal lipids could theoretically mean that the active fumaroles would leave a more detectable signature of past life than other locations. The possibility for a lack of correlation between microbial biomass, diversity and preservation potential emphasises the need for multiple samples from diverse hydrological regimens.

From the point of view of Mars sample collection, however, the lesson is clear. Apparently intuitive ideas may not be borne out by microbial ecology and therefore sample collection on Mars should be informed by terrestrial ecology.

Predicting how biomass and diversity would be influenced by a series of diverse hydrological regimes in any given local environment is problematic. Although the paradigm of collecting from more hydrologically active environments is merited, the results from the NASA BASALT project show that ideally a large number of samples from different hydrological regimens should be collected to maximise the chance of finding rock materials that could have potentially hosted the highest biomass and diversity.

These observations on microbial diversity in volcanic environments on Earth suggest that rather than 'follow the water' the overall strategy should be the more cumbrous, but more accurate: 'follow the water to large-scale environments that have been exposed to long-lived hydrological cycles, but collect across all hydrological gradients within that area'.

7 Short-Lived or Recently Formed Habitats May Contain Low Biomass and Diversity

There are very few studies that examine the microbial communities associated with newly formed volcanic materials that can be considered analogous to young materials on Mars or volcanic materials that experienced transient interactions with liquid water. In general, microbial biomass and diversity increases over time in the first years after colonisation of a lava flow, but most studies have focused on lava flows many tens or even hundreds of years old. Over these time periods, it is known that volcanic terrains on Earth make felicitous places for microbial colonisation. It has been known for a long time, particularly from studies before the use of molecular biology, that volcanic materials are colonised early by oxygenic photosynthesisers such as algae and cyanobacteria (Kristinsson 1970, 1974; Schwabe 1970; Schwabe and Behre 1972; Brock 1973; Englund 1976; Henriksson and Henriksson 1982; Adamo and Violante 1991; Adamo et al. 1993; Carson and Brown 1978; Fermani et al. 2007; Stefansdottir et al. 2014). Other studies have also shown lithotrophs as early components of volcanic communities, such as organisms that oxidise volcanically-derived CO and H_2 (King 2003; Dunfield and King 2005; Nanba et al. 2004). Dunfield and King (2005) showed that the diversity of CO-oxidisers (based on amplification of *coxL* genes) was low in a recent (42 years old) lava flow in Hawai'i, but lava between 108 and 300 years old had similar diversity. In volcanic environments where CO and sulphur gases (e.g., H_2S, SO_2) are actively emitted, chemoautotrophic primary production in newly formed volcanic rocks may ultimately generate organic carbon sufficient to sustain heterotrophs.

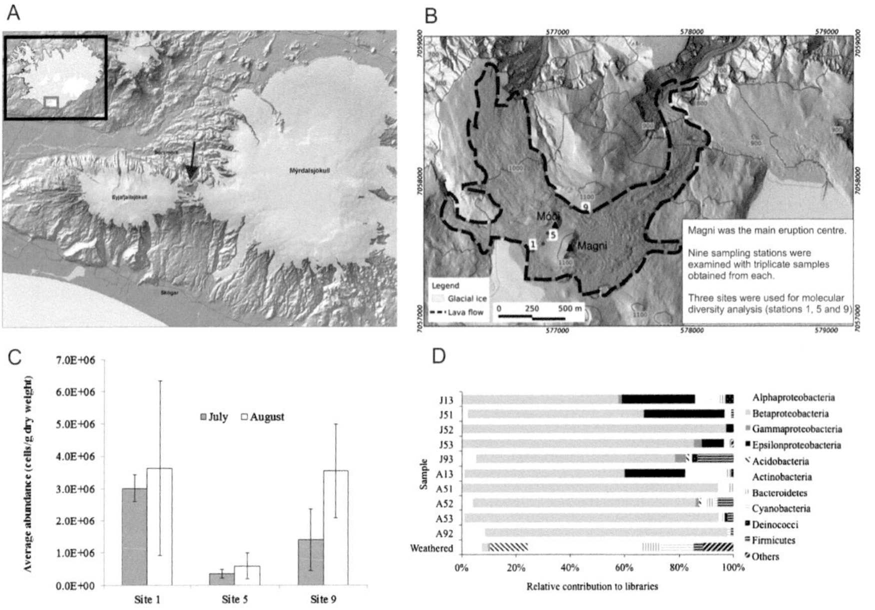

Fig. 4 Newly formed materials and those exposed to transient habitability contain highly heterogeneous, low-diversity populations. The newly formed lava flows of the Fimmvörðuháls lava flow, Eyjafjallajökull, Iceland in 2010 were colonised within 3 to 5 months by a low diversity biota. (**A**) Location of lava flows in Iceland (arrow), (**B**) Map of lava flows showing sample sites 1, 5 and 9 from which diversity analysis was undertaken, (**C**) Microbial abundance measured as direct cell counts at sample sites (error bars are standard deviation) taken on July 5 and August 31, 2010 showing heterogeneity. (**D**) Microbial diversity at phylum level shows communities dominated by Proteobacteria (specifically Betaproteobacteria). 'J' samples are samples collected in July; 'A' samples are samples collected in August. 'Weathered' samples are samples from the basaltic Hnausahraun lava flow, erupted circa 150–300 A.D. and is given for comparison. 'Others' phylum refers to other phyla, each represented by a maximum of two clones in any Fimmvörðuháls library (data from Kelly et al. 2014)

Where studies on the early (less than a year old) colonisation of volcanic terrains have been undertaken, they have shown the extremely low diversity that is often sustained by new volcanic materials, even on Earth, where atmospheric inoculation occurs and microbially-accessible organic carbon is available. A study of the early colonisation of Icelandic lava flows showed that within three months after the eruption of Eyjafjallajökull in 2010, colonisation had occurred (Fig. 4) (Kelly et al. 2014). Biomass in some sites was low, but at others exceeded 10^6 cells/g dry weight (Fig. 4) in samples of $\sim$50 cm^3. Nevertheless, the diversity was low and was dominated by *Betaproteobacteria*, in contrast to older lava flows that typically contain an order of magnitude more biomass and have higher abundance of *Acidobacteria* and *Actinobacteria* (Kelly et al. 2010, 2011; Byloos et al. 2018). Although some of the organisms on the fresh lava may have used organic material atmospherically delivered to the surface of the lava flows, the presence of chemoautotrophic members of the *Thiobacillus* genus suggested that some of the population may have used iron and sulphur from the primary volcanic materials (Kelly et al. 2014).

It is a significant point that arguments about the evolutionary likelihood of particular metabolisms on Mars (i.e. whether it is relevant to discuss searching for biomarkers of the relatively biochemically and evolutionarily advanced oxygenic photosynthesis compared to chemolithotrophic metabolisms) are irrelevant to the points made in this paper. All metabolic and taxonomic groups of microorganisms examined on early lava flows on Earth have shown a heterogeneous distribution with a low biomass and diversity. These observations emphasise the points in Sects. 4–6. If volcanic materials have not been exposed to long-lived hydrological regimens and mature microbial communities have not been established, then the heterogeneous distribution of biomass and diversity may be exacerbated by overall low biomass, emphasising the need for multiple samples and suitable sample size.

8 Is It Possible for Samples to Be Habitable, but Devoid of Life?

Is there a possibility of collecting samples on Mars that fully satisfy the conditions for habitability but are devoid of a biota? If Mars never hosted an origin of life or a transfer of life to that planet then all environments that have energy sources, CHNOPS elements and liquid water and are habitable by terrestrial standards would be devoid of life. However, even if Mars has hosted life, transient habitats such as near-surface water melted by impacts could be devoid of life if that habitat remained disconnected from inhabited regions (Cockell 2011). If sample collection is undertaken within a confined area (for example, a single impact crater or outflow channel), the possibility of sampling a once habitable, but never inhabited palaeoenvironment is conceivable. As we cannot quantify this likelihood in the absence of a knowledge of life on Mars, this theoretical possibility would suggest that collecting samples from different environments, not merely within the same local environment but across much wider areas, should be a priority to maximise the chances of collecting samples containing life (Cockell and McMahon 2019). It is important to note that if Mars was lifeless, this strategy will also strengthen the conclusion that the planet was lifeless by multiplying the number of sampled habitable environments that can be shown to be devoid of life.

An important distinction is whether uncolonised regions in a sample are devoid of life because they lack some fundamental requirement for life making them uninhabitable, or whether they are habitable zones that can support life, but they have not yet been colonised. For example, uncolonised regions of obsidian studied by Herrera et al. (2008) are likely uncolonised because of the lack of freely available iron. In other words, they may be uninhabitable.

On Earth, habitable samples devoid of life at the centimetre scale do seem to be rare. As organisms reproduce, fluid flow tends to carry them into new habitable spaces so that at micron-scales, some part of that habitable space is colonised. Although not all habitable space will be colonised, the movement of microbes does lead to the ecological reality that when habitable samples are examined at centimetre scales, they usually contain life.

Real examples on Earth of pervasive centimetre-scale habitable samples devoid of life are recently formed lava flows in which hours to days after cooling to below the upper temperature limit for life, microbes are so heterogeneous and sparse that at centimetre scales some rocks remain uncolonised, particularly those beneath the surface of the flow that have not been subject to atmospheric colonisation (Cockell 2014).

These ecological considerations lead to relevant observations. In general, where life has accessed an environment, fluid flow ensures that even though space can be habitable but unoccupied, at macroscopic centimetre scales samples generally contain examples of life. In environments where habitable conditions were extremely transient, this assumption may break down and habitable samples devoid of life at centimetre scales may be prevalent. In the specific case of Mars, if the planet was lifeless or the biomass extremely localised to certain regions, entire regions or environments may be habitable but devoid of life. Large samples suites are required to test these hypotheses.

An important question is how many samples are minimally required to determine that a given material is devoid of life in all examples of that material. On the Earth, a high level of biological confidence might be obtained by showing that life was not present in three sets of three triplicates of such a material across one location with this sample set repeated across three geographically separated regions of that material, whether on the surface or subsurface.

It may never be possible to demonstrate conclusively that Mars has always been lifeless, since short-lived, localised pockets of life could have left scant or undetectable evidence, or no evidence at all. However, if the sampling regimen described in the previous paragraph was deployed across the geological records of diverse palaeohabitats with good potential to preserve biosignatures (e.g., lake sediments, evaporitic deposits, and hydrothermal sinter deposits) and no evidence of life was found, the balance of evidence would shift in favour of the hypothesis that Mars has never been inhabited. In principle, a Bayesian approach could be developed to model this shift; these statistical issues merit greater attention.

9 A Strategy for Sample Collection on Mars

In this section, we summarise six recommendations for sample collection on Mars based on observations of the microbial ecology of volcanic materials on Earth. We provide a brief rationale for each. The strategy does not take into account cost, mass, or space restrictions associated with specific missions such as robotic Mars sample return missions (Sherwood et al. 2003; MEPAG 2008; iMOST 2018), but rather provides an insight into a biologically ideal sample suite, whether it be collected by robot or human explorers.

9.1 Collect Sufficiently Large Samples

Heterogeneity in microbial biomass and diversity has been observed in all volcanic materials. These effects are likely to be caused by physical, chemical and biological influences that will vary in different samples. 'Sufficient' size cannot be easily quantified, but observations on volcanic rocks discussed here suggest that typical sample sizes should be a minimum 10 cm^3 to retain millimetre-scale vesicles, fractures and mineral grains and their associated biological targets. Defining ideal samples based on mass is less important than volume. For example, a lower density sample can contain more pore space for potential microbial colonisation than a higher density sample, counterintuitively making a lighter sample of comparable volume potentially more valuable to a biologist than a more massive sample. Less is sometimes more.

9.2 Collect Sufficiently Many Samples

Heterogeneity in samples, particularly in those where habitability is short-lived, results in high standard deviations in biomass and diversity. Statistically, to derive meaningful statements on the distribution of life, multiple samples should be collected. Three samples is a typical statistical minimum. However, microbial ecological analysis yields suggestions on preferred numbers. For example, based on sample strategies used to investigate heterogeneities in low biomass volcanic rocks (Kelly et al. 2014) sample collection numbers such as nine provide for a more comprehensive analysis, made up of three independent sets of three triplicates from the same material at regional scale. Additional samples add to the robustness of the study, but might be limited by field time and cost. If environments are devoid of life, large sample sizes can increase the robustness of the conclusion that they were never inhabited. These sample number recommendations do not take into account separate samples used in other studies such as geology, atmospheric history and investigations for future resource use that might be collected in the limited sample sets envisaged for robotic Mars sample return (e.g., Sherwood et al. 2003; MEPAG 2008; iMOST 2018).

9.3 Collect Samples of the Same Material from Different Locations

Following directly from point (2) is the recognition that the heterogeneity in the biomass and diversity of life is related to large-scale biogeographical influences. It is therefore important to collect samples of the same material, but from different locations separated by ~kilometre distances. This allows for stronger statistical statements about the distribution of life over large scales. If true replication is to be achieved, avoiding the pitfall of pseudo-replication by sampling the same region, then the sample suite discussed in point (2) must be repeated in three completely geographically separated regions of the same material, yielding 27 samples in total of a given sample type. If a habitable environment was transient, this strategy maximises the chances of obtaining life-bearing rocks or being able to conclude with greater confidence that all such materials in all geographical locations are devoid of life.

9.4 Collect Across Varied Hydrological Regimes

'Follow the water' is broadly a correct view with respect to the fundamental requirement that all known life has used liquid water as a solvent. However, microbial ecology shows that locations with a high abundance of liquid water do not always harbour the most biomass and diversity. As we have discussed here, in volcanic hydrothermal regimens, high temperatures can limit life to lower biomass and diversity than drier locations at more mesophilic temperatures. It is well known that in some locations liquid water can even be uninhabitable. For example, on the Earth, magnesium chloride brines can have water activities so low as to be uninhabitable (Hallsworth et al. 2007) and such brines may exist on Mars. The lessons from volcanic environments show that sampling should be carried out across hydrological gradients and not merely in locations that appear to have hosted the most vigorous hydrological regimes. Transects capture the variation in the interaction between geochemistry and physical conditions that can influence habitability in a way independent of the quantity of liquid water.

9.5 Maintain Sample Integrity

Cells are distributed heterogeneously and their locations provide information on the factors that control life (e.g. presence of iron-bearing crystals, alteration rinds, fracture networks). It is therefore essential to maintain sample integrity. Where possible they should be collected with minimal damage caused by drilling, excavation etc. Sample integrity also includes the need to collect aseptically to avoid external biological contamination.

9.6 Minimise Post-Collection Modifications

Once collected, samples should receive minimal alteration. This includes irradiation and exposure to other physical and chemical agents. Where samples are disrupted for examination, information on the orientation of the sample fragments and their locations should be maintained so that the geological factors that control microbial distribution within the material can be reconstructed.

10 Conclusions

Mars is a planet of volcanically derived materials. The search for life on Mars and the study of the distribution of habitable conditions is therefore greatly advanced by investigating the microbial ecology of similar environments on Earth where biomass and diversity can be directly quantified and related to environmental variables with a statistically high level of confidence. Keeping in mind environmental differences between present-day Earth and Mars, one purpose of this paper has been to provide a review of well-studied volcanic materials and their enclosed biota. We have used volcanic environments on Earth to extract general patterns of microbial growth that are applicable across many types of volcanic material. Using these patterns, we suggest principles of sample collection that, combined with appropriate taphonomic considerations, will yield maximum information for astrobiologists and planetary scientists. These principles apply to robotic exploration and to the human exploration of the planet. Ultimately, to be able to test hypotheses with high statistical confidence, sample sets with sufficient coverage and diversity, exploration across large scales, and multiple replicates are required.

Acknowledgements This paper was made possible by support from the Science and Technology Facilities Council (STFC). Grant: ST/R000875/1.

References

P. Adamo, P. Violante, Weathering of volcanic rocks from Mt. Vesuvius associated with the lichen *Stereocaulum vesuvianum*. Pedobiologia **35**, 209–217 (1991)

P. Adamo, A. Marchetiello, P. Violante, The weathering of mafic rocks by lichens. Lichenologist **25**, 285–297 (1993)

E.S. Amador, M.L. Cable, N. Chaudry, T. Cullen, D. Gentry, M.B. Jacobsen, G. Murukesan, E.W. Schwieterman, A.H. Stevens, A. Stockton, C. Yin, D.C. Cullen, W. Geppert, Synchronous in-field application of life-detection techniques in planetary analog missions. Planet. Space Sci. **106**, 1–10 (2015)

A.M. Anesio, S. Lutz, N.A.M. Chismas, L.G. Benning, The microbiome of glaciers and ice sheets. NPJ Biofilms Microbiomes **3**, 10 (2017)

O. Arnalds, Volcanic soils of Iceland. Catena **56**, 3–20 (2004)

O. Arnalds, P. Dagsson-Waldhauserova, H. Olafsson, The Icelandic volcanic aeolian environment: Processes and impacts—A review. Aeolian Res. **20**, 176–195 (2016)

W. Bach, K.J. Edwards, Iron and sulphide oxidation within the basaltic ocean crust: Implications for chemolithoautotrophic microbial biomass production. Geochim. Cosmochim. Acta **67**, 3871–3887 (2003)

R.A. Bell, Cryptoendolithic algae of hot semi-arid lands and deserts. J. Phycol. **29**, 133–139 (1993)

V.P. Belobrov, S.V. Ovechkin, Soils and soil cover patterns of volcanic plateaus in Indochina, Eurasian. Soil Sci. **38**, 1065–1074 (2005)

C.A. Benson, R.W. Bizzoco, D.A. Lipson, S.T. Kelley, Microbial diversity in nonsulfur, sulfur and iron geothermal steam vents. FEMS Microbiol. Ecol. **76**, 74–88 (2011)
J.-P. Bibring, Y. Langevin, J.F. Mustard, F. Poulet, R. Arvidson, A. Gendrin, B. Gondet, N. Mangold, P. Pinet, F. Forget (the OMEGA team), Global mineralogical and aqueous Mars history derived from OMEGA/Mars Express data. Science **312**, 400–404 (2006)
A.L. Brady, G.F. Slater, E. Gibbons, S. Kobs Nawotniak, S.S. Hughes, S. Payler, A. Stevens, C.S. Cockell, C. Haberle, A. Sehlke, R.C. Elphic, D.S.S. Lim, in *Detection of Microbial Organic Biomarkers in Terrestrial Basalts in a Mars Analogue Environment*, AbSciCon, Mesa, AZ (2017)
T.F. Bristow, E.B. Rampe, C.N. Achilles, D.F. Blake, S.J. Chipera, P. Craig, J.A. Crisp, D.J. Des Marais, R.T. Downs, R. Gellert, J.P. Grotzinger, S. Gupta, R.M. Hazen, B. Horgan, J.V. Hogancamp, N. Mangold, P.R. Mhaffy, A.C. McAdam, D.W. Ming, J.M. Morookian, R.V. Morris, S.M. Morrison, A.H. Treiman, D.T. Vaniman, A.R. Vasavada, A.S. Yen, Clay mineral diversity and abundance in sedimentary rocks of Gale Crater, Mars. Sci. Adv. **4**, eaar3330 (2018)
T.D. Brock, Primary colonization of Surtsey, with special reference to the blue-green algae. Oikos **24**, 239–243 (1973)
B. Byloos, P. Monsiers, M. Mysara, N. Leys, N. Boon, R. Van Houdt, Characterisation of the bacterial communities on recent Icelandic volcanic deposits of different ages. BMC Microbiol. **18**, 122 (2018)
N.A. Cabrol, E.A. Grin, The evolution of lacustrine environments on Mars: (is Mars only hydrologically dormant?). Icarus **149**, 291–328 (2001)
N.A. Cabrol, D. Wettergreen, K. Warren-Rhodes, E.A. Grin, J. Moersch, G. Chong Diaz, C.S. Cockell, P. Coppin, C. Demergasso, J.M. Dohm, L. Ernst, G. Fisher, J. Glasgow, C. Hardgrove, A.N. Hock, D. Jonak, L. Marinangeli, E. Minkley, G. Gabriele Ori, J. Piatek, E. Pudenz, T. Smith, K. Stubbs, G. Thomas, D. Thompson, A. Waggoner, M. Wagner, S. Weinstein, M. Wyatt, Life in the Atacama: Searching for life with rovers (science overview). J. Geophys. Res. **112**, G04 (2007)
E.J. Carpenter, S. Lin, D.G. Capone, Bacterial activity in South pole snow. Appl. Environ. Microbiol. **66**, 4514–4517 (2000)
J.L. Carson, R.M. Brown, Studies of Hawaiian freshwater and soil algae. 2. Algal colonization and succession on a dated volcanic substrate. J. Phycol. **14**, 171–178 (1978)
C.S. Cockell, Vacant habitats in the Universe. Trends Ecol. Evol. **26**, 73–80 (2011)
C.S. Cockell, Types of habitat in the Universe. Int. J. Astrobiol. **13**, 158–164 (2014)
C.S. Cockell, S. McMahon, Lifeless Martian samples and their significance. Nat. Astron. **3**, 468–470 (2019)
C.S. Cockell, D. Catling, W.L. Davis, R.N. Kepner, P.C. Lee, K. Snook, C.P. McKay, The ultraviolet environment of Mars: Biological implications past, present and future. Icarus **146**, 343–359 (2000)
C.S. Cockell, K. Olsson-Francis, F. Knowles, L.C. Kelly, A. Herrera, T. Thorsteinsson, V. Marteinsson, Bacteria in weathered basaltic glass, Iceland. Geomicrobiol. J. **26**, 491–507 (2009)
C.S. Cockell, L.C. Kelly, S. Summers, V. Marteinsson, Following the kinetics: Iron-oxidising microbial mats in cold Icelandic volcanic habitats and their rock-associated carbonaceous signature. Astrobiology **11**, 679–694 (2011)
C.S. Cockell, L.C. Kelly, V. Marteinsson, Actinobacteria: An ancient phylum active in volcanic rock weathering. Geomicrobiol. J. **30**, 706–720 (2013)
C.S. Cockell, J.P. Harrison, A. Stevens, S.J. Payler, S.S. Hughes, S.E. Kobs Nawotniak, A.L. Brady, R.C. Elphic, C.W. Haberle, A. Sehlke, K. Beaton, A. Abercromby, P. Schwendner, J. Wadsworth, H. Landenmark, R. Cane, A.W. Dickinson, N. Nicholson, L. Perera, D.S.S. Lim, A low diversity microbiota inhabits extreme terrestrial basaltic terrains and their fumaroles: Implications for the exploration of Mars. Astrobiology **19**, 284–299 (2019)
E.K. Costello, S.R.P. Halloy, S.C. Reed, P. Sowell, S.K. Schmidt, Fumarole-supported islands of biodiversity within a hyperarid, high-elevation landscape on Socompa Volcano, Puna de Atacama, Andes. Appl. Environ. Microbiol. **75**, 735–747 (2009)
C.R. Cousins, I.A. Crawford, Volcano-ice interaction as a microbial habitat on Earth and Mars. Astrobiology **11**, 695–710 (2011)
C.R. Cousins, M. Fogel, R. Bowden, I. Crawford, A. Boyce, C.S. Cockell, M. Gunn, Biogeochemical probing of microbial communities in a basalt-hosted hot spring at Kverkfjöll volcano, Iceland. Geobiology **16**, 507–521 (2018)
K.V. Dahigaonkar, P.N. Chavan, Diversity of microorganisms associated with pristine and extreme environment of mud volcanoes. Int. J. Sci. Res. Publ. **8**, 157–164 (2018)
L.R. Dartnell, L. Desorgher, J. Ward, A. Coates, Modelling the surface and subsurface Martian radiation environment: Implications for astrobiology. Geophys. Res. Lett. **34**, L02207 (2007)
C.J. Daughney, J.-P. Rioux, D. Fortin, T. Pichler, Laboratory investigation of the role of bacteria in the weathering of basalt near deep sea hydrothermal vents. Geomicrobiol. J. **21**, 21–31 (2004)
J.R. de la Torre, B.M. Goebel, E.I. Friedmann, N.R. Pace, Microbial diversity of cryptoendolithic communities from the McMurdo Dry Valleys, Antarctica. Appl. Environ. Microbiol. **69**, 3858–3867 (2003)

W.W. Dickinson, M.R. Rosen, Antarctic permafrost: An analogue for water and diagenetic minerals on Mars. Geology **31**, 199–202 (2003)
K.E. Dunfield, G.M. King, Analysis of the distribution and diversity in recent Hawaiian volcanic deposits of a putative carbon monoxide dehydrogenase large subunit gene. Environ. Microbiol. **7**, 1405–1412 (2005)
K.J. Edwards, D.R. Rogers, C.O. Wirsen, T.M. McCollom, Isolation and characterisation of novel psychrophilic, neutrophilic, Fe-oxidising, chemolithoautotrophic α- and γ-Proteobacteria from the deep sea. Appl. Environ. Microbiol. **69**, 2906–2913 (2003)
B. Ehlmann, J.F. Mustard, G.A. Swayze, R.N. Clark, J.L. Bishop, F. Poulet, D.J. Des Marais, L.H. Roach, R.E. Milliken, J.J. Wray, O. Barnouin-Jha, S.L. Murchie, Identification of hydrated silicate minerals on Mars using MRO-CRISM: Geologic context near Nili Fossae and implications for aqueous alteration. J. Geophys. Res. **144**, E00D08 (2009)
B. Ehlmann, J.F. Mustard, S.L. Murchie, J.-P. Bibring, A. Meunier, A.A. Fraeman, Y. Langevin, Subsurface water and clay mineral formation during the early history of Mars. Nature **479**, 53–60 (2011)
B. Englund, Nitrogen fixation by free-living microorganisms on the lava field of Heimaey, Iceland. Oikos **27**, 428–432 (1976)
K.A. Farley, C. Malespin, P. Mahaffy, J.P. Grotzinger, P.M. Vasconcelos, R.E. Milliken et al., In situ radiometric and exposure age dating of the Martian surface. Science **343**, 1247166 (2014)
P. Fermani, G. Mataloni, B. Van de Vijver, Soil microalgal communities on an Antarctic active volcano (Deception Iceland, South Shetlands). Polar Biol. **30**, 1381–1393 (2007)
C.B. Field, M.J. Behrenfeld, J.T. Randerson, P. Falkowski, Primary production in the biosphere: Integrating terrestrial and oceanic components. Science **281**, 237–240 (1998)
N. Fierer, R.B. Jackson, The diversity and biogeography of soil bacterial communities. Proc. Natl. Acad. Sci. **103**, 626–631 (2006)
F. Garcia-Pichel, S.L. Johnson, D. Youngkin, J. Belnap, Small-scale vertical distribution of bacterial biomass and diversity in biological soil crusts from arid lands in the Colorado Plateau. Microb. Ecol. **46**, 312–321 (2003)
D. Gat, Y. Mazar, E. Cytryn, Y. Rudich, Origin-dependent variations in the atmospheric microbiome community in Eastern Mediterranean dust storms. Environ. Sci. Technol. **51**, 6709–6718 (2017)
P.M. Gaylarde, A. Jungblut, C.C. Gaylarde, B.A. Neilan, Endolithic phototrophs from an active geothermal region in New Zealand. Geomicrobiol. J. **23**, 579–587 (2006)
A. Gendrin, N. Mangold, J.-P. Bibring, Y. Langevin, B. Gondet, F. Poulet, G. Bonello, C. Quantin, J. Mustard, R. Arvidson, S. LeMouélic, Sulfates in Martian layered terrains: The OMEGA/Mars Express view. Science **307**, 1587–1591 (2005)
D.M. Gentry, E.S. Amador, M.L. Cable, N. Chaudry, T. Cullen, M.B. Jacobsen, G. Murukesan, E.W. Schwieterman, A.H. Stevens, A. Stockton, G. Tan, C. Yin, D.C. Cullen, W. Geppert, Correlations between life-detection techniques and implications for sampling site selection in planetary analog missions. Astrobiology **17**, 1009–1021 (2017)
V. Gomez-Alvarez, G.M. King, K. Nüsslein, Comparative bacterial in recent Hawaiian volcanic deposits of different ages. FEMS Microbiol. Ecol. **60**, 60–73 (2006)
P.M. Grindrod, M. West, N.H. Warner, S. Gupta, Formation of an Hesperian-aged sedimentary basin containing phyllosilicates and sulfates in Coprates Catena, Mars. Icarus **218**, 178–195 (2012)
W. Gross, J. Küver, G. Tischendorf, N. Bouchaala, W. Büsch, Cryptoendolithic growth of the red alga Galdieria sulphuraria in volcanic areas. Eur. J. Phycol. **33**, 25–31 (1998)
J.P. Grotzinger, D.Y. Sumner, L.C. Kah, K. Stack, S. Gupta, L. Edgar, D. Rubin, K. Lewis, J. Schieber, N. Mangold, R. Milliken, P.G. Conrad, D. DesMarais, J. Farmer, K. Siebach, F. Calef, J. Hurowitz, S.M. McLennan, D. Ming, D. Vaniman, J. Crisp, A. Vasavada, K.S. Edgett, M. Malin, D. Blake, R. Geliert, P. Mahaffy, R.C. Wiens, S. Maurice, J.A. Grant, S. Wilson, R.C. Anderson, L. Beegle, R. Arvidson, B. Hallet, R.S. Sletten, M. Rice, J. Bell, J. Griffes, B. Ehlmann, R.B. Anderson, T.F. Bristow, W.E. Dietrich, G. Dromart, J. Eigenbrode, A. Fraeman, C. Hardgrove, K. Herkenhoff, L. Jandura, G. Kocurek, S. Lee, L.A. Leshin, R. Leveille, D. Limonadi, J. Maki, S. McCloskey, M. Meyer, M. Minitti, H. Newsom, D. Oehler, A. Okon, M. Palucis, T. Parker, S. Rowland, M. Schmidt, S. Squyres, A. Steele, E. Stolper, R. Summons, A. Treiman, R. Williams, A. Yingst (MSL Science Team), A habitable fluvio-lacustrine environment at Yellowknife Bay, Gale Crater, Mars. Science **343**, 1242777 (2013)
J.E. Hallsworth, M.M. Yakimov, P.N. Golyshin, J.L. Gillion, G. D'Auria, F. de Lima Alves, V. La Cono, M. Genovese, B.A. McKew, S.L. Hayes, G. Harris, L. Giuliano, Limits of life in $MgCl_2$-containing environments: Chaotropicity defines the window. Environ. Microbiol. **9**, 801–813 (2007)
J.J. Halvorson, E.H. Franz, J.L. Smith, R.A. Black, Nitrogenase activity, nitrogen fixation, and nitrogen inputs by lupines at Mount St. Helens. Ecology **73**, 87–98 (1992)
H.R. Harvey, R.D. Fallon, J.S. Patton, The effect of organic matter and oxygen on the degradation of bacterial membrane lipids in marine sediments. Geochim. Cosmochim. Acta **50**, 795–804 (1986)

L.E. Hays, H.V. Graham, D.J. Des Marais, E.M. Hausrath, B. Horgan, T.M. McCollom, M.N. Parenteau, S.L. Potter-McIntrye, A.J. Williams, K.L.K. Lynch, Biosignature preservation and detection in Mars analog environments. Astrobiology **17**, 363–400 (2017)

J.W. Head, H. Hiesinger, M.A. Ivanov, M.A. Kreslavsky, S. Pratt, B.J. Thomson, Possible ancient oceans on Mars: Evidence from Mars Orbiter Laser Altimeter Data. Science **286**, 2134–2137 (1999)

J.W. Head, J.F. Mustard, M.A. Kreslavsky, R.E. Milliken, D.R. Marchant, Recent ice ages on Mars. Nature **426**, 797–802 (2003)

M.H. Hecht, S.P. Kounaves, R.C. Quinn, S.J. West, S.M. Young, D.W. Ming, D.C. Catling, B.C. Clark, W.V. Boynton, J. Hoffman, L.P. Deflores, K. Gospodinova, J. Kapit, P.H. Smith, Detection of perchlorate and the soluble chemistry of martian soil at the Phoenix lander site. Science **324**, 64–67 (2009)

L.E. Henriksson, E. Henriksson, Concerning the biological nitrogen fixation on Surtsey. Surtsey Res. Prog. Rep. IX, 9–12 (1982)

A. Herrera, C.S. Cockell, S. Self, M. Blaxter, J. Reitner, G. Arp, W. Dröse, A. Tindle, Bacterial colonization and weathering of terrestrial obsidian rock in Iceland. Geomicrobiol. J. **25**, 25–37 (2008)

A. Herrera, C.S. Cockell, S. Self, M. Blaxter, J. Reitner, T. Thorsteinsson, G. Arp, W. Dröse, A. Tindle, A cryptoendolithic community in volcanic glass. Astrobiology **9**, 369–381 (2009)

D.W. Hopkins, L. Badalucco, L.C. English, S.M. Meli, J.A. Chudek, A. Ioppolo, Plant litter decomposition and microbial characteristics in volcanic soils (Mt Etna, Sicily) at different stages of development. Biol. Fertil. Soils **43**, 461–469 (2007)

B. Horgan, J.F. Bell, Widespread weathered glass on the surface of Mars. Geology **40**, 391–394 (2012)

Z. Huang, S. Xu, Y. Han, Y. Gao, Y. Wang, C. Song, Comparative study of Dushanzi and Baiyanggou mud volcano microbial communities in Junggar Basin in Xinjiang, China. Int. Res. J. Public Environ. Health **3**, 244–256 (2016)

G.S. Hubbard, F.M. Naderi, J.B. Garvin, Following the water, the new program for Mars exploration. Acta Astronaut. **51**, 337–350 (2002)

J.A. Hurowitz, J.P. Grotzinger, W.W. Fischer, S.M. McLennan, R.E. Milliken, N. Stein, A.R. Vasavada, D.F. Blake, E. Dehouck, J.L. Eigenbrode, A.G. Fairén, J. Frydenvang, R. Gellert, J.A. Grant, S. Gupta, K.E. Herkenhoff, D.W. Ming, E.B. Rampe, M.E. Schmidt, K.L. Siebach, K. Stack-Morgan, D.Y. Sumner, R.C. Wiens, Redox stratification of an ancient lake in Gale Crater, Mars. Science **356**, 6341 (2017)

B.M. Hynek, K.L. Rogers, M. Antunovich, G. Avard, G.E. Alvarado, Lack of microbial diversity in an extreme Mars analog setting: Poás Volcano, Costa Rica. Astrobiology **18**, 923–933 (2018)

A.M. Ibekwe, A.C. Kennedy, J.J. Halvorson, C.H. Yang, Characterization of developing microbial communities in Mount St. Helens pyroclastic substrate. Soil Biol. Biochem. **39**, 2496–2507 (2007)

iMOST, The potential science and engineering value of samples delivered to Earth by Mars sample return, (co-chairs D.W. Beaty, M.M. Grady, H.Y. McSween, E. Sefton-Nash; documentarian B.L. Carrier; plus 66 co-authors), 186 p. white paper (2018)

L.C. Kelly, C.S. Cockell, Y.M. Piceno, G. Andersen, T. Thorsteinsson, V. Marteinsson, Bacterial diversity of weathered terrestrial Icelandic volcanic glasses. Microb. Ecol. **60**, 740–752 (2010)

L.C. Kelly, C.S. Cockell, A. Herrera-Belaroussi, Y. Piceno, G. Andersen, T. DeSantis, E. Brodie, T. Thorsteinsson, V. Marteinsson, F. Poly, X. LeRoux, Bacterial diversity of terrestrial crystalline volcanic rocks, Iceland. Microb. Ecol. **62**, 69–79 (2011)

L.C. Kelly, C.S. Cockell, T. Thorsteinsson, V. Marteinsson, J. Stevenson, Pioneer microbial communities of the Fimmvörðuháls Lava Flow, Eyjafjallajökull, Iceland. Microb. Ecol. **68**, 504–518 (2014)

J.M. Kimble, C.L. Ping, M.E. Sumner, L.P. Wilding, Andosols, in *Handbook of Soil Science*, ed. by M.E. Sumner (CRC Press, Boca Raton, 2000), pp. E209–E224

G.M. King, Contributions of atmospheric CO and hydrogen uptake to microbial dynamics on recent Hawaiian volcanic deposits. Appl. Environ. Microbiol. **69**, 4067–4075 (2003)

H. Kristinsson, Report on the lichenological work on Surtsey and in Iceland. Surtsey Research Progress Report V, 52 (1970)

H. Kristinsson, Lichen colonization in Surtsey 1971–1973. Surtsey Research Progress Report VII, 9–16 (1974)

J. Lasue, N. Mangold, E. Hauber, S.M. Clifford, W. Feldman, O. Gasnault, C. Grima, S. Maurice, O. Mousis, Quantitative assessments of the Martian hydrosphere. Space Sci. Rev. **174**, 155–212 (2013)

J. Lasue, A. Cousin, P.-Y. Meslin, N. Mangold, R.C. Wiens, G. Berger, R.C. Wiens, E. Dehouck, O. Forni, W. Goetz, O. Gasnault, W. Rapin, S. Schroeder, A. Ollila, J. Johnson, S. Le Mouélic, S. Maurice, R. Anderson, D. Blaney, B. Clark, S.M. Clegg, C. d'Uston, C. Fabre, N. Lanza, M.B. Madsen, J. Martin-Torres, N. Melikechi, H. Newsom, V. Sautter, M.P. Zorzano, Martian eolian dust probed by ChemCam. Geophys. Res. Lett. **45**, 10968–10977 (2018). https://doi.org/10.1029/2018GL079210

M.J. Le Bas, R.W. Le Maitre, A.R. Woolley, The construction of the total alkali-silica chemical classification of volcanic rocks. Mineral. Petrol. **46**, 1–22 (1992)

J.E. Lee, H.L. Buckley, R.S. Rampal, G. Lear, Both species sorting and neutral processes drive assembly of bacterial communities in aquatic microcosms. FEMS Microbiol. Ecol. **86**, 288–302 (2013)
D.S.S. Lim, A.J.F. Abercromby, S. Kobs Nawotniak, D.S. Lees, M.J. Miller, BASALT Team, The BASALT Research Program: Designing and developing mission elements in support of human scientific exploration of Mars. Astrobiology **16**, 245–259 (2019)
Q. Liu, Y.-G. Zhou, Y.-H. Xin, High diversity and distinctive community structure of bacteria on glaciers in China revealed by 454 pyrosequencing. Syst. Appl. Microbiol. **38**, 578–585 (2015)
J. Logemann, J. Graue, K. Köster, B. Engelen, J. Rullkötter, H. Cypionka, A laboratory experiment of intact polar lipid degradation in sandy sediments. Biogeosciences **8**, 2547–2560 (2011)
H.P. Lukito, K. Kouno, T. Ando, Phosphorus requirements of microbial biomass in a regosol and an andosol. Soil Biol. Biochem. **30**, 865–872 (1998)
K. Lysnes, I.H. Thorseth, B.O. Steinbu, L. Øvreås, T. Torsvik, R.B. Pedersen, Microbial community diversity in seafloor basalt from the Arctic spreading ridges. FEMS Microb. Ecol. **50**, 213–230 (2004)
L. Maturrano, F. Santos, R. Rosselló-Mora, J. Antón, Microbial diversity in Maras Salterns, a hypersaline environment in the Peruvian Andres. Appl. Environ. Microbiol. **72**, 3887–3895 (2006)
S. McMahon, T. Boask, J.P. Grotzinger, R.E. Milliken, R.E. Summons, M. Daye, S.A. Newman, A. Fraeman, K.H. Williford, D.E.G. Briggs, A field guide to finding fossils on Mars. J. Geophys. Res. **123**, 1012–1040 (2018)
H.Y. McSween, G.J. Taylor, M.B. Wyatt, Elemental composition of the Martian crust. Science **324**, 736–739 (2009)
MEPAG, Science priorities for Mars sample return. Astrobiology **8**, 491–521 (2008)
R.E. Milliken, D.L. Bish, Sources and sinks of clay minerals on Mars. Philos. Mag. **90**, 2293–2308 (2010)
K. Nanba, G.M. King, K. Dunfield, Analysis of facultative lithotroph distribution and diversity of volcanic deposits by use of the large subunit of Ribulose 1,5-bisphosphate carboxylase/oxygenase. Appl. Environ. Microbiol. **70**, 2245–2253 (2004)
R. Navarro-González, F.A. Rainey, P. Molina, D.R. Bagaley, B.J. Hollen, J. de la Rosa, A.M. Small, R.C. Quinn, F.J. Grunthaner, L. Cáceres, B. Gomez-Silva, C.P. McKay, Mars-like soils in the Atacama Desert, Chile, and the dry limit of microbial life. Science **302**, 1018–1021 (2003)
K. Nüsslein, J.M. Tiedje, Characterization of the dominant and rare members of a young Hawaiian soil bacterial community with small-subunit ribosomal DNA amplified from DNA fractionated on the basis of its guanine and cytosine composition. Appl. Environ. Microbiol. **64**, 1283–1289 (1998)
S.L. O'Brien, S.M. Gibbons, S.M. Owens, J. Hampton-Marcell, E.R. Johnston, J.D. Jastrow, J.A. Gilbert, F. Meyer, D.A. Antonopoulos, Spatial scale drives patterns in soil bacterial diversity. Environ. Microbiol. **18**, 2039–2051 (2016)
B. Orcutt, B. Bailey, H. Staudigel, B.M. Tebo, K.J. Edwards, An interlaboratory comparison of 16S-rRNA gene-based terminal restriction fragment length polymorphism and sequencing methods for assessing microbial diversity of seafloor basalts. Environ. Microbiol. **11**, 1728–1738 (2009)
A. Oren, Microbial life at high salt concentrations: Phylogenetic and metabolic diversity. Saline Syst. **4**, 2 (2008)
H. Oskarsson, O. Arnalds, J. Gudmundsson, G. Gudbergsson, Organic carbon in Icelandic Andosols: Geographical variation and impact of erosion. Catena **56**, 225–238 (2004)
M.M. Osterloo, V.E. Hamilton, J.L. Bandfield, T.D. Glotch, A.M. Baldridge, P.R. Christensen, L.L. Tornabene, F.S. Anderson, Chloride-bearing materials in the southern highlands of Mars. Science **319**, 1651–1654 (2008)
M.M. Osterloo, F.S. Anderson, V.E. Hamilton, B.M. Hynek, Geologic context of proposed chloride-bearing materials on Mars. J. Geophys. Res. **115**, E10012 (2010)
R.L. Parfitt, J.M. Kimble, Conditions for formation of allophane in soils. Soil Sci. Soc. Am. J. **53**, 971–977 (1989)
F. Poulet, J.-P. Bibring, J.F. Mustard, A. Gendrin, N. Mangold, Y. Langevin, R.E. Arvidson, B. Gondet, C. Gomez, M. Berthé, S. Erard, O. Forni, N. Manaud, G. Poulleau, A. Soufflot, M. Combes, P. Drossart, T. Encrenaz, T. Fouchet, R. Melchiorri, G. Belluci, F. Altieri, V. Formisano, S. Fonti, F. Capaccioni, P. Cerroni, A. Coradini, O. Korablev, V. Kottsoy, N. Ignatiey, D. Titov, L. Zasova, P. Pinet, V.B. Schmitt, C. Sotin, E. Hauber, H. Hoffmann, R. Jaumann, U. Keller, F. Forget, Phyllosilicates on Mars and implications for early Martian climate. Nature **438**, 623–627 (2005)
L. Preston, L.R. Dartnell, Planetary habitability: Lessons learned from terrestrial analogues. Int. J. Astrobiol. **13**, 81–98 (2014)
A.E. Richardson, Prospects for using soil microorganisms to improve the acquisition of phosphorus by plants. Funct. Plant Biol. **28**, 897–906 (2001)
C.K. Robinson, J. Wierzchos, C. Black, A. Crits-Christoph, B. Ma, J. Ravel, C. Ascaso, O. Artieda, S. Valea, M. Roldán, B. Gómez-Silva, J. DiRuggiero, Microbial diversity and the presence of algae in halite endolithic communities are correlated to atmospheric moisture in the hyper-arid zone of the Atacama Desert. Environ. Microbiol. **17**, 299–315 (2015)

J.C.M. Santelli, B.N. Orcutt, E. Banning, W. Bach, C.L. Moyer, M.L. Sogin, H. Staudigel, K.J. Edwards, Abundance and diversity of microbial life in ocean crust. Nature **453**, 643–657 (2008)

G.H. Schwabe, On the algal settlement in craters on Surtsey during summer 1968. Surtsey Research Progress Report V, 68–69 (1970)

G.H. Schwabe, K. Behre, Algae on Surtsey in 1969–1970. Surtsey Research Progress Report VI, 85–89 (1972)

C.E. Sharp, A.L. Brady, G.H. Sharp, S.E. Grasby, M.B. Stott, P.F. Dunfield, Humboldt's spa: Microbial diversity is controlled by temperature in geothermal environments. ISME J. **8**, 1166–1174 (2014)

B. Sherwood, D.B. Smith, R. Greeley, W. Whittaker, G.R. Woodcock, G. Barton, D.W. Pearson, W. Siegfried, Mars sample return: Architecture and mission design. Acta Astronaut. **53**, 353–364 (2003)

A.J. Solon, L. Vimercati, J.L. Darcy, P. Arán, D. Porazinska, C. Dorador, M.E. Farias, S.K. Schmidt, Microbial communities of high-elevation fumaroles, penitentes, and dry tephra "soils" of the Puna de Atacama volcanic zone. Microb. Ecol. **76**, 340–351 (2018). https://doi.org/10.1007/s00248-017-1129-1

R.M. Soo, S.A. Wood, J.J. Grzymski, I.R. McDonald, S.C. Cary, Microbial biodiversity of thermophilic communities in hot mineral soils of Tramway Ridge, Mount Erebus, Antarctica. Environ. Microbiol. **11**, 715–728 (2009)

H. Staudigel, R.A. Chastain, A.Y.W. Boucier, Biologically mediated dissolution of glass. Chem. Geol. **126**, 147–154 (1995)

H. Staudigel, A. Yayanos, R. Chastain, G. Davies, E.A. Th Verdurmen, P. Schiffman, R. Boucier, H. de Baar, Biologically mediated dissolution of volcanic glass in seawater. Earth Planet. Sci. Lett. **164**, 233–244 (1998)

H. Staudigel, H. Furnes, N.R. Banerjee, Y. Dilek, K. Muehlenbach, Microbes and volcanoes: A tale from the oceans, ophiolites, and greenstone belts. GSA Today **16**, 4–102 (2006)

G. Stefansdottir, A.L. Aradottir, B.D. Sigurdsson, Accumulation of nitrogen and organic matter during primary succession of Leymus arenarius dunes on the volcanic island Surtsey, Iceland. Biogeosciences **11**, 5763–5771 (2014)

N.A. Stroncik, H.U. Schminke, Evolution of palagonite: Crystallization, chemical changes, and element budget. Geochem. Geophys. Geosyst. **2**, 2001 (2001). https://doi.org/10.1029/2000GC000102

N.A. Stroncik, H.U. Schminke, Palagonite—A review. Int. J. Earth Sci. **91**, 680–697 (2002)

B.M. Tebo, H.A. Johnson, J.K. McCarthy, A.S. Templeton, Geomicrobiology of manganese(II) oxidation. Trends Microbiol. **13**, 421–428 (2005)

A.S. Templeton, H. Staudigel, B.M. Tebo, Diverse Mn(II)-oxidising bacteria isolated from submarine basalts at Loihi Seamount. Geomicrobiol. J. **22**, 127–139 (2005)

I.H. Thorseth, H. Furnes, O. Tumyr, A textural and chemical study of Icelandic palagonite of varied composition and its bearing on the mechanism of the glass-palagonite transformation. Geochim. Cosmochim. Acta **55**, 731–749 (1991)

I.H. Thorseth, T. Torsvik, V. Torsvik, V. Torsvik, F.L. Daae, R.B. Pedersen, Diversity of life in ocean floor basalt. Earth Planet. Sci. Lett. **194**, 31–37 (2001)

I.H. Thorseth, R.B. Pedersen, D.M. Christie, Microbial alteration of 0-30-Ma seafloor and sub-seafloor basaltic glasses from the Australian Antarctic Discordance. Earth Planet. Sci. Lett. **215**, 237–247 (2003)

T. Torsvik, H. Furnes, K. Muehlenbachs, I.H. Thorseth, O. Tumyr, Evidence for microbial activity at the glass-alteration interface in oceanic basalts. Earth Planet. Sci. Lett. **162**, 165–176 (1998)

M. Vos, A.B. Wolf, S.J. Jennings, G.A. Kowalchuk, Micro-scale determinants of bacterial diversity in soil. FEMS Microbiol. Rev. **37**, 936–954 (2013)

J. Wadsworth, C.S. Cockell, Perchlorates on Mars enhance the bacteriocidal effects of UV light. Sci. Rep. **7**, 4662 (2017)

J.J. Walker, J.R. Spear, N.R. Pace, Geobiology of a microbial endolithic community in the Yellowstone geothermal environment. Nature **434**, 1011–1014 (2005)

K. Wall, J. Cornell, R.W. Bizzoco, S.T. Kelley, Biodiversity hot spot on a hot spot: Novel extremophile diversity in Hawaiian fumaroles. Microbiol. Open **4**, 267–281 (2015)

C.M. Weitz, R.C. Anderson, J.F. Bell III., W.H. Farrand, K.E. Herkenhoff, J.R. Johnson, B.L. Jolliff, R.V. Morris, S.W. Squyres, R.J. Sullivan, Soil grain analyses at Meridiani Planum, Mars. J. Geophys. Res. **111**, E12S04 (2006)

R.J. Whitaker, D.W. Grogan, J.W. Taylor, Geographical barriers isolate endemic populations of hyperthermophilic archaea. Science **301**, 976–978 (2003)

D. Wolff-Boenisch, S.R. Gíslason, E.H. Oelkers, C.V. Putnis, The dissolution rates of natural glasses as a function of their composition at pH 4 and 10, and temperatures from 25 to 74°C. Geochim. Cosmochim. Acta **68**, 4843–4858 (2004)

D. Wolff-Boenisch, S.R. Gíslason, E.H. Oelkers, The effect of crystallinity on dissolution rates and CO_2 consumption capacity of silicates. Geochim. Cosmochim. Acta **70**, 858–870 (2006)

A. Yin, Structural analysis of the Valles Marineris fault zone: Possible evidence for large-scale strike-slip faulting on Mars. Lithosphere **4**, 286–330 (2012)

Space Sci Rev (2020) 216:56
https://doi.org/10.1007/s11214-020-00678-7

Concerns of Organic Contamination for Sample Return Space Missions

Queenie Hoi Shan Chan[1,2] · **Rhonda Stroud**[3] ·
Zita Martins[4] · **Hikaru Yabuta**[5]

Received: 27 June 2019 / Accepted: 23 April 2020 / Published online: 12 May 2020

Abstract Analysis of organic matter has been one of the major motivations behind solar system exploration missions. It addresses questions related to the organic inventory of our solar system and its implication for the origin of life on Earth. Sample return missions aim at returning scientifically valuable samples from target celestial bodies to Earth. By analysing the samples with the use of state-of-the-art analytical techniques in laboratories here on Earth, researchers can address extremely complicated aspects of extra-terrestrial organic matter. This level of detailed sample characterisation provides the range and depth in organic analysis that are restricted in spacecraft-based exploration missions, due to the limitations of the on-board *in-situ* instrumentation capabilities. So far, there are four completed and in-process sample return missions with an explicit mandate to collect organic matter: Stardust and OSIRIS-REx missions of NASA, and Hayabusa and Hayabusa2 missions of JAXA. Regardless of the target body, all sample return missions dedicate to minimise terrestrial organic contamination of the returned samples, by applying various degrees or strategies of organic contamination mitigation methods. Despite the dedicated efforts in the design and execution of contamination control, it is impossible to completely eliminate sources of organic contamination. This paper aims at providing an overview of the successes and

Role of Sample Return in Addressing Major Questions in Planetary Sciences
Edited by Mahesh Anand, Sara Russell, Yangting Lin, Meenakshi Wadhwa, Kuljeet Kaur Marhas and Shogo Tachibana

✉ Q.H.S. Chan
Queenie.Chan@open.ac.uk; Queenie.Chan@rhul.ac.uk

1 Planetary and Space Sciences, School of Physical Sciences, The Open University, Walton Hall, Milton Keynes MK7 6AA, UK

2 Present address: Department of Earth Sciences, Royal Holloway University of London, Egham Surrey TW20 0EX, UK

3 Code 6360, Naval Research Laboratory, Washington, DC 20375, USA

4 Centro de Química Estrutural, Departamento de Engenharia Química, Instituto Superior Técnico (IST), Universidade de Lisboa, Avenida Rovisco Pais 1, 1049-001 Lisbon, Portugal

5 Department of Earth and Planetary Systems Science, Hiroshima University, 1-3-1 Kagamiyama, Hiroshima 739-8526, Japan

lessons learned with regards to the identification of indigenous organic matter of the returned samples vs terrestrial contamination.

Keywords Sample return · Contamination · Organic matter · Comets · Asteroids · Extra-terrestrial samples

1 Introduction

Better understanding of the types of organic matter in the early solar system, and its subsequent evolution, is key to addressing the origin of life on Earth, and to its potential detection elsewhere in the solar system or beyond. Among the fundamental questions that remain only partially answered or controversial are: What is the inventory of organic compounds on comets and asteroids? How much variation among comets and asteroids is there, and how much between comets and asteroids? What is the formation and processing history for the individual components? Are there identifiable features associated with organics from different locations in protoplanetary disks? Are there identifiable features that can be linked to interstellar medium/molecular cloud chemistry? What organics exist in asteroids and comets that may have been delivered to the early Earth? What mixture of asteroidal and cometary material is responsible for the development of life on Earth? Which organic matter is present on other evolved bodies, such as Mars and its moons (Phobos and Deimos), Titan or Enceladus, and how does it relate to those present in small bodies?

Sample return missions are essential to fully answer these questions (Fig. 1). Spacecraft-based missions alone, while extremely valuable scientifically in addressing these topics, can only provide partial answers. This is largely because limitations on the *in-situ* instrumentation capabilities, based on payload weight and size, and operating power constraints, are inevitable. With the right returned samples, and the analytical capabilities of laboratories world-wide, researchers can answer extremely complicated, and subtle aspects regarding organic matter, inaccessible with on-board instrumentation. With careful curation, returned samples also remain available in pristine condition for decades as analytical instrumentation continues to improve, new contextual information from other investigations becomes available, and as new questions emerge over time.

Determination of the "right" solar system samples, and how to capture, return, curate and perform organic analysis on them is no simple matter. So far, there have been two completed sample return missions with an explicit mandate to collect organic matter: the NASA Stardust mission to comet 81P/Wild 2; and the JAXA Hayabusa mission to asteroid 25143 Itokawa. Samples from two more asteroids, 162173 Ryugu and 101955 Bennu, are scheduled for return by the JAXA Hayabusa2 and NASA OSIRIS-REx spacecrafts by 2020 and 2023, respectively. Although the CAESAR mission was not selected under the NASA's New Frontiers Program, it remains as a well-developed comet nucleus sample return mission concept. The first mission that targeted to return up to 200 g of soil from the Martian moon Phobos by 2014 was Russia's Phobos-Grunt mission (Marov et al. 2004), however the mission terminated in 2012 upon a malfunction that stranded the spacecraft in Earth orbit. A JAXA-led sample return from the Martian moons – the Martian Moons eXploration (MMX) mission, is in the planning stages, which targets to return samples to Earth from Phobos, with a possible return in 2029 (Krüger et al. 2019; Kuramoto et al. 2018; Usui et al. 2020). Analysis of organic matter is also a major motivation behind sample return mission concepts for Mars, Enceladus, and Titan, among others (e.g., Barucci et al. 2012; Mattingly and May 2011; Tsou et al. 2012).

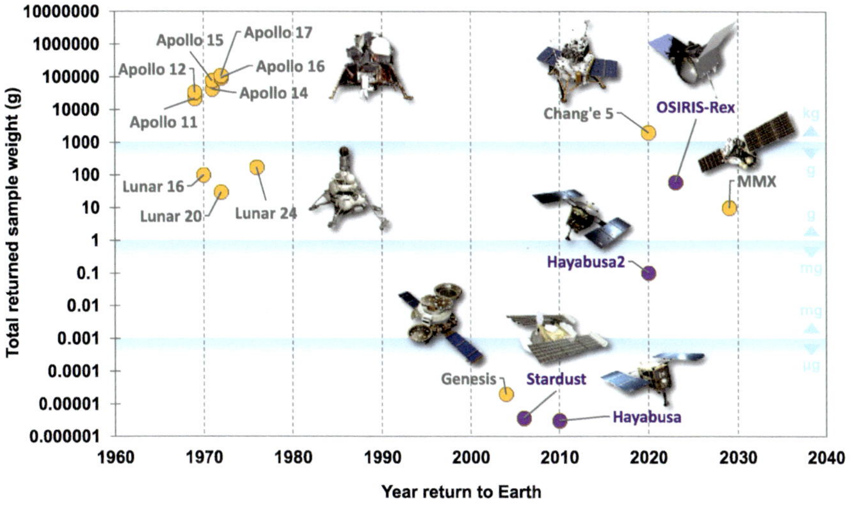

Fig. 1 Sample return missions and the total weight (or target weight) of the returned samples. Missions with an explicit mandate to collect organic matter are shown as purple symbols. Minimum target total weights of returned samples are shown in the figure for Hayabusa2 (target: 0.1–10 g) and OSIRIS-Rex (target: 0.06–2 kg). Image credit: NASA, JAXA, CNSA

Small bodies, e.g., comets and asteroids, were logical target choices for the first organic sample return due to both scientific and logistical reasons. The planetary protection issues associated with introducing extra-terrestrial organic matter to Earth are rendered moot for asteroids and comet targets, due to the large amounts of such material entering the terrestrial atmosphere daily in the form of interplanetary dust particles (IDPs) and meteorites. Furthermore, to date no such samples have shown to contain any traces of life. The degree of risk associated with Martian sample return however, remains a matter of significant debate. The potential risk of contaminating the Earth with potential extra-terrestrial life forms, combined with the cost of both the return phase of the mission and the appropriate curation methods, has so-far prevented realisation of the long-anticipated Mars sample return. Compared to larger outer body sample return missions, many comet and asteroid targets have the advantage of shorter mission timelines, which generally correlates with reduced cost. Most importantly, if the questions of greatest interest concern tracing the earliest history of organic matter in the solar system and the delivery of organic matter to the early Earth, comets and asteroids are the best sources of the relevant materials.

In all cases, regardless of the target body or planetary protection considerations, minimisation of any terrestrial organic contamination of the returned samples is both a top priority and a major challenge. Each of the completed and in-progress missions has contended with organic contamination mitigation to varying degrees and with different strategies, which provide an important knowledge base for future mission planning. The primary goal of this paper is to provide an overview of the successes and lessons learned with regards to returned organic samples.

2 Extra-Terrestrial Organic Matter

The language used to discuss solar system organics varies significantly across the different planetary science communities, so it is important to define explicitly the terms for our discussion. The terms labile, volatile, and refractory for example, take on different meanings depending on the temperature and pressure context. Meteoriticists often draw the distinction between solvent-insoluble organic matter (IOM) and solvent-soluble organic matter (SOM), because the majority of organic matter in meteorites is in the form of complex macromolecules. The IOM is a kerogen-like polymer that is solid at 275K, and thus can be considered non-labile, or refractory (e.g., Alexander et al. 2007; Bonal et al. 2007; Bonal et al. 2006; Busemann et al. 2007; Chan et al. 2017; Clemett et al. 1993; Cody and Alexander 2005; Cody et al. 2008b; Derenne and Robert 2010; Glavin et al. 2018; Kebukawa et al. 2019; Kebukawa et al. 2017; Nakamura 2005; Yabuta et al. 2010; Yabuta et al. 2005; Yabuta et al. 2007). Soluble organic matter in meteorites consists of smaller macromolecules, as well as simpler species such as amino acids, alcohols and sugars. These species may diffuse rapidly through pores in solid samples, and undergo chemical reactions with aqueous fluids, at temperatures as low as 275K, depending on the molecule (e.g., Aponte et al. 2015; Burton et al. 2012; Burton et al. 2013; Burton et al. 2015; Chan et al. 2018b; Elsila et al. 2016a; Elsila et al. 2013; Glavin et al. 2018; Glavin et al. 2010a; Glavin et al. 1999; Glavin et al. 2010b; Glavin and Dworkin 2009; Glavin et al. 2012; Martins 2018; Martins 2019; Martins and Sephton 2009; Pizzarello et al. 2006; Pizzarello and Groy 2011; Pizzarello and Holmes 2009; Pizzarello et al. 2001). In this chapter, we refer to these as labile components, although in the context of comet surfaces, these species are relatively inert. Comets also contain many organic molecules as trapped species in organic and water ices. These we will refer to as volatiles (e.g., methane CH_4, formaldehyde CH_2O, methanol CH_3OH, ammonia NH_3, hydrogen cyanide HCN, hydrogen sulphide H_2S (e.g., Bajt et al. 2009; Bockelée-Morvan et al. 2004; Le Roy et al. 2015; Quirico et al. 2016; Sandford et al. 2006)).

3 Missions with an Explicit Mandate to Collect Organic Matter

In 2006, the Stardust mission returned the first samples unambiguously identified as originating from a specific comet – comet 81P/Wild 2 (Brownlee et al. 2006; Sandford et al. 2006; Zolensky et al. 2006) (Fig. 1). The Stardust cometary dust collector consisted of an aluminium (Al) frame filled with an array of silica aerogel tiles and Al foils. The tray was exposed to the comet 81P/Wild 2 dust stream at an encounter distance of 237 km at closest approach, and relative velocity of 6.1 km/s (Brownlee et al. 2006; Tsou et al. 2003). The aerogel provided a low density, porous target in which the effects of the hypervelocity capture on the cometary materials could be reduced to the extent that some organic matter was preserved. After the initial impact into the aerogel tile face, the captured cometary particles gradually decelerated by impacting into the nanoscale, amorphous-silica-mesoparticle aerogel backbone, leaving a physical track that is visible to inspection with an optical microscope. Three primary track shapes (A, carrot; B, bulb; and C, hedgehog) were produced (Hörz et al. 2006). Type A tracks are formed when the impacting dust grain retains the majority of its original mass in one or two "terminal" particles deposited at the track terminus. Types B and C are associated with the explosive disaggregation of the impacting dust grains, for which the impacting mass is spread across the track walls, and for the Type C's, along and at the termini of side tracks. Dust grain components volatised during the heat of impact (labile organics, noble gasses, and possibly ice or water from hydrated dust components) diffused out from the track through the aerogel pore network, eventually adsorbing onto the surrounding silica mesoparticles, and occasionally reaching the Al foils that separate the individual tiles from the collector tray frame (Bechtel et al. 2014;

Sandford et al. 2006; Sandford et al. 2010). During the Stardust preliminary examination, organic matter was identified that chemically and structurally resembled the organic matter previously analysed in carbonaceous chondrites, and in anhydrous IDPs thought to have a cometary origin. Some of this material was found to have ^{15}N and/or D-rich isotope compositions indicative of formation in cold environment such as the outer solar system, or protosolar molecular cloud (Aikawa and Herbst 2001; Dartois et al. 2005; Messenger et al. 2003; Pizzarello and Holmes 2009; Remusat et al. 2009; Terzieva and Herbst 2000). Subsequent studies have revealed additional important results, such as the presence of glycine (Elsila et al. 2009). The Stardust mission results provide important context for interpreting results from the Rosetta cometary mission to explore and land on comet 67P/Churyumov–Gerasimenko. However, Stardust was not designed to directly sample the comet surface, or to return any volatiles, which remain important goals for future sample return missions.

The Hayabusa mission targeted the near-Earth S-type asteroid 25143 Itokawa, with the goal of providing the first samples from a known asteroid available for direct laboratory analysis. The Hayabusa sample canister contained two separate chambers (A and B), for collection of dust grains during two distinct touch-and-go encounters (Nakamura et al. 2011; Yada et al. 2014b). According the original sampling plan, a bullet was to be fired into the asteroid surface at each touch down site, so that the sampler horn could funnel the cloud of regolith particles accelerated by the bullet impact into the sample canister. Although the bullets did not fire according to plan, and the number of collection events was reduced to one, dust grains were collected on the walls of the sample canister during a touchdown in 2005 (Yada et al. 2014b; Yano et al. 2006). After the return of the samples to Earth in 2010, more than 2000 individual grains were identified and underwent preliminary analyses. The initial mineralogical and isotopic analyses demonstrated that the Itokawa grains showed a direct affinity to LL-type ordinary chondrites (OCs) (Nakamura et al. 2011; Yurimoto et al. 2011). This result provided the first direct link between S-type asteroids and LL chondrites. The majority ($\sim$90%) of the particles exhibited characteristics of a high degree of thermal equilibration (up to 800°C), consistent with silicate minerals from LL5 to LL6 chondrites (Nakamura et al. 2011). The remaining 10%, appear to less equilibrated, comparable to those found in less equilibrated LL4 chondrites. Because Itokawa is an airless body, and the Hayabusa samples were collected from the surface, the returned grains provide an important source of information about space weathering processes in a regime where ion irradiation from solar wind and galactic cosmic rays dominate (Bonal et al. 2015; Nagao et al. 2011; Noguchi et al. 2011; Tsuchiyama et al. 2011). The combination of thermal metamorphism and radiation processing are unfavourable conditions for the survival of all but the most refractory organic matter. Thus, it is completely consistent that the results of the preliminary search for an SOM component with amino acid analysis and time-of-flight secondary ion mass spectrometry (ToF-SIMS) generated a null result, with an upper limit of a concentration of < 1 parts per million (ppm) estimated from analytical limits (Naraoka et al. 2015). Coordinated H, C, and N isotope measurements, scanning transmission x-ray microscopy (STXM) using x-ray absorption near-edge structure (XANES) spectroscopy and transmission electron microscope (TEM) analysis of a few Itokawa particles of unknown origin ("category 3") have been performed (Ito et al. 2014; Kitajima et al. 2015; Uesugi et al. 2014; Yabuta et al. 2014a). Although it was challenging to determine their sources only by organic isotopic and molecular analyses, detection of inorganic inclusions (e.g., NaCl, TiO2) as well as comparison with witness coupons led to a conclusion that the possible origins for the category 3 particles include terrestrial contamination.

Two asteroid sample return missions are currently in progress (Fig. 1), this time targeting carbon-rich asteroids. The JAXA Hayabusa2 mission seeks to return samples from the

C-type asteroid Ryugu, with a planned return of samples to Earth in 2020 (Tachibana et al. 2014). The NASA OSIRIS-REx plans to return samples from type B asteroid Bennu in 2023 (Barnouin et al. 2019; Hamilton et al. 2019; Lauretta et al. 2017; Lauretta et al. 2015; Walsh et al. 2019). Both asteroids are thought to be related to the class of meteorites known as carbonaceous chondrites, which contain up to ∼3.5 wt% organic carbon (Alexander et al. 2017; Alexander et al. 2007), of which ∼90% is in acid-insoluble, macromolecule polymer form, with the remainder mostly in labile, soluble organic form (Martins 2019; Martins and Sephton 2009; Pearson et al. 2006; Sears and Dodd 1988). Thus, returned samples from these two targets should provide ample organic matter for laboratory analyses designed to address questions regarding the diversity of the organic matter delivered to the surface of the early Earth. Hayabusa2 has taken a sampling approach similar to that of Hayabusa (Sawada et al. 2017), during touch-and-go encounters. In order to obtain samples from a depth expected to be shielded from space weathering processes that would heavily alter the distribution of organic material, a more complex manoeuvre was planned to excavate subsurface material of Ryugu by dropping an explosive small carry-on impactor (SCI) (Arakawa et al. 2020). The SCI was detonated after the spacecraft was deployed to the opposite side of Ryugu for protection against the resultant ejecta. The SCI operation has generated an ejecta curtain and left a crater with a diameter of 20 m on the surface of Ryugu. After the literal settling of the dust, the spacecraft returned to sample the freshly exposed sub-surface regolith in July 2019. The samples will be returned to Earth around the end of 2020. OSIRIS-REx also uses a touch-and-go sampling strategy. In this case, the sample collector will make direct contact with the Bennu surface, and pressurised nitrogen gas will be used to disrupt the surface regolith, driving small pebbles and dust into the collection chamber.

4 Determination of Terrestrial Contaminants

Rigorous identification of terrestrial organic contaminants relies on the ability to contrast the types, distribution, structures and isotopic compositions of organic compounds between typical terrestrial contaminants and extra-terrestrial material. A detailed description of the techniques used for studying organic matter in extra-terrestrial material is given in a separate paper of this issue (Martins et al. 2020) so we will not go into details in favour of a brief description here.

The analytical techniques that can be used for studying the organic content of a sample strongly depend on how much of the sample is available for laboratory characterisation. Only very few organic analysis techniques have both the required sensitivity and the high spatial resolution necessary to study the organic matter in small samples, such as the Stardust and Hayabusa missions returned particles. Those that do include two-step laser desorption laser-ionisation mass spectrometry (L^2MS), Fourier transform infrared spectroscopy (FTIR), Raman spectroscopy, XANES, and NanoSIMS, etc., which are functional group specific, rather than providing characterisation of the entire organic molecule. These techniques focus on refractory organic contents, and can be further classified into: (1) analytical techniques that are spatially-resolved, structure-oriented, and generally non-destructive to the sample (e.g., FTIR, Raman, XANES) and (2) others that are also spatially-resolved but molecule- or isotope-oriented, and yet are destructive techniques as they require the sample materials to be "desorbed" or "sputtered" away in order for them to be detected by the instrument (e.g., L^2MS, NanoSIMS). The refractory organic contents of extra-terrestrial components typically are highly heterogeneous in elemental and isotopic distributions (e.g., Aléon et al. 2001; Alexander et al. 2013; Chan et al. 2018b; Flynn 2008; Quirico et al. 2005). Organic matter in extra-terrestrial materials displays significant spatial heterogeneity (elemental distribution and ratio) at micrometre-scale (e.g., Alexander et al. 2017; Keller et al. 2004; Quirico et al. 2005). Isotopically, extra-terrestrial refractory organics often contain micrometre-sized D- and ^{15}N-rich anomalies (hot spots, δD value as high as $\sim$30,000‰ observed for IDPs) (Busemann et al. 2009; Chan et al. 2019b; Keller et al. 2004; Matrajt et al. 2012; Messenger et al. 2008; Messenger et al. 2003; Nakamura-Messenger et al. 2006) mixed with moderately enriched isotopically anomalous organic material. Nevertheless, since these techniques do not provide characterisation of the entire organic molecule, it is impossible to provide a definitive answer as to the direct source of terrestrial contaminant.

Obvious biological contamination in the form of viable fungal and/or bacterial colonies can be readily tested by classical microbiological culture (e.g., Oyama et al. 1970; Regberg et al. 2018). Methodical contamination controls have successfully prevented the returned samples from fungal/bacterial contamination. However, other forms of biological contamination are unavoidable, which can be systematically identified by determining the labile organic components using spectroscopic techniques such as gas chromatography mass spectrometry (e.g., GC-MS, GC/GC-MS, GC-ToF-MS, GC-QqQ-MS etc.) and liquid chromatography (e.g., LC-MS, LC-ToF-MS, LC-FD/QToF-MS etc.). For example, amino acids, monomers of protein, have several characteristics that can be used to identify those that are biotically synthesised: (1) proteins and enzymes are made up of $\sim$20 major proteinogenic amino acids (e.g., glycine, alanine, valine, aspartic acid, glutamic acid, leucine), while several extra-terrestrial amino acids (e.g., isovaline, α-aminoisobutyric acid (α-AIB), β-amino-butyric acid (β-ABA)) are not used in terrestrial proteins (e.g., Botta and Bada 2002; Burton et al. 2012; Chan et al. 2018b; Cronin and Chang 1993; Ehrenfreund et al. 2001; Elsila et al. 2016a; Glavin et al. 2010b; Martins et al. 2015; Martins and Sephton 2009), (2) proteins for most microorganisms are made of only the L-enantiomer of chiral amino acids, leading to L-homochirality or a significant excess in the L-enantiomers (e.g., Bada 1995; Blackmond 2010; Glavin et al. 1999; Glavin et al. 2012; Herd et al. 2011; Myrgorodska et al. 2015; Pizzarello et al. 2012), (3) terrestrial amino acids are depleted in the heavy isotopes of carbon, hydrogen and nitrogen compared to extra-terrestrial amino acids (e.g., Burton et al. 2013; Chan et al. 2016; Cronin and Chang 1993; Elsila et al. 2016a; Elsila et al. 2012; Martins et al. 2007; Pizzarello et al. 1994; Pizzarello et al. 2004).

5 Potential Sources of Contamination

Despite the dedicated efforts in the design and execution of contamination control (Allen et al. 2011; Calaway et al. 2019; Dworkin et al. 2017; McCubbin et al. 2019; Sandford et al. 2010; Yada et al. 2014b), it is impossible to *completely* eliminate sources of organic contamination. The complex nature of a sample return mission elucidates the diversity of potential scenarios where various sources of organic contamination could be introduced into the returned samples (Table 1 and Fig. 2).

Organic contamination is to some degree inevitable. Contaminants can potentially be accrued during the design and construction of the spacecraft and associated components, on the course of the mission, during the sample recovery and curation processes, and/or during sample characterisation after their dissemination to analytical teams. For example, the lunar samples returned by the Apollo missions were found to contain amino acids derived from terrestrial biological contamination as the $\delta^{13}C$ isotopic values of glycine, β-alanine, and

Table 1 List of terrestrial organic contaminants detected in samples returned by space missions

Organic compounds[a]	Potential sources of contamination	Missions[b]	Identification techniques[c]	References
Non-labile/refractory (insoluble organic matter)				
Aliphatic $–CH_2–$ and $–CH_3$	• Aerogel • Organic deposition and/or modification during hard X-ray synchrotron XRF measurements • Thermally degraded refractory organic material	Apollo, Stardust	Typical $–CH_2–/–CH_3$ ratios are about 2 in chondritic material, and 3–4 in IDPs, but contaminants can have higher abundance of $–CH_3$ bonded in the Si–O network	(Bechtel et al. 2014; Sandford et al. 2006; Sandford et al. 2010; Thomas-Keprta et al. 2014)
C–N=C (N-heterocycles), C≡N	• Thermally degraded nitrogen-containing polymer material used in the Hayabusa spacecraft	Hayabusa	Chemical, organic, and isotopic distributions of extra-terrestrial materials are heterogenous, while that of terrestrial contaminants can also be homogenous	(Uesugi et al. 2019; Yabuta et al. 2014a)
C=O, C=C	• Aerogel • Polyamide derived from the Hayabusa sample container • Volatile organic during sample storage	Stardust, Hayabusa	Contaminants can show C=O stretching with no or very weak aliphatic C–H	(Bechtel et al. 2014; Kitajima et al. 2015; Sandford et al. 2006; Sandford et al. 2010; Uesugi et al. 2019; Yabuta et al. 2014b)
C (graphitised)	• Shock amorphisation, devolatilisation, and thermally driven graphitisation of carbonaceous materials upon capture	Stardust	Organic contaminants have a homogenous distribution of graphitised C while extra-terrestrial materials show heterogeneity in spatial distribution	(Fries et al. 2009)
D-enriched globules	• Cyanoacrylate adhesive introduced during sample preparation for TEM	Stardust	Contaminants show D enrichment, but isotopically normal nitrogen composition	(De Gregorio et al. 2011; De Gregorio et al. 2010)

Table 1 (*Continued*)

Organic compounds[a]	Potential sources of contamination	Missions[b]	Identification techniques[c]	References
N-bearing aliphatic and aromatics	• Aerogel • Industrial material • Laboratory contaminants • Nylon • Polyacrylamide • Spacecraft Materials	Stardust	Contaminants have nitrogen isotopic composition indistinguishable from values of terrestrial organic matter ($\delta^{15}N = -10$ to $+20$ ‰)	(De Gregorio et al. 2011; Sandford et al. 2010)
PAHs (78 to 206 amu)	• Hypervelocity impact on aerogel C contaminants	Stardust	Structurally indistinguishable from low mass PAHs found in other extra-terrestrial matter, but extra-terrestrial organics are typically enriched in the heavy isotopes	(Clemett et al. 2010; Sandford et al. 2006; Sandford et al. 2010; Sandford and Brownlee 2007; Spencer et al. 2009; Spencer et al. 2008; Spencer and Zare 2007)
PTFE, phthalates, polyethylene	• Plasticizers • Plastics	Apollo, Hayabusa	The presence of these molecular fragments are typical components of plastics, which can be identified by techniques such as ToF-SIMS	(Naraoka et al. 2012; Steele et al. 2001)
Labile (soluble organic matter)				
Amino acids	• Aerogel • Background contaminant in the curation environment • Biological (e.g., handprint, sweat, urea) • Nylon-6 storage bags • Rocket exhaust • Teflon	Apollo, Stardust, Hayabusa	Biology shows homochirality (amino acids present as predominantly the L-enantiomers) and a depletion in the heavy isotopes	(Burlingame et al. 1970; Chan et al. 2019a; Elsila et al. 2016b; Elsila et al. 2009; Fox et al. 1976; Glavin et al. 2008; Nagy et al. 1971; Naraoka et al. 2012; Simoneit et al. 1969)

Table 1 (*Continued*)

Organic compounds[a]	Potential sources of contamination	Missions[b]	Identification techniques[c]	References
Amines	• Aerogel • Synlube 100	Stardust	Similar relative abundances of MEA to L-alanine across different Stardust samples (preflight unbaked aerogel, preflight baked aerogel, flight aerogel adjacent to impact track, flight aerogel without impact track)	(Glavin et al. 2008)
Volatiles				
Ethanol, methanol, acetonitrile	• Aerogel • Polyimide-Kapton	Stardust		(Bajt et al. 2009; Sandford et al. 2010)

[a]Structures of the organic compounds are given in Fig. 2

[b]Sample return missions where organic contaminants were anticipated and identified

[c]Structures of terrestrial organic materials are different from typical extra-terrestrial organics. To prepare for analysing extra-terrestrial organics, a "contamination knowledge" can be obtained by characterising the terrestrial organic compounds present in the background of control samples

Acronyms: MEA=Ethanolamine; PAHs = Polycyclic aromatic hydrocarbons; PTFE = Polytetrafluoroethylene; TEM = Transmission electron microscopy; ToF-SIMS = Time-of-flight secondary ion mass spectrometry; XRF = X-ray fluorescence

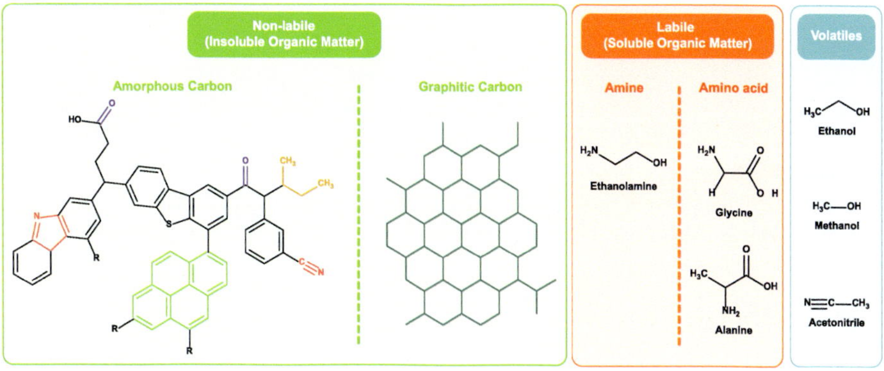

Fig. 2 Examples of structures of the organic contaminants described in Table 1

Fig. 3 Range of values of $\delta^{13}C$ measured for amino acids in the lunar regolith Apollo 70011, and comparison to the values of solar wind, carbonaceous meteorites and terrestrial sources (Elsila et al. 2016b)[1]

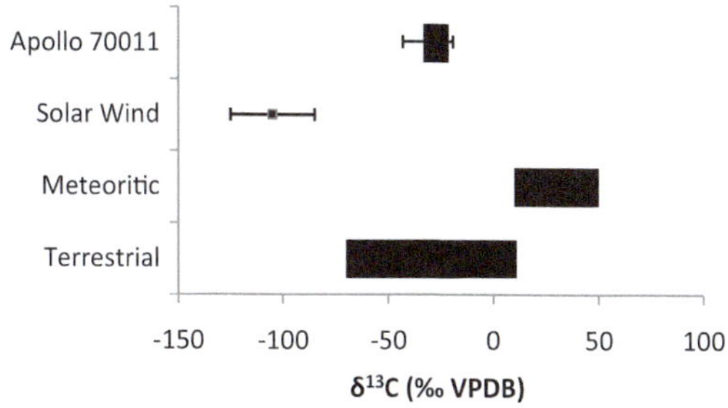

L-alanine were in the range of terrestrial biological sources, ranging from –21‰ to –33‰ (Elsila et al. 2016b) (Fig. 3).

Some Hayabusa category 3 samples contain carbon-rich organic contaminants that overprinted the extra-terrestrial organic signatures detected by scanning electron microscopy/energy dispersive X-ray spectrometry (SEM/EDS) (Uesugi et al. 2019; Uesugi et al. 2014), Raman spectroscopy (Kitajima et al. 2015), XANES (Yabuta et al. 2014a), NanoSIMS (Ito et al. 2014), and ToF-SIMS (Naraoka et al. 2012). Preflight monitoring of the nitrogen-purged sample cabinet desiccator box by the OSIRIS-REx team also showed a steady increase of volatile compounds although at very small amounts (Dworkin et al. 2017). In this section, previous reports of organic contaminants identified in mission returned samples will be addressed and discussed, based on the organic characterisation of samples returned by the Apollo, Stardust and Hayabusa missions.

5.1 Apollo Missions (1969-1972)

Apollo 11 marked the first successful sample return from another solar system body. Following the return of Apollo 11 in July 1969, numerous studies have analysed the returned lunar samples in search of organic compounds (e.g., Brinton and Bada 1996; Elsila et al. 2016b; Fox et al. 1976; Glavin et al. 2010c; Hare 1972; Hare et al. 1970; Nagy et al. 1971;

[1]Reprinted from Geochimica et Cosmochimica Acta, 172, Jamie E. Elsila, Michael P. Callahan, Jason P. Dworkin, Daniel P. Glavin, Hannah L. McLain, Sarah K. Noble, Everett K. Gibson, The origin of amino acids in lunar regolith samples, 357-369, 2016, with permission from Elsevier.

Fig. 4 (Left) Astronaut collects lunar rake samples at Station 1 during the Apollo 17 mission's first spacewalk at the Taurus-Littrow landing site. (Right) During the third Apollo 16 extravehicular activity (EVA), astronaut examines the surface of a boulder at North Ray Crater with a 20-bag dispenser attached to his right wrist (Image credit: NASA)

Steele et al. 2010; Thomas-Keprta et al. 2014). Lunar samples were collected by astronauts while on traverses, with the use of tools such as scoops, rakes, tongs, and electric drill (Fig. 4), and the samples were stored in separate documented sample bags made of Teflon film reinforced by an aluminium band around the rim (Allton 1989). However, the use of Teflon plastic bag could have contributed to the plastic contamination (fluorinated carbon compound detected by ToF-SIMS) identified in lunar samples (Steele et al. 2001).

In parallel to the search for indigenous organics, immense efforts have been made to address the possible sources of terrestrial contamination that might have influenced the interpretation of the organic compositions of the lunar materials. An early study reported that the Apollo 10 and 11 spacecraft were contaminated with terrestrial bacteria (Puleo et al. 1970), which were probably acquired when the spacecraft were tested and assembled in areas that had more environmental and personnel controls. Fortunately, viable bacteria were not detected in the lunar samples when tested by classical microbiological culture (Holland and Simmonds 1973; Oyama et al. 1970).

Without the technology that could determine the chirality and isotopic compositions of organic compounds, earlier studies could only determine the origin of the organics by comparing that to typical biogenic organics and their distributions. However, this led to divergent conclusions, as some determined that the detected organics were of terrestrial origin (e.g., handprint, urea, and Teflon contamination (Nagy et al. 1971), rocket exhaust (Burlingame et al. 1970; Fox et al. 1976; Simoneit et al. 1969)), while others determined the opposite, that the detected organics were indigenous to the lunar samples (e.g., Harada et al. 1971; Mitchell et al. 1971). A variety of other potential sources of contamination were also noted, such as terrestrial water, exhaust products from the lunar descent engine and reaction control system engines, lunar module outgassing, astronaut spacesuit leakage and venting of life support backpack, particulate abraded from spacesuits, venting of lunar module fuel and oxidiser tanks, cabin, and waste systems (Epstein and Taylor 1972; Simoneit and Flory 1970).

With the availability of the more sensitive high performance liquid chromatography fluorescence detection (HPLC/FD) technique, and based on the amino acid enantiomeric ratio, alanine and aspartic acid detected in Apollo 17 mission returned lunar soil were attributed to terrestrial biogenic contamination accrued during collection, transport, and analysis of the lunar soil (Brinton and Bada 1996). Although the quantities of amino acids in lunar soils made compound specific isotopic measurements challenging at that time, the extremely low D/L enantiomeric ratio of amino acids ($D/L < 0.05$) was a strong evidence for a biotic origin for the detected amino acids. 30 years after the return of Apollo 11, Kozar et al. (2001) proposed a null hypothesis, that pristine lunar samples should not contain markers (muramic acid and 3-hydroxy fatty acids) for terrestrial bacteria. This null hypothesis was confirmed by gas chromatography-tandem mass spectrometry (GC-MS/MS) analysis, meaning that similar markers detected in pristine lunar samples in the future should be extra-terrestrial. More recent studies in which highly sensitive modern instruments were used have indicated the presence of organic materials, which are composed predominately of amorphous, structurally disordered kerogen-like organic matter that is distinctive from terrestrial contamination (Thomas-Keprta et al. 2014). However, the detected organics were suggested to be contributed by meteoritic delivery to the surfaces of the Moon, instead of synthesised indigenously on the Moon.

In 2016, with the use of a gas chromatography coupled with mass spectrometry and isotope ratio mass spectrometry (GC-MS/IRMS), Elsila et al. (2016b) determined the carbon isotopic compositions and enantiomeric ratios of the amino acids in seven lunar regolith samples. The strong excess of the L-enantiomer of proteinogenic amino acids (aspartic acid, glutamic acid, serine, threonine, and valine), as well as a terrestrial $\delta^{13}C$ isotopic values, led the authors to conclude that terrestrial contamination was the primary source of the majority of the detected amino acids. In agreement with Fox et al. (1976), lunar module exhaust was not the primary source of the observed amino acids (Elsila et al. 2016b). The level of amino acid contamination was higher in non-curated samples compared to samples stored under NASA curation, reflecting terrestrial contamination accrued in non-curatorial laboratory environment for over 40 years. However, the source of the contamination is unknown, except for a potential contamination from Nylon-6 storage bags, as one of the amino acids detected in the lunar soil – ε-amino-n-caproic acid (EACA) – is a monomer released upon hydrolysis of Nylon-6 (Glavin et al. 2006). The presence of Nylon-6 in lunar samples is also supported by the detection of fibrous materials with a m/z 114 fragment on the ToF-SIMS spectra, which is characteristic of the monomer repeat unit of Nylon-6 (Steele et al. 2001).

The Apollo samples have been stored under strict isolation conditions in an ISO class 6 clean room for the past 50 years, where all pristine Apollo samples are curated and processed inside gloveboxes filled with inert gaseous nitrogen (Allton et al. 1998). Despite the efforts in contamination control (Allen et al. 2011; Flory and Simoneit 1972), it is possible that terrestrial contamination could have occurred during this extended curation period. Hydrocarbons, plasticizers, solvents, silicones, and rubbers have been detected inside Lunar curation gloveboxes (Calaway et al. 2014), some of which are unavoidable as common plasticizers from outside air could have been released into the laboratory through the air handling system which were not effectively filtered. The concern of contamination of lunar samples during long-term curation was further addressed by Fox (2002). Fox (2002) suggested that the Lunar Curation Facility at Johnson Space Center (JSC) was primarily concerned with keeping the collection clean from chemical contamination but not essentially biological contamination. Similar to the Lunar Curation Facility, the Antarctic Curation Facility at JSC is also an ISO class 6 clean room which has been upgraded from ISO class 7 in 1999–2000 (Allen et al. 2011; Calaway et al. 2019) (Fig. 5). Meteorite samples are either processed inside gaseous nitrogen filled gloveboxes (carbonaceous chondrites and martian meteorites),

Fig. 5 (Left) Lunar Curation Facility and (Right) Antarctic Curation Facility at NASA JSC (Image credit: NASA)

or on class 100 laminar flow benches (all other types) from which a small amount of viable fungal and bacterial colonies has been observed (4 colony forming units (CFU)/25 cm^2) (Regberg et al. 2018).

5.2 Stardust Mission (1999-2006)

The Stardust mission has returned to Earth thousands of grains from comet 81P/Wild 2 and interstellar dust. A great effort has been made to identify the possible sources of organic contamination and characterise the potential contaminants by a variety of analytical techniques, such as infrared (IR) microscopy, nuclear magnetic resonance (NMR), STXM-XANES, L^2MS, liquid chromatography with ultraviolet (UV) fluorescence detection and time-of-flight mass spectrometry (LC-FD/ToF-MS) analyses, etc. (e.g., Sandford et al. 2010). These detailed studies indicated the presence of various sources of organic contaminants, such as intrinsic organic compounds found in silica aerogel that was used to capture the Stardust samples, contaminants introduced during sample analyses, organic material being altered during hypervelocity capture, as well as new organic compounds being synthesised by impact heating.

The Stardust samples were captured in aerogel tiles, which were wedged into the sample collection trays and wrapped on all four sides with aluminium foil (1100 aluminium) to capture impacting dust particles as well as facilitate aerogel removal (Tsou et al. 2003) (Fig. 6). The aerogel was produced from tetramethyl and/or tetraethyl orthosilicate and the manufacturing process involved the use of ethanol, water, nitric acid, ammonium hydroxide, and acetonitrile (Tillotson and Hrubesh 1992). A silicone-based mould release (Synlube 100) was sprayed into the aerogel tile moulds in order to prevent aerogel from adhering to the moulds during the gelation process. The aerogel should ideally consist solely of Si and O. However, preliminary studies of the original aerogel collector tiles had identified a variety of organic contaminants present at low abundances (Sandford et al. 2006; Sandford et al. 2010).

IR analyses of the original aerogel collector tiles suggested the presence of "native" refractory organic carbon in the forms of $Si–CH_2$ and $Si–CH_3$ (Bechtel et al. 2014; Sandford et al. 2006; Sandford et al. 2010). The aliphatic $–CH_2–$ and $–CH_3$ groups of the contaminant are covalently bonded to the aerogel network (Tsou et al. 2003), and have IR C–H stretching absorption features in the 3000–2800 cm^{-1} (3.33–3.57 μm) region. The $–CH_3$ absorption is significantly stronger than the $–CH_2–$ absorption, therefore it is easily discernible from

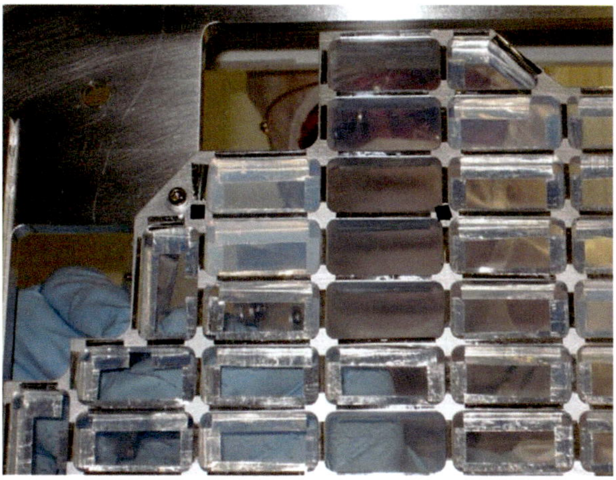

Fig. 6 Portion of Stardust interstellar tray showing the aluminium frame, aerogel tiles (4 cm long), and aluminium foil strips wrapping on all four sides of the aerogel tiles (Allen et al. 2011)[2]

typical comet 81P/Wild 2 organics that have high $-CH_2-/-CH_3$ ratios (Keller et al. 2006; Rotundi et al. 2008; Sandford et al. 2006). However, Si–O has a typical stretching absorption feature at around 1100 cm^{-1} (10 μm) that is characteristic of amorphous silicates. The high abundance of Si–O leads to an intensive absorption feature that has saturated the IR spectra. Therefore, it was impossible to determine the $-CH_3$/Si–O ratio with the IR technique, and thus the absolute abundance of aliphatic $-CH_2$ and $-CH_3$ contaminants is unknown.

The impurities in aerogel can also be reflected by Raman spectroscopy. With a porosity of up to 99 vol%, the intensities of the Raman bands of amorphous silica at low wavenumbers (i.e., in the Si–O–Si bending and Si–O stretching region) are extremely weak, and the Raman spectra of pure silica aerogel should be fluorescence-free. However, Stardust aerogel shows very strong Raman peaks at around 2850 cm^{-1} (symmetric CH_2 stretch), 2908 cm^{-1} (asymmetric CH_2 stretch), and 2965 cm^{-1} (antisymmetric CH_3 stretch), which are typical for alkanes (Rotundi et al. 2008; Wopenka 2012). In addition to the alkane features, the Stardust aerogel also has bands in the at $\sim$ 999 and 1029 cm^{-1} (breathing mode of benzene), as well as in the 1300–1600 cm^{-1} region, which correspond to the presence of CH_2 (1298 cm^{-1}), N=N (1444 cm^{-1}), and C=C (1600 cm^{-1}) functional groups, and overlap with the graphite band of the sp^2-bonded carbon feature of amorphous carbon typically observed for extra terrestrial organics.

Although the absolute abundance of the organic contaminant in the aerogel is unknown, it was possible to estimate the C/Si ratio of the organic carbon in aerogel via NMR analyses (Sandford et al. 2010). NMR analyses indicated that the background C/Si did not exceed 0.12 (or 0.06 in the case of C/O) in the aerogels, which was significantly lower than that observed for organic particles extracted from aerogel coupons (C/Si $\approx$ 1.7; and C/O $\approx$ 0.6) (Cody et al. 2008a). The NMR data also indicated that the organic carbon has a very simple chemistry that consists of just a few simple functional groups, which are readily converted into highly volatile species upon heating, e.g., methanol, ethylene, ethanol. This group of organic contaminants can be reduced by subjecting the Stardust samples to heating. However, even though the degree of C contamination could be reduced by <2 wt% by heating to 300 °C for 72 h in a controlled flow of filtered air at 5–6 psi, it could not be completely

[2]Reprinted from Chemie der Erde - Geochemistry, 71, Carlton Allen, Judith Allton, Gary Lofgren, Kevin Righter, Michael Zolensky, Curating NASA's extraterrestrial samples—Past, present, and future, 1-20, (2011), with permission from Elsevier.

Table 2 List of major amino acids and amines detected in Stardust materials and their potential source(s) (Elsila et al. 2009; Glavin et al. 2008)

Amine detected	Source
Glycine	Comet 81P/Wild 2
BALA	Aerogel
L-alanine	Aerogel
GABA	Aerogel (partial bakeout)
EACA	Nylon-6 (sample curation)
MEA	Synlube 100, Aerogel
EA	Comet 81P/Wild 2
MA	Comet 81P/Wild 2

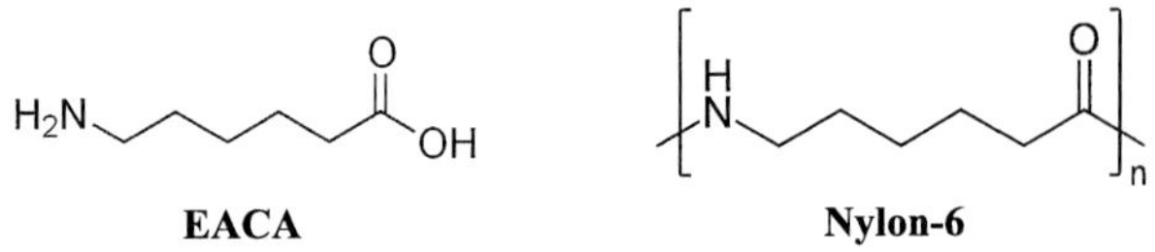

Fig. 7 Chemical structures of EACA and Nylon-6

removed, with the residual organics chiefly in the form of aliphatic $-CH_2-$ and $-CH_3$ (Tsou et al. 2003).

The labile organic contents of Stardust materials (preflight aerogel, Stardust flight aerogel witness tiles, comet-exposed aerogel, comet-exposed foil) have been explored by LC-FD/ToF-MS analysis (Elsila et al. 2009; Glavin et al. 2008; Sandford et al. 2010). Only trace levels of labile organic compounds were identified in the preflight aerogel which included amino acids [L-aspartic acid, L-glutamic acid, L-serine, glycine, β-alanine (BALA), γ-amino-butyric acid (GABA), L-alanine, EACA] and amines [ethanolamine (MEA), methylamine (MA), ethylamine (EA)], with a concentration of 0.04 to 3.4 nmol/g (which translates to $\sim$4 to 340 parts per billion (ppb), with an average molar mass of $\sim$100 g/mol of all amino acids and amines listed above) (Glavin et al. 2008). EACA and glycine are the two most abundant amino acids detected. D-amino acid was not detected in the preflight aerogel above the 0.1 nmol/g ($\sim$10 ppb) level. The low D/L enantiomeric ratios indicate that most of the amino acids are potentially terrestrial contaminants. The sources of amino acids and MEA detected in Stardust aerogel were suggested to be derived from the aerogel, Nylon-6 storage bag during curation, and Synlube 100 (Table 2). Data from a subsequent carbon isotopic analysis of Stardust comet-exposed foils also confirmed that EACA ($\delta^{13}C = -25 \pm 2‰$) was very likely derived from the Nylon-6 storage and shipping bags used during curation (Elsila et al. 2009) (Fig. 7). Nevertheless, glycine, MA and EA were shown to be indigenous to the comet (Table 2) (Elsila et al. 2009; Glavin et al. 2008). Glycine was suggested to have a cometary origin as it was only detected on the comet-exposed side of the Stardust foil sample which indicated that glycine was not a contaminant from the foil (Glavin et al. 2008). Furthermore, the $\delta^{13}C$ value for glycine of $+29 \pm 6‰$ is well outside the terrestrial range for organic carbon of $-6‰$ to $-40‰$ (Elsila et al. 2009), and thus supports an extra-terrestrial origin for glycine. The MA to EA ratio in Stardust comet-exposed materials strongly indicates that these two amines are cometary in origin. The molar ratio of MA to EA in Stardust comet-exposed aerogel and foil (1.0–1.8) is distinct from that in contamination control samples such as preflight aerogel (6.8), SRC heatshield (6.1), and Synlube 100 (0.1) (Glavin et al. 2008). MA and EA are present predominantly in an acid labile bound form, rather than as a free amine, indicating that comet 81P/Wild 2 contains labile amide-rich organic polymer.

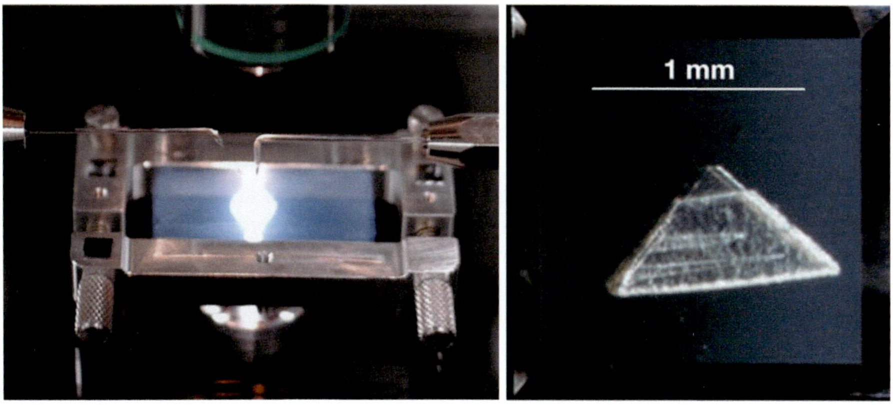

Fig. 8 (Left) An aerogel tile mounted in a custom-designed aluminium clamp, with needles that are being prepared to begin cutting a picokeystone. (Right) A picokeystone is mounted between two Si_3N_4 windows (Frank et al. 2014).[3]

Impact tracks in the aerogel collector were harvested in wedges of aerogel (typical thickness approximately 500 µm) called "keystones" (~70 µm thick for "picokeystones") (Frank et al. 2014; Westphal et al. 2004) (Fig. 8). Characterisation of picokeystones extracted from the Stardust comet and interstellar dust collectors suggested that terrestrial contamination occurred after extraction through the adsorption of volatile organics during sample storage and/or sample transfer (Bechtel et al. 2014). STXM-XANES analysis of 9 picokeystones indicated absorption features around 285 eV and 288.2 eV, which can be assigned to C=C of aromatic carbon and C=O respectively (Sandford et al. 2010). Synchrotron FTIR microscopy analyses revealed organic contamination via beam damage in the form of organic deposition and/or modification during hard X-ray synchrotron X-ray fluorescence (XRF) measurements (Bechtel et al. 2014). In particular, hard XRF analysis can alter the organic content by increasing the $-CH_2-/-CH_3$ ratio significantly.

In addition to the presence of "native" organics in Stardust aerogel, simulation experiments suggested that organic contaminants could be altered, destroyed, or synthesised during the hypervelocity impact (at approximately 6.12 km/s) of the Stardust particles as they were captured in Stardust aerogel. The simulation experiments were conducted by hypervelocity impact of micrometre-size borosilicate glass beads into Stardust-type aerogel, and IR laser pulse heating of Stardust aerogel (Sandford et al. 2006; Sandford et al. 2010; Sandford and Brownlee 2007; Spencer et al. 2009; Spencer et al. 2008; Spencer and Zare 2007). Based on L^2MS examinations of the simulation experiment products (keystone-dissected borosilicate glass bead impact tracks in aerogel), a low-intensity distribution of polycyclic aromatic hydrocarbons (PAHs) at the initial bead impact site was observed. The range of low-abundance artefactual PAHs includes toluene (92 atomic mass unit (amu)), styrene (104 amu), naphthalene (128 amu), phenanthrene (178 amu), and pyrene (202 amu) (Spencer et al. 2009; Spencer et al. 2008). Although the L^2MS technique is particularly well suited for mapping the spatial distribution of PAHs at the micrometre scale and the sub-attomole level (1 amol = 10^{-18} mol) (Clemett et al. 2010;

[3]Reprinted from Meteoritics & Planetary Science, 49, David R. Frank, Andrew J. Westphal, Michael E. Zolensky, et al., Stardust Interstellar Preliminary Examination II: Curating the interstellar dust collector, picokeystones, and sources of impact tracks, 1522–1547, (2013), with permission from John Wiley and Sons.

Clemett and Zare 1997), it was shown that a low-mass envelope of aromatic compounds could be synthesised from the original aliphatic carbon in the aerogel by high laser desorption power, which includes the same PAHs as those detected in the IR laser pulse heating simulation experiment (Sandford et al. 2006; Spencer and Zare 2007). Silica aerogels are particularly prone to PAH contamination due to the adsorption properties of SiO_2 surfaces and their high, density-dependent surface area (Clemett et al. 2010). Fries et al. (2009) has also conducted hypervelocity experiment by firing coal samples into aerogel. It was shown that graphitic materials in direct contact with aerogel could amorphise during aerogel capture, while relatively disordered, heteroatom-rich materials could be devolatilised and then re-condensed within the particle. It is challenging to determine whether the detected PAHs are indigenous to the Stardust samples. Nevertheless, the distributions of the synthesised PAHs in Stardust samples are different from those in the Murchison carbonaceous chondrite and some IDPs (Fig. 9), in particular, the L^2MS spectrum of Stardust sample is complex and contains higher molecular weight PAHs including alkylated (C_0–C_4) phenanthrenes and pyrenes (Sandford et al. 2006).

Transmission electron microscopy analyses of organic matter in capture tracks in the Stardust aerogel collectors have identified extremely N-rich organic matter that is distinctive from typical interstellar organic molecules. Based on C- and N-XANES analyses, this N-rich material was thought to be polymeric contamination, which is aliphatic and dominated by amide (CONHx) functional groups (Cody et al. 2008a), similar to amide polymers such as polyacrylamide and nylon. As these samples also contain nanoscale metal-oxide crystallites, an industrial origin was proposed for their origin (De Gregorio et al. 2011). These N-rich materials could have been present within the aerogel collector during the construction of the Stardust spacecraft, or introduced during particle embedding, although such procedure was conducted in a clean environment and used high-purity sulphur (Cody et al. 2008a; De Gregorio et al. 2011; Wirick et al. 2009).

Other potential sources of contamination to Stardust materials have been explored, which include the landing site soils, the thermal protection system of the sample return capsule (i.e., the pyrolytic C ablator heatshield and backshell), their solid ablation products, and atmospheric vent filter etc. Fortunately, the integrity of the capsule during landing underscores that the collectors were unlikely contaminated by these external sources. The nature of the organic matter detected in these materials differs significantly from the organics seen in the cometary collector, therefore these potential sources did not contribute any significant contaminants to the collectors.

5.3 Hayabusa Mission (2003-2010)

The first mission that brought back to Earth samples from an asteroid was achieved by the JAXA's Hayabusa spacecraft which visited the near-Earth S-type asteroid (25143) Itokawa (Yada et al. 2014b). Since its return to Earth in 2010, 943 particles have been picked up and kept in an ISO 6 clean room at the Planetary Material Sample Curation Facility of JAXA. Despite a lithology related to LL chondrites that typically have low organic contents, 64 particles from the collection are composed predominantly of carbonaceous material, based on field emission SEM/EDS observation at JAXA Extraterrestrial Sample Curation Center (ESCuC) (Uesugi et al. 2014; Yada et al. 2014b). These carbonaceous particles are classified as "category 3" particles.

However, the small sizes ($\sim$20–200 μm) of the carbon-rich category 3 particles make the characterisation of their organic contents and determination of their origins (terrestrial vs extra-terrestrial) challenging (Ito et al. 2014; Kitajima et al. 2015; Uesugi et al. 2014;

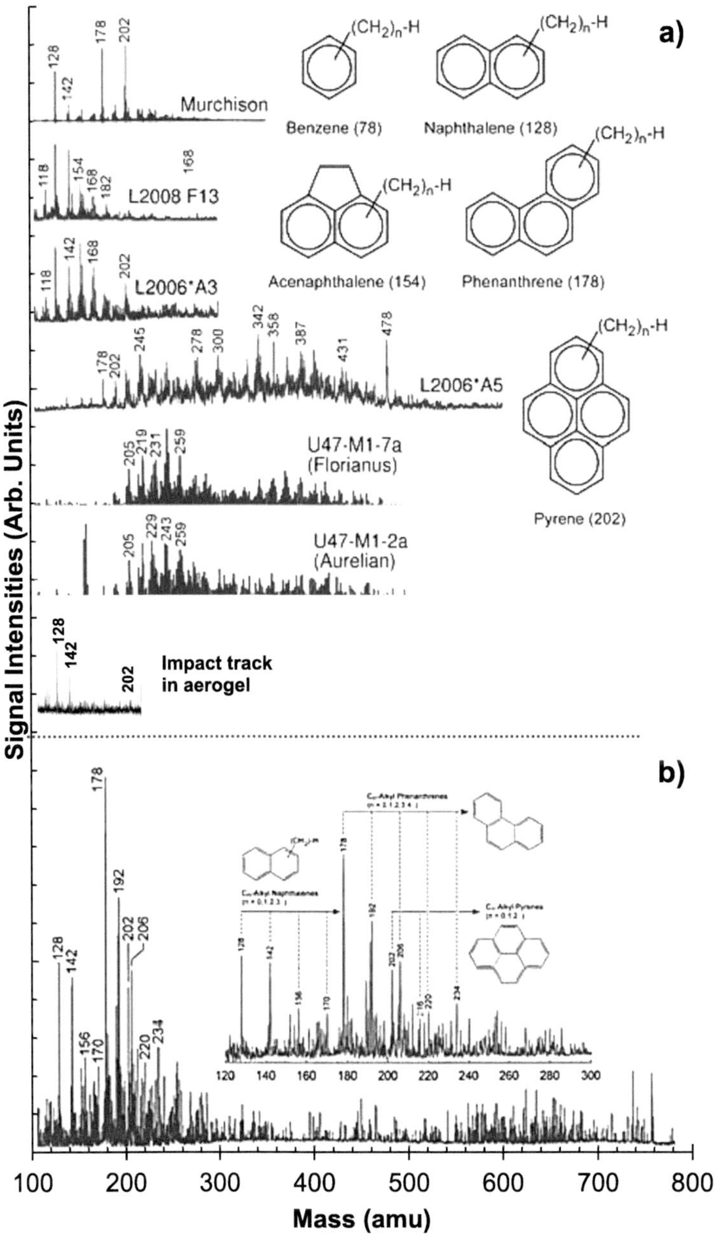

Fig. 9 (a) The L^{2}MS spectra of the Murchison carbonaceous chondrite, stratospheric IDPs, and a keystone-dissected glass bead impact track in aerogel (Sandford et al. 2006; Spencer et al. 2008). (b) The L^{2}MS spectrum of Stardust sample. The PAH distribution of Stardust sample is complex and contains higher molecular weight PAHs, while meteorite, IDPs and impact generated compounds show relatively simple populations, and mostly dominated by small PAHs

Yabuta et al. 2014a) (Fig. 10). Initial examinations of the category 3 particles indicated that the organic material they contained was not detected on the surface of categories 1 and 2 par-

Fig. 10 A category 3 particle (RA-QD02-0012) being picked up using a quartz glass needle with platinum wire inside. The size of the particle is about 100 μm. (Image credit: JAXA)

ticles (Naraoka et al. 2012). The particles were composed predominantly of C, N, O and S, and the elemental distributions of the organic material are homogenous (Ito et al. 2014; Naraoka et al. 2015), which make a clear variation from extra-terrestrial IOM that are typically heterogeneous in elemental and isotope distributions (e.g., Aléon et al. 2001; Alexander et al. 2013; Flynn 2008; Quirico et al. 2005). The preliminary results challenged an indigenous origin for the category 3 carbonaceous material because if they were indigenous to the asteroid, at least some of this organic material would have been found as a component of the categories 1 and 2 silicate particles.

Category 3 particles often take different shapes and appearances. Based on the texture of the category 3 particles, they were subcategorised into 3 types: blocky, fibrous and faint (Naraoka et al. 2015; Uesugi et al. 2014). Type 1 blocky particles are the most abundant among the 3 particle types. They are composed of silicate, iron sulphide, stainless steel, and aluminium. They sometimes exhibit rugged edges with "horn-like" texture that is likely produced by tearing. The less abundant category 3 materials are the type 2 fibrous and type 3 faint particles, which are about equal in abundance. However, type 2 fibrous particles have yet to be characterised in any subsequent analysis, so very little is known about their organic composition.

Subsequent to initial SEM/EDS observations, the organic compositions of selected category 3 particles were further characterised by FTIR, Raman spectroscopy, XANES, ToF-SIMS, and NanoSIMS in order to determine the nature of the organics and their origins. The Raman C parameters of the category 3 particles suggest the presence of disordered organics (i.e. not thermally-altered). However, FTIR spectra of the particles show predominantly C=O stretching with no or very weak aliphatic C–H, which is distinctive from unheated chondritic IOM that often exhibits clear absorption features at around 2960 to 2855 cm^{-1} corresponding to aliphatic C–H in IR spectra (Kitajima et al. 2015). Polyamide – a degradation product of polyimide resin (e.g., derived from polyimide films used in the sample container of the Hayabusa spacecraft) through hydrazinolysis – was suggested to be a possible contaminant that contributed to the C=O, C–F, O–H, and aromatic C=C observed by FTIR for a particle named "white object" (Kitajima et al. 2015; Uesugi et al. 2019; Yabuta et al. 2014b) (Fig. 11).

STXM-XANES analyses of the category 3 particles also came to the same conclusion that the particles displayed typical spectral patterns of heterogeneous organic macromolecules, although they are relatively nitrogen-rich, unlike chondritic IOM (Uesugi et al. 2019; Yabuta et al. 2014a). The results suggested that they are not "fresh/unused" industrial

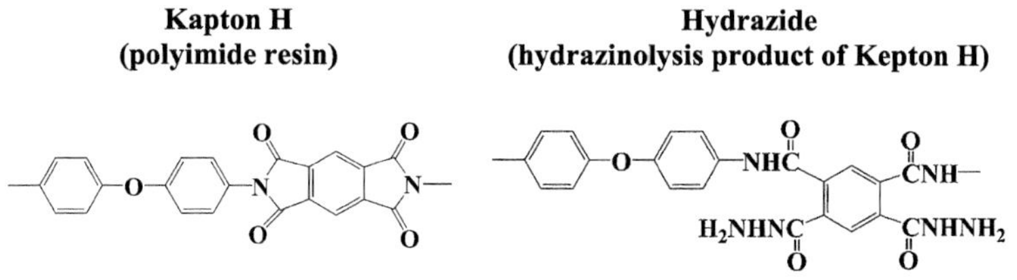

Fig. 11 Chemical structures of the sources of the contaminant "white object" (Kitajima et al. 2015)

or biological polymers (i.e., *not* fluoro-rubber nor silicon rubber). Instead, they are possibility nitrogen-containing polymer material used in the spacecraft that were degraded through chemical and physical alteration processes (e.g., heat, cosmic ray irradiation) during the 7 years of Hayabusa's journey, to form the materials seen in these particles. Further XANES investigations have revealed spectral similarity between one of the category 3 particles and the particles collected from a witness plate in a cleanroom where the Hayabusa2 spacecraft was constructed at ISAS/JAXA, indicating contamination into the sample catcher before the operation of the Hayabusa spacecraft (Uesugi et al. 2019).

As the molecular structures and chemical compositions of the category 3 particle organics did not provide a strong indication of an extra-terrestrial/terrestrial origin, further investigation with the use of isotopic analyses were conducted. The refractory organic components in extra-terrestrial materials, such as meteorites, IDPs, and Stardust cometary particles, often contain large D, ^{13}C, and ^{15}N isotopic enrichments with the presence of hot spots (isotopic anomalies) mixed with moderately enriched isotopically anomalous organic material (e.g., Alexander et al. 2017; Alexander et al. 2007; Busemann et al. 2009; De Gregorio et al. 2013; Messenger et al. 2003; Nakamura-Messenger et al. 2006). However, isotopic analyses of the category 3 particles using NanoSIMS indicated that the organic materials in category 3 particles exhibit terrestrial H, C, and N isotopic compositions, and isotopic anomalies were not observed (Ito et al. 2014; Uesugi et al. 2019). The isotopic compositions of the particles lie in a region that can be represented by either terrestrial or extra-terrestrial organic matter, and since not all extra-terrestrial organic matter contains isotopically anomalous regions, a definitive origin of the category 3 organics could not be established.

The labile organic contents of categories 1 and 2 particles have been explored by Naraoka et al. (2012) using two-dimensional- (2D-) HPLC/FD and ToF-SIMS. Only glycine and alanine were detected in one Itokawa particle at levels slightly higher than that of the procedural blank. As glycine and alanine are common proteic amino acids, while other non-proteic amino acids such as AIB were not detected, Naraoka et al. (2012) concluded that the amino acids present in the samples were likely derived from terrestrial contamination. Similarly, the ToF-SIMS spectra of the Itokawa particles were very similar to that of the procedural blank, with only small differences such as small peaks at $m/z = 55$, 57, 73 and 149 in positive ions and $m/z = 60$, 61 and 77 in negative ions, which were assigned to phthalates. However, phthalates are likely contaminants introduced to the samples in the laboratory as they are common components of plasticizers.

Witness coupons made from aluminium foil that were exposed in a clean chamber in the curation room at the ESCuC sample curation facility of JAXA were analysed in order to examine the type of contaminants and their accumulation rates (Sugahara et al. 2018). Common terrestrial amino acids (glycine, alanine, valine, leucine, isoleucine, proline, aspartic acid and glutamic acid) were detected. Glycine was the most abundant contaminant with an accumulation rate of 1.2 pmol/cm^2/day, which could potentially account for the $\sim$0.1 pmol of glycine detected in the Itokawa particle (Naraoka et al. 2012).

5.4 Extraneous Materials Delivered by Extra-Terrestrial Processes

The process of the delivery of organic matter originating from meteorite infall is not restricted to the planet Earth. Meteorites have been identified on other celestial bodies, such as (i) Bench Crater (McSween 1976; Zolensky et al. 1996) and Hadley Rille (Rubin 1997), which are carbonaceous and enstatite chondrites identified on the Moon, (ii) several iron and stony-iron meteorites (e.g., Meridiani Planum, Lebanon, Littleton) identified on the surface of Mars by Mars Exploration Rovers (MERs) Spirit and Opportunity, and the Curiosity rover of the Mars Science Laboratory (Ashley 2015; Ashley et al. 2011; Connolly et al. 2006; Schröder et al. 2016; Schröder et al. 2008), and (iii) the "Black Boulder", which is a possible xenolith originates from a low-albedo body identified on Itokawa (Nagaoka et al. 2014). Although the acquisition of exogenous materials is a common process as the target body evolves through time, these 'foreign' materials carry information different from the target body and should be carefully identified to avoid confusing their synthetic histories.

Meteoritic impact events can alter the organic records on the asteroidal or planetary body by introducing exogenous organic matter of a different synthetic origin. Impactors are capable of delivering essential exogenous organic molecules (Chyba and Sagan 1992) as well as shock-synthesising organics at the impact zone (Chyba and Sagan 1992; Chyba et al. 1990; Martins et al. 2013). While the shock-synthesised organic molecules are made *in-situ*, organic matter delivered via impact events are synthesised elsewhere and therefore the synthetic process of such organic inventory is unrelated to the local environment. The amount of this exogenous organic is small, for example, only 1–4 wt% of lunar soils was estimated to be derived from carbonaceous chondrite infall, based on the mass balance for iridium in the lunar soils (Haskin and Warren 1991), and approximately 3–6 wt% of carbon at Mercury's surface (in graphite, amorphous, or nanodiamond form) (Syal et al. 2015). However, the amount of the containing organics is sufficient to introduce complexity to the interpretation of the indigenous organic composition of the lunar soil (Brinton and Bada 1996). For examples, the Bench Crater meteorite was suggested to be the source of complex organic material in the Apollo 17 samples (Thomas-Keprta et al. 2014).

Without a thick atmosphere like that of the Earth, impactors arriving at the surfaces of asteroidal bodies and planets with thin or no atmosphere are not sufficiently decelerated to survive the impact. While small impactors such as IDPs can deliver exogenous organics intact (Anders 1989; Chyba and Sagan 1992; Love and Brownlee 1993), high-energy impact events from larger impactors can alter the organics the impactor contains as well as those present at the surface of the target body. The structures of surviving organic materials can also be altered by short-term impact heating. A group of dehydrated meteorites commonly referred as thermally metamorphosed carbonaceous chondrites have been suggested to have experienced short-term heating induced by transient processes such as impact (Chan et al. 2019c; Nakato et al. 2008; Yabuta et al. 2010). Impact heating may have induced aromatic condensation, dehydrogenation and extensive dehydration of organic molecules, which could have influenced the elemental and isotopic compositions by decreasing the H/C ratio and δD isotopic value of the organic material (Yabuta et al. 2010). Any traces of alteration prior to the impact event could have been erased by impact heating, which further complicates the determination of the history of organic processing on the parent body.

Other than the extensive cratering record observed for inner and outer solar system planetary bodies (Strom et al. 2005; Zahnle et al. 2003), material mixing through meteorite delivery or asteroidal collision is also supported by meteoritic evidences as xenoliths hosted within another meteorite of a different origin. Primitive xenolithic clasts are found hosted in ordinary chondrites (e.g. Monahans (1998), Zag, Sharps, Carancas, Dar al Gani [DaG] 139), carbonaceous chondrites (e.g. Kaidun), ureilites (e.g. Almahata Sitta, Nilpena, DaG 165), howardite (e.g. Bholghati, EET 87513), and eucrites (LEW 85300) (Brearley and Prinz 1992; Buchanan et al. 1993; Chan et al. 2018a; Chan et al. 2018b; Goodrich et al. 2004; Ikeda et al. 2003; Ikeda et al. 2000; Kebukawa et al. 2017; Rubin et al. 2002; Zolensky and Ivanov 2003; Zolensky et al. 2009; Zolensky et al. 1999; Zolensky et al. 1992). These xenolithic clasts, often referred to as "dark clasts", contain organic material and mineral components different from the host meteorite (Chan et al. 2018b), and are often classified petrographically, mineralogically, and organically as the primitive types 1 and 2 carbonaceous chondrite matrix-like clasts showing affinities to CM/CI chondrites.

6 Contamination Control Measures

A successful sample return mission necessitates the return of scientifically valuable samples unaltered and uncontaminated by the recovery and curation processes. One goal of the Stardust mission was to investigate whether cometary dust contained complex organic materials, and if so, to study the abundance, chemical, and isotopic compositions of the organic phase(s) (Brownlee et al. 2006; Tsou et al. 2003). Hayabusa2 seeks to investigate the "diversification of organic materials through interactions with minerals and water in a planetesimal" (Tachibana et al. 2014). OSIRIS-REx seeks to "return and analyse a sample of pristine carbonaceous asteroid regolith in an amount sufficient to study the nature, history, and distribution of its constituent minerals and organic material" (Dworkin et al. 2017; Lauretta et al. 2017). A proper, carefully planned contamination control is vital in achieving such mission goals, and a realistically and strategically determined level of "acceptable" contamination of the sample is essential to ensure a meticulous and achievable mission outcome (e.g., Dworkin et al. 2017; Sandford et al. 2010; Tachibana et al. 2014; Yada et al. 2014b). This is particularly challenging for organic analysis of returned samples, as the samples are brought to Earth, a planet where water, life and biogenic materials are ubiquitous. With a top priority of returning pristine extra-terrestrial organic matter, any sample returned from space missions must be isolated and protected against any contact with terrestrial contaminants in order to maintain the sample's pristine nature and scientific importance.

Dworkin et al. (2017) highlighted the essentiality of maintaining the "pristine" state of the returned sample, by underscoring that the "pristine" state can be violated by any alteration to the sample by "changing its inherent states, losing sample components, or adding extraneous components". One key method to monitor the change to the returned sample's pristine state is to monitor and evaluate the added terrestrial extraneous component – terrestrial contaminants – which can be achieved by keeping records of the background contamination that can occur at any time during the span of spacecraft fabrication, operations, and sample curation.

Potential sources of terrestrial contaminants should be strictly avoided via preflight, flight, and postflight controls, including any chemical or solvent required for preparing the sampling system or be in contact with the samples or analytical tools, air contained in and material used to build the sample sampler/container, and air circulated in the curation cabinet. Curation of the returned samples (final curation location, sample form, sample catalogue) should be carefully planned to minimise the handling, relocating, or subsampling of the samples, and when subsampling is needed for a particular sample, influence to other samples will be kept to its minimum. Amino acid-based polymers, such as nylon and latex, should be prohibited, and the use of hydrazine should be limited by spacecraft design and operations. Precision cleaning techniques should also be used to remove terrestrial organic and inorganic contamination to a specific level of cleanliness, and such level of cleanliness is verified to a standard. For example, the sample catcher of Hayabusa2 is cleaned in highly purified 2-propanol and methanol/dichloromethane by an ultrasonic cleaner of 38 kHz frequency, which has been subsequently checked by a series of test analyses (Sawada et al. 2017; Yada et al. 2014a). In the past, NASA JSC used Freon 113 as the final cleaning agent for Apollo, and ultrapure water to decontaminate Genesis solar wind materials after hard landing (McCubbin et al. 2019). The use of ultrapure water as final precision cleaning agent is continued nowadays at NASA JSC. Cleaning spacecraft surfaces would require approaches to also limit recontamination and maintain the cleanliness at high levels. The types of organic solvents used to clean the tools and materials should be carefully chosen as they can become the "contaminant" when left as residues to the targeted organic species. For examples, nitriles found in nitrile containing organic solvents or nitrile gloves commonly used in analytical laboratories can potentially be hydrolysed during typical amino acid extraction protocols to give laboratory-yielded amino acids. Subsequent to precision cleaning, all tools which can be baked at high temperatures should be pyrolysed in air at $\geq$500 °C for at least 3 hours to remove solvent residues and temperature sensitive organic contaminants such as amines, amino, carboxylic, and aromatic acids.

In addition to prudently executing the much-needed contamination control, it is equally important to anticipate the presence of terrestrial contaminants, and to readily identify, monitor, and correct for the organic contamination signatures. Before the return of the samples, the contamination control teams of different missions have made significant research efforts to identify and characterise any intrinsic contaminants in addition to those introduced by the terrestrial environment and sample preparation. To vigorously identify potential contamination sources, "witness coupons", which were introduced into contamination control protocols for various space missions in the past [e.g. Stardust (Sandford et al. 2010)], have been used to track the origins and routes of incorporation of the contaminants found at each stage of the mission, from the construction of the spacecraft to the curation of the returned samples. The use of witness coupons, in the form of aluminium foil, sapphire glass plates, and aerogel tiles, have been incorporated as a key contamination control procedure in future sample return missions including Hayabusa2, OSIRIS-REx and Mars 2020 (Dworkin et al. 2017; Sawada et al. 2017; Summons et al. 2014; Tachibana et al. 2014).

The examination of witness coupons which assessed the level of terrestrial organic contamination in the curation facilities of the Hayabusa2 and OSIRIS-REx missions have been reported recently (Dworkin et al. 2017; Sugahara et al. 2018) (Fig. 12). Seven common proteinogenic amino acids (glycine, alanine, valine, leucine, isoleucine, proline, aspartic acid and glutamic acid) were detected on the witness coupons exposed in the clean chamber in JAXA's curation room. The total abundance of amino acids detected on the witness coupon that had been exposed to the clean chamber environment for 7 days was 0.3–1.2 ng/cm^2. The most abundant terrestrial contaminant was glycine, which was accumulated at a calculated rate of 1.2 pmol ($\sim$0.09 ng) /cm^2/day (Sugahara et al. 2018). This level of terrestrial deposition of glycine is low, compared to the amount of glycine detected in extra-terrestrial samples returned by space missions (e.g. 1.5 ng/cm^2 detected in the comet-exposed side of the Stardust foil sample (Glavin et al. 2008)) and carbonaceous chondrites (e.g. 2000 ng/g in Murchison (Glavin et al. 2006), <20 ng/g in aqueously and thermally altered meteorites (Burton et al. 2014; Chan et al. 2016)). About 0.05–0.8 ng/cm^2 of amino acids and <200 cells/cm^2 of viable cells were detected on the witness coupons exposed to cabinet desiccator boxes located in an ISO 7 clean room at JSC that were reserved for the storage of

Fig. 12 (a, b) The ESCuC sample curation facility of JAXA and the aluminium foil witness coupons (Sugahara et al. 2018).[4] (c, d) Payload Hazardous Servicing Facility of OSIRIS-REx at NASA Kennedy Space Center and the aluminium foil witness plate marked in red circle (Dworkin et al. 2017)[5]

OSIRIS-REx returned samples (Dworkin et al. 2017). Dworkin et al. (2017) suggested that the total abundance of amino acids could not be explained only by the presence of viable cells on the witness plates and thus the amino acids could have been contributed from other unknown sources. The similar total amino acid abundances observed for witness plates that were exposed to the desiccator boxes for variable lengths of time led the authors to conclude that the total amino acid contamination was not a simple accumulation of material over time. Only very small amounts of volatile compounds were detected at the levels equal to or less than the amounts of volatiles found in the analytical laboratory background.

Witness coupons occupy space and mass on missions, therefore an effective number of witness coupons should be used which are to be strategically placed in key locations (spacecraft construction site, testing environment, on board the spacecraft at multiple positions when flown in space, different locations at the curation facility – e.g., inside curation cabinets, inside the gloveboxes where samples are manipulated, on flow benches where other tools and materials are handled, areas in proximity to air circulation systems to monitor the contaminants from filtered air) and for different target contamination species (Table 1). The recommendations for the use of witness coupons have evolved significantly in recent sam-

---

[4]Reprinted from Earth, Planets and Space, 70, Haruna Sugahara, Yoshinori Takano, Yuzuru Karouji, Kazuya Kumagai, Toru Yada, Naohiko Ohkouchi, Masanao Abe and Hayabusa2 project team, Amino acids on witness coupons collected from the ISAS/JAXA curation facility for the assessment and quality control of the Hayabusa2 sampling procedure, 1–10 (2018). This article is distributed under the terms of the Creative Commons Attribution 4.0 International License (http://creativecommons.org/licenses/by/4.0/).

[5]Reprinted by permission from Springer Nature, Space Science Reviews, OSIRIS-REx Contamination Control Strategy and Implementation, J. P. Dworkin, L. A. Adelman, T. Ajluni et al. (2017).

ple return missions and each mission has a detailed plan of the location, types and numbers of witness coupons used to monitor different types of potential contaminants. For example, Hayabusa2 specified to use (i) thirteen witness coupons (plates, dishes, and disks made of aluminium alloy, Pyrex borosilicate glass, and sapphire glass, respectively, on a base plate) to monitor and record the environment during manufacturing of the sampler, its installation to the spacecraft and transportation between facilities including the launch site, (ii) nine coupons on the sampler, and (iii) two inside the sample catcher and container to monitor the inner environment of the entire mission (Sawada et al. 2017). OSIRIS-REx specified that two types of witness plates (sapphire and aluminium) should be flown – ten flown in the sample return capsule and fourteen on the touch-and-go sample acquisition mechanism (TAGSAM), among a well-documented list of other types of coupons used for background facility contaminants and hardware verification (Dworkin et al. 2017).

In addition to the careful plans to monitor the background contamination level from as early as the manufacture of the spacecraft, the flow of mission returned sample analysis should take into consideration that any contaminant accrued by a technique does not influence the interpretation of the data collected by subsequent analytical techniques. Although a coordinated analysis involving multiple analytical techniques can provide a better picture of the organic content of the valuable mission returned samples, the number and types of analytical techniques used in an analytical sequence should be carefully considered, not only to reduce the organic alteration but also to limit the exposure of the sample to potential contaminants. The analytical sequence should always prioritise non- or less-destructive techniques (Table 3). For example, for refractory organic analysis, FTIR should ideally be placed prior to Raman and UV fluorescence spectroscopy, as the organics can be thermally damaged and photo-oxidised by the laser used in the later techniques if the laser intensity is set too high. L^2MS can be used to detect aromatic/conjugated organic molecules at the micrometre scale and the sub-attomole level. However, this technique is considered mildly destructive as it desorbs molecules from the sample surface, and therefore should be conducted subsequent to Raman spectroscopy. SEM is commonly conducted at the beginning of an analytical sequence as it provides images/elemental maps of a sample that help to determine areas of interest. However, SEM also can introduce C contamination during electron irradiation in particular when imaging at high magnifications. The contamination is a function of the accelerating voltage ($C \propto V^{-0.8}$) and current density, duration of beam exposure, conductivity of the sample, and the presence of hydrocarbons, which could reduce the image contrast, and impose artefacts on subsequent C analysis by depositing a layer of disordered C (Chan et al. 2019b). Therefore, it is recommended to place SEM analysis in a later sequence if that is permissible, or verification of the cleanliness of the SEM chamber through high resolution imaging of pre-cleaned metal stubs. STXM-XANES allows chemical mapping (with a high spatial resolution of 30 to 40 nm) and can produce spectra that reveal considerable chemical complexity in the organic macromolecules. However, if focussed ion beam (FIB) extraction of a section is required for the XANES analysis, rather than ultramicrotome sectioning, the sample can suffer from extensive thermal and irradiation damage (Bassim et al. 2012). Finally, SIMS techniques shall be the last of the experimental sequence due to the destructive nature of the ionisation process.

Labile organic analysis often requires the extraction of the target molecules and thus the technique is highly destructive. After the labile organic extraction process, sample residue and refractory materials can be collected, however, they are often extensively altered during the course of the extraction procedure. Therefore, it would be more sensible to allocate samples separately for labile and refractory organic analyses. However, if the availability of samples is extremely limited, labile organic extraction can be performed after the same

Table 3 A list of the most common analytical techniques used for characterisation of organic components in extra-terrestrial samples and the suggested analytical sequence

	Technique[1]	Acronym	Required sample mass[2]	Sample form	Destructiveness[3]	Measurement objective	Contamination or chemical modification
Beginning of sequence ↓	Fourier transform–infrared spectrometry	FTIR	ng	Sample with flat surface	Non-destructive	Functional group compositions	
	Raman spectroscopy		ng-µg	Sample with flat surface	Non-destructive	Molecular vibrations and crystal structures	Laser beam damage at strong laser intensities
	UV fluorescence spectroscopy		ng	Sample with flat surface	Non-destructive	Molecular compositions	
Seq. A[4,5]	Gas chromatography-mass spectrometry, isotope ratio mass spectrometry (or liquid chromatography-fluorescence detection or mass spectrometry)	GC-MS, IRMS (or LC-FD/MS)	µg-g	Powdered sample	Destructive (due to solvent extraction)	Compound-specific molecular composition and/or isotopic compositions	
Seq. B[5]	Scanning electron microscopy/energy dispersive spectrometry	SEM/EDS	ng	Mounted sample	Non-destructive	Elemental compositions, spatial distribution of carbon	C deposition
	Two-step laser desorption/laser ionisation mass spectrometry	L^2MS	ng	Sample with flat surface	Partially	Mainly aromatic compounds	Alteration by high laser desorption power
	Synchrotron Fourier transform–infrared spectrometry	sFTIR	ng	Sample with flat surface/FIB section (100 nm – 1 µm thickness)	Partially	Functional group compositions	
	Synchrotron Scanning transmission X-ray microscopy using X-ray absorption near-edge structure spectroscopy	sSTXM-XANES	ng	Ultra-thin section (100 nm thickness)	Partially	Chemical mapping, functional group compositions	
	Synchrotron X-ray fluorescence spectroscopy	sXRF	ng	Sample with flat surface/FIB section	Partially	Sample compositions	Beam damage in the form of organic deposition

Table 3 (*Continued*)

	Technique[1]	Acronym	Required sample mass[2]	Sample form	Destructiveness[3]	Measurement objective	Contamination or chemical modification
End of sequence	Transmission electron microscopy and Electron Energy Loss Spectroscopy	TEM/EELS	ng	Ultra-thin section (<100 nm)	Partially	Nanoscale composition and morphology	Electron beam damage
	Time-of-flight secondary ion mass spectrometry	ToF-SIMS	ng	Powdered sample/sample with flat surface	Destructive	Spatial distribution of organic molecules	
	Nano-scale secondary ion mass spectrometry	NanoSIMS	ng	Mounted sample/FIB section	Destructive	Spatial distribution of isotopic compositions	Ion beam damage, C, H deposition

[1]The use of C common to sample preparation (e.g. sample mounting, thin section preparation, FIB section preparation) must be completely avoided

[2]Sample mass required for analysis, not taking into account the sample loss during sample preparation, data provided by (Zolensky et al. 2000)

[3]Most analyses as considered partially destructive because of complicated sample preparation

[4]Important note: Since Seq. A analysis requires an exceptionally high level of cleanliness, analytical techniques prior to Seq. A analysis (FTIR, Raman spectroscopy, UV fluorescence spectroscopy) *MUST BE* handled by investigators who (i) have extensive experiences in techniques in Seq. A, (ii) have an in-depth understanding of the level of cleanliness required, and (iii) have a high skill level to work in wet laboratories and cleanrooms in order to keep the contamination at the minimum level required for Seq. A analysis)

[5]Two analytical sequences (A and B) that targets at characterising solvent soluble (labile) and insoluble (non-labile, refractory) organic compounds, respectively. Seq. A typically requires larger amounts of analytical samples (e.g. ~100 mg of meteorite samples for amino acid analysis using GC-MS) and is prone to terrestrial contamination during sample handling. Therefore, only analytical techniques that require minimal sample preparation and shorter analytical duration such as Raman spectroscopy can be incorporated into the analytical sequence. Other analytical techniques that involve further sample preparation (e.g. sample polishing for SEM analysis) should be included in a separate analytical sequence (Seq. B). Sample residues collected after targeted organic compounds have been extracted for Seq. A analysis can subsequently be used for Seq. B analysis, in so far as the investigators are aware of the contaminations that have been introduced to the samples during the extraction protocols, and how the potential alterations could have influenced the interpretation of the analytical results.

Acronym: FIB = focused ion beam

sample is characterised by a few non-destructive analytical techniques, such as FTIR and Raman spectroscopy. Very often, the contamination controls taken during labile and refractory organic analyses are different. For examples, samples targeted for amino acid analysis can only be in contact with sterile materials, or tools that have been sterilised after subjecting to heating at 500°C in air for at least 3 hours, while samples analysed by FTIR and Raman spectroscopy can be in contact with tools that have only been cleaned with organic solvent. If a sample is targeted for amino acid analysis subsequent to characterisation using FTIR and Raman spectroscopy, the same level of cleanliness should be sought for during the entire analytical sequence.

7 Conclusion

Sample return missions are essential in order for us to answer fundamental questions regarding the nature of the organic matter at birth of our solar system, its subsequent evolution, and the implications on the origin of life on Earth. However, proper identification and characterisation of extra-terrestrial organic matter are complicated by potential organic contamination, as the samples are exposed to terrestrial environment subsequent to their arrival on Earth, a planet inhabited by life. Unfortunately, the samples will never be cleaner than the tools and containers used store and handle them. A key lesson learned from past sample return missions is that a certain level of terrestrial contamination is inevitable, despite the best efforts that were made to minimise it. While careful measures of contamination control are planned and implemented, future studies of mission returned samples should be aware of the presence of different levels of terrestrial contamination, and employ state-of-the-art methods in order to distinguish extra-terrestrial organics form the inevitable terrestrial contamination.

Acknowledgements Queenie Chan was supported by a Science and Technology Facilities Council (STFC) rolling grant (ST/P000657/1). Rhonda Stroud was supported by the NASA LARS program, award NNH17AE44I. Zita Martins was financed by Fundo Europeu de Desenvolvimento Regional (FEDER) funds through the COMPETE 2020 - Operational Programme for Competitiveness and Internationalisation (POCI), and by Portuguese funds through FCT - Fundação para a Ciência e a Tecnologia in the framework of the project POCI-01-0145-FEDER-029932 (PTDC/FIS-AST/29932/2017). Centro de Química Estrutural acknowledges the financial support of Fundação para a Ciência e Tecnologia (UIDB/00100/2020). This research is framed within the College on Polar and Extreme Environments (Polar2E) of the University of Lisbon.

Publisher's Note Springer Nature remains neutral with regard to jurisdictional claims in published maps and institutional affiliations.

References

Y. Aikawa, E. Herbst, Two-dimensional distributions and column densities of gaseous molecules in protoplanetary disks. Astron. Astrophys. **371**(3), 1107–1117 (2001)

J. Aléon, C. Engrand, F. Robert, M. Chaussidon, Clues to the origin of interplanetary dust particles from the isotopic study of their hydrogen-bearing phases. Geochim. Cosmochim. Acta **65**(23), 4399–4412 (2001)

C.M.O.D. Alexander, M.L. Fogel, H. Yabuta, G.D. Cody, The origin and evolution of chondrites recorded in the elemental and isotopic compositions of their macromolecular organic matter. Geochim. Cosmochim. Acta **71**, 4380–4403 (2007)

C.M.O.D. Alexander, K.T. Howard, R. Bowden, M.L. Fogel, The classification of CM and CR chondrites using bulk H, C and N abundances and isotopic compositions. Geochim. Cosmochim. Acta **123**, 244–260 (2013)

C.M.O.D. Alexander, G.D. Cody, B.T. De Gregorio, L.R. Nittler, R.M. Stroud, The nature, origin and modification of insoluble organic matter in chondrites, the major source of Earth's C and N. Geochemistry **77**(2), 227–256 (2017)

C. Allen, J. Allton, G. Lofgren, K. Righter, M. Zolensky, Curating NASA's extraterrestrial samples—past, present, and future. Geochemistry **71**(1), 1–20 (2011)

J.H. Allton, *Catalog of Apollo Lunar Surface Geological Sampling Tools and Containers* (NASA Lyndon B. Johnson Space Center, 1989), pp. 1–97

J.H. Allton, J.R. Bagby, P.D. Stabekis, Lessons learned during Apollo lunar sample quanrantine and sample curation. Adv. Space Res. **22**(3), 373–382 (1998)

E. Anders, Pre-biotic organic matter from comets and asteroids. Nature **342**(6247), 255–257 (1989)

J.C. Aponte, J.P. Dworkin, J.E. Elsila, Indigenous aliphatic amines in the aqueously altered Orgueil meteorite. Meteorit. Planet. Sci. **50**(10), 1733–1749 (2015)

M. Arakawa, T. Saiki, K. Wada, K. Ogawa, T. Kadono, K. Shirai, H. Sawada, K. Ishibashi, R. Honda, N. Sakatani, Y. Iijima, C. Okamoto, H. Yano, Y. Takagi, M. Hayakawa, P. Michel, M. Jutzi, Y. Shimaki, S. Kimura, Y. Mimasu, T. Toda, H. Imamura, S. Nakazawa, H. Hayakawa, S. Sugita, T. Morota, S. Kameda, E. Tatsumi, Y. Cho, K. Yoshioka, Y. Yokota, M. Matsuoka, M. Yamada, T. Kouyama, C. Honda, Y. Tsuda, S. Watanabe, M. Yoshikawa, S. Tanaka, F. Terui, S. Kikuchi, T. Yamaguchi, N. Ogawa, G. Ono, K. Yoshikawa, T. Takahashi, Y. Takei, A. Fujii, H. Takeuchi, Y. Yamamoto, T. Okada, C. Hirose, S. Hosoda, O. Mori, T. Shimada, S. Soldini, R. Tsukizaki, T. Iwata, M. Ozaki, M. Abe, N. Namiki, K. Kitazato, S. Tachibana, H. Ikeda, N. Hirata, N. Hirata, R. Noguchi, A. Miura, An artificial impact on the asteroid (162173) Ryugu formed a crater in the gravity-dominated regime. Science **368**(6486), 67–71 (2020)

J. Ashley, The study of exogenic rocks on mars-an evolving subdiscipline in meteoritics. Elements **11**, 10–11 (2015)

J.W. Ashley, M.P. Golombek, P.R. Christensen, S.W. Squyres, T.J. McCoy, C. Schröder, I. Fleischer, J.R. Johnson, K.E. Herkenhoff, T.J. Parker, Evidence for mechanical and chemical alteration of iron-nickel meteorites on Mars: process insights for Meridiani Planum. J. Geophys. Res., Planets **116**(E7), E00F20 (2011)

J.L. Bada, Origins of homochirality. Nature **374**, 594–595 (1995)

S. Bajt, S.A. Sandford, G.J. Flynn, G. Matrajt, C.J. Snead, A.J. Westphal, J.P. Bradley, Infrared spectroscopy of Wild 2 particle hypervelocity tracks in Stardust aerogel: evidence for the presence of volatile organics in cometary dust. Meteorit. Planet. Sci. **44**(4), 471–484 (2009)

O.S. Barnouin, M.G. Daly, E.E. Palmer, R.W. Gaskell, J.R. Weirich, C.L. Johnson, M.M. Al Asad, J.H. Roberts, M.E. Perry, H.C.M. Susorney, R.T. Daly, E.B. Bierhaus, J.A. Seabrook, R.C. Espiritu, A.H. Nair, L. Nguyen, G.A. Neumann, C.M. Ernst, W.V. Boynton, M.C. Nolan, C.D. Adam, M.C. Moreau, B. Rizk, C.Y. Drouet D'Aubigny, E.R. Jawin, K.J. Walsh, P. Michel, S.R. Schwartz, R.L. Ballouz, E.M. Mazarico, D.J. Scheeres, J.W. McMahon, W.F. Bottke, S. Sugita, N. Hirata, N. Hirata, S.I. Watanabe, K.N. Burke, D.N. DellaGiustina, C.A. Bennett, D.S. Lauretta, D.E. Highsmith, J. Small, D. Vokrouhlický, N.E. Bowles, E. Brown, K.L. Donaldson Hanna, T. Warren, C. Brunet, R.A. Chicoine, S. Desjardins, D. Gaudreau, T. Haltigin, S. Millington-Veloza, A. Rubi, J. Aponte, N. Gorius, A. Lunsford, B. Allen, J. Grindlay, D. Guevel, D. Hoak, J. Hong, D.L. Schrader, J. Bayron, O. Golubov, P. Sánchez, J. Stromberg, M. Hirabayashi, C.M. Hartzell, S. Oliver, M. Rascon, A. Harch, J. Joseph, S. Squyres, D. Richardson, J.P. Emery, L. McGraw, R. Ghent, R.P. Binzel, M.M.A. Asad, C.L. Johnson, L. Philpott, H.C.M. Susorney, E.A. Cloutis, R.D. Hanna, H.C. Connolly, F. Ciceri, A.R. Hildebrand, E.M. Ibrahim, L. Breitenfeld, T. Glotch, A.D. Rogers, B.E. Clark, S. Ferrone, C.A. Thomas, H. Campins, Y. Fernandez, W. Chang, A. Cheuvront et al., Shape of (101955) Bennu indicative of a rubble pile with internal stiffness. Nat. Geosci. **12**(4), 247–252 (2019)

M.A. Barucci, A.F. Cheng, P. Michel, L.A.M. Benner, R.P. Binzel, P.A. Bland, H. Böhnhardt, J.R. Brucato, A. Campo Bagatin, P. Cerroni, E. Dotto, A. Fitzsimmons, I.A. Franchi, S.F. Green, L.M. Lara, J. Licandro, B. Marty, K. Muinonen, A. Nathues, J. Oberst, A.S. Rivkin, F. Robert, R. Saladino, J.M. Trigo-Rodriguez, S. Ulamec, M. Zolensky, MarcoPolo-R near Earth asteroid sample return mission. Exp. Astron. **33**(2), 645–684 (2012)

N.D. Bassim, B.T. De Gregorio, A.L.D. Kilcoyne, K. Scott, T. Chou, S. Wirick, G. Cody, R.M. Stroud, Minimizing damage during FIB sample preparation of soft materials. J. Microsc. **245**(3), 288–301 (2012)

H.A. Bechtel, G.J. Flynn, C. Allen, D. Anderson, A. Ansari, S. Bajt, R.K. Bastien, N. Bassim, J. Borg, F.E. Brenker, J. Bridges, D.E. Brownlee, M. Burchell, M. Burghammer, A.L. Butterworth, H. Changela, P. Cloetens, A.M. Davis, R. Doll, C. Floss, D.R. Frank, Z. Gainsforth, E. Grün, P.R. Heck, J.K. Hillier, P. Hoppe, B. Hudson, J. Huth, B. Hvide, A. Kearsley, A.J. King, B. Lai, J. Leitner, L. Lemelle, H. Leroux, A. Leonard, R. Lettieri, W. Marchant, L.R. Nittler, R. Ogliore, W.J. Ong, F. Postberg, M.C. Price, S.A. Sandford, J.-A.S. Tresseras, S. Schmitz, T. Schoonjans, G. Silversmit, A.S. Simionovici, V.A. Solé, R. Srama, F.J. Stadermann, T. Stephan, V.J. Sterken, J. Stodolna, R.M. Stroud, S. Sutton, M. Trieloff, P. Tsou, A. Tsuchiyama, T. Tyliszczak, B. Vekemans, L. Vincze, J. Von Korff, A.J. Westphal, N. Wordsworth, D. Zevin, M.E. Zolensky et al., Stardust interstellar preliminary examination III: infrared spectroscopic analysis of interstellar dust candidates. Meteorit. Planet. Sci. **49**(9), 1548–1561 (2014)

D.G. Blackmond, The origin of biological homochirality. Cold Spring Harb. Perspect. Biol. **2**(5), a002147 (2010)

D. Bockelée-Morvan, J. Crovisier, M. Mumma, H. Weaver, The composition of cometary volatiles. Comets II **1**, 391–423 (2004)

L. Bonal, E. Quirico, M. Bourot-Denise, G. Montagnac, Determination of the petrologic type of CV3 chondrites by Raman spectroscopy of included organic matter. Geochim. Cosmochim. Acta **70**(7), 1849–1863 (2006)

L. Bonal, M. Bourot-Denise, E. Quirico, G. Montagnac, E. Lewin, Organic matter and metamorphic history of CO chondrites. Geochim. Cosmochim. Acta **71**, 1605–1623 (2007)

L. Bonal, R. Brunetto, P. Beck, E. Dartois, Z. Dionnet, Z. Djouadi, J. Duprat, E. Füri, Y. Kakazu, G. Montagnac, P. Oudayer, E. Quirico, C. Engrand, Visible-IR and Raman microspectroscopic investigation of three Itokawa particles collected by Hayabusa: mineralogy and degree of space weathering based on nondestructive analyses. Meteorit. Planet. Sci. **50**(9), 1562–1576 (2015)

O. Botta, J.L. Bada, Extraterrestrial organic compounds in meteorites. Surv. Geophys. **23**, 411–467 (2002)

A.J. Brearley, m. Prinz, CI chondrite-like clasts in the Nilpena polymict ureilite: implications for aqueous alteration processes in CI chondrites. Geochim. Cosmochim. Acta **56**(3), 1373–1386 (1992)

K.L.F. Brinton, J.L. Bada, A reexamination of amino acids in lunar soils: implications for the survival of exogenous organic material during impact delivery. Geochim. Cosmochim. Acta **60**(2), 349–354 (1996)

D. Brownlee, P. Tsou, J. Aléon, C.M.O.D. Alexander, T. Araki, S. Bajt, G.A. Baratta, R. Bastien, P. Bland, P. Bleuet, J. Borg, J.P. Bradley, A. Brearley, F. Brenker, S. Brennan, J.C. Bridges, N.D. Browning, J.R. Brucato, E. Bullock, M.J. Burchell, H. Busemann, A. Butterworth, M. Chaussidon, A. Cheuvront, M. Chi, M.J. Cintala, B.C. Clark, S.J. Clemett, G. Cody, L. Colangeli, G. Cooper, P. Cordier, C. Daghlian, Z. Dai, L. D'Hendecourt, Z. Djouadi, G. Dominguez, T. Duxbury, J.P. Dworkin, D.S. Ebel, T.E. Economou, S. Fakra, S.A.J. Fairey, S. Fallon, G. Ferrini, T. Ferroir, H. Fleckenstein, C. Floss, G. Flynn, I.A. Franchi, M. Fries, Z. Gainsforth, J.-P. Gallien, M. Genge, M.K. Gilles, P. Gillet, J. Gilmour, D.P. Glavin, M. Gounelle, M.M. Grady, G.A. Graham, P.G. Grant, S.F. Green, F. Grossemy, L. Grossman, J.N. Grossman, Y. Guan, K. Hagiya, R. Harvey, P. Heck, G.F. Herzog, P. Hoppe, F. Hörz, J. Huth, I.D. Hutcheon, K. Ignatyev, H. Ishii, M. Ito, D. Jacob, C. Jacobsen, S. Jacobsen, S. Jones, D. Joswiak, A. Jurewicz, A.T. Kearsley, L.P. Keller, H. Khodja, A.L.D. Kilcoyne, J. Kissel, A. Krot, F. Langenhorst, A. Lanzirotti, L. Le, L.A. Leshin, J. Leitner, L. Lemelle, H. Leroux, M.-C. Liu, K. Luening, I. Lyon et al., Comet 81P/Wild 2 under a microscope. Science **314**(5806), 1711–1716 (2006)

P.C. Buchanan, M.E. Zolensky, A.M. Reid, Carbonaceous chondrite clasts in the howardites Bholghati and EET87513. Meteoritics **28**(5), 659–669 (1993)

A.L. Burlingame, M. Calvin, J. Han, W. Henderson, W. Reed, B.R. Simoneit, Lunar organic compounds: search and characterization. Science **167**(3918), 751–752 (1970)

A.S. Burton, J.E. Elsila, M.P. Callahan, M.G. Martin, D.P. Glavin, N.M. Johnson, J.P. Dworkin, A propensity for n-ω-amino acids in thermally altered Antarctic meteorites. Meteorit. Planet. Sci. **47**(3), 374–386 (2012)

A.S. Burton, J.E. Elsila, J.E. Hein, D.P. Glavin, J.P. Dworkin, Extraterrestrial amino acids identified in metal-rich CH and CB carbonaceous chondrites from Antarctica. Meteorit. Planet. Sci. **48**(3), 390–402 (2013)

A.S. Burton, S. Grunsfeld, J.E. Elsila, D.P. Glavin, J.P. Dworkin, The effects of parent-body hydrothermal heating on amino acid abundances in CI-like chondrites. Polar Sci. **8**(3), 255–263 (2014)

A.S. Burton, H. McLain, D.P. Glavin, J.E. Elsila, J. Davidson, K.E. Miller, A.V. Andronikov, D. Lauretta, J.P. Dworkin, Amino acid analyses of R and CK chondrites. Meteorit. Planet. Sci. **50**(3), 470–482 (2015)

H. Busemann, M.O.D. Alexander, L.R. Nittler, Characterization of insoluble organic matter in primitive meteorites by microRaman spectroscopy. Meteorit. Planet. Sci. **42**(7–8), 1387–1416 (2007)

H. Busemann, A.N. Nguyen, G.D. Cody, P. Hoppe, A.L.D. Kilcoyne, R.M. Stroud, T.J. Zega, L.R. Nittler, Ultra-primitive interplanetary dust particles from the comet 26P/Grigg–Skjellerup dust stream collection. Earth Planet. Sci. Lett. **288**(1–2), 44–57 (2009)

M.J. Calaway, C.C. Allen, J.H. Allton, Organic contamination baseline study in NASA Johnson Space Center Astromaterials Curation Laboratories, NASA/TP-2014-217393. NASA Astromaterials Acquisition and Curation Office (2014)

M. Calaway, J. Allton, F. McCubbin, R. Zeigler, The rise of astromaterials curation: NASA's 50 years of preserving pristine samples from the solar system, in *Lunar and Planetary Science Conference* (2019)

Q.H.S. Chan, Y. Chikaraishi, Y. Takano, N.O. Ogawa, N. Ohkouchi, Amino acid compositions in heated carbonaceous chondrites and their compound-specific nitrogen isotopic ratios. Earth Planets Space **68**(1), 7 (2016)

Q.H.S. Chan, M.E. Zolensky, R.J. Bodnar, C. Farley, J.C.H. Cheung, Investigation of organo-carbonate associations in carbonaceous chondrites by Raman spectroscopy. Geochim. Cosmochim. Acta **201**, 392–409 (2017)

Q. Chan, M. Zolensky, Y. Kebukawa, I. Franchi, I. Wright, I. Zhao, Z. Rahman, J. Utas, Primitive oxygen-, nitrogen-, and organic-rich vein preserved in a xenolith hosted in the metamorphosed Carancas meteorite, in *Lunar and Planetary Science Conference* (2018a)

Q.H.S. Chan, M. Zolensky, Y. Kebukawa, M. Fries, M. Ito, A. Steele, Z. Rahman, A. Nakato, A.L.D. Kilcoyne, H. Suga, Y. Takahashi, Y. Takeichi, K. Mase, Organic matter in extraterrestrial water-bearing salt crystals. Sci. Adv. **4**, eaao3521 (2018b)

Q. Chan, M. Zolensky, R. Brunetto, E. Parker, J. Dworkin, C. Sandt, I. Wright, I. Franchi, Coordinated organic analyses of Hayabusa category 3 carbonaceous particles, in *82nd Annual Meeting of the Meteoritical Society* (2019a), p. 6223

Q.H.S. Chan, I.A. Franchi, X. Zhao, A. Stephant, I.P. Wright, C.M.O.D. Alexander, Organics preserved in anhydrous interplanetary dust particles: pristine or not? Meteorit. Planet. Sci. (2019b). https://doi.org/10.1111/maps.13414

Q.H.S. Chan, A. Nakato, Y. Kebukawa, M.E. Zolensky, T. Nakamura, J.A. Maisano, M.W. Colbert, J.E. Martinez, A.L.D. Kilcoyne, H. Suga, Y. Takahashi, Y. Takeichi, K. Mase, I.P. Wright, Heating experiments of the Tagish Lake meteorite: investigation of the effects of short-term heating on chondritic organics. Meteorit. Planet. Sci. **54**(1), 104–125 (2019c)

C.F. Chyba, C. Sagan, Endogenous production, exogenous delivery and impact-shock synthesis of organic molecules: an inventory for the origins of life. Nature **355**, 125–132 (1992)

C.F. Chyba, P.J. Thomas, L. Brookshaw, C. Sagan, Cometary delivery of organic molecules to the early Earth. Science **249**(4967), 366–373 (1990)

S.J. Clemett, R.N. Zare, Microprobe two-step laser mass spectrometry as an analytical tool for meteoritic samples. Symp. - Int. Astron. Union **178**, 305–320 (1997)

S.J. Clemett, C.R. Maechling, R.N. Zare, P.D. Swan, R.M. Walker, Identification of complex aromatic molecules in individual interplanetary dust particles. Science **262**(5134), 721–725 (1993)

S.J. Clemett, S.A. Sandford, K. Nakamura-Messenger, F. HÖRz, D.S. McKay, Complex aromatic hydrocarbons in Stardust samples collected from comet 81P/Wild 2. Meteorit. Planet. Sci. **45**(5), 701–722 (2010)

G.D. Cody, C.M.O.D. Alexander, NMR studies of chemical structural variation of insoluble organic matter from different carbonaceous chondrite groups. Geochim. Cosmochim. Acta **69**(4), 1085–1097 (2005)

G.D. Cody, H. Ade, M.O.D. Alexander, T. Araki, A. Butterworth, H. Fleckenstein, G. Flynn, M.K. Gilles, C. Jacobsen, A.L.D. Kilcoyne, K. Messenger, S.A. Sandford, T. Tyliszczak, A.J. Westphal, S. Wirick, H. Yabuta, Quantitative organic and light-element analysis of comet 81P/Wild 2 particles using C-, N-, and O-μ-XANES. Meteorit. Planet. Sci. **43**(1–2), 353–365 (2008a)

G.D. Cody, C.M.O.D. Alexander, H. Yabuta, A.L.D. Kilcoyne, T. Araki, H. Ade, P. Dera, M. Fogel, B. Militzer, B.O. Mysen, Organic thermometry for chondritic parent bodies. Earth Planet. Sci. Lett. **272**(1–2), 446–455 (2008b)

H.C. Connolly, J. Zipfel, J.N. Grossman, L. Folco, C. Smith, R.H. Jones, K. Righter, M. Zolensky, S.S. Russell, G.K. Benedix, A. Yamaguchi, B.A. Cohen, The Meteoritical Bulletin, No. 90, 2006 September. Meteorit. Planet. Sci. **41**(9), 1383–1418 (2006)

J.R. Cronin, S. Chang, Organic matter in meteorites: molecular and isotopic analyses of the Murchison meteorite, in *The Chemistry of Life's Origins*, ed. by J.M. Greenberg, C.X. Mendoza-Gómez, V. Pirronello (Kluwer Academic, Norwell, 1993), pp. 209–258

E. Dartois, G.M. Muñoz Caro, D. Deboffle, G. Montagnac, L. d'Hendecourt, Ultraviolet photoproduction of ISM dust. Astron. Astrophys. **432**(3), 895–908 (2005)

B.T. De Gregorio, R.M. Stroud, L.R. Nittler, C.M.O.D. Alexander, A.L.D. Kilcoyne, T.J. Zega, Isotopic anomalies in organic nanoglobules from Comet 81P/Wild 2: comparison to Murchison nanoglobules and isotopic anomalies induced in terrestrial organics by electron irradiation. Geochim. Cosmochim. Acta **74**(15), 4454–4470 (2010)

B.T. De Gregorio, R.M. Stroud, G.D. Cody, L.R. Nittler, D.A.L. Kilcoyne, S. Wirick, Correlated microanalysis of cometary organic grains returned by Stardust. Meteorit. Planet. Sci. **46**(9), 1376–1396 (2011)

B.T. De Gregorio, R.M. Stroud, L.R. Nittler, C.M.O.D. Alexander, N.D. Bassim, G.D. Cody, A.L.D. Kilcoyne, S.A. Sandford, S.N. Milam, M. Nuevo, T.J. Zega, Isotopic and chemical variation of organic nanoglobules in primitive meteorites. Meteorit. Planet. Sci. **48**(5), 904–928 (2013)

S. Derenne, F. Robert, Model of molecular structure of the insoluble organic matter isolated from Murchison meteorite. Meteorit. Planet. Sci. **45**(9), 1461–1475 (2010)

J.P. Dworkin, L.A. Adelman, T. Ajluni, A.V. Andronikov, J.C. Aponte, A.E. Bartels, E. Beshore, E.B. Bierhaus, J.R. Brucato, B.H. Bryan, A.S. Burton, M.P. Callahan, S.L. Castro-Wallace, B.C. Clark, S.J. Clemett, H.C. Connolly, W.E. Cutlip, S.M. Daly, V.E. Elliott, J.E. Elsila, H.L. Enos, D.F. Everett, I.A. Franchi, D.P. Glavin, H.V. Graham, J.E. Hendershot, J.W. Harris, S.L. Hill, A.R. Hildebrand, G.O. Jayne, R.W. Jenkens, K.S. Johnson, J.S. Kirsch, D.S. Lauretta, A.S. Lewis, J.J. Loiacono, C.C. Lorentson, J.R. Marshall, M.G. Martin, L.L. Matthias, H.L. McLain, S.R. Messenger, R.G. Mink, J.L. Moore, K. Nakamura-Messenger, J.A. Nuth, C.V. Owens, C.L. Parish, B.D. Perkins, M.S. Pryzby, C.A. Reigle, K. Righter, B. Rizk, J.F. Russell, S.A. Sandford, J.P. Schepis, J. Songer, M.F. Sovinski, S.E. Stahl, K. Thomas-Keprta, J.M. Vellinga, M.S. Walker, OSIRIS-REx contamination control strategy and implementation. Space Sci. Rev. **214**(1), 19 (2017)

P. Ehrenfreund, D.P. Glavin, O. Botta, G. Cooper, J.L. Bada, Extraterrestrial amino acids in Orgueil and Ivuna: tracing the parent body of CI type carbonaceous chondrites. Proc. Natl. Acad. Sci. **98**(5), 2138–2141 (2001)

J.E. Elsila, D.P. Glavin, J.P. Dworkin, Cometary glycine detected in samples returned by Stardust. Meteorit. Planet. Sci. **44**(9), 1323–1330 (2009)

J.E. Elsila, S.B. Charnley, A.S. Burton, D.P. Glavin, J.P. Dworkin, Compound-specific carbon, nitrogen, and hydrogen isotopic ratios for amino acids in CM and CR chondrites and their use in evaluating potential formation pathways. Meteorit. Planet. Sci. **47**(9), 1517–1536 (2012)

J.E. Elsila, S.B. Charnley, A.S. Burton, D.P. Glavin, J.P. Dworkin, Compound-specific isotopic ratios of amino acids in CM and CR chondrites, in *44th Lunar and Planetary Science Conference* (2013), p. 1281

J.E. Elsila, J.C. Aponte, D.G. Blackmond, A.S. Burton, J.P. Dworkin, D.P. Glavin, Meteoritic amino acids: diversity in compositions reflects parent body histories. ACS Central Sci. **2**(6), 370–379 (2016a)

J.E. Elsila, M.P. Callahan, J.P. Dworkin, D.P. Glavin, H.L. McLain, S.K. Noble, E.K. Gibson, The origin of amino acids in lunar regolith samples. Geochim. Cosmochim. Acta **172**, 357–369 (2016b)

S. Epstein, H.P. Taylor Jr., O^{18}/O^{16}, Si^{30}/Si^{28}, C^{13}/C^{12} and D/H studies of Apollo 14 and 15 samples. Proc. Lunar Planet. Sci. Conf. **3**, 1429 (1972)

D.A. Flory, B.R. Simoneit, Terrestrial contamination in Apollo lunar samples. Space Life Sci. **3**(4), 457–468 (1972)

G.J. Flynn, Physical, chemical, and mineralogical properties of comet 81P/Wild 2 particles collected by Stardust, in *Advances in Meteoroid and Meteor Science*, ed. by J.M. Trigo-Rodríguez, F.J.M. Rietmeijer, J. Llorca, D. Janches (Springer, New York, 2008), pp. 447–459

A. Fox, Chemical markers for bacteria in extraterrestrial samples. Anat. Rec. **268**(3), 180–185 (2002)

S.W. Fox, K. Harada, P.E. Hare, Amino acid precursors in lunar fines: limits to the contribution of jet exhaust. Geochim. Cosmochim. Acta **40**(9), 1069–1071 (1976)

D.R. Frank, A.J. Westphal, M.E. Zolensky, Z. Gainsforth, A.L. Butterworth, R.K. Bastien, C. Allen, D. Anderson, A. Ansari, S. Bajt, N. Bassim, H.A. Bechtel, J. Borg, F.E. Brenker, J. Bridges, D.E. Brownlee, M. Burchell, M. Burghammer, H. Changela, P. Cloetens, A.M. Davis, R. Doll, C. Floss, G. Flynn, E. Grün, P.R. Heck, J.K. Hillier, P. Hoppe, B. Hudson, J. Huth, B. Hvide, A. Kearsley, A.J. King, B. Lai, J. Leitner, L. Lemelle, H. Leroux, A. Leonard, R. Lettieri, W. Marchant, L.R. Nittler, R. Ogliore, W.J. Ong, F. Postberg, M.C. Price, S.A. Sandford, J.-A.S. Tresseras, S. Schmitz, T. Schoonjans, G. Silversmit, A.S. Simionovici, V.A. Solé, R. Srama, T. Stephan, V.J. Sterken, J. Stodolna, R.M. Stroud, S. Sutton, M. Trieloff, P. Tsou, A. Tsuchiyama, T. Tyliszczak, B. Vekemans, L. Vincze, J.V. Korff, N. Wordsworth, D. Zevin et al., Stardust interstellar preliminary examination II: curating the interstellar dust collector, picokeystones, and sources of impact tracks. Meteorit. Planet. Sci. **49**(9), 1522–1547 (2014)

M. Fries, M. Burchell, A. Kearsley, A. Steele, Capture effects in carbonaceous material: a Stardust analogue study. Meteorit. Planet. Sci. **44**(10), 1465–1474 (2009)

D.P. Glavin, J.P. Dworkin, Enrichment of the amino acid L-isovaline by aqueous alteration on CI and CM meteorite parent bodies. Proc. Natl. Acad. Sci. **106**(14), 5487–5492 (2009)

D.P. Glavin, J.L. Bada, K.L.F. Brinton, G.D. McDonald, Amino acids in the Martian meteorite Nakhla. Proc. Natl. Acad. Sci. **96**, 8835–8838 (1999)

D.P. Glavin, J.P. Dworkin, A. Aubrey, O. Botta, J.H. Doty III, Z. Martins, J.L. Bada, Amino acid analyses of Antarctic CM2 meteorites using liquid chromatography–time of flight–mass spectrometry. Meteorit. Planet. Sci. **41**(6), 889–902 (2006)

D.P. Glavin, J.P. Dworkin, S.A. Sandford, Detection of cometary amines in samples returned by Stardust. Meteorit. Planet. Sci. **43**(1–2), 399–413 (2008)

D.P. Glavin, A.D. Aubrey, M.P. Callahan, J.P. Dworkin, J.E. Elsila, E.T. Parker, J.L. Bada, P. Jenniskens, M.H. Shaddad, Extraterrestrial amino acids in the Almahata Sitta meteorite. Meteorit. Planet. Sci. **45**(10–11), 1695–1709 (2010a)

D.P. Glavin, M.P. Callahan, J.P. Dworkin, J.E. Elsila, The effects of parent body processes on amino acids in carbonaceous chondrites. Meteorit. Planet. Sci. **45**(12), 1948–1972 (2010b)

D.P. Glavin, D.R. Williams, J.P. Dworkin, G. Kminek, M. Lupisella, J.D. Rummel, In situ biological contamination studies of the moon: implications for planetary protection and life detection missions. Earth Moon Planets **107**(1), 87–93 (2010c)

D.P. Glavin, J.E. Elsila, A.S. Burton, M.P. Callahan, J.P. Dworkin, R.W. Hilts, C.D.K. Herd, Unusual nonterrestrial l-proteinogenic amino acid excesses in the Tagish Lake meteorite. Meteorit. Planet. Sci. **47**(8), 1347–1364 (2012)

D.P. Glavin, C.M.O.D. Alexander, J.C. Aponte, J.P. Dworkin, J.E. Elsila, H. Yabuta, The origin and evolution of organic matter in carbonaceous chondrites and links to their parent bodies, in *Primitive Meteorites and Asteroids*, ed. by N. Abreu (Elsevier, Amsterdam, 2018), pp. 205–271

C.A. Goodrich, E.R.D. Scott, A.M. Fioretti, Ureilitic breccias: clues to the petrologic structure and impact disruption of the ureilite parent asteroid. Chem. Erde **64**(4), 283–327 (2004)

V.E. Hamilton, A.A. Simon, P.R. Christensen, D.C. Reuter, B.E. Clark, M.A. Barucci, N.E. Bowles, W.V. Boynton, J.R. Brucato, E.A. Cloutis, H.C. Connolly, K.L. Donaldson Hanna, J.P. Emery, H.L. Enos, S. Fornasier, C.W. Haberle, R.D. Hanna, E.S. Howell, H.H. Kaplan, L.P. Keller, C. Lantz, J.Y. Li, L.F. Lim, T.J. Mc, C. Merlin, M.C. Nolan, A. Praet, B. Rozitis, S.A. Sandford, D.L. Schrader, C.A. Thomas, X.D. Zou, D.S. Lauretta, D.E. Highsmith, J. Small, D. Vokrouhlick, ý.E. Bowles, E. Brown, K.L. Donaldson Hanna, T. Warren, C. Brunet, R.A. Chicoine, S. Desjardins, D. Gaudreau, T. Haltigin, S. Millington-Veloza, A. Rubi, J. Aponte, N. Gorius, A. Lunsford, B. Allen, J. Grindlay, D. Guevel, D. Hoak, J. Hong, D.L. Schrader, J. Bayron, O. Golubov, P. Sánchez, J. Stromberg, M. Hirabayashi, C.M. Hartzell, S. Oliver, M. Rascon, A. Harch, J. Joseph, S. Squyres, D. Richardson, J.P. Emery, L. Mc, G. Ghent, R.P. Binzel, M.M.A. Asad, C.L. Johnson, L. Philpott, H.C.M. Susorney, E.A. Cloutis, R.D. Hanna, H.C. Connolly, F. Ciceri, A.R. Hildebrand, E.M. Ibrahim, L. Breitenfeld, T. Glotch, A.D. Rogers, B.E. Clark, S. Ferrone, C.A. Thomas, H. Campins, Y. Fernandez, W. Chang, A. Cheuvront, D. Trang, S. Tachibana, H. Yurimoto, J.R. Brucato, G. Poggiali, M. Pajola, E. Dotto, E. Mazzotta Epifani et al., Evidence for widespread hydrated minerals on asteroid (101955) Bennu. Nat. Astron. **3**(4), 332–340 (2019)

K. Harada, P.E. Hare, C.R. Windsor, S.W. Fox, Evidence for compounds hydrolyzable to amino acids in aqueous extracts of Apollo 11 and Apollo 12 lunar fines. Science **173**(3995), 433–435 (1971)

P.E. Hare, Ion-exchange chromatography in lunar organic analysis. Space Life Sci. **3**, 354–359 (1972)

P.E. Hare, K. Harada, S.W. Fox, Analyses for amino acids in lunar fines. Geochim. Cosmochim. Acta, Suppl. **1**, 1799 (1970)

L. Haskin, P. Warren, Lunar chemistry, in *Lunar Sourcebook*, vol. 4(4) (1991), pp. 357–474

C.D.K. Herd, A. Blinova, D.N. Simkus, Y. Huang, R. Tarozo, C.M.O.D. Alexander, F. Gyngard, L.R. Nittler, G.D. Cody, M.L. Fogel, Y. Kebukawa, A.L.D. Kilcoyne, R.W. Hilts, G.F. Slater, D.P. Glavin, J.P. Dworkin, M.P. Callahan, J.E. Elsila, B.T. De Gregorio, R.M. Stroud, Origin and evolution of prebiotic organic matter as inferred from the Tagish Lake meteorite. Science **332**(6035), 1304–1307 (2011)

J.M. Holland, R.C. Simmonds, The mammalian response to lunar particulates. Space Life Sci. **4**(1), 97–109 (1973)

F. Hörz, R. Bastien, J. Borg, J.P. Bradley, J.C. Bridges, D.E. Brownlee, M.J. Burchell, M. Chi, M.J. Cintala, Z.R. Dai, Z. Djouadi, G. Dominguez, T.E. Economou, S.A.J. Fairey, C. Floss, I.A. Franchi, G.A. Graham, S.F. Green, P. Heck, P. Hoppe, J. Huth, H. Ishii, A.T. Kearsley, J. Kissel, J. Leitner, H. Leroux, K. Marhas, K. Messenger, C.S. Schwandt, T.H. See, C. Snead, F.J. Stadermann, T. Stephan, R. Stroud, N. Teslich, J.M. Trigo-Rodríguez, A.J. Tuzzolino, D. Troadec, P. Tsou, J. Warren, A. Westphal, P. Wozniakiewicz, I. Wright, E. Zinner, Impact features on Stardust: implications for comet 81P/Wild 2 dust. Science **314**(5806), 1716–1719 (2006)

Y. Ikeda, M. Prinz, C.E. Nehru, Lithic and mineral clasts in the Dar al Gani (DAG) 319 polymict ureilite. Antarct. Meteor. Res. **13**, 177–221 (2000)

Y. Ikeda, N.T. Kita, Y. Morishita, M.K. Weisberg, Primitive clasts in the Dar al Gani 319 polymict ureilite: precursors of the ureilites. Antarct. Meteor. Res. **16**, 105–127 (2003)

M. Ito, M. Uesugi, H. Naraoka, H. Yabuta, F. Kitajima, H. Mita, Y. Takano, Y. Karouji, T. Yada, Y. Ishibashi, T. Okada, M. Abe, H, C, and N isotopic compositions of Hayabusa category 3 organic samples. Earth Planets Space **66**(1), 91 (2014)

Y. Kebukawa, M.E. Zolensky, Q.H.S. Chan, K. Nagao, A.L.D. Kilcoyne, R.J. Bodnar, C. Farley, Z. Rahman, L. Le, G.D. Cody, Characterization of carbonaceous matter in xenolithic clasts from the Sharps (H3.4) meteorite: constraints on the origin and thermal processing. Geochim. Cosmochim. Acta **196**, 74–101 (2017)

Y. Kebukawa, M. Ito, M.E. Zolensky, R.C. Greenwood, Z. Rahman, H. Suga, A. Nakato, Q.H.S. Chan, M. Fries, Y. Takeichi, Y. Takahashi, K. Mase, K. Kobayashi, A novel organic-rich meteoritic clast from the outer solar system. Sci. Rep. **9**(1), 3169 (2019)

L.P. Keller, S. Messenger, G.J. Flynn, S. Clemett, S. Wirick, C. Jacobsen, The nature of molecular cloud material in interplanetary dust. Geochim. Cosmochim. Acta **68**(11), 2577–2589 (2004)

L.P. Keller, S.A. Bajt, G.A. Baratta, J. Borg, J.P. Bradley, D.E. Brownlee, H. Busemann, J.R. Brucato, M. Burchell, L. Colangeli, L. d'Hendecourt, Z. Djouadi, G. Ferrini, G. Flynn, I.A. Franchi, M. Fries, M.M. Grady, G.A. Graham, F. Grossemy, A. Kearsley, G. Matrajt, K. Nakamura-Messenger, V. Mennella, L. Nittler, M.E. Palumbo, F.J. Stadermann, P. Tsou, A. Rotundi, S.A. Sandford, C. Snead, A. Steele, D. Wooden, M. Zolensky, Infrared spectroscopy of comet 81P/Wild 2 samples returned by Stardust. Science **314**(5806), 1728–1731 (2006)

F. Kitajima, M. Uesugi, Y. Karouji, Y. Ishibashi, T. Yada, H. Naraoka, M. Abe, A. Fujimura, M. Ito, H. Yabuta, H. Mita, Y. Takano, T. Okada, A micro-Raman and infrared study of several Hayabusa category 3 (organic) particles. Earth Planets Space **67**(1), 1–12 (2015)

M.P. Kozar, M.T. Krahmer, A. Fox, L. Larsson, J. Allton, Lunar dust: a negative control for biomarker analyses of extraterrestrial samples? Geochim. Cosmochim. Acta **65**(19), 3307–3317 (2001)

H. Krüger, M. Kobayashi, S. Sasaki, R. Soja, M. Sommer, P. Strub, R. Srama, G. Moragas-Klostermeyer, H. Kimura, T. Hirai, T. Arai, R. Ishimaru, E. Grün, Cometary dust trail simulations for the Martian Moons Exploration (MMX) mission, in *EPSC-DPS Joint Meeting 2019*, Geneva, Switzerland (2019). EPSC-DPS2019-112-1

K. Kuramoto, Y. Kawakatsu, M. Fujimoto, H. Genda, T. Imamura, S. Kameda, K. Matsumoto, H. Miyamoto, T. Morota, H. Nagaoka, T. Nakamura, K. Ogawa, H. Otake, M. Ozaki, S. Sasaki, H. Senshu, S. Tachibana, N. Terada, T. Usui, K. Wada, S. Watanabe (M.S. Team), Martian Moons Exploration (MMX) conceptual study update, in *49th Lunar and Planetary Science Conference* (The Woodlands, Texas, 2018), p. 2143

D.S. Lauretta, A.E. Bartels, M.A. Barucci, E.B. Bierhaus, R.P. Binzel, W.F. Bottke, H. Campins, S.R. Chesley, B.C. Clark, B.E. Clark, E.A. Cloutis, H.C. Connolly, M.K. Crombie, M. Delbó, J.P. Dworkin, J.P. Emery, D.P. Glavin, V.E. Hamilton, C.W. Hergenrother, C.L. Johnson, L.P. Keller, P. Michel, M.C. Nolan, S.A. Sandford, D.J. Scheeres, A.A. Simon, B.M. Sutter, D. Vokrouhlický, K.J. Walsh, The OSIRIS-REx target asteroid (101955) Bennu: constraints on its physical, geological, and dynamical nature from astronomical observations. Meteorit. Planet. Sci. **50**(4), 834–849 (2015)

D.S. Lauretta, S.S. Balram-Knutson, E. Beshore, W.V. Boynton, C. Drouet d'Aubigny, D.N. DellaGiustina, H.L. Enos, D.R. Golish, C.W. Hergenrother, E.S. Howell, C.A. Bennett, E.T. Morton, M.C. Nolan, B. Rizk, H.L. Roper, A.E. Bartels, B.J. Bos, J.P. Dworkin, D.E. Highsmith, D.A. Lorenz, L.F. Lim, R. Mink, M.C. Moreau, J.A. Nuth, D.C. Reuter, A.A. Simon, E.B. Bierhaus, B.H. Bryan, R. Ballouz, O.S. Barnouin, R.P. Binzel, W.F. Bottke, V.E. Hamilton, K.J. Walsh, S.R. Chesley, P.R. Christensen, B.E. Clark, H.C. Connolly, M.K. Crombie, M.G. Daly, J.P. Emery, T.J. McCoy, J.W. McMahon, D.J. Scheeres, S. Messenger, K. Nakamura-Messenger, K. Righter, S.A. Sandford, OSIRIS-REx: sample return from asteroid (101955) Bennu. Space Sci. Rev. **212**(1), 925–984 (2017)

L. Le Roy, K. Altwegg, H. Balsiger, J.-J. Berthelier, A. Bieler, C. Briois, U. Calmonte, M.R. Combi, J. De Keyser, F. Dhooghe, B. Fiethe, S.A. Fuselier, S. Gasc, T.I. Gombosi, M. Hässig, A. Jäckel, M. Rubin, C. Y. Tzou, Inventory of the volatiles on comet 67P/Churyumov-Gerasimenko from Rosetta/ROSINA. Astron. Astrophys. **583**, A1 (2015)

S.G. Love, D.E. Brownlee, A direct measurement of the terrestrial mass accretion rate of cosmic dust. Science **262**(5133), 550–553 (1993)

M.Y. Marov, V.S. Avduevsky, E.L. Akim, T.M. Eneev, R.S. Kremnev, S.D. Kulikov, K.M. Pichkhadze, G.A. Popov, G.N. Rogovsky, Phobos-Grunt: Russian sample return mission. Adv. Space Res. **33**(12), 2276–2280 (2004)

Z. Martins, The nitrogen heterocycle content of meteorites and their significance for the origin of life. Life **8**(3), 28 (2018)

Z. Martins, Organic molecules in meteorites and their astrobiological significance, in *Handbook of Astrobiology* (CRC Press, Boca Raton, 2019), pp. 177–194

Z. Martins, M.A. Sephton, Extraterrestrial amino acids, in *Amino Acids, Peptides and Proteins in Organic Chemistry. Volume 1, Origins and Sybthesis of Amino Acids*, ed. by A.B. Hughes (Wiley-VCH, Weinheim, 2009), p. 701

Z. Martins, C.M.O.D. Alexander, G.E. Orzechowska, M.L. Fogel, P. Ehrenfreund, Indigenous amino acids in primitive CR meteorites. Meteorit. Planet. Sci. **42**(12), 2125–2136 (2007)

Z. Martins, M.C. Price, N. Goldman, M.A. Sephton, M.J. Burchell, Shock synthesis of amino acids from impacting cometary and icy planet surface analogues. Nat. Geosci. **6**(12), 1045–1049 (2013). https://doi.org/10.1038/ngeo1930

Z. Martins, P. Modica, B. Zanda, L.L.S. d'Hendecourt, The amino acid and hydrocarbon contents of the Paris meteorite: insights into the most primitive CM chondrite. Meteorit. Planet. Sci. **50**(5), 926–943 (2015)

Z. Martins, Q.H.S. Chan, L. Bonal, A. King, H. Yabuta, Organic matter in the solar system—implications for future on-site and sample return missions. Space Sci. Rev. (2020, this issue). https://doi.org/10.1007/s11214-020-00679-6

G. Matrajt, S. Messenger, D. Brownlee, D. Joswiak, Diverse forms of primordial organic matter identified in interplanetary dust particles. Meteorit. Planet. Sci. **47**(4), 525–549 (2012)

R. Mattingly, L. May, Mars sample return as a campaign, in *2011 Aerospace Conference* (2011), pp. 1–13

F.M. McCubbin, C.D.K. Herd, T. Yada, A. Hutzler, M.J. Calaway, J.H. Allton, C.M. Corrigan, M.D. Fries, A.D. Harrington, T.J. McCoy, J.L. Mitchell, A.B. Regberg, K. Righter, C.J. Snead, K.T. Tait, M.E. Zolensky, R.A. Zeigler, Advanced curation of astromaterials for planetary science. Space Sci. Rev. **215**(8), 48 (2019)

H.Y. McSween, A new type of chondritic meteorite found in lunar soil. Earth Planet. Sci. Lett. **31**(2), 193–199 (1976)

S. Messenger, F.J. Stadermann, C. Floss, L.R. Nittler, S. Mukhopadhyay, Isotopic signatures of presolar materials in interplanetary dust. Space Sci. Rev. **106**(1), 155–172 (2003)

S. Messenger, K. Nakamura-Messenger, L.P. Keller, ^{15}N-rich organic globules in a cluster IDP and the Bells CM2 chondrite, in *39th Lunar and Planetary Institute Science Conference* (2008), p. 2391

J.M. Mitchell, T.J. Jackson, R.P. Newlin, W.G. Meinschein, E. Cordes, V.J. Shiner Jr., Search for alkanes containing 15 to 30 carbon atoms per molecule in Apollo 12 lunar fines, in *Lunar and Planetary Science Conference Proceedings 2* (1971), p. 1927

I. Myrgorodska, C. Meinert, Z. Martins, L. Le, S. d'Hendecourt, U.J. Meierhenrich, Molecular chirality in meteorites and interstellar ices, and the chirality experiment on board the ESA cometary Rosetta mission. Angew. Chem., Int. Ed. Engl. **54**(5), 1402–1412 (2015)

K. Nagao, R. Okazaki, T. Nakamura, Y.N. Miura, T. Osawa, K-i. Bajo, S. Matsuda, M. Ebihara, T.R. Ireland, F. Kitajima, H. Naraoka, T. Noguchi, A. Tsuchiyama, H. Yurimoto, M.E. Zolensky, M. Uesugi, K. Shirai, M. Abe, T. Yada, Y. Ishibashi, A. Fujimura, T. Mukai, M. Ueno, T. Okada, M. Yoshikawa, J. Kawaguchi, Irradiation history of Itokawa regolith material deduced from noble gases in the Hayabusa samples. Science **333**(6046), 1128–1131 (2011)

H. Nagaoka, S. Takasawa, A.M. Nakamura, K. Sangen, Degree of impactor fragmentation under collision with a regolith surface—laboratory impact experiments of rock projectiles. Meteorit. Planet. Sci. **49**(1), 69–79 (2014)

B. Nagy, J.E. Modzeleski, V.E. Modzeleski, M.A.J. Mohammad, L.A. Nagy, W.M. Scott, C.M. Drew, J.E. Thomas, R. Ward, P.B. Hamilton, H.C. Urey, Carbon compounds in Apollo 12 lunar samples. Nature **232**(5306), 94–98 (1971)

T. Nakamura, Post-hydration thermal metamorphism of carbonaceous chondrites. J. Mineral. Petrol. Sci. **100**(6), 260–272 (2005)

T. Nakamura, T. Noguchi, M. Tanaka, M.E. Zolensky, M. Kimura, A. Tsuchiyama, A. Nakato, T. Ogami, H. Ishida, M. Uesugi, T. Yada, K. Shirai, A. Fujimura, R. Okazaki, S.A. Sandford, Y. Ishibashi, M. Abe, T. Okada, M. Ueno, T. Mukai, M. Yoshikawa, J. Kawaguchi, Itokawa dust particles: a direct link between S-type asteroids and ordinary chondrites. Science **333**(6046), 1113–1116 (2011)

K. Nakamura-Messenger, S. Messenger, L.P. Keller, S.J. Clemett, M.E. Zolensky, Organic globules in the Tagish Lake meteorite: remnants of the protosolar disk. Science **314**(5804), 1439–1442 (2006)

A. Nakato, T. Nakamura, F. Kitajima, T. Noguchi, Evaluation of dehydration mechanism during heating of hydrous asteroids based on mineralogical and chemical analysis of naturally and experimentally heated CM chondrites. Earth Planets Space **60**(8), 855–864 (2008)

H. Naraoka, H. Mita, K. Hamase, M. Mita, H. Yabuta, K. Saito, K. Fukushima, F. Kitajima, S.A. Sandford, T. Nakamura, T. Noguchi, R. Okazaki, K. Nagao, M. Ebihara, H. Yurimoto, A. Tsuchiyama, M. Abe, K. Shirai, M. Ueno, T. Yada, Y. Ishibashi, T. Okada, A. Fujimura, T. Mukai, M. Yoshikawa, J. Kawaguchi, Preliminary organic compound analysis of microparticles returned from Asteroid 25143 Itokawa by the Hayabusa mission. Geochem. J. **46**(1), 61–72 (2012)

H. Naraoka, D. Aoki, K. Fukushima, M. Uesugi, M. Ito, F. Kitajima, H. Mita, H. Yabuta, Y. Takano, T. Yada, Y. Ishibashi, Y. Karouji, T. Okada, M. Abe, ToF-SIMS analysis of carbonaceous particles in the sample catcher of the Hayabusa spacecraft. Earth Planets Space **67**(1), 67 (2015)

T. Noguchi, T. Nakamura, M. Kimura, M.E. Zolensky, M. Tanaka, T. Hashimoto, M. Konno, A. Nakato, T. Ogami, A. Fujimura, M. Abe, T. Yada, T. Mukai, M. Ueno, T. Okada, K. Shirai, Y. Ishibashi, R. Okazaki, Incipient space weathering observed on the surface of Itokawa dust particles. Science **333**(6046), 1121–1125 (2011)

V.I. Oyama, E.L. Merek, M.P. Silverman, A search for viable organisms in a lunar sample. Science **167**(3918), 773–775 (1970)

V.K. Pearson, M.A. Sephton, I.A. Franchi, J.M. Gibson, I. Gilmour, Carbon and nitrogen in carbonaceous chondrites: elemental abundances and stable isotopic compositions. Meteorit. Planet. Sci. **41**(12), 1899–1918 (2006)

S. Pizzarello, T.L. Groy, Molecular asymmetry in extraterrestrial organic chemistry: an analytical perspective. Geochim. Cosmochim. Acta **75**(2), 645–656 (2011)

S. Pizzarello, W. Holmes, Nitrogen-containing compounds in two CR2 meteorites: 15N composition, molecular distribution and precursor molecules. Geochim. Cosmochim. Acta **73**(7), 2150–2162 (2009)

S. Pizzarello, X. Feng, S. Epstein, J.R. Cronin, Isotopic analyses of nitrogenous compounds from the Murchison meteorite: ammonia, amines, amino acids, and polar hydrocarbons. Geochim. Cosmochim. Acta **58**(24), 5579–5587 (1994)

S. Pizzarello, Y. Huang, L. Becker, R.J. Poreda, R.A. Nieman, G. Cooper, M. Williams, The organic content of the Tagish Lake meteorite. Science **293**(5538), 2236–2239 (2001)

S. Pizzarello, Y. Huang, M. Fuller, The carbon isotopic distribution of Murchison amino acids. Geochim. Cosmochim. Acta **68**(23), 4963–4969 (2004)

S. Pizzarello, G. Cooper, G.J. Flynn, The nature and distribution of the organic material in carbonaceous chondrites and interplanetary dust particles, in *Meteorites and the Early Solar System II*, ed. by D.S. Lauretta, H.Y. McSween Jr. (University of Arizona Press, Tucson, 2006), pp. 625–651

S. Pizzarello, D.L. Schrader, A.A. Monroe, D.S. Lauretta, Large enantiomeric excesses in primitive meteorites and the diverse effects of water in cosmochemical evolution. Proc. Natl. Acad. Sci. USA **109**(30), 11949–11954 (2012)

J.R. Puleo, G.S. Oxborrow, N.D. Fields, H.E. Hall, Quantitative and qualitative microbiological profiles of the Apollo 10 and 11 Spacecraft. Appl. Microbiol. **20**(3), 384–389 (1970)

E. Quirico, J. Borg, P.-I. Raynal, G. Montagnac, L. d'Hendecourt, A micro-Raman survey of 10 IDPs and 6 carbonaceous chondrites. Planet. Space Sci. **53**(14–15), 1443–1448 (2005)

E. Quirico, L.V. Moroz, B. Schmitt, G. Arnold, M. Faure, P. Beck, L. Bonal, M. Ciarniello, F. Capaccioni, G. Filacchione, S. Erard, C. Leyrat, D. Bockelée-Morvan, A. Zinzi, E. Palomba, P. Drossart, F. Tosi, M.T. Capria, M.C. De Sanctis, A. Raponi, S. Fonti, F. Mancarella, V. Orofino, A. Barucci, M.I. Blecka, R. Carlson, D. Despan, A. Faure, S. Fornasier, M.S. Gudipati, A. Longobardo, K. Markus, V. Mennella, F. Merlin, G. Piccioni, B. Rousseau, F. Taylor, Refractory and semi-volatile organics at the surface of comet 67P/Churyumov-Gerasimenko: insights from the VIRTIS/Rosetta imaging spectrometer. Icarus **272**, 32–47 (2016)

A. Regberg, A. Burton, C. Castro, S. Stahl, S. Wallace, F. McCubbin, Microbial Ecology of the Johnson Space Center Meteorite Curation Lab and Associated Infrastructure, in *Lunar and Planetary Science Conference* (2018)

L. Remusat, F. Robert, A. Meibom, S. Mostefaoui, O. Delpoux, L. Binet, D. Gourier, S. Derenne, Protoplanetary disk chemistry recorded by D-rich organic radicals in carbonaceous chondrites. Astrophys. J. **698**(2), 2087–2092 (2009)

A. Rotundi, G.A. Baratta, J. Borg, J.R. Brucato, H. Busemann, L. Colangeli, L. D'Hendecourt, Z. Djouadi, G. Ferrini, I.A. Franchi, M. Fries, F. Grossemy, L.P. Keller, V. Mennella, K. Nakamura, L.R. Nittler, M.E. Palumbo, S.A. Sandford, A. Steele, B. Wopenka, Combined micro-Raman, micro-infrared, and field emission scanning electron microscope analyses of comet 81P/Wild 2 particles collected by Stardust. Meteorit. Planet. Sci. **43**(1–2), 367–397 (2008)

A.E. Rubin, The Hadley Rille enstatite chondrite and its agglutinate-like rim: impact melting during accretion to the Moon. Meteorit. Planet. Sci. **32**(1), 135–141 (1997)

A.E. Rubin, M.E. Zolensky, R.J. Bodnar, The halite-bearing Zag and Monahans (1998) meteorite breccias: shock metamorphism, thermal metamorphism and aqueous alteration on the H-chondrite parent body. Meteorit. Planet. Sci. **37**(1), 125–141 (2002)

S.A. Sandford, D.E. Brownlee, Response to comment on "organics captured from comet 81P/Wild 2 by the Stardust spacecraft". Science **317**(5845), 1680 (2007)

S.A. Sandford, J. Aléon, C.M.O.D. Alexander, T. Araki, S. Brucato, G.A. Baratta, J. Borg, J.P. Bradley, D.E. Brownlee, J.R. Brucato, M.J. Burchell, H. Busemann, A. Butterworth, S.J. Clemett, G. Cody, L. Colangeli, G. Cooper, L. D'Hendecourt, Z. Djouadi, J.P. Dworkin, G. Ferrini, H. Fleckenstein, G.J. Flynn, I.A. Franchi, M. Fries, M.K. Gilles, D.P. Glavin, M. Gounelle, F. Grossemy, C. Jacobsen, L.P. Keller, A.L.D. Kilcoyne, J. Leitner, G. Matrajt, A. Meibom, V. Mennella, S. Mostefaoui, L.R. Nittler, M.E. Palumbo, D.A. Papanastassiou, F. Robert, A. Rotundi, C.J. Snead, M.K. Spencer, F.J. Stadermann, A. Steele, T. Stephan, P. Tsou, T. Tyliszczak, A.J. Westphal, S. Wirick, B. Wopenka, H. Yabuta, R.N. Zare, M.E. Zolensky, Organics captured from Comet 81p/Wild 2 by the Stardust Spacecraft. Science **314**(5806), 1720–1724 (2006)

S.A. Sandford, S. Bajt, S.J. Clemett, G.D. Cody, G. Cooper, B.T. Degregorio, V. de Vera, J.P. Dworkin, J.E. Elsila, G.J. Flynn, Assessment and control of organic and other contaminants associated with the Stardust sample return from comet 81P/Wild 2. Meteorit. Planet. Sci. **45**(3), 406–433 (2010)

H. Sawada, R. Okazaki, S. Tachibana, K. Sakamoto, Y. Takano, C. Okamoto, H. Yano, Y. Miura, M. Abe, S. Hasegawa, T. Noguchi, S.M.P.T. Hayabusa, Hayabusa2 sampler: collection of asteroidal surface material. Space Sci. Rev. **208**(1), 81–106 (2017)

C. Schröder, D.S. Rodionov, T.J. McCoy, B.L. Jolliff, R. Gellert, L.R. Nittler, W.H. Farrand, J.R. Johnson, S.W. Ruff, J.W. Ashley, D.W. Mittlefehldt, K.E. Herkenhoff, I. Fleischer, A.F.C. Haldemann, G. Klingelhöfer, D.W. Ming, R.V. Morris, P.A. de Souza, S.W. Squyres, C. Weitz, A.S. Yen, J. Zipfel, T. Economou, Meteorites on Mars observed with the Mars Exploration Rovers. J. Geophys. Res., Planets **113**(E6), E06S22 (2008)

C. Schröder, P.A. Bland, M.P. Golombek, J.W. Ashley, N.H. Warner, J.A. Grant, Amazonian chemical weathering rate derived from stony meteorite finds at Meridiani Planum on Mars. Nat. Commun. **7**(1), 13459 (2016)

D.W.G. Sears, R.T. Dodd, Overview and classification of meteorites, in *Meteorites and the Early Solar System*, ed. by J.F. Kerridge, M.S. Matthews (University of Arizona Press, Tucson, 1988), pp. 3–31

B. Simoneit, D. Flory, Apollo 11, 12 and 13 organic monitoring history. A U. Calif. Berkeley report to NASA (1970)

B.R. Simoneit, A.L. Burlingame, D.A. Flory, I.D. Smith, Apollo lunar module engine exhaust products. Science **166**(3906), 733–738 (1969)

M.K. Spencer, R.N. Zare, Comment on "Organics captured from Comet 81P/Wild 2 by the Stardust spacecraft". Science **317**(5845), 1680 (2007)

M.K. Spencer, M.R. Hammond, R.N. Zare, Laser mass spectrometric detection of extraterrestrial aromatic molecules: mini-review and examination of pulsed heating effects. Proc. Natl. Acad. Sci. **105**(47), 18096–18101 (2008)

M.K. Spencer, S.J. Clemett, S.A. Sandford, D.S. McKay, R.N. Zare, Organic compound alteration during hypervelocity collection of carbonaceous materials in aerogel. Meteorit. Planet. Sci. **44**(1), 15–24 (2009)

A. Steele, J. Toporski, R. Avci, C. Agee, D. McKay, Investigations into the contamination of lunar return material. Part 1. Surface analysis and imaging investigations, in *32nd Lunar and Planetary Science Conference*, Houston, Texas (2001), p. 1674

A. Steele, F.M. McCubbin, M. Fries, M. Glamoclija, L. Kater, H. Nekvasil, Graphite in an Apollo 17 Impact Melt Breccia. Science **329**(5987), 51 (2010)

R.G. Strom, R. Malhotra, T. Ito, F. Yoshida, D.A. Kring, The origin of planetary impactors in the inner solar system. Science **309**(5742), 1847–1850 (2005)

H. Sugahara, Y. Takano, Y. Karouji, K. Kumagai, T. Yada, N. Ohkouchi, M. Abe, H.p. team, Amino acids on witness coupons collected from the ISAS/JAXA curation facility for the assessment and quality control of the Hayabusa2 sampling procedure. Earth Planets Space **70**(1), 194 (2018)

R.E. Summons, A.L. Sessions, A.C. Allwood, H.A. Barton, D.W. Beaty, B. Blakkolb, J. Canham, B.C. Clark, J.P. Dworkin, Y. Lin, R. Mathies, S.M. Milkovich, A. Steele, Planning considerations related to the organic contamination of martian samples and implications for the Mars 2020 rover. Astrobiology **14**(12), 969–1027 (2014)

M.B. Syal, P.H. Schultz, M.A. Riner, Darkening of Mercury's surface by cometary carbon. Nat. Geosci. **8**, 352 (2015)

S. Tachibana, M. Abe, M. Arakawa, M. Fujimoto, Y. Iijima, M. Ishiguro, K. Kitazato, N. Kobayashi, N. Namiki, T. Okada, R. Okazaki, H. Sawada, S. Sugita, Y. Takano, S. Tanaka, S. Watanabe, M. Yoshikawa, H. Kuninaka (The Hayabusa2 Project), Hayabusa2: scientific importance of samples returned from C-type near-Earth asteroid (162173) 1999 JU_3. Geochem. J. **48**(6), 571–587 (2014)

R. Terzieva, E. Herbst, The possibility of nitrogen isotopic fractionation in interstellar clouds. Mon. Not. R. Astron. Soc. **317**(3), 563–568 (2000)

K.L. Thomas-Keprta, S.J. Clemett, S. Messenger, D.K. Ross, L. Le, Z. Rahman, D.S. McKay, E.K. Gibson Jr., C. Gonzalez, W. Peabody, Organic matter on the Earth's Moon. Geochim. Cosmochim. Acta **134**(0), 1–15 (2014)

T.M. Tillotson, L.W. Hrubesh, Transparent ultralow-density silica aerogels prepared by a two-step sol-gel process. J. Non-Cryst. Solids **145**, 44–50 (1992)

P. Tsou, D.E. Brownlee, S.A. Sandford, F. Hörz, M.E. Zolensky, Wild 2 and interstellar sample collection and Earth return. J. Geophys. Res., Planets **108**(E10), 8111 (2003)

P. Tsou, D.E. Brownlee, C.P. McKay, A.D. Anbar, H. Yano, K. Altwegg, L.W. Beegle, R. Dissly, N.J. Strange, I. Kanik, LIFE: life investigation for Enceladus a sample return mission concept in search for evidence of life. Astrobiology **12**(8), 730–742 (2012)

A. Tsuchiyama, M. Uesugi, T. Matsushima, T. Michikami, T. Kadono, T. Nakamura, K. Uesugi, T. Nakano, S.A. Sandford, R. Noguchi, T. Matsumoto, J. Matsuno, T. Nagano, Y. Imai, A. Takeuchi, Y. Suzuki, T. Ogami, J. Katagiri, M. Ebihara, T.R. Ireland, F. Kitajima, K. Nagao, H. Naraoka, T. Noguchi, R. Okazaki, H. Yurimoto, M.E. Zolensky, T. Mukai, M. Abe, T. Yada, A. Fujimura, M. Yoshikawa, J. Kawaguchi, Three-dimensional structure of Hayabusa samples: origin and evolution of Itokawa regolith. Science **333**(6046), 1125–1128 (2011)

M. Uesugi, H. Naraoka, M. Ito, H. Yabuta, F. Kitajima, Y. Takano, H. Mita, I. Ohnishi, Y. Kebukawa, T. Yada, Y. Karouji, Y. Ishibashi, T. Okada, M. Abe, Sequential analysis of carbonaceous materials in Hayabusa-returned samples for the determination of their origin. Earth Planets Space **66**(1), 1–11 (2014)

M. Uesugi, M. Ito, H. Yabuta, H. Naraoka, F. Kitajima, Y. Takano, H. Mita, Y. Kebukawa, A. Nakato, Y. Karouji, Further characterization of carbonaceous materials in Hayabusa-returned samples to understand their origin. Meteorit. Planet. Sci. **54**(3), 638–666 (2019)

T. Usui, W. Fujiya, M. Koike, Y.N. Miura, S. Tachibana, Y. Takano, H. Kato, H. Sawada, Y. Sato, Y. Kawakatsu, K. Kuramoto, H. Otake, The importance of Phobos sample return for understanding the Mars-moon system. Space Sci. Rev. (2020). https://doi.org/10.1007/s11214-020-00668-9

K.J. Walsh, E.R. Jawin, R.L. Ballouz, O.S. Barnouin, E.B. Bierhaus, H.C. Connolly, J.L. Molaro, T.J. McCoy, M. Delbo', C.M. Hartzell, M. Pajola, S.R. Schwartz, D. Trang, E. Asphaug, K.J. Becker, C.B. Beddingfield, C.A. Bennett, W.F. Bottke, K.N. Burke, B.C. Clark, M.G. Daly, D.N. Della Giustina, J.P. Dworkin, C.M. Elder, D.R. Golish, A.R. Hildebrand, R. Malhotra, J. Marshall, P. Michel, M.C. Nolan, M.E. Perry, B. Rizk, A. Ryan, S.A. Sandford, D.J. Scheeres, H.C.M. Susorney, F. Thuillet, D.S. Lauretta, D.E. Highsmith, J. Small, D. Vokrouhlický, N.E. Bowles, E. Brown, K.L. Donaldson, H.T. Warren, C. Brunet, R.A. Chicoine, S. Desjardins, D. Gaudreau, T. Haltigin, S. Millington-Veloza, A. Rubi, J. Aponte, N. Gorius, A. Lunsford, B. Allen, J. Grindlay, D. Guevel, D. Hoak, J. Hong, D.L. Schrader, J. Bayron, O. Golubov, P. Sánchez, J. Stromberg, M. Hirabayashi, C.M. Hartzell, S. Oliver, M. Rascon, A. Harch, J. Joseph, S. Squyres, D. Richardson, J.P. Emery, L. Mc, G. Ghent, R.P. Binzel, M.M. Al Asad, C.L. Johnson, L. Philpott, H.C.M. Susorney, E.A. Cloutis, R.D. Hanna, H.C. Connolly, F. Ciceri, A.R. Hildebrand, E.M. Ibrahim, L. Breitenfeld, T. Glotch, A.D. Rogers, B.E. Clark, S. Ferrone, C.A. Thomas, H. Campins, Y. Fernandez, W. Chang, A. Cheuvront, D. Trang, S. Tachibana, H. Yurimoto et al., Craters, boulders and regolith of (101955) Bennu indicative of an old and dynamic surface. Nat. Geosci. **12**(4), 242–246 (2019)

A.J. Westphal, C. Snead, A. Butterworth, G.A. Graham, J.P. Bradley, S. Bajt, P.G. Grant, G. Bench, S. Brennan, P. Pianetta, Aerogel keystones: extraction of complete hypervelocity impact events from aerogel collectors. Meteorit. Planet. Sci. **39**(8), 1375–1386 (2004)

S. Wirick, G.J. Flynn, L.P. Keller, K. Nakamura-Messenger, C. Peltzer, C. Jacobsen, S. Sandford, M. Zolensky, Organic matter from comet 81P/Wild 2, IDPs, and carbonaceous meteorites; similarities and differences. Meteorit. Planet. Sci. **44**(10), 1611–1626 (2009)

B. Wopenka, Raman spectroscopic investigation of two grains from comet 81P/Wild 2: Information that can be obtained beyond the presence of sp2-bonded carbon. Meteorit. Planet. Sci. **47**(4), 565–584 (2012)

H. Yabuta, H. Naraoka, K. Sakanishi, H. Kawashima, Solid-state 13C NMR characterization of insoluble organic matter from Antarctic CM2 chondrites: evaluation of the meteoritic alteration level. Meteorit. Planet. Sci. **40**, 779–787 (2005)

H. Yabuta, L.B. Williams, G.D. Cody, C.M.O.D. Alexander, S. Pizzarello, The insoluble carbonaceous material of CM chondrites: a possible source of discrete organic compounds under hydrothermal conditions. Meteorit. Planet. Sci. **42**(1), 37–48 (2007)

H. Yabuta, C.M.O.D. Alexander, M.L. Fogel, A.L.D. Kilcoyne, G.D. Cody, A molecular and isotopic study of the macromolecular organic matter of the ungrouped C2 WIS 91600 and its relationship to Tagish Lake and PCA 91008. Meteorit. Planet. Sci. **45**(9), 1446–1460 (2010)

H. Yabuta, M. Uesugi, H. Naraoka, M. Ito, A. Kilcoyne, S. Sandford, F. Kitajima, H. Mita, Y. Takano, T. Yada, Y. Karouji, Y. Ishibashi, T. Okada, M. Abe, X-ray absorption near edge structure spectroscopic study of Hayabusa category 3 carbonaceous particles. Earth Planets Space **66**(1), 1–8 (2014a)

H. Yabuta, M. Uesugi, H. Naraoka, M. Ito, D. Kilcoyne, S. Sandford, F. Kitajima, H. Mita, Y. Takano, T. Yada, Source of carbonaceous materials from Hayabusa-returned samples: evaluation from XANES study, in *NIPR Symposium 2014* (2014b)

T. Yada, M. Abe, M. Uesugi, Y. Karouji, K. Kumagai, W. Satake, Y. Ishibashi, A. Nakato, T. Okada, A nature of particles in the Hayabusa sample catcher and contamination controls for Hayabusa 2 sample containers, in *77th Annual Meeting of the Meteoritical Society* (2014a)

T. Yada, A. Fujimura, M. Abe, T. Nakamura, T. Noguchi, R. Okazaki, K. Nagao, Y. Ishibashi, K. Shirai, M.e. Zolensky, S. Sandford, T. Okada, M. Uesugi, Y. Karouji, M. Ogawa, S. Yakame, M. Ueno, T. Mukai, M. Yoshikawa, J. Kawaguchi, Hayabusa-returned sample curation in the planetary material sample curation facility of JAXA. Meteorit. Planet. Sci. **49**(2), 135–153 (2014b)

H. Yano, T. Kubota, H. Miyamoto, T. Okada, D. Scheeres, Y. Takagi, K. Yoshida, M. Abe, S. Abe, O. Barnouin-Jha, A. Fujiwara, S. Hasegawa, T. Hashimoto, M. Ishiguro, M. Kato, J. Kawaguchi, T. Mukai, J. Saito, S. Sasaki, M. Yoshikawa, Touchdown of the Hayabusa spacecraft at the Muses Sea on Itokawa. Science **312**(5778), 1350–1353 (2006)

H. Yurimoto, K-i. Abe, M. Abe, M. Ebihara, A. Fujimura, M. Hashiguchi, K. Hashizume, T.R. Ireland, S. Itoh, J. Katayama, C. Kato, J. Kawaguchi, N. Kawasaki, F. Kitajima, S. Kobayashi, T. Meike, T. Mukai, K. Nagao, T. Nakamura, H. Naraoka, T. Noguchi, R. Okazaki, C. Park, N. Sakamoto, Y. Seto, M. Takei,

A. Tsuchiyama, M. Uesugi, S. Wakaki, T. Yada, K. Yamamoto, M. Yoshikawa, M.E. Zolensky, Oxygen isotopic compositions of asteroidal materials returned from Itokawa by the Hayabusa mission. Science **333**(6046), 1116–1119 (2011)

K. Zahnle, P. Schenk, H. Levison, L. Dones, Cratering rates in the outer solar system. Icarus **163**(2), 263–289 (2003)

M. Zolensky, A. Ivanov, The Kaidun Microbreccia meteorite: a harvest from the inner and outer asteroid belt. Chem. Erde **63**(3), 185–246 (2003)

M.E. Zolensky, R.H. Hewins, D.W. Mittlefehldt, M.M. Lindstrom, X. Xiao, M.E. Lipschutz, Mineralogy, petrology and geochemistry of carbonaceous chondritic clasts in the LEW 85300 polymict eucrite. Meteoritics **27**(5), 596–604 (1992)

M.E. Zolensky, M.K. Weisberg, P.C. Buchanan, D.W. Mittlefehldt, Mineralogy of carbonaceous chondrite clasts in HED achondrites and the Moon. Meteorit. Planet. Sci. **31**(4), 518–537 (1996)

M.E. Zolensky, R.J. Bodnar, E.K. Gibson, L.E. Nyquist, Y. Reese, C.-Y. Shih, H. Wiesmann, Asteroidal water within fluid inclusion-bearing halite in an H5 hondrite, Monahans (1998). Science **285**(5432), 1377–1379 (1999)

M.E. Zolensky, C. Pieters, B. Clark, J.J. Papike, Small is beautiful: the analysis of nanogram-sized astromaterials. Meteorit. Planet. Sci. **35**(1), 9–29 (2000)

M.E. Zolensky, T.J. Zega, H. Yano, S. Wirick, A.J. Westphal, M.K. Weisberg, I. Weber, J.L. Warren, M.A. Velbel, A. Tsuchiyama, P. Tsou, A. Toppani, N. Tomioka, K. Tomeoka, N. Teslich, M. Taheri, J. Susini, R. Stroud, T. Stephan, F.J. Stadermann, C.J. Snead, S.B. Simon, A. Simionovici, T.H. See, F. Robert, F.J.M. Rietmeijer, W. Rao, M.C. Perronnet, D.A. Papanastassiou, K. Okudaira, K. Ohsumi, I. Ohnishi, K. Nakamura-Messenger, T. Nakamura, S. Mostefaoui, T. Mikouchi, A. Meibom, G. Matrajt, M.A. Marcus, H. Leroux, L. Lemelle, L. Le, A. Lanzirotti, F. Langenhorst, A.N. Krot, L.P. Keller, A.T. Kearsley, D. Joswiak, D. Jacob, H. Ishii, R. Harvey, K. Hagiya, L. Grossman, J.N. Grossman, G.A. Graham, M. Gounelle, P. Gillet, M.J. Genge, G. Flynn, T. Ferroir, S. Fallon, D.S. Ebel, Z.R. Dai, P. Cordier, B. Clark, M. Chi, A.L. Butterworth, D.E. Brownlee, J.C. Bridges, S. Brennan, A. Brearley, J.P. Bradley, P. Bleuet, P.A. Bland, R. Bastien, Mineralogy and petrology of comet 81P/Wild 2 nucleus samples. Science **314**(5806), 1735–1739 (2006)

M. Zolensky, K. Ohsumi, G. Briani, M. Gounelle, T. Mikouchi, W. Satake, T. Kurihara, M. Weisberg, L. Le, *Searching for Chips of Kuiper Belt Objects in Meteorites* (2009)

Space Sci Rev (2019) 215:48
https://doi.org/10.1007/s11214-019-0615-9

Advanced Curation of Astromaterials for Planetary Science

Francis M. McCubbin[1] · Christopher D.K. Herd[2] · Toru Yada[3] · Aurore Hutzler[1] · Michael J. Calaway[4] · Judith H. Allton[1] · Cari M. Corrigan[5] · Marc D. Fries[1] · Andrea D. Harrington[1] · Timothy J. McCoy[5] · Julie L. Mitchell[1] · Aaron B. Regberg[1] · Kevin Righter[1] · Christopher J. Snead[6] · Kimberly T. Tait[7] · Michael E. Zolensky[1] · Ryan A. Zeigler[1]

Received: 11 May 2019 / Accepted: 17 October 2019 / Published online: 7 November 2019

Abstract Just as geological samples from Earth record the natural history of our planet, astromaterials hold the natural history of our solar system and beyond. Astromaterials acquisition and curation practices have direct consequences on the contamination levels of astromaterials and hence the types of questions that can be answered about our solar system and the degree of precision that can be expected of those answers. Advanced curation was developed as a cross-disciplinary field to improve curation and acquisition practices in existing astromaterials collections and for future sample return activities, including meteorite and cosmic dust samples that are collected on Earth. These goals are accomplished through research and development of new innovative technologies and techniques for sample collection, handling, characterization, analysis, and curation of astromaterials. In this contribution, we discuss five broad topics in advanced curation that are critical to improving sample acquisition and curation practices, including (1) best practices for monitoring and testing of curation infrastructure for inorganic, organic, and biological contamination; (2) requirements for storage, processing, and sample handling capabilities for future sample

Role of Sample Return in Addressing Major Questions in Planetary Sciences
Edited by Mahesh Anand, Sara Russell, Yangting Lin, Meenakshi Wadhwa, Kuljeet Kaur Marhas and Shogo Tachibana

✉ F.M. McCubbin

1 NASA Johnson Space Center, Mailcode XI2, 2101 NASA Parkway, Houston, TX 77058, USA

2 Department of Earth and Atmospheric Sciences, University of Alberta, 1-26 Earth Sciences Building, Edmonton, Alberta, T6G 2E3, Canada

3 JAXA, Sagamihara, 252-5210, Japan

4 Jacobs, NASA Johnson Space Center, Mail Code XI2/JETS, 2101 NASA Parkway, Houston, TX 77058, USA

5 Department of Mineral Sciences, National Museum of Natural History, Smithsonian Institution, Washington, DC, USA

6 Texas State University – Jacobs JETS Contract, NASA Johnson Space Center, Mail Code XI2/JETS, 2101 NASA Parkway, Houston, TX 77058, USA

7 Department of Natural History, Royal Ontario Museum, 100 Queen's Park, Toronto, Ontario M5S 2C6, Canada

return missions, along with recent progress in these areas; (3) advancements and improvements in astromaterials acquisition capabilities on Earth (i.e., the collection of meteorites and cosmic dust); (4) the importance of contamination knowledge strategies for maximizing the science returns of sample-return missions; and (5) best practices and emerging capabilities for the basic characterization and preliminary examination of astromaterials. The primary result of advanced curation research is to both reduce and quantify contamination of astromaterials and preserve the scientific integrity of all samples from mission inception to secure delivery of samples to Earth-based laboratories for in-depth scientific analysis. Advanced curation serves as an important science-enabling activity, and the collective lessons learned from previous spacecraft missions and the results of advanced curation research will work in tandem to feed forward into better spacecraft designs and enable more stringent requirements for future sample return missions and Earth-based sample acquisition.

1 Introduction

Human fascination with the night sky and with celestial objects that fall to the Earth from the sky is as old as our species, and use of these astromaterials as a natural resource occurred at least as early as the Bronze Age (Jambon 2017; McCoy 2018; McCoy et al. 2017). However, the initial curation of astromaterials as objects of scientific interest to understand our universe began more recently (Marvin 2006) and in earnest with the curation of meteorite samples in museums starting in the year 1748 at the Natural History Museum Vienna (Brandstätter 2006). Meteorites have remained objects of fascination by scientists and the public alike with the establishment of many meteorite collections across the world. Meteorite recovery and curation practices vary widely and are highly dependent on many factors, including the knowledge and resources of the finder and the financial and technical support available for the collection in which the sample is curated. The scientific importance of the sample can also be a determining factor, but this is predicated on the aforementioned factors. All meteorites, regardless of how they were handled from recovery to curation, have experienced uncontrolled entry and exposure to the terrestrial environment, including, at minimum, the terrestrial atmosphere and the ground. This exposure results in terrestrial contamination, the amount of which is typically dependent on the physicochemical properties of the meteorite, the conditions at the fall site, and the amount of exposure time to the terrestrial environment. Consideration of these factors can also be determining factors in how a meteorite sample is curated. An overview of meteorite collections, their contents, and curation practices is available in McCall et al. (2006).

Until the 1960's, delivery of all astromaterials to Earth were unplanned events that required reactionary responses for recovery and curation. However, with the initiation of the Apollo program, direct return of pristine astromaterials from another body became possible, and with it, established the need to design a facility to keep those samples in a pristine state for an indefinite period of time. Planning for the Lunar Receiving Laboratory (LRL) began in 1964, and the facility was completed in 1967 (Calaway et al. 2017; McLane et al. 1967). As part of this planning, stringent protocols in the handling, storage, and processing of samples were developed. These protocols ensured that portions of the samples remained pristine or as close to an "as returned" state as possible in perpetuity to enable future scientific discoveries from the returned samples. The delivery of Apollo 11 samples to Earth occurred on July 24, 1969 at 12:50 EDT, four days after the first successful human landing on the Moon. This round-trip journey marked a transformative milestone in

Table 1 Planetary sample return missions

Program/Mission	Returned to Earth	Destination	Returned sample
Apollo 11 (USA, NASA)	July 24, 1969	Moon: Mare Tranquillitatis	21.55 kg
Apollo 12 (USA, NASA)	November 24, 1969	Moon: Oceanus Procellarum	34.30 kg
Luna 16 (USSR)	September 24, 1970	Moon: Mare Fecunditatis	101 g
Apollo 14 (USA, NASA)	February 9, 1971	Moon: Fra Mauro Highlands	42.80 kg
Apollo 15 (USA, NASA)	August 7, 1971	Moon: Hadley-Apennine	76.70 kg
Luna 20 (USSR)	February 25, 1972	Moon: Apollonius Highlands	30 g
Apollo 16 (USA, NASA)	April 27, 1972	Moon: Descartes Highlands	95.20 kg
Luna 24 (USSR)	August 22, 1976	Moon: Mare Crisium	170.1 g
Apollo 17 (USA, NASA)	December 19, 1972	Moon: Taurus-Littrow	110.40 kg
Genesis (USA, NASA)	September 8, 2004	Earth-Sun Lagrange 1	Implanted Solar Wind Atoms
Stardust (USA, NASA)	January 15, 2006	Comet Wild 2/Interstellar	Small Particles Captured in Aerogel
Hayabusa (Japan, JAXA)	June 13, 2010	Asteroid 25143 Itokawa	Tens of thousands of recovered small particles
Hayabusa2 (Japan, JAXA)	December 2020 (Planned; in Flight)	Asteroid 162173 Ryugu	0.1 to 10 g Planned
OSIRIS-REx (USA, NASA)	September 2023 (Planned; in Flight)	Asteroid 101955 Bennu	0.06–2 kg Planned

human history and as the first sample return mission, provided the initial fuel to drive the burgeoning field of planetary sample science.

The planning process for curation prior to the return of the Apollo 11 samples set the precedent that curation involvement and planning begins at the inception of a sample return mission, and this founding principle has guided sample return missions subsequent to Apollo (e.g., Allen et al. 2011; Yada et al. 2014). There have been a total of 13 successful sample return missions, including six manned Apollo missions from NASA, three unmanned lunar sample return missions from the Union of Soviet Socialist Republics (USSR), the NASA Long Duration Exposure Facility that exposed various materials to the low-Earth orbit environment for approximately 6 years, the NASA Genesis mission that returned solar wind from the Earth-Sun Lagrange point 1 (L1), the NASA Stardust mission that returned particles embedded in aerogel from the coma of Comet Wild 2 and from interstellar space, and JAXA's Hayabusa mission that returned material from the surface of asteroid Itokawa (Table 1). In addition, there are two sample return missions in flight, including JAXA's Hayabusa2 mission that will return samples from the asteroid Ryugu and NASA's OSIRIS-REx mission that will return material from the asteroid Bennu (Table 1). Further details about each of these missions are provided in Table 1. With each successive sample return mission comes with it an important set of lessons learned that are used to inform subsequent sample return missions, and these lessons learned extend to curation standards and practices.

The Apollo program offered the first set of lessons learned and set forth the modern era of curation practices for astromaterials from the solar system. With the exception of USSR Luna missions, all sample return missions in the last two decades have built upon the legacy of Apollo. While recent missions have contributed to lessons learned, the majority of lessons learned and established practices can be linked to Apollo. The Apollo program actively sought out a wide range of scientists and eventually levied the scientific community at large to influence mission conception and design. Mission decisions and laboratory research on returned samples, at least peripherally, were focused substantially in maximizing science obtained from samples in laboratory research. The majority of these sample scientists were found in the field of geological sciences. The management environment was an integration of human spaceflight mission objectives, engineering constraints, sample scientists, and those responsible to prevent back contamination of the Earth. In this management structure, conflicts routinely arose and were not only turf battles, but were rooted in basic technical conflicts to balance crew safety, lunar sample preservation, and potential hazard containment for unknown biological pathogens. Since Apollo was a series of missions, it was possible to improve sampling hardware and laboratory handling devices using experience and samples from the lunar surface. For example, regolith drive tube function was greatly improved through redesign for Apollo missions 15–17 to allow deep penetration with minimal distortion of stratigraphy. Knowledge gained from examination of the first samples (Apollo missions 11–14) allowed the switch from a high-vacuum gloved handling environment to pure gaseous nitrogen positive pressure gloveboxes, which better preserved sample cleanliness and ease of use. A mission series like Apollo allows fine tuning of sample collection and returned sample handling as knowledge is acquired. Building upon Apollo and later sample return missions, a series of lessons learned and best practices for future sample return missions were developed and listed as follows:

- World-class scientific expertise: Integration of planetary sample scientists as advisors on science issues through formal organizations such as the historical Lunar Sample Analysis Planning Team (LSAPT) and Lunar and Planetary Science Team (LAPST) and as well as today's Curation and Analysis Planning Team for Extraterrestrial Materials (CAPTEM). CAPTEM currently presents findings to NASA on sample allocations ensuring best science and fair access to samples, current curation facilities, and inspection of laboratory operations, capabilities, capacity needs, and staffing. CAPTEM also provides findings for publicizing sample characterization information and service to the community. In addition, CAPTEM provides NASA with findings on design review of sample receiving and curation facilities as well as material restrictions/suggestions to preserve science value of samples.
- The integration of planetary science and geology training for astronauts, mission managers, and engineers involved in sample return missions.
- The integration of sample scientists into landing site selection, traverse planning, and sample acquisition.
- The integration of sample scientists into mission control operations and advisors during missions.
- The integration of Earth receiving and curation operations personnel into mission conception and engineering spacecraft design is critical for any sample return mission.
- Selection of materials that have low to zero particulate shedding mechanical properties for spacecraft, primary sample containment, handling, and storage equipment to preserve sample integrity.
- Selection of materials that have low to zero outgassing mechanical properties for spacecraft, primary sample containment, handling, and storage equipment to preserve sample integrity.

- Selection of a diversity of materials for primary sample containment, handling, and storage equipment to enable scientific investigations of the entire periodic table, organic compounds, and biological matter.
- Sample return missions should establish a concept of sample segregation for primary mission goals (e.g., segregation of samples in different containment/isolation used for inorganic, organic, and biological investigations as well as focused goals of the mission). Sample acquisition and containment must always focus on prohibiting cross-contamination and preservation of the scientific integrity of each sample.
- The integration of curation, proper material selection, and cleaning into mission contamination control requirements and implementation during Assembly, Test, and Launch Operations (ATLO) is critical for sample return.
- Use of inert and/or vacuum environments or environments close to native collection environments for processing and storage of astromaterials. Develop standard practices to mitigate contamination from terrestrial atmosphere, pressures, and temperatures.
- Use of environmental monitoring methods, cleanroom technology, and biological safety isolation to maintain desired processing and storage environments.

Lessons learned not only inform our best practices, but they also help to identify strategic knowledge gaps that require new research to fill. Furthermore, if we look only at improving upon our current curation capabilities, we will not be prepared when returned samples require care that is very different from those within our current collections. At present, most returned samples are geological in nature, with the exception of the Genesis solar wind atoms that are implanted within a number of high purity material substrates. Most of the samples are kept close to room temperature and, when kept in the pristine environments of a clean laboratory, will maintain their fidelity indefinitely. However, future sample return missions could bring back samples that require storage and handling conditions outside of current capabilities, including gases, liquids, ices, or biological materials. To successfully curate these sensitive materials also requires new research, and we describe here a field of research that we refer to as advanced curation.

Advanced Curation is a cross-disciplinary field that seeks to improve curation practices in existing astromaterials collections, including meteorite and cosmic dust samples that are collected on Earth. Specifically, advanced curation has two primary goals that include (1) expansion of the sample processing and storage capabilities of astromaterials facilities to prepare for future sample return missions and Earth-based collection of astromaterials and (2) to maximize the science returns of existing astromaterials sample collections. These goals are accomplished through research and development of new innovative technologies and techniques for sample collection, handling, characterization, analysis, and curation of astromaterials. In addition, advanced curation includes testing and evaluation of new technologies and operational procedures for future sample return missions through human and robotic analog studies. Here we outline best practices and procedures and highlight new results, capabilities, and ongoing activities in the field of advanced curation of astromaterials. In particular, we outline (1) the best practices for monitoring and testing of curation infrastructure for contamination, (2) the development of new storage, processing, and sample handling capabilities, (3) the development and improvement of new astromaterials acquisition capabilities on Earth (i.e., the collection of meteorites and cosmic dust), (4) the importance of contamination knowledge strategies for maximizing the science returns of sample-return missions, (5) best practices and emerging capabilities for the preliminary examination and initial characterization of astromaterials, and finally (6) a summary of the biggest challenges that lie ahead as we look toward future sample-return initiatives.

2 Monitoring and Testing of Curation Infrastructure

All sample return curation facilities are designed and built to meet specific controlled environment and cleanliness standards for the curated samples. Curation infrastructure is defined as all engineering systems that control the sample's storage and processing environment. This definition incorporates brick and mortar, temporary, modular, and mobile facilities. In addition, specialized equipment is included such as isolation chambers, gloveboxes, and desiccators that have the ability to alter the atmospheric chemistry, temperature, and pressure of the environment. During the Apollo program, curation infrastructure borrowed many innovative technologies from handling radioactive materials and biological quarantine practices. Today, curation infrastructure is derived from many industries including the nuclear, biotechnology, pharmaceutical, and semiconductor industries (USP 2013; Whyte 2001; Ramstorp 2000). Methods and techniques are either borrowed, augmented, or invented to maintain the controlled environment to mitigate terrestrial cross-contamination. Contamination covers any element that could compromise sample integrity. To quote the definition of pristine from Dworkin et al. (2018), it means that "no foreign material is introduced to the sample in an amount that hampers the ability to analyze the chemistry and mineralogy of the sample". While sample return missions designate contamination limits on specific elements and compounds at time of launch with focused science goals, samples are effectively allocated over time to study everything on the periodic table. Therefore, the implementation of curation infrastructure should be mindful that everything could be a contaminant to some research group. Modern cleanroom facilities have substantial infrastructure footprints that require continual monitoring to ensure they operate within the defined strict contamination control guidelines. This requires continuous monitoring and testing of the labs to verify that the sample processing environments remain clean from the standpoint of inorganic, organic, and biological contamination. As it is unrealistic to eliminate all contamination, careful monitoring and contamination knowledge must be conducted. To this end, curation laboratories that house astromaterials have developed numerous protocols and methods to monitor curation facilities and we outline those practices below.

2.1 Real-Time Continuous Monitoring and Testing of Curation Cleanroom Laboratories

Cleanrooms are a specialized controlled environment that must be continually monitored to verify whether they are working to defined parameters and specifications. The international standards organization (ISO) have developed fundamental standards for cleanrooms, namely ISO 14644. Curation cleanroom laboratories follow this standard as well as many adopted recommend practices from several industries (e.g., IEST, SEMI, GSA, etc.). For curation facilities, cleanroom measurements are regularly made to ensure that the heating ventilation, and air conditioning (HVAC) system is creating the appropriate cascade of positive or negative pressure and that Fan Filter Units (FFU), in conjunction with the HVAC system, are delivering the proper level of airborne particles to accepted limits for the planned ISO class. Temperature and humidity are also kept within pre-specified limits within the intended operational parameters of the HVAC system.

Ideally, real-time remote monitoring can track airborne particulates, room-to-room differential pressures, temperature, humidity, and HVAC operations. Remote airborne particle counters have either internal or external pumps with a flow rate of 0.1 CFM (2.83 LPM) or 1.0 CFM (28.3 LPM) dependent on ISO Class and desired statistics. Many of them can output up to 6 channels of simultaneous data within the range of 0.3–25.0 μm. For ISO

Class 4 and below, a dedicated 0.1 μm particle counter is desired to improve particle count statistics. While real-time remote monitoring is ideal, hand-held manual particle counters are sometimes used for spot checking spaces and annual ISO Class audits. For ISO Class 5 and above, these handheld particle counting instruments are typically set-up for a 2 minute measurement with a total sampling volume of 5.68 L and particle channels set at 0.3, 0.5, 0.7, 1.0, 5.0, and 10.0 μm. In lieu of real-time continuous remote monitoring, weekly particle counts of all curation labs are desirable of key areas with a full ISO audits conducted annually or bi-annually.

Curation cleanroom laboratories primarily use a positive pressure differential barrier to reduce contamination. A pressure differential barrier is based on the concept of using positive pressure air flow cascade to create a cleaner zone towards a less clean zone as a first line of defense to prevent cross-contamination between two adjacent spaces. The pressure differential should be of significant magnitude and stability to prevent any reversal of air flow between barriers including when barrier thresholds are crossed and/or doors are opened. However, the pressure differential should not be too high as to create turbulent air flow that could compromise the clean zone. In addition, too high of pressure between zones can also prevent doors from opening. For example, at 0.10 inH_2O (inches of water), a 3×7 ft. door requires 11 lbs. of force to open and close. Furthermore, this pressure results in unwanted turbulent air flow. ISO 14644-4, the design, construction, and start-up of cleanrooms and associated controlled environment, contains the international standard for cleanroom air-flow monitoring. ISO 14644-4 Section A.5.3 states that the pressure between clean zones should be set at: $\Delta P = 0.02$ to 0.08 inH_2O (5 to 20 Pa).

The cleanroom technology literature generally recommends a pressure differential of 0.04 inH_2O (10 Pa) between two cleanrooms and a pressure differential of 0.06 inH_2O (15 Pa) between the cleanroom and an unclassified room (Sakraida 2008; Whyte 2001). Whyte (2010) discusses the reason for ISO 14644-4 acceptable minimum of 0.02 inH_2O (5 Pa) pressure between adjacent rooms. This acceptable minimum was established for processing facilities that handle products that can be adversely affected from greater pressures. These low pressure differentials can sometimes be found in long tunnels between processing cleanrooms that contain air flow sensitive products. Whyte (2010) further discusses if 0.02 inH_2O (5 Pa) must be used; confirmation of the air flow direction must be verifiable with routine observable smoke flow tests (assuming such tests would not be a source of contamination). Sakraida (2008) discusses recent experimental studies that have tested the optimal pressure differential between clean zones. Pressure differentials between 0.03 to 0.05 inH_2O were determined to be optimal for mitigating cross-contamination. The study further suggested that clean zones with pressures above 0.05 inH_2O showed little increased benefit to mitigate contamination compared to increased energy costs of operating the air handling unit.

Based on ISO 14644-4 standards and available cleanroom technology literature, astromaterials curation laboratories should ideally maintain ≥ 0.05 inH_2O between interior "dirty" hallways to laboratory anterooms and a minimum of 0.03 to 0.05 inH_2O in most adjacent rooms between anteroom and main laboratory. For primary astromaterials storage areas and processing laboratories, ideally 0.05 to 0.08 inH_2O should be maintained to mitigate the long-term infiltration of contaminates. However, it is important to note that higher pressures may be desired to create a buffer to mitigate the risk of dropping below 0.05 inH_2O based on air flow stability from the HVAC and laboratory layout.

Real-time continuous remote monitoring is common for modern cleanrooms with a desired differential pressure accuracy of about ± 0.001 inH_2O or better. For older cleanroom laboratories, manual magnehelic differential pressure gauges are sometimes still used for monitoring differential pressures. In addition, annual or biannual differential pressure audits are conducted between each room doorway threshold with a handheld manometer and data placed on a building map to verify proper cascade of pressures.

HVAC ON/OFF and velocity (m/s) are continually monitored in real-time. The data displayed are also used to check air changes per hour towards the as-built of the cleanroom and ISO standards. The FFUs are biannually or annually checked to be running at 90 ± 10 fpm. While FFUs are typically not monitored in real-time, this is an important routine check to assess failing blower motors and the efficiency of the ULPA or HEPA filters to determine when they need to be replaced.

Electrostatic charging and discharging in curation laboratories has the potential to cause damage to samples and equipment. In addition, electrostatic discharges are a serious safety hazard to laboratory personnel. Most curation cleanrooms maintain a temperature between 24 to 15 °C $\pm$ 1.0 °C and relative humidity (RH%) of <65% to >35% $\pm$ 1.0% RH. These ranges are based on ISO 14644-4 and ISO 14644-5 standards and are only for laboratory environments and do not reflect the environment of containment, such as in gloveboxes where moisture (H_2O) is commonly measured below 1 ppm.

A deviation in any of these parameters or over a certain threshold (per curation protocol specific to the collection) triggers an investigation to understand the source of the problem and mitigate any faults. In case the issue cannot be resolved in a timely manner, samples are securely placed into storage and work stops in the lab, especially for samples processed outside of gloveboxes.

2.2 Real-Time Continuous Monitoring of Curation Infrastructure Systems

2.2.1 Inert Environments

Most pristine astromaterials benefit from not being stored and processed in terrestrial atmosphere. Since Earth's atmosphere is an oxidizing environment, preservation of astromaterials are preferred to be placed in an indigenous, vacuum, or inert environment. Most astromaterials on Earth are stored and processed in an inert gas such as nitrogen, argon, or helium, with the exception of JAXA's vacuum receiving glovebox used for the Hayabusa mission. Of these three inert gases, nitrogen is the most cost effective and is often chosen over argon and helium for routine storage. However, nitrogen analysis of astromaterials samples are compromised by processing in nitrogen, so nitrogen is not used exclusively.

At NASA Johnson Space Center (JSC), building 31 and 31N has a dedicated 15000 gallon liquid nitrogen (LN_2) tank and tank farm that converts high purity LN_2 to gaseous nitrogen (GN_2) for the entire building infrastructure. This nitrogen gas system provides an inert environment for processing and storing all NASA extraterrestrial sample collections where gloveboxes and desiccators consume $\sim$3500 scfh of GN_2. After gas production, the GN_2 is filtered for particulates by the use of sintered 316 stainless steel filters (1 micron filtration at the tank farm and 3 nm point-of-use filters connected to all devices). In addition to 3 nm particulate filtration, the Genesis lab uses point of use Pall gas purifiers that reduces any H_2O, CO_2, O_2, and CO in the GN_2 to < 1 ppb. The LN_2 is a modified Grade C per MIL-PRF-27401G [LN_2 purity 99.995%; H_2O <10 ppm; Total Hydrocarbons as CH_4 <1.0 ppm; O_2 < 10 ppm; H_2 < 10 ppm; Ar <20 ppm; CO_2 <10 ppm; CO <10 ppm; and particulates <1.0 mg/L]. LN_2 is delivered to JSC weekly and the Curation Office periodically tests the purity of the liquid nitrogen beyond the NASA contract audits. For periodic sampling of the LN_2, a cryogenic liquid sampler is connected directly to the LN_2 tanker truck with the sampler hose. The LN_2 sample is taken to an outside laboratory for analysis. The boil-off of this LN_2 at the tank farm produces high purity gaseous nitrogen (GN_2). JSC currently tests the purity of the delivered GN_2 by conducting airborne molecular organic sampling and SEM triage of inorganic particulates captured in 3 nm sintered stainless steel filters. Adsorbent sample tubes are used for sample collection and sent outside to Balazs Nanoanalysis for TD-GC-MS analysis. GN_2 results routinely show no infiltration particulates past the filters and organic compounds and all hydrocarbon loads are below the reporting limit of < 0.1 ng/L for >C7. The GN_2 is also tested monthly for the nitrogen isotopic ratio in a Finnigan MAT 253 IR-MS to ensure that no fractionation occurs over time or within the line. K-bottles of GN_2, Ar, and He are also supplied at high purity research grade when required for certain processing activities or experiments. For example, the Subzero Facility for Curation of Astromaterials at the University of Alberta (see Cold Curation section) uses high-purity (99.998%) Ar as a source, which is then further refined using a purification system to bring oxygen (O_2) and moisture (H_2O) levels to <0.1 ppmv (Herd et al. 2016).

2.2.2 *High Purity Cleaning Agents*

Cleaning curation sample handling tools, containers, and other equipment (such as glove-boxes, isolation chambers, and desiccators) is required for the curation of astromaterials. Precision cleaning is typically required where equipment is cleaned to a specified cleanliness and the cleanliness is measured and verified to a standard. These precision cleaning facilities are not a small foot-print and use substantial consumables and equipment for operations. During final precision cleaning, specialized equipment is needed to purify the aqueous cleaning solutions. Historically, Apollo used Freon 113 as the final cleaning agent. The Freon 113 recycled in-house by distillation to achieve the required high purity. Today, NASA JSC uses ultrapure water (UPW) as the final cleaning agent and requires substantial initial investment (>$3M USD) and monthly maintenance cost. For JSC, UPW is not only used for precision cleaning, but is also used to decontaminate Genesis solar wind materials contaminated by macro particles during the hard landing (see Genesis section).

The UPW purity is maintained and monitored in continuously flowing production lines. The JSC UPW plant produces 10 gallons/minute of UPW serving 5 laboratories throughout the building within a continuous flowing final loop connected to a 1000 gallon supply tank. Future upgrades to the system will increase the capacity to a 5000 gallon tank producing 15 gallons/minute serving 7 laboratories. Once UPW leaves the final flowing loop, within <5 seconds, CO_2 and other compounds in the air quickly dissolve into the highly deionized water and resistivity is immediately lowered from ~18.18 MΩ-cm to <1.0 MΩ-cm. Therefore, UPW cannot be stored or transported in containers for use and UPW must be used directly from the flowing final loop for the maximum cleaning effectiveness. The UPW system is outfitted with a continuous real-time monitoring of critical components of the system as well as final water quality. The system monitors flow rate, pressure, resistivity, conductivity, temperature, particulates, total organic carbon, and tank levels. The UPW system conforms to ASTM D 5127-13, Standard Guide for Ultra-Pure Water Used in the Electronics and Semiconductor Industries and produces E-1.1 or better quality of water with a resistivity of 18.18 MΩ-cm and total organic carbon (TOC) between 1 to 3 ppb. The quality of the water is routinely tested at least once a year or more for the following: (1) Anions by IC ranging from > 0.05 to 0.02 ppb (µg/L) of Fluoride (F^-), Chloride (Cl^-), Nitrite (NO^{2-}), Bromide (Br^-), Nitrate (NO^{3-}), Phosphate (HPO_4^{2-}), and Sulfate (SO_4^{2-}); (2) Monovalent & Divalent Cations by IC ranging from > 0.02 to 0.01 ppb (µg/L) of Lithium (Li^+), Sodium (Na^+), Ammonium (NH_4^+), Potassium (K^+), Magnesium (Mg^{2+}), and Calcium (Ca^{2+}); (3) 30 elements Ultra Low Level in UPW by ICP-MS ranging from > 10 to 0.02 ppt (ng/L) of Aluminum (Al), Antimony (Sb), Arsenic (As), Barium (Ba), Bismuth (Bi), Boron (B), Cadmium

(Cd), Calcium (Ca), Chromium (Cr), Cobalt (Co), Copper (Cu), Gallium (Ga), Germanium (Ge), Iron (Fe), Lead (Pb), Lithium (Li), Magnesium (Mg), Manganese (Mn), Mercury (Hg), Molybdenum (Mo), Nickel (Ni), Potassium (K), Silver (Ag), Sodium (Na), Strontium (Sr), Tin (Sn), Titanium (Ti), Tungsten (W), Vanadium (V), and Zinc (Zn); (4) Low-level Dissolved Silica at > 0.1 ppb (μg/L); (5) Bacteria-ASTM Method-F1094—87 48 Hr Incubation reported in > 1 Bacteria per 100 mL cfu.

2.2.3 Gloveboxes and Desiccators

The inert environments of gloveboxes and desiccators that house astromaterials in storage or during processing should also be monitored. These environments are typically monitored continuously in real-time for their pressure, temperature, and known contaminates. For Apollo lunar material stored and processed in inert GN_2, as well as the Subzero Facility used for processing Tagish Lake and other pristine astromaterials (Herd et al. 2016), gloveboxes are continuously monitored for O_2 and H_2O at a resolution of ± 1 ppm. For the lunar sample collection at JSC, these environments are required to be at 1 inH_2O positive pressure, room temperature, <25 ppm of O_2, and <50 ppm of H_2O; but actual achievable can be <1.0 ppmv for H_2O and <15 ppmv for O_2 with the current system. It should be noted that the vast majority of H_2O and O_2 levels in gloveboxes do not originate from the GN_2 supply lines, but from the isolator gaskets and gloves through molecular infiltration of terrestrial atmosphere even under 1.0 inH_2O positive pressure.

2.3 Inorganic and Organic Testing of Curation Clean Labs

Since 1998, the NASA JSC Curation Office has contracted *Air Liquide Balazs Nanoanalysis* to analyze airborne molecular inorganic and organic contaminates in cleanrooms and laboratory suites (Calaway et al. 2014). Following sampling protocols developed for the semiconductor industry, vertical exposure of 8″ and 6″ diameter high purity silicon semiconductor wafers are exposed for 24-hours on a work surface or inside gloveboxes to better understand the airborne molecular contamination (AMC). The AMC data is also used to calculate the rate of deposition of surface molecular contamination (SMC). The inorganic and organic AMC for cleanroom monitoring is reported using *ISO 14644-8 Classification of Air Cleanliness by Chemical Contamination* (*ACC*) and the SMC for *ISO 14644-10 Classification of Surface Cleanliness by Chemical Concentration.*

For routine inorganic lab and glovebox monitoring, pre-cleaned 8″ silicon wafers are packaged in two separate polypropylene wafer carriers; one for sample exposure and one for control, which is not opened. After a 24 or 48 hour of vertical exposure, Vapor Phase Decomposition Inductively Coupled Plasma Mass Spectrometry (VPD ICP-MS) is conducted at Balazs laboratories in Freemont, CA. The VDP-ICP-MS analyses report 35 elements (Al, As, B, Ba, Be, Ca, Cd, Ce, Co, Cr, Cu, Fe, Ga, Ge, Hf, In, K, La, Li, Mg, Mn, Mo, Na, Ni, Pb, Sb, Sn, Sr, Ta, Ti, W, V, Y, Zn, and Zr) with reporting limits ranging from 10^8 to 10^{10} atoms/cm^2.

For routine organic lab and glovebox monitoring, two sets of prebaked 8″ silicon wafers are sandwiched together and triple-wrapped in baked-out aluminum foil; two for sample exposure and two for control, which are not opened. After a 24 or 48 hour vertical exposure on an aluminum stand, Thermal Desorption Gas Chromatography Mass Spectroscopy (TD-GC-MS) is conducted at Balazs laboratories. The TD-GC-MS measures organic compounds from C6 to C28 with a reporting limit of 0.1 ng/cm^2.

In addition to organic wafer exposure, which collect airborne molecular and particulate contaminants well, proprietary air absorbent tests are routinely conducted to better understand hydrocarbon and volatile organic compound (VOC) load in cleanroom air or glovebox gaseous nitrogen environments. This test is implemented with an adsorbent tube with a pump running at 100 mL/min, for 6 hours that is exposed to the cleanroom or glovebox. The adsorbent tube is analyzed using the same TD-GC-MS method as the organic wafer, but with a reporting limit of 0.1 ng/L.

Besides these traditional methods of monitoring, the JSC Curation Office also employs the use of optical microscopy and Scanning Electron Microscopy (SEM) as a basic method of direct analysis for inorganic and organic contaminates for the cleanroom laboratory and infrastructure. Cleanroom construction materials, surfaces, sample handling tools, containers, and unknown visible material are analyzed directly or with tape-pulls or polyester wipes. Optical microcopy and SEM typically are used as an initial screening before using other methods of analysis. The following methods have been used in the past at the NASA JSC Curation office, on an as-needed basis, on witness plates, test coupons, millipore filters, and other material samples: (1) Optical Stereomicroscopy/Microscopy for macro particulate/other contamination, (2) FEG-SEM/EDX for micro particulate identification, (3) FT-IR and Raman Spectroscopy for surface contamination, (4) XPS for complete surface/thin-films/oxidation, (5) LA-HR-ICP-MS for gross surface inorganics, (6) VPD-HR-ICP-MS for molecular airborne inorganics, (7) TD-GC-MS with GL Sciences SWA-256 wafer analyzer for molecular airborne organics/outgassing, (8) DART-qTOF-MS for gross surface organics, (9) LC-MS for amino acids, and (10) AFM (Atomic Force Microcopy) for surface roughness/thin-films/cleaning changes. Although not continuously monitored, the Subzero Facility for the Curation of Astromaterials used solid phase microextraction (SPME) fiber GC-MS methods to characterize the glovebox atmosphere during commissioning (Herd et al. 2016); this method shows potential for use in continuous monitoring, although its use requires the assessment and selection of appropriate SPME fibers for the airborne organic compounds of interest.

2.4 Biological Testing of Curation Cleanlabs

Biological testing of clean labs is important in many commercial and academic settings, and biological testing in aerospace and medical settings, like spacecraft assembly facilities, hospital cleanrooms, and pharmaceutical production labs are discussed here in the context of the best practices for monitoring astromaterials curation facilities. The monitoring methods differ among these labs, but the overall goal, to reduce or eliminate contamination, is always the same. A key difference for curation facilities is the need to identify contaminants. Identification is not always a monitoring plan requirement in other industries. Microorganisms like bacteria and fungi are capable of physically and chemically altering astromaterials (Toporski and Steele 2007). Since the nutrient levels in cleanrooms are purposely kept at very low levels, it is likely that microorganisms will seek out nutrient bearing phases in the astromaterials themselves (e.g. phosphorous rich minerals, organic carbon). Therefore, it is important to identify organisms in cleanrooms and understand how they might affect samples stored within the cleanrooms. The most common monitoring method for any cleanroom is cultivation of viable microorganisms like bacteria and fungi. The implementation of a variety of culture-independent analysis techniques that are employed more sporadically are also discussed.

In the aerospace industry, biological testing is most commonly performed to meet planetary protection requirements for individual pieces of hardware and entire missions. The goals are defined by Article IX of the 1967 United Nations Treaty on, "Principles Governing the Activities of States in the Exploration and Use of Outer Space, Including the Moon and Other Bodies". More detailed policies are outlined by COSPAR (Committee on Space Research) (COSPAR 2011). The sampling and testing methods are implemented by NASA (NASA 2010, 2017) and/or ESA (ECSS 2008).

NASA requirements for sampling an aerospace cleanroom to meet planetary protection requirements are described in the "Handbook for the Microbial Examination of Space Hardware" (NASA 2010). Briefly, samples are collected with sterile swabs or wipes made of cotton or preferably a synthetic material like polyester. The samples are exposed to 80 °C for 15 minutes and any surviving microorganisms are transferred to Petri dishes filled with Tryptic Soy Agar (TSA) and incubated at 32 °C for 72 hours. Cultured organisms are counted but not necessarily identified. ESA requirements are similar, but require cultivation on Reasoners 2 Agar (R2A) for oligotrophic bacteria, Thioglycolate Agar (TGA) for anaerobic bacteria, and Potato Dextrose Agar (PDA) for fungi in addition to TSA (ECSS 2008). Only one set of R2A plates are heat shocked, while the remaining samples are incubated without being exposed to heat. The ESA standards also include provisions for collecting air samples with an impactor style sampling device. Sampling to meet planetary protection requirements is conducted with the assumption that all of the hardware will be exposed to DHMR (dry heat microbial reduction) or an equivalent process to sterilize the spacecraft. Organisms that survive the heat-shock treatment are counted as a proxy for what might be capable of surviving DHMR. These are fit for purpose assays that are not designed or intended to capture the total diversity of the cleanroom environment. While some facilities and/or missions do identify and archive isolates, this is not required or routine in every instance. Similar culture-based assays have been used to monitor Chinese and Russian space craft assembly facilities as well (Novikova 2004; Zhang et al. 2018).

Cleanrooms used to manufacture pharmaceuticals and package food are also monitored for biological contamination. There are no detailed methods for how to monitor these types of cleanrooms, but cultivation-based techniques are generally the norm. ISO 14698-1 sets out very general principles and methods for biocontamination control in cleanrooms. The document states, "The appropriate sampling method and related procedures shall be selected and performed to reflect the complexity and variety of situations. Sampling shall be carried out using a device and method selected in accordance with the written procedure and in accordance with the instructions provided by the device manufacturer," (ISO14698 2003). The United States Pharmacopeial Convention also relies on cultivation based methods without specifying a particular set of sampling tools, growth conditions, or nutrients (USP 2013). For example, air samples can be collected with a variety of tools, including: slit to agar samplers, centrifugal samplers, gelatin filter samplers, sieve impactors, impingers, and settle plates (USP 2013). However, the USP document does make several important points regarding sampling methods and data analysis: (1) Total particulate counts from air sampling do not correlate to microbial abundance, although this is an area of open research (Raval et al. 2012). (2) Microbial monitoring is semi-quantitative at best. (3) Colony counts (i.e. the number of culturable organisms) are highly variable from sample to sample and from day to day. Recovery rate is a more reliable statistic for defining a microbial baseline. Recovery rate is defined as: $\frac{\#\ of\ samples\ with{>}0\ CFU}{\#\ total\ number\ of\ samples\ collected\ during\ a\ sampling\ event}$ where CFU is a colony forming unit. For example, an aseptic ISO 7 cleanroom should have a baseline recovery rate $<10\%$ (USP 2013). The USP document also emphasizes the importance of identifying cleanroom isolates and taking action when new isolates appear and or when an individual sample contains >15 CFU. In general, sampling of pharmaceutical cleanrooms is focused on cultivating mesophilic organisms from surface swabs, air samples, and cleanroom personnel (Sandle 2012; Whyte 2010). A variety of media types and growth conditions are acceptable as long as they are suitable for enumerating the organisms of concern.

Standard efforts to monitor microbial contamination in cleanrooms rely on cultivation based techniques across all industries. Cultivation based techniques are relatively cheap and easy to perform on a regular basis. However, they can be highly variable and even the most comprehensive culture-based sampling campaign is guaranteed to under-sample the environment (e.g., Hug et al. 2016; Lynch and Neufeld 2015; Rappe and Giovannoni 2003). The community recognizes the need to assess these, "unculturable," organisms and has employed a variety of techniques to do so.

Next generation DNA sequencing is the most common culture-independent method. Amplification and sequencing of marker genes (tag or amplicon sequencing) like the ribosomal 16S gene for bacteria and archaea and the ITS region for fungi is one promising method for monitoring unculturable organisms in the cleanroom environment. This measurement has changed as the sequencing platforms have improved. Initial tag sequencing was performed using clone libraries and Sanger type sequencing, which only generates about 1000 base pairs of data for a single organism at a time (Shokralla et al. 2012). The 454 platform generates 10^2–10^4 sequences per sample and has allowed researchers to identify hundreds of OTU's (operational taxonomic unit) or organisms per sample (La Duc et al. 2014; Moissl-Eichinger et al. 2015; Vaishampayan et al. 2013). Using this technology, archaea were found to be persistent, viable (Moissl-Eichinger 2011) members of some cleanroom communities (Moissl-Eichinger 2011; Moissl-Eichinger et al. 2015; Moissl et al. 2008). The Illumina sequencing platforms are the current standard for tag sequencing (Mahnert et al. 2015; Minich et al. 2018; Mora et al. 2016). These sequencers can generate 10^5–10^6 sequences per sample, allowing researchers to identify even more organisms. A recent tag sequencing survey of the SAF (spacecraft assembly facility) at JPL identified > 16000 OTU's. Tag sequencing is a powerful monitoring tool, but it does have several important biases. PCR (polymerase chain reaction) based amplification of DNA is required for most low biomass samples. This amplification step does not amplify DNA from every organism equally. For example Moissl-Eichinger et al. (2015) were able to cultivate organisms that they did not detect using, "universal" PCR primers for amplification and subsequent tag sequencing. Secondly, sequencing of any type cannot distinguish DNA from viable organisms from relict environmental DNA inside dead organisms. Several researchers have started treating their samples with compounds like PMA (propidium monoazide) to destroy DNA from non-viable organisms prior to sequencing (e.g., Mahnert et al. 2015; Moissl-Eichinger et al. 2015; Mora et al. 2016; Weinmaier et al. 2015; Zhang et al. 2018). Due to variations in primer choice, sequence length, error rate, and total number of sequences produced, it is very difficult to quantitatively compare data generated by different sequencing platforms (Tremblay et al. 2015). Care should be taken to keep these variables as consistent as possible during monitoring. When changes are made, they should be directly compared to previous methods.

Rather than amplifying and sequencing specific marker genes, it is also possible to sequence all of the DNA in a sample (with or without amplification) using the same types of DNA sequencers discussed above. This sequencing technique is commonly referred to as shotgun metagenomics (e.g., Bashir et al. 2016; Minich et al. 2018; Moissl-Eichinger et al. 2015; Weinmaier et al. 2015). Shotgun metagenomics provides more information about the function of abundant organisms in the environment but often fails to detect rare members of the community (Tessler et al. 2017). Additionally, this technique generates large amounts of data that can be very challenging and time-consuming to interpret. At present, metagenomics is a powerful research tool, but it is probably not yet suitable for routine monitoring. New DNA sequencers, like the MinIon platform (Reuter et al. 2015), that generate longer reads may eventually be able to generate metagenomic data that are easier to assemble and interpret, but they are still being developed and improved.

DNA sequencing can be used to inform the design of more rapid assays for biological monitoring. qPCR (quantitative polymerase chain reaction) can be used to assess the number of copies of genes in a sample that directly correlates to microbial abundance (Cooper et al. 2011; Hubad and Lapanje 2013; Kwan et al. 2011; Mahnert et al. 2015; Moissl-Eichinger 2011; Moissl-Eichinger et al. 2015; Schwendner et al. 2013; Vaishampayan et al. 2013; Zhang et al. 2018). When interpreting qPCR data, care must be taken to account for organisms that have multiple copies of the 16S or other marker gene (Větrovský and Baldrian 2013). DNA microarrays like the Phylochip have also been investigated as potential monitoring tools (Cooper et al. 2011; Jimenez 2011; La Duc et al. 2009, 2014; Probst et al. 2010; Vaishampayan et al. 2013). Both of these techniques show promise as monitoring solutions, but they probably require initial investigation with culturing and DNA sequencing in order to ensure that probes and primers are designed to capture the communities present inside the cleanroom in question.

Techniques that do not involve sequencing DNA are also being tested in cleanroom settings. All living organisms on Earth produce a compound called ATP (adenosine triphosphate) for energy storage. Measuring the concentration of ATP in a cleanroom sample provides information about the total number of viable cells (Benardini and Venkateswaran 2016; La Duc et al. 2007; Mahnert et al. 2015; Venkateswaran et al. 2003), but it is not useful for identifying what organisms are present. MALDI-TOF (Matrix assisted laser desorption time of flight) mass spectrometry is now commonly used in the medical field to identify organisms, and it is being applied in aerospace cleanrooms as well (Andrade et al. 2018; Moissl-Eichinger et al. 2015). However, this technique is still dependent on culturing organisms. Fluorescence based monitoring systems can detect airborne cells but cannot identify them (Hallworth 2012).

Biological testing of clean labs suffers from a lack of repetition. Outside the NASA standard assay, very few measurements are routinely replicated between labs. In some respects, this is good and appropriate. Monitoring methods should be modified to suit the environment and the questions being asked. The microbial profile of cleanrooms will be different in different environments (e.g., La Duc et al. 2009). For example, cold curation facilities should explicitly test for the presence of psychrophilic organisms (Sandle and Skinner 2013). It would be a waste of resources to look for psychrophiles in labs maintained at room temperature. However, variations in sample collection, DNA extraction, DNA sequencing, and data processing methods make inter-lab comparisons very difficult. Testing new methods and techniques is an important area of research, but more effort should be made to relate these new measurements to previously generated data.

Curation labs should design a monitoring plan that is capable of quantifying and identifying the microbes present therein. Unfortunately, there is no single measurement or technique capable of thoroughly describing a microbial community. Each method discussed above has its strengths and weaknesses. Therefore, a selection of culture-based and culture-independent techniques should be used to monitor cleanroom ecology. Samples should be collected from the air and from surfaces regularly and as frequently as daily when critical operations are being conducted. Special care should be taken to avoid organic or inorganic contamination during sampling. For example, agar filled contact plates used in the pharmaceutical industry are inappropriate for curation labs since they would introduce bioavailable organic compounds and trace metals into the lab. Most importantly, sample collection and analysis methods should be as consistent as possible in order to generate a baseline dataset that can be used as a basis of comparison for new techniques. Regular and consistent sampling is the most important feature of any environmental monitoring program.

3 Development of New Storage, Processing, and Sample Handling Capabilities

As technological advancements and new ideas expand the variety and scope of scientific questions that can be asked with astromaterials samples, so expands the need for better storage, processing and sample handling capabilities of curation laboratories that house and process astromaterials samples. Here we summarize a number of important advancements and areas of growth in sample storage, processing, and handling techniques that will be important in the coming decades for maximizing science returns on astromaterials samples.

3.1 Cold Curation of Astromaterials and Associated Gases, Biological Samples, and Hardware

The ever-expanding plans for the return of samples from volatile-rich solar system targets and/or targets of astrobiological significance necessitates the development of curation at temperatures below that of typical curation facilities (20 °C). Temperature requirements depend primarily on which volatiles are expected within the returned sample, which in turn relate to the conditions under which the material formed and has since been preserved. The term "cryogenic" is defined as relating to temperatures below −183 °C; the normal boiling points of the noble gases, oxygen, nitrogen, and air lie below this temperature. More generally, "cryogenic" refers to temperatures below approximately −150 °C (https://www.nist.gov/mml/acmd/cryogenics/aboutcryogenics). The term "high temperature cryogenic" is used to refer to temperatures from the boiling point of liquid nitrogen, −196 °C, up to −50 °C, the generally defined limit of cryogenics (e.g., Zohuri 2017). The curatorial temperatures for terrestrial materials, including tissue samples and ice cores, include: ≤−20 °C (the temperature of typical walk-in freezers in which physical processing and documentation takes place); ≤−40 °C for archival storage (e.g., of ice cores); and −80 to −196 °C (liquid nitrogen) for biological samples (e.g., Anchordoquy and Molina 2007; Rissanen et al. 2010). Thus, with the exception of biological tissue storage, the field of Earth materials curation has not yet entered the realm of cryogenics.

3.1.1 Past and Present Practices in Cold Curation of Astromaterials

The expected range of temperatures required to preserve solar system materials spans from those needed for (water) ice cores to cryogenic. Cold curation and sample handling of astromaterials has been done to a limited extent at NASA-JSC over several decades. Several Apollo 17 samples were initially processed under GN_2 in a processing cabinet at room temperature for about a month before being transported to cold storage (−20 °C) where they have remained. Furthermore, the US Antarctic meteorite collection utilizes cold storage of new Antarctic meteorites, and initially used cold stages in a nitrogen glovebox for cold sample handling. According to Annexstad and Cassidy (1980) "The specimens are transferred from a small staging freezer to the processing cabinet. A specially constructed stage, cooled by liquid nitrogen, is used to keep the sample frozen while an initial cold chip is removed from the meteorite. This chip is immediately returned to freezer storage for future experiments when a frozen piece may be required. The parent meteorite is then allowed to warm to ambient temperature naturally in the cabinet's dry GN_2 environment." In the first few years of Antarctic meteorite handling at JSC, the staff gained experience with storing and handling samples frozen, using a cold processing plate in a cabinet and using a cold storage room. Although some hardware was assembled to do this, it became clear after detailed tests that this was not an effective way to handle samples due to the difficulty of keeping samples cold while still allowing dexterity of the sample processor, length of time required to process individual samples, and overall expense. The cold processing approach was abandoned at JSC in 1979, after review and discussion by the Meteorite working Group (MWG) (Righter et al. 2014).

More recently, insights into the benefits of curation and processing under cold conditions have been gained from the collection, curation, and study of the Tagish Lake meteorite (Herd et al. 2016 and references therein). Tagish Lake is a unique carbonaceous chondrite that fell January 18, 2000 onto the frozen surface of the eponymous lake in northern British Columbia, Canada. The meteorite was collected about a week after the fall, and collection was done without direct hand contact; more significantly, the meteorite specimens were kept below 0 °C after collection and during subsequent transport to curation facilities (Herd et al. 2016). The cold ambient temperatures at the location of the fall, coupled with the care with which the collection and subsequent curation were carried out places Tagish Lake among the most pristine meteorites ever collected (Herd et al. 2016). The meteorite is a type 2 carbonaceous chondrite with affinities to CM and CI meteorites (Blinova et al. 2014; Zolensky et al. 2002), and contains among the highest concentrations of organic matter measured in meteorites (Alexander et al. 2014; Grady et al. 2002; Herd et al. 2011). The pristine nature of the meteorite, coupled with the curation methods used to preserve it, have yielded new insights into the formation of nanoscale organic globules in the coldest regions of the protoplanetary disk (e.g., Nakamura-Messenger et al. 2006) as well as the role of asteroid parent-body aqueous alteration in the modification and synthesis of organic molecules (Herd et al. 2011; Hilts et al. 2014).

The majority of the Tagish Lake meteorite specimens are stored at −30 °C and processed within the Subzero Curation Facility for Astromaterials at the University of Alberta; this facility houses an Ar glovebox within a walk-in freezer maintained at temperatures of −10 to −15 °C (Herd et al. 2016). While there are no indications that the Tagish Lake meteorite contains water ice or other such volatiles, these conditions of storage and handling are justified by the discovery of especially volatile and/or reactive organic species (e.g., formic acid, naphthalene, and styrene; Hilts et al. 2014). Challenges and limitations of the Subzero Curation Facility include: mitigation of glovebox leaks, user comfort, and the extreme dryness of the Ar atmosphere, which would result in the sublimation of water or other ices from the samples (Herd et al. 2016). However, the facility achieves the goal of enabling documentation and processing of pristine astromaterials under low temperature in an inert atmosphere. The low-temperature curation of the Tagish Lake specimens reduces reaction rates, preserves intrinsic (volatile) organic compounds, and discourages microbial activity (Herd et al. 2016)—requirements that are desirable for returned samples from organic-rich asteroids, cometary nuclei, Mars, or other volatile-rich returned sample targets (lunar poles, icy moons, etc.) as discussed below.

3.1.2 Volatile-Rich Samples from the Lunar Poles

The lunar poles are high-priority targets for sample return due to the possibility of significant quantities of water-ice and other volatiles in permanently shadowed regions (PSRs). Remote sensing data indicates that volatiles comprise up to several weight percent of materials in PSRs; the composition of the volatiles in the crater Cabeus' PSR included H_2O, CO_2, CO, H_2S, CH_4, OH, SO_2, NH_3, C_2H_4, and CH_3OH (Colaprete et al. 2010; Gladstone et al. 2010). This mix of compounds present a complex curatorial challenge, and even more so in the presence of local regolith/silicates (largely anorthosite or basalt). The volatiles detected by LCROSS have a range of condensation temperatures, and a subset are highly reactive in the presence of silicate minerals. If the solid and volatile components of a lunar PSR sample are stored together, therefore, a mixed-phase, highly reactive sample will likely result. The preservation of a lunar polar sample would therefore be maximized by separating the solid and volatile components and storing them in that configuration for the long-term.

The presence of numerous reactive species presents several additional challenges. First, the corrosive nature of H_2S limits the materials to which the sample can be exposed without alteration. Materials will therefore need to be selected that accommodate the curatorial requirements for isolation (materials should not significantly contract under cold temperatures), durability during sample processing/preliminary examination, and particulate contamination. Second, volatile-rich samples often contain gases that are hazardous to humans, even at low concentrations (e.g., CO, H_2S, SO_2, NH_3). This additional risk—on top of the existing particulate exposure risk from solid samples—may require the use of respirators or special masks during preliminary examination and curatorial operations. The need to minimize leakage from curatorial hardware (gas containers, analytical equipment, gloveboxes, etc.) will be significantly higher for volatile-rich samples; because they will operate at cold temperatures, proper materials selection from the component to the system level will be a top priority.

3.1.3 Cometary Nucleus Samples

The preservation of a cometary nucleus sample lies at the extreme end of cold-curation storage requirements because the sample would contain hypervolatiles including noble gases, nitrogen, and oxygen (Bieler et al. 2015; Le Roy et al. 2015), although the retention of these gases would only likely be achieved if they are trapped within solid ices of primarily H_2O, CO, and CO_2. Insights from the ROsetta Spectrometer for Ion and Neutral Analysis (ROSINA) instrument, which measured volatiles in the coma of comet 67P/Churyumov-Gerasimenko demonstrate that, while dominated by water, the nucleus of 67P includes a wide range of volatile compounds, including molecular oxygen, CO, CO_2, HCN, H_2S, CH_4, and many others (Bieler et al. 2015; Le Roy et al. 2015). Curation of these ices, which would almost certainly be intimately mixed with non-volatile, fine-grained silicate, oxide, sulfide and more refractory organic materials, would require significant technological development for cryogenic curation—assuming that the sample could be collected and returned to Earth under cryogenic conditions in the first place. Various options for the return of cometary nucleus volatiles have been studied, including cryogenic sample return, for which significant technical challenges exist (Veverka 2010b). Allowing volatile components to be released by warming a comet nucleus sample and capturing them in a separate container removes the need for cryogenic handling (Veverka 2010a), which was the approach proposed for the CAESAR mission concept to comet 67P. No truly cryogenic sample return missions are planned at the time of this writing.

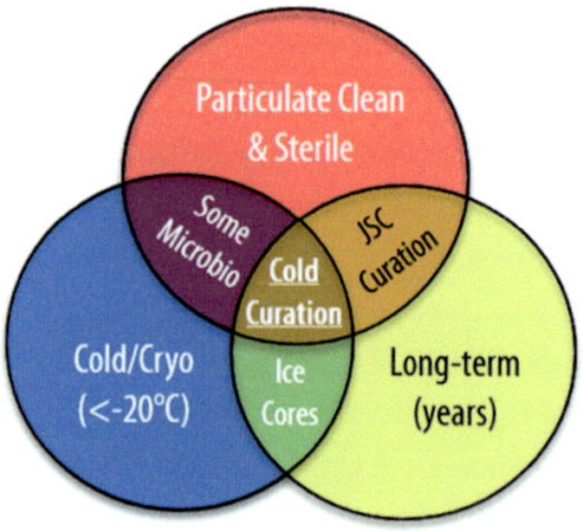

Fig. 1 The overlapping challenges of cold curation and biological containment present unique challenges for future astromaterials curation efforts

3.1.4 Biological Samples

Curation facilities may be required to curate biological samples as part of a contamination knowledge collection from the spacecraft build and sampling of flight hardware. This requirement is currently in place for the Mars2020 mission, which may be the first leg in a Mars Sample Return Campaign. Although there are no requirements that the martian samples be kept cold, biological sampling during the spacecraft build and of the flight hardware includes microbiological samples, including swab samples, liquids, isolated pure cultures of bacteria and fungi, and DNA samples. The requirements for long term preservation of these biological samples varies with sample type and intended use. We will discuss two broad sample types: (1) Samples preserved for later growth and (2) samples intended for molecular analysis like DNA sequencing.

The guidelines for preserving bacteria and fungi for later cultivation are well established (CABRI 1998). Bacteria should be placed in a protective solution of 15–50% glycerol by volume and frozen at -80 °C. Commercially available products like cryobeads should be used to improve long-term viability. If viability needs to be maintained for > 5 years, samples should be frozen at -130 °C. Some species of bacteria and fungi can be freeze dried with liquid nitrogen and stored at 4–8 °C. It is important to test the survivability of each strain prior to committing to a preservation method.

Preservation of swabs, liquids, witness plates, or extracted DNA for later analysis is less straightforward. As a general rule, colder is better, but there is little consensus on what temperature is best. There is some evidence that storing samples at too low a temperature can cause more damage than it prevents (Anchordoquy and Molina 2007; Vaught and Henderson 2011). Rapid changes in DNA sequencing technology make it very difficult to predict how samples will be handled in the future (Reuter et al. 2015). DNA extraction techniques are also evolving and can have significant effects on sample quality (Dauphin et al. 2009; Mitchell and Takacs-Vesbach 2008; Rose et al. 2011; Zielińska et al. 2017). Barring additional research, the best strategy is to store unprocessed samples alongside extracted DNA so that future researchers have options for what to analyze.

Future sample return missions from icy moons will incorporate both the biological sensitivity of a Mars Sample Return (MSR) campaign and the temperature sensitivity of lunar or cometary samples (Fig. 1). Therefore, even if MSR does not have a low-temperature storage requirement, it is inevitable that biological containment and cold curation will eventually be needed concurrently. The challenges of operating a bio-safety level 4 (BSL-4) facility at cold temperatures are unique to astromaterials curation, and will need to be addressed in the coming years. Many materials suitable for biological containment (e.g., many plastics) become brittle at temperatures at or below the freezing point of water. The additional presence

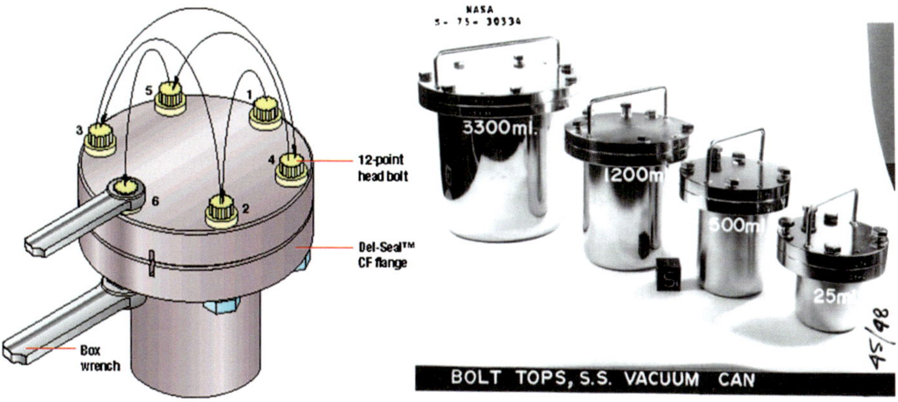

Fig. 2 Sketch of typical assembled conflate flange (left). This design is very similar to the specialty "bolt top containers" used for the Apollo program sample storage in 1969 to present (right), except that these are now commercially available, and in various sizes and materials, as explained above

of salts (chlorides, sulfates, etc.) may pose challenges to the selection/durability of metal components. The overlapping requirements for sterility, particulate cleanliness, temperature control, leak prevention/sample isolation, gas safety, and curator comfort will need to be met in the coming years as exploration efforts at Europa, Enceladus, and other icy moons intensify.

3.2 Curation of Organics and Organic-Rich Materials at Room Temperature

Experience with the curation of the organic-rich Tagish Lake meteorite (Sect. 3.1), has provided ample evidence for the value of cold curation in the preservation of organics and organic-rich materials; namely, the retention of volatile organics, the mitigation of volatile organic contaminants, and the suppression of metabolism by any microorganisms in the curation facility (Herd et al. 2016). However, cold curation is not a requirement for the storage of organic-rich materials. For example, curation planning for OSIRIS-REx turned instead towards hermetically sealed storage of samples to preserve organics (Dworkin et al. 2018). This approach had some precedent with the Apollo missions samples (Fig. 2), avoids the potential contamination and time and dexterity intense processing issues associated with cold curation, and is cost effective utilizing known commercially available and tested hardware and approaches.

When handling organic-rich or organic sensitive materials the use of plastics should be extremely limited. PTFE or Teflon is acceptable in some situations, but glass or metal is preferable. Prior to use, tools and sample containers should be combusted at 500 °C to remove organic contaminants. Long-term storage of organically sensitive samples should use well-characterized glass baked at ≥ 500 °C wherever possible (e.g., Grosjean and Logan 2007; Peters et al. 2005; Sherman et al. 2007). Furthermore, frequent microbial monitoring of labs where organic-rich samples are stored and processed is critically important. Moreover, metagenomics studies of any microbes recovered from curation labs that house and process organic-rich samples will be important, particularly for microbes that can metabolize under anaerobic conditions. The primary goal of these metagenomic studies would be to characterize the metabolic function of these anaerobic microbes to understand how they might alter the samples if they are inadvertently introduced to the samples. This is particularly important for organic-rich sample collections stored at room temperature.

3.3 Future Restricted Earth Return Missions

In the 50 years since the Apollo 11 launch, advancements in knowledge and technology allow for not only unprecedented scientific investigations of extraterrestrial samples but also a greater understanding of the potential hazards of sample exposure or release into the environment (e.g. extraterrestrial life). However, in the case of a biological health hazard, more precautions are required, not only to protect the samples from Earth, but also to protect Earth from the samples. Under the UN Space Treaty of 1967, the Committee on Space Research (COSPAR) maintains a planetary protection policy at the international level for all space faring nations. The policy provides "international standard on procedures to avoid organic-constituent and biological contamination in space exploration" (COSPAR Planetary Protection Policy March 2017). The policy also promotes the prevention of *"adverse changes in the environment of the Earth resulting from the introduction of extraterrestrial matter"* as stated in the UN Space Treaty.

For the United States, the NASA Planetary Protection Office in the Office of Safety and Mission Assurance provides the policies and requirements for all NASA exploration missions regarding forward and backward control of biological contamination. NASA Policy Directive (NPD) 8020.7G, *Biological Contamination Control for Outbound and Inbound Planetary Spacecraft*, complies with the UN Space Treaty and COSPAR planetary protection policy stating *"the Earth must be protected from the potential hazard posed by extraterrestrial matter carried by a spacecraft returning from another planet or other extraterrestrial sources"*. NASA Procedural Requirements (NPR) 8020.12D, *Planetary Protection Provisions for Robotic Extraterrestrial Missions*, outlines requirements for meeting the NPD 8020.7G as well as specifies planning documents and reviews for Category V Restricted Earth Returns. The Planetary Protection Office classifies any "samples from solar system bodies that may harbor indigenous life" as Category V: Restricted Earth Return. Although there are currently three bodies with this designation (Mars, Europa, and Enceladus), this number can change in either direction as more information about any particular planetary body is gained. For example, during the first three Apollo sample return missions (Apollo 11, 12, 13, and 14), the Moon was considered Restricted. Consequently, the Apollo 11, 12, and 14 samples and astronauts were quarantined upon arrival while health assessments and biohazard tests were performed. However by Apollo 15, which launched just over two year after Apollo 11, the Moon was reclassified as Unrestricted and the final three Apollo missions (Apollo 15, 16, and 17) proceeded without the same level of biohazard Planetary Protection precautions.

The scientific community has identified Mars Sample Return (MSR) as a high priority sample return activity for many years, and support for such an endeavor has waxed and waned over the last few decades. Current efforts relating to MSR are focused on a multi-mission campaign, the first of which is the Mars 2020 rover mission to Jezero Crater. At the time of writing, no space agency has fully committed to returning the samples that will be collected by Mars 2020, but NASA and ESA are discussing the possibility of forming a partnership to complete the campaign and decisions are anticipated to be made in the year 2020. Due to both Planetary Protection and Science requirements, the Mars 2020 rover mission has the most stringent inorganic, organic, and biological contamination control requirements of any sample return mission in history. Strategies for satisfying these and other requirements related to MSR and Restricted Earth Return in general are described below.

3.3.1 *Facility Preparation*

Infrastructure Samples returned from any planetary body designated as Restricted must be contained within a Biosafety Level 4 (BSL-4) facility until it can be demonstrated that either (1) the samples do not pose a threat to life on Earth or (2) the samples have been adequately sterilized for release (Rummel et al. 2002, NASA technical publication 211842). The requirements and processes associated with biohazard testing and/or sterilization are developed specifically for each mission and each set of samples. International space treaties with the United States, COSPAR planetary protection policies, and NASA planetary protection policy directives and requirements do not impose any specific design requirements on a biocontainment architecture or BSL-4 facility. The policies simply state that the Earth must be protected from the potential hazard posed by extraterrestrial matter and microbial containment is required on Category V (sample return) Restricted Earth Returns. The U.S. Dept. of Health and Human Services traditionally has jurisdiction of design and operating requirements for a BSL-4 facility in the United States. The "Biosafety in Microbiological and Biomedical Laboratories", 5th Edition (Dec. 2009) authored by the U.S. Department of Health and Human Services: Public Health Service, Centers for Disease Control and Prevention, and the National Institutes of Health; HHS Publication No. (CDC) 21-1112 (hereafter BMBL, 2009), houses the primary recommendations, standards, and design requirements for all BSL labs. Under this regulation, any related agents with unknown risk of transmission are classified to be under BSL-4 containment. Presumably, an extraterrestrial or unknown pathogen would require, at minimum, a BSL-4 containment. At this time, we cannot predict what other federal or international agencies may wish to impose additional guidelines and requirements and/or request jurisdiction of a NASA BSL-4 sample return lab. For example, the National Institute of Health (NIH) imposed additional design requirements at the Galveston National Lab beyond the BMBL requirements. The World Health Organization also has guidelines and requirements for BSL-4 laboratories and the Dept. of Agriculture has claimed some jurisdiction of extraterrestrial soils. For the Apollo Program in January 1966, the Interagency Committee on Back Contamination (ICBC) was established to include the CDC with Dr. David Sencer of the CDC as chairman, Department of Agriculture, Department of the Interior, Department of Health, Education, and Welfare, National Academy of Sciences, and NASA, which imposed strict requirements on the construction of the Lunar Receiving Laboratory (LRL), JSC Bldg. 37. Therefore, historically, other agencies have been involved in the construction and operations of such a BSL-4 type lab.

One of the major challenges in designing a facility for Restricted Earth Return Missions is the integration of the Contamination Control (CC) requirements necessary to protect the samples from terrestrial contamination and Planetary Protection (PP) requirements necessary to protect the Earth and its inhabitants (all life: from humans to animals to plants, etc.) from a possible extraterrestrial pathogen (e.g., microbes, viruses, or prions). While walls can act as physical barriers for protections, developing the proper pressure differentials inside and outside the laboratory is vital (Fig. 3). Unlike the curation of traditional unrestricted samples, which utilizes positive pressure gradients to protect the samples from contamination, BSL-4 facilities rely on negative pressure gradients to protect the scientists and general public from the samples. Although there have been a number of possible iterations demonstrated in the Draft Test Protocol (Rummel et al. 2002; Fig. 3), the presumed baseline requirement is that the samples must be contained within a BSL-4 Facility (BMBL 2009). In order to best protect the samples and Earth, redundancies are built into the design schematic (Fig. 3). Specifically, not only will the entire cleanroom laboratory be constructed within a BSL-4 facility, but the use of a Biosafety Cabinet (BSC) III double walled isolator for

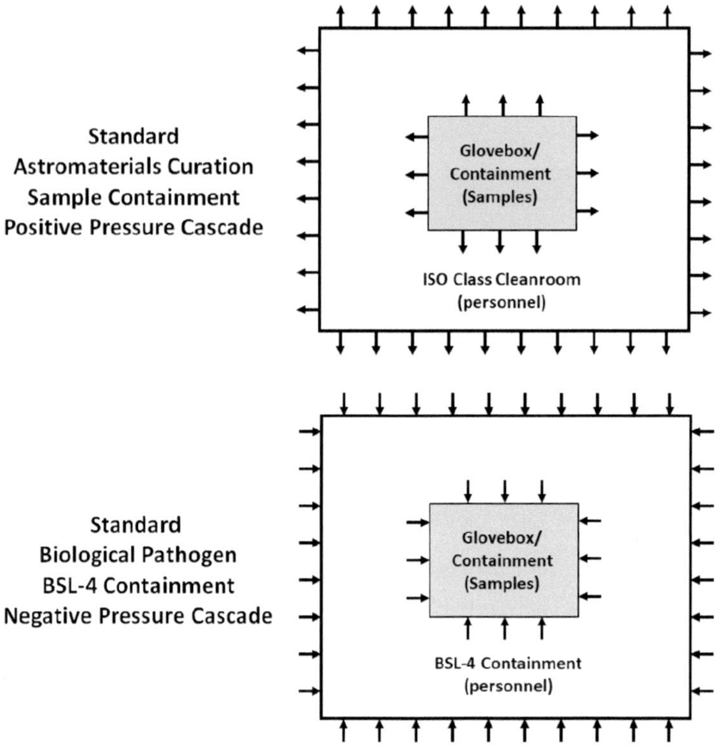

Fig. 3 Comparison of differential pressure gradients used for containment of astromaterials samples from bodies designated as restricted Earth return (adapted from Rummel et al. 2002: 6, Fig. 1): (1) The top schematic represents a standard sample containment design for unrestricted sample return missions. In this configuration, the highest pressure is situated inside a positive pressure glovebox or containment isolation chamber and pushes out to a lower pressurized cleanroom. The cleanroom is also under positive pressure relative to the outside of the lab. This positive pressure cascade is designed to mitigate the infiltration of outside or laboratory contamination to the astromaterials samples. (2) The middle schematic is a standard BSL-4 containment design for working with hazardous biological pathogens. In this particular configuration, the glovebox or containment isolation chamber that houses the biohazard is under negative pressure relative to the laboratory space. In addition, the laboratory is also under negative pressure relative to outside of the lab. This negative pressure cascade is engineered to protect the outside environment from a release of any biohazard. (3) The bottom schematic is the current design concept for a restricted Earth return sample containment that combines both of these technologies. The containment isolation chamber that houses the samples is designed with a double wall and the interstitial space is filled with a high purity gas at a higher pressure relative to the contained isolated samples and the cleanroom. The pressure between the cleanroom and containment is still under a negative pressure differential to maintain BSL-4 containment standards, but any leakage would be the high purity gas that pushes into the containment and out to the laboratory cleanroom. In addition, the cleanroom laboratory would be a positive pressure cascade with a negative differential pressure plenum barrier to maintain BSL-4 containment to the outside environment of the facility

sample processing within the cleanroom will add an additional level of protection with the corresponding differential pressure scheme. For current BSL facilities in the U.S., a Class III BSC glovebox gastight (leak rate) criterion is $< 1 \times 10–5$ cc/s with 100% He tracer gas under 3 inH_2O pressure in the cabinet (Stuart et al. 2012). Dependent on mission science requirements, specialized double walled glovebox or containment seals could be required for maintaining nitrogen or other inert gas environment purity under negative pressure. Non-glove storage isolators can achieve a He leak rate of $< 1 \times 10–7$ cc/s. However, achieving

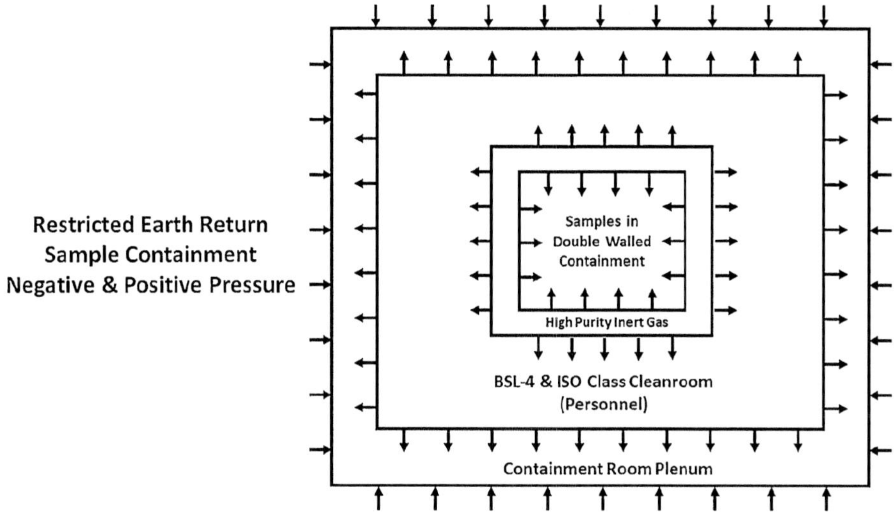

Fig. 3 (*Continued*)

a better leak rate on double-walled isolated containment may require additional engineering development and challenges. While there have been some studies exploring how these requirements could be implemented (Beaty et al. 2009), these studies need to be updated to reflect some significant shifts in possible facility usage (e.g. no animal studies, long-term use, multi-mission use).

Functional Laboratory Design The verification of extinct or extant life within a sample may require the examination of organic compounds within the samples. As such, not only is it vital to ensure no terrestrial biological contamination occurs during sample handling and storage, but the amount of organic contamination must also be minimized. One of the main ways to do this is through the selection of manufacturing equipment and laboratory space with proper materials that have little to no potential to outgas or shed particles. This will require the use of mainly 300 series stainless steel and Teflon in areas with intimate contact with the samples. For this reason (as well as offering additional differential pressure gradients), double walled isolator cabinets are the likely choice for the handling of these materials since the suits utilized in BSL-4 facilities and the glovebox gloves could contaminate the samples with organic matter and make life detection more difficult (Vrublevskis et al. 2018a; Holt et al. 2019). Although the utilization of a double-walled isolator (DWI) helps to mitigate the contamination risk, it requires a significant advancement in remote or robotically assisted manipulation since manual manipulation via a glove port could compromise the organic CC requirements due to material outgassing. Although there is work being done in Advanced Curation related to small particle handling, there are other groups investigating hepatic feedback remote handling, specifically for Mars Sample Return (Vrublevskis et al. 2018b). In addition to robotic sample manipulators, and any analytical equipment should be developed to allow scientists to manipulate samples and run analyses remotely.

A further complication of returning samples from a restricted planetary body is the unknown long-term space requirements of the collection. Although Apollo samples were eventually deemed safe and released (two years after the first samples were returned), this will not necessarily be the case for future restricted Earth return missions. If a potential or actual health hazard is found, or if there are too many concerns about unknown unknowns, samples may never be released from containment. Therefore, multiple facilities and sample use strategies would need to be developed to conduct science in containment. One way to approach this is to construct a modular BSL-4 facility that has walls that can be shifted to accommodate the addition of new analytical instrument suites and other long-term curation/scientific needs.

Integration of Cleaning and Sterilization Techniques The safety of the samples and the technicians and environment at large will require not only a well-designed facility, but also the integration of cleanliness and sterilization protocols. While there are standard sterilization techniques (e.g. heat, peroxide, formaldehyde) and cleaning procedures (e.g. IPA, UPW), there is not one standard procedure that can do both simultaneously. In the case of MSR, where a major part of "life-detection" will rely on DNA extraction and not viability, any unviable biological matter remaining will compromise the samples and studies. Due to these considerations, strategies for integrating these procedures are underway. A similar strategy as that taken for flight hardware will be employed, clean the materials first and then sterilize without generating contamination. Although the Mars 2020 Mission has integrated the use of vapor phase hydrogen peroxide (VHP), they are using heat sterilization on the sample intimate hardware. However, given the specs of isolator cabinets (e.g. size and mechanics) and the systematic sterilizations needed to avoid cross-contamination and ensure safety, heat sterilization is not a viable option.

The need to integrate the isolator cabinets into the facility's infrastructure will mean that they will have to be cleaned and sterilized in place. For this we can draw upon best practices used in Curation Glovebox Laboratories and BSL-4 Cabinet Laboratories. For initial final cleaning/sterilization, a standard UPW/IPA cleaning procedure would likely be utilized (see Sect. 3.7) followed by sterilization utilizing an ultra-pure hydrogen peroxide solution. Given the harshness of the high concentration peroxide required (35%), amount of residual moisture after sterilization is complete, and its limitations in sterilizing instrumentation due to unexposed surfaces, a new technique is being considered. Not only does ionized hydrogen peroxide only require a solution of 8% hydrogen peroxide, it more easily permeates instrumentation, and does not leave a liquid residue (Webb 2011; Grimaldo 2017). Although more research is required to confirm sufficient sterilization without generating long-term corrosion or systemic contamination, the outlook is promising.

3.3.2 Collection of Baselines for Science and Planetary Protection

The concurrent requirements for sterility, particulate and organic volatile cleanliness, temperature control, leak prevention/sample isolation, and gas safety will need to be met in the coming years as exploration efforts at Mars, Europa, and Enceladus come to fruition. The classification of a mission as Category V Restricted Earth Return not only adds more CC and PP constraints (https://planetaryprotection.nasa.gov/categories), it also broadens the scope of required CK (Fig. 4) to include biological witness materials. Not only does this require more rigorous sets of samples, unlike other collections, which only require storage in an inert ultra-pure gaseous environments (e.g. nitrogen), biological CK will also require samples to be frozen (see Sect. 3.1).

While all scientific investigations of returned samples are highly sensitive to terrestrial contamination, contamination is especially detrimental where studies of extant or extinct extraterrestrial life are concerned. The proper collection, storage, and cataloging of Contamination Knowledge (CK) associated with the production and assembly of the spacecraft, rover, lander, orbiter, and/or sampling system will be vital to these investigations. Without

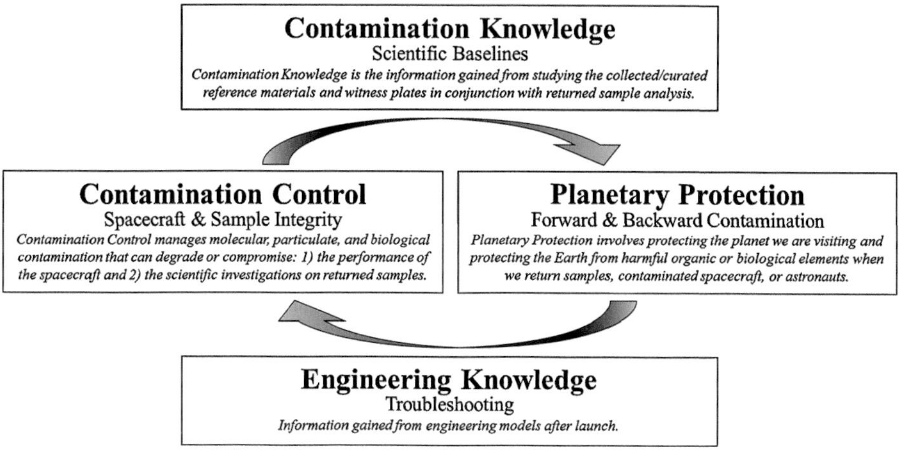

Fig. 4 Cartoon illustrating the categories of samples needed for testing and verification of spaceflight missions. Each type of sample serves a different purpose and hence requirements for each sample collection related to these categories must be defined

a well-constructed and curated CK collection, the baseline for contamination within the returned samples cannot be established. Therefore, after mission inception and design, the development of the CK collection as part of a mission's curation plan (CP) should occur in conjunction with the mission's CC and PP plans. The CP, CC, and PP plans and implementation of these plans during ATLO and Earth receiving operations are paramount to the ultimate value of the returned samples.

Technological advancements to instrumentation are continually progressing with greater precision and accuracy for sample measurements, especially in the field of microbiology. An array of CK samples must be made available to scientists once restricted samples are returned to Earth, and those samples should be preserved in such a manner that they can be analyzed by instrumentation that was not invented at the time of their initial storage. Collecting and curating unanalyzed/unaltered samples will minimize the possibility that current analysis and extraction techniques destroy or alter the samples or otherwise inhibit yet to be developed measurements. Some of the types of biological CK samples the NASA Curation Office requires for restricted Earth return missions include:

(1) Un-analyzed/Un-altered dry swabs and wipes in sterile containers stored at ≤ -80 °C.
(2) All recirculation filters from the clean rooms used for spacecraft and spacecraft hardware assembly and all filters from the laminar flow benches used to assemble sample intimate hardware, packaged in sterile Teflon bags and frozen at -80 °C.
(3) Witness plates collecting airborne contamination within the assembly cleanrooms stored at ≤ -80 °C.

3.4 Sample Processing Cabinets Under Vacuum

Historically, vacuum processing of samples was employed for primary processing of Apollo lunar materials. First envisioned in 1964, the 3 story "High Vacuum Complex" integrated into JSC bldg. 37 Lunar Receiving Laboratory between 1967–1968 was a series of connected glovebox isolation chambers operating at 10^{-6} torr vacuum environment to decontaminate, process, and store samples (Calaway et al. 2014; White 1976). Although vacuum processing takes a very direct approach to minimizing sample contamination by avoiding sample

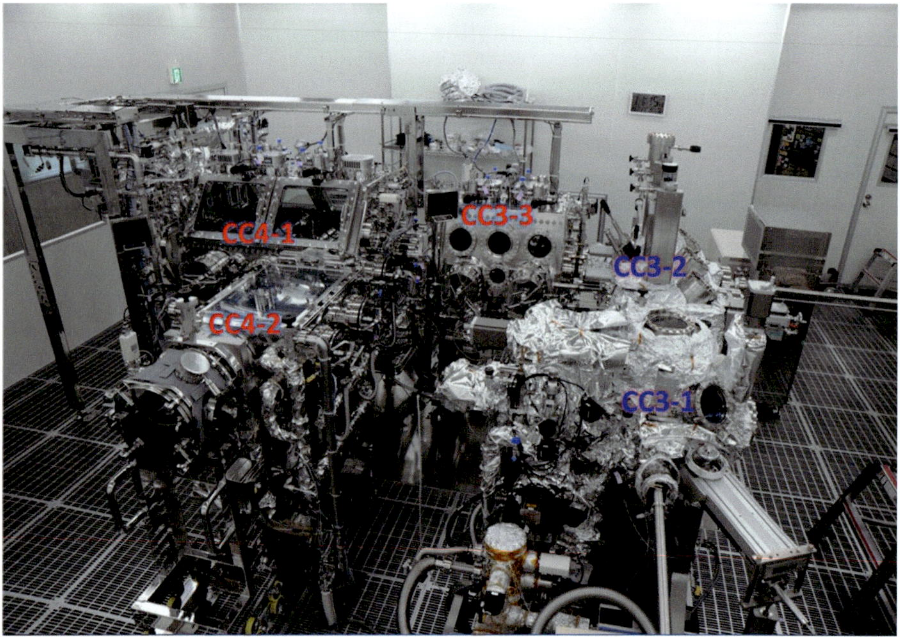

Fig. 5 A bird's-view image of the sample-handling system for Ryugu samples. The system consists of five chambers—CC3-1, CC3-2 CC3-3, CC4-1 and CC4-2

contact with gases, the process is inherently difficult. Maintaining vacuum in large processing cabinets requires constant pumping and the use of cold traps to remove unwanted pumping oils and other contaminants, which renders gloves stiff with attendant processor fatigue, and any leaks in the system tend to introduce relatively humid ambient air. There is also the danger of rapid pressure loss through mechanical failure, which would introduce significant contamination and poses a physical risk to processors. Unfortunately, the F-201 processing glovebox chamber was prone to leaks and glove failures as well as difficulties in using vacuum hardware with an increasingly large volume of lunar samples, which ultimately drove the high vacuum complex to be used only for Apollo 11 and 12. For Apollo 14 onward, the high vacuum complex was replaced by a series of positive pressure gaseous nitrogen gloveboxes called the Sterile Nitrogen Atmospheric Processing (SNAP) Line and Nonsterile Nitrogen Processing Line (NNPL) (Reynolds et al. 1973; Simoneit et al. 1973).

The JAXA Extraterrestrial Sample Curation Center (ESCuC) in Sagamihara, Japan has successfully employed vacuum storage for samples returned from the Hayabusa mission to asteroid Itokawa and is planned for JAXA's Hayabusa2 mission to carbonaceous asteroid Ryugu that is currently in flight (Okazaki et al. 2017; Sawada et al. 2017; Yada et al. 2014). In the case of the Hayabusa samples, all the sample handling processes occurred in purified gaseous nitrogen following the sample container opening process under vacuum conditions (Yada et al. 2014). Installation of a newly-developed sample-handling vacuum processing clean chamber (CC) was completed in October, 2018 at ESCuC (Okazaki et al. 2017; Sawada et al. 2017; Fig. 5) two years prior to sample return. The entire sample-handling system in ESCuC consists of five chambers—CC3-1, CC3-2, CC3-3, CC4-1 and CC4-2. The returned sample container will be first connected to CC3-1 for opening in a vacuum, and will be transferred to CC3-2 for vacuum-handling of samples. The container will be then transferred into CC3-3, where the sample handling environment will be changed from vacuum to purified gaseous nitrogen. Further handling of samples will be done in purified nitrogen in CC4-1 and CC4-2.

3.5 Sample Processing Cabinets with Remote Participation of Scientists

It is rather common for scientific investigations with astromaterials to require a specific sample or specific portion of a larger sample for subsampling. Principle investigators will occasionally travel to a curation facility to pick-out samples, provide input on sample subdivision, and/or sample preparation. This requires that the PI arrange a visit to the facility to communicate directly with the curatorial processors on which samples to pull and ultimately choose for study. In some cases, travel can take a significant amount of time and cost. Today, communication technology can almost eliminate the need for the PI to travel to the facility with the integration of "Live" real-time video conferencing sessions with remote scientists.

NASA JSC has experimented with this technology during the Stardust preliminary examination in 2006 and again in 2014 with the retrofit of one of the Apollo 16 Lunar processing cabinets (Calaway 2015). In this most recent technology demonstration, a Leica DMS1000 Macroscope and Axis Pan-Tilt-Zoom (18× optical zoom) IP camera system was integrated into the Lunar processing glovebox. The Axis camera and macroscope were mounted on the outside of the glovebox and focused through glass. This was done to eliminate the concern for cross-contamination from the two systems. The Axis camera was mounted to the top of the glovebox looking through the lighting window and could be used for situational awareness of processing or zoomed in to look at sample splits. The Leica Macroscope was mounted above the PI observation window at the end of the glovebox. A sample was placed onto a jack-stand and the macrosope could focus on the sample through the glass window.

While both of these commercial-off-the-shelf (COTS) products offer video streaming capabilities, the video integration was complicated by the JSC firewall and mandated government IT security requirements. In addition, remote wireless connections were hampered by the thick walls of the curation facility. For both the systems, the video needed to be securely accessed outside of JSC. Therefore, the video stream was required to push through the JSC internal firewall to the JSC public zone and then pass through another firewall to get to the internet for public access. The Axis camera browser software is capable of secure viewing with passwords, and the IP address would be routed accordingly by our internal IT group. For the macroscope video stream, a streaming service like YouTube or USTREAM from the DMZ could be used. For the tech demo, a dedicated USTREAM account was set-up. At the March 2015 Lunar and Planetary Science Conference, we demonstrated this system. We successfully live-streamed the Leica Macroscope and Axis camera real-time images from the NASA JSC Lunar laboratory to the Marriot Hotel in The Woodlands, Texas. However, for this test we used the VPN network access to simplify the test due to time constraints. Lunar curation now has all the equipment and tools needed to set up a permanent video conferencing with external PIs during a video or teleconference. In the future, more collections could integrate this type of COTS technology to reduce time and travel costs where appropriate.

3.6 Small Particle Handling

One particular objective of advanced curation efforts is the development of new methods for the collection, storage, handling and characterization of small particles. In this context, "small" refers to microscale particles, typically between one and several hundred microns in diameter (though submicron interstellar particles will be analysis targets in the future). Particles in this size range include dust derived from comets and asteroids that is continuously accreted by the Earth, as well as material collected by robotic sampling missions and by astronauts during Apollo missions. The curation of small particles includes the unambiguous identification of particles in/on collection substrates, the transfer of particles between collection, analysis, and storage substrates, sample characterization, sample preparation/subdivision, and the preservation and documentation of samples in a publicly available catalog. Astromaterials Curation facilities in the United States, Russia, and Japan currently maintain several small particle collections (Allen et al. 2011): Lunar regolith fine-grained samples returned by Apollo astronauts and by Soviet Luna robotic spacecraft, Cosmic Dust that has been collected in Earth's stratosphere by ER2 and WB-57 aircraft, Comet 81P/Wild 2 dust returned by NASA's Stardust spacecraft, interstellar dust returned by Stardust, and asteroid Itokawa particles that were returned by the Hayabusa spacecraft. NASA and JAXA Curation offices are currently preparing for the anticipated return of two new astromaterials collections—asteroid Ryugu regolith collected by Hayabusa2 spacecraft in 2019 and returned to Earth in 2020, and asteroid Bennu regolith to be collected by the OSIRIS-REx spacecraft and returned in 2023 (Lauretta et al. 2017). A substantial portion of these anticipated returned samples are expected to consist of small particle components, and mission requirements necessitate the development of new processing tools and methods in order to maximize the scientific yield from these valuable acquisitions.

There are several aspects of microscale astromaterials curation that present challenges that are distinct from macroscopic sample curation. At scales of less than 100 μm, electrostatic and intermolecular forces dominate the behavior of particles. Particles adhere weakly to glass or tungsten needles via Van der Waals intermolecular forces, usually enabling transfer between analysis and storage substrates. These transfer operations are hindered by transient electrostatic forces. Triboelectric charging due to contact, separation, and frictional electrification (Matsusaka et al. 2010) is the primary mechanism by which particles are lost during transfer operations (although environmental and instrumental vibrations also contribute to sample loss during transfers), and these triboelectric effects become more severe in low-humidity environments. Hayabusa2 and OSIRIS-REx collections will be curated in sample processing cabinets purged with dry GN_2, and developing methods for suppressing triboelectric charge accumulation in these dry environments will be critical for successful sample processing.

Sample characterization at the microscale also presents unique challenges. Typically, optical images of submicron to micron-sized particles do not provide sufficient information to investigators to make informed sample selections. Microscale particles are often imaged and characterized in scanning electron microscopes equipped with an energy-dispersive spectroscopic (EDS) detector for elemental characterization; such analyses are useful for investigators requesting samples with desired mineralogy and are necessary to distinguish true extraterrestrial material from terrestrial contamination for samples that are collected in the stratosphere. However, SEM analysis of microscale particles introduces an additional risk of loss due to sample charging from electron beam bombardment; additionally, some fragile organic and mineral phases may be potentially modified by e-beam characterization. In some instances (especially with rare samples), it may be necessary to subdivide a particle via ultramicrotomy or by focused ion beam (FIB) sample preparation. These methods must be carefully considered in order to avoid compromising the scientific integrity of the sample. The objective of advanced microscale astromaterials curation research is to better understand these challenges and to investigate tools, equipment, and methods that facilitate microscale sample processing.

3.6.1 Description of Tools and Equipment Used for Small Particle Handling

Tweezers Commercially available, low-Ni stainless-steel tweezers (e.g. Dumont No. 5 Dumoxel) can be utilized to reliably manipulate samples as small as several hundred microns by hand. Smaller particles ($\geq$ 50 μm) may be manipulated by tweezers that are fixed to devices that enable mechanical or electrical actuation, especially when mounted on a micromanipulator. The NASA Curation Office at JSC has acquired two such devices manufactured by Micro Support Co., Ltd.; these devices are being used to investigate methods of particle removal from OSIRIS-REx contact pads. Challenges remain with small particle manipulation via tweezers (either by hand or by electrical/mechanical actuatable devices) due to lack of force feedback, and risk of deforming or fracturing particles with low tensile strength remains significant.

The use of micro-electro-mechanical systems (MEMS) microtweezers for particle manipulation (Keller and Howe 1997) have also been investigated. Initial experiments with these devices revealed similar force feedback limitations to stainless steel tweezers. In addition, silicon devices were more brittle, making removal and placement of particles on rigid substrates precarious operations that often resulted in the shattering of the microtweezers and the loss of the sample. Finally, the low-cost benefits due to mass production on a single wafer have so far not been realized, and the cost of these microtweezers has remained significantly more expensive than their stainless-steel counterparts.

Needles Manipulation of particles by fine-tipped needles is a technique that has been utilized by curators since NASA initiated the cosmic dust collection in 1981. Particles smaller than 20 μm are typically transferred from a collection medium to an analytical substrate (e.g. beryllium disk or epoxy bullet) or to a storage container (e.g. concavity slide) using a microneedle made from glass or tungsten. With skill and practice, curation personnel can transfer particles as small as 5 μm between substrates by hand using a glass or tungsten needle attached to a pin vise. However, involuntary hand motions on the order of 100 μm make routine and reliable transfer tedious and precarious operations. For critical transfers, needles are attached to mechanical or motorized micromanipulators to improve transfer reliability and precision, while minimizing user fatigue. The intermolecular forces between the needles and the particles in this size range are typically sufficient to overcome repulsion due to triboelectric charge accumulation. Larger ($\geq$ 20 μm) particles have been more challenging to manipulate. When using the same glass and tungsten microneedles for particles larger than 20 μm, triboelectric charging effects significantly hinder the reliable manipulation of particles. We have recently observed that, by utilizing tungsten carbide needles with low taper ratios ($\sim$3:1), particles as large as 200 μm can be manipulated successfully. We speculate that these needles present greater contact surface area for intermolecular forces to capture particles, and that the needle shape may aid in the rapid redistribution of accumulated triboelectric charge; however, more tests are needed.

Glass and quartz needles are fabricated using micropipette pullers that concentrate heat at the center of a solid core glass rod or capillary tube while applying force to each end; this action creates two needles with submicron tips. When a capillary tube is pulled in such a manner, and the tip is carefully broken off, a micropipette is created. A vacuum can be applied to this tube, creating a microscopic version of vacuum tweezers. We have investigated utilizing such a system to manipulate particles that are too large to be handled by Van der Waals adhesion. Our initial results are that, while the vacuum tweezer system is very efficient at securing larger microscale particles, releasing the particle by removing the vacuum frequently results in sample loss. We speculate that the vacuum action induces strong triboelectric charging effects.

Microscopes Stereomicroscopes possess imaging characteristics such as long working distance and laterally correct viewing that make them extremely well-suited for freehand and mechanically assisted manual manipulation of microscale particles. Stereomicroscopes from manufacturers such as Nikon, Leica, and Olympus utilize two main optical designs—the Greenough design, which has two optically independent light paths, and the common main objective (CMO) design, in which optically parallel light paths share a common objective (Zimmer 1998). Greenough designs are preferable in environments in which size and weight must be minimized, and where high magnification is not necessary or desired (e.g., suspended over a collector during cosmic dust harvesting). CMO optical designs afford increased magnification compared with Greenough type microscopes, and are utilized in more critical sample operations such as mounting particles onto analysis substrates.

For the manipulation of very small (<10 μm) samples, upright microscopes equipped with geared XY stages are utilized. These microscopes are equipped with long-working distance objectives capable of providing up to 500X magnification. The geared, manual XY stage is coupled to the Z-focus mechanism that raises and lowers beneath a stationary objective; this enables movement in X, Y, and Z independent of objective position. By placing a needle at the focal point of the objective, it is possible to transfer microparticles between substrates by moving the stage rather than by moving the needle.

Digital microscopes are particularly useful in processing environments where stereo or upright microscopes would be inconvenient—for instance, in a N_2 sample cabinet or a temperature-regulated environment. They also introduce the potential to perform curation activities remotely, reducing contamination risks and operator fatigue. JAXA's Hayabusa sample processing cabinet uses three digital microscopes—two mounted inside the cabinet, and one mounted externally—to image particles during transfer operations. Digital microscopes are best utilized with micromanipulator assisted particle transfers, especially if the microscope suffers from image lag.

Micromanipulators Micromanipulators are mechanical, hydraulic, and motorized/electrical devices that enable the precise handling of microscale samples. Most commercially available micromanipulators have three axes of motion, with motorized versions often providing a virtual fourth axis of motion (which is desirable for performing microfluid injections). Mechanical micromanipulators often use a combination fine-pitched screw mechanisms and linear guide rails to achieve microscale positioning. Singer instruments manufactures a mechanical micromanipulator with a 3D pantograph design (Robert 1951); the user holds a pencil-grip stylus and, through the pantograph mechanism, manipulates a probe with a 4:1 reduction of hand motions.

Motorized micromanipulators employ a combination of precision stepper motors and worm gear mechanisms to achieve microscale positioning and motion. These have the advantage of being able to be operated remotely and, in some cases, can be programmed for autonomous operation. A variety of input mechanisms can be utilized, including joysticks and rotary optical encoders. Motorized manipulators can also be computer controlled.

Integrated Systems In order to achieve reproducible, robust, and reliable particle transfers and processing, a combination of microscopes, micromanipulators, and XY/XYZ stages are required. Motorized micromanipulators and XY stages require bulky power supplies and motor drivers (sometimes one per axis of motion), and microscopes with digital cameras often require desktop computers to operate the imaging software. Such equipment occupies large footprints in cleanrooms with limited space, compromises pristine environments with instrument cooling fans, and leads to unsightly tangled masses of cables and wires. The

NASA Curation Office has recently obtained an integrated system that includes dual motorized micromanipulators, a motorized XYZ stage, and high-resolution digital microscope. The MicroSupport AxisPro system utilizes a graphic user interface control system, allowing all electromechanical components to be operated independently or simultaneously via computer mouse. A number of manipulation and sampling tools are available for the AxisPro, including an ultrasonic milling tool and a device that enables the electrical actuation of stainless-steel tweezers. The compact, integrated design of the system enables the possibility of placing the AxisPro in a N_2 sample cabinet with an operator performing sample processing activities remotely. So far, the AxisPro system has been used extensively for microsample handling technique development (e.g. implanting and extracting particles into polyurethane foam collectors).

In order to minimize the risk of sample contamination (especially for collections that have been returned from extraterrestrial sources via spacecraft), materials restrictions are placed on tools and equipment used in sample processing cabinets.

JAXA manipulates samples within its Hayabusa processing cabinet using an integrated mechanical manipulation system manufactured by Hitachi (Yada et al. 2014). The system consists of an XYZ stage, a left and right micromanipulator, and a sealed digital camera; the system is constructed from T6061 aluminum, 304/316 stainless steel, Teflon, and quartz. No lubrication is used for the bearings, and the entire manipulator is disassembled and serviced annually to maintain performance.

Six-Axis Compact Robot Arms While 3-axis micromanipulators have been extremely successful for activities involving the transfer of isolated particles in the 5–20 μm range (e.g. from microscope slide to epoxy bullet tip, beryllium SEM disk), their limited ranges of motion and lack of yaw, pitch, and roll degrees of freedom restrict their utility in other applications. For instance, curation personnel removing particles from cosmic dust collectors by hand often employ scooping and rotating motions to successfully free trapped particles from the silicone oil coatings. Similar scooping and rotating motions are also employed when isolating a specific particle of interest from an aliquot of crushed meteorite. While cosmic dust curators routinely perform with these kinds of manipulations using handheld tools, operator fatigue limits the number of particles that can be removed during a given extraction session. The challenges for curation of small particles will be exacerbated by mission requirements that samples be processed in N_2 sample cabinets. We have been investigating the use of compact robot arms to facilitate sample handling within gloveboxes. Six-axis robot arms potentially have applications beyond small particle manipulation. For instance, future sample return missions may involve biologically sensitive astromaterials that can be easily compromised by physical interaction with a curator; other potential future returned samples may require cryogenic curation (Calaway and Allen 2013). Robot arms may be combined with high resolution cameras within a sample cabinet and controlled remotely by curation personnel. Sophisticated robot arm and hand combination systems can be programmed to mimic the movements of a curator wearing a data glove; successful implementation of such a system may ultimately allow a curator to virtually operate in a nitrogen, cryogenic, or biologically sensitive environment with dexterity comparable to that of a curator physically handling samples in a glovebox.

3.6.2 Methods for Mitigating Triboelectric Charging

Developing tools and methods for mitigating the effects of triboelectric charging during small particle processing activities is a major objective of microscale advanced curation research. Triboelectric charging results from contact, separation, and frictional charge induction (Matsusaka et al. 2010). Examples include friction between storage substrates and instrument support stages or friction between manipulation tools and particles. Many storage substrates currently in use for small particle curation are fabricated out of glass, quartz, corundum, or other optically transparent material that enables the utilization of transmitted illumination; for example, interplanetary dust particles have traditionally been stored and distributed to investigators between a flat glass slide and a glass concavity slide. However, most transparent materials possess poor electron mobility, and any local accumulated charges are unable to easily redistribute. We have identified friction between these slides, particles, manipulation tools, and instrument support stages as a major source of sample electrification. In cases where substrate transparency is not a curation requirement, the glass support slide may be replaced with a silicon wafer. Particles retain a high level of visibility on such substrates (especially under coaxial illumination), and triboelectric charging is significantly reduced such that particles between 40 and 100 μm can be reliably manipulated and arranged in arrays without additional charge-mitigation devices. Recently, we have experimented with producing storage receptacles in silicon using focused ion beam (FIB) milling (Fig. 6). We used an FEI Quanta 3D-FEG Focused Ion Beam (FIB) to mill several shallow (<20 μm) depressions between 30 μm^2 and 80 μm^2 into the surface of a silicon chip; material was sputtered using a 65 nA Ga^+ beam at 30 kV. A 10 μm particle of CM2 meteorite was placed into one of the FIB-produced wells using a pantograph mechanical micromanipulator. The charge-dissipative nature of the Si chip enabled us to successfully acquire a secondary electron image of the stored particle using a 190 pA beam current at 10 kV. Storage substrates that also enable electron beam imaging and characterization are desirable, as they minimize the need for high-risk microscale particle transfers between storage and analysis substrates. We are currently investigating this technique to produce storage wells in other charge-dissipative substrates that could enable in-situ elemental analyses.

For instances where the use of transparent substrates is unavoidable, other steps may be taken to minimize frictional contact between insulating materials. For instance, we have constructed slide support frames out of conducting and electrically grounded materials (e.g. miniature aluminum extrusion framing systems). One effective method for minimizing triboelectric charging effects is to restrict small particle processing to times when the ambient humidity is above 60% (Guardiola et al. 1995); however, this method is not viable for microscale samples that are processed in dry GN_2 sample cabinets.

Another extremely effective method for mitigating triboelectric charging effects is the use of a ^{210}Po alpha ionizing source. Companies such as NRD® manufacture commercially available devices designed to reduce static charge via alpha particle emission. Because alpha particles have a short penetration range in air, the sources are most effective when placed within 25 mm of the sample. Tools, substrates, and samples can be periodically exposed to the Po-210 source as sample electrification is observed to worsen; alternatively, if working distance permits, the source can be left in place during particle transfer operations to remove any transient charges as they are produced. Due to the short half-life of Po-210 (138 days), sources must be replaced annually to remain effective. Also, the use of radioactive sources may be prohibited in certain facilities and typically requires specific safety training and security protocols.

JAXA has developed an electrostatically controlled particle manipulation system to handle Itokawa particles in an ultrapure GN_2 sample cabinet (Yada et al. 2014). Instead of attempting to neutralize the charge that has accumulated on the particle, they use it to an advantage by attracting the particle with an oppositely charged needle. The system utilizes a quartz needle with an embedded platinum wire; the samples rest on a grounded conductive

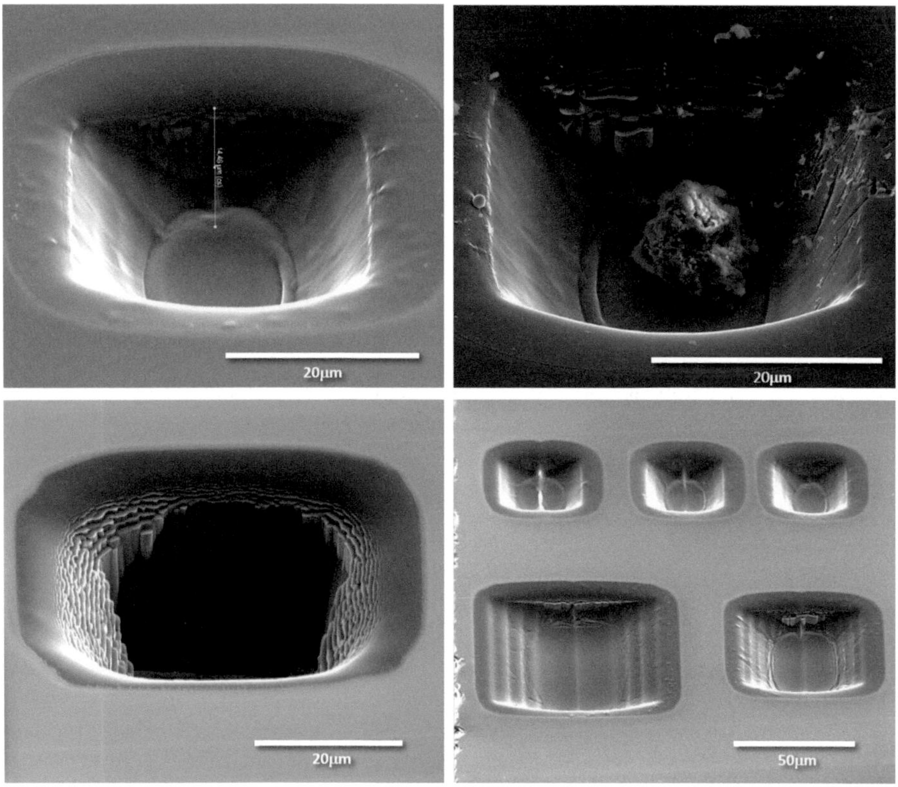

Fig. 6 Secondary electron image of Focus Ion Beam (FIB)-produced wells in Si chip for particle storage and manipulation

surface. When voltage is applied to the system, a charge is induced to the needle by applying a voltage to the platinum wire; this charge is used to attract and release particles and transfer them to custom gold SEM mounts for characterization or to storage wells in quartz slides for allocation and archiving. The NASA Curation Office has reproduced the electromagnetic manipulation system (using needles fabricated at JAXA as part of an ongoing international collaboration between NASA and JAXA curation facilities) and is currently investigating applications for the system for its microscale particle collections.

The tools and methods described here represent only a fraction of the techniques and instrumentation currently utilized and under development for microscale astromaterials sample processing and analysis. An international collective of curators and small-particle scientists at curation facilities, research institutions, and commercial industries continue to collaborate to improve our ability to extract high-quality science from these valuable and unique micro-sized samples.

3.7 Advanced Precision Cleaning for Storing and Handling Astromaterials

Precision cleaning of isolation chambers (e.g., gloveboxes and desiccator cabinets), sample containers, and processing tools is important for mitigating terrestrial cross-contamination to pristine astromaterials. Once samples arrive on Earth, the sample environment and how it will be handled will begin to alter the pristine nature of the sample. As stated before, the term "precision cleaning" simply means cleaning materials to a prescribed level of cleanliness, which is measured and verified. Aerospace, semiconductor, pharmaceutical, and optics industries are historically concerned with precision cleaning. Standards for precision cleaning are widespread across industrial processes through trade organizations like the Institute of Environmental Sciences and Technology (IEST) (https://www.iest.org), ASTM International (formerly American Society for Testing and Materials; https://www.astm.org), SEMI (https://www.semi.org) and others. Since curatorial precision cleaning does not directly align with a single industry's cleaning standard, curation precision cleaning procedures and protocols for handling astromaterials are derived from many of these established industry standards. NASA also has its own flight hardware precision cleaning standards and are often different and dependent on program and mission. For the NASA Curation Office, precision cleaning standards were mainly derived from the Apollo program and early cleaning recipes and history have been discussed in Calaway et al. (2014).

Currently, precision cleaning at the NASA Curation Office is divided into three categories: PreClean, Final Clean, and Advanced Clean. PreClean is considered gross cleaning when parts arrive from fabrication/machining and/or procurements from a vendor. Final Clean is typically linked to the use of a final cleaning agent, drying, and packaging of the part. During Final Clean, the hardware cleanliness is also measured and verified to meet a certain standard of cleanliness for use. Advanced Clean is a term used for non-routine avenues of cleaning and/or testing of new cleaning methods and techniques. Advanced cleaning is typically done after a routine PreClean and Final Clean process has been completed. This might include techniques that require advanced particulate removal, organic-free cleaning or sterility. Advanced Clean may also use standard or advanced cleanliness verification processes to assess surface cleanliness using a variety of state-of-the-art instrumentation.

In the NASA Curation Office, PreClean and Final Clean support all collections by cleaning the sample processing tools and containers. However, each collection has its own tools and containers, which are cleaned in entirely separate cleaning sessions. All hardware items are put away with the exception of items from the collection being cleaned. The cleaning tanks are cleaned and refilled before starting a new collection. All of this careful effort is taken to mitigate the potential for cross-contamination between the different astromaterials collections. Before attempting gross cleaning, it is important to understand and verify the cleaning chemical compatibility with the material that is to be cleaned. In addition, complex equipment and tools are routinely disassembled, then cleaned and then reassembled in a cleanroom after the precision cleaning is completed. PreClean typically consists of removing any visible grease, dirt, adhesives, or other marks with the use of polyester wipes saturated with isopropanol alcohol (IPA) (70% IPA and 30% UPW), if compatible with IPA. If IPA wiping does not work or is not compatible with the material, other cleaners that will not contaminate the material will be used to remove the visible dirt (e.g., citrus-based solvents to remove silicones, ammonia-based solutions, hexane, and household dish liquid have been used for initial gross cleaning). In addition, mechanical gross cleaning may also be necessary in conjunction with cleaning chemicals, such as razor blades and scrubbers (e.g., Scotch Brite pads and nylon brushes). After all visible material is gone, PreClean uses a gross degreasing procedure to remove any machining oils and grease from manufacturing. This is done by soaking and sometimes sonicating the part in a degreasing detergent or chemical. Brulin 815GD (at 5 to 30% concentration with UPW) is commonly used for stainless steel, aluminum, and titanium metal parts. Freon 113 replacements are also sometimes used for degreasing; for example Honeywell Solstice Precision Fluid, DuPont Vertrel specialty fluids (e.g., Vertrel XF), or 3M HFE 7100-DL. Dilute nitric acid is also routinely used to remove trace metal contaminants, such as lead, from newly fabricated items. After degreasing, the hardware is then cleaned with a surfactant. Mechanical scrubbing with polyester wipes or soft brushes are used in a surfactant bath and then placed into a sonication bath for 5 to 15 minutes. Afterwards, parts are removed and spray rinsed with UPW and air dried.

Final Cleaning of equipment and tools are typically centered on high purity chemicals or cleaning agents. Final Clean also incorporates a verification step to evaluate the cleanliness of the part and certify the level of cleanliness and qualification for use. From 1966 to 1994, the final cleaning agent at JSC was Freon 113. Established during Apollo, Freon 113 was an excellent degreaser and final cleaning solvent. The United States government environmental policies on ozone depleting chemicals phased out chlorofluorocarbon production from 1992–1995 which forced the NASA Curation Office to change degreasers and final cleaning agent. After an in-depth research process, the final cleaning agent was changed to UPW in 1994 and is used currently (see Sect. 2.2.2 for details on current UPW purity and system). Final Clean can sometimes redo surfactant cleaning or use of a pre-degreaser, however, most of the time, parts are rinsed with UPW and placed into a UPW cascade bath and sonicated for 5 to 15 minutes. The UPW is often heated to 40 to 70 °C to provide better cleaning. GN_2 is also used during the end of sonication to remove particulates out of the bath. Parts are then removed from the bath and thoroughly spray rinsed with UPW. If another high purity chemical is used (such as IPA or a Freon 113 replacement), this would be applied at this stage, then spray rinsed again with UPW and dried by GN_2 (sometimes heated GN_2) to remove all visible water. During the final rinse, run-off aliquots of UPW may be taken for optical particle counts, liquid particle counts, or TOC analyses. It should be noted that traditional non-volatile residue (NVR) mass balance measurements and black-lights used to be used, however, these methods were eliminated since the Final Clean often showed cleanliness below detection limit of those techniques. The part is then either left to continue to air dry or placed into an oven to remove any more water. After drying, parts are triple bagged in FEP Teflon or nylon bags depending on the collection and its material restrictions. Before bagging, precision cleaning verification often entails an optical inspection of the parts. If parts are shown not to be cleaned to the specified cleanliness standard, parts are sent through the process again.

In the NASA Curation Office, the verification of cleanliness reference standard is frequently IEST-STD-CC1246E, *Product Cleanliness Levels—Applications, Requirements, and Determination* (IEST-STD-CC1246E 2013). This is a derivative of the historical military discontinued standard MIL-STD-1246 (MIL-STD-1246C 1994). It is in IEST-STD-CC1246E that hardware surface cleanliness levels are specified per unit measure both for particles and non-volatile residue (NVR). Particle counts are measured by optical microscopy and/or liquid particle counts. Most NASA astromaterials collections use Level 50 cleanliness standard from IEST-STD-CC1246E. However, the Genesis collection has a dedicated precision cleaning lab, and hardware is generally cleaned to Level 25. For example, cleaning for flight of the Genesis mission involved most surfaces being cleaned to level 25 (no particles $> 50\ \mu m/0.1\ m^2$) for particulates per surface area (a function of particle abundance vs. particle size). NVR is traditional measured by means of gravity mass calculation (a as function of mass vs. surface area) but are often limited to technique sensitivity. As an example, R1E-4 is a designation for NVR indicating $< 100\ ng/0.1\ m^2$. More recent techniques have relied on bench-top total organic carbon (TOC) analyzers (as a function of ppb vs. surface area) and more time-consuming analytical instrumentation. This standard is useful because it is frequently cited when cleaning hardware for spaceflight, including sample

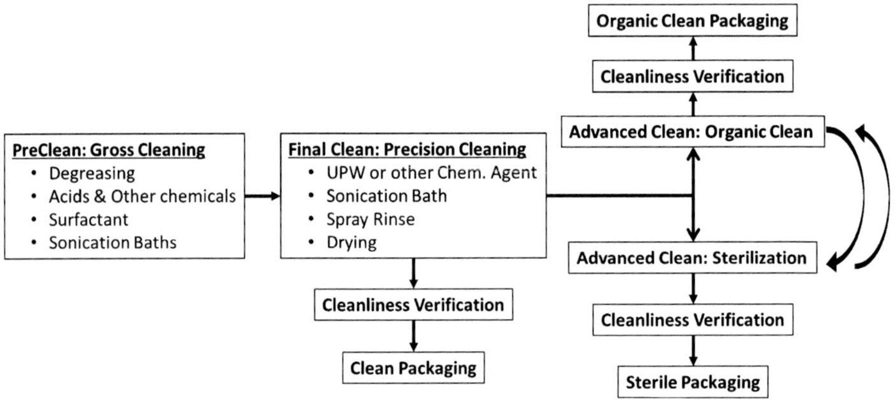

Fig. 7 Basic cleaning process for hardware and tools used in astromaterials curation laboratories. Items to be cleaned are first introduced to a PreClean process that is configured for gross cleaning. Afterwards, items are introduced to a Final Clean process where they are precision cleaned, cleanliness level verified, and packaged for use in the lab. If specialized advanced cleaning is desired, the item(s) are further processed after the routine Final Clean process. This diagram shows the hypothetical process path for organic and sterilization of hardware and tools. These processes could be a single advanced cleaning process or a combination of several advanced cleaning processes. In addition, cleanliness verification can occur at multiple points during the cleaning process or after the process is completed before packaging. Hardware and tools, coupons, and/or final cleaning agent aliquot are commonly used to verify cleanliness to a set standard

collection devices on spacecraft. However, new missions are generally citing total organic carbon (TOC) and the TOC is not always transferable to the NVR level.

3.7.1 Technical Tensions Among Cleaning for Particulate, Molecular Organics, and Sterility

UPW cleaning to remove inorganic particles is a mature process for curation of astromaterials. However, the expanding diversity of requirements for high level organic cleaning and sterility invokes conflicts among cleaning techniques that need to be managed. Management options for reconciling these tensions requires attention to detail, perhaps a complex handling sequence, or subdivision of samples into separate handling tracks such as organic-cleaned vs. metal-free (e.g. metal vs. plastic handling tools and containers).

A simple model of a cleaning process line (Fig. 7) starts with pre-cleaning followed by UPW cleaning for particle removal. This particle-cleaned product is suitable for inorganic usage or for further cleaning. The product could be further cleaned to high organic cleanliness or to sterility by separate tracks. Selection of a direct separate track may be most efficient and adequate. Greater cleanliness or sterility might be achieved by cleaning for both low organics and sterility. However, due to conflicts in materials, environment, process chemicals, and packaging for high level organic cleaning and sterility, adaptability for specific situations is required.

The revived need for both high-level organic cleanliness and sterility, driven by recent missions such as OSIRIS-REx, Hayabusa2, and possibly Mars Sample Return, have resulted in investigations into advanced techniques and methods for cleaning and for verification of cleanliness. Traditional methods to achieve organic cleanliness, such as solvent extraction and bake-out, can be augmented by UV-ozone cleaning, plasma cleaning, supercritical fluid cleaning, and CO_2 SNOW cleaning (Mickelson 2002a, 2002b; Calaway et al.

2007, 2009; King et al. 2010; Schmeling et al. 2013; Kuhlman et al. 2013). Assessment of surface cleanliness has included the use of instrumentation such as SEM, TEM, FT-IR, Raman, XPS, SIMS, AMF, LC-MS, DART-MS, TD-GC-MS, LA-ICP-MS, and VDP-ICP-MS. Protocols to remove adventitious carbon must be followed by surface passivation and extreme control of environment, which includes packaging. Packaging materials must not off-gas organics. Thus, maintaining an organic-clean surface after cleaning is challenging. Traditional heat-sealing will not work for organic samples. For example, the Mars 2020 CK collection has used hermetically-sealing bag clips in-place of traditional heat sealing. Techniques for achieving sterility of containers and tools include standard autoclave, dry heat, UV, hydrogen peroxide vapor and gamma irradiation and electron beam (Allen et al. 1999; Clark 2004). NASA JSC is currently constructing an advanced precision cleaning lab to further study some of these techniques for future missions.

4 Development of New Astromaterials Acquisition Capabilities on Earth

The study of astromaterials in the laboratory allows direct analysis of material arising from the full breadth of the history of our Solar System. The continuing advance of technology has improved not only our technological capacity to make measurements but also our ability to minimize contamination and sample modification during collection of freshly fallen meteorites and cosmic dust. Some of these improvements are related to educating the public on proper methods of handling meteorites, but many of the technological advances have focused on improving our ability to track and find materials. In many cases, these improvements result in reduction of exposure time to uncontrolled conditions, which could reduce terrestrial contamination, especially if clean-collection practices are used during recovery operations. Collection of material has been steadily improved by advances in ground-based and satellite sensors, dissemination of information with the growth of the internet and various social media platforms, and especially by the flow of data through freely available data sources. Although the collection of astromaterials may seem, at first, to be a prerequisite of curation, one of the primary goals of advanced curation is aimed at maximizing the science returns of astromaterials samples, and improvements in sample collection techniques has a direct benefit to science. We are in the midst of an exciting period of growth of truly innovative astromaterials sample collection techniques on Earth, and this section illustrates a current snapshot of a rapidly evolving field.

4.1 New Astromaterials Collection Capabilities for Cosmic Dust

Since inception in 1981, the NASA Cosmic Dust collection has collected interplanetary dust from Earth's stratosphere using flat-plate, oil-coated collectors. The oil is a high-viscosity silicone (polydimethylsiloxane, $(C_2H_6OSi)_n$) that is mechanically stiff at collection altitude and temperature but engulfs and protects collected material on return to room temperature. This oil has a long track record of successful recovery of cosmic dust, and recent tests have shown that even oil used for sampling in 1981 retains its original viscosity properties. This is heartening for long-term storage of cosmic dust, but silicone oil has a significant drawback in that it is a contaminant for important studies such as oxygen isotopes, organic species, and amorphous silicates. The community has expressed a desire for at least a subset of samples collected "dry", without the contaminating oil.

Foam collectors are a promising means to accomplishing dry collection. A small number of foam collectors have been flown as a test of concept, which yielded a few cosmic dust samples (Messenger et al. 2015). The results were generally acknowledged as positive, and comments to the NASA CD Curator indicate that the scientific community has a strong interest in additional dry collection. To this end, NASA will fly foam collectors as a subset of future collection flights. Collection in foam appears to be straightforward and Messenger et al. (2015) claims that the collection rate is comparable with silicone oil (albeit with a short total collection time for reference), but extraction of collected material from foam is a nontrivial exercise. Foam features relatively deep pits in its surface from which the particles must be extracted. This problem is exacerbated by the observation by Messenger et al. (2015) that "20–50%" of individual foam cells were broken when observed post-flight, probably by aerodynamic pressure. In order to test identification and extraction methods, NASA Cosmic Dust personnel fabricated an analog foam collector by adhering a $1/8''$ thick sheet of white polyurethane foam to a surplus Lexan IDP flag using double-stick tape; excess foam was trimmed to match the profile of the flag. Small (<20 mm) particles of Bells CM2 meteorite were transferred from a concavity slide into individual foam cells using a MicroSupport AxisPro micromanipulation system and bent glass needles. ^{210}Po sources were utilized to minimize triboelectric charging effects. Ten particles were implanted into the experimental foam collector apparatus using this technique. The transfer process was then reversed to remove four of the particles from the collector onto cleaned glass slides. None of the particles were lost due to vibration or triboelectric charging effects. Ultimately, foam may be replaced by a more rigid material, but this awaits future work. For the near term, the use of a computer-controlled micromanipulator shows promise in removing material from foam collectors using standard pulled-glass needles.

In addition to dry foam collectors, NASA Cosmic Dust is conducting a project with undergraduate students at Texas A&M to develop a prototype, high altitude balloon-based dust collection platform. The intent of this new system is to supplement existing aircraft-based collection for two major reasons. One, expanding into a new collection platform adds programmatic depth and resilience to CD collection efforts. Should the existing ER-2 and/or WB-57 aircraft become unavailable (either temporarily or permanently), dust collection will continue with a balloon-based platform. NASA balloon flights also regularly operate from both the northern and southern hemispheres, opening up the possibility of CD flights intended to collect material from cometary debris streams, which preferentially impinge on the southern hemisphere. The second reason is to offer new ways to improve CD "timed collection" efforts (Dermott and Liou 1994; Messenger 2002). "Timed collection" is collection of material in the stratosphere timed to coincide with the settling time of material sourced from a specific meteor shower and thus with a specific parent body. Collection by aircraft is possible and has been demonstrated (e.g. timed collection of comet Grigg-Skjellerup in 2003 and of comet Giacobini-Zinner in 2012), but significant flight time constraints can impede both the total flight time and timing of "timed collection" attempts. Collection by high-altitude balloons may be more accommodating, as NASA long-duration balloons (LDBs) feature flight times of up to 100 days. Cosmic dust collector(s) could fly on long-duration missions and deploy for collection only during the settling time for material from a specific meteor shower, maximizing collection time at precisely defined intervals to maximize the chances of collecting material from a specific parent body.

A cosmic dust collector prototype for use on high altitude balloons was tested in mid-2019 at the NASA balloon research center in Fort Sumner, NM. The prototype is called Cometary and Asteroidal Research of Dust in Near-space Atmospheric Levels (CARDINAL), a name chosen by the students. CARDINAL uses a swing arm with collectors at each end as a low-power means to move air over the collectors, spinning the arm with a small electric motor. Collector size, rotation size, and swing arm length were chosen to produce an estimated one cosmic dust particle per day per collector. While this collection rate is about 1/48th that of aircraft-based collection, LDB flights last up to over 100 days as opposed to the typical tens-of-hours collection time of aircraft-based collection. The lower airspeed of balloon-based collection may also be advantageous for collection using collectors that are free of silicone oil, or "dry" collection. CARDINAL seals the swing arm within its body to protect the collectors from contamination, and remains sealed at all times below a set altitude. The lid is movable to expose the collectors at high altitude. CARDINAL draws power from the balloon gondola, and both housekeeping and operations data are stored by the onboard microprocessor for post-flight analysis. CARDINAL is self-contained with respect to communications, with all flight functions programmed prior to flight and no communications needed with a ground station or other external controller. Testing at Fort Sumner revealed significant performance shortcomings, which prevented the first test flight. A second design, which draws heavily on lessons learned from CARDINAL, is currently in early development. This design will simplify the collector encapsulation design, reduce weight considerably, and feature its own battery and solar power module that will perform double duty as flight trim ballast. Ultimately, the intent of this project is to collect cosmic dust during precisely defined periods when specific meteor showers are active, to attempt to collect material from known cometary parent bodies. Both oil-based and dry-foam collectors will be used as part of the continuing development of dry-foam cosmic dust collection.

4.2 New Astromaterials Collection Capabilities for Meteorites

Meteorite falls represent opportunities for the recovery of samples that have been briefly exposed to the oxidative, organic-, and moisture-rich environment at the Earth's surface, which is also teeming with life. Substantial contamination of freshly fallen meteorites by such exposure can occur in a matter of days to weeks (Burton et al. 2014; Hilts et al. 2014; Kebukawa et al. 2009). At the same time, advances in curation and contamination knowledge in support of sample return (e.g., Allen et al. 2011; Dworkin et al. 2018; Yada et al. 2014) have resulted in the establishment of curation facilities that prevent or mitigate against such contamination (Herd et al. 2016). Therefore, the present-day challenge is to reduce the amount of time that meteorite samples spend in the field and apply the best methods for their field collection and laboratory curation and handling. The faster the meteorite is recovered and removed to a curation facility, the more scientifically valuable such a sample will be. Rapid collection of meteorites can be seen as strongly complementary to established efforts to collect meteorites from dry deserts such as the Sahara (Grady 2000) and Antarctica (Harvey 2003). These collection efforts produce the large number of meteorites necessary to routinely produce weathered but unusual types such as martian, lunar, ungrouped achondrites, and many others, while rapid collection of fresh falls produces a relatively small number of unweathered meteorites. The combination of both approaches provides a comprehensive approach to meteorite collection, which facilitates study of a wide range of meteorite types with unweathered examples for studying weathering effects, pristine organics and fragile mineral phases (e.g., Haberle and Garvie 2017), and as ground truth for quantifying alteration due to terrestrial weathering.

Significant technological advances have been made in recent years that enable more precise observations and more accurate modeling of meteorite falls, and thus the more rapid potential recovery of the meteorites. Advances fall into two main categories: improved observation of meteor/fireball phenomena, and characterization and modeling of meteorites in dark flight.

4.2.1 Fireball Detection and Tracking

One of the most important "sample return spacecraft" may be the Earth itself. As our planet orbits the Sun, it collects around 40000 tons of extraterrestrial material each year (Flynn et al. 2004; Zolensky et al. 2006a; Zook 2001), ranging from microns in size to occasional large meteorites (and less frequent large impactors). This material originates from a wide range of parent bodies and so has the potential to inform us about the status and histories of a great many parent bodies. Historically, efforts such as the Prairie Fireball Network (Wetherill and Revelle 1981), the Meteorite Recovery and Observation Project (MORP) in Canada (Halliday et al. 1978), the European Fireball Network (Oberst et al. 1998), and others (Bland 2004; Colas et al. 2015; Cooke and Moser 2011; Gritsevich et al. 2014; Hindley and Houlden 1977; Kokhirova and Borovička 2011; Shiba et al. 1997; Sullivan and Klebe 2004; Trigo-Rodriguez et al. 2006; Watson 2009; Weryk et al. 2008; Wiśniewski et al. 2017) surveyed meteors using networks of cameras and recovered small numbers of meteorites. Perhaps the most scientifically significant outcome of these efforts was their recovery of meteorites paired with calculations of their original orbits. This allowed something new in meteoritics—laboratory studies of the recovered meteorites were given a precise "home" in the Solar System, at least immediately prior to the meteorites' fall to Earth. The large number of fireballs observed by these networks and need for accurate fall location calculation encouraged the development of meteor dynamics (e.g., Ceplecha et al. 1998) and strewn field modeling. Today, improvements in digital imagery, geographical information systems, computational capability, and the growing power of the internet to quickly collate data submitted by the public, have driven the development of new fireball reporting networks (Venton 2017). A hallmark of these new networks is their emphasis on including the general public in reporting and other forms of participation (Day et al. 2018a). Another is regular calculation of fireball orbits, such that today the orbits of more than two dozen meteorite falls are known. The Desert Fireball Network (DFN) has used a network of 52 automated digital camera devices to recover four meteorite falls in the period of 2007–2015 (Bland et al. 2012; Sansom et al. 2015). All of the falls are associated with pre-atmosphere orbits, as are a larger number of fireballs that have not yet yielded meteorites. Recent funding from the Australian Research Council expands the Australian-based Desert Fireball Network into a Global Fireball Observatory (GFO), bringing the total area covered by all-sky cameras to over 12 million km^2, and enabling 24-hour all-sky observation from both hemispheres with 156 state-of-the-art camera stations. The GFO is expected to see 800 bright fireballs every 6 months and track 5 meteorite falls per month globally.

4.2.2 Eyewitness Reporting via the Internet

Historically, most meteorite falls have been recovered based on eyewitness accounts and many of these recoveries were facilitated by eyewitnesses close enough to observe a meteorite striking the ground. Many eyewitnesses doubtlessly observed meteorite falls from a distance, but the fragmented nature of eyewitness reports and lack of widespread public understanding inhibited collection of fallen material. With the growth of the internet, however, several organizations have begun collating eyewitness accounts into meaningful bodies of data capable of rapidly constraining the site of potential meteorite falls. The non-profit American Meteor Society (AMS) stood up a webpage with this purpose in 2006, allowing individuals to log meteor sighting reports at no cost to the user (https://www.amsmeteors.org). The sightings are collated into reports of individual fireballs, and automated calculations,

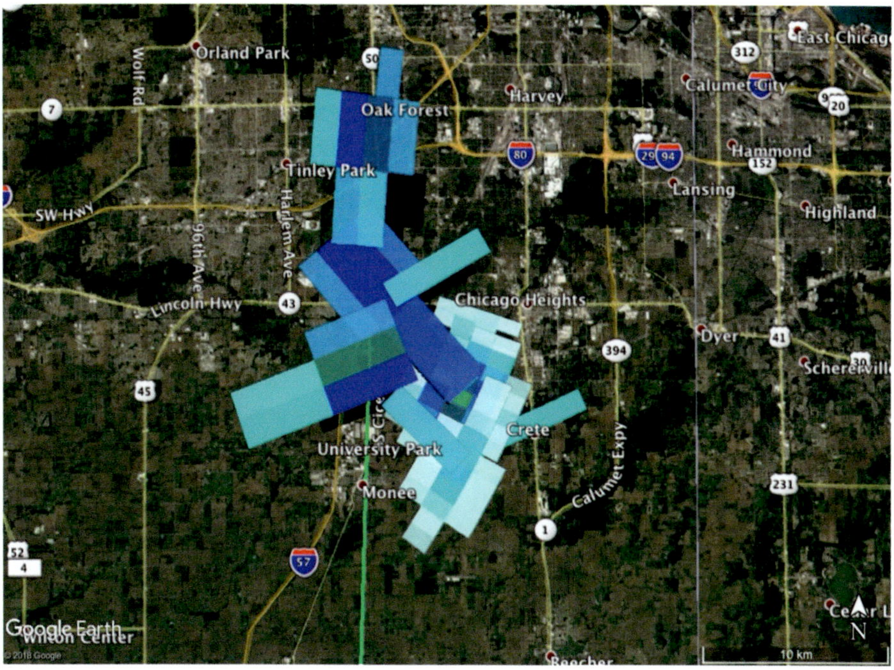

Fig. 8 Top-down composite view of weather radar signatures of falling meteorites. This shows meteorites falling during the Park Forest, IL meteorite fall (26 March 2003). The meteorites are size sorted during free fall to the ground with the most massive stones landing first, and individual radar "sweeps" record a cross-section of the resulting curtain of falling material. Signatures here record meteorites from 15.5 km down to 5.0 km altitude, and the altitude range can change for different falls based on the observation geometry of nearby radars. Typical detection timing for a meteorite fall ranges from radar detection of the fireball itself to observation of the last material (larger than dust) to reach the ground ~10–12 minutes later. Data is provided via website access by NOAA, and this image was composed in Google Earth

based on sighting azimuths, produce an estimated ground track for each event. Similar websites are operated by the International Meteor Organization (IMO) (https://www.imo.net), Fireballs in the Sky in Australia (http://fireballsinthesky.com.au), the UK Meteor Network (https://ukmeteornetwork.co.uk), and EXOSS Citizen Science in Brazil (https://exoss.org). The AMS, for example, recorded 5473 separate fireballs in 2017 and has produced eyewitness-based reports for meteorite falls to include Park Forest, IL (27 March 2003), Battle Mountain, NV (22 August 2012), Sutter's Mill, CA (22 April 2012) and others. With the increase in public access to the internet, these organizations have seen significant growth and can rapidly provide public notice of new meteorite falls (Fig. 8).

4.2.3 Weather Radar Detection of Meteorite Falls

Weather radars are commonly used for weather observation and forecasting worldwide, but the cost of building and maintaining a nationwide weather radar system demands that most weather radars are operated as nation-wide, government-run networks. The upside of this is that data is standardized within a national network and many countries maintain archives of their radar imagery. The downside is that public access to weather radar imagery is scant, with most data available in a short-lived image download format, often in the form of a calculated product such as rainfall rate. The United States has, arguably, led the way in free public dissemination of weather radar imagery, with the nationwide NEXRAD network operated by the National Oceanic and Atmospheric Administration (NOAA). NEXRAD data are freely available in all formats, from lightly formatted data to compiled images and calculated products (Fig. 9). All NEXRAD data are available online at no cost and are updated in near real time. NEXRAD data are made available on the internet at such a rapid rate that, for a typical meteorite fall lasting 10–15 minutes, at least one set of radar data showing falling meteorites is available online before all of the meteorites have reached the ground. Finding a meteorite fall in weather radar imagery currently requires manual data processing and analysis, but falls have been identified within hours of the event via weather radar imagery (Jenniskens et al. 2012) and the possibility exists for more rapid, automated identification of meteorite falls.

Weather radar detection of meteorite falls was first demonstrated with the Ash Creek, TX fall of 15 Feb 2009 (Fries and Fries 2010). Two researchers, Drs. Marc Fries and Robert Matson, had been investigating weather radar for this purpose without knowledge of the others' efforts and independently noted the ability of NEXRAD radars to detect falling meteorites and provide their fall location with great accuracy. Ash Creek was also noted by a National Weather Service office in Dallas, TX, who serendipitously noted the fall in the imagery of the KFWS radar. Ash Creek was a perfect test case in that it was a sizable fall that occurred in otherwise clear skies and stands out clearly in radar imagery. Since then, twenty-four recovered meteorite falls have been identified in NEXRAD imagery from 1997 to the present, with an additional thirteen identified that have not been recovered for various reasons (Fries et al. 2017). This is an average of 1.8 meteorite falls found per year, with 1.1 per year recovered in the United States and Canada (Fig. 10). Falls are identified by a combination of factors. Falls occur at the time and location identified by eyewitnesses, progress from high altitude to low over a period of 10–15 minutes, may include short-range turbulence caused by supersonic meteorites, and progress from moderate to high spectral width (a measure of the range of sizes of reflectors in a given image pixel) to low spectral width as the meteorites size-sort on the way to the ground. By contrast, nearly everything else that appears in weather radar imagery—such as weather, birds, insects, and aircraft—move laterally and do not show the other features listed above. Work is currently underway to include quantitative measurements of meteorite fall mass, and it may be possible to suggest meteorite types from radar data based on the fragmentation behavior of the bolide.

Weather radars are not limited to the U.S., of course. According to the World Meteorological Organization (a United Nations Specialized Agency), weather radars around the world comprise approximately $3.6\times$ the total areal coverage of the NOAA NEXRAD network. In other words, if all the world's weather radars could be put to use finding meteorite falls, the number of falls observed on radar could conceivably increase by an additional $\approx 3.6\times$. Such an increase should provide freshly-fallen meteorites for study and constitutes significant public outreach potential. There are significant obstacles to using weather radar to search for meteorite falls to include (1) some national radar networks choose to limit public access to data, (2) many countries do not archive sufficient data to be useful, and/or (3) some countries use proprietary data formats that are difficult for the public to utilize. These problems must be overcome within individual nations in order to realize the full potential of weather radar data to identify and locate meteorite falls.

4.2.4 Seismic Data

Another important asset for locating meteorite falls is seismic data. Seismometers monitor ground motion over most of the Earth's surface, and they are capable of detecting sonic

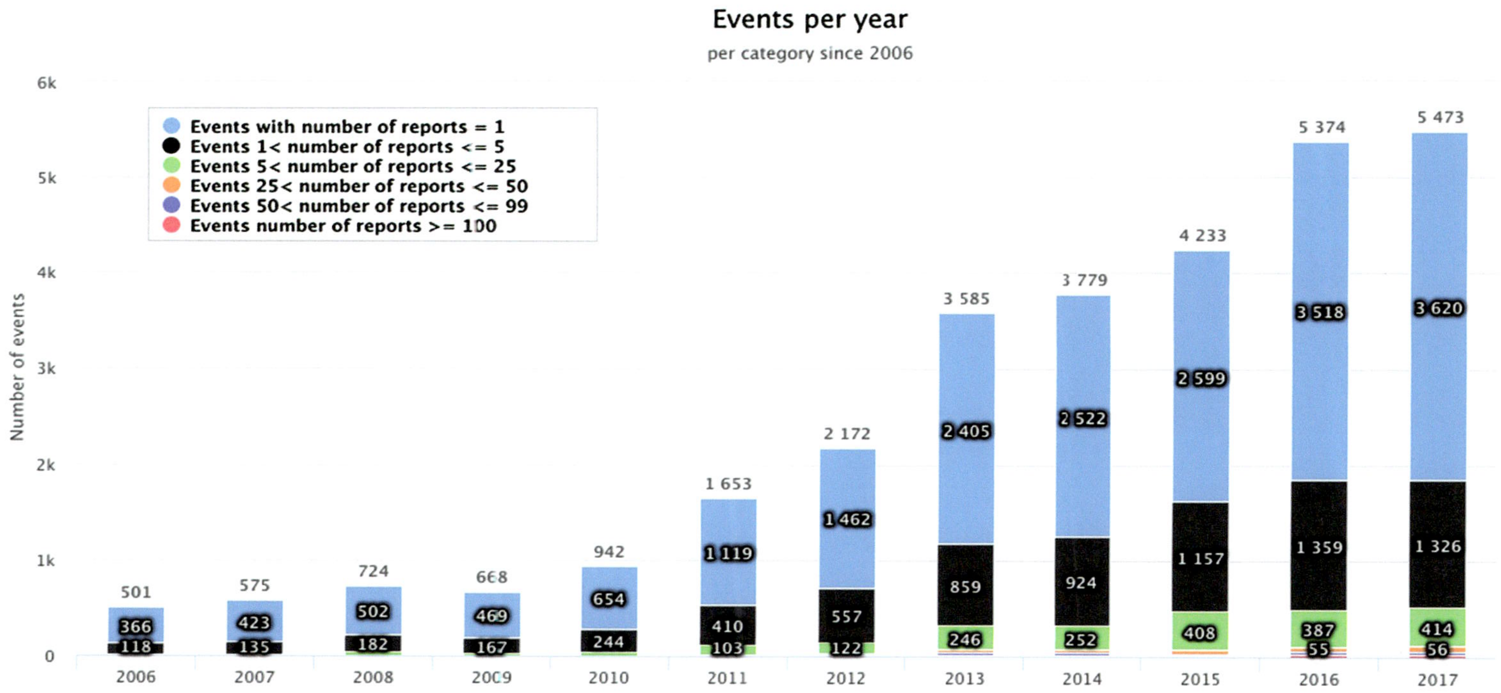

Fig. 9 Eyewitness accounts of individual meteor events collected by the American Meteor Society (AMS), shown by year. The rise in observed events tracks with the growth of the internet and the popularity of the AMS website

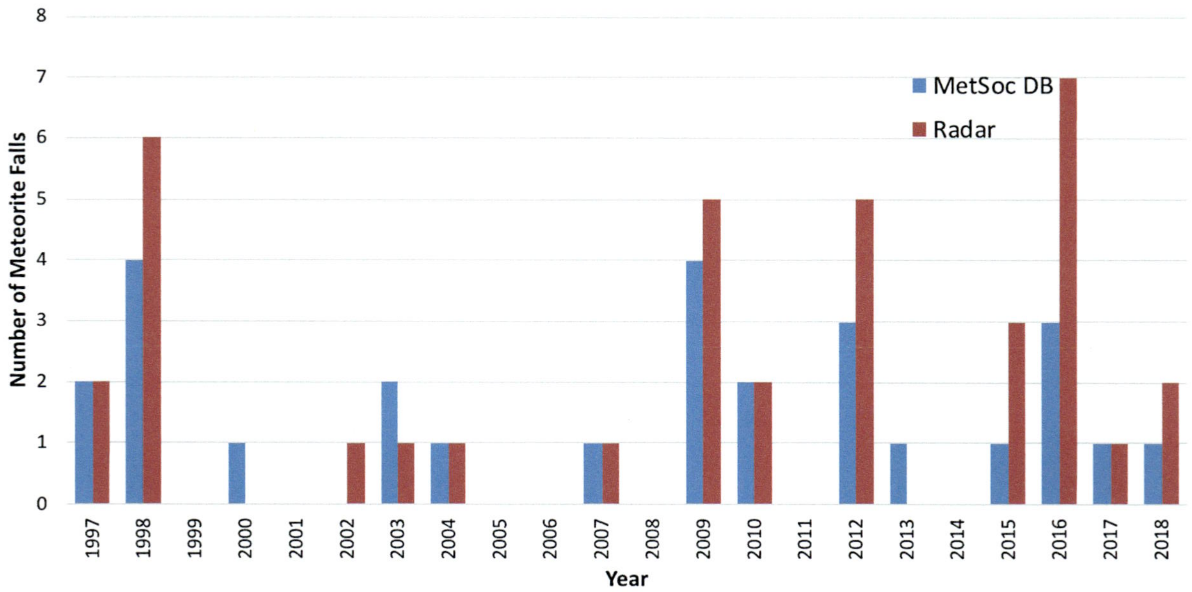

Fig. 10 Comparison between the number of meteorite falls seen in the Meteoritical Society's database of approved falls (blue) and those that appear in NOAA weather radar imagery (red). MetSoc falls are shown only for the area scanned by the NOAA radar network. Most years are in parity between the two, with more falls overall seen by the radar network than appear in the database. This is because some falls occur into water or inhospitable terrain and are not recovered. On average, 1.8 meteorite falls are noted per year since 1997 with 1.1 recovered per year

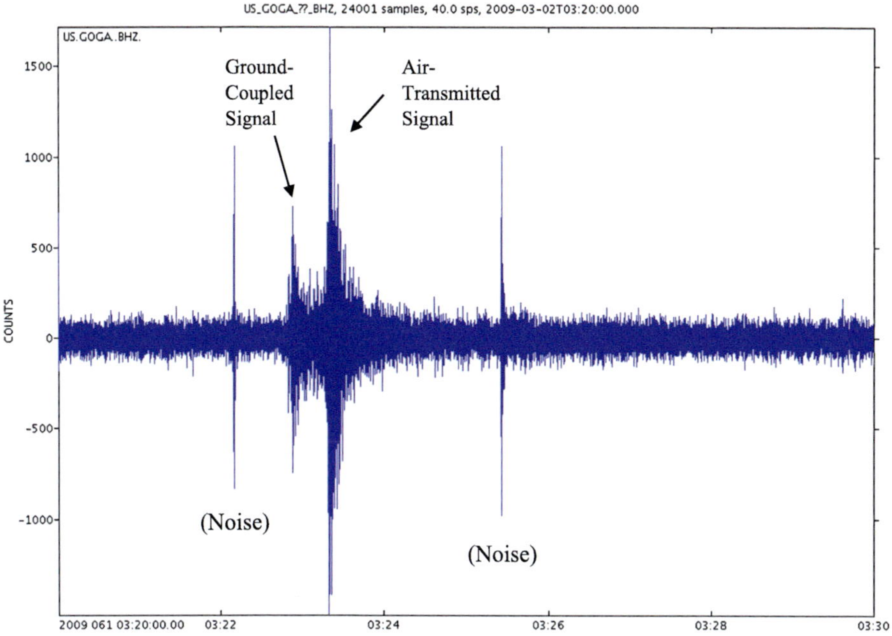

Fig. 11 Seismometer trace showing a sonic boom from a meteorite fall. Seismometers often record a pair of intensity "spikes" like the ones shown here, one from a ground-coupled signal and another transmitted directly through the air (see text). This signature is from the Cartersville meteorite fall (02 Mar 2009), which resulted in a single recovered stone. This illustrates how seismometers provide useful information for even small meteorite falls. Seismometer data provides precise timing of a bolide and can locate the event through triangulation

booms from both high-altitude deceleration of meteoroids (Edwards et al. 2008) and relatively closer low-altitude passage of falling meteorites in "dark flight", or the portion of a meteorite fall after luminous flight. Seismometers are typically operated in networks, and those networks are usually regional or local suites of instruments operated by government entities, universities, or other research facilities. The Incorporated Research Institutions for Seismology (IRIS) database collates publicly-available seismometer data into a common format and makes it available online.

Sonic booms from meteorite falls often appear as a pair of peaks in signal intensity graphs, or seismograms (Fig. 11). The main signature seen in most seismometer data comes from the point of maximum sonic boom generation during the fall, between ~20–30 km altitude (Ceplecha et al. 1998). The first peak comes from coupling of the air-transmitted pulse with the ground underneath the sonic boom. The resulting ground-transmitted pulse moves approximately 6 km/s and outpaces the air-transmitted pulse, usually arriving at the seismometer first. The air-transmitted sonic boom pulse moves approximately 340 m/s but is a stronger signal, producing a stronger pulse shortly after the ground-transmitted pulse arrives. Additional sonic booms may arrive afterwards, from individual falling meteorites in dark flight. Such signals are usually less intense than the initial signature. The arrival time of each of these signals is strongly dependent on the distance between the source and the seismometer, permitting triangulation of the source location.

4.2.5 Meteorite Recovery from Bodies of Water

Approximately 70% of the Earth's surface is covered by water, and that same percentage of meteorite falls disappear into the oceans and other bodies of water. Most of these have historically been lost, although there have been several meteorite falls wherein significant meteorite masses were recovered by people diving after an observed fall. These include Angra dos Reis (the namesake of the angrite clan of meteorites) (Prinz et al. 1977), Peña Blanca Spring where the recovered meteorite landed in a pond in front of a group of ranch hands (Lonsdale 1947), the main mass of Björbole which was recovered from icy water in 1899 (Martin and Mills 1976), and the main mass of Chelyabinsk which was recorded on video landing in a lake (Popova et al. 2013). These were fortuitous recoveries that all feature relatively shallow water and eyewitness(es) who saw the actual point of entry into the water. Recently, attempts have been made to expand this capability by recovering probable meteorite falls from deeper water using Remotely Operated Vehicles (ROVs) and other modern deep-sea technologies.

The Aquarius Project (Bresky and Fries 2018) is a student project to recover meteorites from Lake Michigan. On 06 Feb 2017, a bright green fireball accompanied by sonic booms heralded a large meteorite fall into the lake. The fall was observed by four radars in the NEXRAD national weather radar network. The meteorites now lie on the lake floor at a depth of around 100 meters. A consortium of Lake Michigan-area institutions was formed, including the Field Museum of Chicago, the Shedd Aquarium, and the Adler Planetarium with assistance from NASA and NOAA. Teenage students from Chicago public schools have worked with the project and assist in the design and testing of original devices used to collect meteorites from the lake floor. Aquarius Project updates are archived online (https://openexplorer.nationalgeographic.com/expedition/rovmeteoritehunt). In July of 2018, the students deployed their equipment to attempt to recover meteorites from the lake bed, recovering small rocks and sediment. This material is currently under examination by Aquarius Project institutions, and additional trips to collect material are planned for 2019.

Another attempt to retrieve meteorite material from an observed fall occurred in early July 2018, targeting a very large meteorite fall that occurred in the Pacific Ocean on 07 March 2018 about 20 km off of the coast of Washington state. Calculations of total meteorite mass based on radar reflectivity indicate that this fall was the most massive fall seen to date by the NEXRAD system since its inception in the mid-1990's. More importantly, the distribution of meteorite mass is unlike any of the two dozen recovered meteorite falls recorded by NEXRAD. The Washington coast fall features significantly more large, surviving meteorites proportional to smaller size fractions than previous events. This implies a stronger than typical mechanical toughness for this fall, which may in turn come from a slow infall velocity or a meteorite type that is inherently stronger than typical ordinary chondrites. The fall itself occurred during region-wide cloud cover and so was not observed directly, although at least two videos record bright flashes through the clouds. It is scientifically important to understand the reason why this fall produced an atypically large number of large meteorites, both as a planetary defense issue and to allow identification of future meteorite falls of the same type from radar data. For these reasons, a one-day effort was mounted by the Ocean Exploration Trust (OET) exploration vessel E/V *Nautilus* to map the fall site with multibeam sonar and attempt retrieval of meteorite material sufficient to identify the meteorite type. Sonar revealed that the ~100 m-deep seafloor was flat and featureless. The ROV pair Argus and Hercules performed a seafloor examination/sampling transect along a ~1.6 km track, collecting one sample with a magnetic rake, five more with a water-jet sediment sampler and one with a scoop. These samples were examined by optical microscopy, Raman spectroscopy, and electron beam analysis at NASA JSC. In June

2019, a follow-on team returned to the site aboard the research vessel R/V Falkor, operated by the Schmidt Ocean Institute (SOI). This second expedition focused specifically on retrieving ∼1 cm-sized meteorite fragments from the area of the meteorite strewn field where that size of meteorite should predominate. Over the course of three continuous days of ROV operations, an SOI-developed pair of sediment samplers sifted a large volume of ocean floor sediment but did not recover cm-sized meteorites. A suite of sediment samples was collected, washed, sieved, and searched in a manner identical to the Nautilus samples, in an iterative search for progressively smaller meteorite fragments. This effort was ultimately successful in the size fractions below ∼ 2 mm in diameter. At the time of this writing, over 100 small melt spherules and other fragments have been recovered from a combination of the Nautilus and Falkor samples. Work continues on this project to identify any meteorite type(s) among the spherules and whether they can be definitively linked to the fall event on March 07, 2019.

The public outreach aspect of this effort was a dramatic success, with a large global audience watching in real time via the OET webpage as the Argus and Hercules scanned the seafloor and collected samples. Through the course of eight hours with the ROVs on the seafloor, the crew took questions and narrated the effort for a pan-global audience. Overall, both the Aquarius Project and the Washington coast fall recovery show that modern oceanographic surveying and sampling techniques have made water-borne recovery of meteorite falls a real possibility. These pioneering efforts are yielding a first trip up the learning curve towards optimizing the techniques needed, and have set up a powerful new means of engaging students and the public for meteorite research and recovery. At present, the majority of all meteorite falls are lost to science because they fall into the oceans, which cover ∼70% of the Earth. In the future, development of the meteorite recovery techniques explored in this effort could be used to identify and sample meteorite falls for any event that terminates into water. Costs and effort requirements would naturally limit that number, but seaborne meteorite fall recovery could be employed for extraordinary events such as the recent (18 Dec 2018) 173 ktonne-TNT event over the Bering Sea. Another example of a worthy recovery target is the infall of 2019 MO, which was observed while still in space and fell into the Caribbean Sea on 22 June 2019.

4.2.6 *The Geostationary Lightning Mapper on GOES Satellites—A New Meteorite Fall Detector*

NOAA recently launched the GOES-16 and -17 satellites, a new design of geostationary weather surveillance satellite. These satellites provide weather imagery services for most of the United States and parts of contiguous countries. For the first time, the GOES-16/17 satellites feature a lightning mapper instrument—the Geostationary Lightning Mapper (GLM). GLM "stares" at a large ground footprint area, collecting imagery at a rate of up to 500 frames/second. The instrument features sufficient dynamic range to detect lightning during local daytime on the ground, and can discern altitude sufficiently to differentiate cloud-to-ground from cloud-to-cloud lightning. Inadvertently, NOAA built a superb meteorite fall detector with the GLM sensors. While visible and IR-wavelength weather satellite imagery can detect meteorite falls (e.g. Almahata Sitta as described in Borovička and Charvát 2009 and Chelyabinsk as seen by ESA's Meteosat-9 and reported in Miller et al. 2013), considerable luck is required because the image collection rate is very low. Previous GOES satellites, for example, only collected images once every few minutes and a meteor would have to happen exactly at the moment when the camera was operating to be recorded. GLM's rapid imaging and sensitivity to bright flashes renders it a very capable meteor detector, as

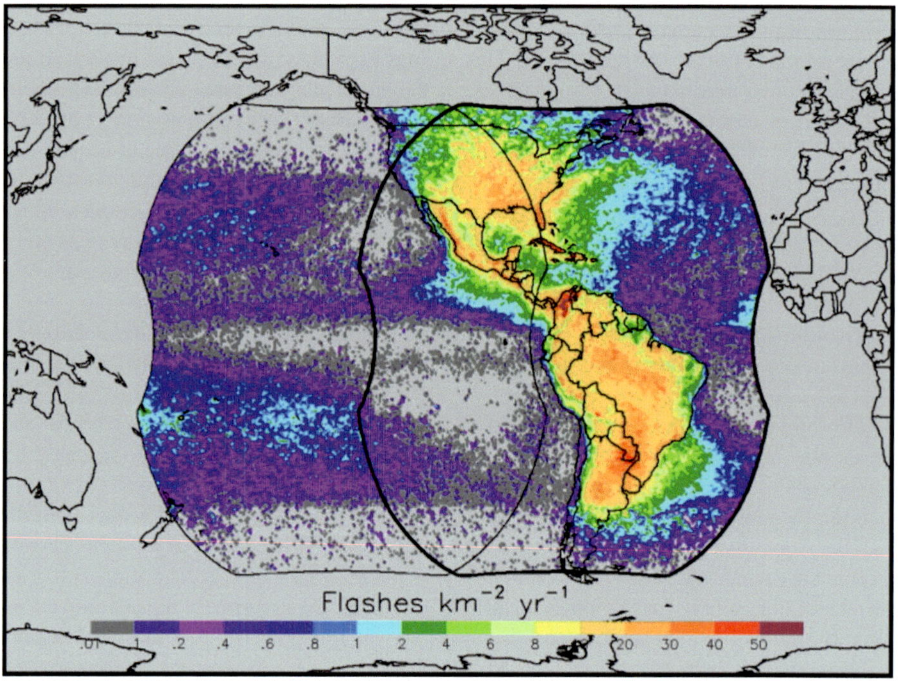

Fig. 12 NOAA graphic indicating the overlapping imaging footprints of GLM sensors on the two GOES satellites, one located off of the US east coast and another off of the west coast. Colors indicate the expected lightning frequency per year. Note that the imaging footprints of the two satellites overlap over most of the contiguous United States and Central America, where stereo imaging of bolides over land is theoretically possible

described recently (Jenniskens et al. 2018). Jenniskens et al. (2018) describe ten separate bolides detected by GLM to include one meteorite fall (British Columbia, Canada, 05 Sep 2017) and has since detected another (Hamburg, MI, 16 Jan 2018). GLM provides location, fireball luminosity, timing, and light curve data for bolides. Presently, both of NOAA's new GOES satellites are in place (GOES-16 and GOES-17) and their GLM sensors are operating and data is available online (https://www.class.noaa.gov). The two satellites cover the continental United States from locations near the east and west coasts, which may allow stereo observation of some meteorite falls (Fig. 12). This feature may be used to rapidly identify bolides among the lightning flashes that GLM is intended to monitor. Lightning usually occurs below 15 km altitude (e.g., Mecikalski and Carey 2018) while meteors produce optically bright signatures between ~20–90 km (Ceplecha et al. 1998). Stereo observation of bolides from the two satellites should yield rapid calculations of the altitude of various events, and finding bolides would be a matter of identifying bright flashes that occur at higher altitudes than that of lightning. The possibility exists that streaming GLM data can be used to identify the location, timing, and potential for producing meteorites from bolides in near real-time.

4.2.7 Putting It All Together—Dedicated Collection of Fresh Meteorite Falls

The march of technology has availed the astromaterials community with the new technologies and capabilities described here. Historically, even though meteorite falls are far outnumbered by meteorite finds, the light terrestrial alteration inherent in recent falls makes them highly scientifically valuable. Regular and rapid collection of freshly fallen meteorites is possible to an extent that a new possibility exists—dedicated recovery of freshly fallen meteorites by a dedicated recovery team. The nature of such a team can take many forms, to include enlisting the assistance of the general public. This approach takes advantage of the spirit of public inclusivity, which created such endeavors as the American Meteor Society's eyewitness reporting program and is a powerful opportunity for outreach. A nationwide program could stand up to mobilize meteorite falls as they are detected, using eyewitness accounts and GLM to identify a new meteorite fall, seismic and radar data to calculate the strewn field, and astromaterials experts encouraging and guiding meteorite recovery at the fall site. Impromptu public lectures and media contact shortly after a meteorite fall have a history of enthusiastic reception from the public and tend to promote recovery of meteorites. The possibility exists that a dedicated collection(s) of freshly fallen meteorites could be founded and sustained in this way.

5 Importance of Contamination Knowledge Strategies for Sample Return Missions to Maximize Science Returns from Samples

The scientific value of the returned samples for all previous sample return missions have benefitted from having an archive of contamination knowledge (CK) materials that could include (1) spacecraft hardware, spares, and flight-like coupons, (2) materials used in the fabrication of spacecraft hardware or construction of a curation lab, and (3) witness materials deployed during ATLO or during construction of the curation lab. The information gained from studying the collected reference materials and witness plates is defined as the CK of a sample collection, and the CK is crucial for verifying and validating scientific results. As part of a sample return mission, these CK samples are archived along with the returned samples and the CK samples are made available for allocation and analysis by the scientific stakeholders of a sample collection. These flown, flight-like, and non-flight reference materials and witness plates provide the scientific community investigating astromaterials with the fundamental ability to reconstruct the contamination history of a sample collection. Furthermore, they serve as a baseline from which to compare tantalizing results attained from the analysis of astromaterials. CK collections are a requirement for sample return missions because contamination control efforts cannot anticipate all possible contamination vectors that can occur during the dynamic activity that is a sample return mission. In fact, all sample return missions have developed non-nominal contamination to some degree, which are highlighted in this section as important lessons learned. In every case, the CK samples have helped to mitigate these unplanned contamination events, improving and, in some cases, enabling scientific returns on the returned astromaterials. The NASA Curation Office has allocated hundreds of samples to scientists in support of gaining contamination knowledge for their respective collections (mainly Apollo, Genesis, and Stardust, and a newly constructed OSIRIS-REx collection). We outline the CK methodology and lessons learned for many completed and ongoing sample return missions and provide insights into best practices for collecting CK when biological contamination from indigenous or exogenous extant life is a possibility.

5.1 Apollo Program

For Apollo, contamination knowledge has a broader definition and overlaps with contamination control, compared with recent sample return missions, principally because Apollo was a series of missions in which in-lab monitoring of returned samples and hardware resulted in improvements for subsequent missions. Here we cite a 1965 formal report wherein three scientific discipline groups made early recommendations concerning sample collecting materials and procedures relevant to contamination. This is followed by examples of organic and inorganic contamination of samples being measured, by both curatorial staff and by researchers using samples, which resulted in improvements both to lunar surface sampling procedures and hardware and curation handling practices. Thus, contamination knowledge is an ongoing process for returned sample collections.

Early advice on contamination issues was captured through 3 of the 7 discipline working groups convened in 1965 for the purpose of advising NASA on science for a 10-year period (NASA 1965). Geochemistry, Bioscience, and Geology Group reports were most concerned with contamination knowledge. For example, top level equipment requirements included: (1) "Sample containers should keep samples sterile and chemically clean. Stainless steel is acceptable. More studies should be completed relative to the use of Teflon in the lunar environment." It was the Geochemistry Group that emphasized the value of sample analysis and foresaw many of the curation guidelines that are still followed today. The report calls for CK studies to determine the amounts and effects of outgassing of the astronaut suits, the escape of atmosphere from the Lunar Excursion Module (LEM), analyses of possible contaminants in LEM fuel and effects of those contaminants on samples. The Geochemistry Group specified acceptable materials to touch the samples (materials that would not interfere with scientific measurements). In general, this meant use of materials of known and simple chemistry, easily distinguishable from lunar material.

5.1.1 Apollo Organic Contamination Knowledge

Apollo organic contamination monitoring conducted in the Lunar Receiving Laboratory (LRL) offers an opportunity to compare the monitoring effort on sample handling equipment to the actual detection of organic compounds in lunar samples by investigators. Apollo planners conducted extensive organic contamination monitoring of the containers, tools, and sample handling facilities. Simoneit et al. (1973) summarized the potential sources of organic contamination: (1) surface contamination of the lunar-bound rock box and its contents; (2) surface contamination on the Apollo lunar hand tools used to obtain samples on the lunar surface; (3) exhaust products from the lunar descent engine and reaction control system engines (both using unsymmetrical dimethyl hydrazine and nitrogen tetraoxide); (4) lunar module outgassing; (5) astronaut spacesuit leakage; (6) particulate material abraded from spacesuits or other sources during EVA; (7) venting of lunar module fuel and oxidizer tanks, cabin, and waste systems; (8) venting of spacesuit life support back packs; (9) exposure to LRL vacuum or nitrogen processing chambers; (10) surface contamination of sample processing tools and containers; (11) surface contamination of containers sent to PIs. Items 1, 2, 3, 9, 10, and 11 were considered most serious. Simulations, modeling, and engineering data were used to estimate the contamination contributed by flight items 3, 4, 5, 7, and 8 (Aronowitz et al. 1966a, 1966b). Virtually all rocket exhaust products were low molecular weight and rapidly diffused over large areas. Because of their low concentration, this was not predicted to be a major contaminant. The varied organic products included acetylene, HCN, ethylene, formaldehyde, methyl amines, and others.

For laboratory handling operations, measurements of contamination via "monitors" or witness plates were used. Clean coupons of a woven aluminum alloy called York mesh (2024 aluminum alloy) or aluminum foil were processed along with the lunar-bound tools or placed inside the rock boxes bound for the Moon. Upon return, these coupons were analyzed by solvent extraction and subsequent gas chromatography and mass spectrometry.

Aliquots of clean Ottawa sand, exposed inside sample processing cabinets, were analyzed by direct pyrolysis and mass spectrometry. The solvent rinsings from tool, container, and cabinet cleaning were also analyzed. Some of the most frequently encountered contaminants were hydrocarbons from pump oils and fatty acids. Detected in the vacuum chamber, some of the fatty acids were thought to be from the polishing compound used on the rock boxes. Dioctylphthalate, a common plasticizer additive for polyethylene, was ubiquitous in cabinets and bags. Simoneit and Flory (1971), Flory and Simoneit (1972), and Simoneit et al. (1973) provide an extensive list.

York mesh and aluminum foil monitored organic contamination levels of about 1 $\mu g/cm^2$ inside the rock boxes. Bakeout of the Apollo 11 rock box actually added organic contamination, but as a result of the monitoring, cleaning improvements were made which produced flight hardware for Apollo 12 through Apollo 15 with only 10–100 ng/cm^2 contamination (Simoneit et al. 1973). With exceptional care, curatorial cleaning procedures during Apollo could produce 1–10 ng/cm^2 contamination ranges for polished, planar surfaces. Flory and Simoneit (1972) concluded that organic contamination to lunar samples during Apollo 11 was in the 1 µg/g (ppm) range, but improved to 0.1 µg/g (ppm) for Apollo 12.

The actual analyses of lunar samples were consistent with the estimated contamination levels. Burlingame et al. (1970) concluded, based on analyses of their allocated samples, existence of systematic organic contamination of about 5 µg/g for Apollo 11 samples, except for those samples processed in the organic reserve cabinet. Reports of organic compounds in lunar fines from other investigators were mixed, ranging from no detection at the ng/g levels (Abell et al. 1970; Lipsky et al. 1970; Meinschein et al. 1970) to ng level detection of various organics (Henderson et al. 1971; Murphy et al. 1970; Preti et al. 1971) and 0.5 ppm via pyrolysis (Oro et al. 1970). Porphyrin-like pigments were detected at the trace ng to pg level by Kvenvolden et al. (1970) and Hodgson and co-workers (1970, 1971), but not by Rho et al. (1970, 1971, 1972). Porphyrins as a possible rocket exhaust contaminant were discussed. Amino acids were detected at the 50 ng/g level by Hare et al. (1970) and Gehrke et al. (1970) after aqueous or other processing to the sub-nanogram level (Murphy et al. 1970) and below detection by Gehrke et al. (1972). No viable organisms were detected in Apollo 11 and 12 samples (Oyama et al. 1970, 1971; Taylor et al. 1971). (For estimates of indigenous lunar carbon in soils and breccias, Vaniman et al. 1991 present values of about 100 ppm selected from analyses less likely, but still possibly, containing terrestrial contamination.) The presence of diamond and polishing compounds on the surfaces of Apollo thin sections have also been documented by Raman spectroscopy, and these contaminants are likely a common result of the thin-section making process (e.g., Steele et al. 2010).

Keeping samples organically clean in the LRL proved difficult. Thus, a small facility to analyze and repackage a lunar sample collected in a special container was constructed at the University of California, Berkeley (Burlingame et al. 1971). The organically clean area consisted of two gloveboxes in tandem preceded by a vacuum entry chamber. The atmospheric nitrogen gas was scrubbed to remove oxygen, water, and organics. The glovebox was equipped with a liquid nitrogen cold finger to remove water generated by the glove operator.

In summary, Apollo organic contaminants were greatly reduced by institution of (1) restrictions on materials allowed contact with, or proximity to samples; (2) isolation of samples in controlled environments; (3) procedures to clean all surfaces in proximity to or contact with samples; and (4) controls on fabrication, processing, and handling of lunar sample hardware.

5.1.2 Inorganic Contamination Knowledge

As is the case with organic contaminants, feedback from investigators on inorganic contaminants was essential in improvement. Some materials were selected for use as lubricants or seals because those materials were not predicted to interfere with scientific measurements. Thus, it made engineering sense to use an alloy of 90% indium and 10% silver for the metal knife-edge seal on the Apollo rock boxes. It made engineering sense to use molybdenum disulfide as thread lubricant in curation sample containers. However, these materials were not of sufficient purity for high precision scientific research. Only a few unopened lunar samples remain in indium-silver sealed containers. The use of MoS_2 was discontinued and all containers removed from service. Molybdenum disulfide was replaced with a Teflon thread lubricant Xylan. However, Xylan was not pure Teflon and contained a binding agent, so Xylan was also removed.

Attention to detail is required and compromises made, especially for the fabrication processes, and this is illustrated by engineering selection of surface treatments for the Apollo drive tube cores and drill cores. The materials used for fabrication of the large diameter Apollo drive tubes, anodized aluminum, and the drill core tubes, canadized (a proprietary passivation technique) titanium were selected for engineering reasons. A list of materials would not normally raise contamination flags, unless the reviewer understood the details of the process (and in many cases an engineering requirement might be the best solution anyway). Lead content in amounts compromising science results was found in the anodized aluminum and in the canadized titanium by investigators making extremely low level measurements. The solution in the drive tube case was to physically remove the outer 1 mm of regolith during core dissection, before sample allocation.

One excellent example of institutionalized CK in Apollo sample curation was the construction, completed in 1979, of Building 31N at JSC to house the Apollo lunar collection. The entire design and construction was reviewed in real time by a facility subcommittee of the Lunar and Planetary Sample Team. The committee, comprised of planetary petrologists and geochemists—users of the lunar samples for research—requested chemical analysis for most of the material selections and reviewed the data in detail. Examples are chemistry of paint, floor coverings, adhesives, electrical cords, etc.

5.1.3 Ongoing Contamination Knowledge Efforts

Collection of CK for the Apollo curation lab continued after construction of the lab and continues to this day. This process is often a joint effort between curation personnel and members of the scientific community that develop in response to interesting, novel, or unexpected results stemming from scientific analysis of the samples. We provide here two notable examples of successful collaboration between curation personnel and the scientific community to better understand the contamination environment within curation facilities at NASA JSC. As mentioned above, Xylan was used on screw threads in the Apollo processing cabinets, tools, and containers to prevent galling. Over time, however, it became evident that the Xylan did not adhere well to the screws because it was flaking off into the processing cabinets and hence served as a potential source of contamination to the Apollo and Antarctic Meteorite samples (Xylan has 45 wt.% C and 4 wt.% N, and it is not removed by step combustion until samples are heated above 600 °C; Wright et al. 1992). In response to the characterization of Xylan as a potential contaminant that could affect the analysis of H, C, N, and O in astromaterials samples, Xylan was banned from use in any of the curatorial sample-handling hardware at NASA JSC. More recently, a concern was raised that the stainless steel tools used during the processing of Apollo samples may contribute highly siderophile elements (HSE) to the processed samples (Papanastassiou et al. 2015; Tikoo et al. 2014), given the low abundances of HSE that occur naturally in lunar samples (Walker et al. 2004). Consequently, the stainless steel tools and sample containers used in the Apollo curation labs were subsampled and analyzed to determine a contamination threshold that would affect HSE analyses of Apollo samples (Day et al. 2018b). Day et al. (2018b) reported that the potential for HSE contamination from the stainless steel containers and tools was low.

5.2 Genesis

Genesis was the first sample return mission since the Apollo program and ended a 32 year hiatus for sample return missions. The Genesis mission launched on August 8, 2001 and traveled to the Earth-Sun Lagrange 1 (L1) point. The Genesis spacecraft was held in halo orbit at L1 outside Earth's magnetosphere for 2.3 years, and the mission collected solar wind plasma that was implanted into several arrays of high purity materials and subsequently returned to Earth for analysis. Unfortunately, due to an inversion design mistake of the sample return capsule's (SRC) drogue parachute gravity switch, the SRC experienced a terminal velocity hard landing at the Utah Test and Training Range (UTTR) on September 8, 2004. The impact into the lacustrine sediment breached the science canister and littered thousands of broken high purity collectors throughout the science canister and SRC. Despite this set-back, curation contingency plans were invoked and after much effort decontaminating samples at UTTR and JSC's curation laboratory, all primary science mission goals were achieved and Genesis is marked as a successful mission. The ability of Genesis to rise from the ashes was in-part due to the fact that at mission inception, the mission had a well-orchestrated CK plan that led to a remarkable CK collection.

Contamination knowledge for Genesis solar wind sample return is captured in archived reference materials and associated documentation of six types, from pre-launch to post landing events: (a) flight-like collector substrates; (b) science canister duplicate components; (c) assembly environment material coupons and process witness plates; (d) post-landing UTTR soil samples; (e) post-landing science canister and sample return capsule hardware; (f) recovery processing tools and containers. Thus, Genesis is an example that contamination knowledge is an ongoing effort and that post-recovery contamination knowledge is important.

5.2.1 Flight-Like Collector Substrates

Genesis had ambitious goals for determining the elemental and isotopic composition of the solar nebula. Desired elemental accuracy was 2σ limits of $\pm 10\%$ of the number of atoms/cm^2 on the collector substrates. Desired isotopic precision for many elements was $\pm 1\%$ compared to terrestrial standards. Given that the estimated 2-year fluence (atoms per cm^2) for the more abundant elements is in the 10^8 to 10^{12} range, requirements for bulk purity and surface cleanliness of collectors were very stringent (Burnett et al. 2003). Much effort was expended by the science team in verifying purity and surface cleanliness of candidate batches of collector substrates. Fifteen types of ultra-pure materials were flown as collector substrates. Three hundred passive collectors, mounted in 5 arrays, individually consisted of single crystal silicon (FZ and CZ), sapphire, germanium, and sapphire coated with aluminum, silicon, diamond-like-carbon, or gold. Targets in a concentrator for O, N, and C were comprised of silicon carbide, isotopically enriched polycrystalline diamond and diamond-like-carbon coated on silicon. Additionally, special collectors of metallic glass, gold foil, polished aluminum alloy and molybdenum coated foils were deployed (Jurewicz et al. 2003). The diversity of the collector materials on these arrays not only provided multiple analytical background choices for optimum specific analyses, but also multiple material choices for the various surface cleaning processes that the hard landing subsequently required. The samples archived for CK served as reference pieces for purity and surface cleanliness or may have been implanted to make calibration pieces during analysis. These materials were also widely used to test surface cleaning protocols before cleaning Genesis-flown samples. It was evident in some cases of solvent application that reactivity of Genesis-flown pieces was different than non-flight reference pieces, presumably due to solar irradiation. To date, allocations of 600 Genesis-flown samples were accompanied by allocations of 300 collector reference materials. The base inventory supporting these allocations is 5000 Genesis-flown collector samples and 2000 non-flown collector reference substrates.

5.2.2 Science Canister Duplicate Components

The canister containing the payload of samples for return was cleaned using ultrapure water (> 18 MΩ-cm resistivity) inside of an ISO Class 4 cleanroom. Re-assembly, including installation of ultraclean collectors, was performed by staff completely enclosed in powered HEPA filtered Dryden suits. More than 200 duplicate canister components, with many cleaned exactly like flight components, are archived for comparison of contaminants. Some cleaning fluids and manufacturing fluids (e.g. electric discharge machining oil) were also archived.

5.2.3 Assembly Environment Material Coupons and Process Witness Plates

More than 100 environment material coupons of the assembly room are archived. Examples include samples of wall construction material, flooring materials, adhesives, paint, fire retardant and subsamples of air handler intake filters. Process witness plates for particle chemistry and airborne molecular contaminants were periodically set out, but only the data were saved (Allton et al. 2016).

5.2.4 Post-Landing UTTR Soil Samples

Just prior to Genesis capsule return, 8 UTTR soil samples from 5 sites were collected by the helicopter recovery crews as they practiced. These samples are archived and have been allocated for contamination studies on solar wind collectors. After the capsule hard landing and recovery of the spacecraft components, eighteen 5-gallon buckets of UTTR soil and spacecraft materials from the impact site were collected and archived. Subsequently, collector fragments were high-graded and removed from buckets for permanent archive.

5.2.5 Post-Landing Science Canister and Sample Return Capsule Hardware

Many of the collectors were broken, but most remained confined in the science canister. The field crew was able to gather and transport to a nearby cleanroom the entire science canister containing most of the collectors and the sample return capsule major components within 8 hours of the crash. All of this material is archived for CK, except for pyrotechnic devices and batteries, which were deaccessioned after 5 years (Stansbery and Team 2005). Ellipsometry was used to measure molecular contamination on some of the collector plates (McNamara and Stansbery 2005; Stansbery and McNamara 2005).

5.2.6 Recovery Processing Tools and Containers

Select recovery and UTTR processing tools and containers are archived for CK. Examples include polystyrene containers, fine brushes, and cleanroom post-it paper used for securing small fragments. The post-it adhesive remains under investigation and these CK samples have been helpful. The lesson here is awareness that any material added to the handling stream at the last minute should have batch specific material archived.

5.3 Long Duration Exposure Facility

The Long Duration Exposure Facility (LDEF), was a school bus-sized cylindrical facility designed to provide long-term data on low-earth orbit (~300 miles altitude) environment and its effects on space systems, materials, and operations (Kinard and O'Neal 1991). Originally intended to be the cargo for the first space shuttle mission in 1981, it was finally placed in low-Earth orbit by Space Shuttle *Challenger* in April 1984. Fifty-seven science and technology experiments from nine countries flew on the satellite. The original plan called for the LDEF to be retrieved in March 1985, but because of the destruction of Challenger (the only shuttle with a sufficiently large cargo bay to accommodate LDEF) it was eventually returned to Earth by the newly built *Columbia* in January 1990. LDEF was an early test bed for ideas on micrometeoroid capture. It was carefully placed in orbit and gravity stabilized such that its orientation relative to Earth remained constant. Thus, the trailing side of the satellite would capture micrometeoroids but no space debris and see no secondary impacts from satellite surfaces—completely eliminating the most significant sources of contamination to captured astromaterials. No subsequent spacecraft has repeated this feat.

Unfortunately, shortly after recovery into the still open Columbia cargo bay the shuttle began to rotate end over end, destroying many experiments and causing severe sample contamination. NASA Mission Control chose not to waken the sleeping astronauts and halt the rotations, despite pleas from the LDEF science team. Thus, recovery of LDEF by the Space Shuttle was non-nominal, resulting in contamination and thereby degrading many mission goals. An important lesson is to carefully consider the possible deleterious consequences of using an astronaut crewed platform for sample recovery operations.

Despite these problems, the LDEF mission was successful in guiding scientists to the design of the capture media for the subsequent Stardust Mission. However, one LDEF lesson was not properly learned. Outgassing of silicone-based adhesives and lubricants coated most of the exterior of the satellite with a Ca- and Si-containing coating, which was baked to a brown color by solar radiation (Whitaker and Dooling 1995). This "brown stain" was to reappear with a vengeance in the Genesis, and to a lesser extent Stardust Missions.

5.4 Stardust Mission

For the Stardust Mission, contamination control procedures were integral to flow of spacecraft manufacture, assembly, testing, flight, and recovery, and the science team took a very active role in planning and implementing contamination control measures, monitoring contamination through numerous witness materials (Sandford et al. 2010; Zolensky and Girard 1997). However, despite these precautions, the captured comet Wild 2 coma dust grains experienced significant contamination from several sources, including the presence of indigenous organic and inorganic material in the silica aerogel capture media, spacecraft outgassing, and an unfortunate sample return capsule (SRC) recovery procedure.

5.4.1 Preflight Contamination

The flight aerogel used in Stardust was marveled, but cleaning it was not a sufficiently high priority for the mission. There were alternative sources for the silica aerogel, which were known to be cleaner than the material manufactured by the Jet Propulsion Laboratory in Pasadena, CA. In addition, recommended work on improving aerogel cleanliness were not adequately performed. Synthesis of the aerogel employed a tetraethyl orthosilicate precursor in a solvent that included ethanol, methanol, acetonitrile, and/or other organic liquids, and Synlube 1000 was used as a mold release agent (Sandford et al. 2010). In the end, the severely contaminated aerogel was baked to reduce the volatile organic content, but several weight percent of carbon in the form of organics remained strongly bonded to the aerogel. To be fair, most persons believed at the time that organics could not be adequately captured by aerogel at the mission capture velocity of 6.2 km/s. The comet Wild 2 coma grains entered the Stardust aerogel at ~6.1 km/sec. Such collisions are sufficiently energetic that they could alter any organic compounds originally present in both the impacting particles and the aerogel collector material (Sandford et al. 2010; Sandford and Brownlee 2007; Spencer et al. 2009; Spencer and Zare 2007). Thus, it came as a surprise when relatively intact cometary organics were recovered from a few captured coma grains. Had we known this was possible, we would have undoubtedly made greater efforts to fly organically-clean aerogel.

It is now clear that some fraction of the impacting comet Wild 2 coma particles survived with little or no alteration, while other portions of the samples were severely heated (Brownlee et al. 2006; Elsila et al. 2009; Sandford et al. 2006, 2010; Zolensky et al. 2006b). Conversion of carbon original to the aerogel, and in the impacting cometary particles, into new forms likely occurred in a similarly variable manner. Thus, before one can assign organics seen in Stardust samples to a cometary origin, it is necessary to consider the possibility that they are either altered cometary materials or materials formed from carbon original to the aerogel. IR absorption difference-maps of individual tracks suggest that impacting Stardust particles do not convert the majority of the original carbon in the aerogel tiles into new chemical forms that remain in the aerogel. However, laser ablation, laser ionization mass spectrometry (L^2MS) studies demonstrate that at least a small amount of the original aliphatic carbon in the aerogel is converted into aromatic materials in the form of lightweight polycyclic aromatic hydrocarbons (PAHs). Thus, while most of the original aerogel carbon appears to be unaffected by the impact process, the issue of the possible presence of impact converted organics must be considered on a case-by-case basis whenever specific organics are being sought in Stardust aerogel samples.

Additional contaminants found their way into the aerogel during flight. These include materials outgassed from nearby spacecraft components, propellant byproducts, and secondary materials from dust impacts on other parts of the spacecraft, particularly the Whipple shields and solar panels. Cometary particles impacted on the aerogel tiles in the collector tray perpendicular to the forward direction. Thus, any tracks seen with oblique orientations must be either due to strikes by random interplanetary dust particles or to secondary materials from impacts on other parts of the spacecraft. Oblique tracks have, in fact, been found in the flight aerogel tiles, most of which fall in non-random spatial distributions on the cometary collector (Westphal et al. 2008). The materials in these tracks could include components from both the original impactor and from the spacecraft. Many of these tracks originated from a grazing impact on the central Whipple shield of the spacecraft as the origin of clustered low-angle oblique tracks. In these tracks, the most likely contaminant would be the Mylar thermal protection material that wrapped the edge of the Whipple shields. A second population of high-angle oblique tracks unambiguously originate from a non-cometary impact on the spacecraft bus just forward of the collector. The exact location of this strike on the spacecraft bus is not known, but possible contaminants include materials used for the sides of the spacecraft bus—highly ordered graphite embedded in an epoxy matrix.

In summary, it is clear that the Stardust cometary collector tray was struck by a limited number of secondary particles resulting from impacts on other parts of the spacecraft. Materials in these oblique tracks should be viewed with considerable caution before interpreting their significance as possible cometary materials. Fortunately, the most likely contaminants, Mylar wrap on the Whipple shields and carbon composites from the body of the spacecraft, have distinctive C X-ray absorption near-edge spectra (XANES) that make them relatively easy to recognize. At present, there is no evidence that this process has introduced contamination outside the domain of the oblique tracks themselves.

5.4.2 Contamination During Flight

It is possible that contaminants could have been introduced to the Stardust sampling trays directly from the spacecraft during its nearly 7-year flight. This is of special concern for the aerogel collectors since aerogel, with its very large surface area to mass ratio, is an excellent 'sponge' for adsorbing contaminants. To assess the extent of on-flight contamination, several 'witness coupons' were enclosed in the Stardust SRC (Tsou et al. 2003). These coupons included 1 cm diameter disks of aluminum and sapphire, and one 'interstellar' aerogel tile (2 cm wide × 4 cm long × 1 cm deep). These coupons were located on the arm that deployed the aerogel collector array and were placed low enough that they resided in the shadow of the main Whipple shield. Thus, these coupons were exposed to the same flight environment as the aerogel collectors for the entire mission, but were never directly exposed to the cometary influx.

Examination of the aerogel witness coupon showed no visible signs of adhering materials or stains. Although we know that silicone-based adhesives and lubricants outgassed during the mission (famously coating the cold camera optic surface), none of the exposed surfaces in the Stardust sample return canister showed any signs of the 'brown stain' seen on many of the surfaces of the LDEF (Fred Hörz, personal communication, 1990) and Genesis hardware (Burnett 2013). Analyses of the aerogel witness coupon (i.e., aerogel that was exposed to all environmental conditions as the collector aerogel except the comet) shows similarities to collector aerogel, although the levels of contaminants, when detected, are generally lower and some components (for example, the carrier of the 1700 cm-1 IR C=O feature) are dramatically less abundant. This suggests that contamination associated with the operational environment of the spacecraft during flight was not a major source of sample contamination.

5.4.3 Contamination from SRC Recovery and Curation Operations

The accumulation of local soils (mud) from the recovery site was a major issue of concern for the Stardust Science Team prior to recovery of the SRC. Fortunately, integrity of the SRC during landing, the relative inability of the mud at the recovery site to stick to the SRC, and the fact that none of the bouncing impacts occurred at the locations of the two backshell vents greatly decreased the magnitude of this concern. Thus far, there is no indication that soils from the recovery site infiltrated the sample canister or in any way contaminated the returned samples. However, an unfortunate decision was made to place the recovered SRC into a polypropylene bag during the brief helicopter transport from the landing site to a hangar for preliminary deintegration operations. Subsequent detailed analysis revealed that the aerogel soaked up outgassed organics from this bag, providing an additional source of organic contamination to the comet coma grains (Hope Ishii, personal communication, 2007). One important lesson learned from the stardust mission is that recovery operations for the SRC significantly suffered from the lack of a hermetic seal for the samples, probably in many additional ways that will only become apparent in the future. Mission engineers should be pushed to provide truly hermetic seals for future returned samples. Contamination from curatorial operations has been carefully mitigated, although there is some evidence that the Stardust Interstellar aerogel has collected some contamination during handling in various labs during analyses (Bechtel et al. 2011).

5.5 Hayabusa Mission

The curation and contamination knowledge of the samples recovered from asteroid Itokawa by the Hayabusa mission (2004–2010) are very well described by Yada et al. (2014). In order to limit contamination to the recovered samples, the constituents of the Hayabusa sampler were limited to A6061 aluminum alloy coated with pure aluminum, stainless steel (304), Viton, aluminum oxide glass, and Teflon. Before launch, every part of the sample container was cleaned in 2-propanol using an ultrasonic cleaner, installed in an ISO Class 7 cleanroom. A contamination coupon made of aluminum oxide glass was installed inside the sample catcher to monitor contamination during the mission. In order to minimize ground contamination following Earth return, the sample container was designed to seal the samples, though the mission budget did not permit a hermetic seal. However, terrestrial atmosphere permeating through the double O-rings seal was estimated to be <1 Pa. During atmospheric entry, the sample container was designed to experience less than 80 °C by using carbon fiber-reinforced plastic (CFRP) capsule ablators. The actual temperature from recovery on the Australian desert until introduction to the JAXA cleanroom was monitored with a temperature logger attached to a transportation box for the reentry capsule. The data of the logger showed that the sample container had been kept under 30 °C. The JAXA cleanrooms are maintained under 26 °C. The magnetic condition of Hayabusa-returned samples should have been disturbed during a return trip to Earth due to the Hayabusa ion engine operation. Additional electric disturbance and shock from atmospheric entry, landing, and transportation, which might affect the samples, are still poorly understood.

5.5.1 JAXA Hayabusa Lab

The JAXA Hayabusa curation laboratory (hereafter "curation lab") consists of four cleanrooms of different clean levels: a planetary sample handling room (ISO Class 5 to 6), an electron microscope room (ISO Class 6), a sample preparation room (ISO Class 6), and a manufacturing and cleaning room (ISO Class 7). These have vertical air flow from the ceiling into a raised, perforated stainless steel floor. All filters used in fan filter units are polytetrafluoroethylene (PTFE), and an additional chemical filter absorbs acid gases such as halogen, sulfate, nitrate, elemental boron, and borate. Design of this lab made effective use of decades of knowledge from NASA's curation labs, and then significantly improved upon them.

Four special-use rooms exhaust their air independently to outside of the cleanrooms to protect the other cleanrooms from chemical and particle contamination. Additionally, there is a basement for equipment that cannot be set in the cleanrooms, such as roughing pumps for vacuum systems, a compressed air supply system, an ultra-pure water supply system, and nitrogen purifiers.

The curation lab has two clean chambers for initial sample handling, Nos. 1 and 2. These are constructed mainly of 304 stainless steel, and their inside walls were electrochemically polished. They were baked in vacuum to at least 120 °C before and after the installation to reduce residual contamination. Both chambers are equipped with turbo molecular pumps (TMPs) and dry scroll pumps. Clean nitrogen supplied by a cyclic type nitrogen purifier and a flow type nitrogen purifier. The former is directly connected to each of the chambers to exclude H_2O, O_2, and hydrocarbon from the circulating nitrogen. The chambers operate at positive pressure to exclude ambient air. Chamber 1, where the Hayabusa capsule was initially opened, can be operated at conditions of ultrahigh vacuum or purified nitrogen. Residual gas expanding into the chamber from the container was collected in bottles made of stainless steel. The lower part of the sample container was maintained in cabinet No. 1 under a vacuum. Chamber No. 1 was also equipped with Viton gloves through gate valves permitting the sample container and catcher to be manipulated with special tools.

After opening, the sample catcher was sent to chamber 2 (in nitrogen) for the extraction of captured Itokawa grains. Both chambers 1 and 2 contain ultraviolet (UV) neutralization lamps to compensate electrostatic charge, which should occur in the pure nitrogen condition. Also, an alpha-ray neutralizer containing a grain of ^{210}Po is employed for the same purpose—use of this neutralizer was pioneered in NASA's Stardust and Cosmic Dust Labs. The clean chambers were constructed using stainless steel (304 and 316), aluminum, and A6061 aluminum alloy, quartz glass, PTFE, and Viton. Gold and copper are used for materials in Itokawa grain sample holders, borosilicate glass for containers of less important items, and polyetheretherketone (PEEK) for electric connectors.

5.5.2 Measurement of Gas in the Capture Cell upon Initial Opening

Residual gas sampling bottles connected to clean chamber 1 were prepared to capture gas released from the container at the time of its initial opening (on June 23, 2010, 13 days after capsule recovery). As it expected that the container could contain some terrestrial atmosphere, O_2 and ^{40}Ar could be used to identify any leaks into the "sealed" sample container.

Noble gases sampled in the gas bottles were analyzed at the University of Tokyo (Okazaki et al. 2011). Elemental ratios of the noble gases collected from the sample container were essentially identical to the terrestrial atmosphere. The inner pressure of the sample container was much higher than expected. Possible causes are a small leak of air through the double Viton O-rings seal, larger-than-expected permeability of the Viton O-rings, or a temporary leak of air accidentally happening during deintegration of the sample return capsule.

5.5.3 Sample Removal from the Container

On 24 June, 2010, the sample container was transferred to the transportation chamber from clean chamber 1 and the inner lid and the sample catcher were set into a catcher handling container. The inner lid, which was connected to the cover of sample catcher room A (there was also a room B), was removed, and its inner surface was observed and photographed. The catcher was placed into the catcher handling container, and these were transferred to clean chamber 2. The inner surface of catcher room A was observed in detail by optical microscopy, and very few particles larger than a few hundred micrometers were observed inside room A.

Initial attempts to remove Itokawa grains from the aluminum sample catcher were unsuccessful. Without exception, every suspected grain proved to be a small piece of protruding Al metal. Next, a special PTFE spatula was used to sweep the interior surface of catcher room A. Observation of the spatula in the FESEM showed the presence of hundreds of rocky particles 1–30 μm in size, half of which proved to be Itokawa regolith grains (based on EDX spectra, Nakamura et al. 2011). Finally, the inverted sample catcher rooms A and B were tapped with a screw driver, causing a rain of Itokawa and Al grains to fall onto specially-prepared quartz glass disks. The particles were moved from the quartz disks to copper SEM mounts for examination by FESEM-EDS at low kV and no conductive coating. It was thought that particles analyzed by FESEM might be contaminated by vacuum pump hydrocarbons. However, Naraoka et al. (2012) showed that no measurable hydrocarbons could be detected on identically-treated witness surfaces by time of flight-secondary ion mass spectrometry.

Because of differences in the requirements for proposed analyses, sample containers used in the initial analyses of Itokawa grains varied. The samples analyzed for the mainstream of initial analyses, including synchrotron X-ray computed tomography and diffraction, FE-SEM and FE electron microprobe analysis, and a secondary ion mass spectrometry were embedded in epoxy resin and mounted onto glass fibers. They were transported within a stainless steel container filled with nitrogen. Samples for transmission electron microscopy were mounted in epoxy resin in a special nitrogen glovebox and transported within the same containers. In this case, they were processed without exposure to air. Samples for noble gas analyses were set in holes in a stainless steel base in a special flange of stainless steel, which had been baked beforehand to decrease contamination. This process was performed in nitrogen. Samples for organic analyses and instrumental neutron activation analysis were set in holes in a diamond plate with a diamond cover. Each investigation required special sample handling and encapsulation procedures.

5.5.4 NASA's Hayabusa Curation Laboratory

Ultimately, 10% of the captured Itokawa grains will be transferred to NASA, and curated in the Hayabusa Curation Laboratory at JSC, although at the time of this writing less than 100 grains have been transferred to NASA's care. Sample containers for the NASA samples distribution consists of a pair of vacuum flanges of stainless steel as an outer container and a pair of synthetic quartz glass plates as a case to enclose the samples. All parts of the containers are separately cleaned by JAXA. The samples are placed into the quartz dimple slides using an electrostatically-controlled micromanipulator in clean chamber 2 at JAXA. The flanges are sealed with six screw bolts and oxygen-free copper gasket coated by gold. The sealing was also performed in the clean chamber, so the inside of the container was filled with atmospheric pressure nitrogen in the clean chamber. Only three materials are used for these sample containers: synthetic quartz, gold, and stainless steel. The NASA Hayabusa Cleanroom is a single room, containing at its core a stainless steel and glass cabinet for sample storage. No special sample handling is performed in this lab, as of now. Samples are merely stored, and allocated as needed.

5.6 OSIRIS-REx

Contamination knowledge for the OSIRIS-REx mission was covered in detail by Dworkin et al. (2018). This section will be a broad overview of the approaches and implementation of CK for OSIRIS-REx. Because curation scientists were involved in mission planning from the start, level 1, 2 and 3 mission requirements address aspects of CK and are integrated into the Contamination Control and Contamination Knowledge Plans. Because the mission is focused on amino acids and in general a carbonaceous asteroid, contamination efforts included attention to organics. An amino acid baseline of 180 ng/cm^2 for OSIRIS-REx was based partly on analysis of Stardust foils, most of which was from a known contaminant called epsilon amino caproci acid (EACA), which is derived from hydrolyzation of nylon (Elsila et al. 2009). The planning for CC/CK involved identification of restricted materials and assessment of hydrazine contamination (from monoprop thrusters), ATLO cleanroom and payload faring monitoring, coupon and material archiving, flight system witness plates, and sample container air filter system. All of these activities and categories have led to a detailed understanding of potential contaminants for the collected sample.

5.6.1 Materials Restrictions

The OSIRIS-REx team was already aware of several specific and classes of compounds that would need to be avoided or restricted due to contamination concerns such as nylon and organic polymers (e.g., silicones, lubricants, adhesives). However, through open communication channels with engineers, additional materials or components were identified in advance, allowing ample time for identification of substitutes. For example, one process required diamond abrasives while another used a coating that included amorphous silica; both nanodiamonds and amorphous silica may also be present in primitive asteroid materials. The diamond-abraded surface was cleaned and verified diamond-free at JSC via FTIR, and the silica-containing material was removed. Galling (i.e., wear resulting from adhesion between sliding surfaces) is a frequent problem in spacecraft assembly and can be mitigated using various lubricants. This unavoidable use of lubricants is an example where material archiving can be helpful, and indeed the Braycote lubricant was archived for every use on the spacecraft during ATLO. Open communication between the subdiscipline engineers also led to the chemical investigation of products whose chemical makeups were unclear and/or proprietary. Analyses at GSFC and JSC allowed materials of concern to be tested in more detail, including couplants and adhesives, in several cases helping to identify replacement products (Dworkin et al. 2018). Detailed reports of materials testing were shared with the mission contamination knowledge scientists and placed on the internal science team website for review.

5.6.2 Hydrazine

OSIRIS-REx thruster propellant, hydrazine, is known to react with organics via a Wolff-Kishner reduction (Dworkin et al. 2018). The mission team conducted tests of the reactivity of various organic compounds with anhydrous hydrazine, and decided that spacecraft thrusters should be canted away from the sampling site, which would result in <180 ng/cm^2 hydrazine to be deposited on TAGSAM surfaces. Even this hydrazine will rapidly evaporate from bare metal at sampling temperatures, but traces might be adsorbed by minerals or react with free carbonyls. In addition, drawing on experience from the Mars Phoenix lander mission and carrying out new calculations specific to OSIRIS-REx, mission engineers were able to estimate the amount of unreacted hydrazine in a thruster plume seeing the sample would be <120 ng/cm^2 for a single collection event. The only times when the spacecraft thrusters could deposit hydrazine onto the TAGSAM head are when the head is in the sampling configuration. This occurs during initial deployment and checkout, baseline sample-mass measurements, the TAG rehearsals, and the TAG event(s). All these considerations led to a much better understanding of potential for hydrazine interaction with the samples, and alleviated concerns for contamination.

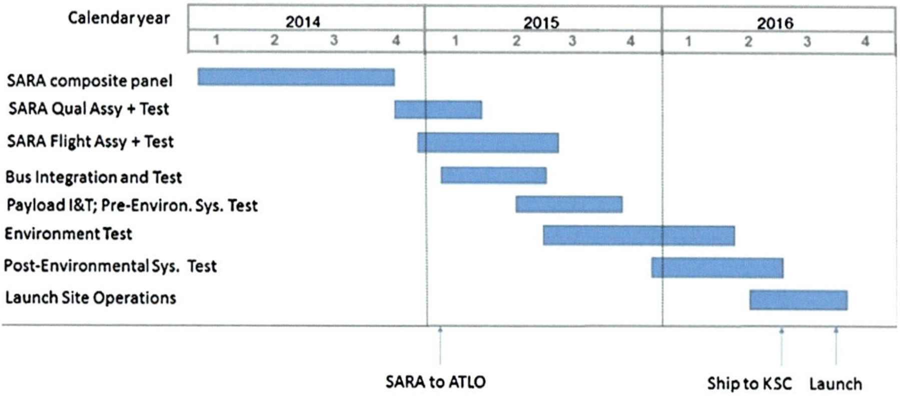

Fig. 13 Simplified schedule for SARA development, fabrication and assembly, and OSIRIS-REx Assembly, Testing and Launch Operations (ATLO)

5.6.3 Materials and Coupon Archiving

Archiving of materials identified to be of potential concern included those associated with the construction of the spacecraft, launch vehicle, SRC, TAGSAM (and associated hardware), science instruments, and materials used for packaging, containment, and processing of samples. During spacecraft assembly, the science and curation teams worked with the OSIRIS-REx mission engineers and ATLO personnel to archive materials from the spacecraft. Archiving began in February 2014, with work on the sample acquisition and retention assembly (SARA) composite panel in Denver, peaked near launch, and was completed by January 2017. Additionally, as the instruments were assembled and readied for integration, members of the instruments team packaged materials to send to JSC. As instruments and sub-assemblies of the spacecraft were tested and integrated, material coupons and items were sent to JSC through integration at KSC, with the last items having arrived at JSC in early 2017. A total of 406 items were received for the non-flight contamination knowledge collection. Data archived for each item also includes photos, its location on the spacecraft, physical description, the company that made the item along with its webpage or other contact information, the archiving location, archiver, and date. The materials fall into general categories including metals (stainless steel, aluminum, titanium alloys), epoxies, paints, polymers, lubricants, non-volatile-residue samples (NVR), sapphire, and various miscellaneous materials (a detailed list of items is found in Dworkin et al. 2018). The collection of CK materials, including witness plates, is archived and stored at NASA JSC in an ISO 7 cleanroom in dedicated stainless steel desiccator cabinets, with separately supplied dry nitrogen lines.

5.6.4 Monitoring ATLO Cleanrooms: LM, KSC, and Payload Faring

Witness plates were deployed to monitor cleanliness levels in the spacecraft assembly cleanrooms, high bay, environmental test facilities, spacecraft transport containers, Kennedy Space Center (KSC) Payload Hazardous Servicing Facility (PHSF), and the interior of the Atlas V launch vehicle fairing up until one day prior to launch of O-REx. All through the ATLO process (from March 2015 until late August 2016; Fig. 13) Si wafer and Al foil witness plates were deployed in these areas at LM and KSC to provide a record of particle counts and volatiles for current and future scientific studies. These plates were deployed in roughly monthly increments for 16 months with each unit containing 4 Si wafers and 4 Al foils, for a total of 128 individual witness plates (64 Si wafers and 64 Al foils). One of each witness plate (Si and Al) was analyzed immediately, while the remaining three are archived at JSC with the materials archive (described in previous section) for future analysis. In addition to the witness plates, gas samples were collected in selected ATLO locations such as cleanrooms and testing facilities, with the goal of identifying any unexpected or problematic species, although no unexpected species were identified.

5.6.5 *Flight Witness Plates and Air Filter CK*

To witness the environment experienced by the sample collection system on the spacecraft, a series of witness plates were designed and implemented in three different areas in the sample canister—the top of the TAGSAM head, the TAGSAM wrist joint assembly, and the inside of the sample canister. The aluminum and sapphire witness plates are designed to be deployed on TAGSAM and in canister recording three different exposure timeframe—always, pre-stow, and post-stow. These plates will be removed immediately upon return to JSC in the cleanroom and then stored in dedicated cabinets for contamination assessment studies. In addition to the witness plates, the sample canister has a two-way air filter in its lid which protects the sample from external contaminants, but the lid can also trap any volatile or particulate material from leaving the canister after TAGSAM is stowed. The filter performance was tested for moisture, particulate, and organic trapping efficiency during prelaunch activities and also drew heavily on similar filters used for the Stardust mission.

Additional contamination knowledge activities will continue during Phase E of the mission (Sept. 2016 to September 2023) with archiving of materials associated with the cleanroom construction at JSC and materials from the UTTR recovery site (might include soil, air, and other environmental background materials that could pose a contamination risk to OSIRIS-REx samples).

5.7 Mars 2020 Rover Mission

Although the Mars 2020 rover mission is not a sample return mission itself, it will be collecting and caching samples from the martian surface that could be picked up by a future mission, and hence Mars 2020 may represent the first mission in an overall Mars sample return campaign. Apart from in situ surface science, the goal of this first mission is to assemble a collection of rigorously documented and returnable cached samples. Contamination knowledge samples from Mars 2020 are in the process of being collected and curated to be part of the overall Mars sample collection if the cached samples are eventually returned.

While there are many aspects of the MSR contamination knowledge samples that will be similar to previous sample return missions, this sample return campaign has presented a number of unique opportunities and challenges. Perhaps the most notable is that due to the possibility of extinct or extant life on the surface of Mars, this sample return mission is designated as restricted by NASA's Planetary Protection Office. This designation requires more stringent organic contamination control requirements as well as the addition of biological contamination control and CK. Furthermore, this designation requires that the returned martian samples are curated in a containment facility in order to protect Earth from possible martian biohazards. The second notable difference is that the Mars 2020 mission would be part of a MSR campaign. Therefore, not only will the CK collection need to be coordinated between multiple missions, but significant point sources of possible contamination will remain on the martian surface. This mission architecture could make it more challenging to track a contaminant to its source. To minimize this knowledge gap, the CK samples collected will have to be more extensive than traditional sample return mission architectures. Finally, the Mars 2020 mission was not officially considered a sample return mission from its inception, so interaction with curation personnel and/or appropriate curation expertise was delayed and did not occur during the early stages of mission design, which has added an extra complication given the additional costs and time constraints of assembling a comprehensive CK collection.

5.7.1 Contamination Knowledge Samples

As with other sample return missions, an array of CK samples will be collected throughout ATLO. These samples will range from high fidelity flight reference materials to airfall witness samples. The flight and non-flight hardware and spacecraft components considered for CK are those items that are a potential contamination risk to the sample intimate hardware during launch, cruise, Mars entry/decent, landing, and rover surface operations. These "line-of-site" items are determined based on the Master Equipment List (MEL) and the Master Material List (MML). The MML will also be utilized to determine materials that have any potential to shed particulates and/or outgas molecular organics, directly or indirectly impacting the sample and caching subsystems. These samples can range from paint samples to flight spares or flight spare equivalents. Given the expanded scope of CK samples, an array of non-flight reference materials could help in tracing contamination. Some of these non-flight reference materials include: facility tools, equipment, and environmental components used during the fabrication, part processing, precision cleaning, and assembly that are considered a potential contamination risk. Witness items (e.g. plates and wipe samples) will be deployed within cleanrooms and on flow-benches during the assembly of the sample and caching subsystem. These witness items are duplicates of the witness items deployed and taken for CC and PP verification. All data collected during CC and PP verification will be tied to the respective CK sample, integrated into the CK database, and made available to the scientific community. Finally, unlike all other CK collections stored in nitrogen at ambient laboratory temperatures, the introduction of biological CK samples add the requirement to curate frozen samples (<-80 °C).

5.7.2 Innovation in Sample Storage

Due to the stringent organic, inorganic, and biological CC and PP requirements, new ways to secure and store CK samples were developed for Mars 2020. Due to organic contamination concerns, all storage bags utilized for organic CK cannot be heat-sealed, so we developed customized bag clips. These two piece bag clips are constructed from 300 series stainless steel and Teflon. These clips will provide a strong seal that will ensure sample safety during shipment and long-term curation, and they will be used for future missions.

For the long term storage of CK samples, a few different storage containers needed to be upgraded or designed from scratch. Due to size constraints, this most high fidelity sample intimate hardware will be stored in a custom 316 stainless steel bolt-top canister with a Teflon seal, plumbed to accommodate an inert atmosphere to ensure sample safety in case of a breach in the primary containment. However, the bulk of the inorganic and organic reference and witness items will be stored within highly customized desiccators. As with the bolt-top containers, this highly customized desiccator is a modified version of the desiccators utilized for the other Astromaterials collections. However, due to stringent contamination control, the construction materials are highly limited. For example, unlike other desiccators that can utilize traditional flexible and highly compressible multi-use gasket material, the organic contamination requirements for Mars 2020 preclude their use. Therefore, a new door gasket design was required. Due to its low outgassing properties, a Teflon gasket was the preferred material. However, the trade-off for low outgassing is low flexibility and low compressibility, which means a whole new door gasket design was required. The new design leverages the slight compressibility of the Teflon gasket material while also utilizing it as a barrier material between a possible contaminant and the samples.

The collection of Mars 2020 CK is an ongoing activity that will continue until the spacecraft is launched from Kennedy Space Center in July of 2020. Full details of the contamination knowledge collection and the laboratories that support these samples will be described in subsequent publications through joint efforts between the Mars 2020 science team and JSC curation personnel after a decision is made regarding the overall MSR campaign, which is expected sometime in 2020.

6 Preliminary Examination of Samples

The preliminary examination (PE) of returned samples, be they collected from an extraterrestrial body and returned by spacecraft or recovered from a frozen lake or someone's back yard on Earth, is arguably the first step in their curation. The importance of this step cannot be overstated with respect to either the early identification of the sample type or the careful preservation of planetary materials for the future. Preliminary examination affects both of these, and many activities in between. Simply defined, preliminary examination is the process by which returned samples are documented and characterized to the point that the appropriate scientific research community is provided with enough information to select and request the samples for their individual, PI-led scientific studies. The results of preliminary investigations are typically presented in catalogs or online databases that are publicly available, and preliminary examination is considered to be a science-enabling activity.

6.1 Steps Involved in Preliminary Examination

The very first steps of preliminary examination of samples may take place before they are even collected. For all returned samples, the astronauts and/or instruments onboard the spacecraft (orbiter, lander, or rover) will gather as much data as possible about the surface of the body from where the samples are gathered. These data will likely include photographs, spectroscopic data, and other possible measurements given the available instruments/crew. For samples collected on the ground, such as meteorites, ideally, a photograph including a scale bar with an indication of compass orientation, a general description of the sample (e.g., percentage of fusion crust, possible rock type, notable physical characteristics), and a description of the site where the sample was found or any other noteworthy features. Care should be taken during these steps to minimize exposure to contamination sources, and any potential contamination should be documented in the collection notes; a list of recommended procedures and materials for collection of freshly-fallen meteorites is provided in Herd et al. (2016). Photographs of sample containment vessels may occur during and after collection. At the very least, the type of sample containment/transport vessel should be documented, to inform future sampling and curation documentation as well as subsequent scientific analysis. In all cases, the utmost care should be taken to ensure curation best practices are implemented to minimize forward contamination.

Once samples are received in a curatorial facility, documentation of their current state should be made. These details may include information about the type of materials that the astromaterials samples were transported in, the state of those materials after landing (e.g., are seals intact? are they covered in dust?), and the documentation should include photographs. Any sampling of head gas in sample containment vessels that would be required should be completed and documented before sample containers are opened. Once the sample vessels are breached, information that may be of interest to research will be lost, along with the opportunity to ever gather it on those particular samples again. In addition, any tomographic scanning (e.g., X-ray or neutron) that is required to occur before samples are opened should be completed, and all processes involved in those analyses should be carefully documented.

Once samples are opened and, if applicable, removed from their sample containment vessels, the initial, curatorial steps involved in basic characterization of these materials take place. The main purpose of these efforts is to document exactly how the sample existed when it was opened. A documentation should be made as to the sample mass and its appearance at the time of opening. There should be a written description of the state of the material (intact rock, crumbled rock or sediments, powder, microscopic grains in a gel, etc.), its general attributes (color, grain size, physical appearance) and any notable features that are present (veins, fractures, fusion crust, metal content, etc.). Of utmost importance is photographic documentation (possibly with video, depending on the type of samples). Sample numbers should be assigned during this phase if they were not designated during collection. In short—the initial characterization of samples includes any process that can take place without making changes to the state of the samples other than opening their sample containment device, which is obviously unavoidable. Ideally, these steps should be made without touching the samples with anything other than curatorial tools comprised of materials determined to make contact with the samples without compromising their pristine nature. Determination of appropriate materials to use are based on several factors and typically represents a compromise between functionality and contamination risk, but foremost those materials must not compromise the ability of the samples to be used to answer the primary science requirements for a mission.

After basic characterization is completed, the curatorial phase of preliminary examination of the samples begins. Preliminary examination is distinct from science activities, and its goal is to produce a sample catalog with a level of detail about each sample that is sufficient for members of the scientific community to make informed requests of materials to conduct their PI-led scientific investigations. Preliminary examination of materials can occur for each representative portion of a sample, if needed to produce a meaningful and informative sample catalog. For Preliminary examinations on samples that also have a mission science team, preliminary examination can happen in parallel with science activities, but the goals of these two activities remains distinct. The methods and analytical techniques used during preliminary examination of a sample will be tailored to the primary science requirements for that sample. Furthermore, these processes will be determined based on sample size, sample form, sample vessel, and the need to prevent either forward or backward contamination. Large samples such as meteorites need to be touched with tools/gloves to be weighed, photographed, and described, and they need to be broken with tools to provide material for their classification. The next steps of preliminary examination of meteorites require the classification chip to be processed even further. As part of the U.S. Antarctic Meteorite Program (AMP) the smaller meteorite chip is weighed, placed into a sample container, and sent from NASA JSC to the Smithsonian National Museum of Natural History for further visual examination via binocular microscope, chipping, insertion into a sample holder grid, and polishing for energy dispersive spectroscopic (EDS) analyses, and/or made into a petrographic thin or thick section (both of which involve exposure to epoxy and polishing grit). These samples are coated with carbon and analyzed with a scanning electron microscope with EDS and/or an electron microprobe to determine their mineral compositions (namely olivine, pyroxene, and in the case of iron meteorites, FeNi metal). Other collections contain much smaller specimens and require much more careful micromanipulation of materials as discussed previously in the small particle handling section.

Preliminary examination of gas and volatile-rich samples present unique challenges to the curatorial preliminary examination process. In the gas phase, samples cannot be photo-documented or weighed. Condensed volatiles can be weighed, but only as a supplement to total or partial-pressure measurements of the quantity of sample in the gas phase. Whether the sample is condensed or not, total and partial pressures of major species (e.g. H_2O, CO_2, etc., depending on the sample origin) should be monitored using techniques that consume little to no sample. Spectroscopy-based techniques (e.g., FTIR, Raman, cavity ring-down spectroscopy) provide possible non-destructive means by which an initial characterization of the compounds in a sample can be ascertained; however, care must be taken to ensure that the techniques (and wavelengths) used do not pose a risk of altering the sample composition or isotopic distribution. High-sensitivity gas chromatography-mass spectrometry (GC-MS) or other high-sensitivity gas analysis techniques can be used to supplement spectroscopy using small quantities (<1 g) of sample.

The concept of a "representative sample" for allocation purposes requires further development for gas and volatile-rich samples. Depending on the temperature at which preliminary examination takes place, some species may be condensed while others remain in the gas phase. Additionally, thorough mixing of a gas-phase sample may be impossible to guarantee, especially while the sample remains sealed in the flight sample container. Therefore, when producing aliquots of gas samples for distribution to the scientific community, homogeneity between aliquots should not be expected. One possible solution is to separate gas-phase compounds by their condensation temperatures, freezing out compounds in a sequence that allows them to be separated by composition. This would limit (and possibly prohibit) bulk compositional analyses, but it would separate the sample into known compositions from which aliquots could be obtained. Regardless of how representative samples are defined, significant development is still required to address this capability gap for future volatile-rich sample return missions. An alternative to conducting preliminary examination of gas samples is to immediately conduct analysis of the gases for scientific purposes prior to preliminary examination. This option would be desirable in any instance where the primary science goals for a sample could be compromised through any preliminary examination processes.

Once the data are collected during preliminary examination, they are compiled and released to the scientific community in various online/digital formats or sample catalogs in order to allow researchers to request them for study. For U.S. meteorites, this is the Antarctic Meteorite Newsletter published in February and September each year (for example, https://curator.jsc.nasa.gov/antmet/amn/amn.cfm#n412). For other collections, different mechanisms of reporting of available materials are put in place. All NASA curated samples are detailed in the ARES website (https://curator.jsc.nasa.gov/), and new samples announced biennially in the Astormaterials Newsletter (https://curator.jsc.nasa.gov/newsletter/#n0101). And meteorites from around the world are detailed in the Meteoritical Bulletin (https://www.lpi.usra.edu/meteor/metbull.php).

Preliminary examination does not end with the process of reporting data in a newsletter. The curation process regularly involves the subsampling of materials. Each time this occurs, the subsampling process is ideally well documented with photographs, diagrams of subsampling, the weights of both removed and remaining materials, and descriptions of those materials if noteworthy. The process of subsampling larger samples involves exposing new material to the sample surface and to the curatorial environment. These newly exposed materials may require description, and in extreme cases, if they reveal something extraordinary, may necessitate announcement to the scientific community of new sample opportunities. Preliminary examination, unlike initial characterization, will go on as long as there is sample remaining in our collections, and hence sample catalogs are living documents.

6.2 Who Does PE?

Who should be involved in the preliminary examination of samples depends on the mechanism by which the samples were recovered. For spacecraft missions to planetary bodies within our Solar System, there are numerous stakeholders. These stakeholders include the mission science team that orchestrated and successfully executed the sample return mission, the scientific community at large that will also want to study the returned samples, and future generations of scientists that will want to study these samples with technology that has not yet been invented. It is the responsibility of a collection curator to defend all of these stakeholders and to find the right balance between sample consumption and sample conservation that maximizes science returns on the samples over multi-decade timescales. Given the important role of a collection curator for the safety and long-term viability of returned samples, it only makes sense for astromaterials returned from spacecraft missions to be received, opened, processed, and characterized within a sample curation facility by specially trained personnel under the management of a curatorial authority. That said, the people involved in the preliminary examination of returned materials will not be limited to curation personnel. The people involved in the preliminary examination should include some combination of curatorial processors trained to process and document miniscule samples, collection curators, and members of the sample science team. There may be a healthy tension that develops between the curator and members of a sample science team (i.e., conservation vs. consumption, respectively) because the science team has scientific mission requirements to achieve within a fixed period of time, and the curator must think beyond that time frame to the long-term viability and availability of the samples. To minimize such tensions, it is important to have policies in place prior to the samples being returned that outline how much sample can be consumed by the mission science team to achieve the primary science goals.

With less controlled sample collection, such as meteorite falls recovered anywhere in the world, the very first steps of preliminary examination (i.e., basic characterization) may take place in the field by trained meteorite hunting/recovery programs such as those run by the US (ANSMET), Japan, Belgium, China, South Korea, and the UK. The ANSMET program, for example, documents the field location of each meteorite recovered with a GPS position, a photograph next to a field number, the percentage of fusion crust visible on the sample, an educated guess at the meteorite type, a general description of any other notable features (i.e., it was found in liquid water, it was found half buried in ice, it was broken in half), and anything else that may require comment (i.e., it was accidentally touched with a glove, a bare hand, a snowmobile) and could possibly affect future analyses. Once the field documentation is completed and the samples are sent to a curation facility, preliminary examination largely occurs by curation personnel, and the samples are made available for request without first being analyzed by a mission science team.

6.3 Where do We Draw the Line Between Preliminary Examination for Curation and Science?

Basic characterization and preliminary examination are, as stated above, the processes by which returned samples are initially documented and sufficiently characterized to provide the appropriate scientific research community with enough information to select and request them for scientific study. A long standing discussion is carried on by curators worldwide as to what constitutes "too much characterization" and where the line is drawn between performing that characterization and conducting research that should be PI-led. The line is particularly difficult to delineate for small particle collections where the entire particle may be $< 10\ \mu m$ and almost any observation with an electron beam or laser could alter the sample (e.g., interplanetary dust particles in NASA's Cosmic Dust collection). However, there are also some examples of disagreement among curators as to which instrumentation and analyses are appropriate for preliminary examination on large samples where subsampling does not negatively impact the availability of the material (e.g., oxygen isotopic measurements, X-ray computed tomography, etc.). Nonetheless, the line between characterization and science is going to be different for each sample type, and that line should be optimally placed such that wasted sample consumption resulting from insufficient information about the samples in a catalog (i.e., the consumed sample did not have the phase of interest) is minimized whilst serendipitous discoveries within the samples during scientific investigations can still occur.

One of the questions meteorite curators commonly ask, for example, is whether or not oxygen isotopic compositions should be determined for meteorite samples returned from Antarctica or elsewhere. These measurements are very much in the realm of PI-led research, but with meteorites coming from a variety of different Solar System bodies, providing the $\Delta^{17}O$ composition gives the curator and requesting scientist the background information needed to identify potentially unique samples or those thought to be from Mars or other potentially unknown sources. Generally, isotope labs in the U.S. have provided the U.S. Antarctic Meteorite Program with these data as needed, but O-isotope analysis should only be used as a characterization tool when other, less destructive, methods cannot be used to uniquely classify the material. Another example is the use of X-ray computed tomography (XCT) as a characterization tool. Having three-dimensional context of samples provides curation personnel with invaluable information about the contents of samples, particularly highly heterogeneous samples like regolith breccias. This information can be used to identify clasts that are not exposed at the surface, and it can be used to make informed decisions about cuts or sample splits during sample processing. However, XCT exposes samples to radiation doses that can have lasting effects on the samples. Thermoluminescence in particular is negatively impacted by XCT analysis of samples (Sears et al. 2016, 2018), and studies are underway to better characterize its effect on organic compounds in samples (Hanna and Ketcham 2017; Friedrich et al. 2019).

Most curated astromaterials samples today are room-temperature solids; however, as we advance to collecting materials that require cryogenic storage or gas-phase samples, further questions arise as to how to conduct a preliminary examination, especially if the "shelf-life" of the samples prohibit waiting for the technology of tomorrow to analyze the samples. An alternative approach to a preliminary examination for materials that may have a short shelf-life is to organize a large scale consortium study where all stakeholders in the scientific community are given the opportunity to join, either through an "all are welcome" approach like Stardust or competitively through a proposal process if there is a need to limit the number of investigators. Conservation for the sake of conservation is not a meaningful philosophical approach to astromaterials curation. It is important that the goals of maximizing science returns on samples over their viable lifetimes are an integral part of a long-term sample conservation plan.

7 Looking Forward

Advanced curation is a critical function to the success of sample return missions and Earth-based sample acquisition and plays an integral part in enabling the high precision measurements that are often done on astromaterials samples. Looking forward, advanced curation must prepare for sample return missions from any celestial body within the solar system, including planets, moons, asteroids, and/or comets. The direction and scope of advanced curation research is driven by (1) existing strategic knowledge gaps identified through lessons learned from previous sample return missions and Earth-based programs that collect astromaterials; (2) the emerging needs of the scientific community that study astromaterials samples; and (3) the selection of new targets for sample return missions and the associated curation and sample handling requirements of those missions (e.g., Beaty et al. 2019; Haltigin et al. 2018; McLennan et al. 2011; Vander Kaaden et al. 2019). The primary result of advanced curation is to both reduce and quantify contamination to astromaterials and preserve the scientific integrity of all samples from mission inception and through ATLO, sample collection, curation/preliminary examination on Earth, curation/storage, and secure delivery of the samples to Earth-based laboratories for in-depth scientific analysis. Advanced curation is an interdisciplinary field of research and development and also serves as an important science-enabling activity. The collective lessons learned from previous spacecraft missions and the results of advanced curation research will work in tandem to feed forward into better spacecraft designs and enable more stringent requirements for future sample return missions.

Acknowledgements This work was supported, in part, by NASA's Science Mission Directorate. We are grateful for the editorial handling of this manuscript by Sara Russell, and this work was improved on the basis of input from the editor as well as two anonymous reviewers. This work is dedicated to the women and men that have worked tirelessly to make sample return missions possible, to the countless astromaterials curation personnel that have processed and cared for astromaterials collections, and to the members of the scientific community that study these carefully curated astromaterials.

References

P.I. Abell, C.H. Draffan, G. Eglinton, J.M. Hayes, J.R. Maxwell, C.T. Pillinger, Organic analysis of the returned Apollo 11 lunar sample, in *Apollo 11 Lunar Science Conference, Proc.*, vol. 2 (Pergamon, New York, 1970), pp. 1757–1773

C.M.O.D. Alexander, G.D. Cody, Y. Kebukawa, R. Bowden, M.L. Fogel, A.L.D. Kilcoyne, L.R. Nittler, C.D.K. Herd, Elemental, isotopic, and structural changes in Tagish Lake insoluble organic matter produced by parent body processes. Meteorit. Planet. Sci. **49**, 503–525 (2014)

C.C. Allen, F.G. Albert, J. Combie, A. Banin, Y. Yablekovitch, I. Kan, R.J. Bodnar, V.E. Hamilton, B.L. Jolliff, K. Kuebler, A. Wang, D.J. Lindstrom, P.A. Morris, R.V. Morris, R.W. Murray, L.E. Nyquist, P.D. Simpson, A. Steele, S.J. Symes, Effects of sterilizing doses of gamma radiation on Mars analog rocks and minerals. J. Geophys. Res., Planets **104**, 27043–27066 (1999)

C. Allen, J. Allton, G.E. Lofgren, K. Righter, M. Zolensky, Curating NASA's extraterrestrial samples—Past, present, and future. Chem. Erde **71**, 1–20 (2011)

J.H. Allton, J.D. Hittle, E.T. Mickelson, E.K. Stansbery, *Cleaning Genesis Sample Return Canister for Flight: Lessons for Planetary Sample Return* (NASA Johnson Space Center, Houston, 2016), p. 41

T.J. Anchordoquy, M.C. Molina, Preservation of DNA. Cell Preserv. Technol. **5**, 180–188 (2007)

L.D. Andrade, R. Awasthi, K. Dua, T.D.A. Pinto, Matrix-assisted laser desorption ionization-time of flight mass spectrometry for identification of bacteria isolated from pharmaceutical cleanrooms. Interv. Med. Appl. Sci. **10**, 45–53 (2018)

J.O. Annexstad, W.A. Cassidy, Collecting and processing Victoria Land meteorites, in *Memoirs of National Institute of Polar Research*, Special Issue (1980), pp. 14–20

L. Aronowitz, C. Baulknight, J. Berkowitz-Mattuck, A. Buchler, P. Glaser, F. Koch, T. Luzzi, N. Milford, S. Penn, F. Pomilla, Investigation of lunar surface chemical contamination by LEM descent engine and associated equipment Final report. Grumman Aircraft Engineering Corp.; Research Dept.; Bethpage, NY, United States (1966a) p. 219

L. Aronowitz, C. Baulknight, A. Buchler, F. Koch, T. Luzzi, S. Penn, A. Wechsler, D. Weiss, Investigation of Lunar Surface Chemical Contamination by LEM Descent Engine and Associated Equipment. Grumman Aircraft Engineering Corp.; Research Dept.; Bethpage, NY, United States (1966b), p. 55

M. Bashir, M. Ahmed, T. Weinmaier, D. Ciobanu, N. Ivanova, T.R. Pieber, P.A. Vaishampayan, Functional metagenomics of spacecraft assembly cleanrooms: Presence of virulence factors associated with human pathogens. Front. Microbiol. **7**, 12 (2016)

D. Beaty, C. Allen, D. Bass, K. Buxbaum, J. Cambell, D. Lindstrom, S. Miller, D. Papanastassiou, Planning considerations for a Mars sample receiving facility: Summary and interpretation of three design studies. Astrobiology **9** (2009). https://doi.org/10.1089/ast.2009.0339

D.W. Beaty, M.M. Grady, H.Y. McSween, E. Sefton-Nash, B.L. Carrier, F. Altieri, Y. Amelin, E. Ammannito, M. Anand, L.G. Benning, J.L. Bishop, L.E. Borg, D. Boucher, J.R. Brucato, H. Busemann, K.A. Campbell, A.D. Czaja, V. Debaille, D.J. Des Marais, M. Dixon, B.L. Ehlmann, J.D. Farmer, D.C. Fernandez-Remolar, J. Filiberto, J. Fogarty, D.P. Glavin, Y.S. Goreva, L.J. Hallis, A.D. Harrington, E.M. Hausrath, C.D.K. Herd, B. Horgan, M. Humayun, T. Kleine, J. Kleinhenz, R. Mackelprang, N. Mangold, L.E. Mayhew, J.T. McCoy, F.M. McCubbin, S.M. McLennan, D.E. Moser, F. Moynier, J.F. Mustard, P.B. Niles, G.G. Ori, F. Raulin, P. Rettberg, M.A. Rucker, N. Schmitz, S.P. Schwenzer, M.A. Sephton, R. Shaheen, Z.D. Sharp, D.L. Shuster, S. Siljeström, C.L. Smith, J.A. Spry, A. Steele, T.D. Swindle, I.L. ten Kate, N.J. Tosca, T. Usui, M.J. Van Kranendonk, M. Wadhwa, B.P. Weiss, S.C. Werner, F. Westall, R.M. Wheeler, J. Zipfel, M.P. Zorzano, The potential science and engineering value of samples delivered to Earth by Mars sample return. Meteorit. Planet. Sci. **54**, S3–S152 (2019)

H.A. Bechtel, C. Allen, S. Bajt, J. Borg, F. Brenker, J. Bridges, D.E. Brownlee, M. Burchell, M. Burghammer, A.L. Butterworth, P. Cloetens, A.M. Davis, C. Floss, G.J. Flynn, D. Frank, Z. Gainsforth, E. Grun, P.R. Heck, J.K. Hillier, P. Hoppe, L. Howard, G.R. Huss, J. Huth, A. Kearsley, A.J. King, B. Lai, J. Leitner, L. Lemelle, H. Leroux, L.R. Nittler, R.C. Ogliore, F. Postberg, M.C. Price, S.A. Sandford, J.A. Sans Tresseras, S. Schmitz, T. Schoonjans, G. Silversmit, A. Simionovici, R. Srama, F.J. Stadermann, T. Stephan, J. Stodolna, R.M. Stroud, S.R. Sutton, R. Toucoulou, M. Trieloff, P. Tsou, A. Tsuchiyama, T. Tyliczszak, B. Vekemans, L. Vincze, A.J. Westphal, M.E. Zolensky et al., FTIR analysis of aerogel keystones from the stardust interstellar dust collector: Assessment of terrestrial organic contamination and X-ray microprobe beam damage, in *42nd Lunar and Planetary Science Conference*, Abstract #1971 (2011)

J.N. Benardini, K. Venkateswaran, Application of the ATP assay to rapidly assess cleanliness of spacecraft surfaces: A path to set a standard for future missions. AMB Express **6**, 8 (2016)

A. Bieler, K. Altwegg, H. Balsiger, A. Bar-Nun, J.J. Berthelier, P. Bochsler, C. Briois, U. Calmonte, M. Combi, J. De Keyser, E.F. van Dishoeck, B. Fiethe, S.A. Fuselier, S. Gasc, T.I. Gombosi, K.C. Hansen, M. Hässig, A. Jäckel, E. Kopp, A. Korth, L. Le Roy, U. Mall, R. Maggiolo, B. Marty, O. Mousis, T. Owen, H. Rème, M. Rubin, T. Sémon, C.Y. Tzou, J.H. Waite, C. Walsh, P. Wurz, Abundant molecular oxygen in the coma of comet 67P/Churyumov–Gerasimenko. Nature **526**, 678 (2015)

P.A. Bland, The Desert Fireball Network. Astron. Geophys. **45**, 5.20–5.23 (2004)

P.A. Bland, P. Spurny, A.W.R. Bevan, K.T. Howard, M.C. Towner, G.K. Benedix, R.C. Greenwood, L. Shrbeny, I.A. Franchi, G. Deacon, J. Borovicka, Z. Ceplecha, D. Vaughan, R.M. Hough, The Australian Desert Fireball Network: A new era for planetary science. Aust. J. Earth Sci. **59**, 177–187 (2012)

A.I. Blinova, C.D.K. Herd, M.J.M. Duke, Testing variations within the Tagish Lake meteorite—II: Whole-rock geochemistry of pristine samples. Meteorit. Planet. Sci. **49**, 1100–1118 (2014)

J. Borovička, Z. Charvát, Meteosat observation of the atmospheric entry of 2008 TC over Sudan and the associated dust cloud. Astron. Astrophys. **507**, 1015–1022 (2009)

F. Brandstätter, History of the meteorite collection of the Natural History Museum of Vienna. Geol. Soc. (Lond.) Spec. Publ. **256**, 123–133 (2006)

C.E. Bresky, M. Fries, The Aquarius project: The first student-driven attempt to retrieve meteorites from underwater, in *49th Lunar and Planetary Science Conference*, Abstract #3004 (2018)

D. Brownlee, P. Tsou, J. Aleon, C.M.O. Alexander, T. Araki, S. Bajt, G.A. Baratta, R. Bastien, P. Bland, P. Bleuet, J. Borg, J.P. Bradley, A. Brearley, F. Brenker, S. Brennan, J.C. Bridges, N.D. Browning, J.R. Brucato, E. Bullock, M.J. Burchell, H. Busemann, A. Butterworth, M. Chaussidon, A. Cheuvront, M.F. Chi, M.J. Cintala, B.C. Clark, S.J. Clemett, G. Cody, L. Colangeli, G. Cooper, P. Cordier, C. Daghlian, Z.R. Dai, L. D'Hendecourt, Z. Djouadi, G. Dominguez, T. Duxbury, J.P. Dworkin, D.S. Ebel, T.E. Economou, S. Fakra, S.A.J. Fairey, S. Fallon, G. Ferrini, T. Ferroir, H. Fleckenstein, C. Floss, G. Flynn, I.A. Franchi, M. Fries, Z. Gainsforth, J.P. Gallien, M. Genge, M.K. Gilles, P. Gillet, J. Gilmour, D.P. Glavin, M. Gounelle, M.M. Grady, G.A. Graham, P.G. Grant, S.F. Green, F. Grossemy, L. Grossman, J.N. Grossman, Y. Guan, K. Hagiya, R. Harvey, P. Heck, G.F. Herzog, P. Hoppe, F. Horz, J. Huth, I.D. Hutcheon, K. Ignatyev, H. Ishii, M. Ito, D. Jacob, C. Jacobsen, S. Jacobsen, S. Jones, D. Joswiak, A. Jurewicz, A.T. Kearsley, L.P. Keller, H. Khodja, A.L.D. Kilcoyne, J. Kissel, A. Krot, F. Langenhorst, A. Lanzirotti, L. Le, L.A. Leshin, J. Leitner, L. Lemelle, H. Leroux, M.C. Liu, K. Luening, I. Lyon, G. MacPherson, M.A. Marcus, K. Marhas, B. Marty, G. Matrajt, K. McKeegan, A. Meibom, V. Mennella, K. Messenger, S. Messenger, T. Mikouchi, S. Mostefaoui, T. Nakamura, T. Nakano, M. Newville, L.R. Nittler, I. Ohnishi, K. Ohsumi, K. Okudaira, D.A. Papanastassiou, R. Palma, M.E. Palumbo, R.O. Pepin, D. Perkins, M. Perronnet, P. Pianetta, W. Rao, F.J.M. Rietmeijer, F. Robert, D. Rost, A. Rotundi, R. Ryan, S.A. Sandford, C.S. Schwandt, T.H. See, D. Schlutter, J. Sheffield-Parker, A. Simionovici, S. Simon, I. Sitnitsky, C.J. Snead, M.K. Spencer, F.J. Stadermann, A. Steele, T. Stephan, R. Stroud, J. Susini, S.R. Sutton, Y. Suzuki, M. Taheri, S. Taylor, N. Teslich, K. Tomeoka, N. Tomioka, A. Toppani, J.M. Trigo-Rodriguez, D. Troadec, A. Tsuchiyama, A.J. Tuzzolino, T. Tyliszczak, K. Uesugi, M. Velbel, J. Vellenga, E. Vicenzi, L. Vincze, J. Warren, I. Weber, M. Weisberg, A.J. Westphal, S. Wirick, D. Wooden, B. Wopenka, P. Wozniakiewicz, I. Wright, H. Yabuta, H. Yano, E.D. Young, R.N. Zare, T. Zega, K. Ziegler, L. Zimmerman, E. Zinner, M. Zolensky, Comet 81P/Wild 2 under a microscope. Science **314**, 1711–1716 (2006)

A.L. Burlingame, M. Calvin, J. Han, W. Henderson, W. Reed, B.R. Simoneit, Study of carbon compounds in Apollo 11 lunar samples, in *Apollo 11 Lunar Science Conference, Proc.*, vol. 2 (Pergamon, New York, 1970), pp. 1779–1791

A.L. Burlingame, P.T. Holland, W.H. McFadden, B.R. Simoneit, J.T. Wilder, P.C. Wszolek UCB Space Sciences Laboratory Organic Cleanroom and Lunar Material Transfer Facilities. Space Sciences Laboratory Report, University of California, Berkeley (1971)

D.S. Burnett, The Genesis solar wind sample return mission: Past, present, and future. Meteorit. Planet. Sci. **48**, 2351–2370 (2013)

D.S. Burnett, B.L. Barraclough, R. Bennett, M. Neugebauer, L.P. Oldham, C.N. Sasaki, D. Sevilla, N. Smith, E. Stansbery, D. Sweetnam, R.C. Wiens, The Genesis Discovery mission: Return of solar matter to Earth. Space Sci. Rev. **105**, 509–534 (2003)

A.S. Burton, D.P. Glavin, J.E. Elsila, J.P. Dworkin, P. Jenniskens, Q.Z. Yin, The amino acid composition of the Sutter's Mill CM2 carbonaceous chondrite. Meteorit. Planet. Sci. **49**, 2074–2086 (2014)

CABRI, Laboratory procedures for microorganisms: Survey of examples of detailed protocols for the different types of preservation methods, in *Common Access to Biotechnological Resources and Information Consortium* (1998) http://www.cabri.org/guidelines/micro-organisms/M300Ap500.html. Accessed April 24, 2019

M.J. Calaway, Lunar processing cabinet 2.0: Retrofitting gloveboxes into the 21st century, in *Lunar and Planetary Science Conference XLVI*, March 16–20, The Woodlands, TX, Abstract #1492 (2015)

M.J. Calaway, C.C. Allen, Cryogenic curation: Isolated technology and mission operational requirements for sample return, in *76th Annual Meteoritical Society Meeting*, Edmonton Alberta, Canada (2013)

M.J. Calaway, D.S. Burnett, M.C. Rodriguez, S. Sestak, J.H. Allton, E.K. Stansbery, Decontamination of genesis array materials by UV ozone cleaning, in *Lunar and Planetary Science Conference XXXVIII*, Houston, TX, Abstract #1627 (2007)

M.J. Calaway, M.C. Rodriguez, J.H. Allton, E.K. Stansbery, Decontaminating solar wind samples with the genesis ultra-pure water megasonic cleaner, in *Lunar and Planetary Science Conference XL*, Woodlands, TX, Abstract #1183 (2009)

M.J. Calaway, C.C. Allen, J.H. Allton, Organic Contamination Baseline Study in NASA Johnson Space Center Astromaterials Curation Laboratories, NASA TP-2014-217393, Lyndon B. Johnson Space Center, Houston (2014), p. 108

M.J. Calaway, J.H. Allton, R.A. Zeigler, F.M. McCubbin, 50th anniversary of the world's first extraterrestrial sample receiving laboratory: The Apollo program's lunar receiving laboratory, in *48th Lunar and Planetary Science Conference*, Abstract #1224 (2017)

Z. Ceplecha, J. Borovicka, W.G. Elford, D.O. Revelle, R.L. Hawkes, V. Porubcan, M. Simek, Meteor phenomena and bodies. Space Sci. Rev. **84**, 327–471 (1998)

B.C. Clark, Temperature–time issues in bioburden control for planetary protection. Adv. Space Res. **34**, 2314–2319 (2004)

A. Colaprete, P. Schultz, J. Heldmann, D. Wooden, M. Shirley, K. Ennico, B. Hermalyn, W. Marshall, A. Ricco, R.C. Elphic, D. Goldstein, D. Summy, G.D. Bart, E. Asphaug, D. Korycansky, D. Landis, L. Sollitt, Detection of water in the LCROSS ejecta plume. Science **330**, 463–468 (2010)

F. Colas, B. Zanda, J. Vaubaillon, S. Bouley, C. Marmo, Y. Audureau, M.K. Kwon, J.-L. Rault, S. Caminade, P. Vernazza, French fireball network FRIPON, in *Proceedings of the International Meteor Conference*, Mistelbach, Austria (2015), pp. 27–30

W. Cooke, D. Moser, The status of the NASA all sky fireball network, NASA report, in *Proceedings of the International Meteor Conference, 30th IMC*, Sibiu, Romania (2011)

M. Cooper, M.T. La Duc, A. Probst, P. Vaishampayan, C. Stam, J.N. Benardini, Y.M. Piceno, G.L. Andersen, K. Venkateswaran, Comparison of innovative molecular approaches and standard spore assays for assessment of surface cleanliness. Appl. Environ. Microbiol. **77**, 5438–5444 (2011)

COSPAR, COSPAR planetary protection policy (20 October 2002, as amended to 24 March 2011), in *COSPAR/IAU Workshop on Planetary Protection, COSPAR* (2011)

L.A. Dauphin, B.D. Moser, M.D. Bowen, Evaluation of five commercial nucleic acid extraction kits for their ability to inactivate Bacillus anthracis spores and comparison of DNA yields from spores and spiked environmental samples. J. Microbiol. Methods **76**, 30–37 (2009)

B.H. Day, P.A. Bland, R. Sayers, Fireballs in the sky: Citizen science with the desert fireball network, in *49th Lunar and Planetary Science Conference*, Abstract #2229 (2018a)

J.M.D. Day, J. Maria-Benavides, F.M. McCubbin, R.A. Zeigler, The potential for metal contamination during Apollo lunar sample curation. Meteorit. Planet. Sci. **53**, 1283–1291 (2018b)

S.F. Dermott, J.C. Liou, Detection of asteroidal dust particles from known families in near-Earth orbits, in *AIP Conference Proceedings, vol. 3, no. 1* (AIP, New York, 1994)

J.P. Dworkin, L.A. Adelman, T. Ajluni, A.V. Andronikov, J.C. Aponte, A.E. Bartels, E. Beshore, E.B. Bierhaus, J.R. Brucato, B.H. Bryan, A.S. Burton, M.P. Callahan, S.L. Castro-Wallace, B.C. Clark, S.J. Clemett, H.C. Connolly, W.E. Cutlip, S.M. Daly, V.E. Elliott, J.E. Elsila, H.L. Enos, D.F. Everett, I.A. Franchi, D.P. Glavin, H.V. Graham, J.E. Hendershot, J.W. Harris, S.L. Hill, A.R. Hildebrand, G.O. Jayne, R.W. Jenkens, K.S. Johnson, J.S. Kirsch, D.S. Lauretta, A.S. Lewis, J.J. Loiacono, C.C. Lorentson, J.R. Marshall, M.G. Martin, L.L. Matthias, H.L. McLain, S.R. Messenger, R.G. Mink, J.L. Moore, K. Nakamura-Messenger, J.A. Nuth, C.V. Owens, C.L. Parish, B.D. Perkins, M.S. Pryzby, C.A. Reigle, K. Righter, B. Rizk, J.F. Russell, S.A. Sandford, J.P. Schepis, J. Songer, M.F. Sovinski, S.E. Stahl, K. Thomas-Keprta, J.M. Vellinga, M.S. Walker, OSIRIS-REx contamination control strategy and implementation. Space Sci. Rev. **214**, 53 (2018)

ECSS, Space Product Assurance: Microbial Examination of Flight Hardware and Cleanrooms. European Cooperation for Space Standardization, ECSS-Q-ST-70-55C, Noordwijk, The Netherlands (2008)

W.N. Edwards, D.W. Eaton, P.G. Brown, Seismic observations of meteors: Coupling theory and observations. Rev. Geophys. **46**, 21 (2008)

J.E. Elsila, D.P. Glavin, J.P. Dworkin, Cometary glycine detected in samples returned by Stardust. Meteorit. Planet. Sci. **44**, 1323–1330 (2009)

D.A. Flory, B.R. Simoneit, Terrestrial contamination in Apollo lunar samples. Space Life Sci. **3**, 457–468 (1972)

G.J. Flynn, L.P. Keller, C. Jacobsen, S. Wirick, An assessment of the amount and types of organic matter contributed to the Earth by interplanetary dust, in *Space Life Sciences: Steps Toward Origin(S) of Life*, ed. by M.P. Bernstein, M. Kress, R. NavarroGonzalez (Pergamon, Kidlington, 2004), pp. 57–66

J.M. Friedrich, H.L. McLain, J.P. Dworkin, D.P. Glavin, W.H. Towbin, M. Hill, D.S. Ebel, Effect of polychromatic X-ray microtomography imaging on the amino acid content of the Murchison CM chondrite. Meteorit. Planet. Sci. **54**, 220–228 (2019)

M. Fries, J. Fries, Doppler weather radar as a meteorite recovery tool. Meteorit. Planet. Sci. **45**, 1476–1487 (2010)

M. Fries, C. Laird, M. Hankey, J. Fries, R. Matson, V. Reddy, Estimation of meteorite fall mass and other properties from weather radar data, in *80th Annual Meeting of the Meteoritical Society*, Abstract #6251 (2017)

C.W. Gehrke, R.W. Zumwalt, W.A. Aue, D.L. Stalling, A. Duffield, K.A. Kvenvolden, C. Ponnamperuma, Carbon compounds in lunar fines from Mare Tranquillitatis; III, Organosiloxanes in hydrochloric acid hydrolysates, in *Apollo 11 Lunar Science Conference, Proc.*, vol. 2 (Pergamon, New York, 1970), pp. 1845–1856

C.W. Gehrke, R.W. Zumwalt, K. Kuo, W.A. Aue, D.L. Stalling, K.A. Kvenvolden, C. Ponnamperuma, Amino acid analyses of Apollo 14 samples, in *Lunar Science Conference, 3rd, Proc.*, vol. 3 (Pergamon, Oxford, 1972), pp. 2119–2129

G.R. Gladstone, D.M. Hurley, K.D. Retherford, P.D. Feldman, W.R. Pryor, J.-Y. Chaufray, M. Versteeg, T.K. Greathouse, A.J. Steffl, H. Throop, J.W. Parker, D.E. Kaufmann, A.F. Egan, M.W. Davis, D.C. Slater, J. Mukherjee, P.F. Miles, A.R. Hendrix, A. Colaprete, S.A. Stern, LRO-LAMP observations of the LCROSS impact plume. Science **330**, 472–476 (2010)

M.M. Grady, Meteorites from cold and hot deserts: How many, how big, and what sort, in *Workshop on Extraterrestrial Materials from Cold and Hot Deserts. Lunar and Planetary Institute, Houston, Workshop on Extraterrestrial Materials from Cold and Hot Deserts*, ed. by L.F.I.A. Schultz, A.M. Reid, M.E. Zolensky (2000), pp. 36–40

M.M. Grady, A.B. Verchovsky, I.A. Franchi, I.P. Wright, C.T. Pillinger, Light element geochemistry of the Tagish Lake CI2 chondrite; comparison with CI1 and CM2 meteorites. Meteorit. Planet. Sci. **37**, 713–735 (2002)

M. Grimaldo, Decontamination with cold plasma activated ionized hydrogen peroxide: Does it behave like a gas? in *60th Annual Biological Safety Conference, ABSA 1717B* (2017)

M. Gritsevich, E. Lyytinen, J. Moilanen, T. Kohout, V. Dmitriev, V. Lupovka, V. Midtskogen, N. Kruglikov, A. Ischenko, G. Yakovlev, V. Grokhovsky, J. Haloda, P. Halodova, J. Peltoniemi, A. Aikkila, A. Taavitsainen, J. Lauanne, M. Pekkola, P. Kokko, P. Lahtinen, M. Larionov, First meteorite recovery based on observations by the Finnish Fireball Network, in *Proceedings of the International Meteor Conference*, Giron, France, 18–21 September 2014 (2014), pp. 162–169

E. Grosjean, G.A. Logan, Incorporation of organic contaminants into geochemical samples and an assessment of potential sources: Examples from Geoscience Australia marine survey S282. Org. Geochem. **38**, 853–869 (2007)

J. Guardiola, V. Rojo, G. Ramos, Influence of particle size, fluidization velocity and relative humidity on fluidized bed electrostatics. J. Electrost. **37**, 1–20 (1995)

C.W. Haberle, L.A.J. Garvie, Extraterrestrial formation of oldhamite and portlandite through thermal metamorphism of calcite in the Sutter's Mill carbonaceous chondrite. Am. Mineral. **102**, 2415–2421 (2017)

I. Halliday, A.T. Blackwell, A.A. Griffin, The Innisfree meteorite and the Canadian camera network. J. R. Astron. Soc. Can. **72**, 15–39 (1978)

M. Hallworth, Rapid microbiological monitoring in pharmaceutical environments, in *Environmental Monitoring: A Comprehensive Handbook*, vol. 6, ed. by J. Moldenhaer (PDA, Bethesda, 2012), p. 93

T. Haltigin, C. Lange, R. Mugnolo, C. Smith (co-chairs), H. Amundsen, P. Bousquet, C. Conley, A. Debus, J. Dias, P. Falkner, V. Gass, A-M. Harri, E. Hauber, A.B. Ivanov, A.O. Ivanov, G. Kminek, O. Korablev, D. Koschny, J. Larranaga, B. Marty, S. McLennan, M. Meyer, E. Nilsen, P. Orleanski, R. Orosei, D. Rebuffat, F. Safa, N. Schmitz, S. Siljeström, N. Thomas, J. Vago, A-C. Vandaele, T. Voirin, C. Whetsel, Astrobiology (2018). https://doi.org/10.1089/ast.2018.29027.mars

R.D. Hanna, R.A. Ketcham, X-ray computed tomography of planetary materials: A primer and review of recent studies. Chem. Erde **77**, 547–572 (2017)

P.E. Hare, K. Harada, S.W. Fox, Analyses for amino acids in lunar fines, in *Apollo 11 Lunar Science Conference, Proc.*, vol. 2 (Pergamon, New York, 1970), pp. 1799–1803

R. Harvey, The origin and significance of Antarctic meteorites. Chem. Erde **63**, 93–147 (2003)

W. Henderson, W.C. Kray, W.A. Newman, W.E. Reed, B.R. Simoneit, M. Calvin, Study of carbon compounds in Apollo 11 and Apollo 12 returned lunar samples, in *Lunar Science Conference, 2nd, Proc.*, vol. 2 (Pergamon, Oxford, 1971), pp. 1901–1912

C.D.K. Herd, A. Blinova, D.N. Simkus, Y. Huang, R. Tarozo, C.M.O. Alexander, F. Gyngard, L.R. Nittler, G.D. Cody, M.L. Fogel, Y. Kebukawa, A.L.D. Kilcoyne, R.W. Hilts, G.F. Slater, D.P. Glavin, J.P. Dworkin, M.P. Callahan, J.E. Elsila, B. De Gregorio, R.M. Stroud, Origin and evolution of prebiotic organic matter as inferred from the Tagish Lake meteorite. Science **332**, 1304–1307 (2011)

C.D.K. Herd, R.W. Hilts, A.W. Skelhorne, D.N. Simkus, Cold curation of pristine astromaterials: Insights from the Tagish Lake meteorite. Meteorit. Planet. Sci. **51**, 499–519 (2016)

R.W. Hilts, C.D.K. Herd, D.N. Simkus, G.F. Slater, Soluble organic compounds in the Tagish Lake meteorite. Meteorit. Planet. Sci. **49**, 526–549 (2014)

K.B. Hindley, M.A. Houlden, The British Fireball Network. Meteoritics **12**, 257 (1977)

G.W. Hodgson, E. Bunnenberg, B. Halpern, E. Peterson, K.A. Kvenvolden, C. Ponnamperuma, Carbon compounds in lunar fines from Mare Tranquillitatis; II, Search for porphyrins, in *Apollo 11 Lunar Science Conference, Proc.*, vol. 2 (Pergamon, New York, 1970), pp. 1829–1844

G.W. Hodgson, E. Bunnenberg, B. Halpern, E. Peterson, K.A. Kvenvolden, C. Ponnamperuma, Lunar pigments; porphyrin-like compounds from an Apollo 12 sample, in *Lunar Science Conference, 2nd, Proc.*, vol. 2 (Pergamon, Oxford, 1971), pp. 1865–1874

J.M.C. Holt, J.C. Bridges, J. Vrublevskis, F. Gaubert, Double walled isolator technology for Mars sample return facilities, in *Proceedings of the 50th Lunar and Planetary Science Conference*, Woodlands, TX, #2408 (2019)
B. Hubad, A. Lapanje, The efficient method for simultaneous monitoring of the culturable as well as nonculturable airborne microorganisms. PLoS ONE **8**, 9 (2013)
L.A. Hug, B.J. Baker, K. Anantharaman, C.T. Brown, A.J. Probst, C.J. Castelle, C.N. Butterfield, A.W. Hernsdorf, Y. Amano, K. Ise, Y. Suzuki, N. Dudek, D.A. Relman, K.M. Finstad, R. Amundson, B.C. Thomas, J.F. Banfield, A new view of the tree of life. Nat. Microbiol. **1**, 6 (2016)
IEST-STD-CC1246E, Product cleanliness levels—Applications, requirements, and determination. Institute of Environmental Sciences and Technology, Schaumburg, IL, USA (2013)
ISO14698, Cleanrooms and associated controlled environments—Biocontamination control—Part 1: General Principles and Methods, ISO 14698-1:2003 (2003)
A. Jambon, Bronze Age iron: Meteoritic or not? A chemical strategy. J. Archaeol. Sci. **88**, 47–53 (2017)
P. Jenniskens, M.D. Fries, Q.Z. Yin, M. Zolensky, A.N. Krot, S.A. Sandford, D. Sears, R. Beauford, D.S. Ebel, J.M. Friedrich, K. Nagashima, J. Wimpenny, A. Yamakawa, K. Nishiizumi, Y. Hamajima, M.W. Caffee, K.C. Welten, M. Laubenstein, A.M. Davis, S.B. Simon, P.R. Heck, E.D. Young, I.E. Kohl, M.H. Thiemens, M.H. Nunn, T. Mikouchi, K. Hagiya, K. Ohsumi, T.A. Cahill, J.A. Lawton, D. Barnes, A. Steele, P. Rochette, K.L. Verosub, J. Gattacceca, G. Cooper, D.P. Glavin, A.S. Burton, J.P. Dworkin, J.E. Elsila, S. Pizzarello, R. Ogliore, P. Schmitt-Kopplin, M. Harir, N. Hertkorn, A. Verchovsky, M. Grady, K. Nagao, R. Okazaki, H. Takechi, T. Hiroi, K. Smith, E.A. Silber, P.G. Brown, J. Albers, D. Klotz, M. Hankey, R. Matson, J.A. Fries, R.J. Walker, I. Puchtel, C.T.A. Lee, M.E. Erdman, G.R. Eppich, S. Roeske, Z. Gabelica, M. Lerche, M. Nuevo, B. Girten, S.P. Worden (Sutter's Mill Meteorite C.), Radar-enabled recovery of the Sutter's Mill meteorite, a carbonaceous chondrite regolith breccia. Science **338**, 1583–1587 (2012)
P. Jenniskens, J. Albers, C.E. Tillier, S.F. Edgington, R.S. Longenbaugh, S.J. Goodman, S.D. Rudlosky, A.R. Hildebrand, L. Hanton, F. Ciceri, R. Nowell, E. Lyytinen, D. Hladiuk, D. Free, N. Moskovitz, L. Bright, C.O. Johnston, E. Stern, Detection of meteoroid impacts by the Geostationary Lightning Mapper on the GOES-16 satellite. Meteorit. Planet. Sci. **53**, 2445–2469 (2018)
L. Jimenez, Molecular applications to pharmaceutical processes and cleanroom environments. PDA J. Pharm. Sci. Technol. **65**, 242–253 (2011)
A.J.G. Jurewicz, D.S. Burnett, R.C. Wiens, T.A. Friedmann, C.C. Hays, R.J. Hohlfelder, K. Nishiizumi, J.A. Stone, D.S. Woolum, R. Becker, A.L. Butterworth, A.J. Campbell, M. Ebihara, I.A. Franchi, V. Heber, C.M. Hohenberg, M. Humayun, K.D. McKeegan, K. McNamara, A. Meshik, R.O. Pepin, D. Schlutter, R. Wieler, The Genesis solar-wind collector materials. Space Sci. Rev. **105**, 535–560 (2003)
Y. Kebukawa, S. Nakashima, T. Otsuka, K. Nakamura-Messenger, M.E. Zolensky, Rapid contamination during storage of carbonaceous chondrites prepared for micro FTIR measurements. Meteorit. Planet. Sci. **44**, 545–557 (2009)
C.G. Keller, R.T. Howe, Hexsil tweezers for teleoperated micro-assembly, in *Proceedings IEEE the Tenth Annual International Workshop on Micro Electro Mechanical Systems. An Investigation of Micro Structures, Sensors, Actuators, Machines and Robots* (IEEE, Nagoya, 1997), pp. 72–77
W. Kinard, R. O'Neal, Long Duration Exposure Facility (LDEF) results, in *29th Aerospace Sciences Meeting*, Abstract #1971 (American Institute of Aeronautics and Astronautics, Washington, 1991)
B.V. King, I.V. Veryovkin, A.V. Zinovev, C.E. Tripa, M.J. Pellin, N. Toyoda, M. Schmeling, Ion beam removal of surface contamination in Genesis samples, in *Lunar and Planetary Science Conference XXXXI*, Abstract #1975 (2010)
G.I. Kokhirova, J. Borovička, Observations of the 2009 Leonid activity by the Tajikistan fireball network. Astron. Astrophys. **533**, 6 (2011)
K.R. Kuhlman, M.C. Rodriquez, C.P. Gonzalez, J.H. Allton, D.S. Burnett, Cleaning study of genesis sample 60487, in *Lunar and Planetary Science Conference XXXXIV*, Abstract #2930 (2013)
K.A. Kvenvolden, S. Chang, J.W. Smith, J. Flores, K. Pering, C. Saxinger, F. Woeller, K. Keil, I. Breger, C. Ponnamperuma, Carbon compounds in lunar fines from Mare Tranquillitatis; I, Search for molecules of biological significance, in *Apollo 11 Lunar Science Conference, Proc.*, vol. 2 (Pergamon, New York, 1970), pp. 1813–1828
K. Kwan, M. Cooper, M.T. La Duc, P. Vaishampayan, C. Stam, J.N. Benardini, G. Scalzi, C. Moissl-Eichinger, K. Venkateswaran, Evaluation of procedures for the collection, processing, and analysis of biomolecules from low-biomass surfaces. Appl. Environ. Microbiol. **77**, 2943–2953 (2011)
M.T. La Duc, A. Dekas, S. Osman, C. Moissl, D. Newcombe, K. Venkateswaran, Isolation and characterization of bacteria capable of tolerating the extreme conditions of cleanroom environments. Appl. Environ. Microbiol. **73**, 2600–2611 (2007)
M.T. La Duc, S. Osman, P. Vaishampayan, Y. Piceno, G. Andersen, J.A. Spry, K. Venkateswaran, Comprehensive census of bacteria in cleanrooms by using DNA microarray and cloning methods. Appl. Environ. Microbiol. **75**, 6559–6567 (2009)

M.T. La Duc, K. Venkateswaran, C.A. Conley, A genetic inventory of spacecraft and associated surfaces. Astrobiology **14**, 15–23 (2014)

D.S. Lauretta, S.S. Balram-Knutson, E. Beshore, W.V. Boynton, C. Drouet d'Aubigny, D.N. DellaGiustina, H.L. Enos, D.R. Golish, C.W. Hergenrother, E.S. Howell, C.A. Bennett, E.T. Morton, M.C. Nolan, B. Rizk, H.L. Roper, A.E. Bartels, B.J. Bos, J.P. Dworkin, D.E. Highsmith, D.A. Lorenz, L.F. Lim, R. Mink, M.C. Moreau, J.A. Nuth, D.C. Reuter, A.A. Simon, E.B. Bierhaus, B.H. Bryan, R. Ballouz, O.S. Barnouin, R.P. Binzel, W.F. Bottke, V.E. Hamilton, K.J. Walsh, S.R. Chesley, P.R. Christensen, B.E. Clark, H.C. Connolly, M.K. Crombie, M.G. Daly, J.P. Emery, T.J. McCoy, J.W. McMahon, D.J. Scheeres, S. Messenger, K. Nakamura-Messenger, K. Righter, S.A. Sandford, OSIRIS-REx: Sample return from Asteroid (101955) Bennu. Space Sci. Rev. **212**, 925–984 (2017)

L. Le Roy, K. Altwegg, H. Balsiger, J.-J. Berthelier, A. Bieler, C. Briois, U. Calmonte, M.R. Combi, J. De Keyser, F. Dhooghe, B. Fiethe, S.A. Fuselier, S. Gasc, T.I. Gombosi, M. Hässig, A. Jäckel, M. Rubin, C.-Y. Tzou, Inventory of the volatiles on comet 67P/Churyumov-Gerasimenko from Rosetta/ROSINA. Astron. Astrophys. **583**, A1 (2015)

S.R. Lipsky, R.J. Cushley, C.G. Horvath, W.J. McMurray, Analysis of lunar material for organic compounds, in *Apollo 11 Lunar Science Conference, Proc.*, vol. 2 (Pergamon, New York, 1970), pp. 1871–1873

J.T. Lonsdale, The Pena Blanca Spring meteorite, Brewster County, Texas. Am. Mineral. **32**, 354–364 (1947)

M.D.J. Lynch, J.D. Neufeld, Ecology and exploration of the rare biosphere. Nat. Rev. Microbiol. **13**, 217–229 (2015)

A. Mahnert, P. Vaishampayan, A.J. Probst, A. Auerbach, C. Moissl-Eichinger, K. Venkateswaran, G. Berg, Cleanroom maintenance significantly reduces abundance but not diversity of indoor microbiomes. PLoS ONE **10**, 20 (2015)

P.M. Martin, A.A. Mills, Size and shape of chondrules in the Bjurböle and Chainpur meteorites. Earth Planet. Sci. Lett. **33**, 239–248 (1976)

U.B. Marvin, Meteorites in history: An overview from the Renaissance to the 20th century. Geol. Soc. (Lond.) Spec. Publ. **256**, 15–71 (2006)

S. Matsusaka, H. Maruyama, T. Matsuyama, M. Ghadiri, Triboelectric charging of powders: A review. Chem. Eng. Sci. **65**, 5781–5807 (2010)

G.J.H. McCall, A.J. Bowden, R.J. Howarth, The history of meteoritics—Overview. Geol. Soc. (Lond.) Spec. Publ. **256**, 1–13 (2006)

T.J. McCoy, Hopewell meteoritic metal beads: Clues to trade 2,000 years ago. Elements **14**, 360–361 (2018)

T.J. McCoy, A.E. Marquardt, J.T. Wasson, R.D. Ash, E.P. Vicenzi, The Anoka, Minnesota iron meteorite as parent to Hopewell meteoritic metal beads from Havana, Illinois. J. Archaeol. Sci. **81**, 13–22 (2017)

J.C. McLane, E.A. King, D.A. Flory, K.A. Richardson, J.P. Dawson, W.W. Kemmerer, Lunar receiving laboratory. Science **155**, 525–529 (1967)

S.M. McLennan, M.A. Sephton, C. Allen, A.C. Allwood, R. Barbieri, D.W. Beaty, P. Boston, M. Carr, M. Grady, J. Grant, V.S. Heber, C.D.K. Herd, B. Hofmann, P. King, N. Mangold, G.G. Ori, A.P. Rossi, F. Raulin, S.W. Ruff, B. Sherwood Lollar, S. Symes, M.G. Wilson, Planning for Mars returned sample science: Final report of the MSR End-to-End International Science Analysis Group (E2E-iSAG). Astrobiology **12**, 175–230 (2011)

K.M. McNamara, E.K. Stansbery, Analysis of molecular contamination on genesis collectors through spectroscopic ellipsometry, in *36th Lunar and Planetary Science Conference*, Abstract #2402 (2005)

R.M. Mecikalski, L.D. Carey, Radar reflectivity and altitude distributions of lightning flashes as a function of three main storm types. J. Geophys. Res., Atmos. **123**, 12814–12828 (2018)

W.G. Meinschein, T.J. Jackson, J.M. Mitchell, E. Cordes, V.J. Shiner Jr., Search for alkanes of 15–30 carbon atom length in lunar fines, in *Apollo 11 Lunar Science Conference, Proc.*, vol. 2 (Pergamon, New York, 1970), pp. 1875–1877

S. Messenger, Opportunities for the stratospheric collection of dust from short-period comets. Meteorit. Planet. Sci. **37**, 1491–1505 (2002)

S. Messenger, K. Nakamura-Messenger, L.P. Keller, S.J. Clemett, Pristine stratospheric collection of interplanetary dust on an oil-free polyurethane foam substrate. Meteorit. Planet. Sci. **50**, 1468–1485 (2015)

E.T. Mickelson, Cleaning and Cleanliness Verification Techniques for Mars Returned Sample Handling. An MRSH Document, NASA JSC 29738 (2002a)

E.T. Mickelson, Molecular Contamination Control: A Unified Cleaning and Verification Strategy for a Sample Receiving Facility. An MRSH Document, NASA JSC 29689 (2002b)

MIL-STD-1246C, Military Standard: Product Cleanliness Levels and Contamination Control Program. Department of Defense (1994)

S.D. Miller, W.C. Straka, A.S. Bachmeier, T.J. Schmit, P.T. Partain, Y.J. Noh, Earth-viewing satellite perspectives on the Chelyabinsk meteor event. Proc. Natl. Acad. Sci. **110**, 18092–18097 (2013)

J.J. Minich, Q.Y. Zhu, S. Janssen, R. Hendrickson, A. Amir, R. Vetter, J. Hyde, M.M. Doty, K. Stillwell, J. Benardini, J.H. Kim, E.E. Allen, K. Venkateswaran, R. Knight, KatharoSeq enables high-throughput microbiome analysis from low-biomass samples. mSystems **3**, 16 (2018)

K.R. Mitchell, C.D. Takacs-Vesbach, A comparison of methods for total community DNA preservation and extraction from various thermal environments. J. Ind. Microbiol. Biotech. **35**, 1139–1147 (2008)

C. Moissl, J.C. Bruckner, K. Venkateswaran, Archaeal diversity analysis of spacecraft assembly cleanrooms. ISME J. **2**, 115–119 (2008)

C. Moissl-Eichinger, Archaea in artificial environments: Their presence in global spacecraft cleanrooms and impact on planetary protection. ISME J. **5**, 209–219 (2011)

C. Moissl-Eichinger, A.K. Auerbach, A.J. Probst, A. Mahnert, L. Tom, Y. Piceno, G.L. Andersen, K. Venkateswaran, P. Rettberg, S. Barczyk, R. Pukall, G. Berg, Quo vadis? Microbial profiling revealed strong effects of cleanroom maintenance and routes of contamination in indoor environments. Sci. Rep. **5**, 13 (2015)

M. Mora, A. Mahnert, K. Koskinen, M.R. Pausan, L. Oberauner-Wappis, R. Krause, A.K. Perras, G. Gorkiewicz, G. Berg, C. Moissi-Eichinger, Microorganisms in confined habitats: Microbial monitoring and control of intensive care units, operating rooms, cleanrooms and the international space station. Front. Microbiol. **7**, 20 (2016)

M.E. Murphy, V.E. Modzeleski, B. Nagy, W.M. Scott, M. Young, C.M. Drew, P.B. Hamilton, H.C. Urey, Analysis of Apollo 11 lunar samples by chromatography and mass spectrometry; pyrolysis products, hydrocarbons, sulfur, amino acids, in *Apollo 11 Lunar Science Conference, Proc.*, vol. 2 (Pergamon, New York, 1970), pp. 1879–1890

T. Nakamura, T. Noguchi, A. Tsuchiyama, H. Yurimoto, K. Nagao, M. Ebihara, R. Okazaki, F. Kitajima, H. Naraoka, M. Abe, A. Fujimura, T. Yada, T. Okada, Y. Ishibashi, K. Shirai, T. Mukai, H. Yano, T. Yamada, H. Kuninaka, M. Yoshikawa, J. Kawaguchi, The first recovery of asteroidal samples by the Hayabusa mission, in *AGU Fall Meeting Abstracts* (2011)

K. Nakamura-Messenger, S. Messenger, L.P. Keller, S.J. Clemett, M.E. Zolensky, Organic globules in the Tagish Lake meteorite: Remnants of the protosolar disk. Science **314**, 1439–1442 (2006)

H. Naraoka, H. Mita, K. Hamase, M. Mita, H. Yabuta, K. Saito, K. Fukushima, F. Kitajima, S.A. Sandford, T. Nakamura, T. Noguchi, R. Okazaki, K. Nagao, M. Ebihara, H. Yurimoto, A. Tsuchiyama, M. Abe, K. Shirai, M. Ueno, T. Yada, Y. Ishibashi, T. Okada, A. Fujimura, T. Mukai, M. Yoshikawa, J. Kawaguchi, Preliminary organic compound analysis of microparticles returned from Asteroid 25143 Itokawa by the Hayabusa mission. Geochem. J. **46**, 61–72 (2012)

NASA, in *Summer Conference on Lunar Exploration and Science*, Falmouth, Massachusetts, July 19–31, 1965 (Scientific and Technical Information Division, National Aeronautics and Space Administration, Washington, 1965), 421 pp.

NASA, in *Handbook for the Microbial Examination of Space Hardware. NASA Technical Handbook, NASA-HDBK-6022* (2010), 52 pp.

NASA, in *Planetary Protection Provisions for Robotic Extraterrestrial Missions, NASA Interim Directive, NID 8020.109A* (2017), 49 pp.

N.D. Novikova, Review of the knowledge of microbial contamination of the Russian manned spacecraft. Microb. Ecol. **47**, 127–132 (2004)

J. Oberst, S. Molau, D. Heinlein, C. Gritzner, M. Schindler, P. Spurny, Z. Ceplecha, J. Rendtel, H. Betlem, The "European Fireball Network": Current status and future prospects. Meteorit. Planet. Sci. **33**, 49–56 (1998)

R. Okazaki, K. Nagao, Y.N. Miura, T. Osawa, K. Bajo, S. Matsuda, T. Nakamura, K. Shirai, M. Abe, T. Yada, T. Noguchi, Y. Ishibashi, A. Fujimura, T. Mukai, M. Ueno, T. Okada, M. Yoshikawa, J. Kawaguchi, Noble gases recovered from the Hayabusa sample container, in *Lunar and Planetary Science Conference* (2011) p. 1653

R. Okazaki, H. Sawada, S. Yamanouchi, S. Tachibana, Y. Miura, K. Sakamoto, Y. Takano, M. Abe, S. Itoh, K. Yamada, H. Yabuta, C. Okamoto, H. Yano, T. Noguchi, T. Nakamura, K. Nagao, Hayabusa2 sample catcher and container: Metal-seal system for vacuum encapsulation of returned samples with volatiles and organic compounds recovered from C-type asteroid Ryugu. Space Sci. Rev. **208**, 107–124 (2017)

J. Oro, W.S. Updegrove, J. Gibert, J. McReynolds, E. Gil-Av, J. Ibanez, A. Zlatkis, D.A. Flory, R.L. Levy, C.J. Wolf, Organogenic elements and compounds in type C and D lunar samples from Apollo 11, in *Apollo 11 Lunar Science Conference, Proc.*, vol. 2 (Pergamon, New York, 1970), pp. 1901–1920

V.I. Oyama, E.L. Merek, M.P. Silverman, A search for viable organisms in a lunar sample, in *Apollo 11 Lunar Science Conference, Proc.*, vol. 2 (Pergamon, New York, 1970), pp. 1921–1927

V.I. Oyama, E.L. Merek, M.P. Silverman, C.W. Boylen, Search for viable organisms in lunar samples; further biological studies on Apollo 11 core, Apollo 12 bulk, and Apollo 12 core samples, in *Lunar Science Conference, 2nd, Proc.*, vol. 2 (Pergamon, Oxford, 1971), pp. 1931–1937

D. Papanastassiou, J.M.D. Day, D.P. Glavin, G.R. Huss, R.L. Korotev, L.E. Nyquist, M. Wadhwa, T.J. Zega, Lunar Curation Task Force Report. NASA report to the Curation and Analysis Planning Team for Extraterrestrial Materials (2015), p. 14

K.E. Peters, C.C. Walters, J.M. Moldowan, *The Biomarker Guide: Volume 1. Biomarkers and Isotopes in the Environment and Human History* (Cambridge University Press, Cambridge, 2005), 492 pp.

O.P. Popova, P. Jenniskens, V. Emel'yanenko, A. Kartashova, E. Biryukov, S. Khaibrakhmanov, V. Shuvalov, Y. Rybnov, A. Dudorov, V.I. Grokhovsky, D.D. Badyukov, Q.Z. Yin, P.S. Gural, J. Albers, M. Granvik, L.G. Evers, J. Kuiper, V. Kharlamov, A. Solovyov, Y.S. Rusakov, S. Korotkiy, I. Serdyuk, A.V. Korochantsev, M.Y. Larionov, D. Glazachev, A.E. Mayer, G. Gisler, S.V. Gladkovsky, J. Wimpenny, M.E. Sanborn, A. Yamakawa, K.L. Verosub, D.J. Rowland, S. Roeske, N.W. Botto, J.M. Friedrich, M.E. Zolensky, L. Le, D. Ross, K. Ziegler, T. Nakamura, I. Ahn, J.I. Lee, Q. Zhou, X.H. Li, Q.L. Li, Y. Liu, G.Q. Tang, T. Hiroi, D. Sears, I.A. Weinstein, A.S. Vokhmintsev, A.V. Ishchenko, P. Schmitt-Kopplin, N. Hertkorn, K. Nagao, M.K. Haba, M. Komatsu, T. Mikouchi, C. Chelyabinsk Airburst, Chelyabinsk airburst, damage assessment, meteorite recovery, and characterization. Science **342**, 1069–1073 (2013)

G. Preti, R.C. Murphy, K. Biemann, The search for organic compounds in various Apollo 12 samples by mass spectrometry, in *Lunar Science Conference, 2nd, Proc.*, vol. 2 (Pergamon, Oxford, 1971), pp. 1879–1889

M. Prinz, K. Keil, P.F. Hlava, J.L. Berkley, C.B. Gomes, W.S. Curvello, Studies of Brazilian Meteorites: 3. Origin and history of Angra-Dos-Reis achondrite. Earth Planet. Sci. Lett. **35**, 317–330 (1977)

A. Probst, P. Vaishampayan, S. Osman, C. Moissl-Eichinger, G.L. Andersen, K. Venkateswaran, Diversity of anaerobic microbes in spacecraft assembly cleanrooms. Appl. Environ. Microbiol. **76**, 2837–2845 (2010)

M. Ramstorp, *Introduction to Contamination Control and Cleanroom Technology* (Wiley, New York, 2000)

M.S. Rappe, S.J. Giovannoni, The uncultured microbial majority. Annu. Rev. Microbiol. **57**, 369–394 (2003)

J.S. Raval, E. Koch, A.D. Donnenberg, Real-time monitoring of non-viable airborne particles correlates with airborne colonies and represents an acceptable surrogate for daily assessment of cell-processing cleanroom performance. Cytotherapy **14**, 1144–1150 (2012)

J.A. Reuter, D.V. Spacek, M.P. Snyder, High-throughput sequencing technologies. Mol. Cell **58**, 586–597 (2015)

M.A. Reynolds, N.L. Turner, J.C. Hurgeton, M.F. Barbee, D.A. Flory, B.R. Simoneit, Environmental control of lunar samples in the lunar receiving laboratory, in *Analytical Methods Developed for Application to Lunar Samples Analyses, American Society for Testing and Materials, ASTM STP*, vol. 539 (1973), pp. 3–15

J.H. Rho, A.J. Bauman, T.F. Yen, J. Bonner, Fluorometric examination of the returned lunar fines from Apollo 11, in *Apollo 11 Lunar Science Conference, Proc.*, vol. 2 (Pergamon, New York, 1970), pp. 1929–1932

J.H. Rho, A.J. Bauman, J. Bonner, Absence of porphyrins in an Apollo 12 lunar surface sample, in *Lunar Science Conference, 2nd, Proc.*, vol. 2 (Pergamon, Oxford, 1971), pp. 1875–1877

J.H. Rho, E.A. Cohen, A.J. Bauman, Spectrofluorometric search for porphyrins in Apollo 14 surface fines, in *Lunar Science Conference, 3rd, Proc.*, vol. 3 (Pergamon, Oxford, 1972), pp. 2149–2155

K. Righter, C.M. Corrigan, T.J. McCoy, R.P. Harvey, *35 Seasons of US Antarctic Meteorites* (Am. Geophys. Union, Washington, 2014)

A.J. Rissanen, E. Kurhela, T. Aho, T. Oittinen, M. Tiirola, Storage of environmental samples for guaranteeing nucleic acid yields for molecular microbiological studies. Appl. Microbiol. Biotechnol. **88**, 977–984 (2010)

B. Robert, A low-power micromanipulator and microdissector. J. Sci. Instrum. **28**, 65 (1951)

H.L. Rose, C.A. Dewey, M.S. Ely, S.L. Willoughby, T.M. Parsons, V. Cox, P.M. Spencer, S.A. Weller, Comparison of eight methods for the extraction of bacillus atrophaeus spore DNA from eleven common interferents and a common swab. PLoS ONE **6**, e22668 (2011)

J. Rummel, M. Race, D. DeVincenzi, P. Schad, P. Stabekis, M. Viso, S. Acevedo, A Draft Test Protocol for Detecting Possible Biohazards in Martian Samples Returned to Earth (2002). NASA technical publication 211842

V.E. Sakraida, Cleanroom Design in 10 Easy Steps, Engineered Systems Magazine, Troy, MI (2008)

S.A. Sandford, D.E. Brownlee, Response to comment on "Organics captured from Comet 81P/Wild 2 by the Stardust spacecraft". Science **317**, 1 (2007)

S.A. Sandford, J. Aleon, C.M.O. Alexander, T. Araki, S. Bajt, G.A. Baratta, J. Borg, J.P. Bradley, D.E. Brownlee, J.R. Brucato, M.J. Burchell, H. Busemann, A. Butterworth, S.J. Clemett, G. Cody, L. Colangeli, G. Cooper, L. D'Hendecourt, Z. Djouadi, J.P. Dworkin, G. Ferrini, H. Fleckenstein, G.J. Flynn, I.A. Franchi, M. Fries, M.K. Gilles, D.P. Glavin, M. Gounelle, F. Grossemy, C. Jacobsen, L.P. Keller, A.L.D. Kilcoyne, J. Leitner, G. Matrajt, A. Meibom, V. Mennella, S. Mostefaoui, L.R. Nittler, M.E. Palumbo, D.A. Papanastassiou, F. Robert, A. Rotundi, C.J. Snead, M.K. Spencer, F.J. Stadermann, A. Steele, T. Stephan, P. Tsou, T. Tyliszczak, A.J. Westphal, S. Wirick, B. Wopenka, H. Yabuta, R.N. Zare, M.E. Zolensky, Organics captured from comet 81P/Wild 2 by the Stardust spacecraft. Science **314**, 1720–1724 (2006)

S.A. Sandford, S. Bajt, S.J. Clemett, G.D. Cody, G. Cooper, B.T. Degregorio, V. de Vera, J.P. Dworkin, J.E. Elsila, G.J. Flynn, D.P. Glavin, A. Lanzirotti, T. Limero, M.P. Martin, C.J. Snead, M.K. Spencer, T. Stephan, A. Westphal, S. Wirick, R.N. Zare, M.E. Zolensky, Assessment and control of organic and other contaminants associated with the Stardust sample return from comet 81P/Wild 2. Meteorit. Planet. Sci. **45**, 406–433 (2010)
T. Sandle, Environmental monitoring: A practical approach, in *Environmental Monitoring: A Comprehensive Handbook*, vol. 6 (2012), pp. 29–54
T. Sandle, K. Skinner, Study of psychrophilic and psychrotolerant micro-organisms isolated in cold rooms used for pharmaceutical processing. J. Appl. Microbiol. **114**, 1166–1174 (2013)
E.K. Sansom, P. Bland, J. Paxman, M. Towner, A novel approach to fireball modeling: The observable and the calculated. Meteorit. Planet. Sci. **50**, 1423–1435 (2015)
H. Sawada, R. Okazaki, S. Tachibana, K. Sakamoto, Y. Takano, C. Okamoto, H. Yano, Y. Miura, M. Abe, S. Hasegawa, T. Noguchi, Hayabusa2 sampler: Collection of asteroidal surface material. Space Sci. Rev. **208**, 81–106 (2017)
M. Schmeling, D.S. Burnett, J.H. Allton, M. Rodriquez, C.E. Tripa, I.V. Veryovkin, Application of CO2 snow jet cleaning in conjunction with laboratory based total reflection X-ray fluorescence, in *Lunar and Planetary Science Conference XXXXIV*, Abstract #2465 (2013)
P. Schwendner, C. Moissl-Eichinger, S. Barczyk, M. Bohmeier, R. Pukall, P. Rettberg, Insights into the microbial diversity and bioburden in a South American spacecraft assembly cleanroom. Astrobiology **13**, 1140–1154 (2013)
D.W.G. Sears, H. Sears, D.S. Ebel, S. Wallace, J.M. Friedrich, X-ray computed tomography imaging: A not-so-nondestructive technique. Meteorit. Planet. Sci. **51**, 833–838 (2016)
D.W.G. Sears, A. Sehlke, J.M. Friedrich, M.L. Rivers, D.S. Ebel, X-ray computed tomography of extraterrestrial rocks eradicates their natural radiation record and the information it contains. Meteorit. Planet. Sci. **53**, 2624–2631 (2018)
L.S. Sherman, J.R. Waldbauer, R.E. Summons, Improved methods for isolating and validating indigenous biomarkers in Precambrian rocks. Org. Geochem. **38**, 1987–2000 (2007)
Y. Shiba, C. Shimoda, T. Maruyama, S. Okumura, M. Tomita, A. Murasawa, K. Ohtsuka, H. Tomioka, E. Hidaka, Photographic observations of the 1996 Leonid fireballs in Japan. Earth Moon Planets **77**, 47–54 (1997)
S. Shokralla, J.L. Spall, J.F. Gibson, M. Hajibabaei, Next-generation sequencing technologies for environmental DNA research. Mol. Ecol. **21**, 1794–1805 (2012)
B.R. Simoneit, D.A. Flory, Apollo 11, 12, and 13 Organic Contamination Monitoring History. Lunar and Earth Sciences Division Internal Note, NASA Document MSC-04350 (1971)
B.R. Simoneit, D.A. Flory, M.A. Reynolds, Organic contamination monitoring and control in the lunar receiving laboratory, in *Analytical Methods Developed for Application to Lunar Samples Analyses, American Society for Testing and Materials, ASTM STP*, vol. 539 (1973), pp. 16–34
M.K. Spencer, R.N. Zare, Comment on "Organics Captured from Comet 81P/Wild 2 by the Stardust Spacecraft". Science **317**, 1680 (2007)
M.K. Spencer, S.J. Clemett, S.A. Sandford, D.S. McKay, R.N. Zare, Organic compound alteration during hypervelocity collection of carbonaceous materials in aerogel. Meteorit. Planet. Sci. **44**, 15–24 (2009)
E.K. Stansbery, K.M. McNamara, Genesis preliminary examination—Ellipsometry overview, in *42nd Lunar and Planetary Science Conference*, Abstract #2145 (2005)
E.K. Stansbery, G.R.P. Team, Genesis recovery processing, in *36th Lunar and Planetary Science Conference*, Abstract #2179 (2005)
A. Steele, F.M. McCubbin, M. Fries, M. Glamoclija, L. Kater, H. Nekvasil, Graphite in an Apollo 17 impact melt breccia. Science **329**, 51 (2010)
D. Stuart, D. Eagleson, R. Lloyd, C. Hersey, D. Eagleson, Analysis of the class III biological safety cabinet integrity test. Appl. Biosafety **17**, 128–131 (2012)
G.M. Sullivan, D.I. Klebe, Searching for fireballs, in *American Astronomical Society Meeting*, Abstracts #204 (2004)
G.R. Taylor, W. Ellis, P.H. Johnson, K. Kropp, T. Groves, Microbial assay of lunar samples, in *Lunar Science Conference, 2nd, Proc.*, vol. 2 (Pergamon, Oxford, 1971), pp. 1939–1948
M. Tessler, J.S. Neumann, E. Afshinnekoo, M. Pineda, R. Hersch, L.F.M. Velho, B.T. Segovia, F.A. Lansac-Toha, M. Lemke, R. DeSalle, C.E. Mason, M.R. Brugler, Large-scale differences in microbial biodiversity discovery between 16S amplicon and shotgun sequencing. Sci. Rep. **7**, 14 (2017)
S.M. Tikoo, B.P. Weiss, W.S. Cassata, D.L. Shuster, J. Gattacceca, E.A. Lima, C. Suavet, F. Nimmo, M.D. Fuller, Decline of the lunar core dynamo. Earth Planet. Sci. Lett. **404**, 89–97 (2014)
J. Toporski, A. Steele, Observations from a 4-year contamination study of a sample depth profile through the martian meteorite Nakhla. Astrobiology **7**, 389–401 (2007)

J. Tremblay, K. Singh, A. Fern, E.S. Kirton, S.M. He, T. Woyke, J. Lee, F. Chen, J.L. Dangl, S.G. Tringe, Primer and platform effects on 16S rRNA tag sequencing. Front. Microbiol. **6** (2015)

J.M. Trigo-Rodriguez, J. Llorca, A.J. Castro-Tirado, J.L. Ortiz, J.A. Docobo, J. Fabregat, The Spanish fireball network. Astron. Geophys. **47**, 26–28 (2006)

P. Tsou, D.E. Brownlee, S.A. Sandford, F. Horz, M.E. Zolensky, Wild 2 and interstellar sample collection and Earth return. J. Geophys. Res., Planets **108**, 21 (2003)

USP, Microbiological Evaluation of Cleanrooms and Other Controlled Environments. USP (United States Pharmacopeia) General Chapter <1116>, in USP 36 (United States Pharmacopeia Convention, North Bethesda, 2013) (2013), http://ftp.uspbpep.com/v29240/usp29nf24s0_c1116.html. Accessed 24 April 2019

P. Vaishampayan, A.J. Probst, M.T. La Duc, E. Bargoma, J.N. Benardini, G.L. Andersen, K. Venkateswaran, New perspectives on viable microbial communities in low-biomass cleanroom environments. ISME J. **7**, 312–324 (2013)

K.E. Vander Kaaden, F.M. McCubbin, P.K. Byrne, N.L. Chabot, C.M. Ernst, C.L. Johnson, M.S. Thompson, Revolutionizing our understanding of the Solar System via sample return from Mercury. Space Sci. Rev. (2019, in press). https://doi.org/10.1007/s11214-019-0614-x

D. Vaniman, B.M. French, H.H. Schmitt, G.H. Heiken, *Lunar Sourcebook; A User's Guide to the Moon* (Cambridge University Press, Cambridge, 1991)

J.B. Vaught, M.K. Henderson, in *Biological Sample Collection, Processing, Storage and Information Management*. IARC Sci. Publ., vol. 163 (2011), pp. 23–42

K. Venkateswaran, N. Hattori, M.T. La Duc, R. Kern, ATP as a biomarker of viable microorganisms in clean-room facilities. J. Microbiol. Methods **52**, 367–377 (2003)

D. Venton, Inner workings: Networks of cameras are tracking meteorites with unprecedented precision. Proc. Natl. Acad. Sci. **114**, 7472 (2017)

T. Větrovský, P. Baldrian, The variability of the 16S rRNA gene in bacterial genomes and its consequences for bacterial community analyses. PLoS ONE **8**, 10 (2013)

J. Veverka, Comet Surface Sample Return (CSSR) Mission Concept Study (2010a), pp. 1–33

J. Veverka, Cryogenic Comet Nucleus Sample Return (CNSR) Mission Technology Study (2010b), pp. 1–45

J. Vrublevskis, L. Berthoud, Y. McColluch, P. Bowman, J. Holt, J. Bridges, A. Bennett, F. Gaubert, L. Duvet, Description of European Space Agency (ESA) double walled isolator (DWI) breadboard currently under development for demonstration of critical technology foreseen to be used in the Mars sample receiving facility (MSRF), in *Proceedings of the Second International Mars Sample Return Conference*, Berlin, Germany, #6009 (2018a)

J. Vrublevskis, S. Duncan, L. Berthoud, P. Bowman, R. Hills, Y. McCulloch, D. Pisla, C. Vaida, B. Gherman, M. Hofbaur, B. Dieber, N. Neythalath, C. Smith, M. van Winnendael, L. Duvet, Description of European Space Agency (ESA) remote manipulator (RM) system breadboard currently under development for demonstration of critical technology foreseen to be used in the Mars sample receiving facility (MSRF), in *Proceedings of the Second International Mars Sample Return Conference*, Berlin, Germany, #6010 (2018b)

R.J. Walker, M.F. Horan, C.K. Shearer, J.J. Papike, Low abundances of highly siderophile elements in the lunar mantle: Evidence for prolonged late accretion. Earth Planet. Sci. Lett. **224**, 399–413 (2004)

W. Watson, Enhancements to the Sentinel fireball network video software, in *Society for Astronomical Sciences Annual Symposium*, vol. 28 (2009), p. 43

R. Webb, A fast track to zero environmental pathogens using novel ionized hydrogen peroxide technology. Infection Control Today. February 1 issue (2011)

T. Weinmaier, A.J. Probst, M.T. La Duc, D. Ciobanu, J.F. Cheng, N. Ivanova, T. Rattei, P. Vaishampayan, A viability-linked metagenomic analysis of cleanroom environments: Eukarya, prokaryotes, and viruses. Microbiome **3**, 14 (2015)

R.J. Weryk, P.G. Brown, A. Domokos, W.N. Edwards, Z. Krzeminski, S.H. Nudds, D.L. Welch, The southern Ontario all-sky meteor camera network, in *Advances in Meteoroid and Meteor Science*, ed. by J.M. Trigo-Rodríguez, F.J.M. Rietmeijer, J. Llorca, D. Janches (Springer, New York, 2008), pp. 241–246

A.J. Westphal, R.K. Bastien, J. Borg, J.C. Bridges, D.E. Brownlee, M.J. Burchell, A.F. Cheng, B.C. Clark, Z. Djouadi, C. Floss, I. Franchi, Z. Gainsforth, G.A. Graham, S.F. Green, P.R. Heck, M. Horanyi, P. Hoppe, F.P. Hoerz, J. Huth, A.T. Kearsley, H. Leroux, K. Marhas, K. Nakamura-Messenger, S.A. Sandford, T.H. See, F.J. Stadermann, N.E. Teslich, S. Tsitrin, J.L. Warren, P.J. Wozniakiewicz, M.E. Zolensky, Discovery of non-random spatial distribution of impacts in the Stardust cometary collector. Meteorit. Planet. Sci. **43**, 415–429 (2008)

G.W. Wetherill, D.O. Revelle, Which fireballs are meteorites? A study of the Prairie Network photographic meteor data. Icarus **48**, 308–328 (1981)

A.F. Whitaker, D. Dooling, LDEF materials results for spacecraft applications: Executive summary. NASA STI/Recon Technical Report N (1995), p. 23549

D.R. White, Lunar sample Processing in the Lunar Receiving Laboratory High-Vacuum Complex. Apollo Experience Report, NASA Technical Note D-8298, Johnson Space Center, Houston, TX (1976)

W. Whyte, *Cleanroom Design*, 2nd edn. (Wiley, New York, 2001)

W. Whyte, *Cleanroom Technology: Fundamentals of Design, Testing and Operation*, 2nd edn. (Wiley, New York, 2010)

M. Wiśniewski, P. Zoladek, A. Olech, Z. Tyminski, M. Maciejewski, K. Fietkiewicz, R. Rudawska, M. Gozdalski, M.P. Gawronski, T. Suchodolski, M. Myszkiewicz, M. Stolarz, K. Polakowski, Current status of Polish Fireball Network. Planet. Space Sci. **143**, 12–20 (2017)

I.P. Wright, S.S. Russell, S.R. Boyd, C. Meyer, C.T. Pillinger, Xylan: A potential contaminant for lunar samples and Antarctic meteorites, in *Proceedings of the 22nd Lunar and Planetary Science Conference*, vol. 22 (1992), pp. 449–458

T. Yada, A. Fujimura, M. Abe, T. Nakamura, T. Noguchi, R. Okazaki, K. Nagao, Y. Ishibashi, K. Shirai, M.E. Zolensky, S. Sandford, T. Okada, M. Uesugi, Y. Karouji, M. Ogawa, S. Yakame, M. Ueno, T. Mukai, M. Yoshikawa, J. Kawaguchi, Hayabusa-returned sample curation in the planetary material sample curation facility of JAXA. Meteorit. Planet. Sci. **49**, 135–153 (2014)

Y. Zhang, C-x. Xin, X. Wang, Y.-L. Deng, Detection of microorganism from China's spacecraft assembly cleanroom. Acta Astronaut. (2018, in press). https://doi.org/10.1016/j.actaastro.2018.08.024

S. Zielińska, P. Radkowski, A. Blendowska, A. Ludwig-Galezowska, J.M. Los, M. Los, The choice of the DNA extraction method may influence the outcome of the soil microbial community structure analysis. MicrobiologyOpen **6**, 11 (2017)

K.-P. Zimmer, Optical designs for stereomicroscopes, in *International Optical Design Conference* (SPIE, Bellingham, 1998), p. 8

B. Zohuri, *Physics of Cryogenics: An Ultralow Temperature Phenomenon* (Elsevier, Amsterdam, 2017)

M.E. Zolensky, T. Girard, Stardust Spacecraft Program Contamination Control Plan. Astromaterials Research and Exploration Science, NASA Johnson Space Center, JSC-27954 (1997), p. 26

M.E. Zolensky, K. Nakamura, M. Gounelle, T. Mikouchi, T. Kasama, O. Tachikawa, E. Tonui, Mineralogy of Tagish Lake; an ungrouped type 2 carbonaceous chondrite. Meteorit. Planet. Sci. **37**, 737–761 (2002)

M. Zolensky, P. Bland, P. Brown, I. Halliday, Flux of extraterrestrial materials, in *Meteorites and the Early Solar System II*, ed. by D.S. Lauretta, H.Y. McSween (University of Arizona Press, Tucson, 2006a), pp. 869–888

M.E. Zolensky, T.J. Zega, H. Yano, S. Wirick, A.J. Westphal, M.K. Weisberg, I. Weber, J.L. Warren, M.A. Velbel, A. Tsuchiyama, P. Tsou, A. Toppani, N. Tomioka, K. Tomeoka, N. Teslich, M. Taheri, J. Susini, R. Stroud, T. Stephan, F.J. Stadermann, C.J. Snead, S.B. Simon, A. Simionovici, T.H. See, F. Robert, F.J.M. Rietmeijer, W. Rao, M.C. Perronnet, D.A. Papanastassiou, K. Okudaira, K. Ohsumi, I. Ohnishi, K. Nakamura-Messenger, T. Nakamura, S. Mostefaoui, T. Mikouchi, A. Meibom, G. Matrajt, M.A. Marcus, H. Leroux, L. Lemelle, L. Le, A. Lanzirotti, F. Langenhorst, A.N. Krot, L.P. Keller, A.T. Kearsley, D. Joswiak, D. Jacob, H. Ishii, R. Harvey, K. Hagiya, L. Grossman, J.N. Grossman, G.A. Graham, M. Gounelle, P. Gillet, M.J. Genge, G. Flynn, T. Ferroir, S. Fallon, D.S. Ebel, Z.R. Dai, P. Cordier, B. Clark, M.F. Chi, A.L. Butterworth, D.E. Brownlee, J.C. Bridges, S. Brennan, A. Brearley, J.P. Bradley, P. Bleuet, P.A. Bland, R. Bastien, Report—Mineralogy and petrology of comet 81P/Wild 2 nucleus samples. Science **314**, 1735–1739 (2006b)

H.A. Zook, Spacecraft measurements of the cosmic dust flux, in *Accretion of Extraterrestrial Matter Throughout Earth's History*, ed. by B. Peucker-Ehrenbrink, B. Schmitz (Springer, Boston, 2001), pp. 75–92

Printed by Printforce, the Netherlands